W9-AGA-438

GENETICS

Harper & Row, Publishers, Inc.
New York, Hagerstown, San Francisco, London

Project Editor: David Nickol
Designer: T. R. Funderburk
Production Supervisor: Kewal K. Sharma
Photo Researcher: Myra Schachne
Compositor: Bi-Comp, Incorporated
Printer and binder: The Murray Printing Company
Art Studio: e h Technical Services

GENETICS

Library of Congress Cataloging in Publication Data

Farnsworth, Marjorie Whyte, Date-
Genetics.

Includes bibliographical references and index.
1. Genetics. I. Title.
QH430.F37 575.1 77-18020
ISBN 0-06-042003-0

CONTENTS

PREFACE

Genetics now occupies a central position in biology, for it provides a set of concepts that serve to unify all branches of the life sciences from biochemistry, cell biology, development and morphogenesis to ecology, medicine, and agriculture. The basic concepts of genetics were initially derived from the study of higher organisms. The elucidation of the structure of DNA and the exploitation of viruses and bacteria as experimental subjects has led to immense advances in our understanding of the molecular basis of gene action. The techniques and ideas generated by these remarkable discoveries are now being intensively applied to higher forms, including humans. Thus it appears that the thrust of the science is coming full circle from plants and animals to microorganisms and back to higher forms once more. Accordingly, an artificial division of the science into the classical versus the molecular is unrealistic, for an appreciation of the genetic mechanisms of microorganisms depends on prior knowledge gained from higher forms, and the basis for gene action in higher forms is derived to a large extent from studies of viruses and bacteria. With this interdependence in mind, an attempt has been made to produce a text that is neither classical nor molecular, but a balanced presentation of both areas. In addition, because of the great strides being made in human genetics and because of the interests of many students, human examples and applications have been included wherever appropriate.

This book is intended for a one-semester course in genetics for students majoring in the biological sciences who have had some work in college level biology and chemistry. It employs the historical approach not only because the science of genetics has a well-defined history, but also because this approach best illustrates our evolving understanding of the nature of the gene and its activities. Students are naturally more successful when they proceed from the simple to the more complex, and molecular concepts have more meaning and are more readily grasped once the basic rules of inheritance are thoroughly understood. Also helpful is some experience with genetics as a way of thinking, a way of drawing inferences and logical deducatons from observations and data. This kind of thinking is clearly demonstrated in early experiments, and the initial material on classical genetics is presented without reference to chromosomes in order to illustrate how much information can be deduced from the results of simple experiments and numerical relationships. Instructors who prefer to begin with chromosomes and meiosis can, of course, do so by changing the order of subject matter and beginning their course with Chapter 6. Similarly, those who prefer to begin with DNA can do so with little difficulty by starting with Chapter 11.

Since genetics is an experimental, problem-solving science, problems that permit the student to apply genetic principles have been included with each chapter. The answers to these problems, as well as the means of their solution, are given at the end of the book. These problems have been taken from those used at SUNY/Buffalo over a number of years, as well as those contributed by colleagues. Insofar as known, they have not been published elsewhere. If any have indeed previously appeared in print, my thanks are gratefully extended to the authors.

For motivational purposes, references are cited at the end of each chapter. Those who actually explore such references are primarily interested in the current status of a field and in an entré into the literature. For this reason, most of the references are to recent reviews that survey a topic and provide numerous citations to the original literature that can then be followed up within the context of the field. This has been a worthwhile and useful approach for undergraduates faced with the preparation of papers and for graduate students refreshing their background.

I am grateful to the Literary Executor of the late Sir Ronald A. Fisher, F.R.S., to Dr. Frank Yates, F.R.S. and to Longman Group Ltd., London, for permission to reprint Table IV from their book *Statistical Tables for Biological, Agricultural and Medical Research.* (6th edition, 1974).

I am deeply indebted to the many authors who have most graciously provided original illustrations. Many thanks are also due to those who have read and criticized all or part of the manuscript and have provided numerous kind suggestions for its improvement. In particular, I wish to thank Dr. Rosemary Elliott for her critical reading of the sections on viral and microbial genetics and Dr. William F. Duggleby whose invaluable assistance in the area of population genetics was so generously given. Most of all, I am indebted to fellow P_1 WEF for unflagging assistance and to F_1's SBF and MWF for their forbearance and encouragement. In a project of this size, some inaccuracies are unavoidable, and for these I must assume sole responsibility. I would be grateful if any such were brought to my attention.

M.W.F.

GENETICS

Chapter 1

THE BEGINNING OF GENETICS: MENDEL'S FIRST LAW

COMPARED to astronomy, mathematics, chemistry, or natural history, genetics is a young science, originating only in 1900, with the rediscovery and independent proof of Mendel's Laws. In the past 75 years it has developed almost explosively to such a level of refinement and sophisticated technology that serious questions are being raised about the possible manipulation of human genes. The more recent studies at the molecular level which give rise to these apprehensions can best be understood and appreciated when viewed in perspective and within the context of the science as a whole. Indeed, they owe their development and much of their meaning to concepts which had been firmly established before the term DNA became a household word.

In tracing the development of genetics from its inception to its current status, we will endeavor to interweave the converging contributions of many persons and many disciplines so as to illuminate a constantly evolving central theme: the nature and activities of genetic material.

The observations of Gregor Mendel on heredity in the edible garden pea represent the first disciplined, analytical inquiry into the mechanism of inheritance. Mendel was a clergyman and a teacher, in later life becoming prelate of the Augustinian monastery of St. Thomas in the town of Brünn, in what is now Czechoslovakia. His interest in botany began early in life, for farming and the development of new varieties of apples were his family's chief occupation. This natural interest was further stimulated by his formal education which centered around mathematics, physics, and the botanical and zoological sciences. Mendel entered the monastery in 1845 and was ordained in 1848, becoming a parish priest and, subsequently, a teacher in a nearby high school. The monastery of St. Thomas provided a stimulating environment. It was a center of cultural, intellectual, and religious life, and its members and visitors included many notable scholars and scientists of the period. In 1851, at the encouragement of the prelate, Mendel entered the University of Vienna, and upon completion of his course of studies in 1854, he returned to his teaching responsibilities at Brünn. His experiments in plant hybridization were carried out in the monastery garden over several years, beginning in 1856.

By this period of the nineteenth century, botany had become a flourishing discipline. Sexuality in plants had been described by Rudolf Camerarius in 1694 and by Nehemiah Grew in 1782. It was known that pollen functioned as the male element and that the ovules of flowers contained the female element (eggs), both being necessary for fertilization and the formation of a new individual. Artificial pollination had long been practiced, and the first experiments in plant hybridization had been successfully carried out by Joseph Kölreuter in 1760. A number of botanists contemporary to Mendel were industriously crossing pure breeding varieties of plants in an effort to produce improved strains of economic importance. Detailed descriptions of the hybrids resulting from these experiments had been published, and it was known that these plants did not breed true, but instead produced offspring which varied with respect to the original parental characters. Actually, the phenomena we now identify as dominance or as segregation were clearly evident in several of these investigations, but the significance of the observations was unrecognized, and in none of these studies was systematic, quantitative analysis employed.

Although Mendel was undoubtedly influenced by the work of his predecessors and contemporaries, he approached his experiments in an innovative way. First, he deliberately restricted his attention to the single character whose inheritance was under consideration in any experiment. Second, he kept accurate pedigree records for each plant. And, finally, he counted the different kinds of individuals that resulted from his experimental crosses. This last procedure was crucial, for more than anything else, it was the numerical data that permitted Mendel to formulate his rules of inheritance. Since his methods of analysis are as applicable and useful today as they were over 100 years ago, it is worthwhile to examine his findings in some detail.

EXPERIMENTS WITH MONOHYBRIDS AND DERIVATION OF THE LAW OF SEGREGATION

For his studies Mendel obtained a number of pure breeding varieties of the garden pea *(Pisum sativum)* (Figure 1–1), a plant previously used for a similar purpose by Alexander Seton in 1822. This plant species offers the advantages of being easy to grow and fertilizing itself before the flower opens. The male (♂) sex cells, or pollen, from a flower fertilize the female (♀) sex cells, or eggs, of that same flower. Self-fertilization, or selfing, is a type of natural inbreeding found in many plants. Cross fertilization is easily accomplished by removing the pollen-producing anthers of the flower and introducing pollen taken from a different plant. A cross between two varieties can be performed by this method, a procedure generally known to botanists as hybridization.

Mendel worked with 14 pure breeding varieties which could be arranged as 7 pairs, the members of each such pair exhibiting alternative inherited states of some single aspect of plant structure or color. For example, one such aspect of structure was the height of the plant and the alternative states of that aspect were represented by two varieties, one tall, the other short. We can refer to tallness and shortness as contrasting *phenotypes,* the term phenotype referring to appearance or condition with respect to any aspect, or character, under consideration, in this case, height.

Before attempting hybridization, Mendel cultivated his varieties for two generations. This preliminary step was undertaken to verify the purity of each line, that is, to make sure that the phenotype characteristic of each was indeed inherited by all offspring. The purity being established, he then performed crosses between individuals of contrasting phenotypes. Since he had seven pairs of inherited, contrasting characters, he carried out seven separate experiments (Table 1–1), only one of which need be used as a representative exam-

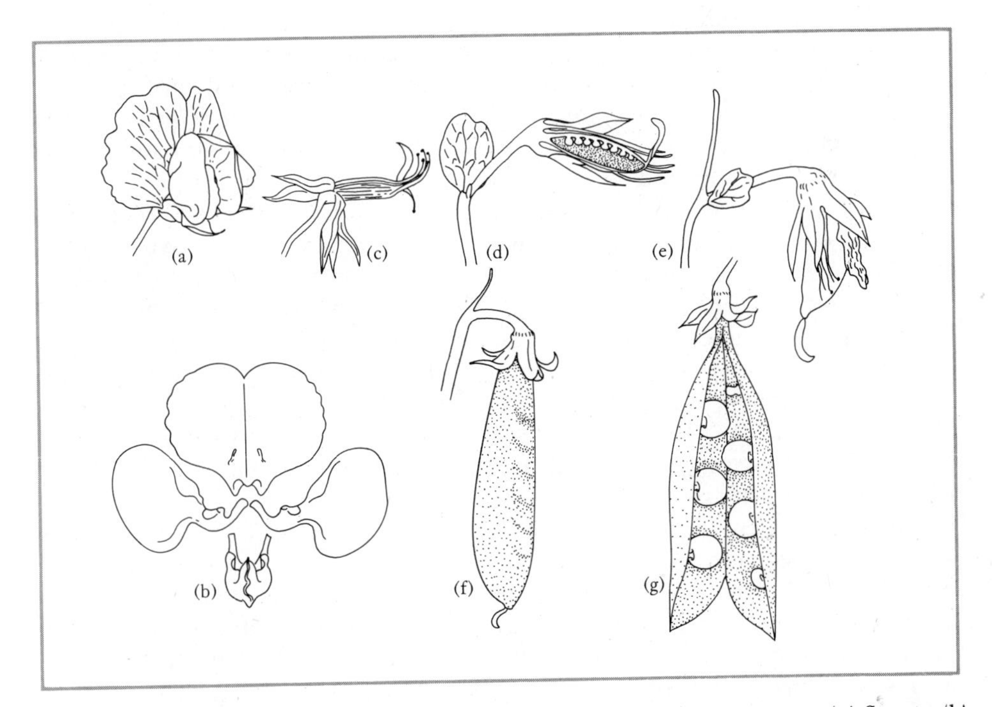

Figure 1–1. Flower, fruit, and seeds of the garden pea, *Pisum sativum:* (a) flower; (b) the five petals; (c) flower with petals removed so that the stamen tube which surrounds the pistil may be seen; (d) arrangement of ovules as seen in a longitudinal section of the pistil; (e, f, g) stages in the development of the fruit from the pistil and of the seeds from ovules. [Figure 153 (p. 355) from *Textbook of Botany,* revised edition, by E. N. Transeau, H. C. Sampson, and L. H. Tiffany. Copyright 1940, 1953 by Harper & Row, Publishers, Inc.]

TABLE 1–1
Data from Mendel's Seven Experiments with Garden Peas

Trait	P_1 crosses between plants of alternative phenotypes	F_1 phenotype
Seed form	round × wrinkled	round
Cotyledon color	yellow × green	yellow
Color of seed coat	gray × white	gray
Form of pod	inflated × constricted	inflated
Color of pod	green × yellow	green
Position of flower	axial × terminal	axial
Length of stem	long × short	long

ple. The best example, because the data are more extensive, is the experiment involving crosses between plants with yellow and green cotyledons. Cotyledons are embryonic leaves containing stored food; they comprise the bulk of the seed and their color can be seen through the seed coat.

Mendel initially carried out *reciprocal crosses* using pollen from the yellow variety to fertilize the eggs of the green variety and vice versa: yellow ♂ × green ♀ and green ♂ × yellow ♀. Since these crosses involved the original parents of the experiment, they can be designated as the *parental* or P_1 crosses.

The mature seeds which formed from each experimental pollination comprised the *first filial* or F_1 generation. These seeds were collected and classified as to color, and it was found that all were yellow and that reciprocal crosses yielded identical results. The green characteristic was not apparent in the seeds of any plant and seemed to have disappeared entirely (Figure 1–2). The seeds were saved and planted the following spring and the resulting F_1 plants, grown to maturity, were allowed to self-fertilize. They produced a *second filial* or F_2 generation totaling 8023 individuals, of which 6022 were yellow and 2001 were green, a phenotypic ratio of approximately 3 yellow : 1 green. Pedigree records indicated that this same 3 : 1 phenotypic ratio occurred in the descendents of either reciprocal P_1 cross. In addition, the green phenotype had not only reappeared, but seemed to be unaltered, for the F_2 green seeds were indistinguishable in color from those of the original P_1 green strain.

Mendel's experiments included one further (and crucial) step, but let us pause for a moment to analyze the results obtained thus far.

We have seen that all members of the F_1 generation were themselves yellow, yet they transmitted the green characteristic to their offspring. From this we can infer that the F_1 plants must have possessed a hereditary factor for green, as well as one for yellow, even though the factor for green was not expressed in the F_1 phenotype. We must therefore conclude that the F_1 individuals were hybrid in nature.

Comparison of the yellow F_1's with the yellow P_1 parental type shows that despite identity in appearance, the genetic constitution of the F_1 clearly differed from that of the yellow parent. The P_1 parental variety, when self-fertilized, yielded only yellow offspring, but the F_1 individuals, when self-crossed, produced both yellow and green progeny. It is apparent that we must make a distinction between the appearance, or phenotype, of an organism and its un-

F_2 phenotypes	F_2 phenotypic ratio
5474 round : 1850 wrinkled	2.96 : 1
6022 yellow : 2001 green	3.01 : 1
705 gray : 224 white	3.15 : 1
882 inflated : 299 constricted	2.95 : 1
428 green : 152 yellow	2.85 : 1
651 axial : 207 terminal	3.14 : 1
787 long : 277 short	2.84 : 1

P_1 Reciprocal Crosses:

Yellow ♂ X Green ♀ → F_1 Offspring: All yellow

Green ♂ X Yellow ♀ → All yellow

F_1 Inbreed:

F_1 Yellow ♂ X F_1 Yellow ♀ → F_2 Offspring: Yellow and Green

F_1 Yellow ♂ X F_1 Yellow ♀ → Yellow and Green

F_2 Phenotypic Ratio:
3 Yellow : 1 Green

3 Yellow : 1 Green

Total of F_2 Progeny: 8023

Total of F_2 Phenotypes: 6022 Yellow : 2001 Green

F_2 Phenotypic Ratio: 3.01 Yellow : 1 Green

Figure 1–2. A summary of the data from Mendel's experiments with green and yellow peas.

derlying genetic constitution, or *genotype,* for although the phenotype may be predicted from the genotype, the reverse does not necessarily hold.

The fact that the hybrid nature of the F_1 individuals was not apparent in their phenotype, Mendel ascribed to a phenomenon he called *dominance*. It was clear that the hereditary unit for green had to be present in the F_1 hybrids in order to appear in their offspring. Yet, when in company with the factor for yellow, only yellow was expressed. Although Mendel could not explain the reasons for this phenomenon, he recognized it and took it into account. The character which was hidden in a hybrid was called the *recessive;* the character which was expressed in the hybrid was called the *dominant*.

Additional inferences with respect to the genotype can also be drawn from Mendel's data. We know that a new individual is formed by the union of male and female sex cells *(gametes),* and we can infer from the data that we cannot assign a greater hereditary contribution to the pollen than to the egg, or vice versa. Both contribute equally to the inheritance of cotyledon color, for no matter which way the original reciprocal crosses were made, the results were the same in that identical F_1 hybrids were formed. Furthermore, if the F_1 individuals can be shown to possess two hereditary factors for cotyledon color, one being received from each parent via fertilization, then it is likely that all individuals arising from the union of male and female sex cells will also possess two hereditary factors for any character under consideration. If so, it follows that the P_1 yellow parent, since it yields only yellow offspring by selfing, must contain two factors for yellow. Similarly, the P_1 green parent, which when inbred yields only green offspring, must contain two factors for green. From this it follows that green color can be expressed in the phenotype only when its genetic determinant is doubly present in the genotype.

Although the 3 : 1 phenotypic ratio obtained in the F_2 generation is yet to be satisfactorily explained, it can be seen from the above analysis that a remarkable amount of information can be derived from Mendel's simple experiments. Our next step must be the formulation of a hypothesis which not only will account for all of the data, but whose validity can be tested experimentally. Before doing so, however, it is useful to introduce some convenient terms. Today, in the functional sense, we call Mendel's hereditary factors *genes,* and we refer to the alternative states of a gene as *allelic states* or *alleles.* As commonly used, the terms dominant and recessive are applied to allelic states themselves, as well as to the phenotypes which they produce. Thus, we can characterize a gene as a dominant or a recessive allele, and we can also describe the character which it produces as a dominant or recessive trait.

In Mendel's experiment the color of the embryonic leaves was determined by a gene present in two allelic states, one allele being dominant and determining yellow color and the other allele being recessive and determining green color. Although we can name this gene cotyledon color or anything else that we wish, it will be even more helpful if we choose symbols to represent the two allelic states. We can follow Mendel's example and use the capital letter A to represent the dominant allele for yellow and the lower case a to represent the recessive allele for green.

An individual whose genotype contains two identical alleles of a gene is called a *homozygote* or is said to be *homozygous* for the allele in question. As examples, we have inferred that Mendel's original P_1 green variety contained two genes for green color. If so, this variety was homozygous for the recessive allele, and we can symbolize its genotype as aa. Similarly, the P_1 yellow variety was presumably homozygous for the dominant allele, and its genotype can be given as AA.

When two different alleles of a gene are present in the genotype, the individual is called a *heterozygote* or hybrid or is said to be *heterozygous* for the alleles in question. In Mendel's experiments the F_1 individuals were heterozygous, and their genotypes can be designated as Aa. We can also call these F_1 individuals *monohybrids,* because, so far as we know, they were heterozygous for only one pair of alleles. We must remember, however, that with dominance, the two genotypes, AA and Aa, will both give rise to the same phenotype. As a result, the only means of distinguishing the homozygote *(AA)* from the heterozygote *(Aa)* is through additional crosses, whereby their nature will be revealed by the types of offspring they produce.

Returning to our analysis of Mendel's data, let us apply the following hypothesis in an attempt to explain the origin of the 3 : 1 phenotypic ratio obtained in the F_2 generation.

If we can infer that the P_1 plants contained two alleles for cotyledon color and yet the gametes which they produced each contained only one, then some mechanism must exist whereby these genes are distributed to different sex cells, each such cell receiving one or the other, but not both, members of the allelic pair. Subsequent union of two such sex cells in the process of fertilization would restore the proposed double allelic condition. Applying this hypothesis to the F_1 hybrids, we would expect that the two alleles, *A* and *a*, present in the genotype would be segregated or separated from each other to pass into different reproductive cells. As a part of our hypothesis, let us also assume that these two classes of reproductive cells are produced in equal numbers in both sexes. Thus, within any one flower, the male sex cells would consist of two classes, one class carrying the allele *A*, the other containing the allele *a*, and equal numbers of the two classes would be formed. Similarly, in the pistillate, or female, structures of the flower, two types of eggs, *A* and *a*, would also be formed in equal numbers. If we assume that fertilization between these gametes is random, then we can expect that all possible combinations between eggs and pollen will occur, as illustrated in Figure 1–3. According to this hypothesis, the resulting F_2 generation should consist of the following classes and proportions:

$\frac{1}{4}$ should be homozygous *AA*, and yellow in phenotype
$\frac{1}{2}$ should be heterozygous *Aa* or *aA*, and yellow in phenotype
$\frac{1}{4}$ should be homozygous *aa*, and green in phenotype

Due to dominance of the *A* allele for yellow over the *a* allele for green, the overall phenotypic ratio should be 3 yellow : 1 green. However, if the hypothesis is wrong, that is, if the two types of eggs or pollen are not produced in equal numbers or fertilization between them is not random, then a phenotypic ratio other than 3 yellow : 1 green should be obtained.

Comparing the expectations based on our hypothesis with Mendel's actual data, we can see that the hypothesis successfully accounts for the results. In addition, the likelihood that the hypothesis is correct is greatly strengthened by the fact that Mendel obtained similar data in experiments which involved six other pairs of contrasting characters (Table 1–1).

However, no matter how attractive or satisfying the explanation, it cannot be unreservedly accepted without additional experimental proof. The hypothesis predicts that the 3 : 1 phenotypic ratio is the result of dominance imposed upon an underlying genotypic ratio of $\frac{1}{4}$ *AA* : $\frac{1}{2}$*Aa* : $\frac{1}{4}$*aa* (1 : 2 : 1). Proof of the hypothesis requires a demonstration that the F_2 green individuals are indeed homozygous for the *a* allele and that the F_2 yellow individuals consist of two genotypic classes: those homozygous for the yellow allele and thus *AA* in genotype and those heterozygous or *Aa*. It must also be shown that there are twice as many yellow heterozygotes as yellow homozygotes. A test of the

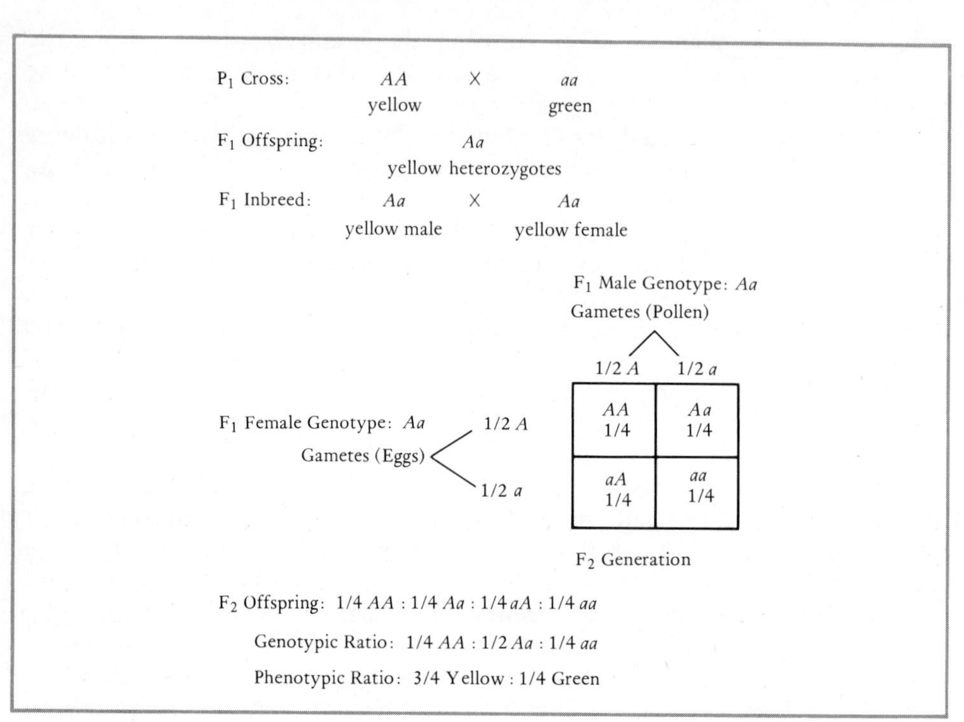

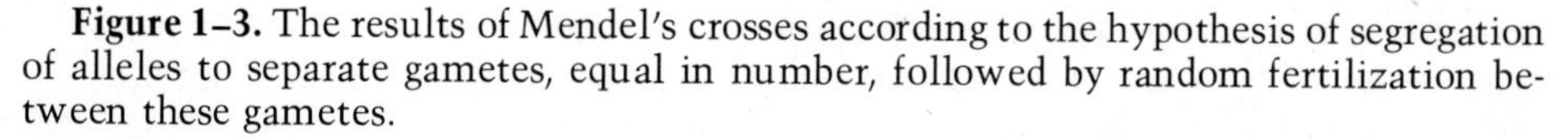

Figure 1–3. The results of Mendel's crosses according to the hypothesis of segregation of alleles to separate gametes, equal in number, followed by random fertilization between these gametes.

hypothesis thus requires that the genotypes of the F_2 individuals be determined. Since these genotypes are revealed only through the types of offspring produced, a further cross becomes necessary.

Mendel, perceiving this need, allowed the F_2 individuals to self-fertilize and produce a *third filial* or F_3 generation. This generation serves to indicate the genotypes present in the preceding F_2 progeny in the following way. Homozygous, self-fertilizing F_2 green plants should produce only green offspring. Similarly, F_2 yellow homozygotes should give rise to only yellow offspring. However, F_2 heterozygotes should be identical in genotype to the first generation F_1 hybrids, and through inbreeding, should yield both yellow and green progeny in a phenotypic ratio of 3 : 1.

Mendel's F_3 generation fully demonstrated the validity of the hypothesis by providing proof of the underlying 1 : 2 : 1 genotypic ratio present in the F_2 generation. F_2 green plants produced only green progeny and therefore must have been *aa* in genotype. Of the F_2 yellow plants, one-third yielded only yellow offspring, indicating that their genotype must have been homozygous *AA*. The other two-thirds produced offspring in the expected phenotypic ratio of 3 yellow : 1 green and therefore must have been heterozygous *Aa* in genotype. Mendel carried his proof further by allowing inbreeding to continue through an F_6 generation. As anticipated, he found that homozygotes continued to produce only homozygotes, while heterozygotes gave rise to progeny

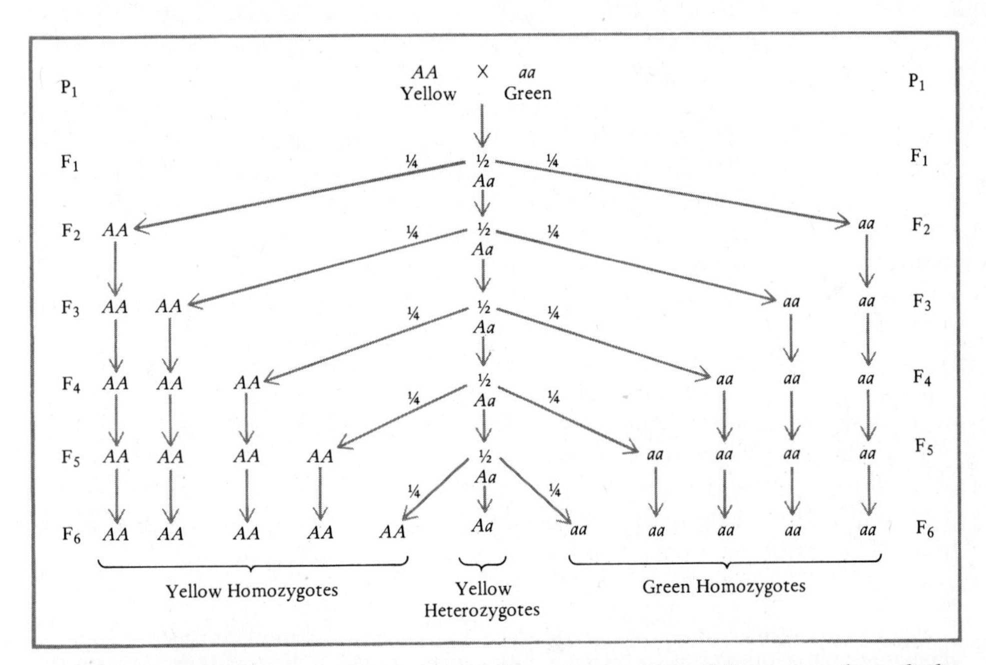

Figure 1–4. Distribution of genotypes in the progeny of Mendel's crosses through the sixth inbred generation. In each generation homozygotes give rise only to homozygotes, while heterozygotes produce offspring in the genotypic ratio of 1/4 *AA* : 1/2 *Aa* : 1/4 *aa*.

in the proportion of 3 yellow : 1 green (Figure 1–4). On the basis of all of these substantiating results, we can accept the hypothesis as proven.

More formally expressed, our hypothesis states that during the process of sex cell (gamete) formation in any individual, the members of a pair of allelic genes are segregated from one another into separate gametes, such gametes being produced in equal numbers, and that subsequent fertilization between these gametes is random. This principle of the separation of allelic genes at sex cell formation is sometimes called Mendel's First Law or the Law of Segregation, and it is the basis for inheritance in all organisms that reproduce by means of the union of sexual cells. Although we tend to take this principle for granted, it rests upon a solid foundation of experimental proof.

That segregation produces two types of gametes, equal in number, can be demonstrated by the examination of pollen phenotypes in corn. Plants heterozygous for the presence or absence of starch produce two types of pollen, and if the pollen is stained with iodine solution, the pollen grains containing starch are colored dark blue, while those without starch remain unstained (Figure 1–5). Counts of the two types of pollen indicate that they are produced in a 1 : 1 ratio.

Another demonstration of Mendel's First Law can be obtained with the unicellular alga *Chlamydomonas,* for during one stage in its life cycle (see Chapter 6) *Chlamydomonas* produces cells that are the equivalent of the gametes of higher organisms. If a parent cell is heterozygous for some biochemical trait, such as the ability or inability to synthesize the amino acid arginine (arg^{+}/arg^{-}), the parent produces two types of gamete-equivalent daughter cells in equal numbers: half are arg^{+} and half are arg^{-}.

We should recognize the astonishing fact that Mendel arrived at his conclusions without the kind of information currently available to us. In the scientific world of the midnineteenth century the function of the cell nucleus was unknown, cell division (mitosis) had not yet been described, and the presence of chromosomes was totally unsuspected. Yet by the application of simple quantitative methods coupled with logical deductions from the data, Mendel was able to formulate this fundamental rule of inheritance, and his data constitute formal genetic evidence for the process, known on the cellular level as meiosis. In addition, Mendel did not consider his factors or determinants, now called genes, to be in any way equivalent to phenotypic characters themselves or to their preformed rudiments, a notion which was popular in the nineteenth century. Instead, he thought of them as concrete particles of some kind whose potential was expressed through purely physiological processes and whose nature was unaltered by residence in a heterozygote in company with an alternate allele. In these concepts alone, he was far in advance of his time.

Figure 1–5. Segregation as demonstrated in pollen from a corn plant heterozygous for genes producing the starchy and nonstarchy phenotypes. Iodine staining reveals the presence of both starchy and nonstarchy pollen grains occurring in approximately equal numbers.

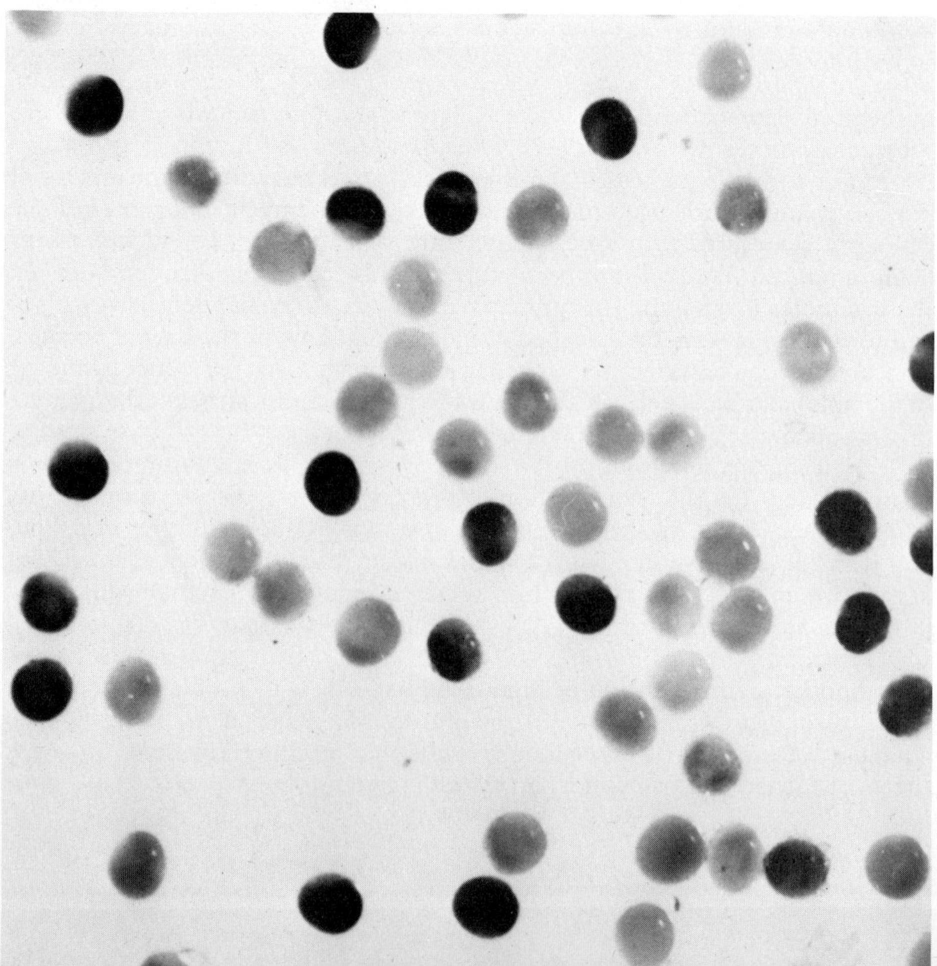

SEGREGATION IN PRACTICE

Although Mendel and the monastery garden may seem remote today, the type of analysis he initiated has immediate practical value. On the basis of the Law of Segregation, we can predict in advance the results of any cross involving two alternative alleles of a given gene. Symbolizing a hypothetical dominant gene as *A* and its recessive allele as *a*, the kinds of crosses that can be made and the results that can be anticipated are the following.

1. cross between homozygotes for the same allele *(AA × AA; aa × aa)*

2. outcross between homozygotes for different alleles *(AA × aa)*

3. cross between heterozygotes for the same alleles *(Aa × Aa)*

4. backcross between a heterozygote and a homozygote *(Aa × AA; Aa × aa)*

In a cross between individuals homozygous for the same allele all gametes formed by both the male and the female parent carry the identical allele, and fertilization between these gametes must again produce the homozygous condition (*AA* or *aa*).

In an outcross between two different genetic lines, each homozygous for one or the other allele, all gametes of one parent contain the allele *A*, while those of the other parent carry the allele *a*. Subsequent fertilization must yield the heterozygote *Aa*.

Individuals heterozygous for the same pair of alleles *(Aa × Aa)* produce gametes with the allele *A* and gametes with the allele *a* in equal numbers. Subsequent random fertilization dictates the 1 : 2 : 1 genotypic ratio and dominance imposes the 3 : 1 phenotypic ratio.

In a backcross between a heterozygote and a homozygote, the heterozygous parent will form two kinds of gametes in equal numbers, those with the allele *A* and those with the allele *a*. The homozygous parent can produce only one class of gametes with respect to these alleles. Fertilization yields progeny of two genotypic classes in a 1 : 1 ratio (Figure 1–6).

When the backcross is made between a heterozygote and an individual homozygous for the dominant allele (*Aa × AA*), half of the progeny will be *Aa* and the other half *AA* in genotype. However, all will exhibit the dominant phenotypic state due to the presence of the dominant allele *A*. If such a backcross is performed to ascertain the genotype of a suspected heterozygote, no useful information will result because, with respect to phenotype, the offspring will be indistinguishable from one another or from those that would be obtained from the cross *AA × AA*.

However, when the backcross is made between a heterozygote and an individual homozygous for the recessive allele *(Aa × aa)*, half of the progeny will be genotypically *Aa* and in consequence show the dominant state, while the other half will be *aa* in genotype and exhibit the recessive state. As a result a 1 : 1 phenotypic and a 1 : 1 genotypic ratio will be obtained, and a suspected heterozygote can be positively identified as such by this means (Figure 1–6). If the unknown parent is, in fact, homozygous *AA*, then all offspring will show the dominant condition. Thus, in either circumstance, a backcross employing the homozygous recessive as one parent tests and verifies the genotype of the other parent. For this reason, this type of backcross is called a *testcross*.

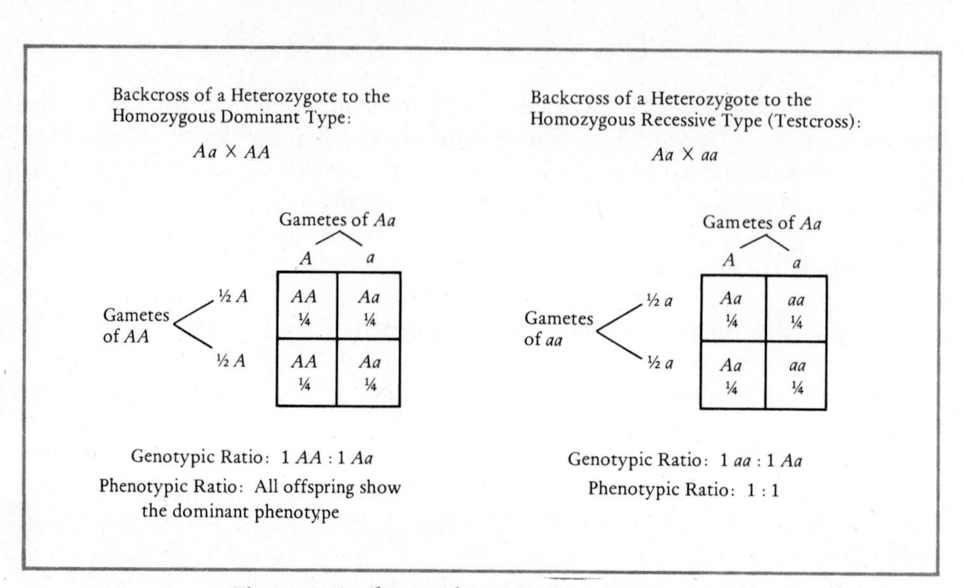

Figure 1–6. The results of backcrossing.

Real life examples of all of these kinds of crosses are commonplace. They may be designed and carried out in a laboratory using various species of plants or animals or they may occur by chance in natural populations in which alternative alleles of a given gene are present. When planned in the laboratory, at least one parental genotype is generally known in advance so that accurate predictions of results can be made. Where controlled matings cannot be carried out, it is often possible to determine probable genotypes and whether a trait is dominant or recessive simply by reasoning from the data according to the principles embodied in Mendel's Law of Segregation. A good example is the application of Mendelian principles to human traits.

HUMAN PEDIGREES

Dominant and recessive traits occur in humans as well as in garden peas, and the only method of distinguishing a dominant from a recessive is by analyzing the pattern of transmission of a trait from one generation to the next. Such a pattern is best identified by use of a pedigree diagram in which the relationships of family members are represented. Pedigree diagrams use the symbols shown at the top of page 13.

The pedigree in Figure 1–7 illustrates the transmission pattern of a dominant trait. Notice that the trait appears in offspring only if it was present in an immediate parent and that if both parents are unaffected, their children are also unaffected. With dominance of the trait established, it then becomes possible to deduce that the male of generation I was a heterozygote, since at least one of his children did not show the character. If he had been homozygous, all of his children would have been affected phenotypically.

The pattern of transmission of a recessive trait differs from that of a dominant in that phenotypically unaffected, but heterozygous parents can pro-

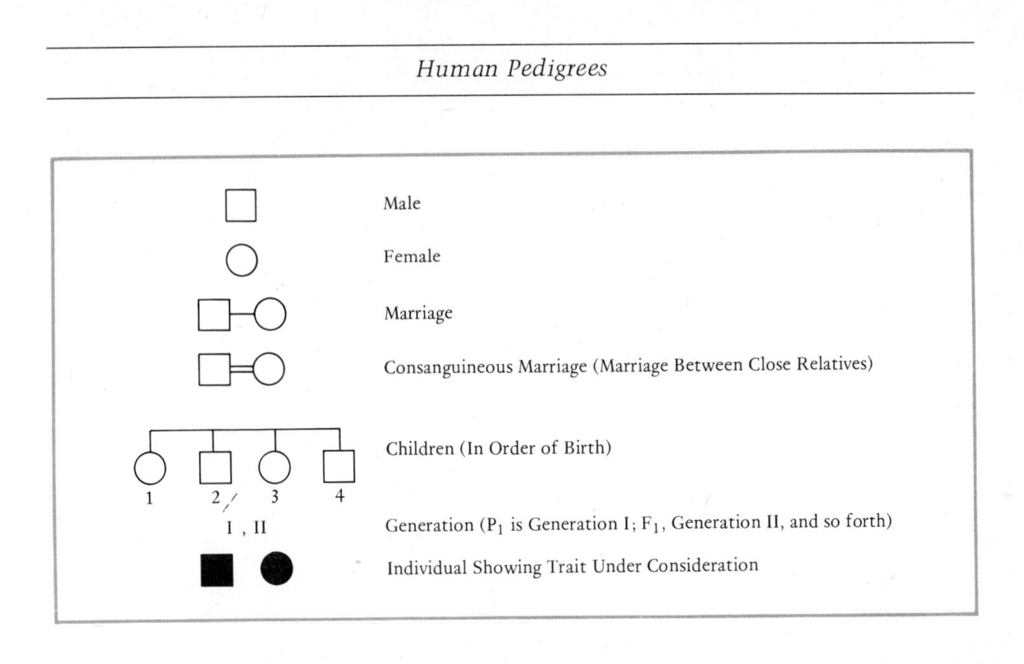

duce homozygous, and therefore phenotypically affected children. Such children are more likely to be born from consanguineous marriages, where the parents are related descendents of a common ancestor who possessed the trait. Chance marriages between unrelated heterozygotes also occur and in some instances may not be exceptionally rare, as in the case of cystic fibrosis where 1 out of every 22 white persons is a carrier, that is, heterozygous for this condition. In addition, sometimes the frequency of a given gene is higher in a particular religious or ethnic group than it is in the population at large. If members of such a group marry within the group, the chance of a marriage between heterozygotes is greatly increased. For example, 1 in 30 Jews of Polish-Lithuanian ancestry (Ashkenazi Jews) is a carrier of Tay-Sachs disease, a condition characterized by severe mental and physical retardation causing death at 2 to 4 years of age. In the general population the frequency of carriers is only about 1 in 300.

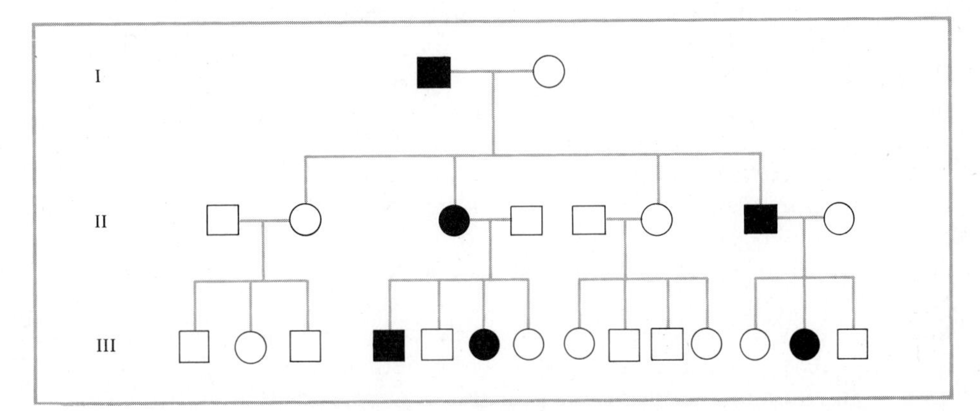

Figure 1–7. Transmission pattern of a dominant trait in humans. Individuals showing the trait are indicated by solid squares and circles.

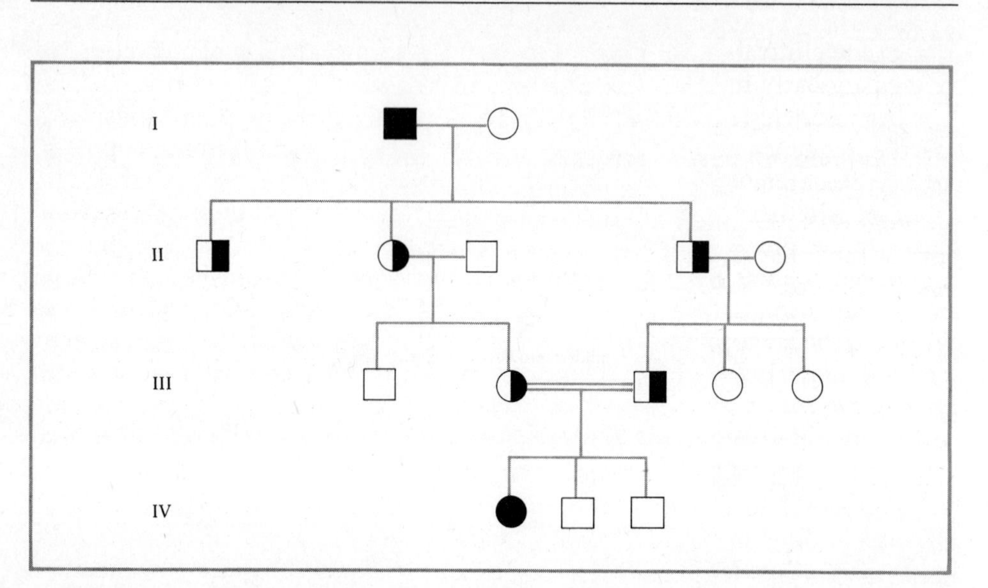

Figure 1–8. Transmission pattern of a recessive trait in humans. Individuals showing the trait are indicated by solid squares or circles; presumed heterozygotes are indicated by half-solid squares or circles; a cousin-cousin marriage is indicated by a double line.

The pedigree shown in Figure 1–8 illustrates the probable line of descent of a recessive gene through four generations. The homozygous individual shown as the solid circle in generation IV is the result of a cousin-cousin marriage. The presence of the recessive phenotype in generation IV automatically indicates that the parents in generation III were heterozygotes (symbolized as half-solid squares and circles). They, in turn, must have inherited the gene from previous parents of generation II, who were brother and sister offspring of generation I.

In the examples of cystic fibrosis and Tay-Sachs disease given above, phenotypically affected offspring are termed abnormal, while those unaffected are termed normal. For many inherited traits, such as eye color, attached or free earlobes, the ability or inability to roll the tongue or to taste certain chemicals, the terms normal and abnormal do not apply. Although one phenotype may be rare and its alternative common, neither will interfere with the health and welfare of the individual, and thus whether or not a given condition is dominant or recessive is of no real concern. However, there are numerous instances in humans where alternative phenotypes are not equally normal or viable, and where the presence of a given allele may result in morphological and metabolic deficiencies. In such cases a medical geneticist is often consulted by parents fearful that a child of their union will be inherently and irreversibly abnormal due to a genetic defect known to occur in the family. By use of the same kind of deductive reasoning illustrated above, the geneticist analyzes the family pedigree in an effort to determine the probable genotypes of the inquiring parents and, on this basis, to predict the "odds" that a defective child will be born. In some instances it is possible to identify heterozygous carriers of the defect by appropriate clinical tests. Where such tests exist, the accuracy of predictions is obviously greatly increased.

In making estimates of the odds, the geneticist must take into consideration the frequency of the different classes of offspring expected from the type of cross represented by the parental genotypes. Of equal importance is a determination of whether the allele causing the defect is dominant or recessive in its phenotypic expression. For example, if the gene was recessive and the parents heterozygous, the expected phenotypic ratio for their offspring would be $\frac{3}{4}$ normal to $\frac{1}{4}$ abnormal. However, since each fertilization is an independent event subject to the laws of chance, the odds would be 1 in 4 that any given child born would be homozygous for the deleterious gene. On the other hand, if the allele causing the defect was dominant and both parents heterozygous, the expected phenotypic ratio would be $\frac{3}{4}$ abnormal to $\frac{1}{4}$ normal, and the odds for a defective child would be 3 chances out of 4. The decision of parents whether or not to undertake the risk might be quite different in these two situations.

THE VALUE OF MUTATIONS

In human pedigrees it sometimes happens that a trait will appear spontaneously in one individual with no evidence of its presence in any other family member of any generation. In such cases it is presumed that an alteration in the structural organization of a gene has occurred and that this change has led to the appearance of the trait. Alterations in gene structure are called *mutations,* and the altered or mutant gene is a new allelic state whose presence gives rise to a phenotype recognizably different from the normal or standard condition. Mutations are valuable tools in genetics because a change in phenotype is the only means by which a gene can be associated with an inherited character. A normal or standard organism presents an integrated phenotype, and it is impossible to distinguish the effects of any one gene from those of another. Only when some process, part, secretion, color, or state is inherited in alternative forms is it possible to assign the function to a specific gene. If a gene never mutated and therefore had no alleles, its identity as the source or controlling agent for any phenotypic character would remain unknown. Although mutation, either spontaneous or induced by radiation or chemicals in the environment, is a relatively rare event, it is the only means by which new alleles arise. Members of existing populations are often heterozygous for alleles of many different genes. When representative organisms from such populations are inbred for a number of generations, varieties homozygous for different alleles can be obtained. Such inbred lines were used by Mendel for his experimental crosses and they have provided much of our knowledge of genetics. Mutant genes and mutant phenotypes should not be regarded as curiosities, but as necessary natural occurrences. They are the building blocks of evolution and are the tools through which the nature of the gene and its activities can best be investigated.

PROBLEMS

1–1. In *Drosophila* ebony is a recessive mutant causing black body color when homozygous, as opposed to the normal tan color.

a. If an ebony colored male is crossed to a homozygous tan female, what will be the color of the offspring?

b. If the F_1 males and females are crossed, what phenotypes can be expected to appear in the F_2 and in what proportions?

c. If an F_1 male is crossed to an ebony female, what kinds of progeny will be produced and in what proportions?

1–2. In tomatoes the flesh can be either red or yellow. The following crosses involving plants of these two phenotypes are made.

Parents	Progeny
red × red	75 red
red × red	63 red, 15 yellow
red × yellow	68 red
yellow × yellow	84 yellow
red × yellow	47 red, 53 yellow

a. What phenotype is dominant?

b. What are the genotypes of the parents and offspring in each case?

1–3. In the cross $AaBBCC \times Aabbcc$ what proportion of the progeny will be homozygous for A?

1–4. Pigs can be solid colored or half colored (the anterior half may be white). A solid-colored pig was crossed to a half-colored pig and the litter consisted entirely of solid-colored piglets. The F_1 offspring were used in the following crosses.

(1) F_1 solid × P_1 solid produced all solid offspring.

(2) F_1 solid × F_1 solid produced 7 solid and 2 half colored.

(3) F_1 solid × P_1 half colored produced 3 solid and 4 half colored.

a. Which phenotype is dominant and which is recessive?

b. Assigning S to the dominant allele and s to the recessive allele, what are the genotypes of the original parents?

c. What genotypes occur in the progeny of cross 2?

d. What are the genotypes of the solid-colored progeny of cross 3?

e. What proportion of the eggs of an F_1 female will carry the gene for half color?

1–5. Deafness in cocker spaniels is inherited. A breeder of champion cocker spaniels observed the following among the normal dogs in his kennel.

female A bred to male C produced all normal puppies.

female B bred to male C produced some deaf puppies.

a. Is the gene for deafness dominant or recessive?

b. If the breeder wished to eliminate the gene for deafness from his kennel and had the opportunity to sell female A, female B, and male C for house pets, which should he sell?

1–6. In cats the hair can be either short or long (angora). A cat fancier made the following crosses.

Parents	Kittens
(1) angora × angora	all angora
(2) angora × short	all short
(3) short × angora	4 short, 5 angora

a. Which phenotype is dominant?

b. What are the genotypes of the parents and offspring of these crosses?

c. If the short-haired offspring from cross 3 were bred to their short parent, what kinds of kittens would be expected and in what proportions?

d. If the short-haired offspring from cross 3 were bred to the short-haired parent of cross 2, what kinds of kittens would be expected and in what proportions?

1–7. In oats the grains can be yellow or white, this condition being determined by a single pair of alleles. The plant is normally self-fertilizing. A yellow-grained plant crossed with a white variety produced all yellow offspring. The F_1 plants were then allowed to self-fertilize and produce an F_2 generation.

a. Which phenotype is dominant?

b. In the F_2 generation what proportion of the plants will have white grains?

c. What proportion of the F_2 plants will yield both yellow- and white-grained offspring in an F_3 generation?

d. What proportion of the F_2 yellow-grained plants will yield only yellow-grained progeny in an F_3 generation?

e. If 5000 pollen grains are produced by an anther of an F_1 plant, how many of these grains would be expected to carry the gene for white?

1–8. In another experiment with oats a yellow-grained plant was crossed with a white-grained plant. The F_1 generation consisted of yellow and white individuals in approximately equal numbers. Using Y for yellow, and y for white,

a. Symbolize the genotypes of the parents.

b. What are the genotypes of the F_1 individuals?

c. If the F_1 plants are allowed to self-fertilize, what will be the appearance of the F_2 plants in each case?

d. What proportion of the F_2 plants in each case will be homozygous for yellow or for white?

1–9. In humans earlobes can be attached or free. A man with attached earlobes marries a woman with free earlobes. Their seven children have free earlobes. One son marries and of his children, half have free and half have attached earlobes.

a. What is the phenotype of the son's wife?

b. Is the phenotype of attached earlobes dominant or recessive?

1–10. The ability to roll the tongue is a dominant trait in humans. When tested, a man and his wife lacked this ability, yet their child was able to roll his tongue. What possible explanation(s) can you give for this circumstance?

1–11. Examine the pedigree shown below.

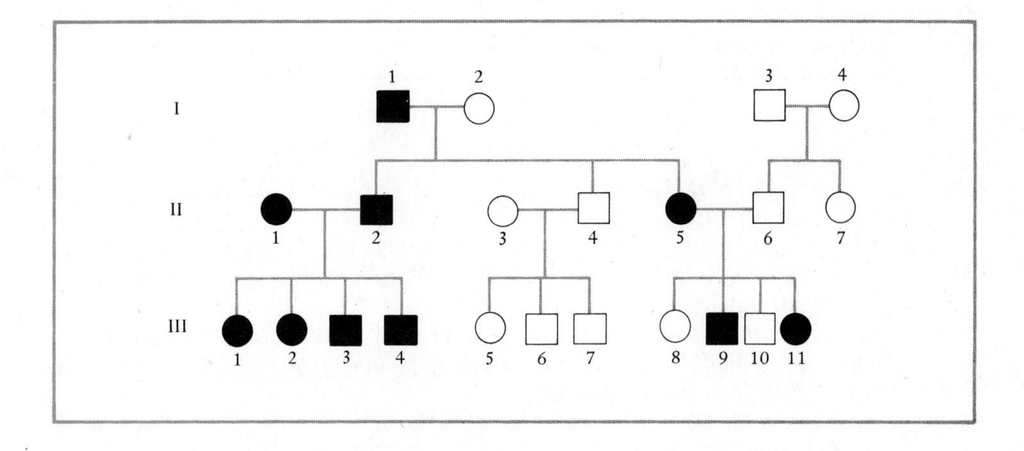

a. Is the trait illustrated in the pedigree dominant or recessive?

b. Using letters A and a as symbols, what are the most probable genotypes for all individuals in generations I and II?

c. What proportion of individuals 1 through 4 of generation III would be expected to be homozygous for the trait?
d. What are the genotypes of individuals 5 through 11 in generation III?
e. What proportion of the eggs of female 11 (generation III) will carry the gene for this trait?
f. If individual 7 of generation III were to marry individual 11 of generation III, what proportion of their children would show the trait?

1–12. Examine the pedigree shown below.

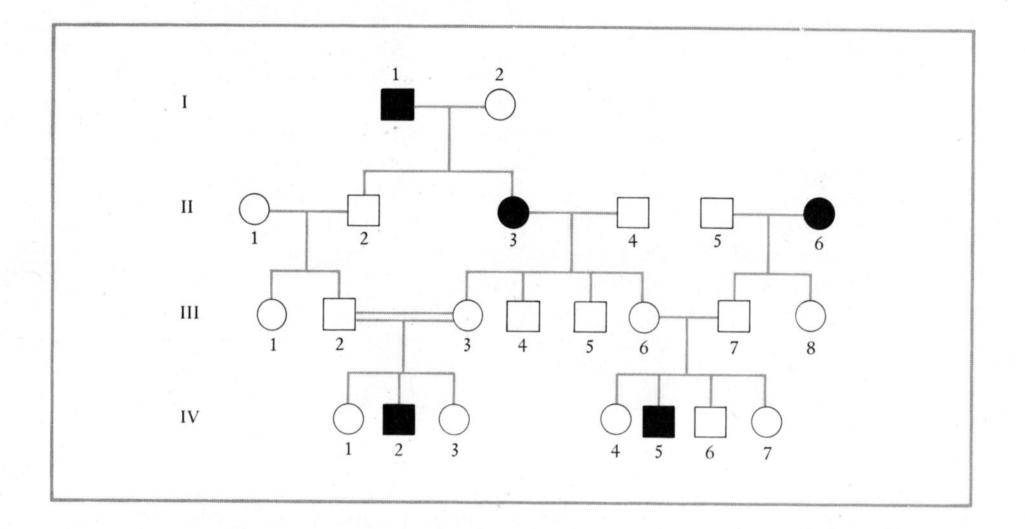

a. Is the trait illustrated in the pedigree dominant or recessive?
b. Using *A* and *a* as gene symbols, what are the genotypes of:
   individuals 1 and 2 of generation I
   individuals 1, 2, 3, and 4 of generation II
   individuals 2, 3, 6, and 7 of generation III
c. What type of cross is represented by the marriage between:
   individuals 1 and 2 of generation I
   individuals 3 and 4 of generation II
   individuals 6 and 7 of generation III

REFERENCES

Dunn, L. C., 1965. *A Short History of Genetics*. McGraw-Hill, New York.

Iltis, H., 1924. *Gregor Mendel: Leben, Werk und Wirkung,* Springer, Berlin. (English translation by E. Paul and C. Paul, 1932. *The Life of Mendel*. Norton, New York.)

McKusick, V., 1969. *Human Genetics,* 2nd ed. Foundations of Modern Genetics Series, Prentice-Hall, Englewood Cliffs, N.J.

Mendel, G., 1866. Versuche über Pflanzen Hybriden. *Verh. naturf. Ver. in Brünn, Abhandlungen, iv*. (English translation in J. A. Peters (Ed.), 1959. *Classic Papers in Genetics*. Prentice-Hall, Englewood Cliffs, N.J.)

Thompson, J. S., and M. W. Thompson, 1973. *Genetics in Medicine,* 2nd ed. Saunders, Philadelphia.

Whitehouse, H. L. K., 1973. *Towards an Understanding of the Mechanism of Heredity,* 3rd ed. St. Martin's Press, New York.

Chapter 2

MENDEL'S SECOND LAW

In the preceding chapter we examined the rules which govern the inheritance of a single pair of contrasting characters. The genotype of any organism contains many genes, however, and if we are to formulate general principles of heredity, we must take into consideration any relationships which govern the combined inheritance of more than one pair of alleles. Mendel was aware of this requirement and his studies did not end with verification of the Law of Segregation. Having discovered the mode of inheritance for a single pair of contrasting characters, he proceeded to apply his first law to the simultaneous inheritance of two pairs of characters and, eventually, even of three pairs of characters.

EXPERIMENTS WITH DIHYBRIDS AND DERIVATION OF THE LAW OF INDEPENDENT ASSORTMENT

For his contrasting phenotypes Mendel chose round versus wrinkled seeds and yellow versus green cotyledons. His previous work had already established that each inbred line was homozygous and that round was dominant over wrinkled and yellow dominant over green. He symbolized the genes responsible for these characters as follows: *A* for round and *a* for wrinkled seeds, *B* for yellow and *b* for green cotyledons. He began his experiments with a P_1 cross of round, yellow plants × wrinkled, green plants or *AABB* × *aabb*. The seeds resulting from this cross-fertilization represented the F_1 generation. They could be classified as to phenotype by observing the nature of the seed coat and the color of the seed. Mendel found that all F_1 individuals were phenotypically round, yellow, showing both dominant characters. He then permitted the F_1 plants to grow to maturity and self-fertilize to produce the seeds of the F_2 generation. These seeds were then examined for form and color. Tabulation of the data revealed an F_2 generation composed of 556 individuals which were distributed in four phenotypic classes according to the following approximate proportions: $\frac{9}{16}$ round, yellow; $\frac{3}{16}$ round, green; $\frac{3}{16}$ wrinkled, yellow; and $\frac{1}{16}$ wrinkled, green. The phenotypic ratio of these four classes was thus 9 : 3 : 3 : 1 (Figure 2–1).

If we analyze these data making use of our knowledge of biparental inheritance, segregation, and dominance, some facts are immediately clear. The round, yellow individuals of the F_1 generation were heterozygous for both pairs of characters, that is, they were *dihybrids*. Their genotype must have been *AaBb,* the presence of the recessive genes *a* and *b* being masked by the dominant alleles *A* and *B*.

If we apply the Law of Segregation to each pair of characters separately, a 3 : 1 phenotypic ratio can be distinguished for the proportion of dominant to recessive phenotypes in the F_2 generation. To illustrate, if we consider only the character of round versus wrinkled seed coat, ignoring cotyledon color, it is clear that the number of round seeds in the F_2 generation is 315 + 108, a total of 423. The number of wrinkled seeds is 101 + 32, a total of 133. This proportion of 423 round : 133 wrinkled is a 3 : 1 phenotypic ratio, and with complete dominance, such a ratio indicates an underlying 1 : 2 : 1 genotypic ratio. The recessive phenotype of wrinkled shows that members of this class are genotypically *aa*. On the basis of a 1 : 2 : 1 genotypic ratio seeds with the dominant round phenotype should fall into two genotypic classes: one-third should be homozygous *(AA)* and two-thirds heterozygous *(Aa)* in genotype.

P_1 Cross: round, yellow × wrinkled, green

F_1 Offspring: all round, yellow

F_1 Inbreed: $F_1 \times F_1$

F_2 Offspring:

Phenotypic Classes	Proportion
315 round, yellow	9/16
108 round, green	3/16
101 wrinkled, yellow	3/16
32 wrinkled, green	1/16
556 Total F_2 progeny	

Phenotypic Ratio: 9 : 3 : 3 : 1

Figure 2–1. A summary of the data from Mendel's experiments with round, yellow and wrinkled, green peas.

We can apply the same reasoning to the other pair of characters, yellow versus green cotyledons. Considering these characters apart from seed coat, it can be seen that in the F_2 generation 416 (315 + 101) yellow individuals and 140 (108 + 32) green individuals were produced. This proportion of 416 yellow : 140 green is also a 3 : 1 phenotypic ratio, from which a genotypic ratio of 1 : 2 : 1 can be inferred. Based on this ratio we can expect that phenotypically yellow seeds will be of two classes: one-third should be homozygous *(BB)* in genotype and two-thirds should be heterozygous *(Bb)* in genotype. The green individuals should be homozygous *bb*.

Consideration of each of these pairs of characters separately indicates that each is obeying the Law of Segregation independently of the other. We can therefore conclude that the overall F_2 phenotypic ratio of 9 : 3 : 3 : 1 must represent a combination of these two separate and independent 3 : 1 ratios.

The source of the F_2 9 : 3 : 3 : 1 phenotypic ratio must reside in the kinds of gametes produced by members of the F_1 generation, as well as in the subsequent fertilizations which occur between these gametes. If so, the types of eggs and pollen formed by the F_1 hybrids *(AaBb)* should be our next consideration. Applying the Law of Segregation, the 3 : 1 phenotypic ratio for round versus wrinkled seeds tells us that the alleles *A* and *a* present in the heterozygote separated from one another at sex cell formation. As a result, two kinds of pollen and two kinds of eggs were produced in equal numbers, that is, those carrying allele *A* and those with allele *a*. Similarly, the 3 : 1 ratio for yellow versus green cotyledon color indicates that male and female gametes carrying the alleles *B* and *b* were also produced in equal numbers. Since biparental inheritance is the rule and members of the F_2 generation inherited both characters equally from their parents, we must assume that each gamete (pollen or egg) contained one member of each of these two pairs of alleles. In addition, random fertilization between these sex cells must also have occurred, as demonstrated by the fact that a 3 : 1 ratio for each pair of characters was obtained in the F_2 generation.

Taking all of these considerations into account, we can formulate the following hypothesis. At sex cell formation in the anthers as well as in the ovaries of any flower, the segregation of alleles *A* and *a* into separate gametes occurs independently of the segregation of the alleles *B* and *b*. As a result, the gametes formed contain all possible combinations of these alleles and consist of the following classes: *AB, Ab, aB,* and *ab*. In addition, these four types of gametes must be produced in equal numbers, as demonstrated by the fact that 3 : 1 ratios for each pair of characters was obtained. Finally, fertilization between these four types of eggs and four types of pollen must be random.

If this hypothesis is correct, we would predict that all possible combinations between the four types of eggs and pollen will occur. The 16 possible genotypic combinations are listed in the checkerboard, or Punnett Square, shown in Figure 2–2.

9 with both dominant alleles *A* and *B*; round, yellow in phenotype
3 with a dominant allele for round *(A)* and homozygous for the recessive allele for green *(bb)*; round, green in phenotype

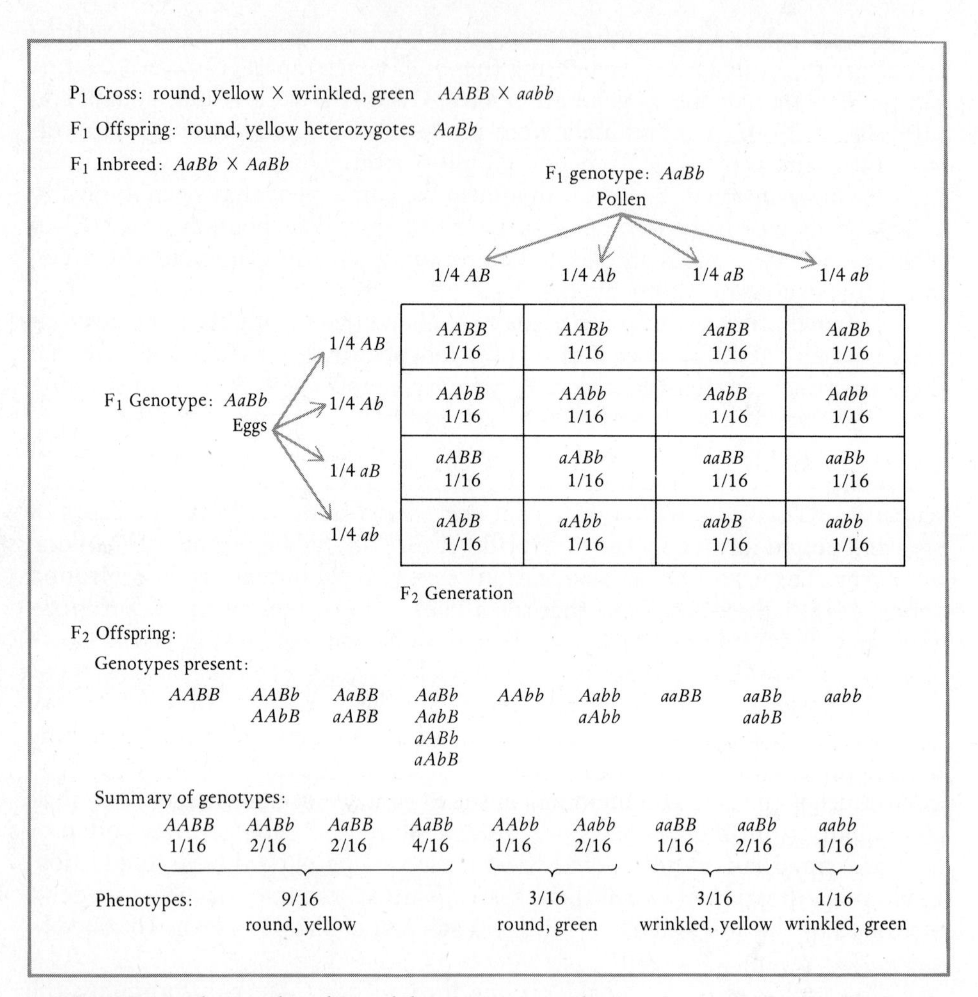

Figure 2–2. The results of Mendel's crosses according to the principles of segregation and independent assortment.

3 homozygous for wrinkled *(aa)* and with a dominant allele for yellow *(B)*; wrinkled, yellow in phenotype

1 homozygous for wrinkled and green (*aa* and *bb*); wrinkled, green in phenotype

Thus, the predictions made according to our hypothesis coincide with the data of Mendel's experiment.

However, all hypotheses must be subjected to experimental proof before they can be accepted. Here, such proof requires that the classes of genotypes actually obtained in the F_2 generation conform to the expectations of the hypothesis. By referring to Figure 2–2, we can count up the kinds of expected F_2 genotypic classes and their proportions. They are: $\frac{1}{16}$ *AABB* : $\frac{2}{16}$ *AABb* : $\frac{2}{16}$ *AaBB* : $\frac{4}{16}$ *AaBb* : $\frac{1}{16}$ *AAbb* : $\frac{2}{16}$ *Aabb* : $\frac{1}{16}$ *aaBB* : $\frac{2}{16}$ *aaBb* : $\frac{1}{16}$ *aabb*. Only one of these genotypic classes, the double recessive *aabb*, can be positively identified by phenotype, and it occurs in the expected frequency according to our

Postulated F_2 Genotypes	Predicted Phenotypes of F_3 Progeny from Self-Fertilization of F_2 Individuals
1/16 *AABB*	all both *A* and *B*
2/16 *AABb*	all *A*; 3 : 1 ratio for *B* and *b*
2/16 *AaBB*	3:1 ratio for *A* and *a*; all *B*
4/16 *AaBb*	9:3:3:1 ratio, as in F_2 generation
1/16 *AAbb*	all *A*; all *b*
2/16 *Aabb*	3:1 ratio for *A* and *a*; all *b*
1/16 *aaBB*	all *a*; all *B*
2/16 *aaBb*	all *a*; 3:1 ratio for *B* and *b*
1/16 *aabb*	all both *a* and *b*

Figure 2–3. Predicted phenotypes of F_3 individuals resulting from self-fertilization of F_2 progeny. The F_3 phenotypes provide proof of genotype of the F_2 plants. In Mendel's experiment the proportion of F_2 plants giving rise to these respective phenotypes agreed with the postulated F_2 genotypic ratio.

hypothesis. The genotypes of the other classes can be only partially verified through phenotype, and thus additional crosses are needed to determine their nature. Mendel obtained this information through the progeny produced by inbreeding and backcrossing.

He allowed F_2 plants to self-fertilize and produce a third generation. The phenotypes of the F_3 seeds indicated the genotypes of their parents (Figure 2–3). For example, an F_2 plant that was genotypically *AaBB* would be expected, by selfing, to produce all yellow progeny in which the round versus wrinkled character assorted in a 3 : 1 ratio. An F_2 plant that was genotypically *AaBb* would be expected to produce progeny in a ratio of 9 : 3 : 3 : 1. An F_2 plant that was *AAbb* should yield offspring all of whom would be round and green in phenotype. On the basis of these kinds of data, Mendel was able to demonstrate that the F_2 generation contained the genotypic frequencies predicted by the hypothesis.

As additional proof, Mendel also performed reciprocal backcrosses between F_1 hybrids and the respective P_1 parental lines. We would predict that a backcross to the homozygous dominant parent (F_1 *AaBb* × *AABB*) would produce offspring all of whom should show both dominant phenotypes. They would, however, be distributed equally among four genotypic classes: *AABB*, *AaBB*, *AABb*, and *AaBb*. Mendel verified the fact that these four genotypes were present in equal numbers by allowing self-fertilization of the backcross progeny and examination of the phenotypes of the resulting offspring (Figure 2–4).

The backcross to the recessive parental type (F_1 *AaBb* × *aabb*) was even more revealing in that, as expected from such a testcross, the phenotypic and genotypic ratios coincided (Figure 2–5). Thus, the cross produced four phenotypic and corresponding genotypic classes in a 1 : 1 : 1 : 1 ratio. These

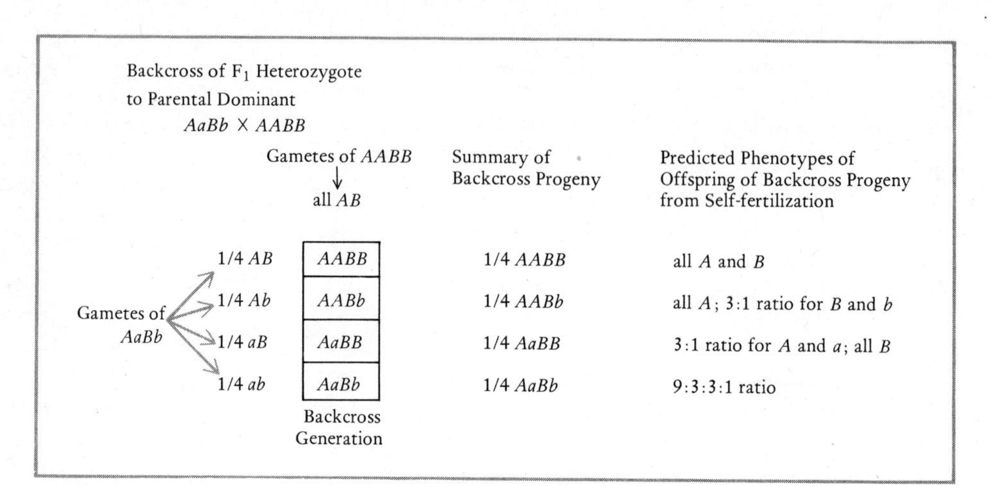

Figure 2–4. Backcross of F_1 heterozygote *(AaBb)* to parental dominant variety *(AABB)* and verification of F_1 genotype. The progeny from this backcross were allowed to self-fertilize. The phenotypes of the resulting offspring provided proof of genotype for the backcross progeny and demonstrated the 1 : 1 : 1 : 1 genotypic ratio arising from the backcross. These results permitted verification of the doubly heterozygous nature of the F_1 individuals.

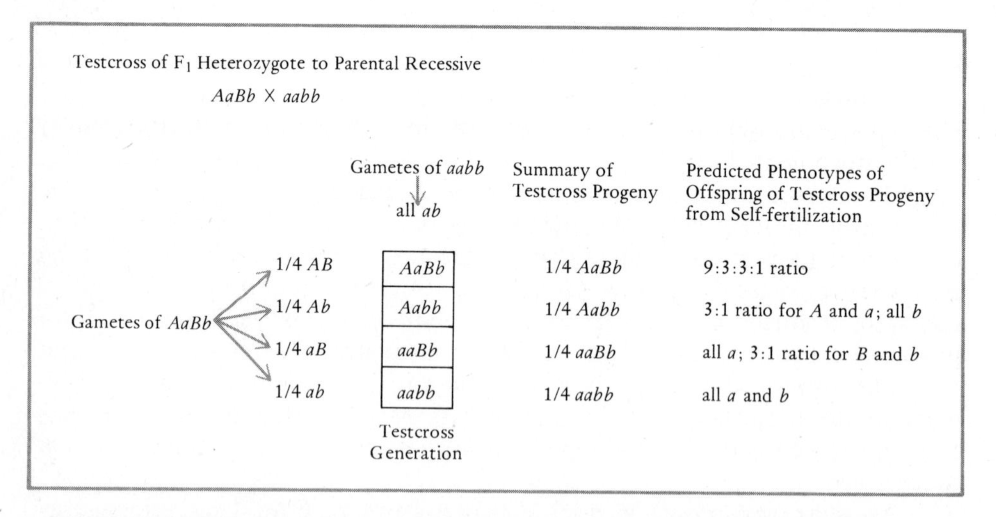

Figure 2–5. Testcross of F_1 heterozygotes *(AaBb)* to the parental double recessive variety *(aabb)* and verification of F_1 genotype. Although the phenotypes of the testcross progeny, as well as the 1 : 1 : 1 : 1 phenotypic ratio, provide verification of the genotype of F_1 individuals, an additional generation obtained by self-fertilization was studied by Mendel. Results from this generation supported the interpretation based on the results of the testcross.

classes were round, yellow *(AaBb)*, wrinkled, yellow *(aaBb)*, round, green *(Aabb)*, and wrinkled, green *(aabb)*. Their presence established the genotype of the F_1 individuals used in the cross. Mendel also subjected these testcross progeny to a further assay of genotype by permitting self-fertilization to produce an additional, verifying generation. Thus, our hypothesis can be accepted in view of the supporting evidence obtained in these confirming experiments.

This hypothesis is known as Mendel's Second Law or the Law of Independent Assortment. It can be simply stated as follows. In sex cell formation the segregation of one pair of alleles is independent of the segregation of any other pair of alleles, and therefore gametes containing all possible combinations of these alleles will be produced in equal numbers. It must be emphasized that Mendel derived this principle entirely through logical deductions from his data and no recourse to chromosomes or to DNA was then, or is now, necessary. We have, of course, since learned that pairs of genes which exhibit this behavior are almost always located on separate and different chromosomes and that exceptions to Mendel's Second Law are the basis for the recognition of linkage, that is, where the alleles in question are located on the same chromosome.

Mendel combined his studies on segregation and independent assortment into a single paper which was published in 1866 in the *Proceedings of the Brünn Natural Science Society*. Unfortunately, his results were not received with enthusiasm nor were they widely read, and it is doubtful that any of his contemporaries appreciated the significance of his work. The use of mathematics in biological studies was an innovation considered to be of at best questionable value. Thus it was that Mendel's work lapsed into obscurity, and it was not until 1900 that his principles of inheritance were rediscovered and confirmed by three independent investigators, Hugo De Vries, Carl Correns, and Erich von Tschermak-Seysenegg. Only then was the science of genetics finally launched.

Beginning in 1900 genetics became an intensely active field in which the principles governing the transmission of traits from one generation to the next were confirmed in a variety of plants as well as in such animals as the fowl and the laboratory mouse. It soon became recognized that Mendel's rules were universally applicable to organisms that reproduced by the union of sex cells.

Inbred lines of plants were the subjects of most of these early studies, and they offered the advantage that many seeds, each representing a new individual, were produced in a generation. As an example, the corn plant *(Zea mays)* has been a favorite object of study for many years not only because of its economic importance, many variable inherited characters, and favorable cytology, but also because an ear of corn contains several hundred kernels, each one an offspring of the cross. Thus, with only a few ears of corn a large population may be scored.

Animals are more difficult subjects in that even the small rodents, such as mice, are expensive to house and feed, have a long life cycle, and produce relatively few offspring. One animal that lacks these difficulties, however, is *Drosophila melanogaster* (Figure 2–6), a small fly introduced as an experimental organism in 1905. *Drosophila* species offer numerous advantages. Very large numbers can be raised cheaply and in a small space, the life cycle from egg to adult spans less than two weeks, and a single female can produce several hundred offspring. Added to these attractions is the fact that various aspects of the body, such as legs, wings, bristles, eyes, and color, are subject to many inherited changes whose transmission can be followed through breeding experiments. As a final and unique advantage, the immature stages of *Drosophila* contain cells with gigantic chromosomes that offer unparalleled opportunities

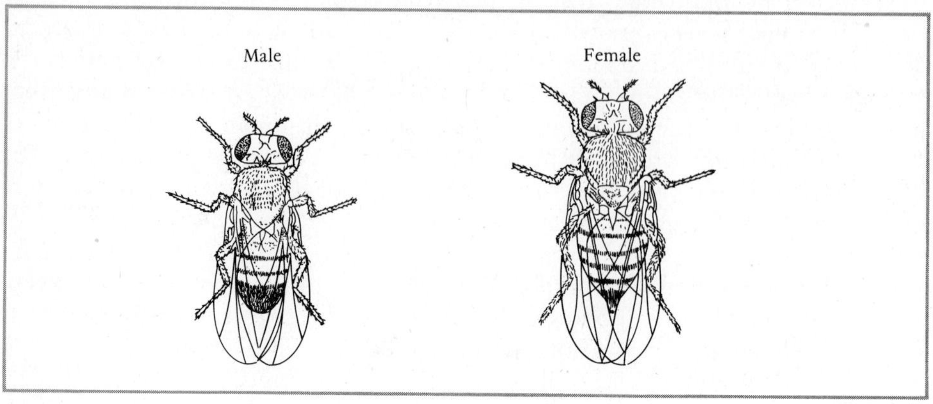

Male Female

Figure 2–6. Adult male and female *Drosophila melanogaster.* [From T. H. Morgan, *The Theory of the Gene,* 2nd ed., Yale University Press, New Haven, Connecticut, 1928.]

to associate hereditary changes with chromosome structure. It is no wonder that this tiny insect has been exploited so thoroughly in studies of inheritance.

As we will see, geneticists have also employed a number of other organisms that offer particular advantages for certain types of investigations. For example, the desire to associate gene action more immediately with cell phenotype has led to the use of yeasts, molds, algae, and other forms that lack the structural complexity of higher plants and animals. In addition, the huge populations obtainable in a very short time with bacteria and viruses have made possible the identification and analysis of rare genetic events.

THREE OR MORE PAIRS OF GENES

Although Mendel's principle of independent assortment has been discussed only in terms of two pairs of characters, it can be applied to three or as many more traits as are inherited in a truly independent fashion. Indeed, Mendel himself tested his rule in crosses in which three pairs of characters were assorting. We will use three such pairs of independent alleles to illustrate some shortcuts that can be employed when several traits must be considered at once.

When three or more pairs of alleles are involved, the number of kinds of gametes that can be produced by a heterozygote increases, as does the number of phenotypes and genotypes that can be expected from an inbreed or a testcross. As an example, let us consider three hypothetical dominant genes *A, B,* and *C* and their respective recessive alleles *a, b,* and *c*. To obtain individuals heterozygous for all three pairs of alleles, any one of the following initial P_1 crosses can be made.

P_1 Crosses	F_1 Trihybrids
$AABBCC \times aabbcc$	$AaBbCc$
$AABBcc \times aabbCC$	$AaBbCc$
$AAbbcc \times aaBBCC$	$AaBbCc$
$AAbbCC \times aaBBcc$	$AaBbCc$

An identical F_1 heterozygote will be produced by each of these crosses, and it is not necessary that all dominant or all recessive alleles be present in the same parental individual. Because biparental inheritance is the rule, the dominant or the recessive allele of any pair of genes can be inherited from either parent.

To determine the kinds of gametes that would be produced by one of these F_1 heterozygotes, or trihybrids, we must apply the principles of segregation and independent assortment. We already know that an individual heterozygous for 1 pair of alleles forms 2 types of sex cells in equal numbers and that an individual heterozygous for 2 pairs of alleles produces 2^2 or 4 kinds of gametes, again in equal numbers. Our present heterozygote has 3 pairs of contrasting alleles and thus forms 2^3 or 8 different kinds of gametes in equal numbers. Using n to stand for the number of independent gene pairs involved, we can employ the simple expression 2^n to determine the number of gamete types that any heterozygote can produce (Table 2–1). Accordingly, the 8 types of male or female gametes that would be formed by our trihybrid are:

1. *ABC*
2. *ABc*
3. *Abc*
4. *AbC*
5. *abc*
6. *abC*
7. *aBC*
8. *aBc*

If an inbreed between two such trihybrids is performed, random fertilization between these male and female sex cells will produce 64 possible combinations. Although these could be derived from a checkerboard of 64 squares,

TABLE 2-1
Relationship Between the Number of Gene Pairs and the Number of Kinds of Gametes That Can Be Produced by a Heterozygote*

Number of gene pairs	Number of kinds of gametes formed by each sex	Number of gamete combinations produced by random fertilization	Number of phenotypic classes in the progeny	Number of genotypic classes in the progeny
1	2	4	2	3
2	4	16	4	9
3	8	64	8	27
4	16	256	16	81
5	32	1,024	32	243
6	64	4,096	64	729
7	128	16,384	128	2,187
8	256	65,536	256	6,561
9	512	262,144	512	19,683
10	1,024	1,048,576	1,024	59,043
n	2^n	4^n	2^n	3^n

* The number of gamete combinations possible through fertilization and the number of phenotypic and genotypic classes which will appear as the result of an inbreed between heterozygotes are also listed. The number of possible phenotypic classes is based on the assumption that complete dominance occurs for all gene pairs.

the total number of possible gamete combinations can be determined more easily by applying the simple expression 4^n, where n equals the number of pairs of alleles involved in the cross. The expression is derived from the fact that a monohybrid inbreed yields 4 combinations: *AA, Aa, aA,* and *aa* (four squares in the checkerboard). When 2 independent pairs of alleles are involved, the number of possible gamete combinations is 4^2 or 16, and when 3 pairs of alleles are involved, it is 4^3 or 64. We could, of course, also calculate the total number of gamete combinations by squaring the number of kinds of gametes, once these have been determined (for example, in a trihybrid 8^2 or 64).

The number of genotypes resulting from such a cross could be obtained either by preparation of the checkerboard or, preferably, by applying another simple formula. As an illustration, a monohybrid inbreed produces 3 genotypes (*AA, Aa,* and *aa*) from the random union of gametes. An inbreed between dihybrids produces 2 separate and independent 1 : 2 : 1 genotypic ratios simultaneously in the progeny. The resulting overall F_2 genotypic ratio contains 9 classes which are the product of 2 separate 1 : 2 : 1 ratios composed of 3 terms each (3^2). Similarly, the genotypic classes produced by a trihybrid inbreed represent 3 independent 1 : 2 : 1 genotypic ratios occurring simultaneously, and the number of resulting classes would be 3^3 or 27. We can thus determine the total number of different genotypes produced by any inbreed of heterozygotes where the gene pairs are assorting randomly by the expression 3^n, where n equals the number of allelic pairs involved (Table 2–1).

The number of phenotypic classes expected from an inbreed depends on whether or not the alleles involved exhibit complete or incomplete dominance and also on whether or not they affect the same or different traits. Let us put aside the latter possibility for the moment and consider only dominance relationships. If complete dominance occurs with every gene pair, then the number of phenotypic classes can be predicted by the expression 2^n, where n equals the number of allelic pairs involved. Since with complete dominance, 1 pair of genes yields 2 phenotypic classes (3 : 1 ratio), the simultaneous assortment and random combination of 2 independent pairs of alleles produces 2^2 or 4 different phenotypic classes (9 : 3 : 3 : 1 ratio). Our trihybrid inbreed would thus be expected to yield a phenotypic ratio containing 2^3 or 8 terms in the ratio of 27 : 9 : 9 : 9 : 3 : 3 : 3 : 1.

If all the alleles involved are codominant or incompletely dominant (see Chapter 4), more classes would result, because the heterozygotes would be phenotypically distinguishable, and the phenotypic ratio would be the same as the genotypic ratio. The phenotypic ratio resulting from a cross involving some dominant and some codominant gene pairs would reflect these differences. In general, the number of phenotypic classes produced will always depend on the expression of the particular alleles involved in the cross.

An inspection of Table 2–1 also permits some generalizations as to the number of offspring which might be required to realize any given ratio or to obtain individuals of some specified genotype. As can be seen, the number of different kinds of gametes and their combinations, as well as the possible genotypic and phenotypic classes, rises dramatically with the involvement of increasing numbers of independent gene pairs. With only 4 gene pairs, 256 different combinations of gametes are possible and of these only 1 will be homozygous for all 4 dominant alleles and 1 homozygous for all 4 recessive alleles. With 8 independent pairs of genes, 65,536 gamete combinations are possible and, again, only 1 of these will be homozygous for all recessive alleles and only 1 homozygous for all dominant alleles. With odds such as these, the size of an F_2 population would have to be enormous in order to have the slightest chance of recovering either homozygote.

The example of 4 gene pairs is useful as an illustration. Here, the odds on recovering an individual homozygous for all recessive alleles are 1 in 256. We could increase this chance significantly if we were to increase our F_2 population sixfold to around 1500 individuals. We would be almost sure of recovering the desired homozygote if the F_2 population were even greater, say, 3000 or higher. Although this seems a large number, it is not difficult to achieve when the organism investigated produces many seeds or is as easy to culture as is *Drosophila*. It should be noted, however, that if the purpose of the experiment was to recover the homozygous recessive type, an inbreed would not be the cross of choice. Instead, a testcross would be used. With 4 independent pairs of alleles, a testcross yields 2^4 or 16 phenotypic and genotypic classes, 1 of which is homozygous for all 4 pairs of alleles. As compared to the odds of 1 in 256 found in an inbreed, a testcross increases the chances of recovery to 1 in 16. However, regardless of the type of cross, the difficulties to be encountered in following the inheritance of many gene pairs at the same time are readily apparent. It must be emphasized that the likelihood of realization of any ratio or the recovery of any infrequent class depends upon the size of the sample population.

Although the independent assortment of genes is observed only with allelic pairs that are located on separate and different chromosomes, Table 2–1 is a vivid illustration of the enormous variation in genotype that can be achieved in a population by this means alone. In humans, where 23 pairs of chromosomes occur, independent assortment can involve 23 pairs of alleles. In a marriage between persons heterozygous for only 15 such gene pairs, a child could inherit any 1 of over *14 million* different genotypes. If such persons were heterozygous for 23 pairs of independent genes, over *94 billion* different genotypes would be possible. Truly it can be said that, with the exception of identical twins, each of us is genetically unique.

PROBLEMS

2–1. Most dogs' eyes reflect a yellowish green light in the dark. However, when homozygous for a recessive gene called ruby eye *(e)*, the light reflected is red. A dominant gene *A* results in black hair, while the recessive allele a^t produces a black and tan pattern when homozygous. A black and tan female crossed to a black male produced nine pups, all black. In the light of a flashlight the eyes of the mother and four of her puppies shone red.

a. What is the genotype of the father with respect to reflectance of color?

b. What is his genotype with respect to coat color?

2–2. In barley orange seedlings *(or)* is recessive to green seedlings *(Or)*, and waxy endosperm *(wx)* is recessive to starchy endosperm *(Wx)*. Barley is self-fertilizing.

A cross between a homozygous orange, waxy plant and a homozygous green, starchy plant produced green, starchy F_1 progeny.

a. If some F_1 individuals are allowed to self-fertilize, what phenotypes will be found in the F_2 generation and in what proportions?
b. If other F_1 individuals are crossed to orange, waxy plants, what phenotypes would be found in their progeny? What genotypes would be found? What would be the proportion of each in the progeny?

2–3. A cross between a homozygous orange, starchy strain of barley and a homozygous green, waxy variety produced F_1 progeny that were green, starchy in phenotype. The F_1 were allowed to self-fertilize and produce an F_2 generation. Some of these F_2 progeny self-fertilized to produce the F_3 individuals listed below.

F_2 PHENOTYPE	F_3 PROGENY
(1) green, starchy	42 green, starchy; 14 green, waxy
(2) green, waxy	76 green, waxy
(3) orange, starchy	58 orange, starchy; 20 orange, waxy
(4) orange, starchy	85 orange, starchy
(5) green, starchy	102 green, starchy
(6) green, starchy	70 green, starchy; 21 orange, starchy
(7) green, waxy	47 green, waxy; 16 orange, waxy
(8) green, starchy	63 green, starchy; 19 orange, starchy; 22 green, waxy; and 7 orange, waxy

a. What are the genotypes of each self-fertilizing F_2 plant listed above?
b. What kind(s) of pollen are produced by plant 3 and in what proportions?
c. What kind(s) of eggs are produced by plant 8 and in what proportions?

2–4. In coffee plants the ability to flower continuously *(f)* is recessive to seasonal flowering *(F)*, and noncrinkled leaves *(c)* is recessive to crinkled leaves *(C)*. A continuous flowering variety with normal leaves was crossed to a seasonal flowering strain with crinkled leaves. The F_1 plants all flowered seasonally and all had crinkled leaves. Cross fertilization between the F_1 plants produced the following progeny.

continuous flowering, crinkled
continuous flowering, noncrinkled
seasonal flowering, crinkled
seasonal flowering, noncrinkled

a. What are the genotypes of the P_1 and F_1 individuals?
b. If the F_2 generation consisted of a total of 96 plants, approximately how many plants would be expected to occur in each F_2 phenotypic class?
c. How many plants would be expected to be heterozygous for both characters?
d. How many would be expected to be homozygous for continual flowering and heterozygous for crinkled leaves?
e. How many would be expected to be homozygous for seasonal flowering and homozygous for noncrinkled leaves?

2–5. In horses the trotting gait *(T)* is dominant over the pacing gait *(t)*, and bay color *(B)* is dominant over black *(b)*. A prize bay trotter stallion was crossed to two mares. Mare 1, a black trotter, produced a black trotter colt. Mare 2, a bay trotter, produced a bay pacer colt. What was the genotype of the stallion?

2–6. In doves a white iris of the eye *(w)* is recessive to an orange iris *(W)*, while barless *(b)* is recessive to the presence of transverse black bars *(B)* on the wings. Barred, white iris doves were crossed to nonbarred, orange iris individuals. The F_1 consisted of the following types.

barred, orange
barred, white
nonbarred, orange
nonbarred, white

a. What are the genotypes of the parents?
b. What proportion of the progeny would be expected to occur in each phenotypic class?
c. If the barred, orange offspring were bred together, what proportion of their progeny would be expected to be barred, white in phenotype?
d. What proportion would be barred, orange in phenotype?
e. What proportion would be nonbarred, orange in phenotype?

In Problems 2–7 through 2–10 *A, B, C,* and *D* are dominant over *a, b, c,* and *d,* respectively.

2–7. In the cross *AaBBCcdd* × *AabbccDd*
a. What proportion of the progeny would be homozygous *dd*?
b. What proportion of the progeny would show *B* in their phenotype?

2–8. In the cross *AaBbCcDd* × *AaBbCcDd*
a. What proportion of the progeny would be expected to be *AABBCCDD* in genotype? In phenotype?
b. What proportion of the progeny would be expected to be *aabbccdd* in genotype? In phenotype?
c. How many kinds of gametes would be produced by each parent?
d. How many phenotypes are expected in the progeny?
e. How many genotypes are expected in the progeny?

2–9. In the cross *AaBbCcdd* × *AaBbCcDD*
a. How many phenotypes are expected in the progeny?
b. How many genotypes are expected in the progeny?

2–10. In the cross *AaBbCcDd* × *aabbccdd*
a. What proportion of the progeny will be homozygous for both *a* and *b*?
b. What proportion will be homozygous for *c* and *d*?
c. What proportion will be homozygous for *a, b,* and *d*?
d. How many phenotypes are expected in the progeny?
e. How many genotypes are expected in the progeny?

2–11. In *Drosophila* ebony *(e)* is a recessive mutant causing black body color, as opposed to the normal, dominant tan color *(E)*. A recessive mutant, dumpy *(d)*, causes reduced wings, as opposed to normal long wings *(D)*. A tan female with normal wings is crossed to a tan male with dumpy wings. The progeny consist of:

41 tan, long wing
44 tan, dumpy wing
17 ebony, dumpy wing
15 ebony, long wing

a. What are the genotypes of the parents?
b. What kinds of eggs and in what proportions are produced by the female parent?
c. What kinds of sperm and in what proportions are produced by the male parent?

d. What is the most likely ratio of tan to ebony in the progeny?
e. What is the most likely ratio of long to dumpy wings in the progeny?
f. What genotypes would be found in each class of progeny?

2–12. Of the tan progeny of the preceding question
a. What proportion would be homozygous for tan?
b. What proportion would be heterozygous for tan?
c. If it were necessary to distinguish between these genotypes, how would you go about it?

The following is the key for the human pedigrees of the next three questions.

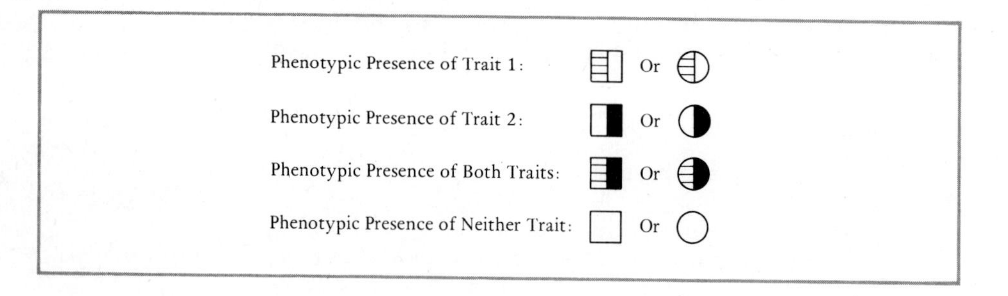

2–13. Examine the pedigree shown below.

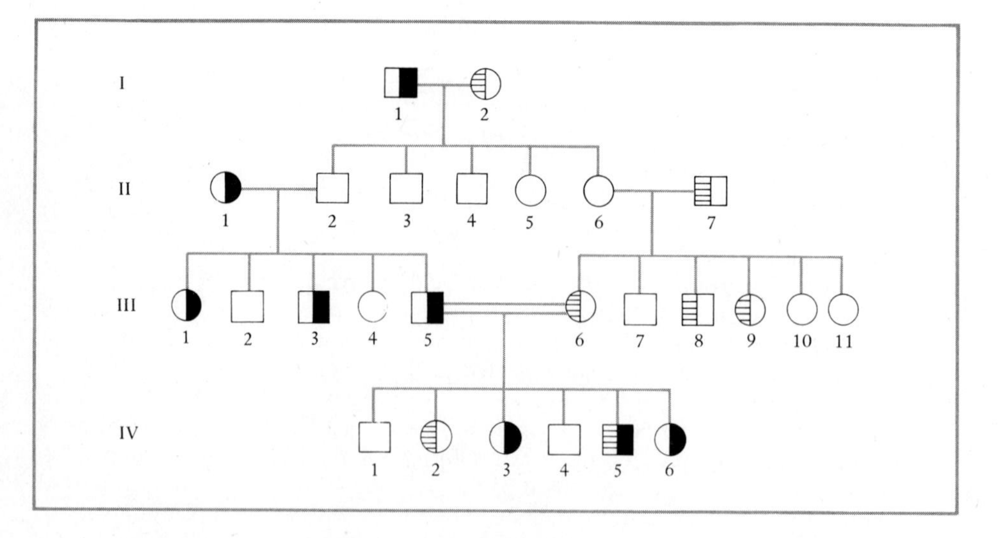

a. Are traits 1 and 2 dominant or recessive?
b. Using *A* and *a* for trait 1 and *B* and *b* for trait 2, what are the genotypes of individuals 1 and 2 of generation I?
c. What are the most probable genotypes of individuals 1, 2, 6, and 7 of generation II, individuals 5 and 6 of generation III, and all the individuals in generation IV?

2–14. Examine the pedigree.

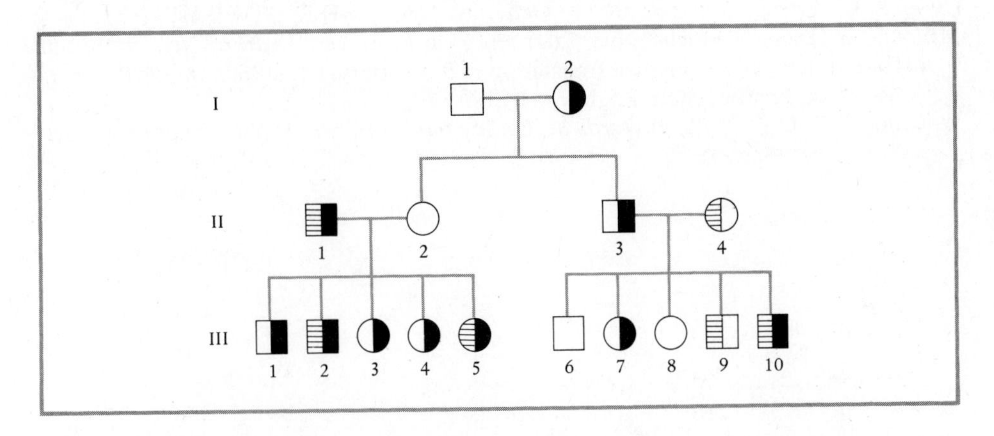

a. Are traits 1 and 2 dominant or recessive?
b. Using A and a for trait 1 and B and b for trait 2, what are the genotypes of all of the individuals in the pedigree?
c. If individual 10 of Generation III marries a woman of identical genotype, what proportion of their children will show trait 1, trait 2, and both traits?
d. If individual 10 of generation III marries a woman showing neither trait, what proportion of the children will show trait 1, trait 2, and both traits?

2–15. Examine the pedigree.

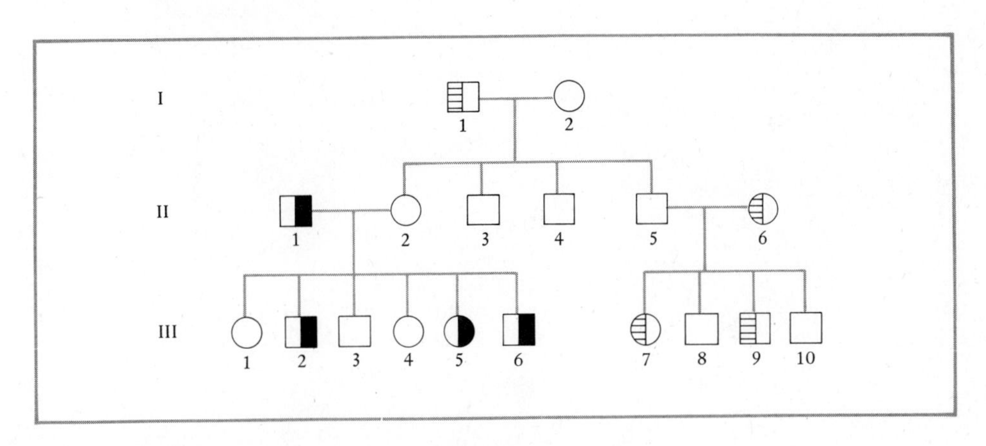

a. Are traits 1 and 2 dominant or recessive?
b. Using A and a for Trait 1 and B and b for trait 2, what are the genotypes of

    individuals 1 and 2 of generation I
    individuals 1 and 2 of generation II
    individuals 5 and 6 of generation II
    individuals 7 and 8 of generation III

c. If a marriage between individuals 6 and 7 of generation III occurred, what proportion of the children would show trait 2?
d. Under what circumstances would their children show trait 1?

REFERENCES

DUNN, L. C., 1965. *A Short History of Genetics*. McGraw-Hill, New York.

MENDEL, G., 1866. Versuche über Pflanzen Hybriden. *Verh. naturf. Ver. in Brünn, Abhandlungen, iv*. (English translation in J. A. Peters (Ed.), 1959. *Classic Papers in Genetics*. Prentice-Hall, Englewood Cliffs, N.J.)

WHITEHOUSE, H. L. K., 1973. *Towards an Understanding of the Mechanism of Heredity*, 3rd ed. St. Martin's Press, New York.

Chapter 3

PREDICTIONS: PROBABILITY, BINOMIAL THEOREM, AND THE CHI-SQUARE TEST

ALTHOUGH checkerboards can be used to determine all genotypes resulting from any cross, their preparation is laborious, time consuming, and subject to inadvertent error, especially when three or more pairs of independent genes are involved. As we have seen, the results of a dihybrid cross are diagramed in a checkerboard of 16 squares, and a cross between individuals heterozygous for three pairs of genes requires 64 squares. Clearly, a convenient, shorthand method for determining the frequency of any genotypic or phenotypic class is desirable. Such a method is offered by the use of probabilities.

USE OF PROBABILITY FORMULATIONS TO PREDICT OUTCOME OF CROSSES

A female of the genotype *Aa* can produce two kinds of eggs in equal numbers, the frequency of these two types being $\frac{1}{2}A$ and $\frac{1}{2}a$. Based on this frequency, the probability that a type *A* egg will be involved in any particular instance of fertilization is $\frac{1}{2}$ and, likewise, the probability that a type *a* egg will be involved is also $\frac{1}{2}$. Similarly, a male of genotype *Aa* forms pollen (or sperm) in the frequencies of $\frac{1}{2}A$ and $\frac{1}{2}a$, and the probability of involvement in a given fertilization event is $\frac{1}{2}$ for each type. Since the encounter between egg and pollen is a random event, the chance that any given egg will be fertilized by any given pollen grain or sperm is the product of their separate probabilities. Thus, the chance that

egg *A* will be fertilized by pollen *A* is $\frac{1}{2} \times \frac{1}{2}$ or $\frac{1}{4}$
egg *A* will be fertilized by pollen *a* is $\frac{1}{2} \times \frac{1}{2}$ or $\frac{1}{4}$
egg *a* will be fertilized by pollen *A* is $\frac{1}{2} \times \frac{1}{2}$ or $\frac{1}{4}$
egg *a* will be fertilized by pollen *a* is $\frac{1}{2} \times \frac{1}{2}$ or $\frac{1}{4}$

In calculating these probabilities, we have once more derived our familiar genotypic ratio of $\frac{1}{4}AA : \frac{1}{4}Aa : \frac{1}{4}aA : \frac{1}{4}$aa or $\frac{1}{4}AA : \frac{1}{2}Aa : \frac{1}{4}aa$. We can also look at the ratio in a somewhat different fashion and conclude that the chance or probability of obtaining a heterozygote from any fertilization is $\frac{1}{2}$, while the chance that an individual will be either homozygous *AA* or *aa* is $\frac{1}{4}$. From the point of view of phenotype, we can also conclude that with complete dominance, the probability that a fertilization will result in an individual showing the dominant trait is $\frac{3}{4}$, while the probability of obtaining the homozygous recessive phenotype is $\frac{1}{4}$.

Such simple probability considerations can be used to make predictions in cases where two or more independent gene pairs are involved. In the cross $AaBb \times AaBb$, where genes *A* and *B* are dominant, we have seen that separate 1 : 2 : 1 genotypic and 3 : 1 phenotypic ratios can be observed in the progeny for either pair of alleles because the alleles are assorting independently of one another. Applying the rule that the chance of two independent events coinciding or occurring simultaneously is the product of their separate probabilities, we can predict the genotypes and phenotypes resulting from this cross without the use of a checkerboard. All that we must remember are the genotypic and phenotypic frequencies that would be expected from a similar cross involving either of the allelic pairs separately. These probabilities are then multiplied together to obtain the overall probability that such separate events will occur at the same time. In this way we can predict that from the cross $AaBb \times AaBb$ the proportion of the progeny that will show

both dominant traits is $\frac{3}{4} \times \frac{3}{4}$ or $\frac{9}{16}$
trait *A* dominant and trait *b* recessive is $\frac{3}{4} \times \frac{1}{4}$ or $\frac{3}{16}$
trait *a* recessive and trait *B* dominant is $\frac{1}{4} \times \frac{3}{4}$ or $\frac{3}{16}$
traits *a* and *b* both recessive is $\frac{1}{4} \times \frac{1}{4}$ or $\frac{1}{16}$

We can also determine the proportion of progeny expected to be of any particular genotype. For example, the proportion of offspring of genotype *AaBb* would be $\frac{1}{2} \times \frac{1}{2}$ or $\frac{1}{4}$; the proportion of genotype *AABb* would be $\frac{1}{4} \times \frac{1}{2}$ or $\frac{1}{8}$.

Probabilities can also be used to predict the outcome of other types of crosses, such as testcrosses or outcrosses. As an example, let us consider the cross $AaBBCc \times Aabbcc$. Considering the *A* and *a* alleles only, the cross is between heterozygotes and we can expect a genotypic ratio of 1 : 2 : 1 and a phenotypic ratio of 3 : 1. Considering the *B* and *b* alleles only, the cross is an outcross, and all progeny will be heterozygous and identical in phenotype and genotype. Considering the *C* and *c* alleles only, the cross is a testcross which will produce a 1 : 1 genotypic and phenotypic ratio. To predict the proportion of offspring from this cross showing the dominant phenotype for *A* and *B* and the recessive phenotype for *c*, our calculation would be: $\frac{3}{4} \times 1 \times \frac{1}{2}$ or $\frac{3}{8}$.

In summary, use of simple probability formulations is a valuable shortcut for predicting the outcome of genetic crosses. However, they can be used only when the genetic events under consideration are truly independent of one another and where the simultaneous occurrence of two or more such events is governed by random chance alone.

USE OF THE BINOMIAL THEOREM TO PREDICT THE OUTCOME OF CROSSES

As we have seen, the calculations of probabilities can be useful in predicting the frequency of different phenotypes or genotypes arising from a cross. However, in previous examples, probabilities were not employed to predict the chance that a particular combination of phenotypes would be the result of a specified number of fertilizations. Such considerations are of little importance in experimental genetics where large populations are often obtained. The experimental investigator is usually more interested in the number of individuals to be found in the different classes of offspring than in what might result from one, two, three, or more particular fertilizations.

However, the latter problem must be faced by the human geneticist confronted by anxious parents concerned about their future offspring. For example, suppose their questions concern the inheritance of the dominant trait polydactyly, *P* (extra fingers and toes), where one parent is heterozygous for the character *(Pp)* and the other parent is homozygous for the normal allele *(pp)*. The cross involved is thus $Pp \times pp$. The probability of a fertilization between gametes carrying the alleles *P* and *p*, respectively, is $\frac{1}{2}$ and the probability of a fertilization between gametes with the alleles *p* and *p*, respectively, is also $\frac{1}{2}$. Consequently, in a large number of fertilizations, we would expect to obtain a 1 : 1 ratio of normal *(pp)* to abnormal *(Pp)* offspring on the basis of chance. If the parents restrict their family to one child, the chances for a normal offspring ($\frac{1}{2}$) and the chances for an abnormal offspring ($\frac{1}{2}$) are equal.

Suppose the parents opt for two children and wish to know the chances that neither, one, or both will be abnormal. In this case we must consider the possible results of two separate fertilizations, each an independent event with no relationship to, or influence upon, preceding or succeeding fertilizations. Accordingly, the rule concerning probabilities can be applied, that is, the chance that two separate events will coincide is the product of their separate probabilities.

If the chance of a fertilization resulting in a normal child is $\frac{1}{2}$, then the probability of two such fertilizations is $\frac{1}{2} \times \frac{1}{2}$ or $\frac{1}{4}$. Likewise, the chance for two abnormal children is also $\frac{1}{2} \times \frac{1}{2}$ or $\frac{1}{4}$. However, there are two sequences of fertilization by which a child of each type could result: first, a normal child followed by an abnormal child or $\frac{1}{2} \times \frac{1}{2} = \frac{1}{4}$ and, second, an abnormal child followed by a normal child or $\frac{1}{2} \times \frac{1}{2} = \frac{1}{4}$. Since both of these sequences are possible, their separate probabilities must be added together to determine the total chance for obtaining, in two separate fertilizations, a child of each type, that is, $\frac{1}{4} + \frac{1}{4}$ or $\frac{1}{2}$.

If we substitute *a* for normal and *b* for abnormal, two normal children would be $a \times a$ or a^2, and two abnormal children would be $b \times b$ or b^2. The two combinations of a normal and an abnormal would be represented by $ab + ba$ or $2ab$. The equation is thus $a^2 + 2ab + b^2 = 1$. This equation is the expansion of the binomial $(a + b)^2$.

If the parents decide to risk three children, three separate and independent fertilizations are involved and the probabilities for all of the possible combinations of normal and abnormal children can be expressed by $(a + b)^3 = 1$. The three children resulting from the three fertilizations can be all normal, two normal and one abnormal, one normal and two abnormal, or all abnormal. The individual probabilities for each such combination of children can be found by expanding the binomial $(a + b)^3$, where a is $\frac{1}{2}$ and b is $\frac{1}{2}$.

If we let a stand for a normal child and b for an abnormal child, the chance that all three children will be normal is $a \times a \times a$ or a^3 and, similarly, the chance that all three will be abnormal is b^3. There are three sequences of births which will produce the combination of two normal and one abnormal child: $a\ a\ b$, $a\ b\ a$, and $b\ a\ a$, the sum of these terms being $3a^2b$. There are also three birth sequences which lead to the combination of one normal and two abnormal children: $a\ b\ b$, $b\ a\ b$, and $b\ b\ a$, their sum being $3ab^2$. When all of these separate terms are combined, the expanded binomial is

$$a^3 + 3a^2b + 3ab^2 + b^3 = 1$$

To obtain the numerical probabilities for each of these terms, we must substitute the respective values for a and b in the equation. Since $a = \frac{1}{2}$, and $b = \frac{1}{2}$, upon substitution, the equation becomes

a^3	$+\ 3a^2b$	$+\ 3ab^2$	$+\ b_3$	$= 1$
$(\frac{1}{2})(\frac{1}{2})(\frac{1}{2})$	$+\ (3)(\frac{1}{2})(\frac{1}{2})(\frac{1}{2})$	$+\ (3)(\frac{1}{2})(\frac{1}{2})(\frac{1}{2})$	$+\ (\frac{1}{2})(\frac{1}{2})(\frac{1}{2})$	$= 1$
$\frac{1}{8}$	$+\ \frac{3}{8}$	$+\ \frac{3}{8}$	$+\ \frac{1}{8}$	$= 1$
(3 normal)	(2 normal, 1 abnormal)	(1 normal, 2 abnormal)	(3 abnormal)	

Having substituted in our equation, we are now able to provide to the inquiring parents specifying three offspring the chances of realizing any particular combination of normal and abnormal children. If the parents choose to have four children, we would base our predictions on the expansion of $(a + b)^4$, and for five children we would use $(a + b)^5$. In summary, the binomial $(a + b)^n$ can be useful in situations where either one of two events can occur by chance, for example, normal versus abnormal, boy versus girl, and so on. The number of individuals specified determines the power to which the binomial is raised. The probability of occurrence of any combination of the two alternatives can be calculated by substitution in an appropriate term of the binomial expansion.

When a series of expanded binomials is examined (Figure 3–1), the following observations can be made.

1. $(a + b)^n$ has $n + 1$ terms, i.e., one more than the power to which the binomial is raised.

2. The first term of each expansion is a^n and the last term is b^n. The exponent of a decreases by 1 in each term, while that of b increases by 1 in each term. The sum of the exponents of a and b in each term is n.

3. If the numerical coefficient of any term is multiplied by the exponent of a in that term and then divided by the number (position) of the term in the series, the result is the coefficient of the next following term.

Binomial	n	$n+1$	Number of Combinations	Expanded Binomial
$(a+b)$	1	2	2	$a+b$
$(a+b)^2$	2	3	4	$a^2+2ab+b^2$
$(a+b)^3$	3	4	8	$a^3+3a^2b+3ab^2+b^3$
$(a+b)^4$	4	5	16	$a^4+4a^3b+6a^2b^2+4ab^3+b^4$
$(a+b)^5$	5	6	32	$a^5+5a^4b+10a^3b^2+10a^2b^3+5ab^4+b^5$
$(a+b)^6$	6	7	64	$a^6+6a^5b+15a^4b^2+20a^3b^3+15a^2b^4+6ab^5+b^6$

Figure 3–1. Expansion of the binomial $(a+b)^n$, where n = the power of the binomial, $n+1$ = the number of terms in the corresponding expansion, and $a=b=1/2$. The number of possible combinations doubles with each successive increase in the power of the binomial. The coefficient of the first and last term is always 1, and the coefficient of the second and next to last term is the same as the power of the binomial in any given case. The coefficient of other terms is the sum of the coefficients above and to either side of the term in question. Note that the values of the coefficients form a symmetrical distribution.

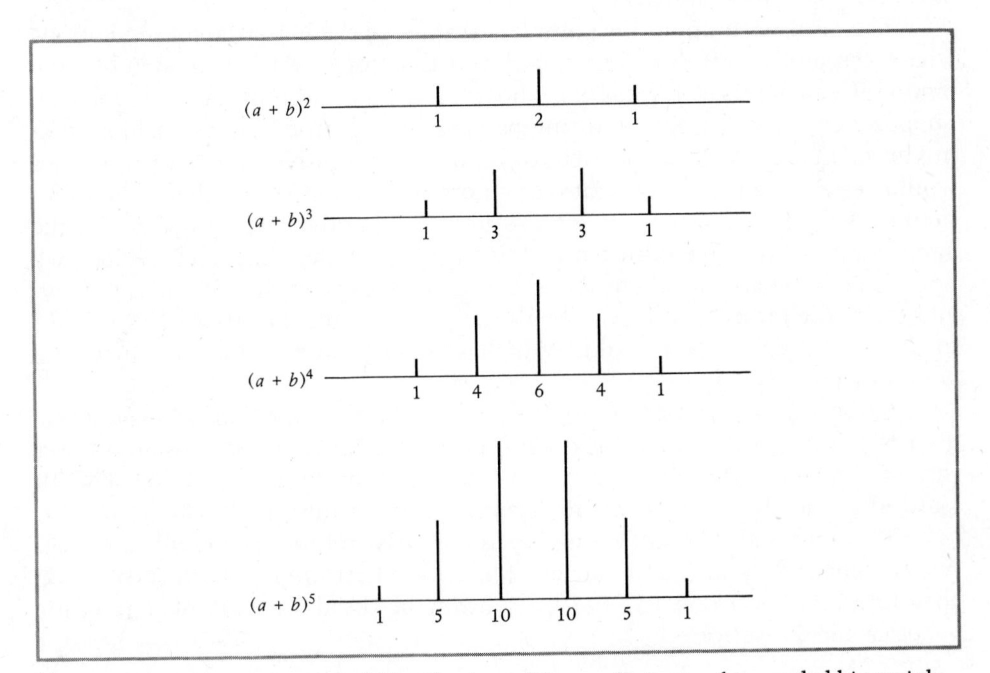

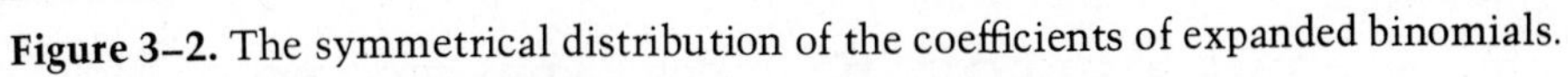
Figure 3–2. The symmetrical distribution of the coefficients of expanded binomials.

4. The numerical coefficients of the terms comprise a symmetrical or normal distribution when $a=b=\frac{1}{2}$ (Figure 3–2).

We should also note that the probability for any specific combination of genotypes or phenotypes can be calculated without the necessity of expanding the binomial to find the requisite term. For example, our parents might specify five children and wish to know the probability that three would be normal and two abnormal. Instead of expanding the binomial $(a + b)^5$ and selecting the proper term, we could apply the following formula to obtain this term.

$$\frac{n!}{x!\,(n - x)!} \times (a)^x (b)^{n-x}$$

For our example, in the above formula n = the number of events (children) specified (5); x = the number of normal children specified (3); and $(n - x)$ = the number of abnormal children specified (2). The symbol ! means factorial, that is, the product of all the integers from 1 to the term specified. Thus, in this example $n! = 5! = 1 \times 2 \times 3 \times 4 \times 5$; $x! = 3! = 1 \times 2 \times 3$; and $(n - x)! = 2! = 1 \times 2$. With the previous assumption that $a = \frac{1}{2}$ and $b = \frac{1}{2}$, we can substitute in the formula as follows.

$$\frac{5!}{(3!)(2!)} \times (\tfrac{1}{2})^3(\tfrac{1}{2})^2 = \frac{120}{12} \times (\tfrac{1}{2})^3(\tfrac{1}{2})^2 = 10(\tfrac{1}{2})^3(\tfrac{1}{2})^2 = \frac{10}{32} = 0.31$$

Comparison of this value with the term $10a^3b^2$ of the expansion of $(a + b)^5$ in Figure 3–1 will demonstrate that we have obtained the identical term and its value by the use of the above formula, and without expanding the entire binomial expression. The formula thus provides a shortcut when the probability for only one specific combination is desired.

The discussion up to this point has employed the binomial for predictions where the probability for event a is $\frac{1}{2}$ and that for event b is also $\frac{1}{2}$, but the binomial can be used for events whose occurrence is not equally probable. Suppose we return to our inquiring parents for a moment and consider what might result if they were each heterozygous for a recessive gene. Here the cross would result in a phenotypic ratio of $\frac{3}{4}$ normal to $\frac{1}{4}$ abnormal. Using the binomial for this situation, we can let a = normal = $\frac{3}{4}$ and b = abnormal = $\frac{1}{4}$. If the parents postulate three children and wish to know the chance of having two normal and one abnormal child, we could again expand the binomial $(a + b)^3$ and select the term $3a^2b$. However, this time we would substitute $\frac{3}{4}$ for a and $\frac{1}{4}$ for b. Accordingly, the probability of having two normal and one abnormal would be $(3)(\frac{3}{4})^2(\frac{1}{4}) = \frac{27}{64}$.

In the expansion of a binomial where a and b are not equal, a symmetrical distribution of coefficient values is not obtained. Thus where $a = \frac{3}{4}$ and $b = \frac{1}{4}$, the expansion is $(\frac{3}{4} + \frac{1}{4})^2 = \frac{9}{16} + \frac{6}{16} + \frac{1}{16}$, and this distribution is lopsided or skewed in the direction of the higher probability values (Figure 3–3).

The binomial theorem is employed not only in human pedigree analysis, but also in studies on the frequency of alleles in natural populations. We shall therefore have occasion to refer to it again in the discussion of population genetics and evolution.

THE CHI-SQUARE TEST APPLIED TO THE OUTCOME OF CROSSES

The ratios which enable us to deduce genotypes are based upon actual counts of the different kinds of offspring produced by any cross. The number of individuals available for counting depends on the species being studied, its fertility, and the environmental conditions under which it is grown or cul-

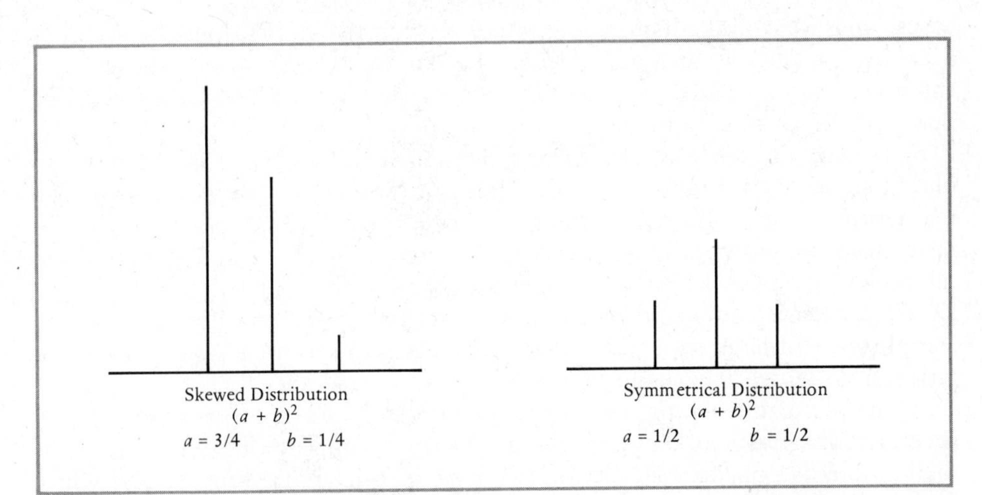

Figure 3–3. Comparison of a skewed distribution with a symmetrical (normal) distribution.

tured. In addition, the sample being counted must be large enough so that on the basis of chance all possible genotypes and phenotypes can be expected to appear in proportions representative of the population from which the sample was derived. The requirement of minimum sample size in realizing a ratio can be illustrated by the following marble game.

Suppose marbles of the same size and weight, but colored either red or white, are uniformly mixed together in a barrel. Suppose also that the barrel is large enough to contain what appears to be an infinite, or at least an enormously large, number of marbles. The rules of the game allow a blindfolded player to take marbles, one at a time, from the barrel. If equal numbers of red and white marbles are present, the chance that a player will draw a red marble from the barrel is $\frac{1}{2}$ and the chance of his selecting a white marble is also $\frac{1}{2}$. Furthermore, these same probabilities would rule in every draw since each selection is an independent event without any relation to preceding or succeeding selections, and the population of marbles is so large that the equal distribution of red and white marbles is essentially unchanged by the withdrawal of samples. Suppose the object of the game is to see how close several players can approach a 1 : 1 ratio of red to white in 10 draws each, that is, how close can they come to the theoretically perfect score of 5 red : 5 white. One player might score 4 red : 6 white; another, 7 red : 3 white; and a third, 2 red : 8 white. How can we describe these scores in terms of the perfect score? The first player with 4 red : 6 white comes close since each term departs from theoretical expectation by only 1 marble. The 7 : 3 ratio for the second player deviates 2 marbles in each direction. The deviation of 3 marbles in the third score of 2 : 8 seems to be so large that it can hardly be considered an example of a 1 : 1 ratio. However, suppose identical deviations from the theoretical expectation were found when the number of draws was increased to 100. Here, a deviation of 1 marble in each direction would produce a score of 49 : 51, a deviation of 2 marbles would give a score of 48 : 52, and a deviation of 3 marbles would yield a score of 47 : 53. With these results we would feel confident that a 1 : 1 ratio was present in all cases, and we would dismiss the small departures from a perfect score as meaningless and due to chance. Thus what appeared to be a large deviation in the first instance becomes a meaningless deviation in the second, the difference between the two being the size of the sample.

To test a hypothesis concerning the results of a cross it is necessary to obtain sufficient progeny so that deviations due to chance do not obscure the real nature of the results. For example, if you performed a cross in which the ratio, according to your hypothesis, was expected to be 1 : 1 and you obtained the results of 2 : 8 in 10 individuals, you might abandon your original hypothesis as incorrect. The abandonment might, in fact, be premature, because the sample was not large enough to reflect accurately the composition of the population from which it was obtained.

Sample size assumes even more importance when ratios composed of several terms must be realized. As an example, suppose the marble game is made more complicated by having two barrels, one containing red and white marbles and the other containing green and brown marbles. A player drawing a marble from each barrel simultaneously could draw the combination of red and brown, red and green, white and brown, or white and green. The probability for a red or a white is $\frac{1}{2}$, the probability for a green or a brown is also $\frac{1}{2}$, and, therefore, the overall probability for any one of the four combinations is $\frac{1}{2} \times \frac{1}{2}$ or $\frac{1}{4}$. The theoretical expectation is that with a sufficient number of draws a 1 : 1 : 1 : 1 ratio for the four combinations will be achieved. In a small number of draws it is very unlikely that any player would realize this ratio, and common sense, rather than calculations can provide the reason: If it is difficult to achieve equal values and a 1 : 1 ratio for two alternatives in a small sample, it will be very much more difficult to obtain equal values for four alternatives in a similar size sample. An analogous situation would exist in the case of a dihybrid testcross where the results are expected to conform to a similar 1 : 1 : 1 : 1 ratio.

Even if the sample of the population being tested is sufficiently large, chance deviations will still occur, and it cannot be expected that the data obtained in any experiment will be a perfect example of any ratio. A small deviation can be regarded as simply due to chance, but a large deviation represents a significant departure from expected results and may indicate that the initial hypothesis was wrong. Thus, the question is what amount of deviation is too small to heed and what amount is too big to ignore? Statisticians have come to a rule of thumb on this point. If the deviation obtained in an experiment could be expected to occur on the basis of chance alone at least once in every 20 trials, then the deviation is not significant, that is, it can be ignored. If a deviation as large or larger than that observed could be expected to occur on the basis of chance alone less often than once out of every 20 trials, it is considered to be significant and should be heeded. Although a significant deviation does not necessarily imply that the hypothesis is wrong, it does indicate that some factor other than chance may be involved. For example, the data may not fit the expected ratio if one class, such as the homozygous recessive, is less viable and therefore more apt to be reduced in numbers or if one phenotype is lethal in early stages of development and therefore is not scored with the rest of the progeny. Although large deviations from the expected can often be attributed to these and other factors, the possibility of an incorrect hypothesis must, of course, be taken into consideration. In general, a probability that a deviation as large or larger than that observed might occur by chance once in 20 trials, or 5 percent of the time ($P = 0.05$) is considered borderline, and additional experiments to test the hypothesis are desirable.

A simple statistical test, called the chi-square test (χ^2), makes use of probability values in the determination of whether or not a given set of data fits a particular ratio. The chi-square test is best explained by example, and for this we can use Mendel's original data on the inheritance of yellow versus green cotyledons. In this experiment 8023 individuals were scored; of these 6022 were yellow and 2001 were green. To use the chi-square test, we must first establish the hypothesis to be tested. Our hypothesis in this case is that yellow and green appear in the ratio of 3 : 1, respectively. The formula for chi-square is: $\chi^2 = \Sigma\,(d^2/e)$, where Σ stands for "the sum of," d represents deviation, and e is expected value. If the total progeny of 8023 conformed perfectly to the 3 : 1 ratio, we would expect that 6017 would be yellow, and 2005.7 (2006) would be green. These are the theoretically expected values *(e)* based on the number of progeny scored. They differ somewhat from the values actually observed. This difference between what was observed and what is expected for each class is the deviation *(d)* for each class. Use of these values in the test is shown below.

χ^2 Test for a Progeny Sample of 8023

Hypothesis	3 yellow	1 green
Observed	6022	2001
Expected (e)	6017	2006
Deviation (d)	5	5
d^2	25	25
d^2/e	$\frac{25}{6017} = 0.004$	$\frac{25}{2006} = 0.012$

$\chi^2 = 0.004 + 0.012 = 0.016$

After the χ^2 value has been calculated as above, reference must be made to a table of chi-squares in order to obtain the corresponding probability value *(P)*. Use of such a table requires that another measurement, the degrees of freedom *(df)*, first be established. For these kinds of data, the number of degrees of freedom is always one less than the number of classes in the ratio because when the expected number *(e)* for all but one class has been calculated, the last class must necessarily contain all remaining individuals of the sample. If there are only two classes, as above, data can be assigned to one class by choice, but once that class is designated, the other class is automatically determined. Hence in the present example, one degree of freedom is available. To determine the probability that our data fit the ratio and that the deviations observed can be attributed to chance, the calculated χ^2 value is compared with those given horizontally in Table 3–1 for 1 degree of freedom (n in Table 3–1). When a reasonable match is found, the heading at the top of the appropriate column of chi-squares is consulted for the corresponding probability value. In the present case, our chi-square of 0.016 corresponds to an approximate probability of 0.90, which means that 9 times out of 10, or 90 times out of 100, a deviation as great as that observed would be expected to occur by chance. Thus, a probability of 0.90 indicates that Mendel's data fit the proposed ratio very well indeed.

TABLE 3-1
Table of Chi-Square

	Probability						
n	.99	.98	.95	.90	.80	.70	.50
1	0.000157	0.000628	0.00393	0.0158	0.0642	0.148	0.455
2	0.0201	0.0404	0.103	0.211	0.446	0.713	1.386
3	0.115	0.185	0.352	0.584	1.005	1.424	2.366
4	0.297	0.429	0.711	1.064	1.649	2.195	3.357
5	0.554	0.752	1.145	1.610	2.343	3.000	4.351
6	0.872	1.134	1.635	2.204	3.070	3.828	5.348
7	1.239	1.564	2.167	2.833	3.822	4.671	6.346
8	1.646	2.032	2.733	3.490	4.594	5.527	7.344
9	2.088	2.532	3.325	4.168	5.380	6.393	8.343
10	2.558	3.059	3.940	4.865	6.179	7.267	9.342

Source: Table IV of Fisher and Yates: *Statistical Tables for Biological, Agricultural and Medical Research,* published by Longman Group Ltd., London (previously published by Oliver and Boyd, Edinburgh). Reprinted by permission of the authors and publishers.

The chi-square test can also be used for ratios containing more than two terms. As an example, we can use the data from Mendel's dihybrid inbreed in which the hypothesis predicts a 9 : 3 : 3 : 1 ratio. Since the ratio consists of four terms, the number of degrees of freedom available is three. The following calculations demonstrate again how closely Mendel's data fit the hypothesis.

χ^2 Test for a Progeny Sample of 556

Hypothesis	9 round, yellow	3 wrinkled, yellow	3 round, green	1 wrinkled, green
Observed	315	101	108	32
Expected (e)	313	104	104	35
Deviation (d)	2	3	4	3
d^2	4	9	16	9
d^2/e	0.013	0.087	0.154	0.257

$\chi^2 = 0.013 + 0.087 + 0.154 + 0.257 = 0.511$
$df = 3$
$P = >0.9$

In summary, chi-square provides a most convenient yardstick by which data can be tested for goodness of fit to a proposed hypothesis. However, this test can be applied only to instances where the smallest class contains at least five, and preferably more, individuals. It is most useful in experimental genetics where controlled matings can be made and population size can be planned in advance. Neither of these is possible in human genetics, and single families are not large enough for proper use of this statistical test. Therefore, in studies of humans chi-square is applied to collections of data or observations made over a period of time on individuals from many different families. When enough data from published cases or hospital records are available, the chi-square test can be used to indicate whether or not a given trait is associated with sex or with some other condition. Obviously such implications are important to medicine.

.30	.20	.10	.05	.02	.01
1.074	1.642	2.706	3.841	5.412	6.635
2.408	3.219	4.605	5.991	7.824	9.210
3.665	4.642	6.251	7.815	9.837	11.345
4.878	5.989	7.779	9.488	11.668	13.277
6.064	7.289	9.236	11.070	13.388	15.086
7.231	8.558	10.645	12.529	15.033	16.812
8.383	9.803	12.017	14.067	16.622	18.475
9.524	11.030	13.362	15.507	18.168	20.090
10.656	12.242	14.684	16.919	19.679	21.666
11.781	13.442	15.987	18.307	21.161	23.209

PROBLEMS

3–1. In an individual of the genotype *AaBbcc*
- a. What proportion of the gametes will carry genes *ABc*?
- b. What proportion of the gametes will be *abc*?
- c. What proportion of the gametes will carry allele *a*?
- d. What proportion of the gametes will carry allele *c*?

3–2. If a cross is made between *AaBbcc* and *aaBbCc*,
- a. What proportion of the offspring will be *C* in phenotype?
- b. What proportion of the offspring will be homozygous for genes *b* and *c*?
- c. What proportion of the offspring will be *ABC* in phenotype?
- d. What proportion of the offspring will be *AaBbCc* in genotype?
- e. What proportion of the offspring will be *aabbcc* in genotype?

3–3. In the cross *AaBBccDdEeff* (parent 1) × *aabbCCddEeFF* (parent 2)
- a. What proportion of the gametes of parent 2 will carry the alleles *a, b, C, d, F*?
- b. What proportion of the gametes of parent 2 will carry the gene *e*?
- c. What proportion of the gametes of parent 1 will carry genes *A, B, c, d, e, f*?
- d. What proportion of the progeny will be heterozygous for all gene pairs?
- e. What proportion of the progeny will show all dominant traits?

3–4. Suppose parents asked you for predictions of their childrens' phenotypes, when four children are postulated and the parents are heterozygous and homozygous, respectively, for a recessive abnormal trait.
- a. What binomial would you use for your predictions?
- b. What terms would you select for all children normal, 2 normal and 2 abnormal children, and 1 normal and 3 abnormal children?
- c. What probabilities would you predict for these three possibilities?

3–5. Two heterozygotes for the dominant character free earlobes plan to have three children.
- a. What probability would you predict for a child with attached earlobes, a child with free earlobes, and a child with attached earlobes, *in that order?*
- b. What probability would you predict for one child with free earlobes and two children with attached earlobes *in any order?*

3–6. Suppose parents asked you to predict the chance of dominant free earlobes occurring in all children resulting from two pregnancies. In this case one parent is heterozygous for free earlobes, the other homozygous for the attached condition.

a. What would be your prediction for the above?
b. If the first pregnancy resulted in identical (one egg) twins and the second pregnancy resulted in a single child, should you revise your predictions? If so, what would they be?
c. If the first pregnancy resulted in fraternal (two egg) twins and the second in a single child, should you revise your predictions? If so, what would they be?

3–7. Four players try the marble game with red and white, and green and brown marbles, described in the text. The players achieve the following scores.

	Red and Green	Red and Brown	White and Green	White and Brown
Player 1	16	25	19	40
Player 2	20	28	24	28
Player 3	36	15	21	28
Player 4	30	16	27	27

Plot the above distributions as bar graphs (histograms) on graph paper and determine χ^2 and P. Which players achieved a good fit to a 1 : 1 : 1 : 1 ratio?

3–8. Five samples of *Drosophila* heterozygous for the recessive genes ebony and reduced eye were inbred and the following data was obtained.

Sample	Normal Color, Normal Eye	Normal Color, Reduced Eye	Ebony Color, Normal Eye	Ebony Color, Reduced Eye
(1)	95	29	17	3
(2)	138	47	42	13
(3)	113	33	37	9
(4)	152	60	55	21
(5)	62	30	21	15

a. Calculate χ^2 and determine P for the samples. Which of the ratios fits a 9:3:3:1 distribution?
b. Does a significant deviation occur in any instance?
c. If data from examples of a good fit to the ratio are pooled, is a better fit obtained?

3–9. Apply the chi-square test to data given in problem 2–11. Do the data fit the ratios?

3–10. A survey of hospital records shows that of 70 patients with an inherited disorder, 55 were males and 15 were females. Is the association of the condition with the male sex statistically significant? Apply the χ^2 test to the data, assuming a 1 : 1 ratio for males versus females.

REFERENCES

Fisher, R. A., 1936. Has Mendel's work been rediscovered? *Ann. Sci., 1,* 115.
Goldstein, A., 1964. *Biostatistics, An Introductory Text.* Macmillan, New York.
Mather, K., and J. L. Jinks, 1971. *Biometrical Genetics,* 2nd ed. Chapman and Hall, London.
Zar, J. H., 1974. *Biostatistical Analysis.* Prentice-Hall, Englewood Cliffs, N.J.

Chapter 4

THE VARIETIES OF GENE EXPRESSION

WITH the rediscovery and application of Mendel's principles, it soon became apparent that genes could be other than dominant or recessive in their phenotypic expression. Some were found to have variable effects, while others exhibited incomplete dominance. It also became evident that the number of allelic states of a gene was not necessarily limited to two. Such findings provoked questions as to the mechanism by which the phenotype is produced. It was clear that the gene was not equivalent to the trait caused by its presence, but was, instead, the initiator of the trait. Furthermore, with proof that genes were contained within chromosomes, this initiating action was properly assigned to the cellular, and ultimately, to the molecular level of organization. Real understanding of the source of the phenotype was achieved only recently, made possible by advances in the science of biochemistry coupled with the use of genetic principles and techniques not yet discussed. Therefore, although familiarity with the kinds of gene expression and allelic states commonly encountered is useful now, an explanation of the actual source of the phenotype will be reserved to later chapters.

INTERMEDIATE OR INCOMPLETE DOMINANCE

Thus far we have considered only the phenomenon of complete dominance which is characterized by two alternative phenotypic states: the phenotype produced by the dominant allele and the opposing phenotype which results from the homozygous presence of the recessive allele. With complete dominance heterozygotes cannot be distinguished from homozygotes for the dominant allele, that is, $AA = Aa$ in phenotype. However, other states of gene expression exist, one of them being *intermediate* or *incomplete dominance*. In this state the dominant allele does not completely mask the presence of the recessive allele and, consequently, the heterozygote possesses a phenotype of its own, one which can be distinguished from that of either homozygote.

A classic and frequently cited example of intermediate dominance is the color of flowers in snapdragons. When homozygous, a dominant gene results in red flowers and the recessive allele in white flowers. The heterozygote, however, is a phenotypic blend, and the flowers are pink. Crosses between red- and white-flowered plants $(AA \times aa)$ produce a uniformly pink F_1 generation (Aa). Upon inbreeding, the F_2 phenotypic ratio is 1 red : 2 pink : 1 white and is thus identical with the genotypic ratio (1 AA : 2 Aa : 1 aa). Intermediate dominance produces three phenotypic classes of progeny in the F_2 generation, the heterozygote being recognizably different from either homozygote.

An excellent example of intermediate dominance in humans is sickle cell anemia, a severe hemolytic disease found in equatorial Africa and in the descendents of persons who migrated from that region. Around 8 percent of American blacks are heterozygous for the sickle cell gene. The disease is caused by the presence of abnormal hemoglobin molecules in the red blood cells (see Chapter 10). Under the conditions of low oxygen tension encountered in the venous circulation, the abnormal hemoglobin tends to precipitate from solution within the cell, forming linear aggregates which distort the cell to a characteristic sicklelike shape from which the name of the disease is taken (Figure 4–1). As a result, homozygotes suffer red cell destruction, anemia, and blockage of small blood vessels with consequent areas of tissue destruction, all of which lead to death at an early age.

Under most conditions of life heterozygotes are superficially normal, but their red cells contain a mixture of normal and abnormal hemoglobin, and blood samples taken from such individuals and subjected to low oxygen tension will exhibit some sickling behavior. This test allows the clinical identification of heterozygotes. The test also indicates that under some special conditions the trait may be revealed. For example, heterozygotes would be well advised not to fly at high altitudes in unpressurized airplanes.

Persons homozygous for the normal allele show no evidence of sickle cell disease. Since their hemoglobin contains unaltered hemoglobin molecules, their blood cells do not sickle, either spontaneously or when subjected to a clinical test.

Data collected on offspring from marriages between heterozygotes show the expected 1 : 2 : 1 phenotypic and genotypic ratios. The trait of sickle cell anemia thus satisfies the criterion for intermediate dominance because, with appropriate tests, the heterozygous phenotype can be distinguished from that

Figure 4–1. Photomicrographs of human red blood cells: (top) cells from a homozygote *(HbA/HbA)* with normal hemoglobin; (middle) cells from a heterozygote *(HbA/HbS)* with sickling of some cells; (bottom) cells from a homozygote *(HbS/HbS)* with complete sickling. [Courtesy of Dr. James V. Neel and the University of Michigan Heredity Clinic.]

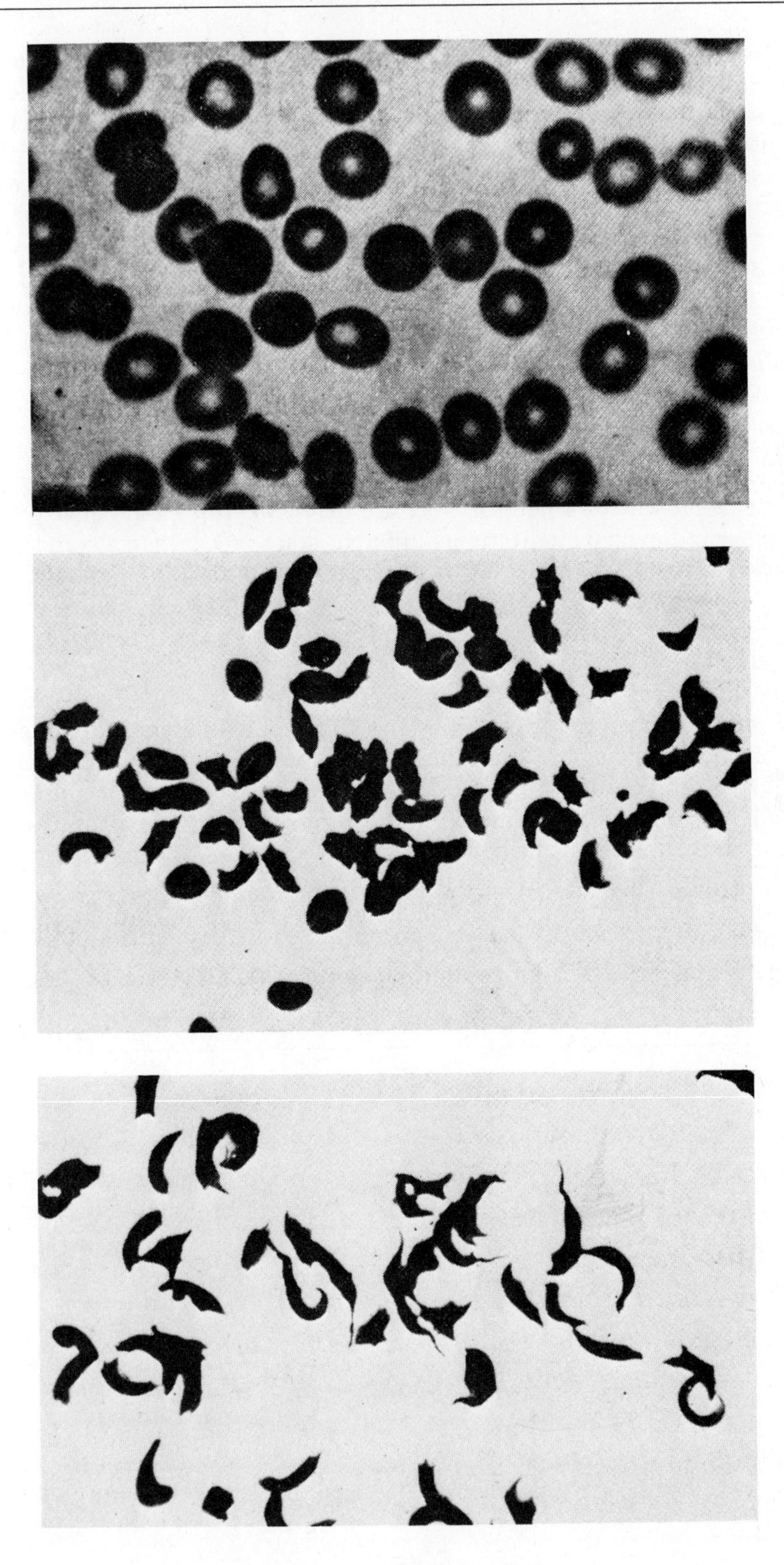

characteristic of either homozygote and can be viewed as an intermediate condition between the two extremes.

It should be noted that the classification of a gene and its effects as dominant, incompletely dominant, or recessive is somewhat arbitrary since all degrees of expression are found. A trait may be labeled a dominant on the basis of superficial appearance, whereas a more careful examination of a heterozygote might reveal some expression of the recessive allele. Thus, although the presence of a dominant normal allele in a heterozygote often compensates for the effects of an inactive or deleterious recessive gene, subtle differences which distinguish the heterozygote can frequently be found if one is willing to make the effort to look for them.

CODOMINANCE

Another type of gene expression is known as *codominance*. Codominant allelic genes present in a heterozygote are expressed fully and equally in the phenotype, producing a qualitatively different state rather than a blend of two extremes. As a good example, let us consider the human ABO blood groups.

On the surface of red blood cells are found substances called *antigens* whose presence or absence is an inherited trait. Such antigens show a coordinated relationship with certain proteins present in the blood serum. These proteins are known as *agglutinins* or, more familiarly, *antibodies*. When red cells of one individual are mixed with the serum of a different individual, one of two reactions can be observed: The red cells may remain freely suspended or they may become clumped together, a phenomenon called *agglutination* (Figure 4–2). Agglutination is caused by the interaction of the antigen on the surface of the cell with an antibody present in the serum. The binding between antigen and antibody is specific, that is, a particular antigen will react with its particular matching antibody and with no other. A test for agglutination makes it possible to classify people into four major groups with respect to the antigen or antigens present. These groups are known as the ABO blood types. They

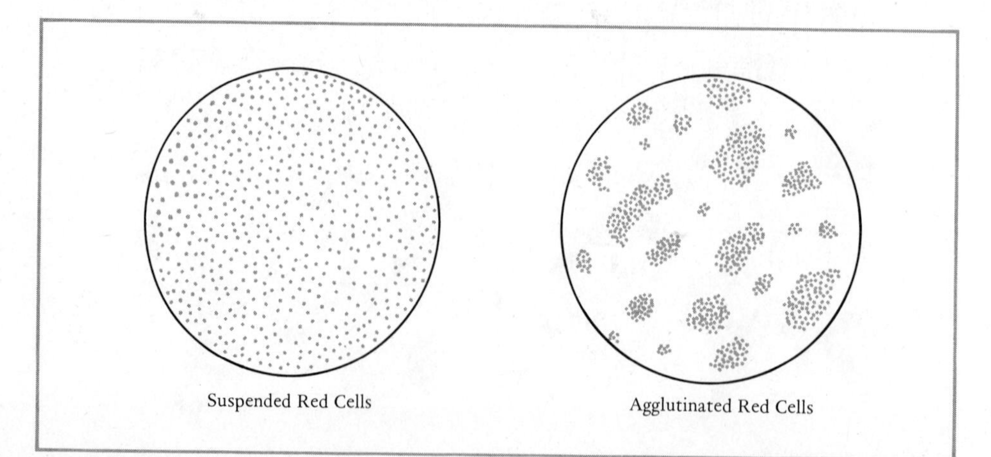

Figure 4–2. Red blood cells clumped by agglutination in an antigen-antibody reaction.

were first reported by Karl Landsteiner in 1901. The four phenotypic classes consist of: (1) persons with antigen A, (2) those with antigen B, (3) those with both antigen A and antigen B, and (4) those with neither antigen.

The four phenotypes are determined by three major alleles: a dominant allele, I^A, the presence of which results in the production of antigen A; a second dominant allele, I^B, which results in the production of antigen B; and a recessive allele, I^O, in whose homozygous presence no antigen at all is produced. Since more than two allelic states are known, these genes can be considered an example of *multiple alleles*. It must be remembered, however, that genes occur in pairs, and therefore only two of these alleles can be present in the genotype of any one individual.

The dominant alleles I^A and I^B are fully expressed in the phenotype by their respective antigens, but since the allele I^O produces no identifying antigen, its presence in a heterozygote cannot be detected. Thus an individual showing antigen A in his blood cells could be either I^AI^A or I^AI^O in genotype (Table 4–1). Codominance is seen in individuals heterozygous for alleles I^A and I^B. In such persons both alleles are fully expressed in the phenotype by their respective antigens. Neither allele causes any diminution in the activity of the other allele, and the phenotype produced is not a blend exhibiting new antigenic properties. Instead, the A and B antigen specificities are distinct, unaltered, and equally apparent, and their presence directly indicates the genotype of such a heterozygote.

Although these antigens and their respective immunoglobulin antibodies are quite different kinds of macromolecules, produced through the action of separate sets of genes, they exhibit the interesting, although not surprising, reciprocal relationship illustrated in Table 4-1. If an individual belongs to blood group A, his serum contains antibodies against antigen B; if he is blood group B, his serum is anti-A; if his cells contain both antigens A and B, his serum contains neither antibody; and if he has neither antigen, his serum contains both antibodies. These relationships are important in preventing self-agglutination and in blood transfusions and organ transplants. Avoidance of agglutination requires that donors and recipients be matched for compatibility.

Other instances of codominance have also been identified in connection with human red cell antigens. For example, antigens symbolized as M and N occur in the presence of a pair of allelic genes, *M* and *N*, respectively. These genes are codominant, so that a heterozygote of the genotype *MN* possesses both antigens. Closely associated with the *M* and *N* genes is another pair of

TABLE 4-1
Relationship Between Blood Group Antigens, Serum Antibodies, and Genotype

Blood group phenotype	Antigen present on red cells	Antibody in serum	Genotype
O	—	anti-A, anti-B	I^OI^O
A	A	anti-B	I^AI^A or I^AI^O
B	B	anti-A	I^BI^B or I^BI^O
AB	A and B	—	I^AI^B

codominant alleles, *S* and *s*, whose presence leads to the appearance of yet another red cell specificity, the corresponding antigens S and s. The *MN* and *Ss* alleles, first discovered by Landsteiner and P. Levine, are inherited together almost as a block and have been interpreted as gene sites which are located closely adjacent to one another on the same chromosome.

No antibodies specific for antigens M, N, S, or s are naturally present in human blood. Therefore, identification of these antigens is accomplished by mixing human red cells with rabbit serum containing the appropriate antibodies. This test uses to advantage the immune response mechanism which functions to produce antibodies against foreign proteins or other complex compounds of invading infectious organisms. Thus, human cells injected into a rabbit stimulate the production of rabbit antibodies specific for the human antigens. A sample of the injected rabbit's blood serum can be used to distinguish the human antigen phenotype. The field of immunology makes extensive use of this technique since antigen-antibody reactions are much more sensitive and specific than chemical methods of identification.

The ABO series, the *MN* and *Ss* alleles, and other inherited antigenic characteristics of red blood cells are often used in medical-legal situations involving identification or disputed paternity. In paternity cases a comparison of the blood groups of mother, child, and alleged father may exclude the man as a possible parent of the child. For example, a child of blood type AB, whose mother is A, could not have as a father a man whose blood group was O. Blood typing does not prove that an individual is the father of a child; it merely indicates whether or not he is a possible parent.

Besides those mentioned here, numerous other human blood cell antigens are known, some of them extremely rare. All are detected through antigen-antibody reactions. Some can be employed as genetic markers for human chromosomes and many are useful with respect to identity or paternity disputes. In addition, they are valuable tools for the study of gene frequencies in human populations.

MULTIPLE ALLELES

We have discussed the effects of alternative alleles, that is, the contrasting phenotypes produced by two allelic states of the same genetic unit, and have also considered the effect of multiple alleles, the phenotypes produced by three or more allelic states of the same gene. As we have seen, human blood groups are determined by such a series of multiple alleles. Only three alleles, I^A, I^B, and I^O, were used to illustrate codominance. However, group A antigen has been found as two common, but different subtypes known as A_1 and A_2, and the presence of these antigens has been attributed to the corresponding alleles I^{A_1} and I^{A_2}. With respect to phenotypic expression, both of these group *A* alleles are codominant with allele I^B, and, together with I^B, exhibit dominance over allele I^O. As an additional complexity, I^{A_1} is dominant over I^{A_2}, so that a heterozygote of the genotype $I^{A_1}I^{A_2}$ is phenotypically A_1 in blood type. Other, but rare variants of the ABO blood group alleles have also been discovered, for example, I^{A_3}, I^{A_4}, and variants of I^B.

The Rh Alleles

A well-known example of multiple allelism in humans, the Rh blood group system, has received much attention because of adverse transfusion reactions and because of its association with a hemolytic disease of the newborn. The hemolytic disease is caused by an incompatibility between Rh antigens of the red blood cells of the fetus and antibodies produced by the mother. Although the placenta forms a barrier between the circulatory systems of mother and fetus, small breaks in the barrier sometimes occur permitting fetal cells to enter the mother's circulation (Figure 4–3). If these cells possess an antigen not found in the mother and against which the mother's immune system can form antibodies, such antibodies can subsequently enter the fetus causing destruction of fetal red blood cells.

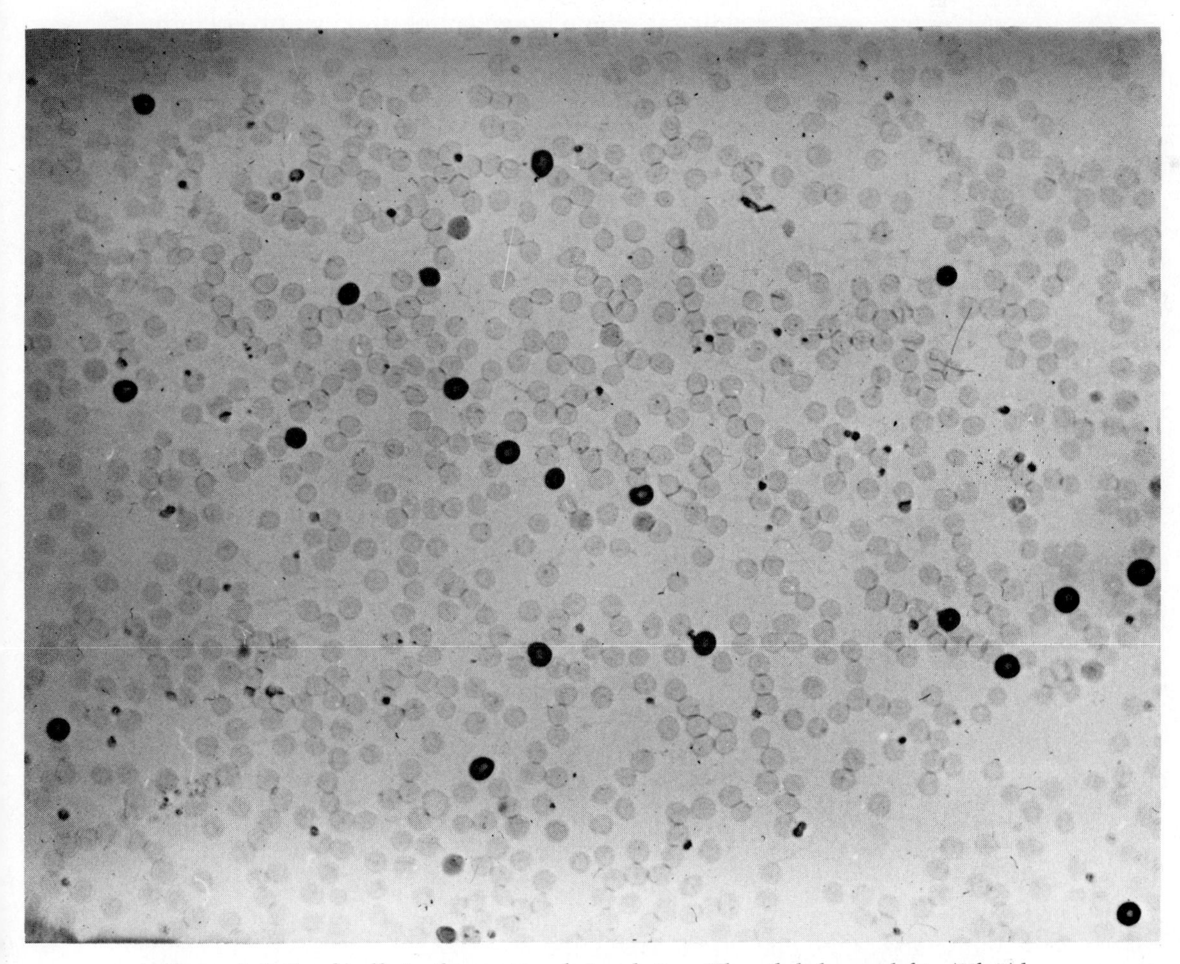

Figure 4–3. Fetal cells in the maternal circulation. The adult hemoglobin (HbA) has been washed out of the mother's red blood cells which therefore appear as "ghosts." Fetal cells stain darkly because fetal hemoglobin (HbF) is not removed by the buffer. [From Cyril A. Clarke and R. B. McConnell, *Prevention of Rh-Hemolytic Disease,* 1972. Courtesy of Cyril A. Clarke and Charles C. Thomas, Publisher, Springfield, Illinois.]

The genes involved in the production of these antigens are collectively called the Rh blood group alleles and were named for the Rhesus monkeys in which the system was first studied by Karl Landsteiner and A. S. Wiener in 1940. These investigators found that around 85 percent of the white population of the United States possess a red cell antigen which is lacking in the other 15 percent of persons tested. The antigen was called the Rhesus factor and symbolized as Rh_0. Since that time other antigens, major and minor, as well as combinations of these antigens have been discovered in human blood. They have all been identified through agglutination reactions with antibodies that have been found in the serum of various patients.

Persons whose blood cells contain an Rh antigen (specifically Rh_0) are known as *Rh positive* (Rh^+) while those whose cells are without this antigen are called *Rh negative* (Rh^-). Antibodies are not naturally present in Rh negative individuals. However, if antigens are introduced via blood transfusion or through entrance of fetal cells into the mother's circulation, antibody formation is initiated. Antibody production can persist for years, and once an Rh^- person has been sensitized, or immunized, against an Rh antigen, a subsequent transfusion with Rh^+ blood or pregnancy with an Rh^+ fetus may call forth an immediate agglutination reaction.

Hemolytic disease of the newborn is characteristically seen in Rh^+ babies whose mothers are Rh^- and whose fathers are Rh^+. In such cases each successive pregnancy carries more risk to an Rh^+ fetus because of the progressive rise in antibody concentration, or titer, in the mother. Recently it has been discovered that a high degree of protection to the fetus is afforded by treatment of Rh^- mothers with anti-Rh_0 gamma globulin. Usually the first Rh^+ baby of such a mother suffers few ill effects, and if the mother receives anti-Rh_0 gamma globulin immediately after delivery of the first Rh^+ child, antibody formation is inhibited.

The ABO blood group alleles have also been implicated in maternal-fetal incompatibilities when the mother belongs to blood group O and the fetus belongs to blood group A, B, or AB. It will be recalled that group O persons possess naturally occurring antibodies against antigens A and B. These antibodies belong to a class of immunoglobulins (IgM) which ordinarily do not diffuse readily through the placenta. Therefore, maternal ABO blood group antibodies are not likely to move across the placenta and into the fetus, although any fetal red blood cells which reach the mother will be destroyed by these antibodies. This destruction is occasionally protective. For example, suppose an Rh^-, blood group O mother had not been sensitized to Rh antigen and therefore had not yet formed Rh antibodies. If any Rh^+ fetal blood cells carrying antigens A or B were to enter the mother's circulation, they might be destroyed before the Rh antigen could stimulate the formation of anti-Rh antibodies. The destruction of the Rh^+ blood cells would be protective to the fetus because the Rh antibodies belong to a different class of agglutinins and, once formed, readily diffuse across the placenta to the fetus.

Initially, the Rh phenotype was thought to be determined by a single pair of alleles, the Rh^+ allele being dominant. However, antigens capable of eliciting antibody formation have been found in supposedly Rh^- individuals so codominance is a more accurate designation for the expression of these genes.

TABLE 4–2
The Rh Multiple Alleles According to the Wiener Hypothesis*

Allele	Agglutinogen	Antigenic properties of agglutinogens					Phenotype based on presence of antigen Rh_0
R^0	Rh_0	Rh_0	—	—	hr′	hr″	Rh positive
R^1	Rh_1	Rh_0	rh′	—	—	hr″	
R^2	Rh_2	Rh_0	—	rh″	hr′	—	
R^z	Rh_z	Rh_0	rh′	rh″	—	—	
r	rh	—	—	—	hr′	hr″	Rh negative
r'	rh′	—	rh′	—	—	hr″	
r''	rh″	—	—	rh″	hr′	—	
r^y	rh^y	—	rh′	rh″	—	—	

* Only the five major antigens are included to represent the differences between agglutinogens. The antigenic properties found for each agglutinogen are given. These properties were discovered over a period of time and thus the symbols used to designate the various antigens are not always consistent.

Although over thirty combinations of antigens can be identified with specific antisera, only five are considered to be major types. These are Rh_0, rh′, rh″, hr′, and hr″. These five differ significantly in their potency to provoke antibody formation. The first such antigen discovered, Rh_0, is the most potent in this respect. Most Rh incompatibilities involve the Rh_0 antigen, although occasionally hr′ or hr″ are implicated. In ordinary clinical practice Rh_0 is the only Rh antigen for which compatibility tests are carried out and any individual possessing this antigen is typed as Rh positive.

The genetics of the Rh blood group system is complicated by the fact that more than one antigen can be identified as the result of the presence of a given Rh gene (Table 4–2). As a result, controversy has arisen over the interpretation and naming of these genes. The original hypothesis, formulated by Wiener, attributes all variations in phenotype to the multiple alleles of a single genetic unit. According to this hypothesis there is a one-to-one relationship between an allele and a macromolecule called an agglutinogen which possesses more than one chemical grouping capable of inducing antibody formation. Any Rh allele would thus result in the presence of a particular agglutinogen, and the several specific chemical groupings present in the macromolecule would account for the several antigen-antibody reactions observed (Figure 4–4). Indirect support for this hypothesis was obtained by Landsteiner who showed that a single, known, artificial antigen could elicit the production of numerous different antibodies. The Rh multiple alleles, their proposed agglutinogens, and the antigenic properties which have been associated with them are illustrated in Table 4–2.

The second hypothesis, developed by R. A. Fisher and R. R. Race, attributes all phenotypes to a complex of three separate genetic units located immediately adjacent to one another. The hypothesis proposes that these three separate genes can each occur in a dominant or a recessive state. These genes have been designated as *D* and *d*, *C* and *c*, and *E* and *e*, with gene *D* presumed the first of the series. Any individual inherits two sets of these three genes, and

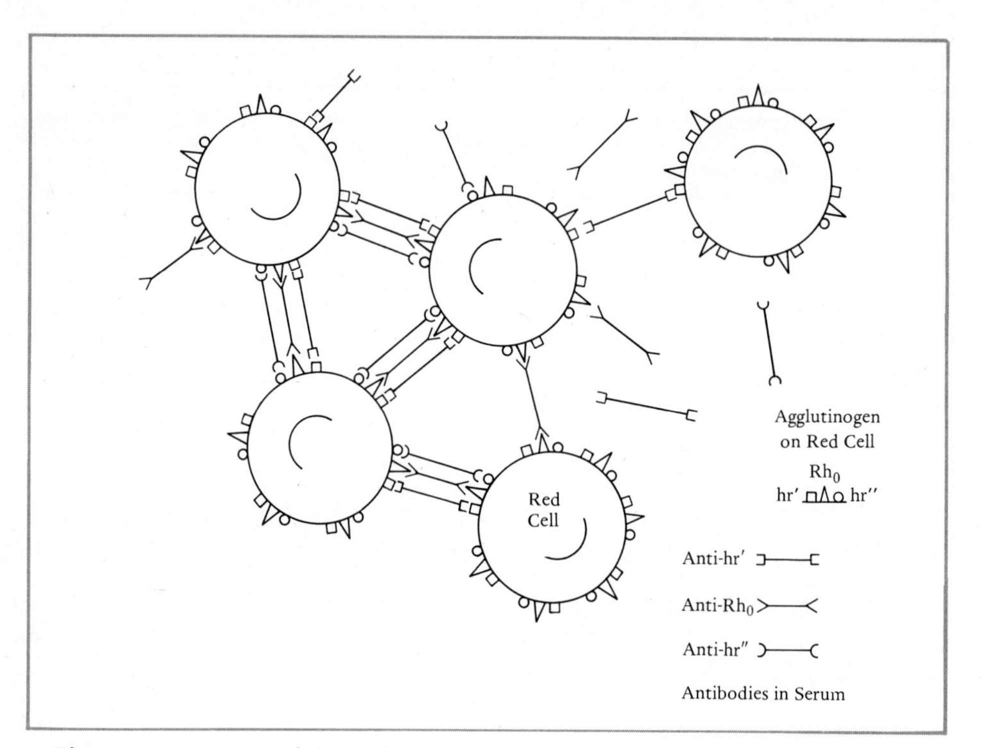

Figure 4–4. Diagram of the agglutination of red blood cells of the genotype R^0/R^0 by anti-Rh antibodies. The genotype R^0/R^0 (or R^0/r) is characterized by three antigenic specificities (anti-Rh_0, anti-hr', and anti-hr''). The three antigenic specificities are represented in the diagram as parts of an agglutinogen, according to the Wiener hypothesis.

the combinations of dominant and recessive alleles results in the particular antigenic specificities observed. According to this hypothesis each gene is represented in the phenotype by a single antigen detectable by use of the corresponding antibody. The five major antigens have been assigned to the proposed genes as follows: Rh_0, gene D; rh', gene C; rh'', gene E; hr', gene c; hr'', gene e. No antigen has been found for gene d (as in the case of I^O of the ABO series).

In summary, the first hypothesis views the Rh genes as a multiple allelic series, while the second views them as a cluster of separate genes. As yet there is no unequivocal way to distinguish between the two alternative explanations. However, most blood group workers prefer the Fisher-Race hypothesis and use its less cumbersome system of notation. Both the DCE system and the Wiener system are given in Table 4–3.

In any particular instance the probable genotype of an individual is estimated by the agglutination reactions of his blood cells with different types of Rh antisera. As an example, a person whose blood cells gave a positive test for antigens Rh_0, hr', and hr'' would have to be R^0/R^0 (or Dce/Dce) or R^0/r (or Dce/dce) in genotype. Analysis of parental and sibling antigens is also of considerable help when the phenotype observed could result from more than one genotype. While the medical practitioner is primarily concerned with potential

TABLE 4–3
The Fisher-Race Concept of the Rh Blood Group Genes Compared to the Wiener Hypothesis of Multiple Alleles.

Antigen and corresponding Fisher-Race gene		Fisher-Race genes, corresponding antigenic phenotype, and equivalent Wiener alleles which produce the phenotype		
Antigen	Fisher-Race gene	Fisher-Race genes present	Antigenic phenotype	Weiner allele
Rh_0	*D*	*Dce*	Rh_0, hr′, hr″	R^0
rh′	*C*	*DCe*	Rh_0, rh′, hr″	R^1
rh″	*E*	*DcE*	Rh_0, rh″, hr′	R^2
—	*d**	*DCE*	Rh_0, rh′, rh″	R^z
hr′	*c*	*dce*	—, hr′, hr″	*r*
hr″	*e*	*dCe*	—, rh′, hr″	*r′*
		dcE	—, rh″, hr′	*r″*
		dCE	—, rh′, rh″	r^y

* Gene *d* has no antigen, indicated by —.

incompatibilities, the population geneticist or physical anthropologist is more interested in genotype, for significantly different frequencies of these alleles have been found in various population subgroups.

Other Multiple Allelic Series

Multiple alleles are commonly encountered in all organisms and their presence results in an increase in the number of genotypes and phenotypes which may occur. The different alleles belonging to any given series are not necessarily codominant with one another, however. In some instances, one allele, that producing the normal or standard phenotype, is dominant over all others, so that any heterozygote containing the dominant allele will exhibit the dominant phenotype.

Relationships among the recessive alleles in such a series vary. Frequently, individuals heterozygous for any two such alleles exhibit an intermediate phenotype. As an example, one of the first multiple allelic series to be discovered involved the phenotype of eye color in *Drosophila*. The wild type or normal eye color of this insect is red, and the pigments present result from the combined action of a number of different, independent genes. The alleles of one such gene, the white gene, produce phenotypes which range in color from white (absence of pigment) through shades named ivory, buff, eosin, cherry, and so on. Using conventional genetic symbols, we can designate the white gene as *w*, the wild type allele for this gene as $+^w$ or, more simply, +, and an intermediate allele, such as eosin, as w^e. The wild type, red-eyed condition is completely dominant over all other phenotypes produced by members of this series, that is, $w/+ = +/+$ in phenotype. However, the other alleles exhibit intermediate dominance when in combination in a heterozygote. For example, a heterozygote (w^e/w) has a distinguishably paler shade of eye color than an individual homozygous for eosin (w^e/w^e). In the homozygous presence of the white allele *w*, no pigment is produced, and thus it has been termed an *amorph*. The other recessive members of the series cause the production of some pigment, but in quantitatively different amounts; they are called *hypomorphs*. An analogous series of hypomorphic alleles, found in the guinea pig, determines the relative amount of melanin in the hair. The probable amorphic allele, when homozygous, results in the absence of melanin or albinism. Another example of a probable amorph is the blood group allele I^O.

In other allelic series phenotypic dominance of one intermediate allele over another may be evident. In one of the first instances of multiple allelism to be studied, R. A. Emerson, in 1911, found in the corn plant a series of alleles which determined the presence of red color in the cob as well as in a maternally derived layer (pericarp) of the seed. An allele in whose presence red color was produced was found to be dominant over a second allele which caused red spotting in these tissues. This allele, in turn, was dominant over a third allele in whose homozygous presence no color at all was formed. A number of additional alleles, some with intermediate effects, have since been identified. In general, it may be said that all variations of dominance, codominance, and incomplete dominance can be found among the multiple allelic series of different organisms.

The identification of the separate allelic states of any gene depends on our ability to distinguish between phenotypes. Sometimes such distinctions are obvious, but on other occasions more than mere observation is required. As an example, seemingly identical alleles may produce the same phenotype under ordinary circumstances. However, when heterozygous with an amorphic allele or when other and modifying genes are present in the genotype or even when environmental conditions are changed, differences in the phenotypes produced by these alleles may become apparent. Genes which exhibit this behavior are called *isoalleles*. Ordinarily they produce the same phenotype, but in some situations they can be shown to differ slightly in expression.

Criteria for Allelism

Since the activities of more than one gene may be involved in the production of a trait, it is frequently necessary to distinguish alleles from nonalleles. To do so merely requires the application of Mendel's principles to plant or animal breeding experiments. For example, suppose a new recessive mutation causing pink eye color, as opposed to wild type red eye color was found in *Drosophila*, and one wished to determine whether or not pink was allelic to another recessive mutant which we can call orange. To test for allelism, we would cross homozygous pink-eyed flies with homozygous orange-eyed flies. If the genes are allelic, the F_1 eye color should be pink or orange or some blend in between, depending on dominance relationships. Upon inbreeding the F_1 flies, a 3 : 1 phenotypic ratio with complete dominance or a 1 : 2 : 1 phenotypic ratio with incomplete dominance should be obtained.

On the other hand, if pink and orange are not alleles, but are independent genes, we would expect the pink-eyed flies to be homozygous for the wild type allele for orange, and the orange-eyed flies to be homozygous for the wild type allele for pink. Therefore, a cross between pink and orange should yield F_1 offspring that are wild type (red-eyed) in phenotype. An inbreed of F_1 flies should produce an F_2 generation composed of $\frac{9}{16}$ red : $\frac{3}{16}$ pink : $\frac{3}{16}$ orange : $\frac{1}{16}$ pink, orange (probably an intermediate color, such as peach). Thus, not only is the phenotype of the F_1 generation indicative, but a 3 : 1 or a 1 : 2 : 1 phenotypic ratio in the F_2 generation is diagnostic of allelism. A 9 : 3 : 3 : 1 phenotypic ratio, variations thereof, or any other significant departure from the 3 : 1 or 1 : 2 : 1 ratios indicate that more than one gene is involved. It should be noted that exceptions to these results may be observed when one or the other of the genes involved is associated with the sex of the organism *(sex linkage)*. However, even in this case similar principles apply and allelism is equally easy to detect (see Chapter 7).

Multiple Alleles and Mutation

Alleles arise by the process of mutation and the presence of a multiple allelic series is evidence of the fact that numerous mutational events involving the same genetic unit have occurred at various times in the past. Studies of mutation indicate that a dominant allele can mutate to other states and that these other states can also change, either back to the original dominant state or to some other allelic condition. For example, the hypothetical gene A can mutate to the recessive allele a^1 or to a^2 or to a^3 and so on, and each of these states can also undergo mutation in either direction. In addition, the frequency with which the respective alleles of a gene mutate varies widely and depends on the allele in question, for example, the rate at which A mutates to a^1 would be significantly different from the rate at which a^1 mutates to A.

These phenomena provoke questions as to the nature of a genetic unit which can exist in such a variety of forms, each of which may lead to a recognizably different phenotype. A concept of the gene as a particle does not allow sufficient scope to deal with the variations observed. It is therefore more reasonable, and indeed easier, to think of the gene as a linear unit of function which offers many sites at which alterations in structure (mutations) could occur, each such structural change altering the function of the gene.

LETHAL GENES

Some allelic states of a gene may exert their effect on superficial characters with negligible consequences to viability. The presence of other genes may affect the functioning of the organism in subtle ways so as to lessen longevity, as well as reproductive and physiological fitness. Still others can result in malformations which may limit the activities and, therefore, the survival of the organism. Finally, some genes produce such drastic alterations in structure or physiology as to render the condition incompatible with life. Genes belonging to this last group are known as *lethals*.

The term lethal gene is loosely used and somewhat confusing, for it suggests that death occurs whenever such an allele is present in the genotype. If this were the case, lethality would be dominant, and all individuals heterozygous for the gene would die. In fact, unless heterozygotes lived long enough to reproduce, a homozygote could not be formed nor could the gene be inherited. Although most lethal genes do not cause death when heterozygous, some genes permitting reproduction, but lethal in a single dose do exist. An example is Huntington's chorea, a dominant inherited disease in humans. The condition is characterized by spasmodic twitchings and by progressive mental deterioration leading inevitably to death. Whether heterozygous or homozygous, all persons who inherit the gene for Huntington's chorea eventually develop the disease, but usually only half of the susceptible individuals show symptoms by age 40. Since this is an age by which families have already been established, the gene can be transmitted to future generations. Because death occurs when the gene is heterozygous, Huntington's chorea can be considered a dominant lethal in the broad sense, even though the age of onset is late.

A more common situation is one where death of the organism occurs only when the allele in question is homozygous, and thus the lethal effect itself is recessive. However, when such a gene is heterozygous, there is often a dominant phenotypic effect, frequently a malformation of some sort. Numerous illustrations of this type of lethal gene exist in both plants and animals. For example, in snapdragons *(Antirrhinum)* a dominant gene called Aurea is known which, when heterozygous, results in the presence of yellow leaves. When homozygous, the gene behaves as a lethal in that young plants fail to develop chlorophyll and in the absence of photosynthesis die of starvation. In mice a factor called Brachyury *(T)* results in the presence of a short tail in heterozygous animals. Homozygotes die as embryos and are characterized by absence of the notochord, reduced and irregular somites, and other abnormalities (Figure 4–5).

Even more numerous are the recessive lethals. A recessive lethal gene produces no obvious effect on the phenotype when present in the heterozygous condition. As in previous examples, the lethal effect is recessive in that death results only when the gene is homozygous. For example, recessive alleles for the Brachyury gene, such as t^0, t^1, t^4, and others, have been described by L. C. Dunn and S. Gluecksohn-Schoenheimer. These genes are lethal when homozygous, but when combined with the wild type allele in a heterozygote, a normal phenotype results. Other examples of recessive lethal genes can be found in all species. They occur with detectable frequency in natural populations, including human populations, and can be present without undue harmful effects in carrier heterozygotes. We must note, however, that such recessive lethals, at least in *Drosophila,* do exert an effect on viability, and if a number of such genes are present in the genotype, a severe viability effect may be observed. In humans, carriers of recessive lethals can sometimes be identified by clinical tests, as in sickle cell anemia or in Tay-Sachs disease. The value to prospective parents of carrier identification is obvious. Unfortunately, the detection of carriers of a number of other recessive lethal conditions in humans is not presently possible.

In addition to genes whose heterozygous or homozygous presence ordinarily results in death, there are a large number which can be classified as semilethal and many more which are deleterious in that viability is lowered to some degree. Viewed overall, the resulting phenotypes form a continuous spectrum ranging from the normal to slight impairment to serious abnormality to death.

When a cross involves an allele having lethal effects, the resulting progeny ratios may be modified depending on the type of cross used. Since those homozygous for a lethal gene ordinarily die without reproducing, only crosses involving heterozygotes need be considered. Ordinarily, a cross between

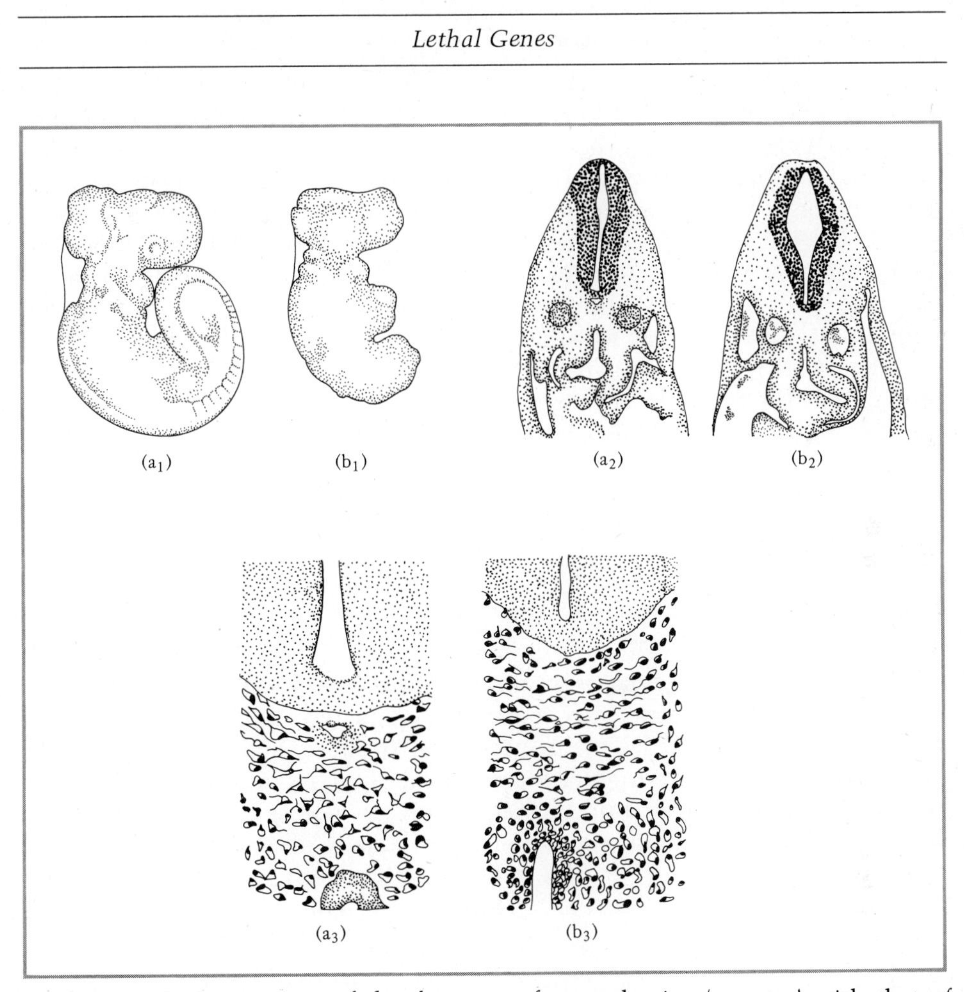

Figure 4–5. Comparison of development of normal mice (a_1 to a_3) with that of homozygous Brachyury *(T/T)* sibs (b_1 to b_3). The pictures illustrate embryos aged 10 days, 9 hours. (a_3) and (b_3) are enlargements of parts of (a_2) and (b_2). In the inviable embryos the notochord does not develop. [From E. Hadorn, *Developmental Genetics and Lethal Factors,* Methuen & Co. Ltd., 1961, originally published by Georg Thieme Verlag, Stuttgart, Germany, 1955.]

heterozygotes is expected to produce a 1 : 2 : 1 genotypic ratio in the progeny, the phenotypic ratio being either 3 : 1 with complete dominance or 1 : 2 : 1 with intermediate or incomplete dominance. When such a cross is made between individuals heterozygous for a lethal gene, $\frac{1}{4}$ of the offspring will be homozygous for the lethal allele. Since death frequently occurs at an early developmental stage, the presence of this class of offspring may not be detected. Thus, recognition that a lethal allele is involved in the cross often depends on phenotypic classification of the viable progeny. Their phenotypes, in turn, depend on whether or not the lethal allele produces a dominant but nonlethal effect when heterozygous. Let us consider two examples as illustrations.

In the following, conventional genetic symbols are used, whereby a plus sign (+) represents the normal, standard, or wild type allele of any gene under consideration. An individual homozygous for this allele can be shown as + / +. As an example of a lethal factor having a dominant phenotypic effect, we will use the Brachyury gene *(T)* of the mouse, which is lethal when homozygous and results in a short tail when heterozygous. The heterozygote is symbolized as *T*/+ and the homozygote as *T*/*T*. A cross between mice heterozygous for Brachyury is illustrated in Figure 4–6. Notice that the true phenotypic ratio is 1 : 2 : 1 and therefore identical with the genotypic ratio. However, since the homozygous *T*/*T* class dies *in utero* and thus does not appear among the viable offspring, the phenotypic ratio actually observed is 2 : 1, that is, 2 short tail : 1 normal tail. To verify that all short-tailed mice are *T*/+, a backcross to +/+ may be carried out as shown below. Since no Brachyury homozygotes are produced by the cross, the two expected classes of progeny appear in the anticipated 1 : 1 phenotypic and genotypic ratios.

	T/+ (short tail)	× ↓	+/+ (normal tail)
	T/+ (short tail)		+/+ (normal tail)
Genotypic ratio	1	:	1
Phenotypic ratio	1	:	1

Congenital hydrocephalus *(ch)* in the mouse can be chosen as an example of a recessive lethal trait. This condition is characterized by abnormalities in the development of cartilage which secondarily affect the growth of the skull and brain and the distribution of cerebrospinal fluid. Homozygous *ch*/*ch* mice are stillborn. As illustrated in the diagram below, a cross between individuals heterozygous for this factor yields a phenotypic ratio of 3 normal : 1 abnormal. Since the presence of the lethal is masked by the dominant wild type allele, heterozygous *ch*/+ individuals cannot be distinguished from homozygous normal +/+ mice.

			ch/+ (normal) × ↓ *ch*/+ (normal)		
	ch/*ch* (stillborn)		*ch*/+ (normal)		+/+ (normal)
Genotypic ratio	1	:	2	:	1
Phenotypic ratio	1	:		3	

In general, the presence of a lethal gene, whether dominant or recessive, is best recognized through the identification of the homozygous lethal class which results from an inbreed of heterozygotes. Where such individuals are eliminated very early in development and therefore do not appear, a marked decrease in the number of offspring produced, as well as the 2 : 1 phenotypic ratio in the case of a dominant allele, are good indications of the presence of such a gene.

The action of genes at the cellular level is unitary, that is, one gene, one action. Sometimes, however, the presence of a given gene results in a whole spectrum of phenotypic changes so that it appears that the gene has multiple actions. This phenomenon is called *pleiotropism,* and it is found primarily in higher organisms where complex and interrelated developmental events occur.

Many lethal genes are pleiotropic, especially those which affect early embryonic stages, for an initial divergence from the normal developmental pattern

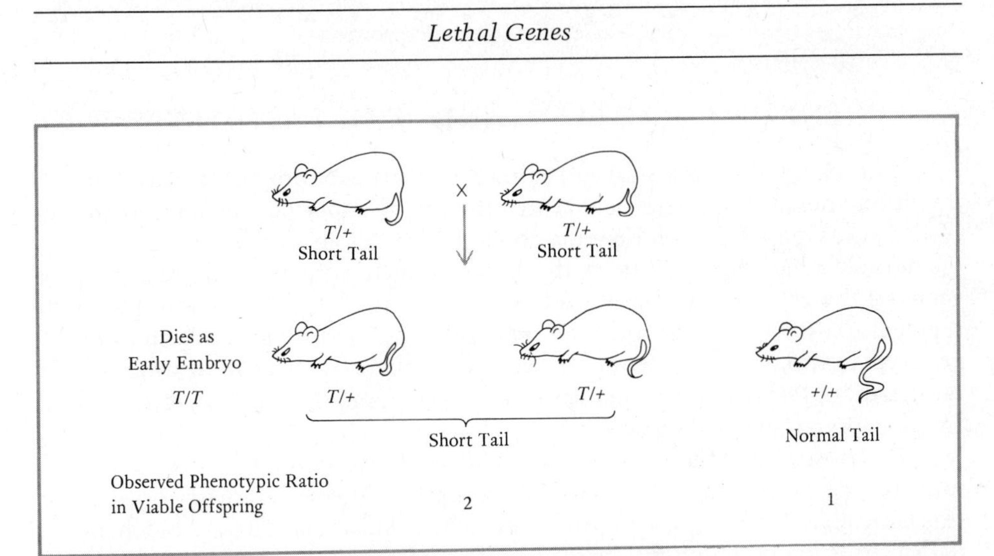

Figure 4–6. Progeny which result from a cross between mice heterozygous for the dominant gene for Brachyury. A phenotypic ratio of 2 : 1 is found in the viable offspring.

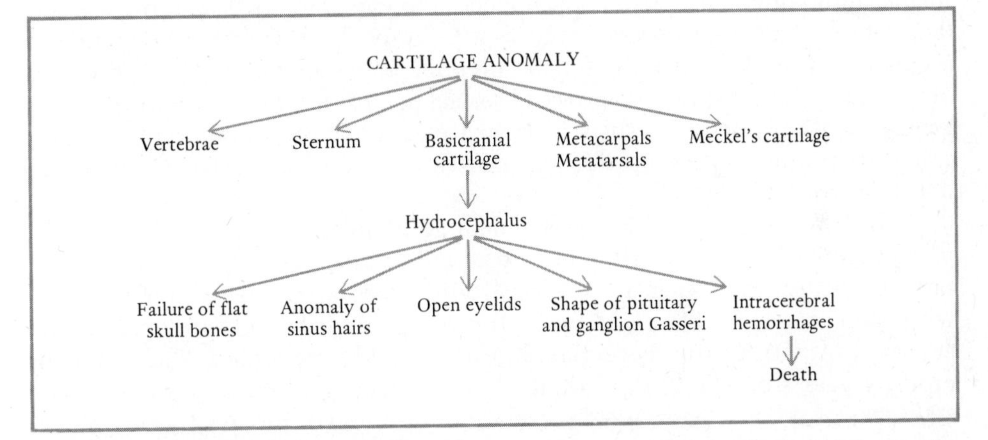

Figure 4–7. Pleiotropism illustrated by the pattern of damage which occurs in mice homozygous for congenital hydrocephalus *(ch)*. [From H. Grüneberg, *J. Genetics, 45,* 1–21, 1943.]

can produce widespread changes in many organ systems. Examples of pleiotropic genes are Brachyury and congenital hydrocephalus in the mouse. In mice homozygous for Brachyury a failure in notochord formation secondarily results in disorganized and missing somites such that the whole posterior portion of the body is malformed. Congenital hydrocephalus is an even more striking example. Here, a fundamental defect in cartilage secondarily affects the skull, ribs, and other skeletal parts. These, in turn, impose changes upon associated structures such as the brain. The final result is an array of abnormalities which involve many organ systems of the body (Figure 4–7).

Attempting to sort out the multiple effects of a pleiotropic gene in an effort to discover a first cause becomes a formidable task. It is much easier to associate a gene with a clear-cut biochemical deficiency than it is to show that a particular gene is the cause of the first of a series of morphological events.

GENE INTERACTION AND THE PHENOTYPE

The Mendelian ratios discussed in preceding sections result from crosses involving independent genes affecting different phenotypic traits, for example, seed coat as opposed to height or color. However, in numerous instances a phenotypic character results from the combined action of several different genes. If the genes involved are independent in all save their action, they will segregate, assort, and recombine to yield the expected and appropriate genotypic ratio, but the phenotypic ratio may represent a modification of that ordinarily observed. Some examples of gene interaction in the production of the phenotype will be considered.

In *Drosophila,* the normal red, wild type eye color (+) results from a mixture of two kinds of pigments, red and brown, whose chemistry does not at this time concern us. These pigments are synthesized via separate biochemical pathways, each of which is controlled by a number of different genes. If a gene active in an early step in the synthesis of red pigment mutates to an inactive state, the entire pathway may be rendered nonfunctional so that no red pigment can be produced. As a consequence, the eye color will be brown. Similarly, if such a mutation occurs in a gene active in the pathway for brown pigment, no brown pigment will be present, and the eye will be a brilliant orange-red. Mutants of this type are observed in *Drosophila;* a recessive mutant gene called brown *(bw)* results in brown eye color when homozygous and a recessive mutant called scarlet *(st)* causes the eyes to be orange-red when homozygous.

The interaction of these independent genes in the production of the final phenotype is best seen in the progeny of a dihybrid inbreed. The phenotypic ratio resulting from such a cross is $\frac{9}{16}$ wild type : $\frac{3}{16}$ brown : $\frac{3}{16}$ orange-red : $\frac{1}{16}$ white (Figure 4–8). The wild type progeny result from the presence in the genotype of both of the respective dominant wild type alleles. The $\frac{3}{16}$ brown progeny possesses a wild type allele for *st,* but are homozygous *bw/bw*. The synthesis of red pigment fails to occur and only brown pigment is present. In the $\frac{3}{16}$ orange-red progeny red pigment is formed, but the production of brown pigment is blocked. Finally, in the $\frac{1}{16}$ of the progeny homozygous for both *bw* and *st,* both synthetic pathways fail. As a consequence, neither red nor brown pigment is produced, and the eye is white. The genes *bw* and *st* are thus related in their action because each contributes to the final phenotype. In this case the 9 : 3 : 3 : 1 phenotypic ratio is not altered, but a new and unexpected phenotype (white eyes) results from the combined effects of the two independent genes.

Other types of gene interaction can give rise to modified ratios. As an example, we can again use eye color mutants of *Drosophila*. In addition to the recessive gene scarlet *(st),* another recessive, but independent gene called cinnabar *(cn)* is also known to produce vivid orange-red eyes when homozygous. The phenotype produced by *st/st* is almost impossible to distinguish from the phenotype produced by *cn/cn,* for in both cases the synthesis of brown pigment is blocked. The brown pigment is the end product of a biochemical pathway composed of a series of steps, each step under the control of a separate gene. Thus, a mutation which changes any one of the genes involved to an inactive state will yield the same result: no brown pigment. Cinnabar and scarlet are

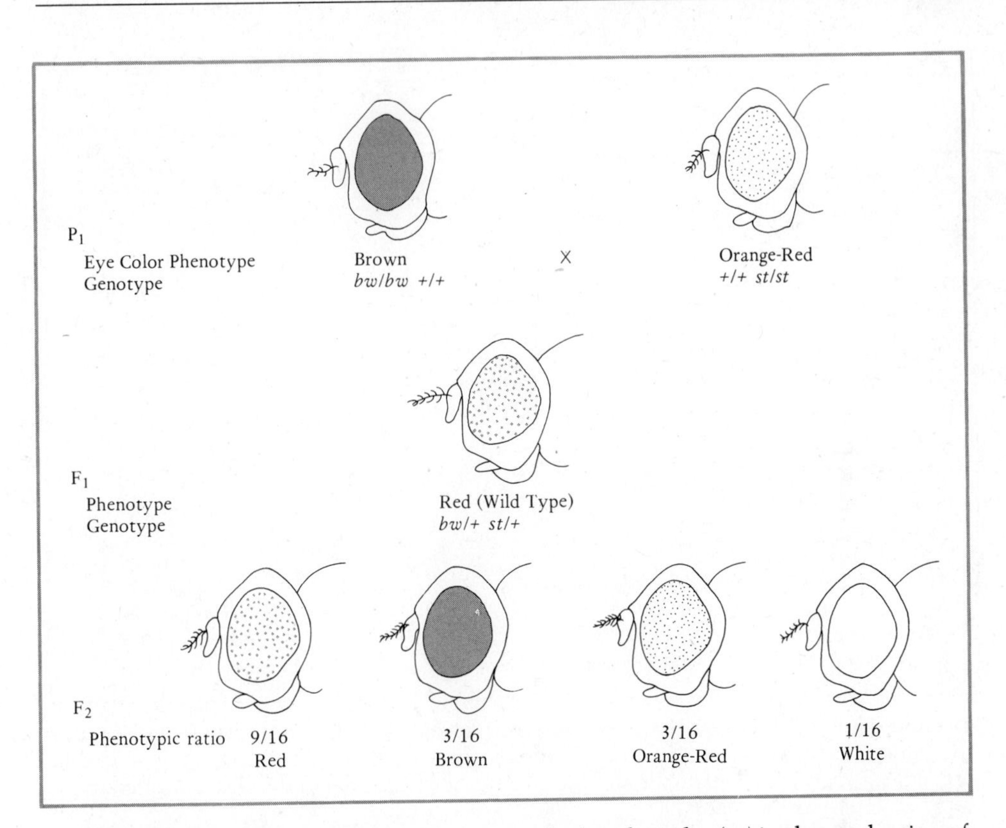

Figure 4–8. The interaction of genes brown *(bw)* and scarlet *(st)* in the production of eye color phenotypes in *Drosophila*.

examples of mutant genes whose actions are exerted at different steps in the same pathway, that is, the pathway which leads to brown pigment. As a result, a dihybrid cross involving these mutants and their respective wild type alleles *(cn/+ st/+ × cn/+ st/+)* produces the modified phenotypic ratio of 9 wild type : 7 orange-red. The basis for the ratio is the familiar 9 : 3 : 3 : 1 phenotypic ratio customarily observed, but in this case the last three terms are added together because individuals homozygous for *cn* or homozygous for *st* or homozygous for both *cn* and *st* cannot be distinguished phenotypically from one another. Notice, however, that the terms of the ratio still sum to 16.

Sometimes an expected ratio may be altered due to a phenomenon called *epistasis*. The term epistasis is applied to situations where the expression of one gene interferes with or prevents the expression of another, independent gene, this relationship being nonreciprocal. For example, eye color mutants such as *st, bw,* or *cn* are not expressed at all in the presence of the gene white *(w),* which blocks any and all eye pigment formation. The gene for white can be called epistatic to other eye color genes in that it prevents their expression. A similar case has been described in the mouse where a gene for albinism prevents the expression of coat color alleles. The recessive albino condition in humans behaves in a like fashion.

When epistatic genes are involved in a dihybrid cross, modified ratios often result. For example, in *Drosophila* the homozygous presence of the re-

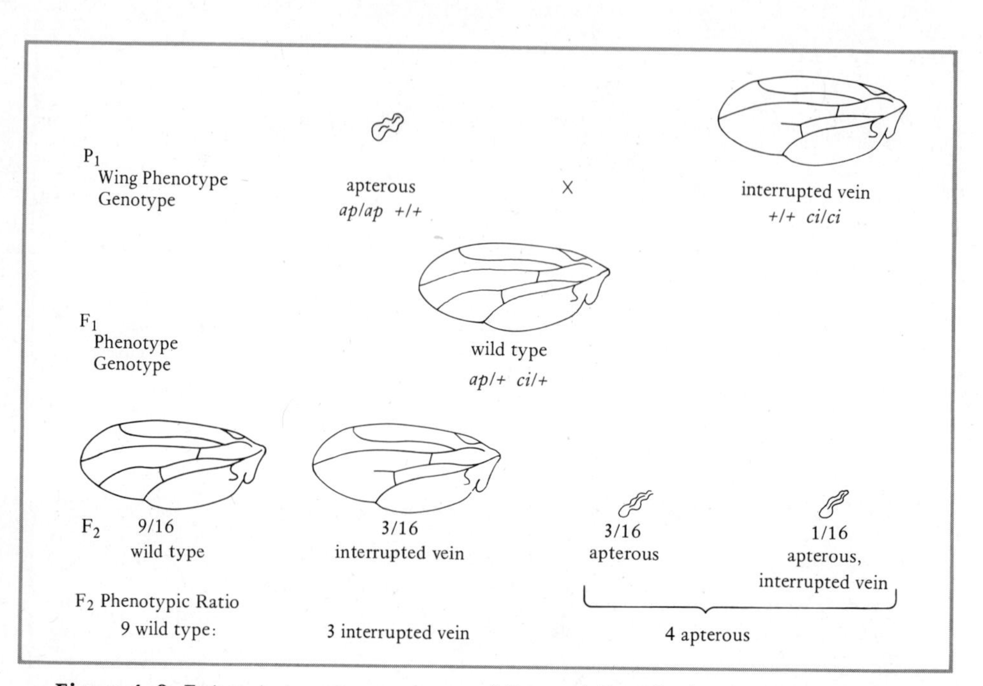

Figure 4–9. Epistasis in wing mutants of *Drosophila.* The homozygous presence of the gene apterous *(ap)* masks the expression of the gene interrupted vein *(ci),* thus preventing identification of the double homozygote as a separate class. A phenotypic ratio of 9 : 3 : 4 is the result.

cessive mutant apterous *(ap)* results in small, crumpled, stubby wings instead of the normal expanded, transparent wings. The homozygous presence of another recessive mutant known as *cubitus interruptus* or interrupted vein *(ci)* causes a small discontinuity to occur in the cubitus vein, the fourth longitudinal vein of the wing. From a dihybrid inbreed involving these mutants, a 9 : 3 : 4 ratio would be observed, $\frac{9}{16}$ wild type : $\frac{3}{16}$ interrupted wing vein : $\frac{4}{16}$ apterous wing (Figure 4–9). This ratio is modified from the classical 9 : 3 : 3 : 1 distribution in that the last two classes are phenotypically indistinguishable from one another and are therefore added together. In each of these two classes *ap* is present in the homozygous condition and the resulting shriveled stublike wing masks the expression of *ci* and prevents the identification of the double recessive as a separate class. Many examples of epistasis are known in a variety of plants and animals. Depending on the interrelationships of the particular genes involved, 15 : 1, 12 : 3 : 1, and other variations of the 9 : 3 : 3 : 1 ratio are often observed.

An interesting case of epistasis occurs in humans. The antigens of the ABO blood groups are found in an alcohol soluble form on the red blood cells. In the majority of persons, called *secretors,* these antigens are also found in water soluble form in secretions such as saliva and gastric juice. The presence of these antigens in saliva is due to a dominant secretor gene *(Se),* and the absence of antigens in secretions is due to homozygosity of the recessive allele *(se).* Marriages between individuals heterozygous for dominant and recessive

P_1 Phenotype: Blood Type B, Secretor × Blood type B, Secretor
P_1 Genotype: $I^B I^O$ *Se/se* × $I^B I^O$ *Se/se*

F_1 offspring:

$I^B I^B$ *Se Se* $I^B I^B$ *Se se* $I^B I^O$ *Se Se* $I^B I^O$ *Se se*	$I^B I^B$ *se se* $I^B I^O$ *se se*	$I^O I^O$ *Se Se* $I^O I^O$ *Se se*	$I^O I^O$ *se se*
9/16 Type B, secretor	3/16 Type B, nonsecretor	3/16 Type O, secretor	1/16 Type O, nonsecretor
Phenotypic Ratio 9		7	

Figure 4–10. A case of epistasis in humans involving blood group and secretor genes. A phenotypic ratio of 9 : 7 is found with respect to the presence of antigen in secretions. The ratio is based on the combined data of many families.

blood group alleles and also heterozygous for the secretor alleles represent typical dihybrid inbreeds, i.e., I^B/I^O *Se/se* × I^B/I^O *Se/se* or I^A/I^O *Se/se* × I^A/I^O *Se/se.* Although the red blood cells will always contain the blood group antigen designated by the genotype, offspring can also be classified according to the presence or absence of antigen in saliva. In this case we would expect that $\frac{9}{16}$ of the offspring would show both dominant phenotypes (antigen B and secretion), while $\frac{7}{16}$ would have no antigen in their saliva. The $\frac{7}{16}$ actually consists of $\frac{3}{16}$ blood group B nonsecretors, $\frac{3}{16}$ blood group O secretors, and $\frac{1}{16}$ blood group O nonsecretors. The members of these three classes contain no antigen in the saliva either because of absence of antigen (blood group O) or absence of the ability to produce an antigen in water soluble cell products (nonsecretors). Epistatic interaction between these genes is thus observed and a 9 : 7, rather than a 9 : 3 : 3 : 1 phenotypic ratio is obtained when the children of many such marriages are scored (Figure 4–10).

Other instances of interactions between gene controlled cell products can be found in humans and other organisms especially with respect to enzymes. Frequently an enzyme is composed of two or more subunits produced through the actions of two or more genes. Mutual interaction between such subunits necessarily occurs and may modify overall enzyme activity. In other cases more than one molecular form of an enzyme may be capable of performing the same catalytic function *(isozymes)*. Examples such as these are more fully discussed in subsequent chapters.

DEGREES OF GENE EXPRESSION

In observing the offspring from a cross, we tend to think of the phenotype in terms of an all or none phenomenon—a trait is expressed or it is not—and to draw conclusions as to genotype based on that form of expression. While it is true that some phenotypes, such as the ABO blood group antigens, conform to this expectation, not all genes are expressed in such an absolute and clear-cut fashion. Instead, the expression of a gene may be variable.

In some cases not all individuals of appropriate genotype actually show the corresponding phenotypic trait. This aspect of gene action, that is, whether or not the gene finds expression in the phenotype, is called *penetrance.* A completely penetrant gene is always expressed; an incompletely penetrant gene may be expressed in some individuals, but not others.

A gene may also vary in the degree to which it is expressed in different individuals or in different parts of the same individual. This *variable expressivity* may result in a range of phenotypes displaying slight to extreme effects of gene action. As an example, in *Drosophila* the recessive gene eyeless, when present in homozygous condition, causes a reduction in the size of the compound eye. However, the extent to which the eye is reduced varies considerably from individual to individual. In some perhaps only a slight decrease in eye size occurs, while in others the eye may be completely absent. Moreover, in any one fly the left eye may be almost normal in appearance, while the right eye may be drastically reduced. A similar range in gene expression is also often encountered in humans, for example, in inherited skeletal abnormalities such as polydactyly. Extra digits may be present on only the feet, only the hands, or on all extremities.

The source of this variation is partly genotypic and partly environmental. The genotype contains thousands of genes and the actions of many are interrelated so as to modify one another's effects. Such modifying genes may be present in a population in different allelic forms so that the genetic background of one individual may differ significantly from that of another, leading to a variation in the expression of the particular gene under consideration.

Modifying genes can act as enhancers such that in their presence the phenotype produced by an entirely different gene will be more strongly evident. *Phenotypic enhancement* is not hard to visualize for almost any trait results from the sequential activities of several to many genes, for example, a biochemical pathway whose several steps are each controlled by a different gene. The presence of a hypomorphic allele for more than one such step could easily cause a phenotype to be expressed more strongly.

A different and more puzzling phenomenon is that of *phenotypic suppression,* where one gene suppresses or seemingly reverses the action of a second gene so that the normal, and not the mutant phenotype is produced. Barring the presence in the genotype of an extra, duplicated wild type allele, phenotypic suppression is difficult to explain. Presumably, it occurs at the time of phenotype initiation, for only at this point can the development of a mutant phenotype be suppressed or reversed. This presumption has been verified by studies in which the cause of gene suppression has been successfully demonstrated in particular instances (see Chapter 17).

ENVIRONMENTAL MODIFICATION OF GENE EXPRESSION

In general, the phenotype can be altered by either of two processes: (1) mutation or (2) adaptation of the individual to environmental conditions. The first of these, mutation, involves a change in the genotype secondarily reflected in the phenotype. Such a change in genotype is inherited. In contrast, phenotypic adaptation involves a noninherited response to a particular environment. Let us consider the changes in phenotype which are induced by the environment, reserving a discussion of mutation to a later chapter (Chapter 17).

The genotype establishes a fixed potential for the expression of any trait, but whether or not and to what extent that potential is realized depends on the environment, both internal and external. For example, it is obvious that a potential for normal growth cannot be achieved under conditions of deprivation. Such factors as light, temperature, moisture, minerals, nutrients or diet, vitamins, hormones, and other aspects of the total environment play a major and permissive role in gene expression. The wild type organism, plant or animal, presents the wild type phenotype under nonlimiting environmental conditions. However, the same organism placed in a harsh and restrictive environment must adapt its physiology for survival. As a consequence, the degree of expression of a given gene may change leading to a corresponding change in some aspect of the phenotype. Thus, although the genotype is fixed, the response of the genotype to varying environmental conditions is flexible to the extent that component genes can vary in their range of expression.

Some genes, such as the blood group alleles, exhibit all-or-none expression regardless of the environment. Others, and especially those with variable penetrance and expressivity, may or may not be expressed. An individual susceptible to a phenotype having variable penetrance may not happen to encounter the environmental conditions necessary for the development of the trait. For example, it has been found that many strains of organisms, from bacteria to mammals, possess temperature-sensitive alleles which are active only within certain defined temperature ranges. One such temperature-sensitive allele is found in rabbits. The coat color of rabbits is determined by a series of multiple alleles. One of these alleles, the recessive gene a^n, determines a striking pigmentation pattern called Himalayan, in which the body is albino and the extremities are black. A Himalayan rabbit (a^n/a^n) thus has white fur over most of its body, but black fur on the nose, ears, paws, and tail (Figure 4–11). The allele

Figure 4–11. The phenotypic appearance of Himalayan rabbits under different temperature conditions: (a) a rabbit raised at a temperature above 30°C; (b) a rabbit raised at a temperature about 25°C; (c) a rabbit which has had the left flank artificially cooled at a temperature below 25°C. [From R. P. Wagner and H. K. Mitchell, *Genetics and Metabolism*, 2nd ed., John Wiley & Sons, New York. After Daneel, *Ergeb. Biol.*, 18, 55–87, 1941.]

a^{h} is temperature sensitive and, when homozygous, is capable of initiating melanin pigment formation at temperatures below 34°C. Since the extremities are ordinarily somewhat cooler than the rest of the body, black fur results. However, if the extremities are artificially warmed or the rabbit is raised at a higher temperature, new hair grows in white. Again, if some region of the body is artificially cooled, new hair is black. Other members of this allelic series, chinchilla, agouti, albino, and others, do not exhibit temperature sensitivity in their phenotypic expression.

Instances of temperature dependent gene expression can also be found in *Drosophila* where, for example, homozygous vestigial winged flies *(vg/vg)* raised at temperatures above 29°C will develop wings that are almost normal in phenotype. Many other similar examples can be cited from among the mutants of *Drosophila* as well as those of other organisms, such as the mold *Neurospora* or various types of bacteria. Studies of such mutants suggest that temperature sensitivity can be attributed to altered forms of cellular proteins, especially enzymes, which may require a higher temperature for their activity or may be unstable in certain temperature ranges.

Efforts to modify the internal and external environment have employed temperature shocks, radiation, poisons, vitamin deficiencies, and a wide range of chemical agents. Application of such treatments to the normal organism has in many cases resulted in phenotypes that mimic those produced by known genes. Such induced phenotypes are called *phenocopies*. It has frequently been

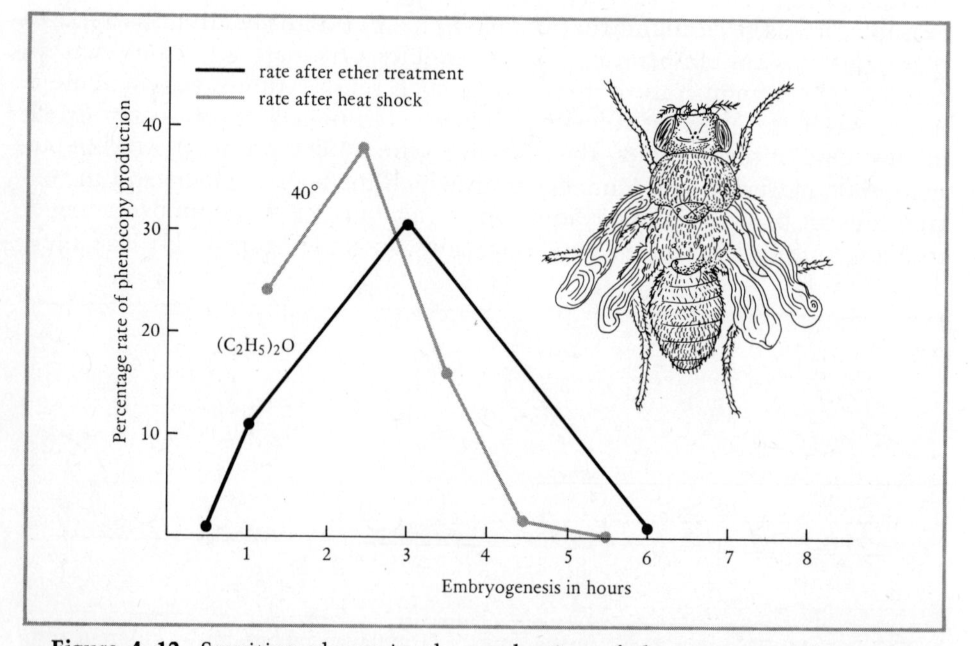

Figure 4–12. Sensitive phases in the production of the tetraptera phenocopy in *Drosophila melanogaster* by means of ether and heat shock. Drawing shows a tetraptera phenocopy (as the fly had to be dissected out of the puparium, the wings have not unfolded). [From E. Hadorn, *Developmental Genetics and Lethal Factors*, Methuen & Co. Ltd., 1961, originally published by Georg Thieme Verlag, Stuttgart, Germany, 1955.]

observed that a phenocopy can be produced only if the treatment is applied during a sensitive stage in development, application at other stages being ineffective. Presumably, such sensitive periods mark a turning point or critical period in embryonic determination or differentiation. The phenotype called tetraptera in *Drosophila* can be used as an illustration. Although most higher insects possess two pairs of wings, members of the Order Diptera, which includes *Drosophila,* have only one pair of wings, the second pair being reduced to small knoblike structures (halteres) which serve as balancing organs. In the homozygous presence of the recessive gene tetraptera, these small balancers develop into true winglike structures. When a wild type embryo is treated by heat shocks or exposure to ether, a phenocopy of the tetraptera condition is produced in the adult fly (Figure 4–12). The sensitive period during which such treatment is effective occurs a few hours after fertilization, although the wing itself does not differentiate until much later, at the time of metamorphosis. These findings suggest that in *Drosophila* the ultimate fate of embryonic cells is decided during very early embryonic stages.

Numerous attempts with various agents to induce phenocopies in vertebrates have produced phenotypes that mimic those caused by lethal genes. Even vitamin deficiencies are effective. For example, hens fed a diet deficient in biotin produce chicks characterized by malformations of the long bones and beak. This phenotype is practically identical with that which results from the homozygous presence of the recessive gene for chondrodystrophy (Figure 4–13). Other modifications of the internal environment also bring about changes in phenotype, for oxygen or carbon dioxide tension, pH, level of nu-

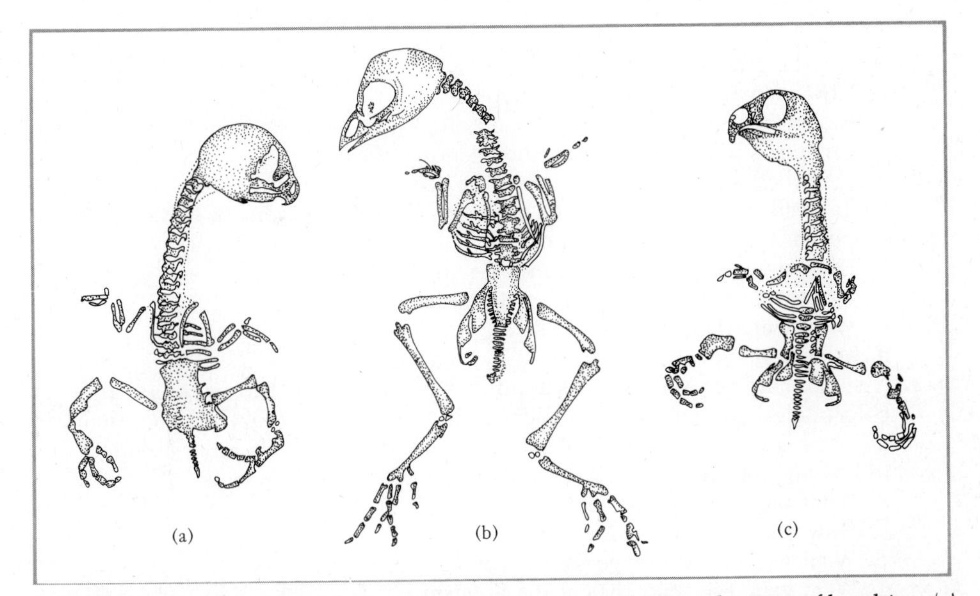

Figure 4–13. The development of the skeleton of the chick at the time of hatching: (a) pattern of damage of the lethal factor chondrodystrophy *(ch);* (b) normal condition on the twenty-first day; (c) effect of biotin deficiency. [From E. Hadorn, *Developmental Genetics and Lethal Factors,* Methuen & Co. Ltd., 1961, originally published by Georg Thieme Verlag, Stuttgart, Germany, 1955.]

trients, minerals, and many other factors can affect the expression of genes. Additional agents obviously capable of inducing phenotypic changes are hormones, whose actions as chemical coordinators and growth stimulators are pervasive, affecting every organ system of the body. Changes in the level of various hormones, brought about by accident or disease, rather than by genotype, can cause profound alterations in metabolism, appearance, and behavior.

In summary, the expression of any gene can rarely be considered as an activity separate and distinct from the environment in which it occurs. Although gene action is usually compared or described under normal conditions, it must be remembered that different degrees of expression may be evoked by significant departures from the standard environment. It is the response of the genotype to such differing environments that makes possible the adaptation and survival of the individual.

PROBLEMS

4–1. The following crosses in fowl produced the results listed.

Cross	F_1
1. black × blue-splashed-white	27 blue
2. blue-splashed-white × blue-splashed-white	25 blue-splashed-white
3. blue-splashed-white × blue	15 blue, 17 blue-splashed-white
4. black × blue	20 black, 18 blue
5. blue × blue	10 black, 18 blue, 9 blue-splashed-white

a. Are the genes involved allelic or nonallelic?
b. How many pairs of genes are involved?
c. Assign symbols to represent a dominant or recessive gene. What are the genotypes of the parents and the offspring of each cross?

4–2. A black cock was crossed to a pure white hen and all offspring were blue in color. An inbreed of the F_1 produced the following classes of progeny: $\frac{3}{16}$ black, $\frac{6}{16}$ blue, $\frac{3}{16}$ blue-splashed-white, and $\frac{4}{16}$ white.
a. How many pairs of genes are involved?
b. What were the genotypes of the P_1 birds?
c. What were the genotypes of the F_1 birds?
d. What genotypes are represented in each F_2 class?

4–3. Assume that in pigs black *(B)* is dominant over the spotted condition *(b)* and that color *(C)* is dominant over noncolor *(c)*. Black pigs crossed to white pigs produced offspring all of which were black. When inbred, these offspring produced the following progeny: 45 black, 16 spotted, and 21 white.
a. What ratio do the data represent?
b. How many gene pairs are involved?
c. What were the genotypes of the parents?
d. What were the genotypes of the F_1 offspring?
e. What genotypes would be found among the F_2 white individuals and in what proportions?

4–4. In a second breeding experiment three of the above F_2 black pigs were crossed to three of the F_2 white pigs with the following results.

Cross	Progeny
1. black × white	$\frac{1}{2}$ black, $\frac{1}{2}$ black spotted
2. black × white	$\frac{1}{2}$ black, $\frac{1}{2}$ white
3. black × white	$\frac{1}{4}$ black, $\frac{1}{4}$ spotted, $\frac{1}{2}$ white

What were the genotypes of the parents in each cross?

4–5. Two strains of tan colored pigs were crossed and all of the F_1 individuals were brown in phenotype. Crosses between these F_1 individuals produced 91 brown, 59 tan, and 10 white offspring.

a. What ratio do the data represent?
b. How many pairs of genes are involved?
c. Assign gene symbols to represent a dominant versus a recessive allele. What were the genotypes of the P_1, the F_1, the F_2 tan individuals, and the F_2 white individuals?

4–6. Whether or not oat grains are red or nonred is determined by the dominant gene *R* and its recessive allele *r*. Nonred grains *(rr)* may be yellow or white, these characteristics being determined by a dominant gene *Y* (yellow) and its recessive allele *y* (white). A cross between plants heterozygous for both pairs of genes produced 101 red, 23 yellow, and 8 white offspring.

a. What ratio do the data represent?
b. What genotypes would be found among the red offspring?
c. What genotypes would be found among the yellow offspring?
d. What proportion of the red individuals will produce only red progeny if allowed to self-fertilize?
e. What proportion of the yellow progeny will produce only yellow individuals if allowed to self-fertilize?

4–7. In cats intense black color is caused by the presence of the dominant gene *C*, while the Siamese pattern results from homozygosity of the recessive allele *c*. However, if the dominant gene *W* (white) is present, no color is formed. A cross between two white cats produced all white kittens. When some of the F_1 kittens were inbred, all of the offspring were white. When others of the F_1 kittens were inbred, progeny in the ratio of $\frac{12}{16}$ white : $\frac{3}{16}$ black : $\frac{1}{16}$ Siamese were produced. Further crosses between F_1 cats showed that no other results could be obtained.

a. What were the genotypes of the original parents?
b. What genotypes were present in F_1 individuals and in what proportions?
c. What F_1 crosses would produce the F_2 data obtained?

4–8. When the gene *W* (dominant white) is present in cats, the eyes are either yellow or blue. When blue eyes are present, they are accompanied by deafness. If only one eye is blue, the cat is deaf on the same side. Which of the following terms would you use to describe this situation and why: epistasis, penetrance, expressivity, pleiotropism.

4–9. In dogs the genes for straight versus curly hair and for black versus pink tongue exhibit incomplete dominance. Straight-haired males with pink tongues were crossed to curly-haired females with black tongues. The puppies had wavy hair and gray tongues. Some F_1 individuals backcrossed to the male parent produced the following offspring.

$\frac{1}{4}$ straight hair with gray tongue
$\frac{1}{4}$ straight hair with pink tongue
$\frac{1}{4}$ wavy hair with gray tongue
$\frac{1}{4}$ wavy hair with pink tongue

Other F_1 individuals backcrossed to the female parent produced the following offspring.

$\frac{1}{4}$ wavy hair and black tongue
$\frac{1}{4}$ wavy hair and gray tongue
$\frac{1}{4}$ curly hair and black tongue
$\frac{1}{4}$ curly hair and gray tongue

a. Assigning symbols to designate the genes involved, what were the genotypes of the parents?
b. Using the same symbols, what were the genotypes of the F_1 individuals?

4–10. If the F_1 individuals of the preceding problem were inbred, what phenotypes and genotypes would be obtained in the F_2 generation and in what proportions?

4–11. Docked tail as opposed to long tail occurs in some breeds of dogs. A long-tailed female crossed to a long-tailed male produced 12 puppies, all with long tails. A long-tailed female crossed to a docked-tail male produced 12 puppies, half with long tails and half with short tails. A docked-tail female crossed to a docked-tail male produced 9 puppies, 6 with docked tail and 3 with long tails. How would you explain the inheritance of the docked tail trait?

4–12. Wavy-haired, docked-tail, pink-tongued dogs are crossed to curly-haired, docked-tail, gray-tongued dogs.
a. What are the genotypes of these parents?
b. What proportion of the fertilizations will give rise to viable puppies?
c. Of the viable puppies, what proportion will be wavy-haired?
d. What proportion will have pink tongues?
e. What proportion will have curly hair and gray tongues?
f. What proportion will have long tails?
g. What proportion will have curly hair, pink tongues, and docked tails?

4–13. In guinea pigs the intensity of coat color varies from black (variety A) to dark sepia (variety B) to medium sepia (variety C) to white (variety D). P_1 and F_1 crosses were made between these varieties with the following results.

P_1	F_1	F_2
A × B	black	3 black : 1 dark sepia
A × C	black	3 black : 1 medium sepia
A × D	black	3 black : 1 white
B × C	dark sepia	3 dark sepia : 1 medium sepia
B × D	dark sepia	3 dark sepia : 1 white
C × D	light sepia	1 medium sepia : 2 light sepia : 1 white

Explain the inheritance of coat color intensity.

4–14. The following P_1 crosses were made between mice of different coat colors. F_1 offspring of the same phenotype were then inbred to produce the F_2 offspring given below.

P_1	F_1	F_2
1. agouti × agouti	all agouti	all agouti
2. agouti × nonagouti	all agouti	3 agouti : 1 nonagouti
3. agouti × yellow	$\frac{1}{2}$ agouti	all agouti
	$\frac{1}{2}$ yellow	2 yellow : 1 agouti
4. yellow × nonagouti	$\frac{1}{2}$ agouti	3 agouti : 1 nonagouti
	$\frac{1}{2}$ yellow	2 yellow : 1 nonagouti

a. What were the genotypes of the parents in each of the four crosses?
b. Which of the genes involved in these crosses are dominant and which recessive?

c. Are the genes involved in these crosses alleles or nonalleles?
d. What is the basis for the ratios obtained in the F_2 progeny of yellow mice?

4–15. In mice the Brachyury gene *(T)* causing short tail is dominant when heterozygous and lethal when homozygous. An allele of *T*, called t^0, is recessive to the wild type long-tailed condition, but is also lethal when homozygous. The heterozygote T/t^0 is viable.
a. If a cross was made between individuals heterozygous for *T* and t^0, what would be the phenotype(s) and genotype(s) of the viable progeny?
b. If the normal litter size is 12, approximately how many living mice would be born?

4–16. In *Drosophila* a mutant, called curly, causes the wings to be curled up over the back, while another mutant, called stubble, results in the presence of short, stubby bristles, as opposed to the normal long, pointed bristles.

A curly female, homozygous for normal bristles, was crossed to a stubble male, homozygous for normal wings. The offspring consisted of:

15 curly wing, long bristle
14 curly wing, stubble bristle
16 normal wing, long bristle
13 normal wing, stubble bristle

a. Are the genes for curled wing and stubble bristle dominant or recessive?
b. What were the genotypes of the parents?
c. What are the genotypes of the offspring?

4–17. When F_1 curly wing, stubble bristle flies (from the preceding question) were inbred, the F_2 hatching flies consisted of four phenotypic classes which appeared in the ratio of 4 : 2 : 2 : 1 as follows.

4 curly wing, stubble bristle
2 curly wing, normal bristle
2 normal wing, stubble bristle
1 normal wing, normal bristle

Explain the basis for this ratio.

4–18. In *Drosophila* a dominant gene, hairless *(H)*, causes the absence of bristles of the head and body. *H/H* is lethal. An independent and also dominant gene, suppressor of hairless *(Su-H)*, reverses the effects of hairless when both are present in the heterozygous condition. *Su-H* has no phenotypic effect by itself when heterozygous, but is lethal when homozygous. Hairless females were crossed to hairy males heterozygous for *Su-H*. The progeny consisted of 11 hairless and 29 hairy flies. What proportion of the hairy progeny carried the gene for hairless?

4–19. If hairy flies heterozygous for both *H* and *Su-H* are crossed, what proportion of the total progeny will be inviable? What proportion of the viable progeny will be hairy in phenotype, hairless in phenotype, heterozygous for both hairless and *Su-H*?

4–20. A woman of blood group A marries a man of blood group O. There are three children in the family. The children's blood types are O, A, and AB. Which child was adopted?

4–21. Parents of blood types AB and O wish to know the probability that of three children two will be blood group A and one will be blood group B. What would you predict?

4–22. What genotypes and their proportions would be produced by the following crosses.

a. $I^AI^O \times I^OI^O$
b. $I^AI^B \times I^AI^O$
c. $I^BI^B \times I^AI^O$
d. $I^AI^B \times I^AI^B$

4–23. A woman whose father was blood group AB and whose mother was A marries a man of blood group B. Of their two children one belongs to blood group O and the other to group A.
a. What is the genotype of the woman?
b. What is the genotype of her husband?
c. What is the genotype of her mother?

4–24. Four persons were found to have the following Rh antigenic specificities.

1. rh′, hr″
2. Rh_0, rh′, hr″
3. Rh_0, rh′, rh″, hr′
4. rh′, rh″, hr′

a. What possible genotype(s) can be assigned in each case?
b. Which phenotypes would be classified clinically as Rh positive?

In problems 4–25, 4–26, 4–27, and 4–28, assume that the Rh phenotype is determined by a single pair of alleles, Rh^+ (dominant) and Rh^- (recessive).

4–25. A woman who suffered hemolytic disease of the newborn when an infant marries an Rh^- man. Would you expect that any of their children would suffer from this disease?

4–26. A woman whose older brother suffered hemolytic disease of the newborn did not herself have the disease when an infant.
a. What are the probable genotypes of the woman, her mother, and her father?
b. If she marries a man heterozygous for Rh^+, what is the probability that any of her children will have this disease?

4–27. Consider the cross: I^A/I^O, MN, Rh^+/Rh^-, $se/se \times I^B/I^O$, MN, Rh^+/Rh^-, Se/se. With respect to phenotype,
a. What proportion of the progeny will be blood group AB?
b. What proportion will be blood group B and M?
c. What proportion will be A, MN, Rh^+?
d. What proportion will be AB, N, Rh^-, Se?

4–28. A woman of phenotype A, MN, S, Rh^+, nonsecretor has a child of phenotype: B, M, Ss, Rh^-, secretor. Which of the following could be excluded as the father and on what basis?
a. AB, M, Ss, Rh^+, secretor
b. O, N, ss, Rh^-, secretor
c. B, M, S, Rh^+, secretor
d. A, M, Ss, Rh^-, nonsecretor
e. AB, MN, Ss, Rh^-, secretor

REFERENCES

CLARKE, C. A., AND R. B. MCCONNELL, 1972. *Prevention of Rh-Hemolytic Disease.* Thomas, Springfield, Ill.

HADORN, E., 1961. *Developmental Genetics and Lethal Factors.* Wiley, New York.

LINDSLEY, D. L., AND E. H. GRELL, 1968. *Genetic Variations of Drosophila melanogaster* (Publ. 627). Carnegie Institution, Washington, D.C.

RACE, R. R., AND R. SANGER, 1968. *Blood Groups in Man,* 5th ed. Blackwell, Oxford.

SEARLE, A. G., 1968. *Comparative Genetics of Coat Colour in Mammals.* Academic Press, New York.

STERN, C., 1973. *Principles of Human Genetics,* 3rd ed. Freeman, San Francisco.

SUTTON, H. E., 1975. *An Introduction to Human Genetics,* 2nd ed. Holt, Rinehart and Winston, New York.

THOMPSON, J. S., AND M. W. THOMPSON, 1973. *Genetics in Medicine,* 2nd ed. Saunders, Philadelphia.

WAGNER, R. P., AND H. K. MITCHELL, 1964. *Genetics and Metabolism,* 2nd ed. Wiley, New York.

Chapter 5

THE MEASUREMENT OF GENE EXPRESSION: QUALITATIVE AND QUANTITATIVE TRAITS

QUALITATIVE traits are those in which alternative phenotypes are sharply distinct and do not overlap. All traits considered thus far have been qualitative in nature. Crosses involving such characters give rise to the clearly distinguishable classes which form the basis for Mendelian ratios. Mendel's crosses between plants with round versus wrinkled seeds can be used as an example. Here, no question arises as to phenotype: a seed is either round or it is wrinkled, and no gradations in between are seen. As a result, one tends to think that members of each class are uniform in phenotype.

However, a closer examination would reveal that of the round seeds, some are less round than others and that although all wrinkled seeds have wrinkles, some have more than others. If the seeds of each respective class were identical in genotype, we would have to ascribe any phenotypic differences occurring among the members of a class to environmental effects. Such factors as the position of the seed in the pod or the varying distribution of fertilizer in the field or the closeness of a plant to nearby buildings or shade trees could readily bring about differences in gene expression sufficient to cause slight variations in phenotype.

APPLICATION OF MEASUREMENTS TO THE EXPRESSION OF TRAITS

If we wished to determine the amount of variation within either group, it would be necessary to classify a random sample of seeds for degree of phenotypic expression. For round seeds, we might ask how round is each seed as compared to a standard, such as a ball bearing. For wrinkled seeds, the number of wrinkles per seed could be counted. Let us use the number of wrinkles for a hypothetical example.

Suppose we have a bag of wrinkled seeds which were produced by a whole field of plants. If we could count the number of wrinkles per seed in a good sized random sample, we should encounter, by chance, seeds with all degrees of phenotypic expression, in numbers corresponding to their proportional representation in the whole. If we classified 500 seeds, the data as to number of wrinkles per seed might be the following.

Class (Number of Wrinkles per Seed)	Frequency (Number of Individuals per Class)
10	1
9	4
8	20
7	58
6	103
5	125
4	101
3	64
2	22
1	2
0	0

When these data are represented by a bar graph or histogram, it can be seen that they cluster around a middle value and fall off equally to either side (Figure 5–1). The bars on the histogram can also be connected by a symmetrical, bell-shaped curve, which for the present data we can call a *normal curve*. If we were to draw a perpendicular line from the exact center of the curve to the baseline, approximately equal numbers of individuals would be included on either side. [It will be recalled that a similar symmetrical distribution is evident for the coefficients of expanded binomials, Chapter 4).

The Mean

Although the nature of the data can more easily be comprehended by such a curve, it might also be useful to know the average value or mean number of wrinkles per seed. The *mean*, represented in statistics by the symbol $\bar{x}$, is determined by summing the values obtained for each seed and dividing by the total number of seeds in the sample. In cases where the data are arranged in classes, as in the present example, the number of wrinkles found for each class can be multiplied by the number of individuals in the class, and these values summed to obtain the total wrinkles found. The mean is obtained by dividing this total by the number of seeds scored. In the present instance, the mean is 4.99 or, for simplicity, 5.0 wrinkles per seed (Table 5–1).

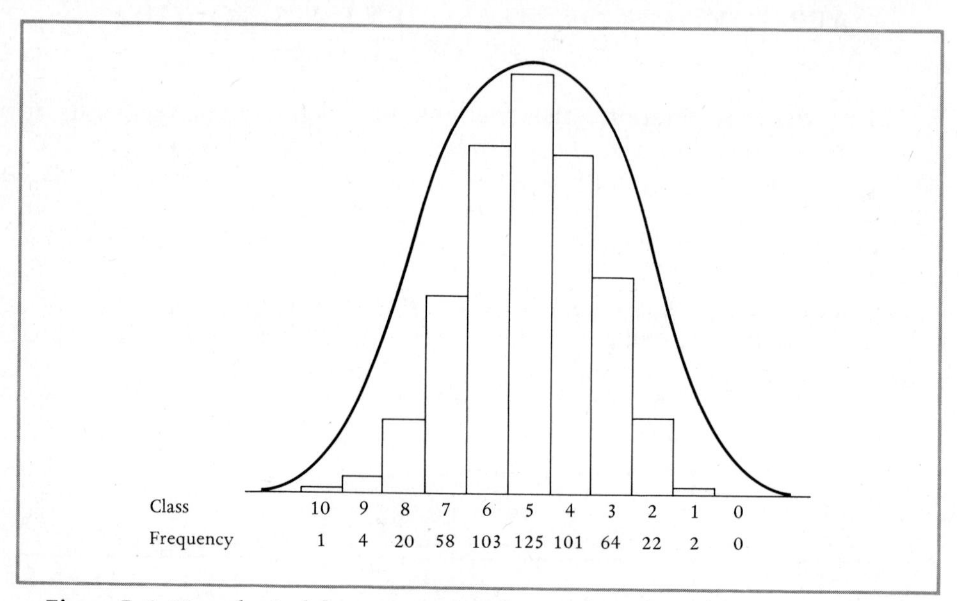

Figure 5–1. Hypothetical data concerning the number of wrinkles per seed arranged as a normal distribution.

TABLE 5-1
Calculation of Mean Number of Wrinkles per Seed

Class (number of wrinkles per seed)	Frequency (number of seeds per class)	Total number of wrinkles per class
10	1	10
9	4	36
8	20	160
7	58	406
6	103	618
5	125	625
4	101	404
3	64	192
2	22	44
1	2	2
0	0	0
	Total seeds = 500	Total wrinkles = 2497

$$\bar{x} = \frac{2497}{500} = 4.99 \cong 5.0 \text{ wrinkles per seed}$$

Variance

Although the mean can be used to represent the sample, it gives no information on the spread of the data, for an entirely different distribution of values might give the same arithmetic mean. Therefore, to characterize our data more completely, another kind of expression is needed to represent the extent of spread of the measurements. One such expression called *variance (V)* is determined as follows.

The deviation of each measurement from the mean value is first determined. Each deviation is then squared, and the sum of these squares is divided by the total number of individuals in the sample or, more properly, by one less than this total number. This calculation can be represented by the following formula, where V stands for variance, Σ means the sum of, d is deviation, and n is the total individuals in the sample.

$$V = \frac{\Sigma d^2}{n - 1}$$

In cases where the data are arranged in classes, the squared deviation from the mean of each class is multiplied by the number of individuals (frequency, f) in the class. The sum of all such calculations would again be divided by $n - 1$ to obtain the variance. The formula for variance when the sample involves classes thus becomes

$$V = \frac{\Sigma f d^2}{n - 1}$$

The calculation of the value of variance for our wrinkled seed sample ($V = 2.38$) is given in Table 5–2. A variance value of 2.38 would mean that the average squared difference from the mean exhibited by our sample of seeds was 2.38 wrinkles.

Standard Deviation

Variance is one way of expressing the spread of the measurements about the mean value and is most frequently used in connection with the quantita-

TABLE 5-2
Calculation of Variance, Standard Deviation, and Standard Error of the Mean

Class	Frequency (f)	Deviation (d)	d^2	fd^2
10	1	5	25	25
9	4	4	16	64
8	20	3	9	180
7	58	2	4	232
6	103	1	1	103
5	125	0	0	0
4	101	1	1	101
3	64	2	4	256
2	22	3	9	198
1	2	4	16	32
0	0	0	0	0
	$n = 500$			1191

$$\bar{x} = 5.0$$

$$V = \frac{\Sigma f d^2}{n - 1} = \frac{1191}{499} = 2.38$$

$$\sigma = \sqrt{\frac{\Sigma f d^2}{n - 1}} = \sqrt{2.38} = 1.5$$

$$s_{\bar{x}} = \frac{\sigma}{\sqrt{n}} - \frac{1.5}{\sqrt{500}} = 0.07$$

tive traits to be discussed in the next section of this chapter. In the present instance, a more useful description of the spread of the data would be one given in the same units of measurement as the sample and the mean, that is, the number of wrinkles per seed. Such an expression, the *standard deviation* (σ), is obtained by taking the square root of the variance.

$$\sigma = \sqrt{\frac{\Sigma fd^2}{n - 1}}$$

For our data the standard deviation is 1.5 (Table 5–2), and we can now more fully describe our sample of wrinkled seeds by stating the mean along with the standard deviation: 5.0 ± 1.5 wrinkles.

We can also apply the mean and standard deviation to the curve which represents our sample. The mean *($\bar{x}$)* can be indicated by a vertical line drawn through the curve to intersect the baseline (abscissa) at the mean value (Figure 5–2). One standard deviation (±1.5 wrinkles) can then be measured off to either side of the mean, with limits marked again by vertical lines. Suppose we next count the actual number of seeds whose phenotypes fall between one standard deviation in either direction from the mean, that is, within the range of 6.5 to 3.5 wrinkles per seed. Our data is arranged in nonoverlapping classes

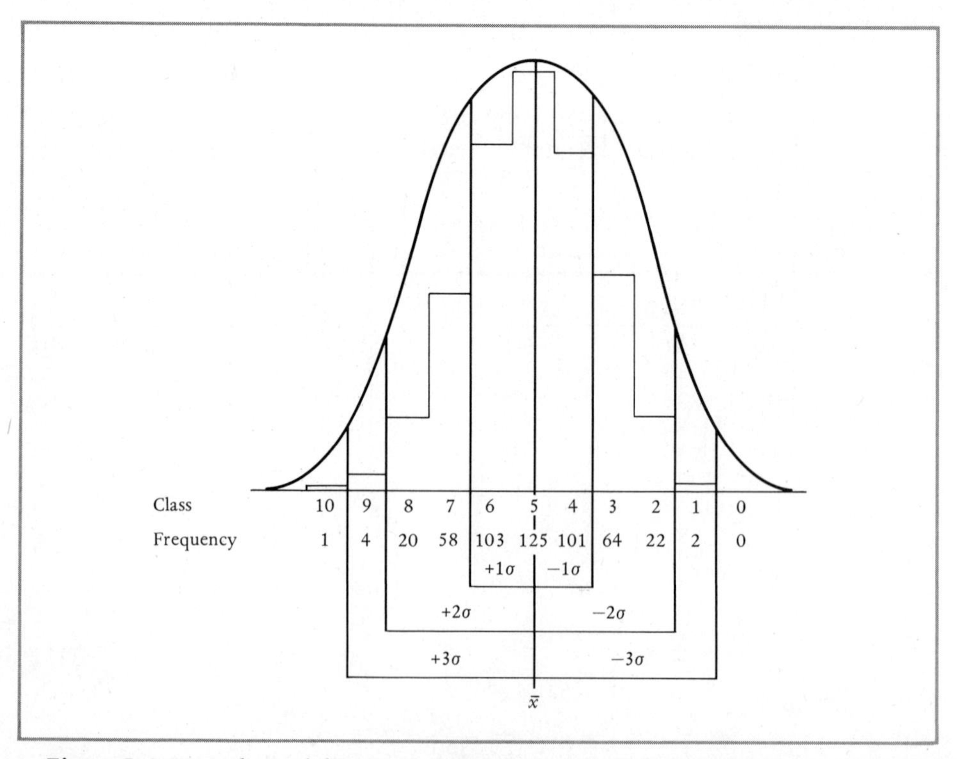

Figure 5–2. Hypothetical data concerning the number of wrinkles per seed arranged as a normal curve. One, two, and three standard deviations (σ) to either side of the mean *($\bar{x}$)* are indicated by vertical lines drawn through the curve. The data for the frequency of seeds in each class shows the number of seeds included within ±1, 2, or 3 σ from the mean.

which presents a problem in determining the number of seeds to be included. Since the range falls short of the classes with 7 and 3 wrinkles, the best we can do is to include the individuals in the classes with 4, 5, and 6 wrinkles. These number 329 or approximately 66 percent of the seeds in the sample.

We can measure off a second standard deviation beyond the first, and count up all of the individuals whose phenotypes fall within these two standard deviations to either side of the mean. The range of two standard deviations would be from 8.0 to 2.0 wrinkles per seed and would include 493 seeds or 98.6 percent of our sample. If we measured off a third standard deviation in either direction and again counted the seeds falling within these limits, 99.8 percent of the sample would be included.

These results vary somewhat from the theoretical, partly due to the data itself and partly due to the presence of discontinuous classes. In a perfect distribution, such as that obtained by expansion of one of the powers of the binomial, one standard deviation to either side of the mean would include 68.26 percent of the area under the curve, two standard deviations would include 95.44 percent of the area, and three standard deviations would include 99.74 percent of the area under the curve. These relationships hold regardless of whether or not the normal curve is compact with steeply sloping sides or is flattened and spread out, because in either case, one, two, or three standard deviations to either side of the mean will directly represent the distribution of the data in any particular case. Obviously, ours is not a perfect sample, although it comes close to being so.

Standard Error of the Mean

Our data for this sample of wrinkled seeds originally consisted of a rather unwieldy collection of numbers. However, by calculating the mean and the standard deviation, we have reduced the data to representative terms which serve to describe the sample in a concise and usable fashion. One would expect that if our sample truly represented the population of wrinkled seeds in the bag, analysis of an additional sample would yield a mean number of wrinkles per seed not identical, but probably very close to that obtained in the initial trial. If we took many such samples we would obtain a whole series of means, all quite similar. These means could themselves be plotted as a histogram or in the form of a normal curve. Although such a curve, or such a calculated mean of means, would represent the population more accurately than data from a single sample, usually it is not possible to perform many trials, especially if each trial is the result of a lengthy experiment. Furthermore, no matter how many trials are made or how many samples are taken, it is impossible to include all of the data because the population from which any sample is taken is theoretically infinite (all those who ever were or will be). Thus, since practical considerations preclude exhaustive sample taking, it becomes useful to estimate how precisely a sample mean represents the theoretical mean of the population. Such an estimate can be made by calculating the *standard error of the mean* ($s_{\bar{x}}$) provided the mean is based on a reasonably large sample, such as the 500 wrinkled peas used here. The standard error of the mean is found by dividing the standard deviation of the mean of a sample by the square root of the total number of individuals in that sample.

$$s_{\bar{x}} = \frac{\sigma}{\sqrt{n}}$$

We characterized our 500 peas by a mean and standard deviation of 5.0 ± 1.5 wrinkles per seed. The calculation of the standard error (shown in Table 5–2) yields a value of 0.067 or, for simplicity, 0.07. We can now characterize our sample by the mean ± the standard error, that is, 5.0 ± 0.07 wrinkles per seed.

The standard error can be considered as the standard deviation of many means, and therefore, it is essentially a standard deviation of the theoretical population mean. In the present example, our standard error of ±0.07 indicates that the population mean lies somewhere in the range of 5.07 wrinkles to 4.93 wrinkles. A standard error of the mean has the same implications as the standard deviation in that we can estimate that the population mean falls in the interval 5.0 ± 0.07, with a probability of 68 percent, and in the interval 5.0 ± 0.14, with a probability of 95 percent.

Standard Error of the Difference in Means

The mean and standard errors of two different samples can also be compared to determine if the populations from which they were taken are the same or different. As a hypothetical illustration, suppose we wished to determine whether or not the addition of fertilizer to a field caused a significant change in gene expression with respect to the number of wrinkles in the seed. In this experiment the genotypes of the plants would be the same as before and only the environment would be altered by the withholding of fertilizer. After harvesting the plants, we can again obtain a bag of wrinkled seeds from which we take 500 as a random sample for examination and classification. If we determine the mean, standard deviation, and standard error, we can compare this second sample grown without fertilizer to the first sample which received fertilizer. Such a comparison is shown in Table 5–3.

Examination of Table 5–3 shows that measurements made for these two samples do not overlap, even when the range of values included within three standard errors from the mean is considered. Accordingly, we would judge that

TABLE 5–3
Comparison of Ranges of Expression Within One, Two, and Three Standard Errors of the Mean

	Trait: number of wrinkles per seed	
	Sample 1 (with fertilizer)	Sample 2 (without fertilizer)
Mean ($\bar{x}$)	5.0	6.5
Standard deviation (σ)	±1.5	±2.0
Standard error ($s_{\bar{x}}$)	±0.07	±0.09
	Range of Expression	
$\bar{x} \pm 1s_{\bar{x}}$	5.07 to 4.93	6.59 to 6.41
$\bar{x} \pm 2s_{\bar{x}}$	5.14 to 4.86	6.68 to 6.32
$\bar{x} \pm 3s_{\bar{x}}$	5.21 to 4.79	6.77 to 6.23

these two samples represent different populations, and in terms of our experiment we would conclude that the withholding of fertilizer results in a measurable and significant difference in gene expression.

In this example inspection alone justifies our conclusion. However, not all cases are so conveniently clear-cut. If the values overlap, it is necessary to determine more precisely whether the difference between the two samples is significant or merely due to chance. To make this determination, we can compare the difference between the means of the two samples with the standard error of this difference. To explain, let us consider the following. Suppose many paired samples are taken from a population and the difference between the means of each pair determined. We would thus obtain a whole series of differences which, by themselves, could be plotted as a normal distribution. The standard deviation and standard error of these differences between the means could be calculated. If the paired samples came from the same population, we would expect that 99.7 percent of all differences between their means would fall within three standard deviations or three standard errors of the mean of our curve. To put it another way, the chance is less than 1 in 100 that the difference between the means in any pair of samples would exceed three standard errors of the differences in means.

This reasoning can be applied as a practical test to determine whether or not there is a significant difference between two experimental samples. For large samples, such as ours, the mean and standard error of each sample is first calculated, and then the difference between the means of the two samples is found. The standard error of the difference in means (s_D) is then obtained by substituting in the equation:

$$s_D = \sqrt{(s_{\bar{x}_1})^2 + (s_{\bar{x}_2})^2}$$

where $s_{\bar{x}_1}$ is the standard error of the mean of sample 1 and
$s_{\bar{x}_2}$ is the standard error of the mean of sample 2.

The standard error of the difference in means is then compared with the actual difference between the means of the two samples. If the actual difference falls within two standard errors of the difference in means, the variation between the two samples can be attributed to chance, and it can be concluded that the two samples were drawn from the same population. However, if the actual difference exceeds two standard errors of the difference in means, the probability that the samples represent two different populations is better than 95 percent.

We can apply this test to our wrinkled peas grown with and without fertilizer. The mean and standard errors of the two samples were 5.0 ± 0.07 (sample 1) and 6.5 ± 0.09 (sample 2). The standard error of the difference in means would be calculated as:

$$\begin{aligned} s_D &= \sqrt{(0.07)^2 + (0.09)^2} \\ &= \sqrt{0.0049 + 0.0081} \\ &= \sqrt{0.013} \\ &= 0.11 \end{aligned}$$

The standard error of the difference in means is thus ±0.11, two such standard errors would be ±0.22, and three such standard errors would be

±0.33. The actual difference between the means of the two samples is ±1.5. If we compare these values, it is clear that the actual difference between the means is much greater than even three standard errors of the difference in means (1.5 versus 0.33), and we can therefore conclude that the two samples represent significantly different populations.

In summary, it can be seen that statistical tools can be of considerable use in reducing numerical values to manageable terms and in permitting comparisons to be made between experimental groups. We have applied these methods to a qualitative trait where the degree of gene expression can be measured quantitatively. In our example an identical genotype was confronted with two different environments, and the use of statistics enabled us to distinguish real differences in response. It should be realized that these methods are just as valuable when the circumstances are reversed, that is, when varying genotypes confront a constant environment. The use of statistics is often helpful in such cases in estimating the number of genes which may be involved in the expression of a trait or in distinguishing the influence of the environment from that of the genotype. Of course, a number of statistical tests other than those described here are commonly applied to experimental data, and the reader is referred to appropriate texts for descriptions of their use.

QUANTITATIVE TRAITS AND POLYGENES

Although the example with wrinkled peas involved measurements of the degree of gene expression, it must be remembered that the distinction between the round and the wrinkled phenotype is qualitative, the two types of seeds represent two different classes, and the two types are clearly different in appearance. The statistical analysis of differences (the quantitative values) we have just discussed is a comparative tool applied to members of a single class defined on the basis of the presence of a specific phenotype. Not all traits are qualitative, however, for some are characterized by phenotypes that vary continuously from one extreme to the other so that no separate classes, such as round versus wrinkled, can be distinguished. Such traits must be analyzed through the use of quantitative measurements. They are often of great economic importance and include such characters as height, intelligence, musical ability, weight of fruits, vegetables, animals, or people, butterfat content of milk, nutritional content of grain, and so on. Not only are many genes involved in the expression of such characters, but the individual effects of each gene are usually slight and therefore frequently obscured by the effects of the environment. Consequently, it is often hard to make distinctions between the contributions of "nature" and those of "nurture." Anywhere from 3 to 40 to 100 or more such genes, called *polygenes,* may be involved in the expression of a quantitative trait. Since usually it is not possible to identify them individually by their effect on the phenotype, informed estimates of the number of genes whose actions are contributing to any given character are often given in terms of a range, such as probably 6, but not more than 10.

The difference between qualitative and quantitative gene expression is immediately obvious if we compare the results of crosses, using normal curves to represent the parental and the F_1 and F_2 progeny populations in each case

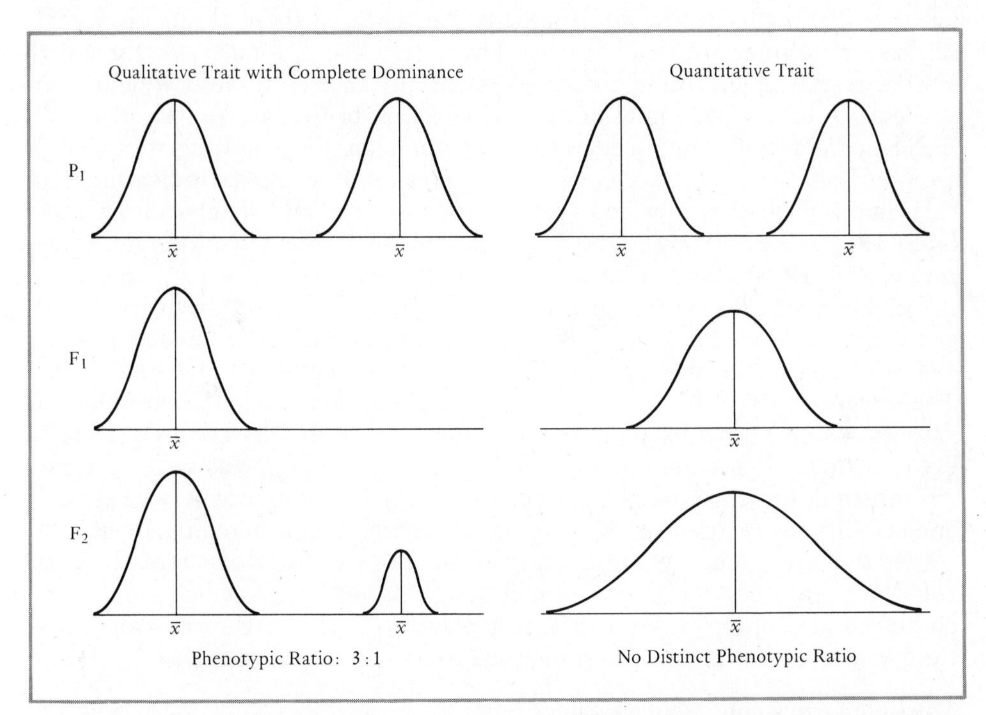

Figure 5–3. The distribution of phenotypes in the P_1, F_1, and F_2 generations for a qualitative versus a quantitative trait. The mean of each group is indicated by a vertical line. The ratios found with qualitative characters result from nonoverlapping classes of progeny. Crosses involving quantitative traits do not yield clear-cut ratios in the F_2 generation, but instead produce offspring which vary continuously from one parental extreme to the other.

(Figure 5–3). It can be seen that the qualitative trait yields clearly separable classes in the F_2 generation. Furthermore, assuming comparable environments, the respective means of the F_1 generation and the dominant class of the F_2 generation are similar to each other and also to that of the dominant parent. Similarly, the mean of the F_2 recessive class corresponds to that of the P_1 parental type.

In contrast, a cross between parental types differing in a quantitative trait produces F_1 offspring whose phenotype is intermediate between those of the parents. The F_2 generation exhibits increased variation (spread) as compared to the F_1, as well as a continuous range of phenotypes from one parental extreme to the other. No separate classes are evident. In addition, the means of the F_1 and F_2 curves are similar.

These clear-cut differences in inheritance were recognized in the 1900s and caused some early workers to postulate that Mendelian principles did not apply to traits showing continuous variation. Some of the confusion regarding such characters was dissipated by Wilhelm L. Johannsen in 1903, who showed that this variation was due to two components: heredity and environment. Using the character of seed weight in beans, Johannsen found that within a pure line seed weight varied closely about a mean which did not differ significantly in successive generations. He ascribed the variation which did occur to the action of the environment. He repeated these studies using 19 separate pure lines which differed in bean weight and in each case obtained similar results. Each line maintained essentially the same mean weight over successive generations. In addition, the selection of the heavier beans in any line as parents for a new generation did not alter the resulting mean bean weight in that generation. We would interpret these results as indicating that each line was homozygous and that variations in size and weight within each line were due exclusively to the environment. Therefore, selection within the line would not be expected to give rise to new phenotypes.

Johannsen also distinguished between environment and heredity. When beans of the 19 pure lines were mixed, they formed a variable population which exhibited a mean weight between the two extremes represented in the 19 lines. The selection of heavier or lighter beans from the mixed population to be the parents of a new generation did produce new lines with different mean values, because such seeds would be expected to differ in genotype as well as in environmental influences on phenotype. To express the difference between inherent heredity as opposed to the external evidence of that heredity, Johannsen coined the terms gene, genotype, and phenotype. His results also provide some rules for selection: Attempts to extract new varieties by inbreeding individuals chosen because of the presence of a given phenotype will be unsuccessful unless that phenotype has a genetic, as opposed to an environmental, basis.

Multiple Genes and Additive Effect

The inheritance of quantitative characters was further clarified by H. Nilsson-Ehle in 1909. Nilsson-Ehle analyzed the results of crosses between varieties of wheat which differed with respect to the color of the seed. From a cross between a dark red and a white variety, medium red F_1 progeny were produced. However, as is the case with most quantitative traits, members of the F_2 generation showed a much wider variation in phenotype, with colors ranging from dark red to white. The proportion of colored to noncolored seeds was 15 : 1 and, furthermore, 1 out of every 16 seeds was as dark in color as the dark red P_1 variety. These findings provoked the hypothesis that two gene pairs were involved. Close examination permitted the separation of the F_2 progeny into five classes based on the depth of color of the seed: dark red, medium dark, medium, light red, and white. Members of these classes appeared in the proportions of 1 : 4 : 6 : 4 : 1, respectively, and these frequencies could be represented by a normal curve (Figure 5–4). The interpretation of these results was somewhat difficult because, although two gene pairs were undoubtedly involved, the F_2 phenotypic classes did not conform to the 9 : 3 : 3 : 1 proportions usually encountered with standard dominance and epistasis did not seem to be involved. To explain the results, Nilsson-Ehle proposed that dominance in the usual sense was absent and, instead, each gene for color might exert an additive "plus" effect. Absence of color could be represented by the genotype *aabb* and dark red (containing four genes for color) by *AABB*. Replacement of an *a* with an *A* or a *b* with a *B* would add one increment of color to the seed. The phenotypes produced by *Aabb* and by *aaBb* would thus be the same since one unit for color would have been introduced in both cases. When the F_2 genotypes are plotted according to the proposed number of plus genes present, it can be

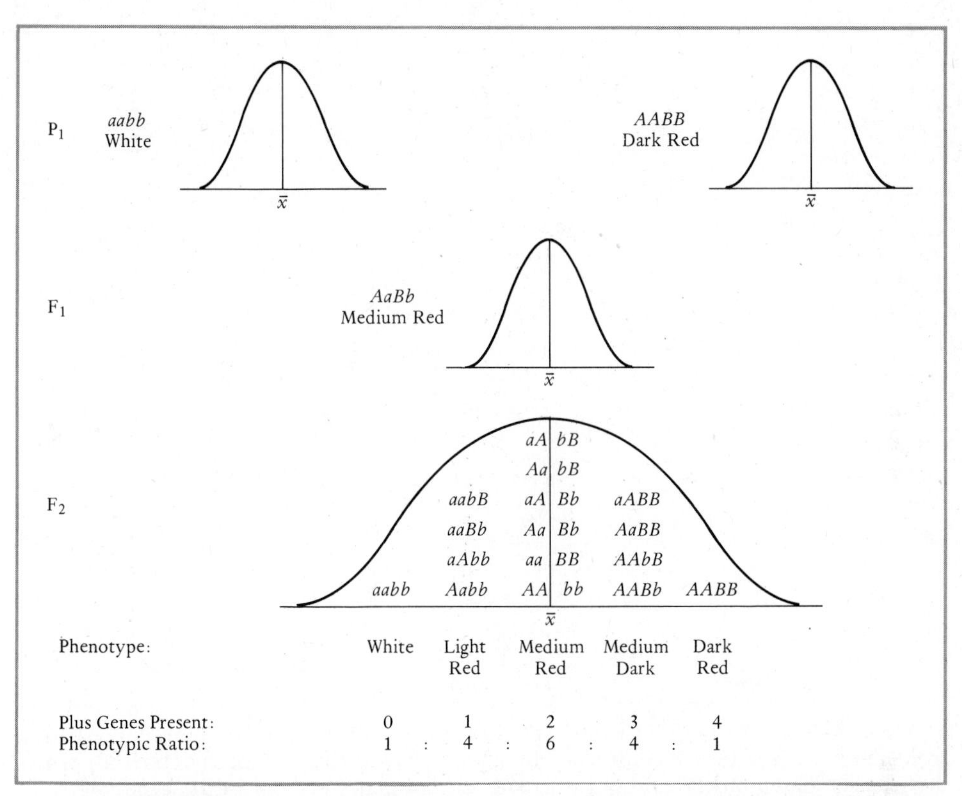

Figure 5–4. Diagram of Nilsson-Ehle's crosses between a red variety *(AABB)* of wheat and a white variety *(aabb)*. The 16 possible gamete combinations of the F_2 generation are shown as a normal distribution based on the number of plus genes present in the genotype. The phenotypic ratio reflects the additive nature of these genes.

seen that the distribution of genotypes is in agreement with the observed 1 : 4 : 6 : 4 : 1 distribution of phenotypes.

In another cross between a red and a white variety Nilsson-Ehle found similar results, except that in this case the F_2 ratio of colored to noncolored seeds was 63 : 1, and only 1 in every 64 seeds was as red as the original P_1 variety. Since 64 different gamete combinations occur in a cross between trihybrids, these results suggested that three gene pairs for color were involved. Nilsson-Ehle successfully separated the F_2 seeds into seven classes which ranged in color from white to dark red. These classes contained the proportions of $\frac{1}{64} : \frac{6}{64} : \frac{15}{64} : \frac{20}{64} : \frac{15}{64} : \frac{6}{64} : \frac{1}{64}$ or a phenotypic ratio of 1 : 6 : 15 : 20 : 15 : 6 : 1, which can be represented by a normal curve (Figure 5–5). Nilsson-Ehle again proposed the hypothesis that genes with equal and additive effects were involved to account for the data. The white phenotype would thus result from the recessive genotype *aabbcc*. The darkest red phenotype would occur when six plus genes were present *(AABBCC)*, each such gene causing the addition of a small increment of color to the seed coat. The various classes whose color phenotypes lay between these two extremes would contain one, two, three, four, or five such additive color genes. When plotted according to the proposed

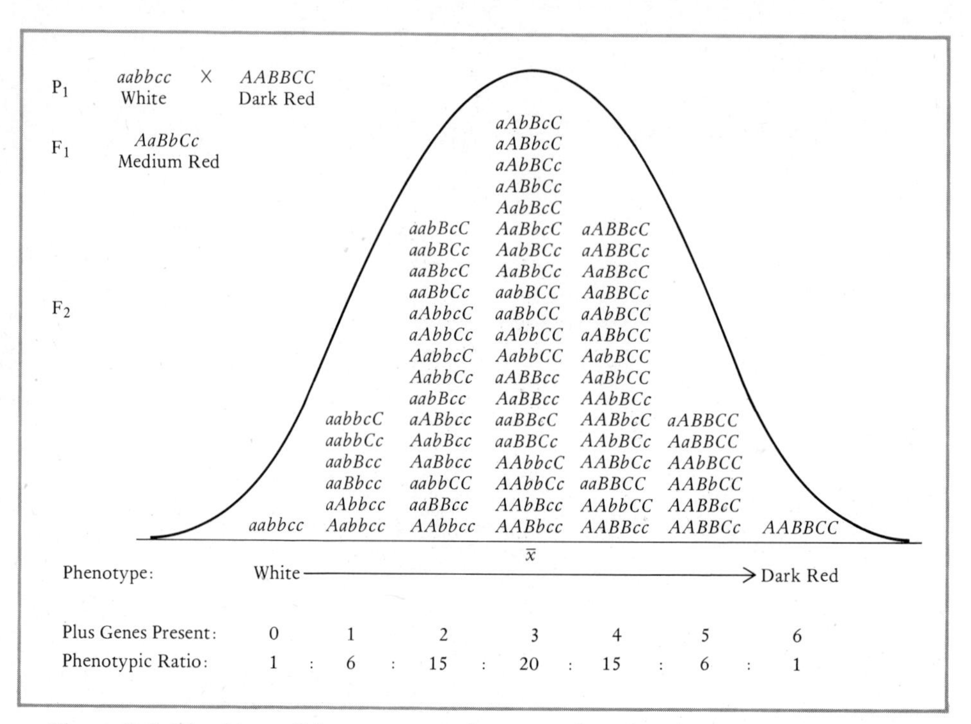

Figure 5–5. The 64 possible gamete combinations found in the F_2 generation resulting from the cross *AaBbCc* × *AaBbCc*. The phenotypes of the F_2 progeny show a normal distribution based on the number of additive color genes present in the genotype.

number of plus genes present, the distribution of genotypes again agreed with the observed proportions of phenotypes.

As a result of these studies, Nilsson-Ehle proposed that other characters showing continuous variation might be inherited in the same fashion as seed color in wheat and in accordance with Mendelian principles. Although in this instance he had, with difficulty, separated the F_2 progeny into distinct phenotypic classes, he suggested that in other cases the slight differences between the many classes could readily be blurred by environmental effects, so that a continuous sequence of phenotypes would be presented to the observer. As a result of Nilsson-Ehle's studies, as well as those of East, Emerson, and others, it has come to be recognized that genes for quantitative traits are inherited in the same way and according to the same rules as genes for qualitative traits. However, the identification of the individual genes contributing to a quantitative character is much more difficult because their separate effects may be masked through the influence of the environment.

It must be recognized that in any cross involving a quantitative trait, it cannot be assumed that the parental varieties are homozygous for all plus genes or for all inert or neutral genes, as was the case in Nilsson-Ehle's experiments. One parental variety may be homozygous for some plus genes, the other variety homozygous for others. When crossed, their progeny will inherit plus genes from both parents, and if these genes are additive in effect, the F_2 generation may contain more extreme phenotypes than those found in either parental line.

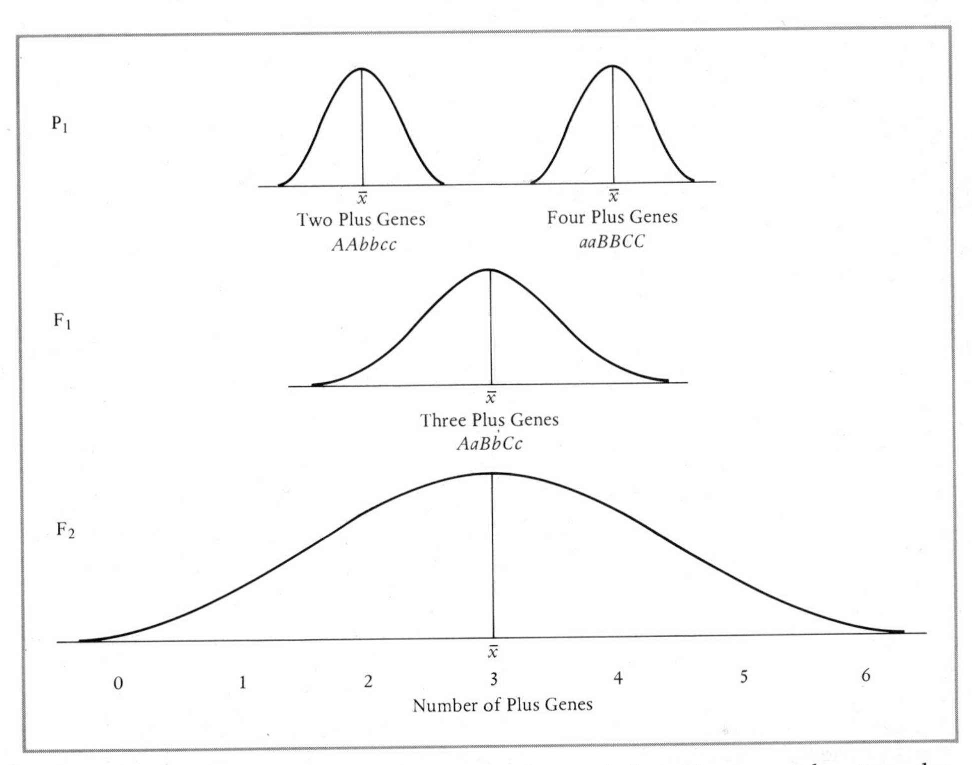

Figure 5–6. Diagram illustrating transgressive variation. One parent has two plus genes, the other has four. F_1 individuals have three such genes and are intermediate in phenotype. Members of the F_2 generation can contain from 0 to 6 plus genes and thus the range of expression can exceed that of either parental type.

Figure 5–6 presents an example. One parent has two plus genes, the other has four. The F_1, having three such genes, is intermediate in phenotype, while the F_2 exhibits the full range of expression, exceeding that of either P_1 variety. Such a phenomenon is called *transgressive variation,* and it is of common occurrence.

Estimating the number of genes that contribute to a quantitative character is difficult. If only 2 pairs of genes are involved, the extreme phenotypes produced by homozygosity can be expected to occur once in every 16 individuals. When 3 pairs of genes are concerned, only $\frac{1}{64}$ will be homozygous for all plus genes and $\frac{1}{64}$ will be homozygous for all inert or neutral genes. With 4 gene pairs the respective phenotypes produced by homozygosity will occur by chance in only 1 out of 256 offspring. Thus, if the extreme phenotypes were to occur in proportions of 1 in 16, 1 in 64, or even 1 in 256, we would have good evidence for the participation of 2, 3, or 4 pairs of genes, respectively.

However, it must be realized that if more than 3 or 4 gene pairs are involved, the chance of recovering in the F_2 generation individuals homozygous for all contributing genes becomes less and less probable. Table 2–1 presents convincing evidence for this fact. For example, with only 5 pairs of genes, there are 1024 different gamete combinations and 243 different genotypes which will occur as the result of a cross between heterozygotes. With 10 pairs of genes, over 59,000 different genotypes are possible, and an estimate that at least 10 pairs of genes are contributing to a quantitative character is not especially unusual.

Calculations of variance are often helpful in making estimates of the number of pairs involved. It will be recalled that variance is a good measure of the spread of data to either side of the mean. When only a few pairs of genes are involved, variance from the mean is greater because more individuals at the extremes are found in the population. However, when many genes are involved and the population is equivalent in size, variance is less because the more extreme phenotypes are either infrequent or absent altogether due to the fact that the probability of their occurrence within a limited population is remote. Thus, with many genes, the data tend to be distributed more closely around the mean.

It should also be recognized that our discussion is based on several assumptions with respect to quantitative characters. We have assumed that genes with quantitative expression are additive and identical in their effects, that they show no dominance and no epistasis, and that they are independent of one another, thereby conforming to Mendel's Law of Independent Assortment. We should now point out that these assumptions are not necessarily valid.

In a model system it is useful to consider one genotype, such as *aabbcc,* as inert or as providing a minimal level of expression to which is added the effects of plus genes. However, although this concept is valid in some instances, such inert genes may actually subtract from expression, rather than being neutral. Furthermore, the presence of plus genes may not merely add an increment to the expression of a trait, but may multiply an effect, thus producing geometric rather than additive increments in expression. In addition, the likelihood that plus genes are identical in their action is remote and contravenes common sense. Considering what is known concerning gene action at the cell and molecular level, it is hardly possible that 20 to 30 or more such genes exert exactly the same effect.

The assumption of no dominance, no gene interaction, and no epistasis is also naive, especially when many genes are involved. Observations of gene expression demonstrate that some degree of dominance is frequent, and it occurs with quantitative as well as with qualitative traits. The presence of dominance can be expected to reduce the number of phenotypic classes, thereby suggesting to an observer that fewer genes are involved than is actually the case and thus compounding the difficulties of analysis. As for gene interaction and epistasis, it is hard to conceive of the development of any multigenic trait where interactions in gene expression would not occur—indeed, they can hardly be avoided.

Finally, not all quantitative genes assort independently of one another. Some may be inherited together because of their location on the same chromosome (linkage).

Heritability

The study of quantitative traits is made even more difficult by the necessity of distinguishing between the effects of numerous genes and the influence of the environment. In doing so, the variance *(V)* of a population sample is often used in statistical analyses. Phenotypic variance (V_P) can be considered to consist of genotypic variance (V_G) combined with environmental variance (V_E), plus any variation in phenotypic expression induced by the interaction of the environment with some specific combination of genes, that is, variance due to a particular gene-environment interaction (V_{GE}). Phenotypic variance can thus be expressed as:

$$V_P = V_G + V_E + V_{GE}$$

The term V_G, which expresses phenotypic variance due to genotype, includes the effects of additive genes, dominant genes, and epistasis. The contributions of each to the phenotype can best be estimated in rigidly controlled experiments and from crosses involving homozygous varieties and their F_1, F_2, and F_1 backcross progeny. Variance due to the environment (V_E) is also most easily determined when homozygous organisms are studied, that is, when V_G is zero. The value of the term V_{GE} is much harder to estimate and therefore must often be ignored.

Animal and plant breeders find useful a value called *heritability (H)* which expresses the degree to which a trait is influenced by the genotype. If the environment is controlled (or roughly the same for all) and one ignores the term V_{GE}, then an estimate of heritability can be obtained by the expression:

$$H = \frac{V_G}{V_P}$$

A heritability of 1.0 would indicate that the trait in question was produced solely by the action of the genotype. An example is the ABO blood group phenotype. A heritability of 0 means that the phenotype is due entirely to the environment, for example, the accidental loss of an ear or a tail. Intermediate values estimate the relative contribution of heredity, as opposed to environment, in the expression of a trait. Thus a heritability of 0.75 is an estimate that 75 percent of the expression of the trait is due to genotype.

Some heritability percentages for traits of economic importance are given in Table 5–4. When plants or animals are to be chosen as breeding stock, heritability estimates for various traits are helpful in forecasting how closely the appearance or performance of offspring will match that of the parents. Also useful to animal breeders is a determination of how closely the presence of one trait is correlated with the presence of a different trait or condition. A positive correlation indicates an association too great to be explained by chance alone and often permits the breeder to select for two traits by testing for only one of them. Whether or not a correlation exists between two traits is estimated by use of a statistical measure called a correlation coefficient whose method of computation can be found in any general text in statistics.

Twin Studies

Quantitative traits in humans have been studied most successfully with twins. Comparisons are made between fraternal twins, identical twins that have been reared together in the same family, and identical twins that have been reared separately in different families. Identical (monozygotic) twins arise from the division of a single embryo during an early stage in development.

TABLE 5–4
Heritability Percentages for Some Traits of Economic Importance

Heritability	Trait
5%	conception rate in cattle
17	ear length in corn
20	egg production in *Drosophila*
20	egg production in poultry
25	yield in corn
30	milk production in cattle
40	fleece length in sheep
60	egg weight in poultry
65	root length in radishes
70	plant height in corn
85	slaughter weight in cattle

Source: James L. Brewbaker, *Agricultural Genetics*, © 1964, p. 12. Reprinted by permission of Prentice-Hall, Inc., Englewood Cliffs, New Jersey.

Since they originate from the same fertilized egg, they have identical genotypes and are always of the same sex. Fraternal (dizygotic) twins arise from two different fertilized eggs which happen by chance to be ovulated at the same time. Fraternal twins thus possess different genotypes and can be of the same or different sexes. Other than being the same age, fraternal twins are no more alike than are other brothers and sisters of the same family.

Studies of identical twins are useful in determining whether the presence of a given trait is due to heredity or to environment or both. The members of a twin pair are each examined for presence of the trait. If both exhibit the condition, they are termed concordant for that trait; if not, they are discordant. Traits expected to show 100 percent concordance are such qualitative aspects as sex, ABO blood group, Rh group, hair and eye color, and others. Traits with variable penetrance may not be expressed to the same extent in each twin. In studies of quantitative inheritance a given trait is measured in selected groups of twins, and the degree to which the members of twin pairs are alike in the expression of the trait is then determined. A low to zero correlation between twinning and the expression of the character implies that the character has primarily an environmental basis. A strong positive correlation suggests a common cause, presumably genotype, for the appearance of the trait, especially since different twin pairs are often raised in different environments with different degrees of access to educational and social advantages. Evidence for the participation of the genotype in human quantitative characters is best seen when identical twins that have been reared apart are compared. Since members of any pair of identical twins have the same genotype, a high positive correlation between such twins points to a strong hereditary component in the expression of a trait. The heritability of different human characters has also been calculated (Table 5–5), but it must be emphasized that heritability is only an estimate, only a good guess based on the limited information currently available.

TABLE 5–5
Correlation and Heritability for Certain Traits in Identical and Fraternal Twins

Trait	Fraternal	Identical reared together	Identical reared apart	Heritability
Stature	0.64	0.93	0.97	0.81
Sitting height	0.50	0.88	0.96	0.76
Weight	0.63	0.92	0.89	0.78
IQ	0.62	0.92	0.73	0.80
Word meaning	0.56	0.86	—	0.68
Spelling	0.73	0.87	—	0.53

Source: Th. Dobzhansky, *Mankind Evolving,* Yale University Press, New Haven, Conn., 1962.

In summary, the analysis of quantitative characters is no simple task. However, despite the obvious complexities of the subject, such analysis has made possible better breeds of livestock and has led as well to the "green revolution," the development of disease- and drought-resistant food crops vastly improved in yield and nutritional value. In a world where famine is an ever present reality, this is no small contribution.

PROBLEMS

5–1. L. A. J. Quetelet was the first person to apply statistical concepts to the measurement of biological characters. In 1871 he measured the height of 1000 French soldiers and compiled the following data

HEIGHT (INCHES)	FREQUENCY
60	2
61	2
62	20
63	48
64	75
65	117
66	134
67	157
68	140
69	121
70	80
71	57
72	26
73	13
74	5
75	2
76	1
	1000

a. Using graph paper, prepare a histogram of the data. Do the data conform to a normal distribution? Can you draw a normal curve?

b. Calculate the mean, standard deviation, and standard error of the mean.

c. Measure off one, two, and three standard deviations to either side of the mean and add up the number of soldiers included within the limits of 1σ, 2σ, and 3σ. What percent of the total number of soldiers is included in each group? Based on the size of the standard error, do you think that Quetelet's data is a representative sample of French soldiers?

5–2. In the following hypothetical case measurements in mm were made of the length of the trumpet of Golden Triumph variety of daffodils and of Spring Delight variety. These two varieties were crossed and similar measurements were made of the flowers of F_1 and F_2 individuals. Data for the parental varieties (P_1 and P_2) and that for the F_1 and F_2 progeny are given below.

Trumpet Length in Spring Delight and Golden Triumph and in Their F_1 and F_2 Progeny

Length in mm	Spring Delight (P_1)	Golden Triumph (P_2)	F_1	F_2
20	1			
21	12			
22	50			
23	77			
24	48			
25	10			1
26	2			1
27				2
28				2
29				2
30			2	6
31			9	18
32			18	57
33			53	115
34			85	145
35			57	108
36			20	55
37			7	13
38			1	6
39				3
40				3
41				2
42				1
43		2		1
44		14		1
45		45		
46		81		
47		47		
48		10		
49		1		

a. Calculate the mean, standard deviation, and standard error of the mean for each.

b. Determine whether the difference between the P_1 and P_2 varieties is statistically significant.

c. Determine if the difference between the F_1 and F_2 is statistically significant.
d. Assuming that each parental variety was homozygous and that the genes involved are additive in effect, how many gene pairs do you estimate are segregating and assorting in the F_2 generation?

5–3. Suppose you have two inbred lines of laboratory rats which differ in size and weight. Suppose also that the difference between the weights of these strains is due to three pairs of genes, where each plus gene adds 50 gm to the weight and each neutral allele adds only 25 gm to the weight.
a. If the mean weight of strain A is 150 gm and that of strain B is 300 gm, what are the genotypes of these two strains with respect to these alleles?
b. If strain A is crossed with strain B, what will be the mean weight of the F_1?
c. What classes of progeny with respect to weight would be expected to occur in the F_2?
d. How many F_2 rats would you have to raise to recover four individuals similar in weight to their 300 gm grandparent?

5–4. In mice suppose the allele spotted 1 (sp^1) causes the presence of small white spots in the coat when homozygous, while the allele spotted 2 (sp^2) causes larger white spots when homozygous. In the heterozygote sp^1/sp^2 the spots are intermediate in size. An experiment is performed in which the mm square area of the spots is measured in these three genotypes at two different temperatures. The mean and standard error of these measurements are given below.

Degrees °C	sp^1/sp^1	sp^1/sp^2	sp^2/sp^2
25°	4 ± 0.51	20 ± 0.40	60 ± 0.34
30°	29 ± 0.46	30 ± 0.62	52 ± 0.37

a. Is there a statistically significant difference in the response of these three genotypes to 30° as compared to 25°?
b. Is there a significant difference in phenotype between sp^1/sp^1 and sp^1/sp^2 at 30°?
c. Is there a significant difference in phenotype between sp^1/sp^2 and sp^2/sp^2 at 30°?
d. What is the effect of temperature on the action of these alleles?
e. What kinds of cell constituents are most susceptible to temperature effects and therefore probably implicated?

REFERENCES

Brewbaker, J. L., 1964. *Agricultural Genetics* (Foundations of Modern Genetics Series). Prentice-Hall, Englewood Cliffs, N.J.

Bulmer, M. G., 1970. *The Biology of Twinning in Man.* Clarendon, Oxford.

Dobzhansky, T., 1963. *Mankind Evolving.* Yale University Press, New Haven, Conn.

East, E. M., 1910. A Mendelian interpretation of variation that is apparently continuous. *Am. Nat., 44,* 65.

Emerson, R. A., and E. M. East, 1913. The inheritance of quantitative characters in maize. *Res. Bull. Neb. Agric. Exp. Sta., 2,* 1.

Falconer, D. S., 1961. *Introduction to Quantitative Genetics.* Ronald Press, New York.

Mather, K., and J. L. Jinks, 1971. *Biometrical Genetics,* 2nd ed. Chapman & Hall, London.

Mather, W. B., 1964. *Principles of Quantitative Genetics.* Burgess, Minneapolis.

Staub, W. J., and M. G. Blase, 1971. Genetic technology and agricultural development. *Science, 173,* 119.

Stern, C., 1973. *Principles of Human Genetics,* 3rd ed. Freeman, San Francisco.

Chapter 6

MITOSIS, CHROMOSOMES, AND MEIOSIS

By the late 1800s cytologists and embryologists were actively engaged in the study of plant and animal cells. The development of techniques for cutting tissues into thin sections for microscopic examination and the use of dyes which selectively stained cell inclusions added impetus to their investigations. Various cell organelles were described and figured in the literature, and although our interpretation of these structures has altered, the drawings and illustrations of these early workers are truly remarkable for their beauty and accuracy (Figure 6–1).

The presence of filamentous bodies within the nucleus was noted, and in 1882 W. Fleming described the longitudinal splitting of these bodies and the distribution of each half to daughter cells during cell division, a process which he named *mitosis.* Almost immediately thereafter, in 1883, Edward van Beneden observed the presence of four such bodies in the fertilized egg *(zygote)* of the roundworm, *Ascaris,* whereas only two of these structures were evident in sperm or in the unfertilized eggs of this species. He reasoned that the presence of four such bodies in the zygote must result from the equal contributions of the male and female gametes. With regard to the number present, the constitution of the gametes was termed *haploid,* as opposed to the double or *diploid* complement found in the fertilized egg.

It was soon recognized that a constant diploid number of these nuclear bodies, termed *chromosomes,* was a characteristic of each species examined

(Table 6–1), and that within somatic cells chromosomes occurred as pairs, or *homologues,* the members of each such pair ordinarily being indistinguishable from one another in size and form. It was also clear that the union of egg and sperm in fertilization established this paired condition of the chromosomes and that one member of each pair was maternally derived and the other paternally derived.

In view of these findings, it was suggested by August Weismann in 1887 that some kind of special process must occur during the maturation of gametes

Figure 6–1. Drawings made by Fleming in 1882 showing various phases of dividing nuclei from human epithelium. [*Archiv Mikroskopic Anatomie,* 1882]

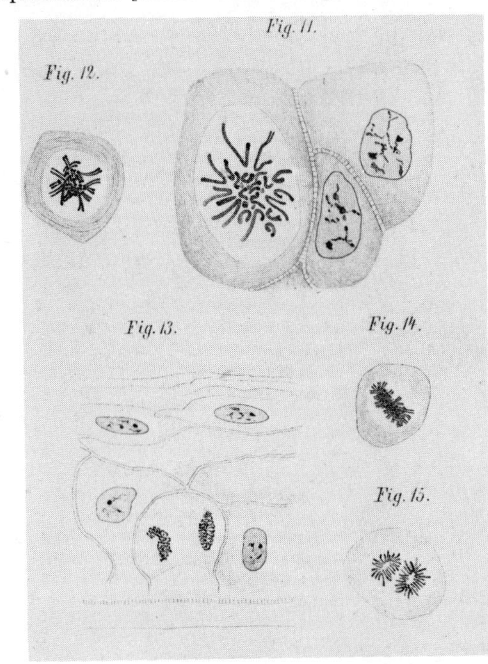

TABLE 6–1.
Somatic Chromosome Numbers of Some Organisms Frequently Used in Biological, Agricultural, and Medical Work

Organism	Chromosome number
Fungi	
Neurospora crassa (pink bread mold)	7 (haploid)
Saccharomyces cerevisiae (brewer's yeast)	probably 17 (haploid)
Higher Plants	
Pinus spp. (pine)	24
Juniperus spp.	22
Zea mays (corn)	20
Hordeum vulgare (barley)	14
Secale cereale (rye)	14
Saccharum officinarum (sugar cane)	80 (octoploid)
Triticum aestivum (bread wheat)	42 (hexaploid)
Musa paradisiaca (banana)	22, 33, 44 (diploid, triploid, tetraploid)

TABLE 6–1. *(Continued)*

Organism	Chromosome number
Higher Plants (Continued)	
Allium cepa (onion)	16
Raphanus sativus (radish)	18
Cucumis sativus (cucumber)	14
Citrillus vulgaris (watermelon)	22
Prunus domestica (plum)	48 (hexaploid)
Malus sylvestris (apple)	34, 51 (diploid, triploid)
Arachis hypogaea (peanut)	40 (tetraploid)
Phaseolus vulgaris (bean)	22
Pisum sativum (garden pea)	14
Trifolium repens (white clover)	32 (tetraploid)
Nicotiana tabacum (tobacco)	48 (tetraploid)
Lycopersicum esculentum (tomato)	24
Solanum tuberosum (potato)	48 (tetraploid)
Camellia sinensis (tea)	30
Coffea arabica (coffee)	44 (tetraploid)
Quercus spp. (oaks)	24
Animals	
Drosophila melanogaster (vinegar fly)	8
Musca domestica (housefly)	12
Apis mellifica (honey bee)	32
Bombyx mori (silkworm)	56
Triturus viridescens (salamander)	22
Rana spp. (frogs)	26
Bufo spp. (toads)	22
Xenopus laevis (clawed toad)	36
Gallus domesticus (chicken)	78
Meleagris gallopavo (turkey)	82
Columba livia (pigeon)	80
Anas platyrhyncha (duck)	80
Mus musculus (mouse)	40
Rattus norvegicus (rat)	42
Mesocricetus auratus (golden hamster)	44
Cricetulus griseus (chinese hamster)	22
Cavia porcellus (guinea pig)	64
Oryctolagus cuniculus (domestic rabbit)	44
Canis familiaris (dog)	78
Felis domestica (cat)	76
Equus caballus (horse)	64
Equus asinus (donkey)	63
Sus scrofa (hog)	38
Ovis aries (sheep)	54
Capra hircus (goat)	60
Bos taurus (cattle)	60
Macaca mulatta (rhesus monkey)	48
Gorilla gorilla	48
Pongo pygmaeus (orang-utan)	48
Pan troglodytes (chimpanzee)	48
Homo sapiens (human)	46

Note: Multiples of the haploid (n) number of chromosomes are called diploid ($2n$), triploid ($3n$), tetraploid ($4n$), hexaploid ($6n$), octoploid ($8n$).

Source: Abridged from M. J. D. White, *The Chromosomes*, 6th ed., 1973, Chapman and Hall Ltd., London.

whereby the number of chromosomes was halved, so that upon fertilization, the diploid number characteristic of the species could be restored. He proposed that such a reduction, followed by restoration to diploidy in each new individual, would permit the maintenance of a constant chromosome number, generation after generation. Subsequently, just such a process, called *meiosis,* was discovered in 1900 by H. von Winiwarter in the developing egg cells of the rabbit. Studies by other cytologists eventually established that during meiosis the members of each pair of chromosomes were separated from one another and by this means haploid gametes were formed.

It should be recalled that 1900 was also the year in which Mendel's work was rediscovered and his findings confirmed. Although Mendel's data were statistical, the behavior of his factors as determined from the results of crosses exactly paralleled the behavior of chromosomes as determined microscopically.

1. Mendel's factors are doubly present in the genotype of an organism; chromosomes are doubly present in each cell as homologous pairs.

2. Mendel's factors undergo segregation at sex cell formation; members of homologous chromosome pairs undergo a physical separation in developing gametes during meiosis.

3. Mendel's factors combine at fertilization so that two alleles for any character are again present in the genotype; the union of haploid gametes restores the diploid chromosome condition to the fertilized egg.

4. Mendel established biparental inheritance through the results of reciprocal crosses; the members of a pair of chromosomes were found to be maternally and paternally derived, one from the egg and the other from the sperm.

5. The results of Mendel's crosses involving two different pairs of hereditary factors indicated that the segregation of one pair was independent of the segregation of the other pair (independent assortment); it was found that the separation of one pair of homologues during meiosis could well be independent of the separation of any other pair.

That these striking parallels in behavior were not coincidental was first recognized in 1903 by Walter S. Sutton, who proposed that the physical location of Mendel's factors must be none other than the chromosomes. Thus, plant hybridization and cytology, two independent lines of investigation employing completely different methods, converged to produce a most powerful and far-reaching generalization: the chromosome theory of heredity.

MITOSIS

The process of mitosis occurs in all organisms whose cells contain nuclei. Such organisms, called *eukaryotes,* include the great bulk of living species from algae, slime molds, fungi, and protozoa to the flowering plants and the animals. Eukaryotes are characterized by the possession of cells of great structural complexity containing nuclei, cell organelles such as plastids and mitochondria, and an intricate cytoplasmic membrane system, the *endoplasmic reticulum* (Figure 6–2). The hereditary material found in the nucleus is embodied in several to many chromosomes of varying size. Each such chromosome is composed of an enormously long molecule of deoxyribonucleic acid,

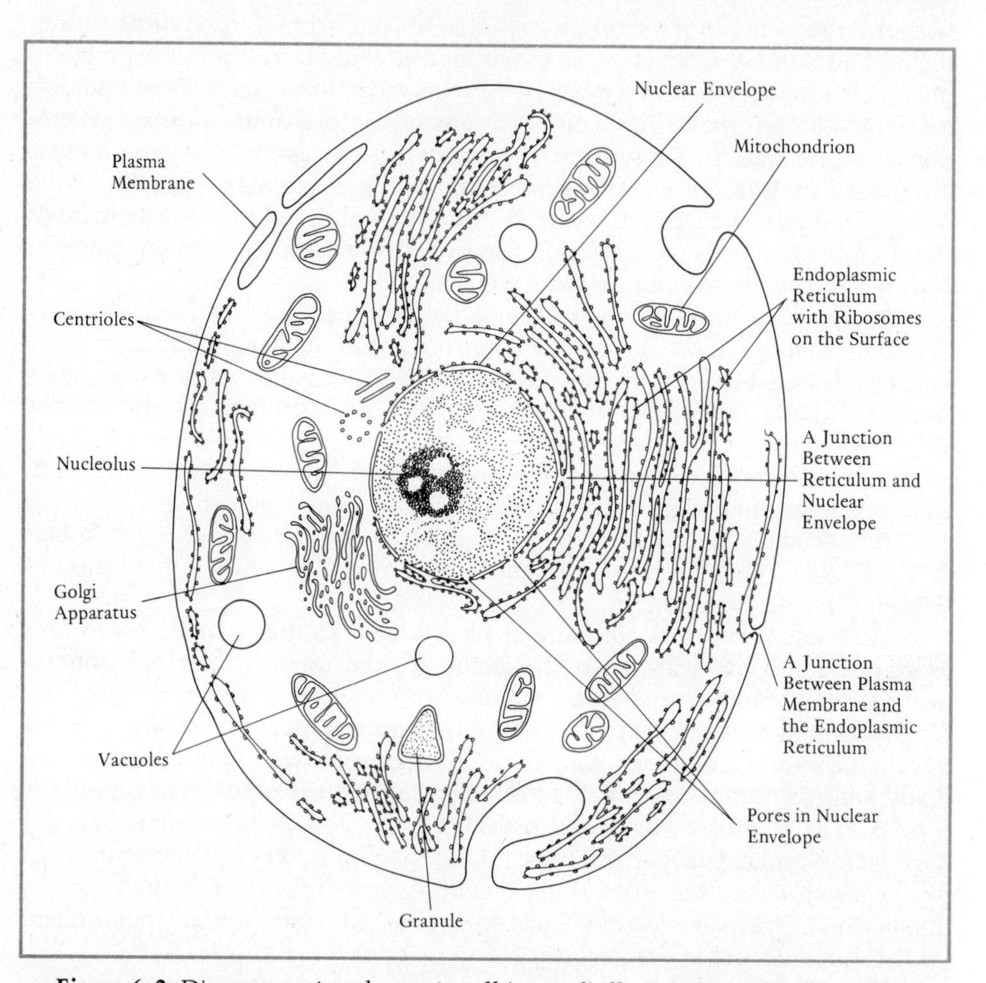

Figure 6–2. Diagrammatic eukaryotic cell (animal) illustrating the various cell components. [From Stephen W. Hurry, *The Microstructure of Cells,* John Murray, London, 1964.]

better known as DNA, to which acidic and basic proteins are firmly bound. The length of this DNA molecule which comprises a single chromosome is truly astounding; for most eukaryotes it has been estimated to be in the *centimeter range*. Yet several to many such chromosomes are contained within a nucleus and cell measured in micrometers (microns).

The few groups of organisms which do not have these features of cell and chromosome structure are known collectively as *prokaryotes*. Prokaryotes include such forms as blue-green algae and bacteria. Though the individuals belonging to any one species are numerous, the number of prokaryotic species is very small compared to the enormous species diversity of the eukaryotes. Characteristically, the cells of prokaryotes are tiny and contain no nuclei, cell organelles, or endoplasmic reticulum. The hereditary material consists of a minute, essentially naked molecule of DNA which is better called a *genophore* or *chromoneme* to distinguish it from the structurally complex chromosomes of eukaryotes. It is interesting to note that mitochondria and plastids also contain a genophore similar in structure to that of bacteria. This discovery has prompted the theory that such organelles represent a symbiotic relationship of great antiquity. A genophore consisting of a single nucleic acid molecule is also characteristic of DNA-containing viruses. The genetic mechanisms of prokaryotes are quite different from those of eukaryotes and involve the transfer of larger or smaller pieces of the naked DNA molecule to other cells. Since mitosis and meiosis do not occur in prokaryotes, cell organelles, or viruses, the following discussion is necessarily restricted to eukaryotic forms.

The Stages of Mitosis

The zygote is the one-cell stage of a new individual and thus it is the progenitor of all of the cells of a multicellular organism. The genotype of the zygote is fixed at the time of fertilization and consists of a particular combination of alleles which have been contributed equally to the new individual by the male and female gametes. Since chromosomes constitute the physical locations of genes, the regular presence of the diploid chromosome number in somatic cells means that the genotype is fully represented in all cells of the

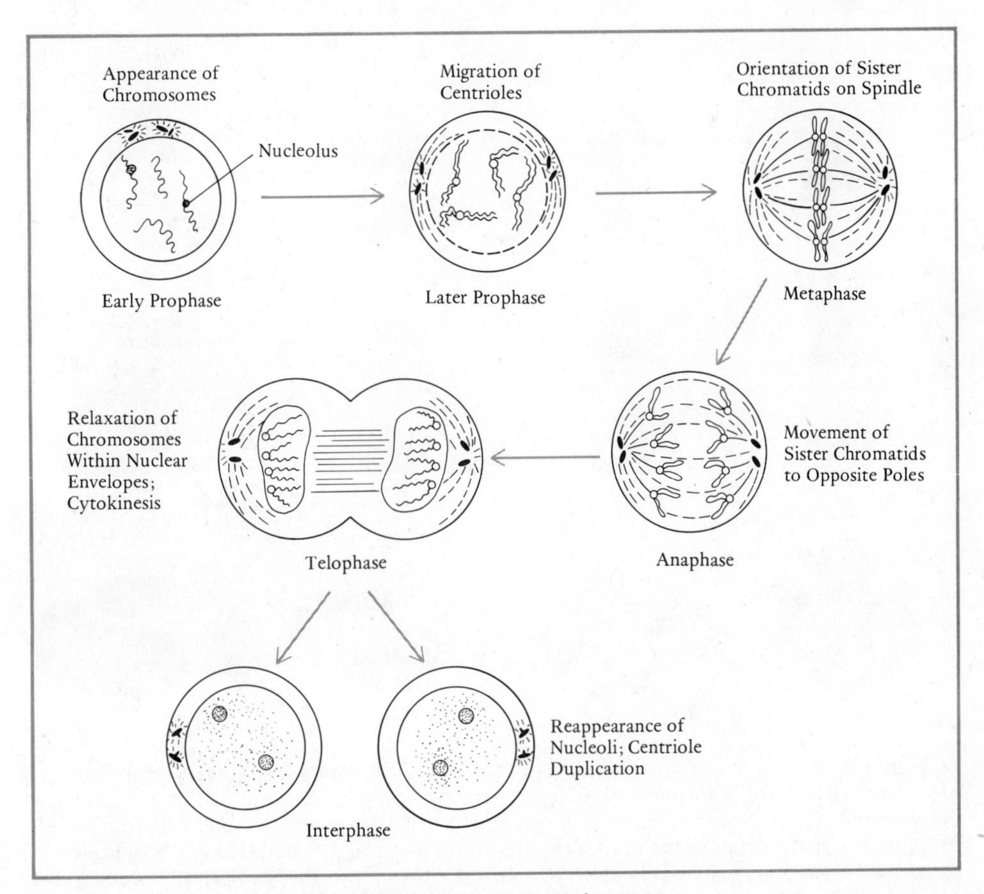

Figure 6–3. Stages of mitosis.

body. This equality of genotype among the cells of an organism requires that in each cycle of cell division every chromosome be replicated exactly, so that an identical set of chromosomes can be distributed to each of the two daughter cells. Such cells, in turn, must also faithfully duplicate each chromosome and ensure its presence in their descendents.

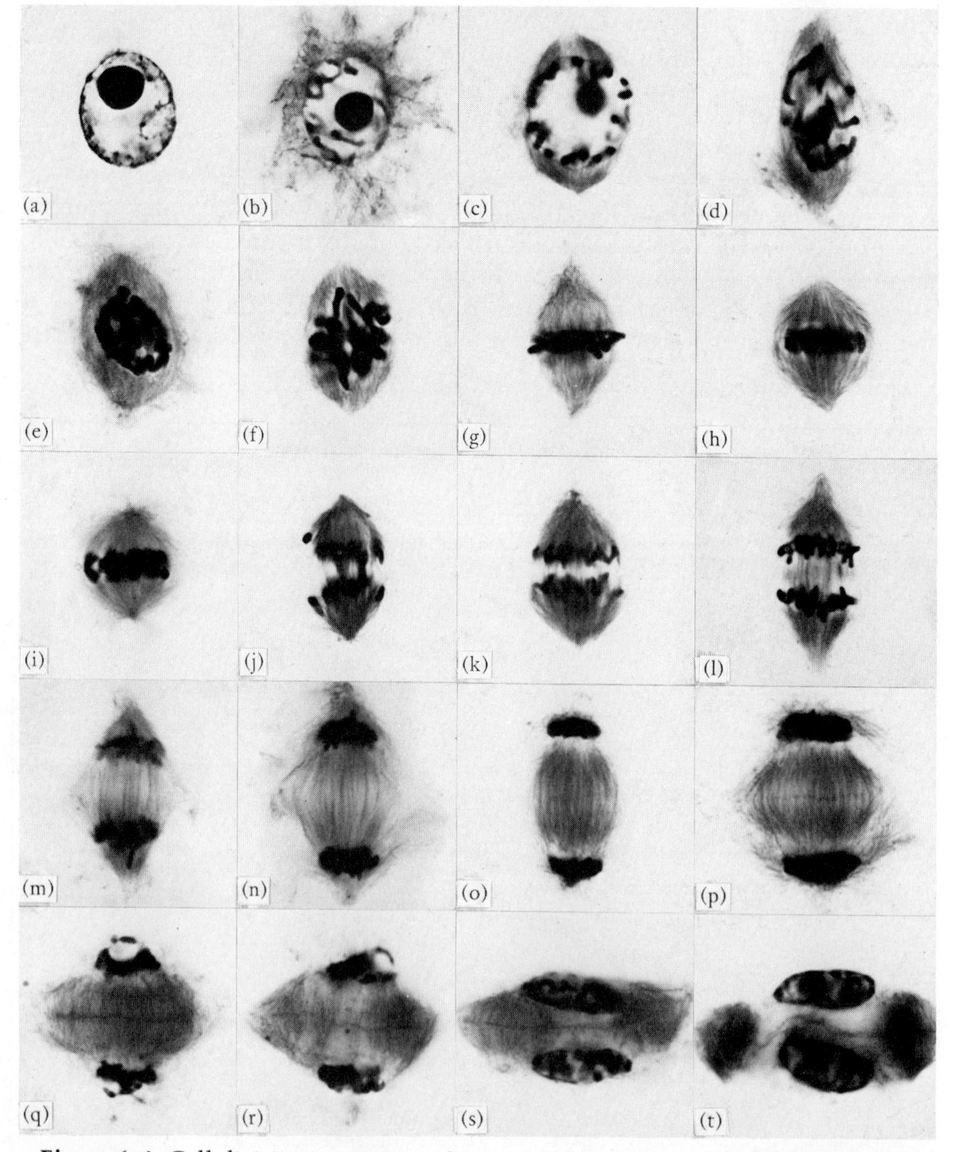

Figure 6–4. Cell division in maize endosperm. (a) Interphase; (b) early prophase; (c) prophase, early spindle formation; (d) prometaphase shortly after nuclear membrane has disappeared; (e, f) prometaphase and orientation of chromosomes on the spindle; (g) metaphase; (h–n) progressive stages in anaphase; (o–t) progressive stages in telophase; (p–t) progressive stages in cell plate formation. (1600×) [From Robert E. Duncan and Maxim D. Persidsky, *Amer. J. Bot., 45:*719–729, 1958.]

The process of cell division of somatic cells consists of several stages. These stages in plant and human cells are illustrated in Figures 6–3, 6–4, and 6–5.

Interphase. In the life of a cell the stage which begins at the completion of one cell division and extends to the onset of the next following division is called *interphase*. During interphase the nuclei of plant and animal cells have a generally similar appearance. The nuclear contents are separated from the cytoplasm of the cell by a nuclear membrane or envelope. The nucleus itself appears to be partially empty and contains small diffuse aggregates of material called *chromatin,* a name assigned by Fleming because of the ability of this material to stain with basic dyes. In addition one or more larger bodies, called

(a) Interphase

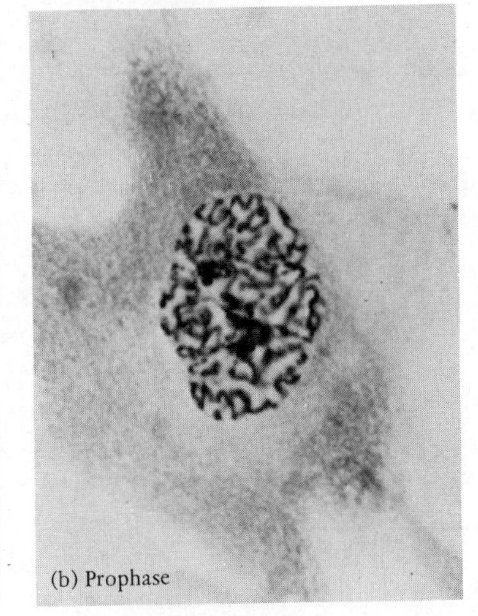

(b) Prophase

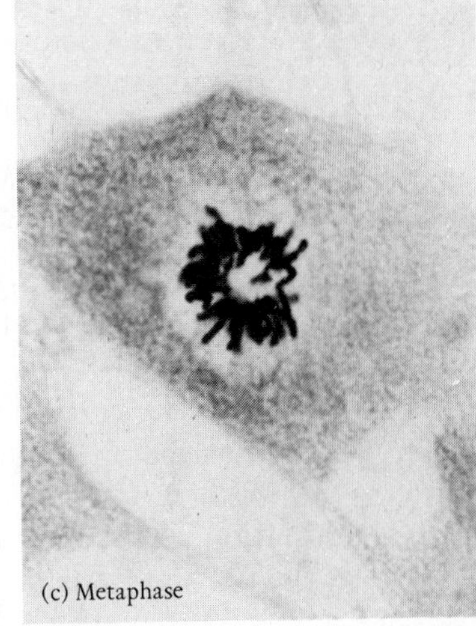

(c) Metaphase

(d) Anaphase

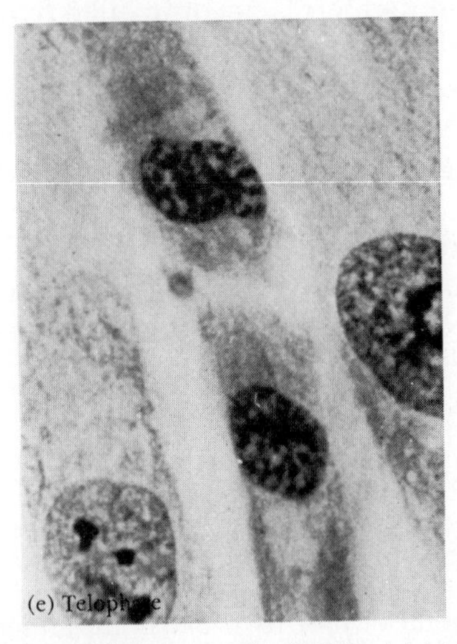

(e) Telophase

Figure 6–5. Mitosis in cultured human cells. (2200×) (a) interphase; (b) prophase; (c) metaphase; (d) anaphase; (e) telophase. [From U. Mittwoch, *Sex Chromosomes,* Academic Press, New York, 1967.]

nucleoli, are also evident. Although nucleoli are attached to particular chromosomes, this association cannot readily be seen in most interphase nuclei.

Prophase. The onset of cell division is marked by the gradual appearance of threadlike bodies within the nucleus. With this occurrence the cell is said to have entered the *prophase* stage of mitosis. During prophase the formerly diffuse substance, which appeared to be isolated granular material, begins to contract and coil to form distinct and separate visible entities. These entities are the chromosomes. With careful observation it can be seen that chromosome replication has been completed, for each chromosome is visibly double and composed of a pair of daughter chromosomes, called *sister chromatids,* that are in close association with one another.

While these events are occurring in the nucleus, concomitant changes take place during prophase in the cytoplasm. In animal cells two pairs of small, densely staining bodies called *centrioles* are located together to one side of the nucleus (Figure 6–6). During prophase these pairs begin to move apart from one another, ultimately taking up positions at opposite sides of the nuclear envelope. As the centrioles begin their migration, the cytoplasm around and between them assumes a fibrillar appearance, the result of the formation of fine microtubules which radiate in all directions from each pair of centrioles, forming the so-called *astral rays*. These fiberlike structures also extend from one pair of centrioles to the other to form the mitotic spindle which gradually begins to impinge upon the nucleus. The astral rays and spindle together constitute the *achromatic figure,* so named because the protein fibers of which it is composed are unstained with basic dyes.

With rare exceptions centrioles are not found in plant cells, but spindle formation occurs in much the same way. Spindle fibers appear at the periphery of the nucleus and extend to opposite sides of the cell, although their termination at these points is not marked by any formed body.

Meanwhile, the chromosomes within the nucleus undergo a further contraction and condensation which is accompanied by the disappearance of the nucleoli. Eventually the nuclear envelope, invaded by rapidly developing spindle fibers, breaks down. At this point the prophase stage is concluded and the next stage in division, *metaphase,* begins.

Metaphase. During metaphase the chromosomes released from the confines of the nuclear membrane take up a central position in the spindle midway between the two ends, or poles. Although some spindle fibers extend in uninterrupted form from pole to pole, others, called half-spindle fibers, become attached to a specialized region on each chromatid. This region is the *centro-*

Figure 6–6. Early prophase illustrating the position of centrioles. (a) Electron micrograph of rat lymphocyte of thymus gland. Chromatin of nucleus is condensing in early prophase and centrioles lie outside and below the nuclear envelope. Arrows indicate bodies similar to centriolar buds. (17,000×) (b) Electron micrograph of two pairs of centrioles. Daughter centrioles grow out at right angles from mother centrioles during early prophase so that a total of four (two pairs) are present at the onset of mitosis. Arrows indicate early microtubule (spindle fiber) formation; PR indicates polyribosomes active in protein synthesis (see Chapter 15). (54,000×) [From R. G. Murray, A. S. Murray, and A. Pizzo, *J. Cell. Biol., 26,* 601–619, 1965.]

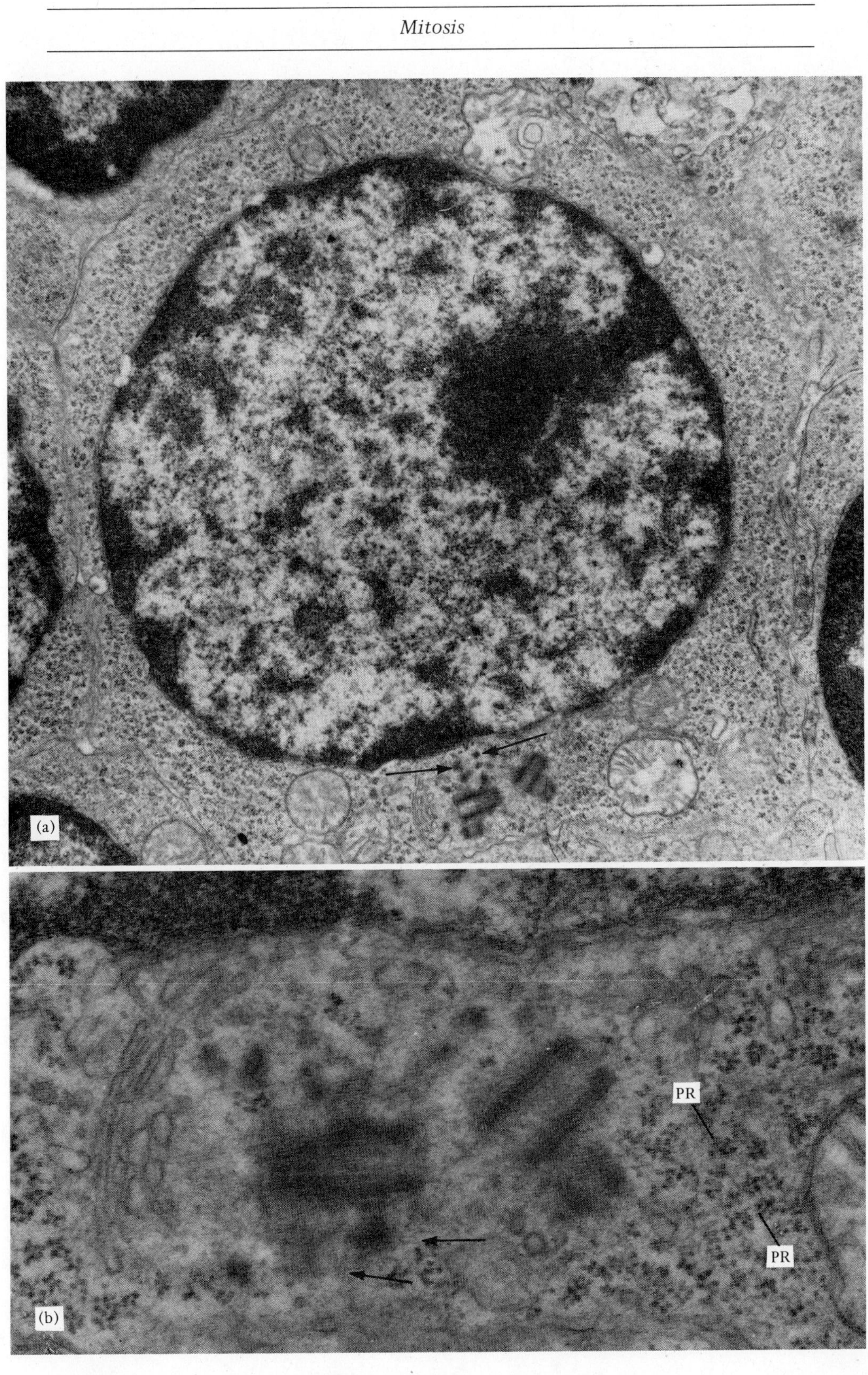
(a)
(b)
PR
PR

mere (kinetochore), or spindle fiber attachment site, and its location is usually marked by a constriction in the body of the chromatid. During early metaphase (prometaphase) orientation of chromosomes is accomplished so that the centromeres of sister chromatids face opposite poles of the spindle. When alignment and spindle fiber attachments are complete, metaphase is concluded and *anaphase* begins.

Anaphase. Anaphase is a stage in mitosis characterized by the separation of sister chromatids from one another and their movement to opposite poles of the spindle. Lengthwise separation of chromatids is first achieved at the centromeres and is followed by the disengagement of more distal regions as the respective sister chromatids move apart and proceed in opposite directions. The movement of these daughter chromosomes ceases when they reach the poles, and the final stage of cell division begins.

Telophase. During *telophase* a new nuclear envelope forms around each group of chromosomes now closely aggregated at opposite poles of the spindle. At the same time division of cytoplasmic components of the cell, *cytokinesis,* commences. In animal cells this is accomplished by the formation of a constriction around the periphery of the cell. This constriction gradually extends inward to divide the cell into two daughter cells, each with its own nucleus. In plant cells cytokinesis is achieved by the formation of a cell plate in the interior of the cell midway between the two poles of the spindle. The cell plate, the forerunner of the cell wall, rapidly grows outward in all directions until separation of daughter cells has been accomplished.

As cytokinesis is completed and the remains of the spindle disappear, the tightly coiled and condensed chromosomes slowly relax, unwind, and become dispersed once more within the confines of the new nuclear envelope. Nucleoli reform and each daughter cell assumes the interphase condition. The length of time required for the entire division cycle from prophase to the succeeding interphase varies greatly depending on cell type. In some early embryos, particularly those of insects, mitosis may occur every 9 to 10 minutes resulting in the rapid build-up of cells for ensuing morphogenesis. In embryos or tissues of other species, however, the process may require anywhere from a few to 24 hours or more.

This elaborate mechanism of chromosome duplication and distribution regularly occurs in dividing cells as a smoothly coordinated and carefully managed activity. It is a marvel of efficiency and is under precise control by the genotype itself. Mitosis accomplishes a purpose of vital importance to the organism, for it ensures that every daughter cell will receive an exact replica of each and every chromosome originally present in the zygote, and by this means the genotype established at fertilization is transmitted to every cell of the organism.

CHROMOSOME MORPHOLOGY AND IDENTIFICATION

The realization of the significance of the chromosomes seen in the stages of cell division stimulated great interest in the structure and organization of these bodies. Initially they were studied in sections or slices of tissue, but somewhat later the introduction of the technique of squashing and staining

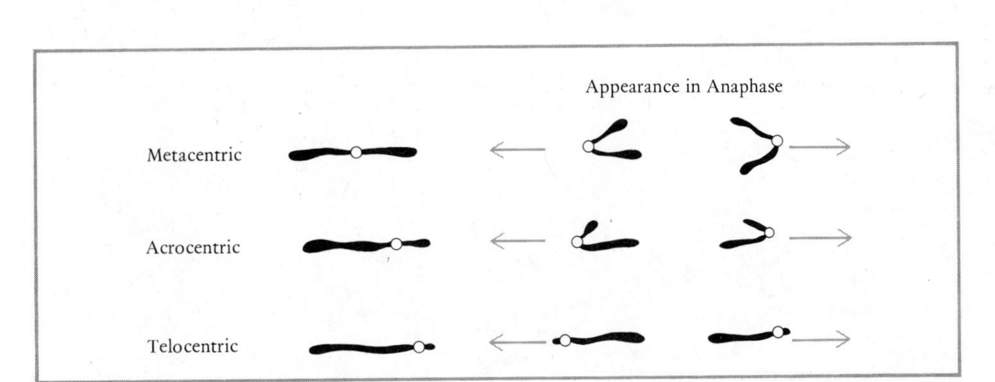

Figure 6–7. The shapes of chromosomes and the position of the centromere.

cells directly on a slide permitted a more accurate determination of chromosome number and morphology. More recently, use of the electron microscope has allowed visualization of the fine structure of the chromosome.

The Chromosomes in Metaphase

As viewed with the light microscope, late prophase to metaphase chromosomes are generally elongate in shape and present the appearance of solid, dense, tightly coiled bodies (see Figure 6–9). Although in a few exceptional cases multiple spindle fiber attachment points have been reported, as a rule, each chromosome (chromatid) contains a single centromere whose location is marked by the presence of the primary constriction. The body of the chromosome to either side of this constriction is designated as the right or left arm. A specialized region associated with the organization of the nucleolus *(nucleolus organizer)* may be evident as an additional constriction located on one or more pairs of chromosomes, depending on species. Other secondary constrictions may also be present.

The position of the centromere is used as a reference point of chromosome classification. For example, a chromosome with a centrally placed centromere is called *metacentric* in type. Its arms are approximately equal in length, and it exhibits a V-shape upon movement toward the pole of the spindle in anaphase. More of a J-shape at this stage denotes a chromosome whose centromere is located closer to one end than the other, a condition known as *acrocentric.* Rod-shaped chromosomes are *telocentric,* that is, their centromeres are located so close to one end that the tiny short arm is not easily, if at all, discernable (Figure 6–7).

With the high magnification made possible by use of the electron microscope, a metaphase chromosome no longer appears solid, but instead it is seen to consist of a mass of tightly compacted fibers (Figure 6–8) upon which the coiling seen at lower magnifications is superimposed (Figure 6–9). The centromere, or kinetochore, appears as a disk-shaped structure from which spindle fibers in the form of microtubules extend (Figure 6–10). Chromosomes are thought to terminate in structures called *telomeres*. Telomeres have been identified in chromosomes undergoing prophase of meiosis and in the specialized salivary chromosomes of *Drosophila* (see Chapter 9). In metaphase chromo-

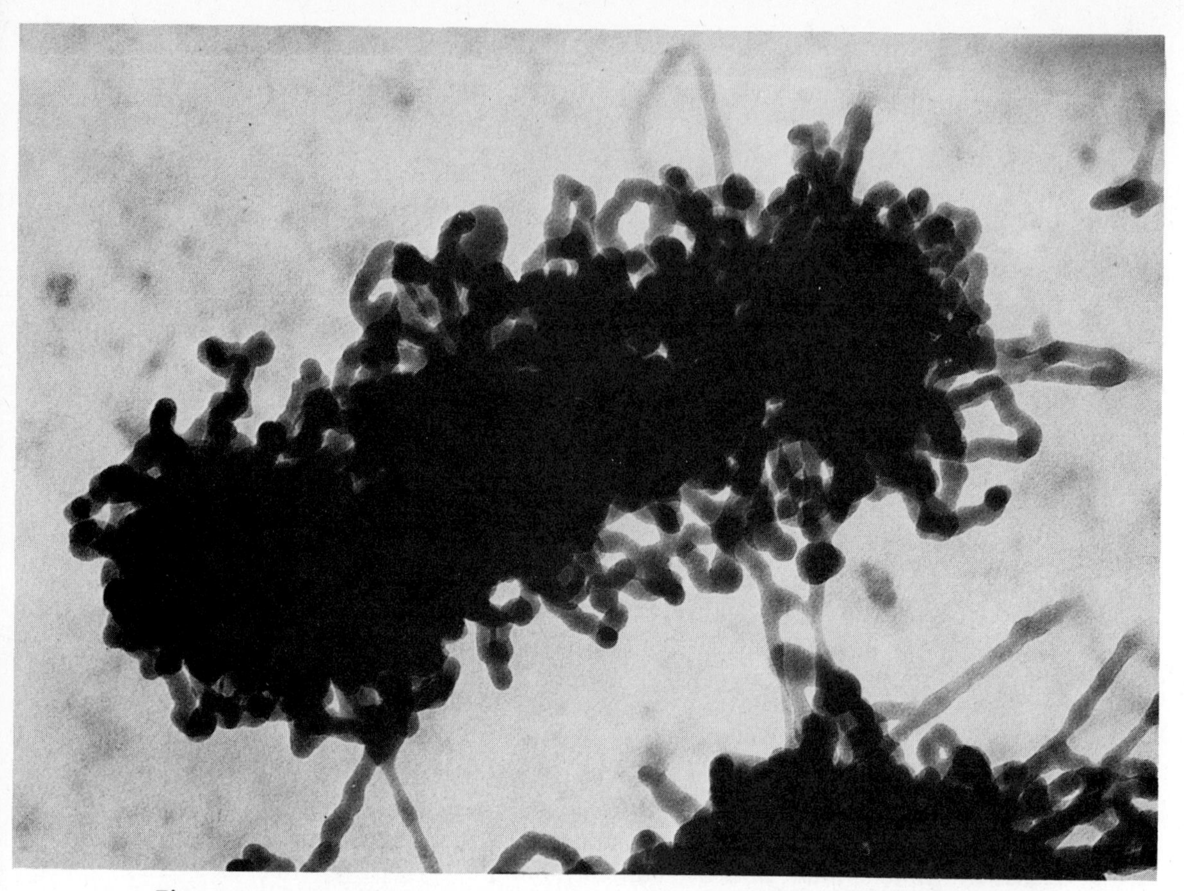

Figure 6–8. A metaphase chromatid from a honey bee embryonic cell. The primary structural component is a tightly folded chromatin fiber. The total length of the fiber is estimated at 300–500 μm, packed into a chromatid 1.36 μm long. (79,560×). [From E. J. DuPraw, *Nature, 206:* 338–343, 1965.]

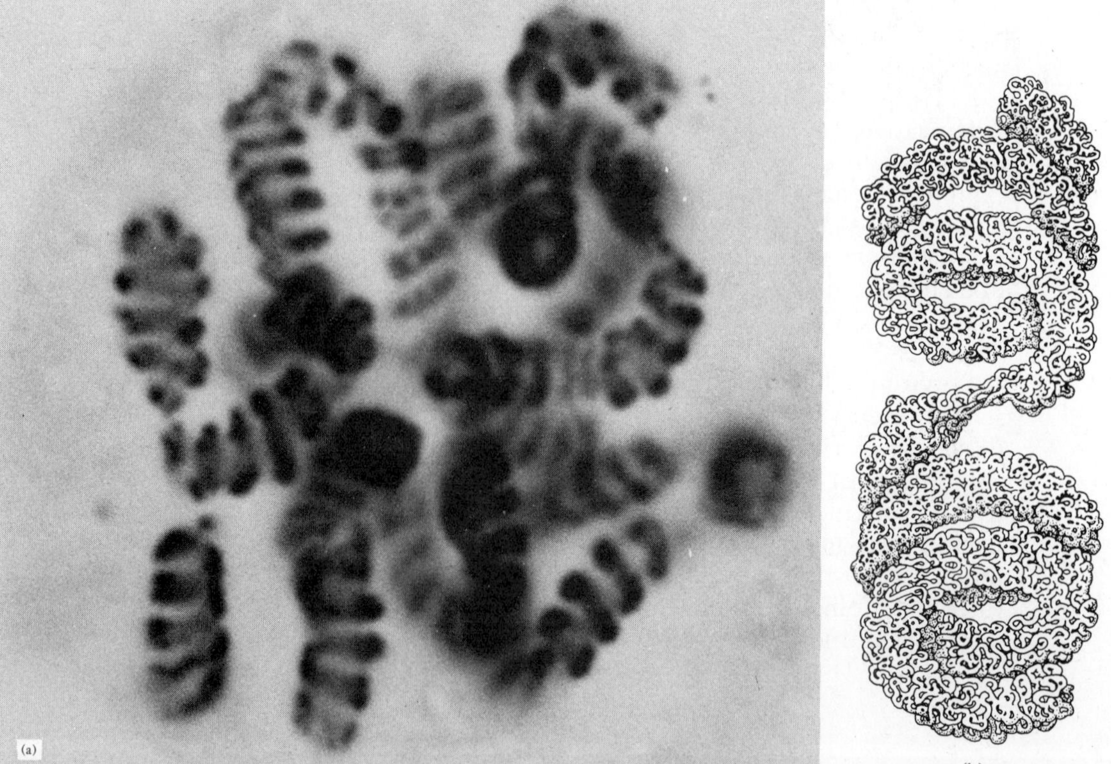

somes, however, no free ends or telomeres are visible, and each chromosome arm appears to terminate in a mass of looped fibrils.

All available evidence supports the concept that each chromosome (chromatid) consists of a single, extremely long and thin molecule of DNA which is stably complexed with protein and that during prophase this delicate filament folds and coils to form the fibrillar mass seen at metaphase. The nature of the mechanism which controls and coordinates this amazing packing process is as yet unknown. We can conclude that it must be precise and

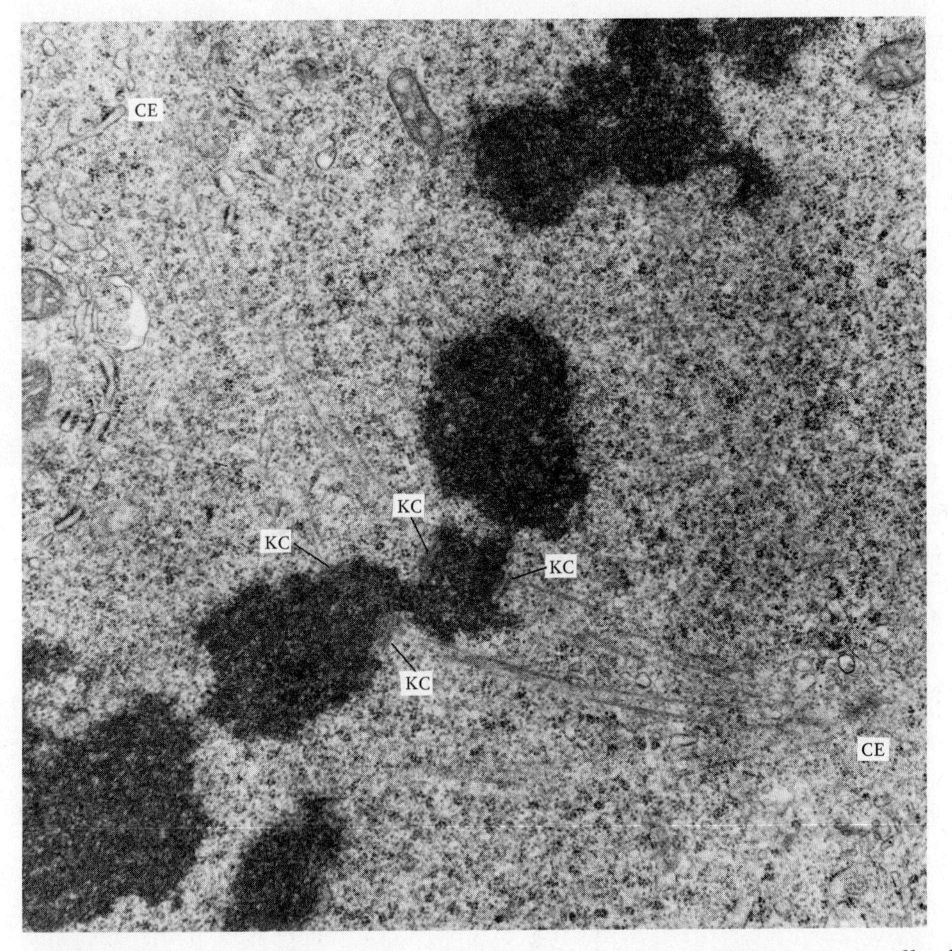

Figure 6–10. Electron micrograph of metaphase kinetochores (KC) in mitotic cells of the rat. Microtubules of the half-spindle fibers extend from each kinetochore (centromere) toward the centrioles (CE). (23,500×) [From P. T. Jokelainen, *J. Ultrastr. Res. 19:* 19–44, 1967.]

Figure 6–9. Chromosome coiling. (a) Spiral structure at first metaphase of meiosis in *Tradescantia virginiana,* after pretreatment with nitric acid vapor before fixation. (4000×) (b) Interpretation to show how coiling might be superimposed on a folded-fiber chromatid. [(a) Reproduced from *The Handling of Chromosomes,* C. D. Darlington and L. F. LaCour, Allen & Unwin, London, VI Edit., 1976; (b) from E. J. DuPraw, *Nature, 209,* 577–581, 1966.]

specific, for it imposes a constant and recognizable morphology upon each kind of chromosome present at metaphase.

To illustrate the complexities involved, consider the following. Human diploid cells contain 23 pairs of chromosomes for a total of 46 chromosomes in all. At metaphase, each chromosome is represented by two chromatids, so that 92 such bodies are present. Since the molecular dimensions of the DNA molecular are known, a determination of the total amount of DNA present in a metaphase cell can be used to estimate the total combined length of all DNA fibers present. In human metaphase cells this total length is approximately 3.8 *meters*. Since 3.8 meters is divided among the 92 chromatids present, the average length of the DNA-protein fiber in a single chromatid can be estimated to measure around 4 *centimeters!* An appreciation of the degree of coiling and packing that is required for the condensation of such a fiber may be gained from the fact that human metaphase chromosomes are actually only 2 to 8 *micrometers* (μ) in length and thus are around 10,000 times shorter than the nucleoprotein fiber of which they are composed. The following equivalents are helpful in visualizing these measurements: 1 millimeter (mm) = 1000 micrometers (μm) or microns (μ); 1 μm = 1000 nanometers (nm) or millimicrons (mμ); 1 nm = 10 Angstroms (Å).

Although models of possible folding patterns such as those illustrated in Figure 6–11 have been proposed, the actual mechanism for folding is not understood.

The Chromosomes in Interphase

As a general rule, during telophase the chromosomes of each daughter cell gradually disappear as formed bodies and appear dispersed within the nucleus. Although early cytologists wondered if this dispersion reflected a dissolution or disintegration of the chromosomes, this interpretation was soon found to be inconsistent with the evidence accumulating from genetic studies. Such studies demonstrated the linear arrangement of genes on the chromosome and the constant association of particular genes with particular locations on chromosomes. The maintenance of these associations clearly required that the individuality and integrity of each chromosome be preserved through succeeding cell divisions, as well as between generations.

The development of cell culture techniques and phase contrast microscopy permitted the observation of living cells as they underwent mitosis. When the same cell was followed through two division cycles, the chromosomes which became essentially invisible in interphase reappeared at the next succeeding prophase in the exact same positions which they had occupied at telophase. Such observations indicated continuity of the chromosomes through interphase and also suggested that these bodies maintained their respective positions within the nucleus during this period, presumably through attachments to the nuclear envelope.

When the fibrillar nature of chromosomes was revealed by the electron microscope, the disappearance of the chromosomes at interphase could be attributed to the gradual relaxation and extension of the tightly wound nucleoprotein fibers. The smallest diameter of such fibers in interphase cells has been measured as 20 Å, although 100 Å, 250 Å, and 500 Å fibers are also found, size

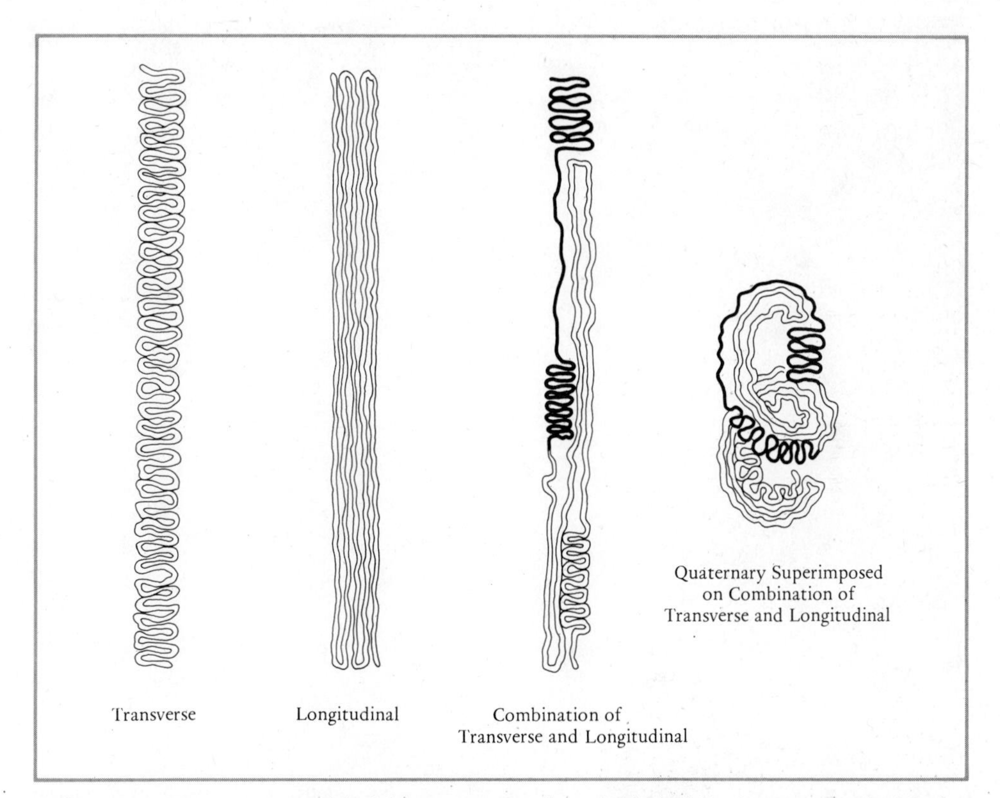

Figure 6–11. Various possible folding patterns for a folded-fiber chromatid. The blackened fiber segment illustrates how the distribution of silver grains seen in autoradiographs depends on the pattern of folding. If during interphase cells are placed for a short time in the presence of radioactive precursors of DNA, these precursors will be incorporated into portions of the DNA molecules. Later, upon contact with the cells, photographic film will be exposed by the radioactivity present. When the film is developed, the presence of black silver grains will indicate the chromosome regions which replicated during exposure to the labeled precursors. Such preparations are called autoradiographs. [From E. J. DuPraw, *Nature, 209,* 577–581, 1966.]

being dependent on the state of coiling. Since the dimensions of an extended fiber are far below the resolving power of the light microscope, the nucleus frequently appears to be lightly staining or empty when viewed with this instrument.

The contents of the interphase nucleus are enclosed by the nuclear envelope which with the electron microscope can be seen to consist of two membranes. The outer member maintains continuity with the endoplasmic reticulum (Figure 6–2), while the inner membrane is adjacent to the chromosomal fibers. This double envelope is pierced by regularly arranged hexagonal pores, called *annuli,* through which particles are described to pass from the nucleus to the cytoplasm. The contents of the interphase nucleus present the appearance of a mass of fibers, some thin, some thick and bumpy, and it is impossible to distinguish the extended fiber of one chromosome from that of another. Collectively, these fibers are called *chromatin*. Attachment of

Figure 6–12. Chromatin fibers of a honey bee embryonic cell firmly attached to the nuclear envelope at the edges of annuli. An unbroken attachment is seen at the arrow and broken fiber ends are visible on other annuli. (94,700×) [From E. J. DuPraw, *Proc. Nat. Acad. Sci.*, *53*, 161–168, 1965.]

chromatin fibers to the nuclear envelope is common and appears to occur adjacent to annuli (Figure 6–12).

The chromosomes or portions of chromosomes which become greatly extended in interphase constitute what is called *euchromatin*. Some chromosome regions, and frequently whole chromosomes, regularly fail to assume this condition at interphase. They remain condensed and are therefore visible as dark staining bodies, collectively called *heterochromatin*. Failure to assume the relaxed and extended state has been correlated with inactivity on the part of genes contained within these regions. Accordingly, the phenomenon reflects the presence of controlling mechanisms which serve to activate or inactivate portions of the complete gene complement *(genome)* at particular times during development or in particular kinds of differentiated cells.

Heterochromatin has been classified into two general types: *constitutive* and *facultative*. Constitutive heterochromatin is associated with centromeric

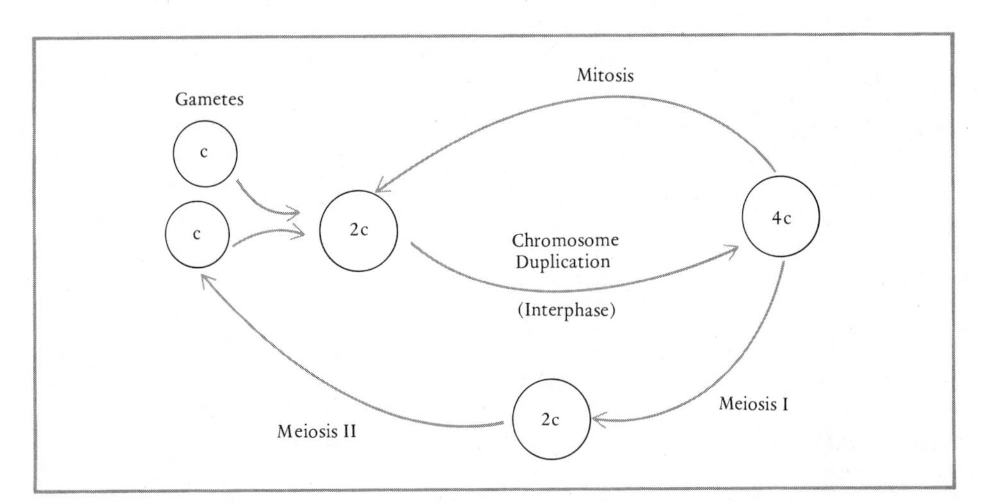

Figure 6–13. DNA content (c) of gametes, somatic cells, and meiotic cells.

regions which remain contracted during interphase, while facultative heterochromatin appears to occur as the result of the regular condensation of whole chromosomes during some stage in development. Such condensation may be reversible. Varying amounts of contracted chromatin may be found in specialized types of tissue cells. This material can also be classified as facultative heterochromatin, and it has been suggested that its presence is somehow associated with maintenance of the differentiated state. It must be noted, however, that the nucleolus cannot be placed in the category of heterochromatin, for its large size and staining capacity reflect intense activity rather than quiescence.

Finally, although the presence of sister chromatids cannot be detected microscopically until the prophase stage of mitosis, the synthesis of sister chromatids actually occurs during interphase when the chromatin fibers are in the extended state. Since DNA is known to constitute the genetic material, the amount of DNA present in the nucleus can be used as an index of whether or not chromosome replication has occurred. DNA can be identified specifically by a technique called Feulgen staining, and the depth of staining can be used to quantitate the amount of DNA present. DNA content can also be assayed by its absorption of ultraviolet light at a wavelength of 260 nm. By these and other methods, it can be demonstrated that the DNA content of the nucleus exactly doubles during the interphase period, well before the onset of prophase. We can therefore infer that the production of daughter chromosomes is an activity which is completed prior to the initiation of mitosis. The DNA content of cells in various stages of the life cycle is summarized in Figure 6–13 (see also Figure 11–29).

Human Metaphase Chromosomes

Although in 1882 Fleming pictured human chromosomes in dividing epithelial cells (Figure 6–1), the diploid number was not known with certainty until 1956, when J. H. Tjio and A. Levan correctly reported the presence of 46 chromosomes in human cells. Until this time studies of human chromosomes lagged far behind those of other organisms because of the difficulty of obtaining favorable material. Great impetus was added to this field, however, by the discovery of J. Lejeune, in 1958, that human mongolism was associated with an abnormal chromosome complement wherein 47, rather than 46 chromosomes were present. By 1960 a number of techniques for the preparation and study of human chromosomes had become available. In addition, it became no longer necessary to obtain tissue samples from surgery or autopsy, for the nucleated leucocytes (especially lymphocytes) present in peripheral blood were found to provide entirely satisfactory material.

The techniques employed then and now utilize the methods of tissue culture. A small sample of blood is cultured for a short time in the presence of phytohemagglutinin, a substance extracted from beans which stimulates chromosome replication and cell division. When colchicine is added to the culture, the cells entering mitosis are held at metaphase, for colchicine is a cell poison causing dissolution of the spindle and the arrest of cell division. Subsequent treatment of the cells with hypotonic saline causes swelling so that the chromosomes become separated and spread apart from one another. Finally, the cells are smeared upon a slide, stained, and examined with the light microscope. For the identification of individual chromosomes, the preparation is photographed through the microscope and an enlarged print is made. The various chromosomes are then cut out of the print and arranged as accurately as possible in pairs to represent the diploid complement. Such an arrangement is called a *karyotype* (see Figure 6–14).

Human chromosomes can be classified into morphological groups according to such criteria as the position of the centromere, the length of the arms, and the presence and location of secondary constrictions (Figure 6–7). An additional distinguishing criterion is the time sequence or pattern of chromosome replication which varies in different chromosomes. Such patterns have been determined by a technique called autoradiography (see Figure 6–11). Although these criteria allow classification into morphological groups and positive identification of a few chromosomes, they do not provide a reliable method for the unambiguous identification of all chromosome pairs.

This goal has now been largely realized through the use of new staining techniques, the first of which was introduced in 1970 by Caspersson (Figure 6–14). This staining method employs a compound called quinacrine (also

Figure 6–14. Banding techniques for human chromosomes. (a) Peripheral blood lymphocyte metaphase from a normal male stained with quinacrine to show Q-banding. Note the bright fluorescence of the distal region of the long arm of the Y chromosome and the small but bright centromeric regions on chromosomes 3 and 4 (indicated by arrows). (b) C-banded metaphase from a male with a deficiency of part of the long arm of one chromosome 9 and a large centric heterochromatic block in one member of chromosome pair 3. The chromosomes with the major C-bands, including the Y, are indicated by arrows. (c) Karyotype from a metaphase cell of a male stained with the acetic acid-saline-Giemsa G-banding technique. It should be noted that both members of a pair show identical banding patterns, but that there are marked differences in pattern among chromosome pairs, so that all pairs are easily distinguishable. X and Y are sex chromosomes. The standard grouping of chromosomes based on morphology is indicated by letters beneath each group. [H. John Evans, *Brit. Med. Bull.*, *29*, 196–202, 1973.]

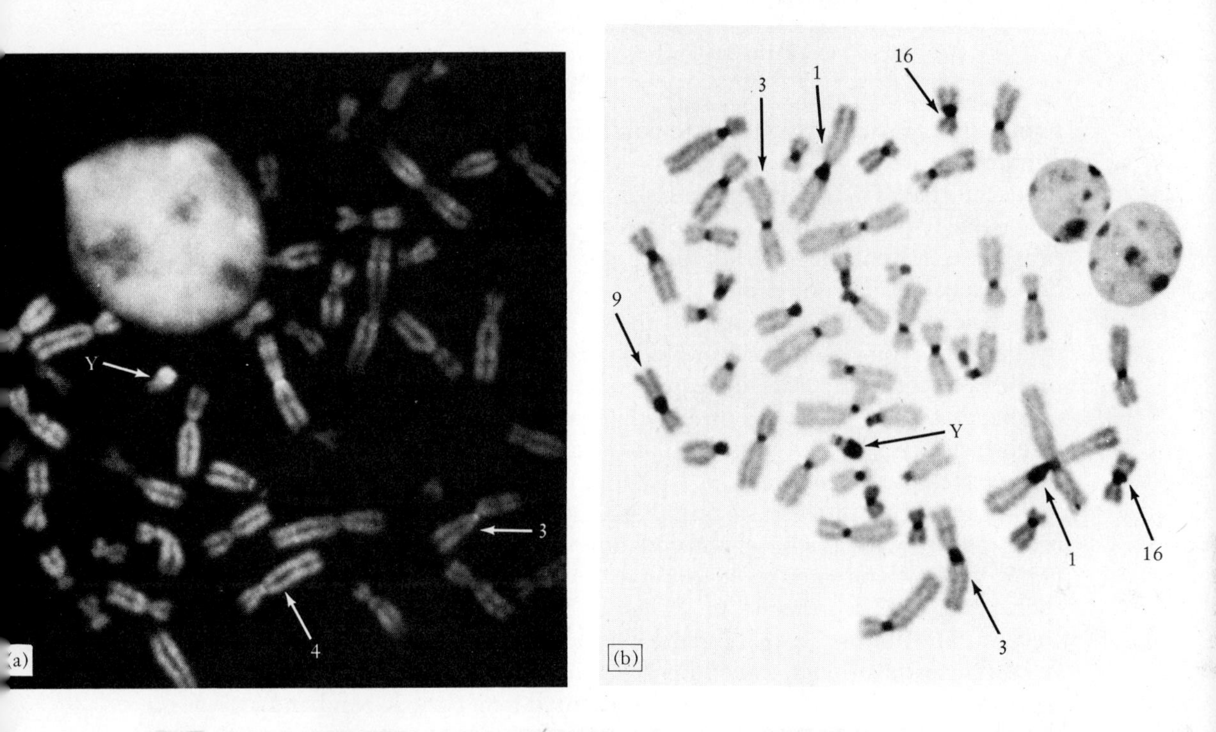
Y
3
4
(a)
16
3
1
9
Y
1
16
3
(b)

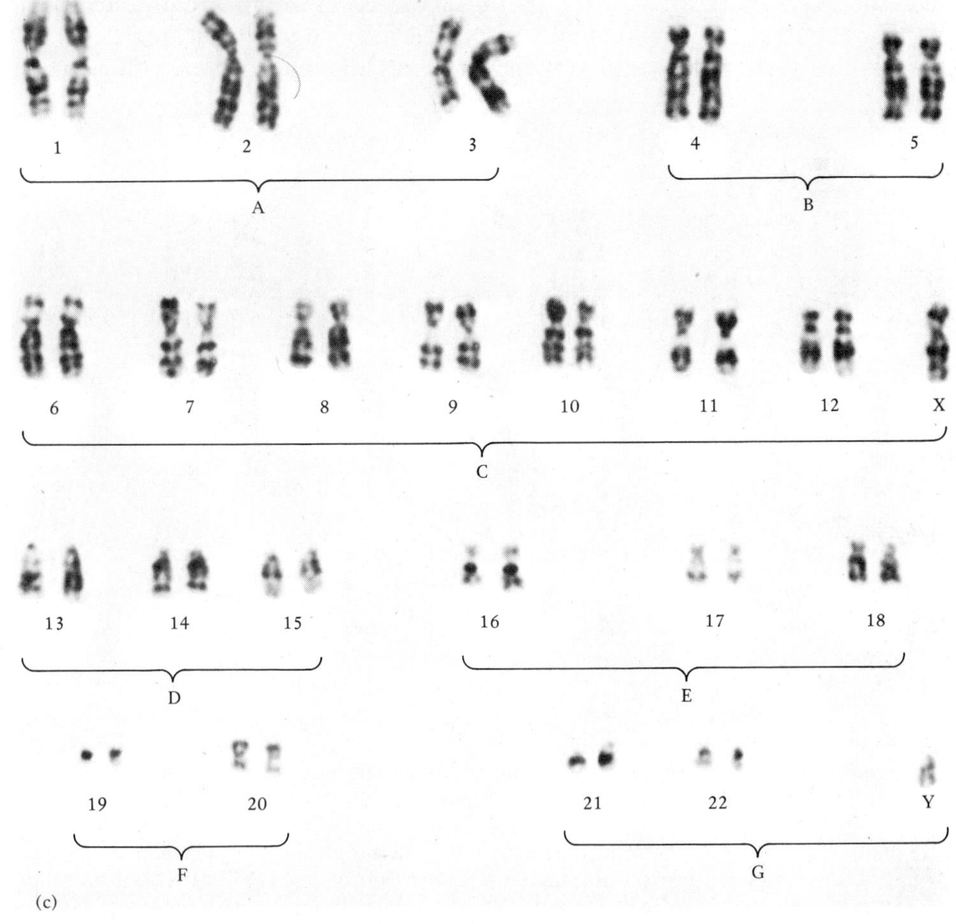
1
2
3
A
4
5
B
6
7
8
9
10
11
12
X
C
13
14
15
D
16
17
18
E
19
20
F
21
22
Y
G
(c)

known as Atabrin), a drug used to combat malaria. Chromosomes treated with quinacrine show transverse bands, called Q-bands, which fluoresce in ultraviolet light. This banding pattern is specific for each chromosome pair and is consistently present in chromosomes obtained from different tissues. In addition, the Y sex chromosome of males can be readily distinguished by the bright fluorescence of its long arm.

A second technique employs the Giemsa stain which is a mixture of four dyes and has long been used for blood smears. When chromosomes are fixed in methanol-acetic acid and subsequently stained with Giemsa, a new and different banding pattern can be observed. These bands are called C-bands because intense staining occurs primarily in the constitutive heterochromatin associated with the centromere.

A third technique also employs fixation in methanol-acetic acid, but before staining with Giemsa, chromosome preparations are specially treated at pH 9.0 or with urea or the proteolytic enzyme trypsin. Subsequent staining then reveals the presence of G-bands whose distribution is again highly specific. The method is called the acetic acid-saline-Giemsa or ASG technique. A modification of this procedure produces R-bands which are the reverse of G-bands, that is, a dark G-band is represented by a light R-band and vice versa. On the basis of these staining methods the chromosomes comprising the human genome can be individually identified and characterized (Figure 6–15).

All of these new techniques and their modifications have revolutionized

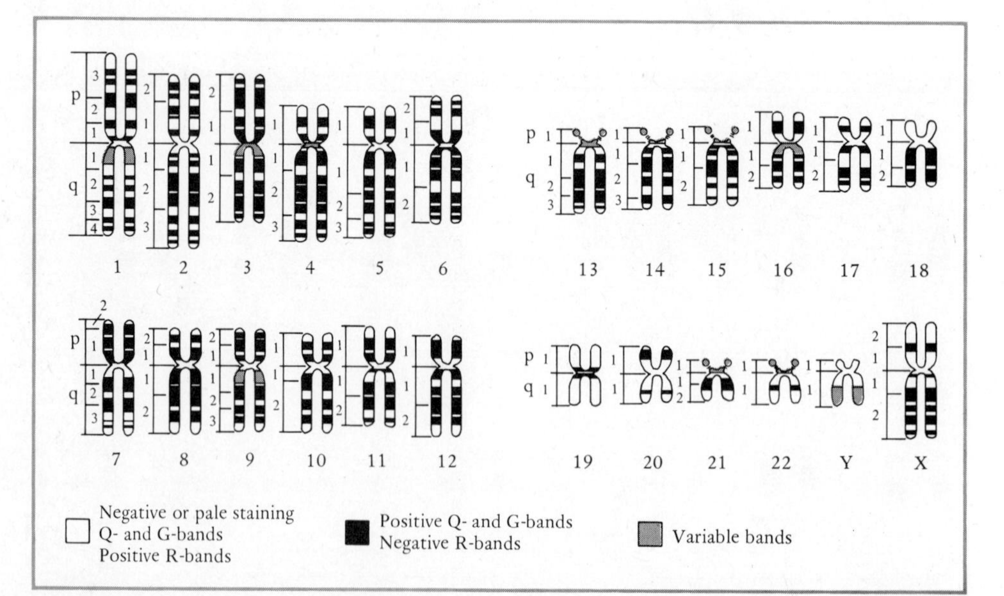

Figure 6–15. Diagrammatic representation of the chromosome bands of human chromosomes as observed with the Q-band, G-band, and R-band staining methods. Centromere represented by Q-band staining only. The position of the centromere is marked by a constriction, and the chromosome arms above and below this constriction are designated as *p* and *q*, *p* being the shorter arm and *q* the longer arm. Within each arm the major banded regions are numbered, and the specific bands of each region are further indicated by smaller numbers (not shown). [*Paris Conference (1971): Standardization in Human Cytogenetics.* Birth Defects: Original Article Series, VIII:7, 1972. The National Foundation, New York.]

the cytogenetics of humans and other mammals. It is now possible to detect the presence of an abnormal chromosome complement through changes in the chromosome banding patterns and to correlate such changes with phenotypic manifestations. In many hospitals newborn infants suspected of carrying chromosomal anomalies are routinely karyotyped, only a few drops of blood being required for cell culture. These procedures are now providing much needed information on the frequencies of various human chromosome aberrations in the population and the correlated consequences of their presence to the individual. Such knowledge is an essential prerequisite for the institution of effective treatment, as well as for genetic counseling.

MEIOSIS

The life cycle of almost all organisms above the level of bacteria and blue-green algae is characterized by an alternation between the diploid and the haploid states. In the diploid ($2n$) phase, a double set of chromosomes occurs in each cell, while in the haploid (n) state, only a single set of chromosomes is present. The portion of the life cycle occupied by each of these phases varies greatly. Algae and fungi are usually haploid in constitution, and their diploid zygotes almost immediately give rise once more to haploid cells. On the other hand, higher plants are diploid through most of the life cycle, the haploid phase being drastically reduced. In animals the haploid phase exists only with respect to gametes. However, in all of these numerous forms of life and regardless of the prominence in the life cycle of either phase, the change from haploidy to diploidy to haploidy regularly occurs.

It is obviously easy to visualize the achievement of diploidy with the paired condition of chromosomes and genes restored by means of the fusion of haploid cells as in fertilization. However, the attainment of haploidy from diploidy requires a reduction in chromosome number to half of that characteristic of the $2n$ state. The process by which this goal is achieved is called *meiosis*. In essence, meiosis in all forms is a mechanism whereby the members of homologous pairs of chromosomes are separated from one another and segregated into different cells. Each such cell thus receives a single representative of each kind of chromosome originally present in the diploid organism. This overall result is accomplished through two cell divisions called the first and second meiotic divisions, or meiosis I and II. Prior to these divisions there occurs but a single replication of the chromosomes (see Figure 6–13). The resulting chromatids are then distributed via the two meiotic divisions into four cells, each of which is haploid. Let us first consider the overall process of meiosis and then examine some of the associated events in more detail.

The Stages of Meiosis

The first and second meiotic divisions are each divided into prophase, metaphase, anaphase, and telophase (Figures 6–16, 6–17). As in mitosis, the movement of chromosomes to opposite poles is facilitated by the presence of a spindle.

Prophase I. The prophase of the first meiotic division is lengthy and is characterized by a number of specific, recognizable stages.

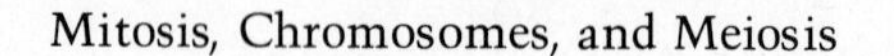

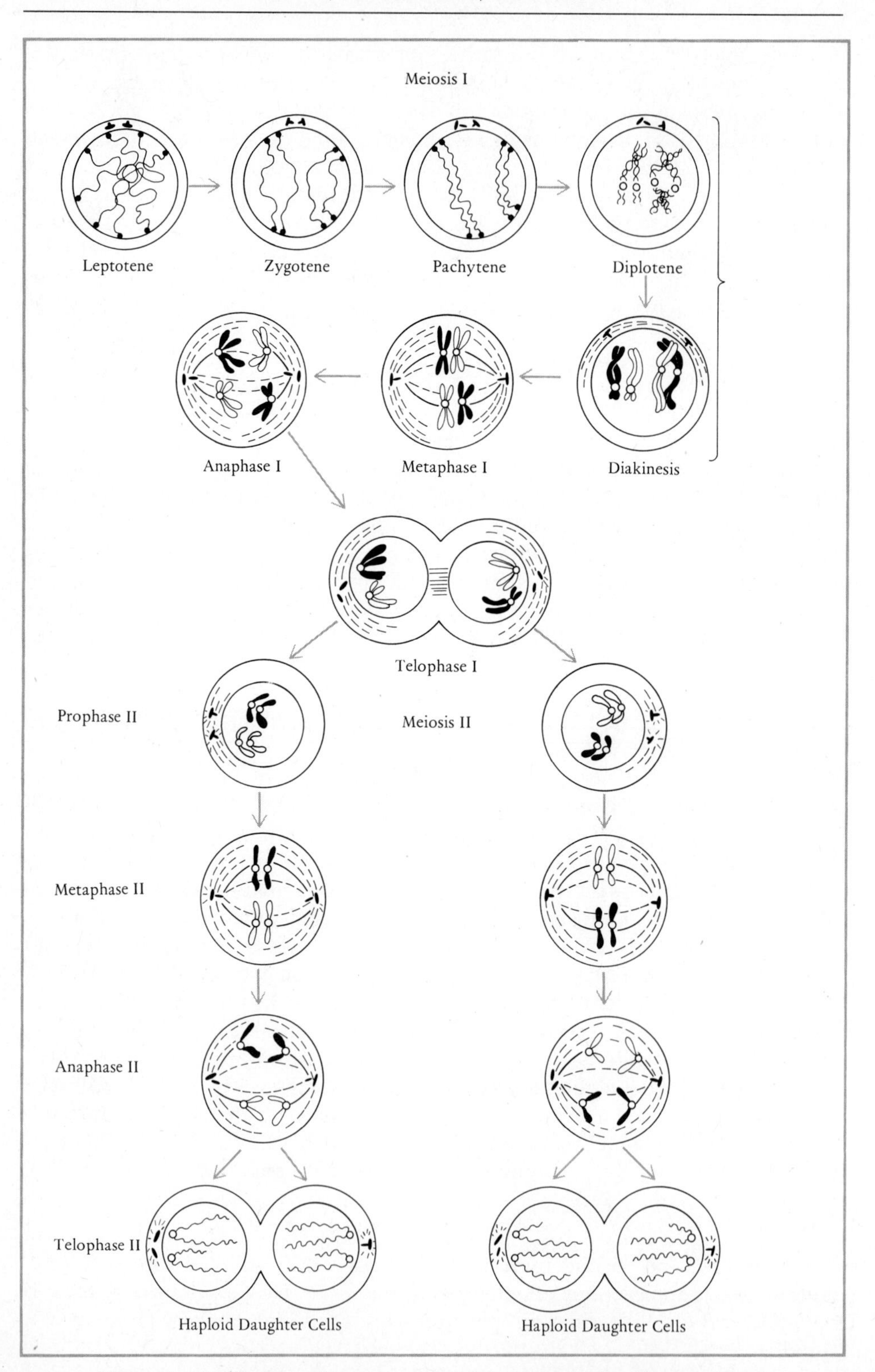

Figure 6–16. Diagram of meiosis I and II in an animal cell where $n = 2$.

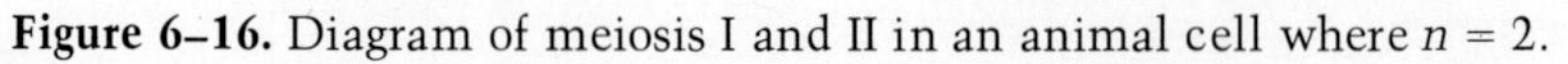

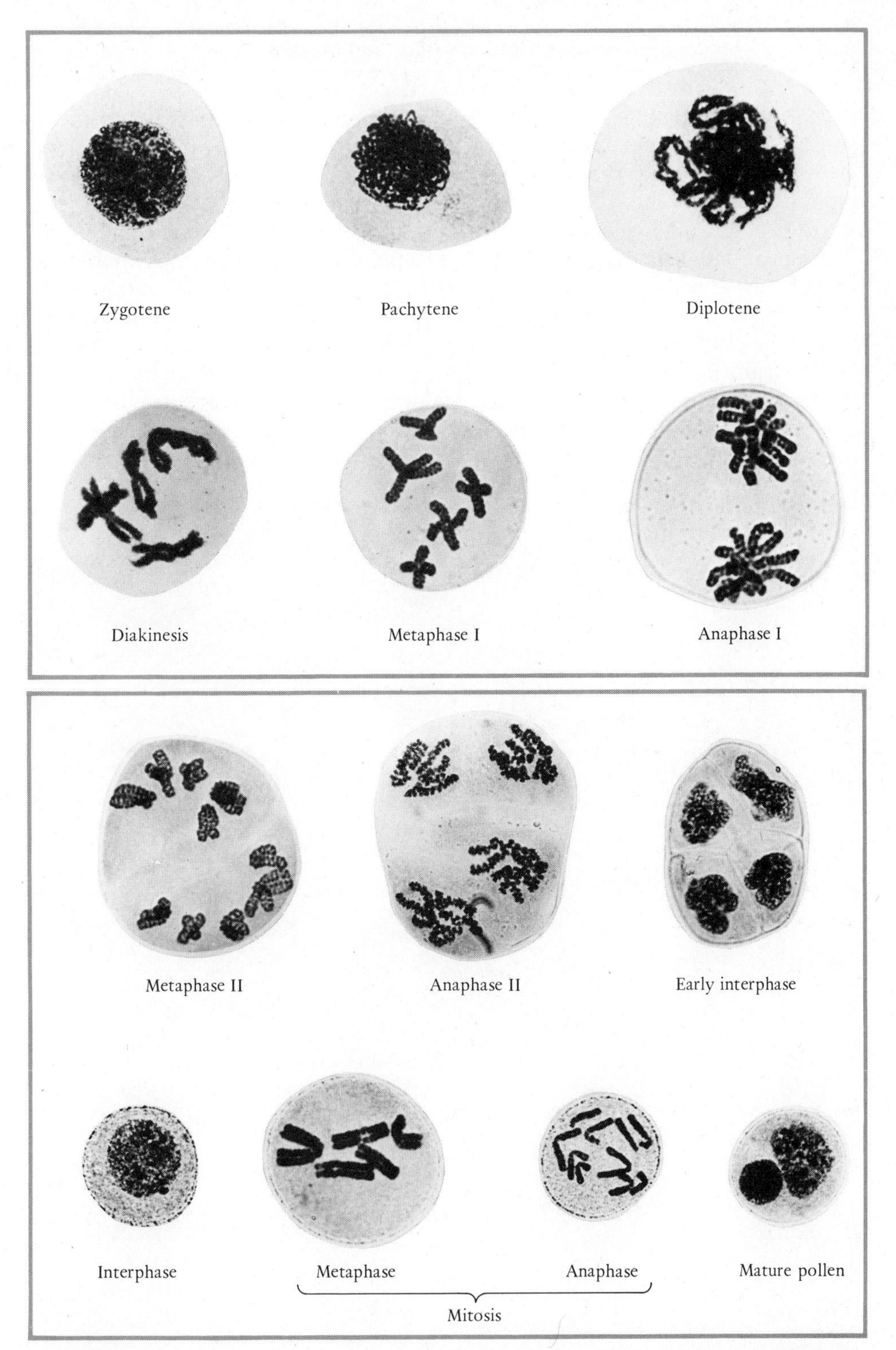

Figure 6–17. Stages in meiosis in *Trillium* ($n = 5$). Each cell resulting from meiosis undergoes a mitotic division to produce a mature pollen grain containing a tube nucleus and a generative nucleus. [From Walter V. Brown, *Textbook of Cytogenetics,* St. Louis, 1972. The C. V. Mosby Co. Courtesy of Dr. A. H. Sparrow and Mr. Robert F. Smith, Biology Department, Brookhaven National Laboratory.]

Leptotene. The first stage of the meiosis I prophase is called the leptotene stage. During leptotene the chromosomes gradually appear within the nucleus as delicate, extended, threadlike structures whose number corresponds to the diploid number of the species. The presence of sister chromatids cannot be seen with the light microscope. However, with close observation one can usually distinguish small bodies, granules, or beads located along each chromosome strand. These bodies are called *chromomeres* (Figure 6–18). In most species they are best seen in prophase of meiosis I and can be interpreted as regions where the chromosome is coiled into small knotlike structures. Since the size and distribution of these chromomeres is a constant feature of the morphology of any given pair of homologues, we can again infer that the lengthwise structural differentiation and coiling of chromosomes is not a random process, but instead reflects the imposition of controlling mechanisms.

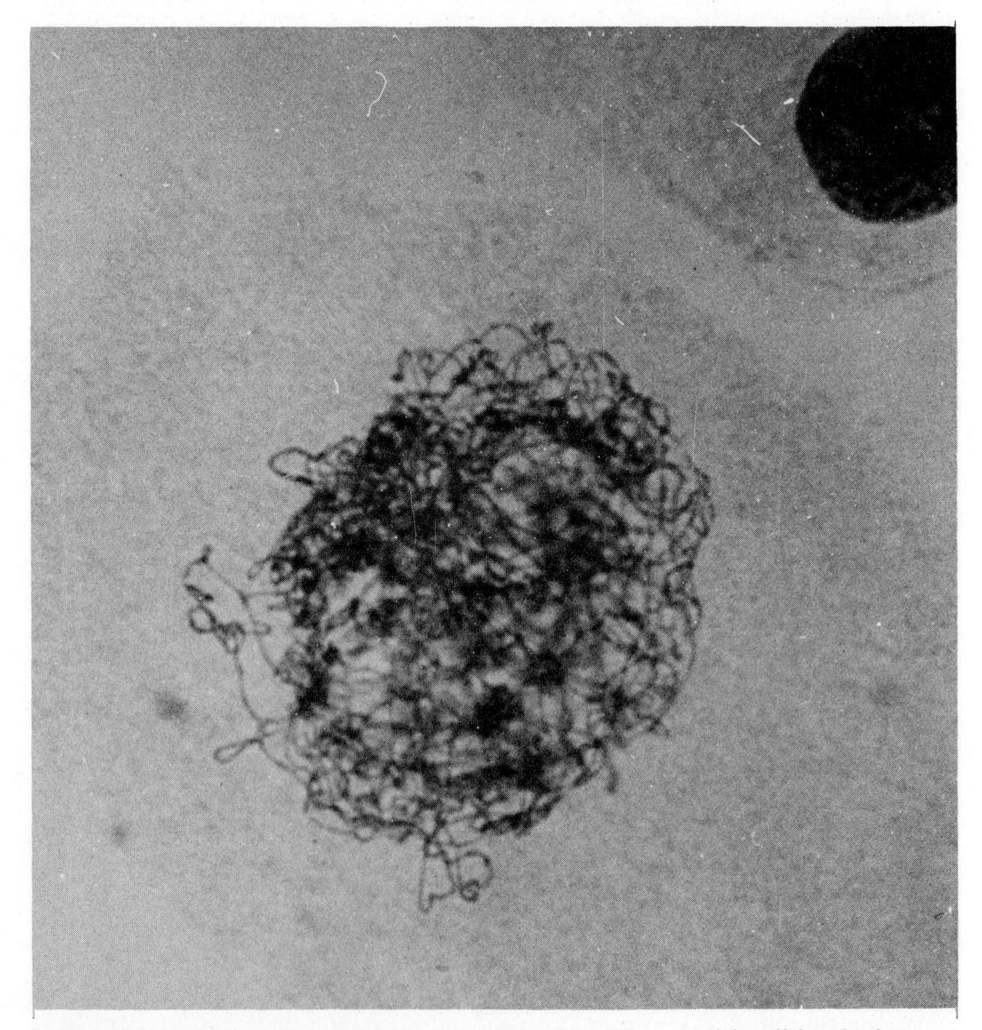

Figure 6–18. Leptotene chromosomes of the lily. The small beadlike thickenings along the chromosome strands are chromomeres. (1000×) [From P. B. Moens, *Chromosoma 23:* 418–451, 1968.]

Zygotene. During the zygotene stage homologous chromosomes begin to pair lengthwise with one another, a process called *synapsis* (Figure 6–19). Synapsis is accompanied by the formation of protein structures called *synaptinemal complexes,* which join together the members of each chromosome pair. These complexes can be seen only with the electron microscope, and an account of their origin and proposed significance will be deferred until our review of the entire meiotic process has been completed.

Pachytene. At the pachytene stage synapsis between homologues is complete and, with few exceptions (notably the X and Y sex chromosomes), each chromosome is in close lengthwise association with its pairing partner, genetic homology being responsible for this intimate pairing. Synapsis between chromosomes sharing only restricted regions of homology is less than complete. This is the case for pairing between the X and Y chromosomes of many species. In humans synapsis between the sex chromosomes involves only the short arms of the X and Y chromosomes (Figures 6–20, 6–25).

Each pair of synapsed homologues is called a *bivalent* by cytologists. The term *tetrad* is used to describe the status and actual composition of the chromosomes involved in such associations. Since each homologue has previously undergone replication, it is present in the form of two chromatids. Thus a pair of synapsed homologous chromosomes is in reality a four-partite structure to which the name tetrad is applicable. At pachytene chromatids are still in a relatively extended state and the four-part nature of the tetrad can be seen with

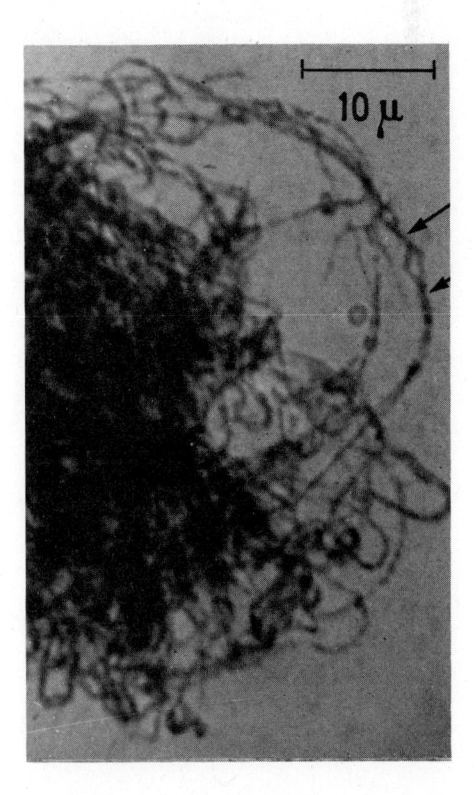

Figure 6–19. Zygotene stage from a lily. Synapsis is in progress with pairing initiated in several segments of one set of homologues (arrows). (1650×) [From P. B. Moens, *Chromosoma 23:* 418–451, 1968.]

the light microscope only in very favorable preparations. With further chromosome condensation in ensuing stages, all four chromatids of the tetrad become readily visible.

Evidence from many sources indicates that during synapsis at pachytene exchanges between the chromatids of homologous chromosomes may occur by a process known as *crossing over*. (This process is discussed in Chapter 8; the occurrence of such a crossover event is illustrated in Figure 6–22.)

Diplotene. At diplotene the bivalents become noticeably shorter and thicker. In addition, the pairing process is terminated. As a result, the homologues tend to fall apart or repulse one another forming characteristic loops.

Complete separation of homologues is prevented by regions called *chiasmata* (*sing*. chiasma) where the arms of the chromosomes are held together (Figure 6–20). Chiasmata are thought to occur as a consequence of preceding crossover events. The number of chiasmata present in any one pair of

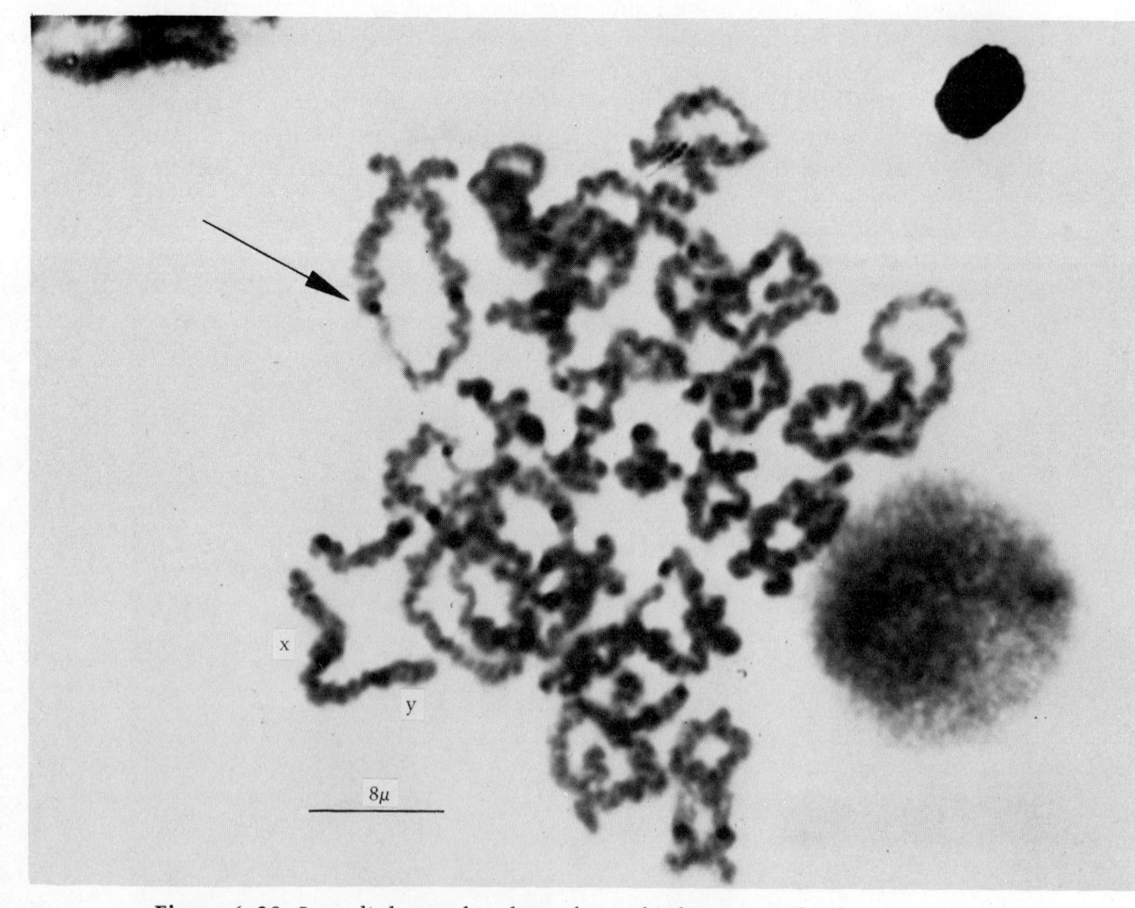

Figure 6–20. Late diplotene bivalents from the human male showing centromere position (arrow). The short arms of the X and Y sex chromosomes are in association with one another. [From Andrew T. L. Chen and A. Falek, *Science, 166,* 1008–1010, 21 November 1969. Copyright © 1969 by The American Association for the Advancement of Science.]

homologues varies with the length of the chromosome, but often two or more can be seen in longer chromosomes. In human males the total number of chiasmata present in a diplotene nucleus is around 55.

Careful examination of the loops of diplotene chromosomes with the light microscope usually reveals the presence of the four chromatids which comprise each tetrad.

Diakinesis. The diakinesis stage is characterized by the continuing condensation and contraction of the chromosomes (Figure 6–21). During condensation the chiasmata tend to slip down the chromosome arms toward their tips, a process called *terminalization* by cytologists. By the end of diakinesis the homologues are strongly contracted into short, densely staining bodies. Spindle fiber formation is initiated at this point, the nuclear membrane disappears, and the tetrads enter metaphase of the first meiotic division.

Metaphase I. During metaphase each tetrad becomes oriented so that the maternally derived centromere faces in one direction and the paternally de-

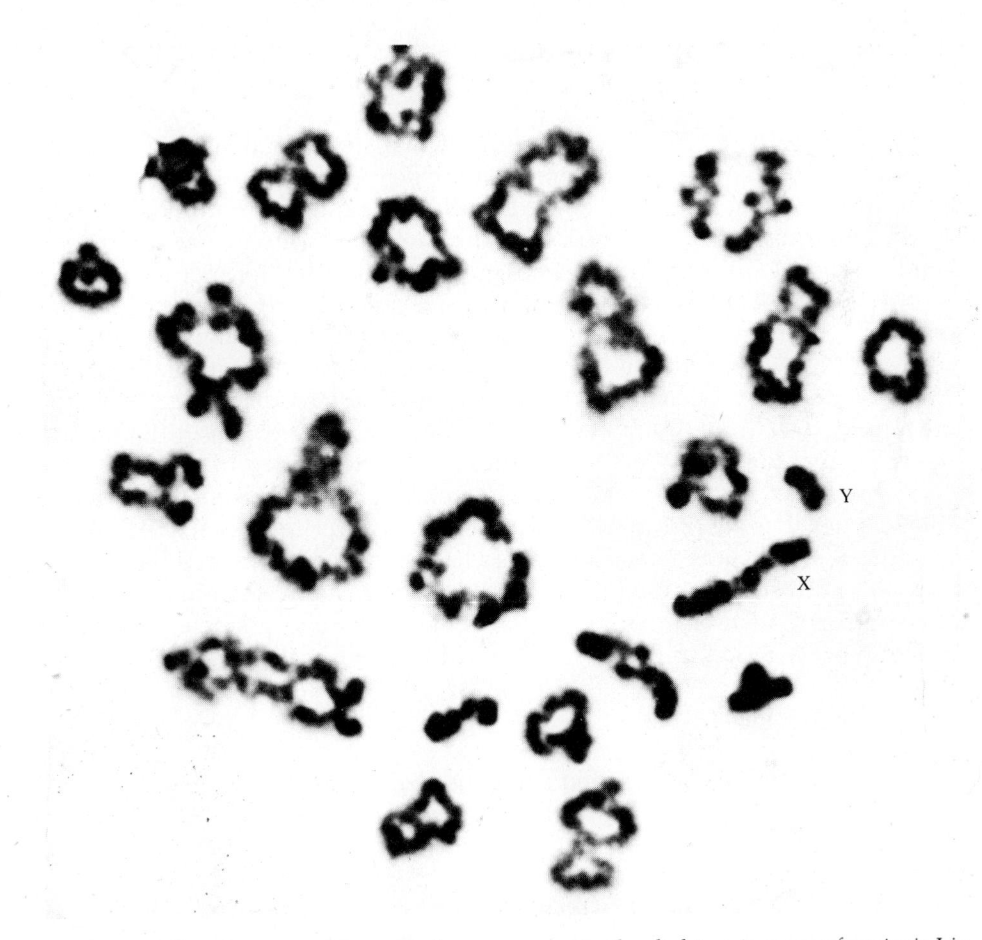

Figure 6–21. Contracted human chromosomes in the diakinesis stage of meiosis I in a spermatocyte from a normal man. The X and Y chromosomes are separated. [Dr. Maj Hultén, Director, Regional Cytogenetics Laboratory, East Birmingham Hospital. *Hereditas, 76*, 55–78, 1974.]

rived centromere faces in the opposite direction (see Figure 6–16). Because of this orientation, both sister chromatids belonging to the same homologue are directed toward the same pole of the spindle (assuming no crossing over).

Anaphase I. During anaphase the homologous chromosomes, each composed of sister chromatids, separate from one another and move to opposite poles of the spindle. Studies with the electron microscope have demonstrated the presence of a centromere within each of the two chromatids of any one homologue. Yet, during the first meiotic division and no other, these paired centromeres behave as if they were single, rather than double entities, with the result that both sister chromatids proceed together to the same pole. For this reason, maternal and paternal centromeres are shown as single entities in Figure 6–16, and others. This behavior on the part of centromeres is very likely crucial to the successful separation of homologous chromosomes and the achievement of haploidy.

Telophase I. During telophase of the first meiotic division, a new nuclear envelope forms around the chromosomes clumped at each pole, and cytokinesis results in the formation of two daughter cells. Since these cells contain only one member of each original pair of homologous chromosomes, they are haploid in constitution.

Interphase. The interphase between the first and second meiotic divisions is often so short that the chromosomes fail to achieve any significant degree of uncoiling. However, regardless of the length of this interval, chromosome replication does not occur. Therefore, the chromosomes which enter the second meiotic division are those which were present at the close of meiosis I.

Meiosis II. The second meiotic division is similar to mitosis. Chromosome condensation occurs during a short prophase, at the end of which the nuclear envelope disappears and the spindle reforms. During metaphase each pair of sister chromatids becomes oriented on the spindle so that their respective centromeres face in opposite directions. Anaphase thus results in the separation of sister chromatids. These move apart from one another to opposite poles of the spindle. During telophase each set of chromosomes becomes enclosed in a nuclear envelope, and cytokinesis results in daughter cells. Since the second meiotic division occurs in each daughter cell produced by meiosis I, a total of four haploid cells is ultimately produced from any one diploid cell which enters meiosis. Each of these four final products contains a single representative from every tetrad originally present at the onset of the meiotic process.

Meiosis and Mendel's Laws

Mendel's Law of Segregation concerns the separation of allelic genes at sex cell formation, but it does not specify the particular time or stage at which segregation occurs, for Mendel's data were derived from the results of crosses and not from direct observation of events occurring at the cytological level. From our review of meiosis it can be seen that separation of homologous chromosomes from one another occurs in the first meiotic division, reducing the chromosome complement from diploid to haploid. However, segregation of maternally and paternally derived alleles can occur at either the first or the second meiotic division depending on whether or not an exchange (crossover) between chromatids of opposing homologues has taken place. These two possibilities are best explained by diagram (Figure 6–22).

Suppose we assume a genotype heterozygous for the alleles *A* and *a*, these genes being located on a pair of homologous chromosomes. At prophase of meiosis I each homologue is represented by two exact replicas or sister chromatids, one pair of sisters with the gene *A* and the other pair with the gene *a*. As previously pointed out, the centromeres of sister chromatids function as one in the first meiotic division and therefore they are shown as single entities in Figure 6–22. Synapsis between these homologues will result in the formation of a tetrad consisting of the four chromatids. First division segregation of the alleles *A* and *a* is shown in Figure 6–22(a). Here, no exchanges between chromatids of opposing homologues have occurred, and therefore the separation of centromeres at anaphase of meiosis I also brings about the separation of alleles *A* and *a* from one another. As a result, these alleles are segregated into different daughter cells at the first meiotic division.

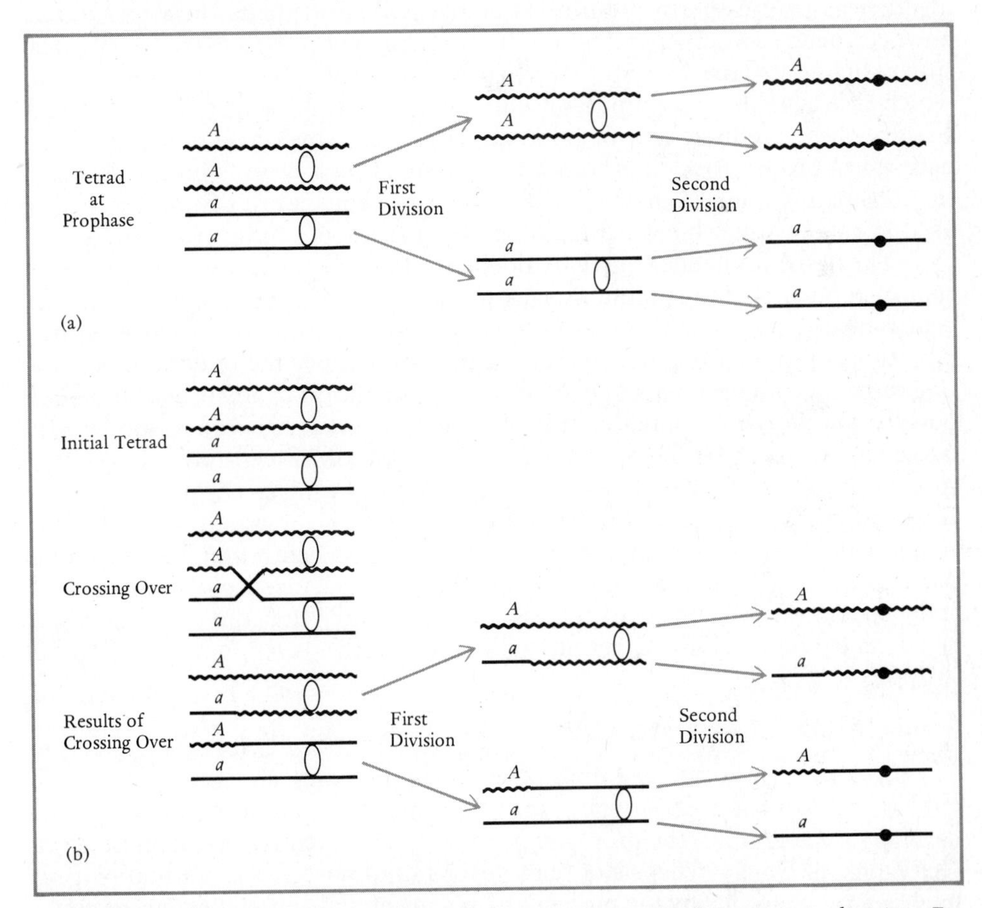

Figure 6–22. (a) Segregation of the alleles *A* and *a* at the first meiotic division. For clarity, a single centromere is shown joining the sister chromatids of each homologue comprising the tetrad. (b) Segregation of the alleles *A* and *a* at the second meiotic division.

Second division segregation is illustrated in Figure 6–22(b). In this case a crossover occurs at a point between the position of the centromere and the position, or *locus,* of the genes *A* and *a*. Note that the two chromatids involved in this exchange are derived from opposing homologues, that is, they are nonsisters. During this crossover event the chromosome segments bearing genes *A* and *a* are exchanged between the two participating chromatids. As a result, the single centromere of each pair of chromatids now carries a representative of each allele, rather than one or the other allele exclusively. Thus, when segregation of centromeres occurs at anaphase of meiosis I, both alleles will be represented in each daughter cell. Only when chromatid separation occurs in the second meiotic division will the opposing alleles *A* and *a* be finally segregated from one another. In summary, second division segregation of a pair of alleles results from the occurrence of a crossover between the gene pair in question and their respective centromeres.

A distinction between first and second division segregation is ordinarily not possible in organisms where the products of meiosis are randomly mixed together as in the sperm of animals or the pollen of plants. In a few fungi, however, such as *Neurospora,* all products of the meiotic process occurring in a single diploid cell are temporarily sequestered together in a structure *(ascus)* which is accessible to manipulation. By dissection it is possible to obtain and analyze these products and to infer the actual crossover events which must have taken place during their formation. In such fungi second division segregation of a pair of alleles can be identified and the frequency of this occurrence is used in establishing the position of a gene with respect to its centromere.

The basis for Mendel's Law of Independent Assortment lies in the orientation of tetrads on the spindle of metaphase I. Whether maternal or paternal centromeres face "north" or "south" is a matter of random chance, either direction being equally probable (see Figure 6–16). Since the orientation of any one tetrad is completely independent of the orientation of any other tetrad, all possible combinations of maternally and paternally derived chromosomes will occur in the gametes. The probability that some or all centromeres derived from a given parent will face the same direction can be calculated. As an illustration, 23 pairs of chromosomes, and therefore 23 tetrads, occur in humans, and the chance that the maternally derived centromeres of any one particular tetrad will face "north" is $\frac{1}{2}$. With the occurrence of random and independent orientation of all tetrads, the probability that the maternally derived centromeres of all 23 tetrads will face "north" is $(\frac{1}{2})^{23}$ or about 1 in 10 million.

Events Associated with Meiosis

Because of its crucial importance to the heredity and life cycle of all eukaryotic organisms, it is worthwhile to examine the process of meiosis in more detail, especially the prophase of the first meiotic division. This stage is characterized by the synapsis of homologous chromosomes, a mutual pairing process which is aided by the presence of synaptinemal complexes. These complexes have been found in over 50 genera ranging from protozoa to higher plants and animals. In all of such varied forms they exhibit an amazingly constant morphology, which suggests that they perform a function of universal and fundamental importance to eukaryotes. The formation of the synaptinemal complexes and the accompanying synapsis of homologues have been described from studies carried out with the electron microscope and are illustrated here by diagram in Figure 6–23 and in the lily plant and in human spermatocytes in Figures 6–24 and 6–25.

The beginning of the complexes can be identified in the leptotene stage. Viewed with the electron microscope, the threadlike leptotene chromosomes appear as elongate fibrillar masses whose ends (telomeres) are attached to the nuclear envelope. The two sister chromatids which represent each homologue are clearly visible. In the split between each pair of sisters, there appears a single structure called a *lateral element* or *chromosome core* which is composed of basic protein. The lateral element, later to comprise part of the synaptinemal complex, extends the full length of each pair of sister chromatids. However, since each chromatid is already partially condensed, the lateral element is actually associated only with those portions of the chromatid (perhaps 2 percent) that are centrally located. The regions of the chromatid not associated with the lateral element lie to the side and appear as masses of fibrils. The reason for the sequestration of most of the chromatid to the side is unknown, but the process is more likely deliberate than random and undoubtedly reflects an underlying pattern of chromosome organization.

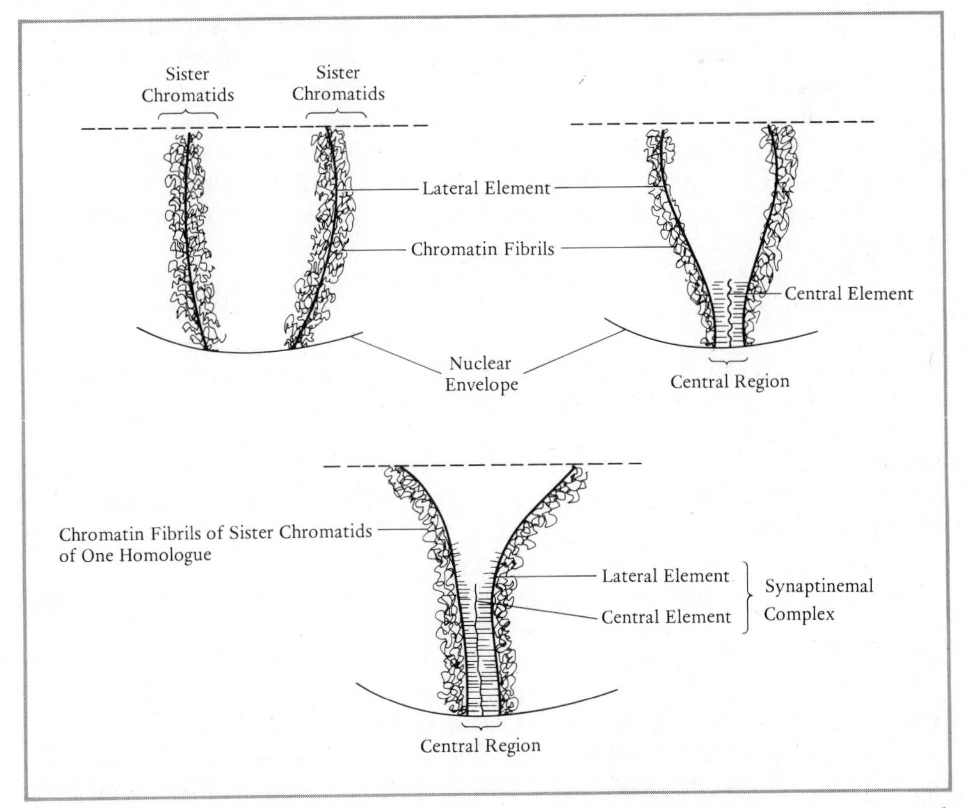

Figure 6–23. Diagrammatic representation of the formation of a synaptinemal complex.

As the cell enters the zygotene stage, synapsis is initiated in the following manner. Each homologue consists of two chromatids conjoined by a single lateral element which, along with the telomeres, is attached at each end to the nuclear envelope. Movement of telomeres on the nuclear envelope then occurs so as to bring the respective lateral elements of the two homologues close to one another in parallel alignment. Transverse filamentous extensions from the two lateral elements then meet and interdigitate to draw the homologues together in a zipperlike fashion. The area thus formed between the lateral elements is called the *central region*. It is composed of the protein filaments in conjunction with an additional protein whose nature has not been well defined.

The two lateral elements together with the central region comprise a synaptinemal complex (Figure 6–24). Running lengthwise through the middle of the central region is a delicate line called the *central element* which marks the midline of each complex. The dimensions of the synaptinemal complex are remarkably uniform. Regardless of species or of the size or morphology of any chromosome pair, each lateral element has been found to measure approximately 500 Å in diameter, while the central region is a uniform 1000 Å in width. Accordingly, as zygotene continues, homologous chromosomes become precisely paired lengthwise at a constant distance of 2000 Å.

In addition to the initiation of pairing, the zygotene stage is also characterized by the synthesis of a small amount of DNA. Ordinarily, chromosome replication is carried out during interphase prior to cell division. However, in the interphase which precedes the first meiotic division chromosome replication is not entirely completed. Quantitative studies on DNA synthesis during the meiotic process have demonstrated that some 0.3 percent of the total required for complete chromosome replication is deferred to the zygotene stage of meiosis I. It has been proposed that this late DNA synthesis is associated with the formation of the synaptinemal complexes and the initiation of pairing.

At the pachytene stage synapsis between homologues is complete and a synaptinemal complex can be seen to extend the full length of each chromosome pair (see Figures 6–24, 6–25). Most evidence suggests that it is during pachytene that exchanges occur between chromatids of opposing homologues. The synaptinemal complex may thus provide a framework essential to this process, for these complexes are present only in the meiotic cells of species and sexes in which crossing over regularly occurs. For example, in *Drosophila* females crossing over is a regular event and synaptinemal complexes are routinely present in egg cells undergoing meiosis. In normal *Drosophila* males, however, crossing over is absent and synaptinemal complexes do not occur. Since crossing over involves the breakage and reunion of chromatids (see Chapters 8 and 14), it is significant that some DNA synthesis occurs during the pachytene stage. This synthesis produces no net increase in the amount of DNA present, and it is therefore thought to be associated with the repair of chromatids broken in the process of genetic exchange.

At pachytene synapsed homologous chromosomes are usually found dispersed through the nucleus with their ends still attached to the nuclear envelope. Since the presence of the synaptinemal complexes parallels that of the

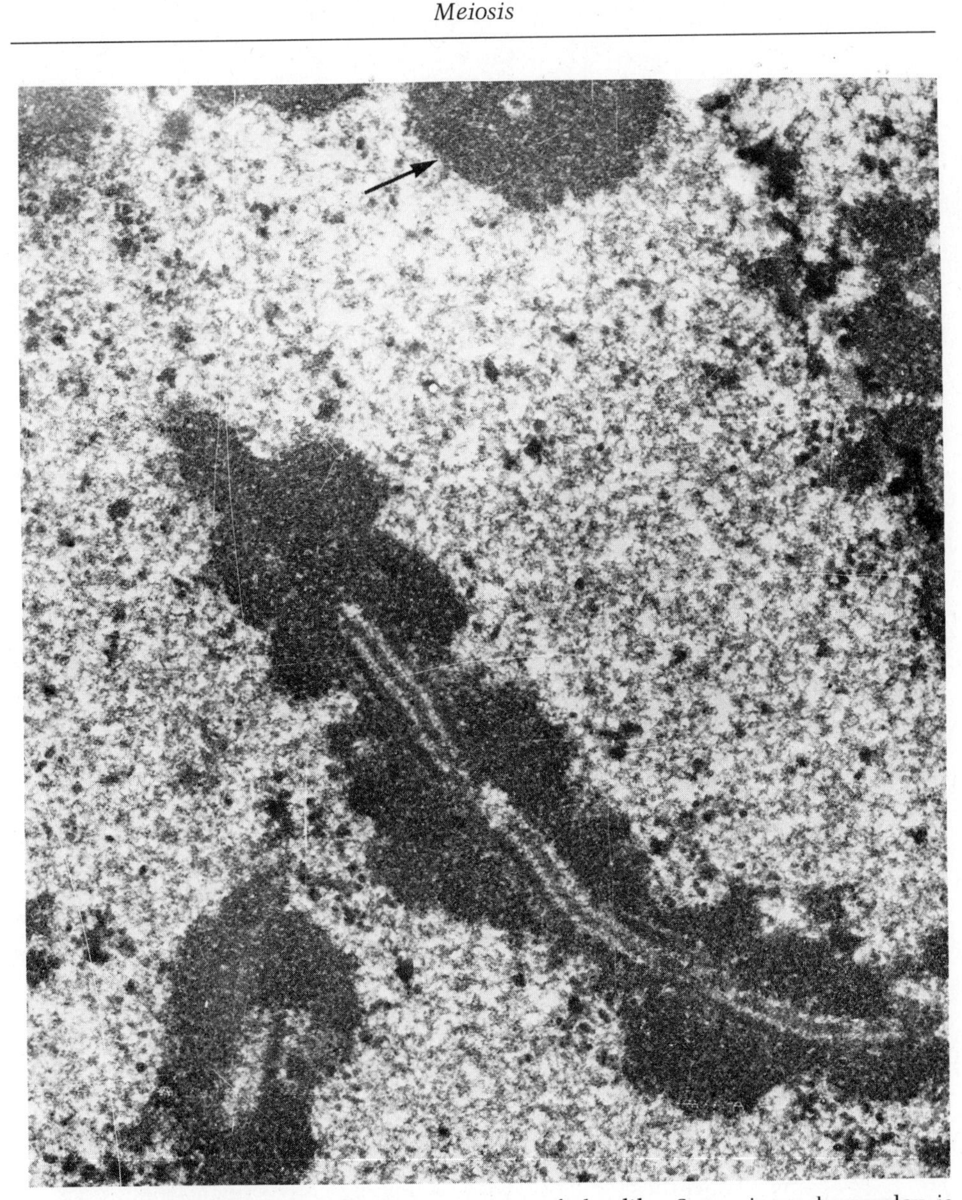

Figure 6–24. Pachytene stage in meiosis I of the lily. Synaptinemal complex is continuous and surrounded by a dense fibrillar mass of chromatin. At top (arrow) is a cross section of the complex and surrounding chromatin. (23,000×) [From P. B. Moens, *Chromosoma 23*:418–451, 1968.]

chromosome pairs, analysis of meiotic events has been aided by selective staining methods which distinguish these protein structures from the chromatin. Such methods have already made possible the visualization of all 23 pairs of human chromosomes at pachytene (Figure 6–25).

At diplotene synapsis is terminated and the central region and lateral elements of the synaptinemal complex are shed or stripped from the

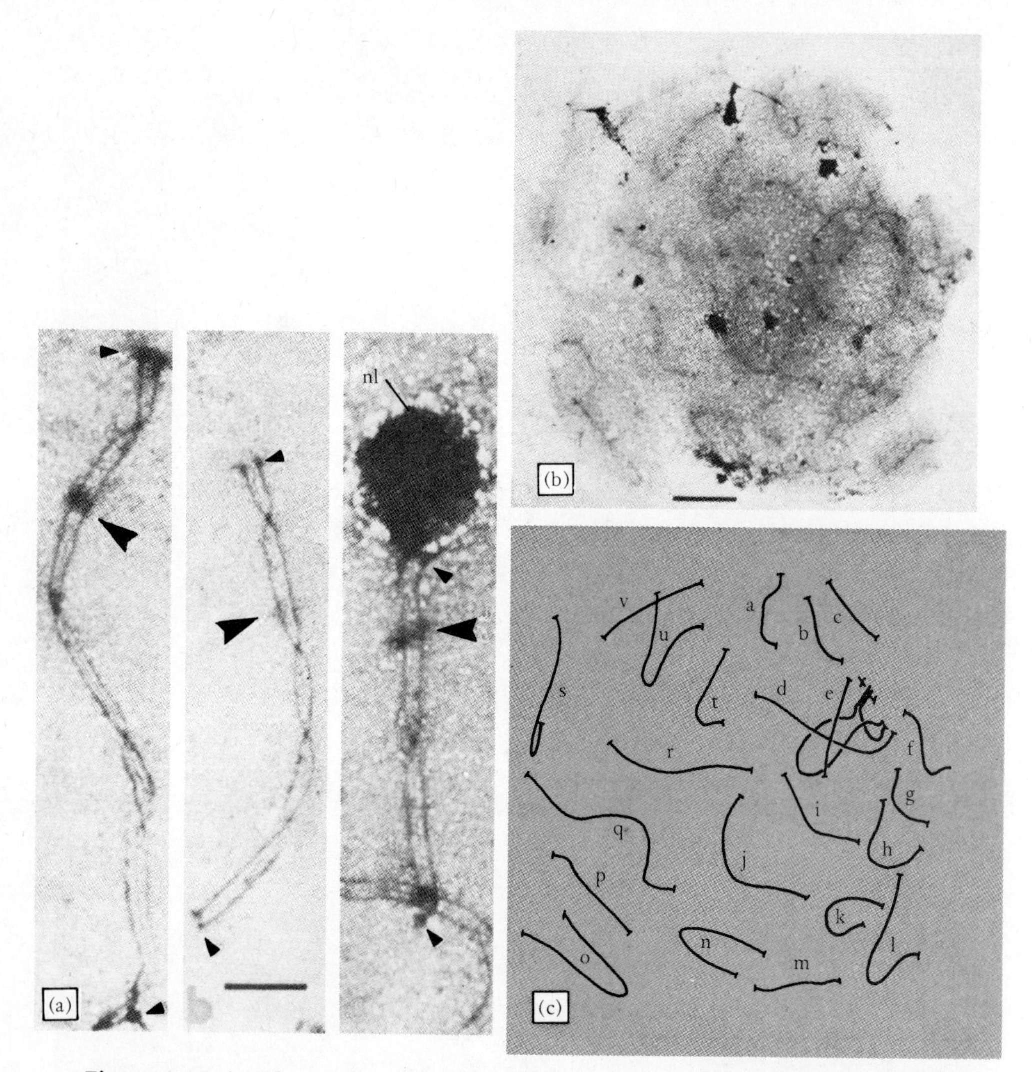

Figure 6–25. (a) Three synaptinemal complexes from spreads of human pachytene chromosomes. Protein of the complex has been selectively stained; chromatin is invisible. The twisting, filamentous synaptinemal complex, seen for its entire length, consists of two parallel lateral elements, each flanking and equidistant from a finer central element to which they are connected by transverse filaments. The lateral elements terminate in dense points of attachment to the nuclear envelope (small arrows). The kinetochore regions (large arrows), one per chromosome, are seen as pronounced thickenings of the lateral elements. A nucleolus (nl) is attached to the synaptinemal complex close to the termination of its short arm. Another complex crosses the long arm attachment region. (Scale 1 μm 9600×) (b) A spread human pachytene nucleus stained to demonstrate the synaptinemal complexes. (c) Tracing of synaptinemal complexes from the nucleus of (b). Positive identification of termini and of the paths of the complexes was made at a higher magnification than that shown. There is a full complement of 22 autosomal bivalents and their XY pair of sex chromosomes. Autosomal bivalents are lettered a through v, and the long X and short Y chromosomes are indicated. (Scale 5 μm, 2350×) [From M. J. Moses, S. J. Counce, and D. F. Paulson, *Science, 187,* 363–365, 31 January 1975. Copyright © 1975 by the American Association for the Advancement of Science.]

chromatids. Short stretches of the complex which remain briefly at regions where chiasmata are present are thought to be associated with the process of chromatid repair which accompanies crossing over. During late diplotene or in the ensuing stage of diakinesis, the homologues lose their attachment to the nuclear envelope. Condensation of the chromosomes in preparation for segregation then commences.

From this brief review it should be evident that the entire meiotic process is a coordinated cell activity of the greatest complexity. Furthermore, meiosis is itself under genetic control, for numerous mutations which alter the sequence of meiotic events in one or more ways have been described. As an example, in *Drosophila* females the homozygous presence of a recessive mutant known as *c(3)G* results in the complete absence of synaptinemal complexes and a drastic reduction in crossing over for all chromosome pairs. In addition, at the first meiotic division homologous chromosomes may fail to segregate from one another, a phenomenon known as *nondisjunction,* and as a consequence the egg cell receives an abnormal chromosome complement. Around 70 percent of the eggs produced by homozygous *c(3)G* females fail to hatch for this reason. Other meiotic mutants of *Drosophila* are known to cause premature cessation of synaptinemal complex formation or precocious division of sister centromeres leading to nondisjunction and chromosome loss in the gametes. Mutants with similar effects are also known in corn, and the genetic control of meiosis and crossing over has been well documented in the fungi. Specific meiotic mutants have not yet been identified in humans, but there is little reason to suppose that they do not occur.

Meiotic Drive

Up to this point it has been assumed that as a result of the meiotic process individuals heterozygous for a pair of alleles will produce two types of gametes with equal frequency and that fertilization between such gametes will be random. Some exceptions to these principles have been observed, and the term *meiotic drive* has been applied to instances where one or another class of progeny consistently appears with a frequency greater than that expected on the basis of random chance. A variety of mechanisms have been implicated in this phenomenon.

In *Drosophila* females some kinds of structurally abnormal chromosomes have been found to segregate from one another in a nonrandom fashion such that one homologue is preferentially included in the functional egg with high frequency. In *Drosophila* males several instances of meiotic drive have been analyzed and found to involve meiotic irregularities, nondisjunction, and the production of abnormal, nonfunctional sperm.

Abnormal segregation ratios are also known in the mouse, where males heterozygous for the tailless alleles transmit the mutant gene to progeny with a frequency greater than 50 percent. In plants similar cases of meiotic drive have been discovered. For example, in corn the presence of a heterochromatic knob on chromosome 10 (or other chromosome) causes that chromosome to be preferentially included in the functional egg.

Preferential inclusion in the functional egg is also a characteristic of the accessory chromosomes found in many plants and in some animals. These chromosomes, called *B chromosomes,* are not homologous to the regularly occurring paired homologues, and their presence or absence is not necessary to life. B chromosomes are usually small, condensed, heterochromatic bodies with seemingly weak effects on the phenotype. When present in large numbers (up to 20 or 30), fertility and vigor may be reduced. Some evidence indicates that an increased frequency of crossing over may occur in their presence. Their origin is obscure and thought to be very ancient. These kinds of chromosomes are found in corn, rye, wheat, and various grasses among the plants and in many insects, especially grasshoppers. In vertebrates such accessory chromosomes are also evident in a few mammals, two examples being the harvest mouse and the fox.

In some plants B chromosomes are preferentially included in the functional egg due to nonrandom orientation on the meiotic spindle. In other species nondisjunction of B chromosomes occurs in the mitotic divisions of the pollen grain and is followed by selective fertilization of the egg by the particular sperm nucleus that contains the supernumerary B chromosomes. In grasshoppers accessory chromosomes undergo nondisjunction in embryonic cells during mitosis and, by this process, become preferentially sequestered in cells of the germ line, that is, cells destined to become gametes.

The occurrence of these and other forms of meiotic drive is apparently widespread in natural populations. Although the primary causes of nonrandom disjunction or of preferential fertilization are presently unknown, it is likely that future studies of these phenomena will greatly increase our understanding of the meiotic process.

POSITION OF MEIOSIS IN THE LIFE CYCLE OF EUKARYOTES

The position of meiosis in the life cycle of eukaryotic organisms varies depending on whether the organism is plant or animal. Both unicellular and multicellular forms of each have been employed extensively as objects of genetic studies, and the life cycles of some representative examples are outlined below.

Primitive Organisms

Chlamydomonas. *Chlamydomonas* is an aquatic, unicellular green alga. Each free-swimming single cell is typically haploid in constitution, contains a single large chloroplast, and moves by means of two flagella. Such cells can be termed *vegetative* in that they reproduce asexually by mitosis.

Under conditions of environmental stress, vegetative cells can assume the properties of gametes. Although identical in appearance, such gametes are physiologically distinct and can be classified as either plus or minus with respect to mating type. Fusion of gametes of opposite mating type (+ with −) occurs to produce a diploid zygote which becomes encased in a cyst wall. The diploid stage is restricted to the zygote which subsequently undergoes meiosis to produce four haploid cells. In some strains these cells divide by mitosis to produce a total of eight haploid cells.

When environmental conditions are favorable, the haploid cells escape from the cyst and assume the vegetative state. Since mating type is inherited through a single pair of alleles, meiosis produces equal numbers of plus and minus vegetative cells (Figure 6–26).

Brewer's Yeast. Brewer's yeast, *Saccharomyces cerevisiae,* is a unicellular fungus found in the vegetative stage in both the haploid and diploid states, since haploid vegetative cells of opposite mating type (+ and −) can fuse to form diploids. Asexual reproduction results from mitosis. During mitosis a daughter nucleus becomes located in a cytoplasmic extension (bud) of the mother cell and such buds eventually pinch off to become separate yeast cells. With the proper environmental conditions diploid cells acquire the ability to undergo meiosis to produce four haploid spores called ascospores, two of which are + and two − in mating type. Each ascospore then assumes the vegetative state (Figure 6–27).

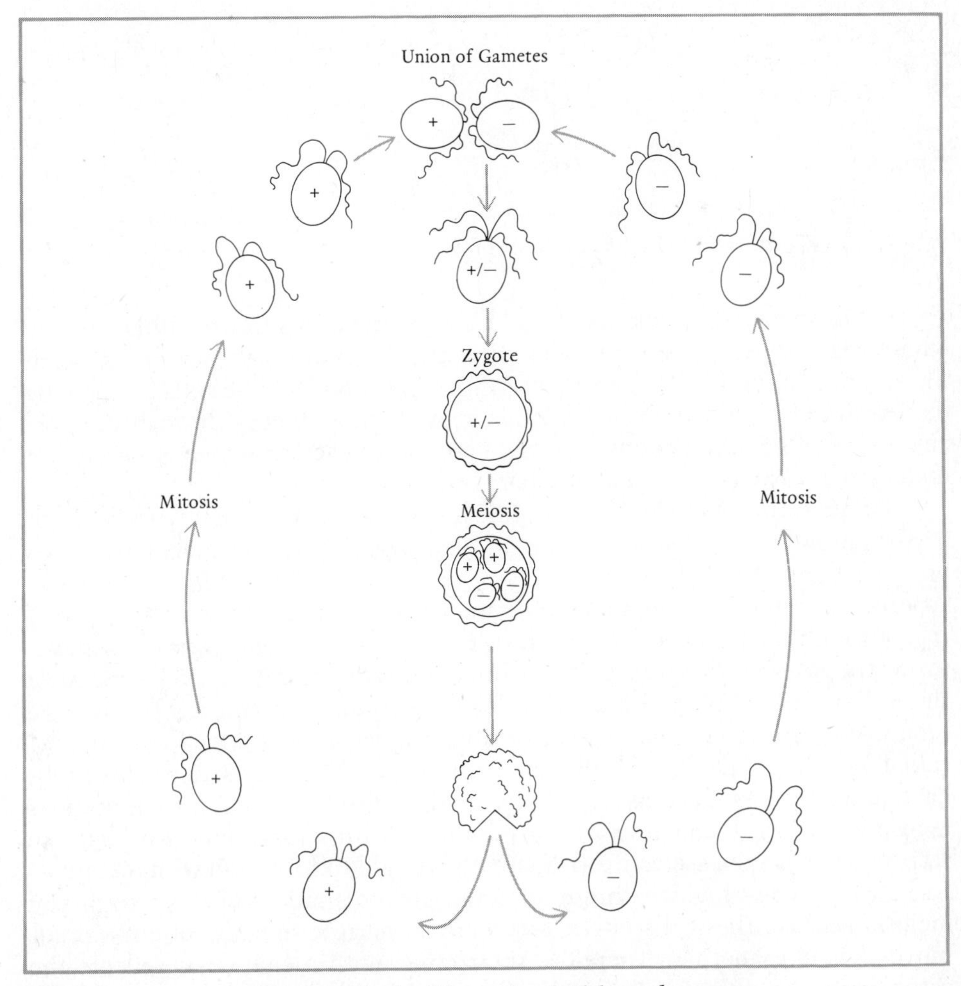

Figure 6–26. The life cycle of *Chlamydomonas.*

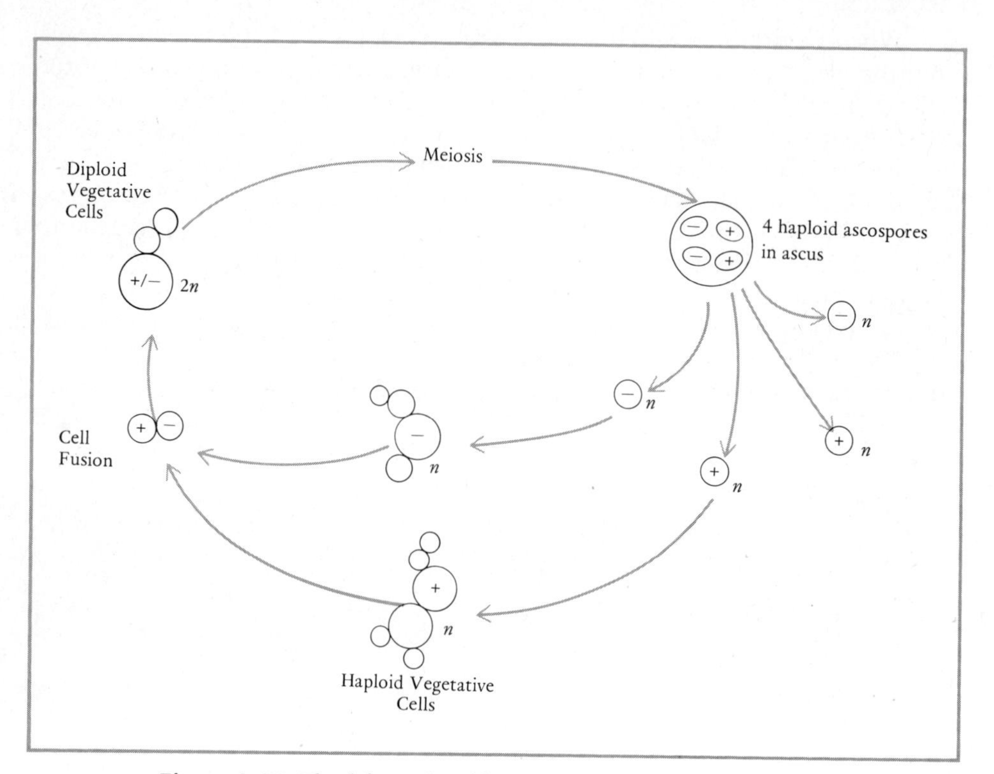

Figure 6–27. The life cycle of brewer's yeast, *Saccharomyces.*

Neurospora. The pink bread mold, *Neurospora,* is a multicellular fungus whose vegetative plant body is haploid and consists of numerous fine, thread-like hyphae collectively called a *mycelium* (Figure 6–28). The cells comprising the hyphae are multinucleate. Asexual reproduction occurs through the production of spores called *conidia* (*sing.* conidium), each of which is capable of undergoing mitosis to form a new haploid mycelium.

Under some conditions the conidia assume a role in sexual reproduction. Two mating types, *A* and *a,* occur in *Neurospora* and the mycelia of each produce immature fruiting bodies called *protoperithecia* (*sing.* protoperithecium) which contain haploid nuclei. When a conidium of one mating type encounters a protoperithecium of the opposite mating type, it is drawn inside the protoperithecium and there undergoes several mitotic divisions. After these divisions individual haploid nuclei of opposite mating types fuse and each resulting diploid nucleus (zygote nucleus) becomes confined within a sac called an ascus. The diploid nucleus then undergoes meiosis to produce four haploid ascospores. Each ascospore divides by mitosis, so that a total of eight ascospores (four *A* and four *a*) are present within each ascus. The haploid ascospores upon release can germinate and divide to form a new mycelium.

Neurospora provides unique material for the analysis of the products of meiosis because the walls of the ascus are so narrow that the nuclei cannot move past one another. Therefore their order within the ascus reflects the actual arrangements of the chromatids at meiosis I and II.

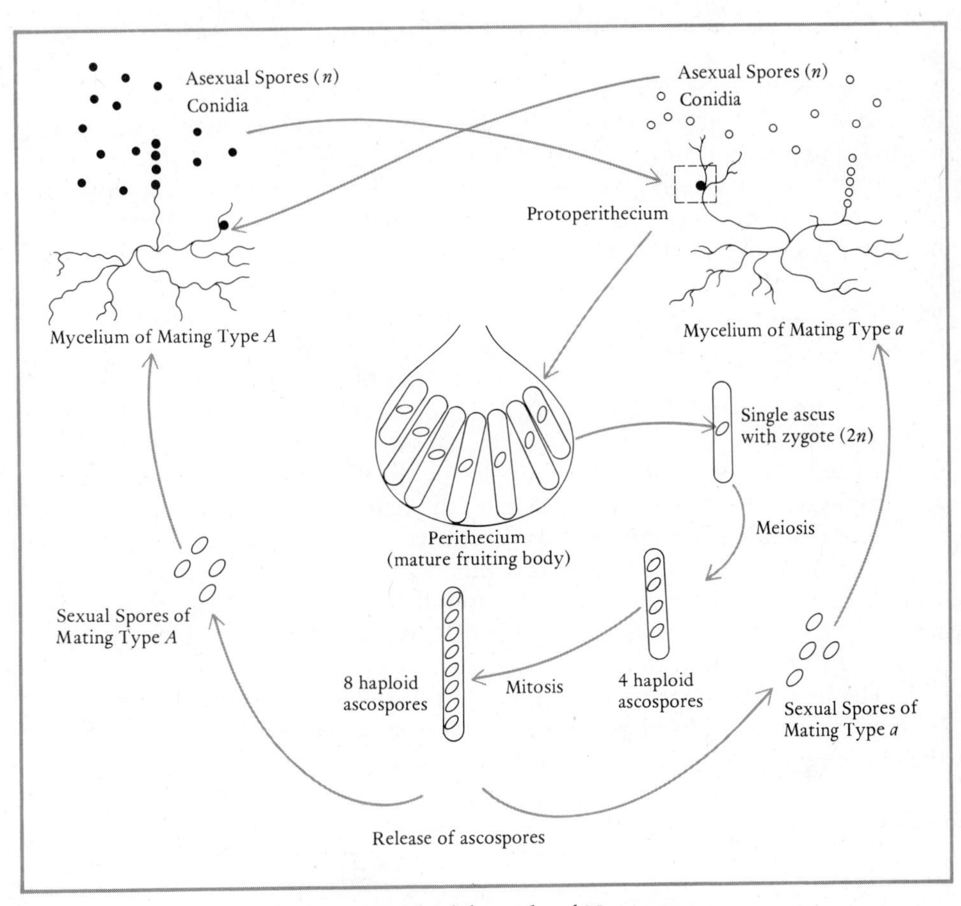

Figure 6–28. The life cycle of *Neurospora.*

Multicellular Green Plants

Primitive organisms, such as *Chlamydomonas* or *Neurospora,* are almost exclusively haploid in constitution, the diploid phase of the life cycle being restricted to the unicellular zygote. At a more advanced evolutionary level, however, both the diploid and haploid phases are represented by multicellular differentiated states.

The haploid sexually reproducing *gametophyte* stage produces haploid male gametes called *microspores* and haploid female gametes called *megaspores*. Since many plants are monoecious, both types may be produced by the same gametophyte, although they may be produced at different times so that cross fertilization is achieved. Union of micro- and megaspore produces a diploid zygote which develops into the diploid plant body *(sporophyte).* Upon maturity specialized cells of the sporophyte undergo meiosis to form haploid spores, each of which can germinate and grow into a new gametophyte.

This generalized life cycle is modified in the higher plants (angiosperms) such that the only obvious stage in the life cycle is that of the diploid sporophyte, the haploid gametophyte stage being greatly reduced and represented by only a few cells. Such a life cycle is characteristic of the corn plant, *Zea mays*.

Corn is monoecious, the male microspores being produced in anthers located in tassels at the tip of the plant and the female megaspores being produced in pistils at the base of the leaves (Figure 6–29). In the anther diploid microspore mother cells (microsporocytes) undergo meiosis, each producing four haploid microspores. Each microspore then divides by mitosis to form two haploid cells, a larger tube cell which surrounds and encloses a smaller generative cell. The generative cell then divides to form two sperm nuclei, both of which are contained within the cytoplasm of the tube cell. Together, these three constitute the male gametophyte or pollen grain.

Pollen is carried by the wind to the end of the long style (silk) which extends from each pistil. The cell wall of the pollen grain fractures and the tube cell containing the two sperm nuclei forms a pollen tube which grows rapidly down through the style to the base of the pistil. Within the pistil is the female gametophyte, or embryo sac, containing the haploid egg cell.

The female gametophyte is formed in the following manner. Each of the female structures (pistils) contains a megaspore mother cell (megasporocyte) which enters meiosis to produce four haploid nuclei. Three of these degenerate while the fourth nucleus divides by mitosis. With additional mitotic divisions, a total of eight haploid nuclei are formed.

The haploid nuclei are contained within the female gametophyte or embryo sac which consists of seven cells. At the end of the embryo sac adjacent to the point of pollen tube entrance is the cell which contains the egg nucleus and

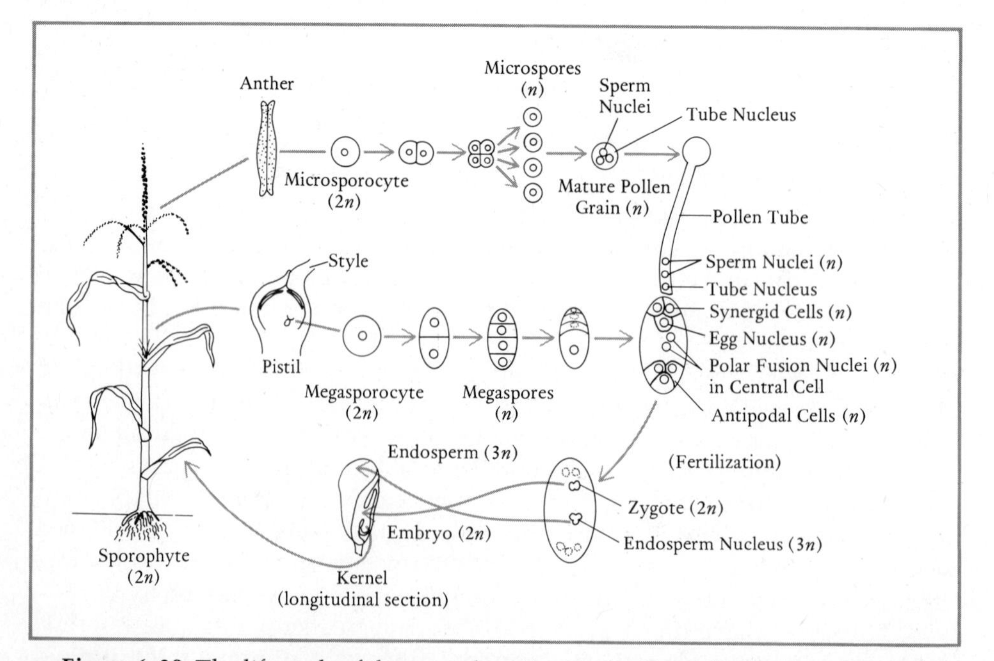

Figure 6–29. The life cycle of the corn plant *Zea mays*. [From A. H. Sturtevant and G. W. Beadle, *An Introduction to Genetics*, Dover, New York, 1962.]

two synergid cells. In the middle of the embryo sac is the central cell, which contains two separate haploid nuclei called polar fusion nuclei. At the outer end of the embryo sac are three antipodal cells each containing a single haploid nucleus.

Upon reaching the embryo sac, the pollen tube discharges the two sperm nuclei, one of which fuses with the egg to form the diploid zygote or new sporophyte. The other fuses with both polar nuclei of the central cell to form a triploid ($3n$) nucleus containing three sets of chromosomes. This $3n$ central cell subsequently develops into the endosperm, a tissue that stores and later provides nutrients to the developing sporophyte embryo. Since both sperm nuclei play a role in the initiation of the new sporophyte generation, the fertilization of the egg and the formation of the triploid endosperm are often referred to as double fertilization.

A mature kernel of corn is most useful for genetic analysis. The outer layers are provided by the parent diploid sporophyte and exhibit the phenotype of that parent. At the same time, the embryo is a representative of the new diploid generation, while the endosperm can be used to determine the phenotypic effects of three doses of a given allele.

Multicellular Animals

In multicellular animals the process of sex cell formation *(gametogenesis)* ordinarily occurs in specialized organs called gonads, the testis producing male gametes (spermatozoa) and the ovary producing eggs (ova).

In the male, sperm formation *(spermatogenesis)* is initiated at the onset of sexual maturity. Within the testis potential sexual cells called spermatogonia may be found. They are diploid and increase in number by mitosis. Eventually, these cells enter a growth phase and enlarge to become recognizable as primary spermatocytes in the prophase of meiosis I. The first meiotic division results in the formation of two haploid daughter cells called secondary spermatocytes, and each of these divides in meiosis II, so that a total of four haploid spermatids is produced (Figure 6–30). Each such spermatid subsequently undergoes a process of differentiation *(spermiogenesis)* through which it is transformed into a flagellated spermatozoon or mature male gamete (Figure 6–31). In the human male the total time required for spermatogenesis from the spermatogonial stage to mature gamete is estimated to be around 74 days. Of this period, approximately 16 days each are required for meiosis I, meiosis II, and spermiogenesis.

In females the undifferentiated sex cells are called oögonia (Figure 6–30). As in the male, they divide by mitosis to produce additional oögonia, some of which enlarge and differentiate into primary oöcytes. The prophase of the first meiotic division is characterized not only by the usual five stages (leptotene, zygotene, and so forth), but also by the cytoplasmic accumulation of yolk. In many animals the yolk must provide the entire nutritional support for the developing embryo from zygote to hatching, whereas in other forms, such as mammals, intrinsic support is required only during early stages prior to placenta formation. However, regardless of the amount of yolk present, the oöcyte is characteristically many to several thousand times larger than the male gamete. The first meiotic division in females produces cells of unequal

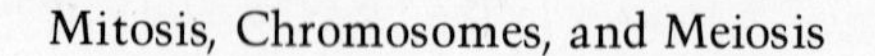

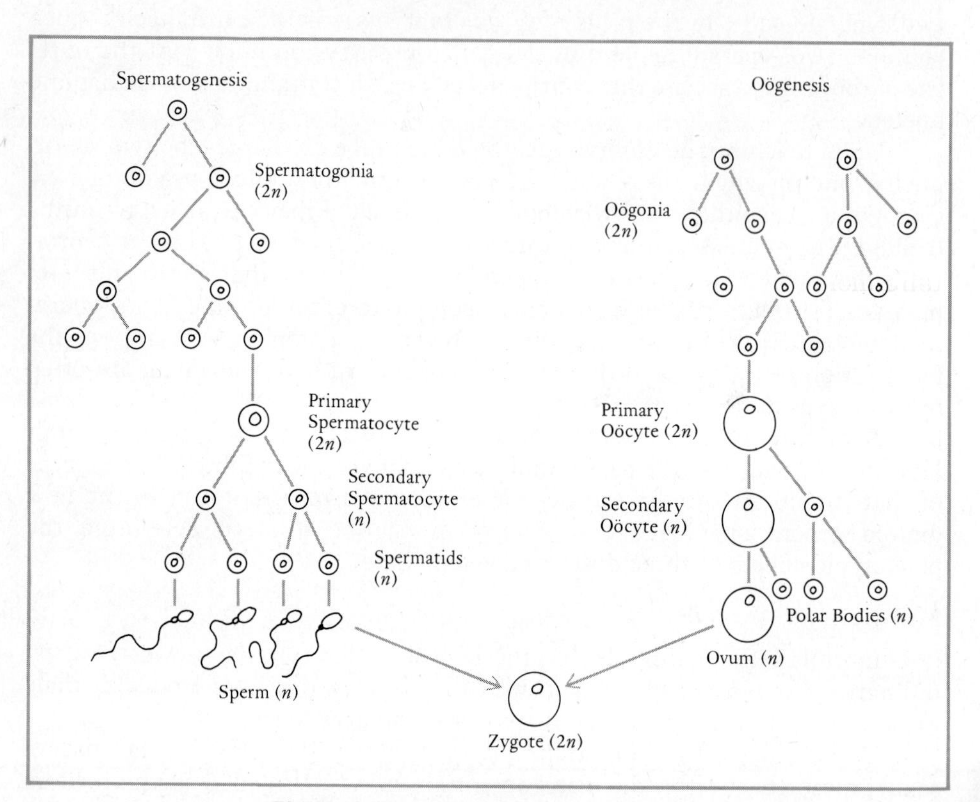

Figure 6–30. Gametogenesis in animals.

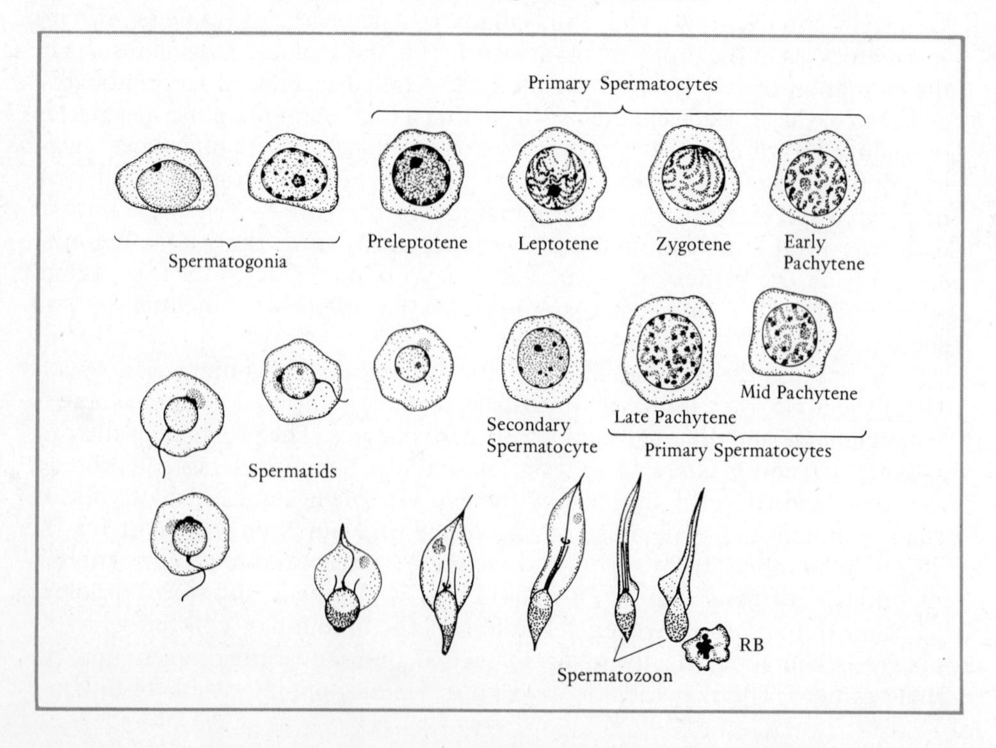

size, presumably for reasons of economy. One cell, the secondary oöcyte, is large and contains all of the stored yolk, while the other cell, the first polar body, is microscopic in size (Figure 6–32). During meiosis II the secondary oöcyte divides again to produce the mature female gamete *(ovum),* as well as a second polar body. In many species the first polar body may also undergo the second meiotic division, so that as a final result, one functional haploid egg cell, and three nonfunctional haploid polar bodies are produced.

In human females the primary oöcytes begin to differentiate during the third month of fetal life, and by birth all oöcytes ever destined for ovulation have reached the diplotene stage of the first meiotic division (Figure 6–33). As a consequence, all genetic exchange via crossing over has already occurred before birth. The primary oöcytes remain suspended at diplotene until the onset of sexual maturity, at which time ovarian follicles begin to develop.

Prior to the ovulation of any oöcyte the first meiotic division recommences and results in the formation of the secondary oöcyte and the first polar body. The oöcyte is then released from the ovary to the Fallopian tube. The second meiotic division then begins immediately and is completed upon

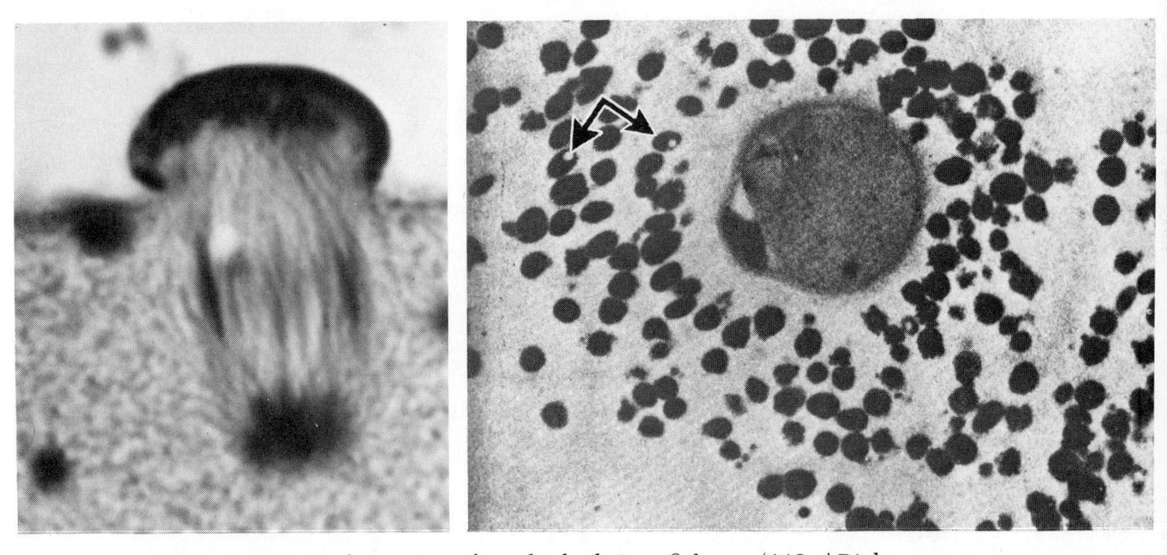

Figure 6–32. Left, formation of a polar body in a fish egg. (440×) Right, mouse ovum in the lumen of the Fallopian tube. The first polar body is seen as a small mass to the left and the second meiotic division is in progress. The metaphase chromosomes are aligned on the equatorial plate of the spindle. Cells which surrounded the egg at ovulation are now degenerating (arrows). (320×) [Photo of mouse ovum from L. Zamboni, *Biology of Reproduction, Suppl. 2,* 44–63, 1970.]

Figure 6–31. Series of drawings illustrating the steps of spermatogenesis in man. Spermatogenesis begins with the spermatogonia at the top left and terminates with a spermatozoon at the bottom right. Spermatids are shown at various steps of spermiogenesis. A spermatozoon is illustrated as seen from its lateral (left) and frontal (right) aspects. The residual body (RB) is cytoplasmic material discarded during maturation of the spermatid. [From Y. Clermont, Chapter II, *The Human Testis,* E. Rosemberg and C. A. Paulson, Eds., Plenum Publishing Corporation, New York, 1970.]

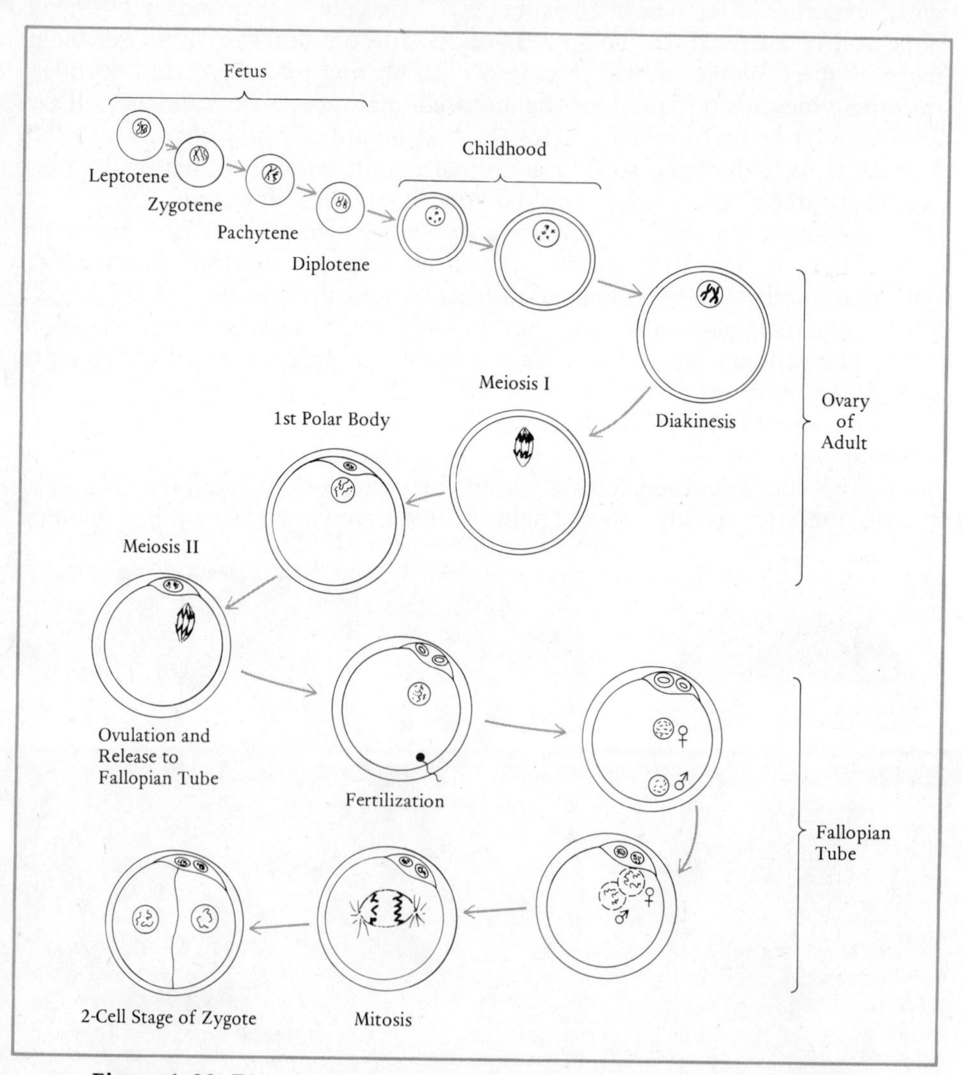

Figure 6–33. Diagram summarizing the stages of oögenesis in humans.

fertilization. Thus, although meiosis in the human female is initiated at the third month of embryonic life, primary oöcytes remain suspended at diplotene anywhere from 12 to 40 years or more before the first meiotic division is actually carried to completion. It should be noted that the risk that some chromosome pairs will fail to segregate normally (nondisjunction) rises with increasing maternal age.

The stages of gametogenesis outlined above are generally found in all multicellular animals. For example, both the mouse and *Drosophila* conform to this pattern. The life cycle of *Drosophila* is of additional interest in that the zygote initially develops into a maggot-type larval form which later, during pupation, becomes transformed into the winged adult. During the larval stage some tissues grow by increase in cell size, rather than by mitosis, and come to contain cells in whose nuclei are chromosomes that have replicated numerous times without dividing. As we shall see, such chromosomes offer unique advantages for cytological analysis.

PROBLEMS

6–1. In a cell of a species where $2n = 6$, diagram the process of meiosis, making sure that you understand chromosome segregation and independent assortment.

6–2. If the diploid chromosome number of a wild strain of wheat *(Triticum)* is 14, how many chromatids (total chromosomes) would be found in:
- a. parenchyma cell of a young leaf at metaphase
- b. vegetative nucleus of the pollen tube
- c. a polar nucleus of the central cell within the embryo sac
- d. a daughter cell forming at telophase in the root tip
- e. a microspore mother cell at the zygotene stage
- f. an egg cell just prior to fertilization
- g. a cell of the endosperm at prophase
- h. a daughter endosperm cell at telophase
- i. a zygote at metaphase of the first cell division after fertilization
- j. each of the two daughter cells forming at telophase in (i), above

6–3. The domestic cat has 38 pairs of chromosomes. How many chromatids (total chromosomes) would be found in:
- a. first polar body
- b. primary oöcyte at diplotene
- c. a liver cell at prophase
- d. a secondary spermatocyte
- e. a zygote at metaphase
- f. a spermatid
- g. a daughter kidney cell at telophase
- h. a flagellated sperm cell

6–4. In a cell with 7 pairs of chromosomes, what is the probability that:
- a. all paternal centromeres will move to the "south" pole
- b. all paternal centromeres will move to the "north" pole
- c. all paternal centromeres will move together to the *same* pole, either north or south

6–5. If an individual inherits centromeres A, B, C, D from the father and centromeres A′, B′, C′, D′ from the mother, what is the probability that any gamete selected at random from such an individual would contain centromeres A, B′, C, D′?

6–6. The interspecific cross horse × donkey produces the mule. How do you account for the fact that mules are almost invariably sterile?

6–7. Assuming $2n = 8$, suppose that in one cell undergoing meiosis one pair of homologues fails to disjoin at the first meiotic division, causing both homologues of the pair to proceed to the same pole.
- a. How many chromatids would be present in each daughter cell resulting from meiosis I?
- b. If the second meiotic division separates all sister chromatids, how many chromatids would be present in each of the four resulting gametes?
- c. Using n to stand for one complete haploid set of chromosomes, how might you designate the chromosome complement of these gametes?

6–8. Assuming $2n = 8$, suppose the first meiotic division of a cell is normal, but in one of the two daughter cells nondisjunction of a pair of sister chromatids occurs during the second meiotic division. How many chromatids would be present in each of the four gametes?

6–9. If a strain of *Neurospora* of genotype p^+ was crossed with a strain carrying the allele p^-, what types of spores would be produced and in what proportions?

6–10. In the preceding cross $(p^+ \times p^-)$ suppose you dissected out the 8 spores, in order, from a single ascus and determined the genotype of each spore.

a. What spore order, with respect to genotype, would be characteristic of first division segregation?

b. Suppose crossing over between any two nonsister chromatids occurs between the centromere and the location of gene *p* resulting in second division segregation. How many different spore orders are possible? What are they?

6–11. Suppose a cross was made between *Neurospora* strains p^-q^+ and p^+q^- in genotype. Assume that the alleles p^+ and p^- are independent of the alleles q^+ and q^- and that crossing over does not occur. What types of spores would be produced and in what proportions?

6–12. Why is it inappropriate to apply the terms dominant and recessive to allelic genes of *Chlamydomonas, Neurospora,* and other similar organisms?

REFERENCES

CASPERSSON, T., AND L. ZECH (Eds.), 1973. *Chromosome Identification, Technique and Applications in Biology and Medicine*. Academic Press, New York.

DUPRAW, E. J., 1970. *DNA and Chromosomes*. Holt, Rinehart and Winston, New York.

HSU, T. C., 1973. Longitudinal differentiation of chromosomes. *Ann. Rev. Genet. 7,* 153.

HSU, T. C., AND K. BENIRSCHKE, 1971. *An Atlas of Mammalian Chromosomes*. Springer-Verlag, New York.

LUYX, P., 1970. Cellular mechanisms of chromosome distribution. *Int. Rev. Cytol. 32* (Suppl. 2).

MOENS, P. B., 1973. Mechanisms of chromosome synapsis at meiotic prophase. *Int. Rev. Cytol., 35,* 117.

MÜNTZING, A., 1974. Accessory chromosomes. *Ann. Rev. Genet., 8,* 243.

NICKLAS, R. B., 1971. Mitosis. *Adv. Cell Biol., 2,* 225.

SHARMA, A. K., AND A. SHARMA, 1972. *Chromosome Techniques, Theory and Practice,* 2nd ed. University Park Press, Baltimore.

WESTERGAARD, M., AND D. VON WETTSTEIN, 1972. The synaptinemal complex. *Ann. Rev. Genet., 6,* 71.

WHITE, M. J. D., 1973. *Animal Cytology and Evolution*. Cambridge University Press, New York.

WHITE, M. J. D., 1973. *The Chromosomes,* 6th ed. Chapman & Hall, London.

ZIMMERING, S., L. SANDLER, AND B. NICOLETTI, 1970. Mechanisms of meiotic drive. *Ann. Rev. Genet., 4,* 409.

Chapter 7

SEX LINKAGE AND SEX DETERMINATION

The proposal by Sutton, in 1903, that Mendelian factors were located in the chromosomes provided great impetus to the field of cytology, and a number of investigators became engaged in intensive study of the chromosomal constitution of a variety of plants and animals. The paired condition of the chromosomes was repeatedly confirmed, but it was observed that homologous chromosomes were sometimes *heteromorphic*, that is, not identical in morphology. The presence of at least two such heteromorphic pairs of homologues provided an opportunity to test Mendel's Law of Independent Assortment. Since the members of each pair could be distinguished from one another, their distribution at meiosis could be followed under the microscope and their independent assortment observed. The results of these investigations not only supported Sutton's proposal, but, by directing attention to heteromorphic pairs of chromosomes, brought about the discovery of sex chromosomes.

PATTERN OF INHERITANCE OF SEX-LINKED GENES

It had been observed by H. von Henking, C. E. McClung, and others that in certain species of insects the chromosome complement of males and females differed in regard to one specific pair of homologues. Males possessed only one chromosome of the pair, while females had two such chromosomes. This chromosome, which occurred singly in males, but doubly in females, was named X, with females designated as XX and males as X0 (X zero).

Studies with other animals showed that the presence of unequal chromosome numbers in the two sexes was the exception rather than the rule, for in the majority of animals investigated the chromosome number of males and females was the same. However, the two sexes were again found to differ with respect to one particular pair of homologues. In females the members of this pair were morphologically identical and were again named X chromosomes. In males this pair of chromosomes was heteromorphic in that a single X was accompanied by a chromosome which differed markedly from the X in size and shape; the second member of the pair was named the Y chromosome. Since the constitution of this pair of homologues, XX for females and XY (or X0) for males, was directly correlated with the sex of an individual, these chromosomes were named *sex chromosomes*. The other chromosome pairs which were present equally in both sexes became known, collectively, as *autosomes*.

It was immediately perceived from the results of cytological studies that the segregation of the sex chromosomes at meiosis and their subsequent reunion as pairs via fertilization provided an obvious means for sex determination. It was reasoned that females, being XX, can produce only X-bearing eggs, that is, they are *homogametic* in that all gametes are alike with regard to sex chromosome constitution. In males, however, segregation of the X from the Y chromosome during meiosis results in the production of two kinds of sperm, one carrying the X chromosome, the other the Y chromosome. Males are thus *heterogametic* in that more than one type of gamete can be produced with respect to sex chromosome content; X- and Y-bearing gametes are regularly produced. Random union between eggs and sperm at fertilization should result in approximately equal numbers of XX (female) and XY (male) zygotes, in agreement with the usually observed 1 : 1 sex ratio. This proposed mechanism could be applied not only to species with the XX-XY type of sex determination, but also to instances where males were X0 in sex type. In such cases two kinds of sperm would also be produced in equal numbers, those with and those without the X chromosome, and random fertilization would again be expected to yield equal numbers of XX (female) and X0 (male) zygotes. The presence of a similar sex-determining mechanism in at least some plants could also be inferred, for it had been shown by Correns, in 1907, that male-determining and female-determining pollen was produced in equal amounts by staminate (male) plants of the dioecious species *Bryonia,* whereas all eggs formed by pistillate (female) plants were alike as regards sex. The extension of cytological investigations to birds and to such insects as butterflies revealed the same sex-determining mechanism, but in reverse. In these groups, females were heterogametic (XY) in sex chromosome constitution and produced equal numbers of X-bearing and Y-bearing eggs, while males were XX and formed sperm, all of which carried a single X chromosome.* (We should note here that the genetic basis for sex determination in *Drosophila* is not based on XX versus XY, but on the ratio of X chromosomes to autosomes.)

* In organisms in which the female is heterogametic, the X chromosome is called Z and the Y chromosome is called W. Since these additional names are confusing, the terms X and Y will be used exclusively in this text.

As a result of these and other studies, most of which were carried out prior to 1910, a general understanding of the chromosomal basis for sex determination was achieved. In addition, it was clear that sex could be viewed as a phenotype whose inheritance and chromosomal basis conformed to the rules of Mendelian genetics, for the 1 : 1 sex ratio obtained in any generation was identical to the results of a testcross between a homozygote and a heterozygote. The resolution of the chromosomal mechanism of sex determination thus provided powerful support for the hypothesis that chromosomes were the physical basis of inheritance. This hypothesis was further strengthened by Thomas H. Morgan in 1910, through the discovery of a recessive mutant in *Drosophila* whose pattern of inheritance paralleled that of the sex chromosomes. The mutant was named white eyes *(w)*, and Morgan analyzed its mode of inheritance by use of the following crosses (Figure 7–1):

1. When white-eyed males were crossed to red-eyed females, all F_1 progeny, both male and female, had red eyes. The results of this cross indicated that the mutant white was recessive to the normal red or wild type allele (w^+).

2. When F_1 red-eyed individuals were inbred, all F_2 females had red eyes, but the F_2 males were of two types, half had red eyes and half had white eyes.

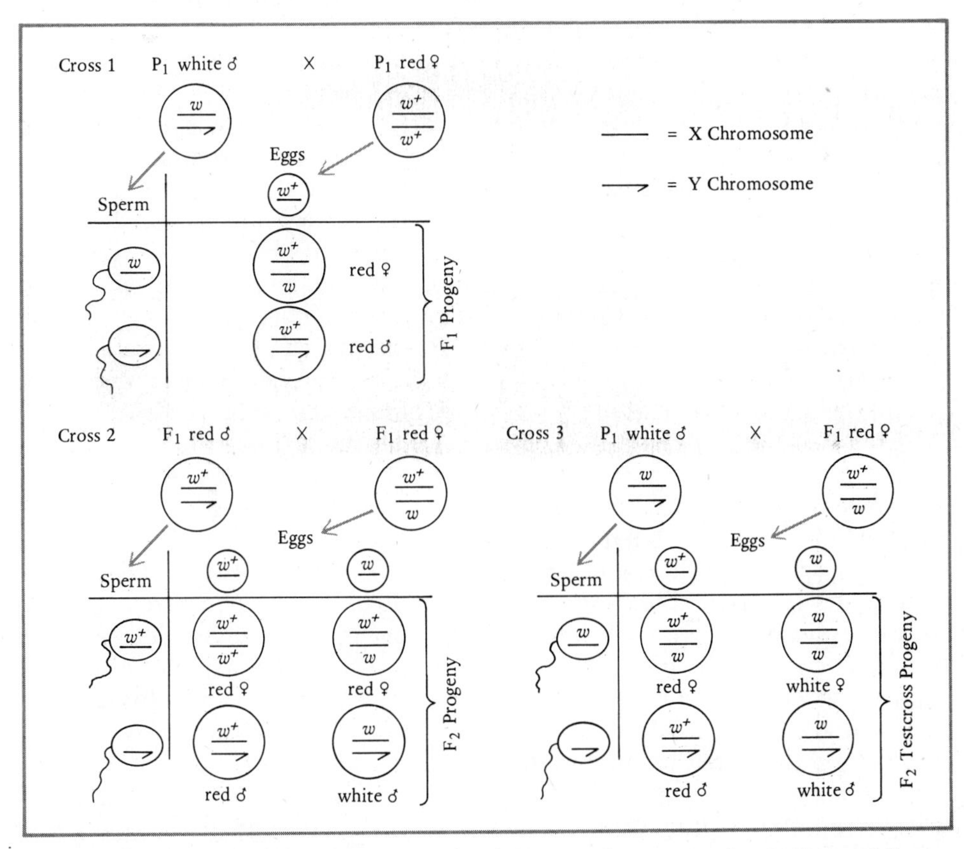

Figure 7–1. Sex-linked inheritance of red versus white eye color in *Drosophila*, as demonstrated by Morgan in 1910.

These results clearly differed from those anticipated. One would expect to obtain a 3 : 1 ratio of red to white as the result of an inbreed between individuals heterozygous for a single pair of alleles, one dominant and the other recessive. An overall ratio of 3 red : 1 white was indeed evident, again indicating the recessive nature of the mutant gene white. However, according to Mendel's principles, the ratio of 3 red : 1 white should be obtained *without regard to sex,* that is, a 3 : 1 ratio of red to white should be evident in both the male *and* the female F_2 progeny. But in Morgan's experiment the recessive phenotype, comprising $\frac{1}{4}$ of the total progeny, was confined exclusively to the male sex and did not appear at all in females.

3. When F_1 red-eyed females were backcrossed to parental white-eyed males, both male and female offspring were of two types: half were red eyed and half were white eyed. These results were typical of a testcross and indicated that the F_1 females used in the cross were heterozygous for red and white, respectively.

The overall results of these three crosses suggested that the inheritance of white eyes followed Mendelian laws, but was, in addition, associated with sex determination. Morgan therefore proposed that the gene white was recessive and located on the X chromosome and that the Y chromosome contained no allele of this gene. According to this hypothesis, the results of these crosses could be explained as follows. The original P_1 male, having only one X chromosome, possessed only the one allele, w. The term *hemizygous* was applied to this state. The allele was expressed in the phenotype because no alternate dominant (wild type) allele was present in the Y chromosome. The P_1 females, possessing two X chromosomes, were homozygous for the dominant allele for red eyes (w^+/w^+). The cross between the P_1 individuals produced F_1 males whose single X chromosome was contributed by the egg and carried the dominant allele, causing F_1 males to be red eyed. F_1 females inherited the recessive allele via an X-bearing sperm and the dominant allele via the egg and, as a consequence, were heterozygous and red eyed in phenotype.

With regard to the inbred progeny, F_2 females would be expected to be uniformly red eyed in phenotype since all females were derived from the union of an X-bearing sperm with the X-bearing egg. In this cross the single X of the F_1 male carried the dominant allele, and therefore the X-bearing sperm contributed this allele to all female zygotes. As a result, all F_2 females were red eyed.

The 1 : 1 ratio between red- and white-eyed F_2 males could also be accounted for by Morgan's hypothesis. Males are produced from the union of a Y-bearing sperm with the X-bearing egg. If, indeed, the white gene was associated only with the X chromosome and no allele of this gene occurred on the Y, then the phenotype of a male with respect to eye color would depend solely on the allele contributed to the zygote by the egg. Since the F_1 females producing these eggs were heterozygous with respect to these alleles, they formed two kinds of gametes, w^+ and w, in equal numbers, thus accounting for the 1 : 1 ratio of red to white observed in the F_2 male progeny.

The results of the backcross of F_1 heterozygous females to parental white-eyed males were also understandable in terms of Morgan's hypothesis. The 1 : 1 ratio between red- and white-eyed male progeny could again be traced to the two kinds of eggs produced by the F_1 heterozygous females. These same two types of eggs also accounted for the 1 : 1 ratio found in the female progeny. Half of the eggs carried the dominant allele w^+ and half the recessive allele w. The X-bearing sperm, required for the formation of a female zygote, all contained the recessive allele w, since all P_1 males were white eyed in phenotype. Fertilization of a w^+ egg by a w-bearing sperm would produce a female heterozygous for these alleles (w^+/w) and therefore red eyed in phenotype. However, union of a w egg with a w-bearing sperm would result in a female homozygous for the recessive allele (w/w) and therefore white eyed in phenotype.

This explanation of the data by Morgan fully accounted for the results of these crosses and was in agreement with the previously established mechanism of sex determination. In addition, these experiments were the first to demonstrate that the inheritance of a specific gene could be associated with, or linked to, the inheritance of a specific chromosome. Since the chromosome involved in this case was a sex chromosome, the phenomenon was named *sex linkage.*

Although Morgan's hypothesis of sex linkage was entirely convincing, the data supporting the proposal were derived entirely from the results of crosses and not from the cytological examination of the sex chromosomes themselves. Critical evidence placing genes on chromosomes required that the pattern of inheritance be correlated with chromosome constitution. Just such evidence was provided in 1916, by Calvin B. Bridges from crosses of white-eyed females with red-eyed males.

All female progeny from such a cross are expected to be red eyed in phenotype due to inheritance of the w^+ allele from the father, but on rare occasions Bridges found a white-eyed daughter whose inheritance of this sex-linked trait appeared to be exclusively maternal in origin *(matroclinous inheritance)*. To account for such an exceptional daughter, he proposed that the two X chromosomes of the mother had failed to segregate from one another during meiosis, the phenomenon known as *nondisjunction,* thereby producing an egg containing both maternal X chromosomes. Because the presence of white eyes in the exceptional female indicated absence of the dominant allele, Bridges reasoned that fertilization of the XX egg by a Y-bearing sperm must have occurred to produce an XXY female whose phenotype resulted from the presence of the recessive allele in each of the maternal X chromosomes.

To substantiate this hypothesis, Bridges crossed exceptional white-eyed daughters to normal red-eyed males. He reasoned that if the daughters were truly X^wX^wY in sex chromosome constitution, chromosome segregation at meiosis should result in four types of eggs: X^wX^w, X^wY, X^w, and Y (Figure 7–2). Fertilization of these eggs by an X^{w+}-bearing sperm should lead, respectively, to $X^{w+}X^wX^w$ females, $X^{w+}X^wY$ females, $X^{w+}X^w$ females, and $X^{w+}Y$ males, all such individuals being red eyed because of the presence of the dominant allele w^+. In contrast to the usual pattern of sex-linked inheritance, males would inherit this trait exclusively from the father *(patroclinous inheritance)*. On the other hand, fertilization of these eggs by a Y-bearing sperm should result, respectively, in X^wX^wY white females, X^wYY white males, X^wY white males, and YY (nullo-X) individuals.

Bridges was able to identify all of these progeny types and to correlate their respective phenotypes with their sex chromosome constitution as ob-

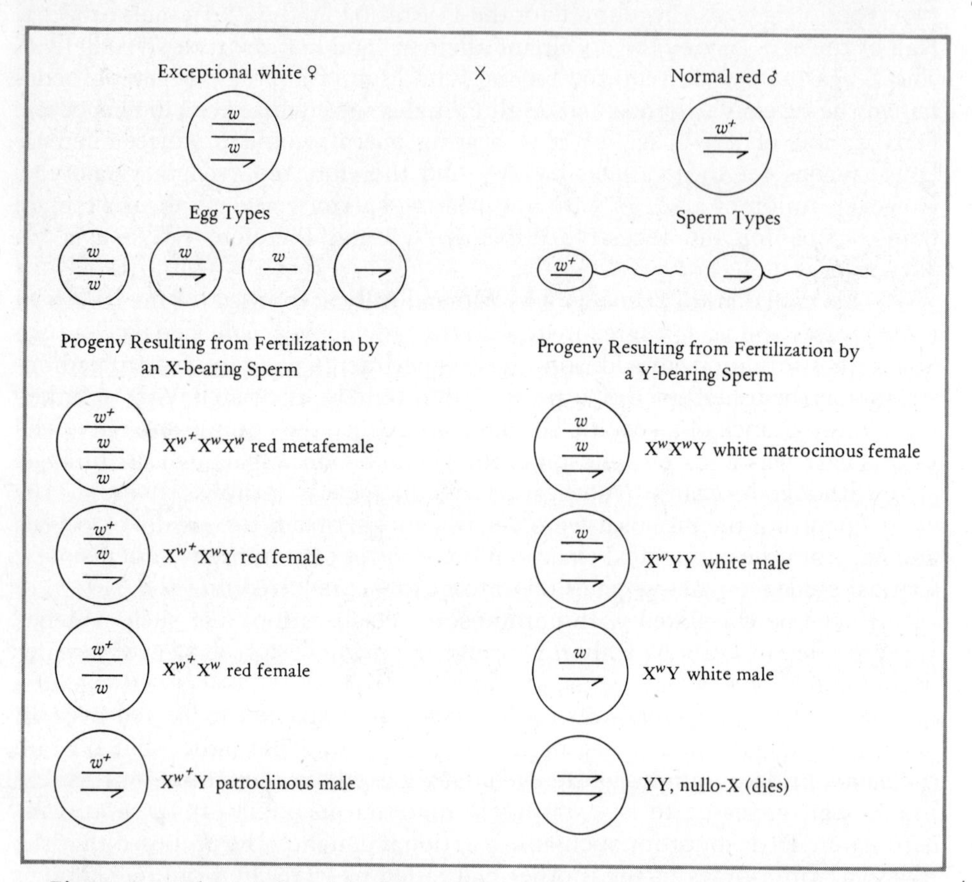

Figure 7–2. The results of Bridges's cross of exceptional white-eyed XXY females with normal red-eyed males.

served under the microscope. He thus provided not only genetic, but also cytological evidence for sex linkage and for the chromosome basis of heredity.

His findings also contributed the following additional information on sex determination in *Drosophila*.

1. The presence of a Y chromosome does not in itself determine maleness, for XXY individuals were normal, fertile females.

2. The XYY constitution is a fertile male.

3. The XXX condition, termed *metafemale* or *superfemale*, is semilethal, the few viable individuals being abnormal female types.

4. The YY constitution is lethal in the egg stage.

It was also noted that the initial occurrence of nondisjunction (primary nondisjunction) could result in either an XX egg or an egg totally lacking the X chromosome. In the latter instance, fertilization by a Y-bearing sperm would result in lethality, whereas fertilization by an X-bearing sperm should produce an X0 male showing patroclinous inheritance of sex-linked traits. Such males have been identified and found to be morphologically normal, although sterile. This has led to the conclusion that the Y chromosome of *Drosophila* bears genes for sperm function.

Bridges's studies also served to emphasize the normal pattern of sex-linked inheritance by which sons inherit their X chromosome exclusively from the maternal parent, while daughters receive an X chromosome from each parent. A change in this pattern such that sons inherit X-linked genes from the father (patroclinous inheritance) or daughters inherit such genes exclusively from the mother (matroclinous inheritance) is an immediate signal that some abnormality, such as nondisjunction, has occurred. One well-known example of such an abnormality is the attached-X condition ($\widehat{XX}$) where both of the X chromosomes have become joined to the same single centromere and are thus necessarily transmitted together to progeny. The consequences of this condition are illustrated in Figure 7–3.

It is also useful to point out a difference in the pattern of inheritance of sex-linked genes versus autosomal genes. Sex-linked traits can be identified most readily through the results of reciprocal crosses or, when this is not possible, by the evident transmission of a trait from mothers to sons. Using white eyes as the example of a recessive sex-linked trait, the cross white-eyed females × red-eyes males will produce progeny consisting of red-eyed females and white-eyed males. However, the reciprocal cross red-eyed females × white-eyed males, will yield all red-eyed offspring. The differing results of such reciprocal crosses permit identification of sex linkage. In contrast, reciprocal crosses involving autosomal genes yield identical results. Using the recessive autosomal mutant brown eyes of *Drosophila*, the cross brown-eyed females × red-eyed males produces all red-eyed offspring due to dominance of the gene for

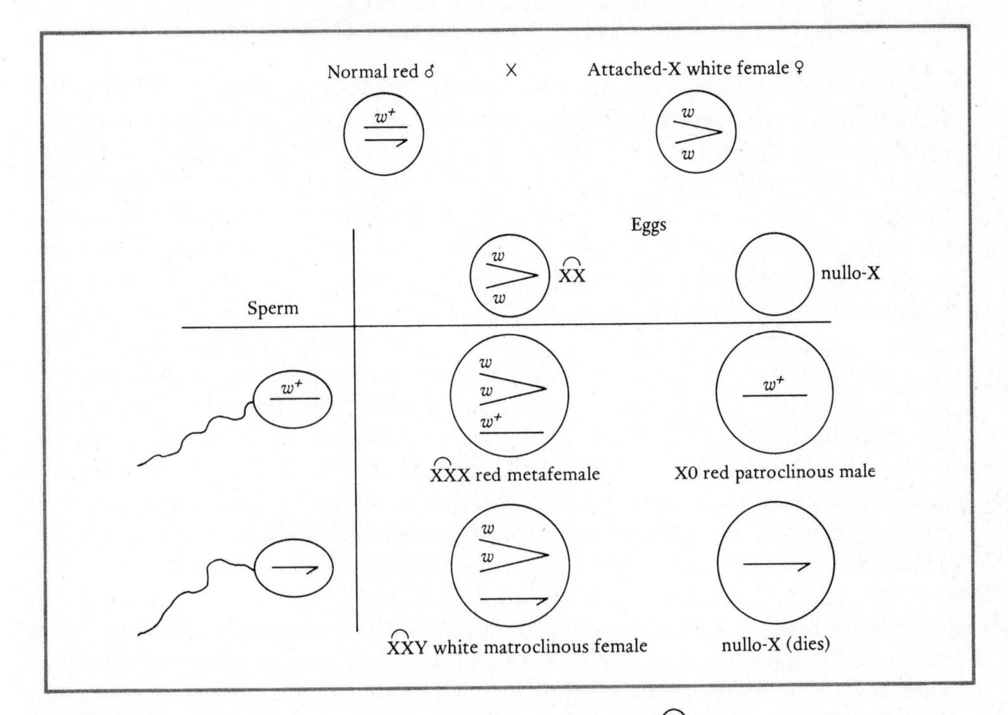

Figure 7–3. Results of a cross between an attached-X ($\widehat{XX}$) white-eyed female and a normal red-eyed XY male.

red over that for brown. The reciprocal cross red-eyed females × brown-eyed males also produces all red-eyed offspring for the same reason.

Additional examples of autosomal inheritance can be found in the results of Mendel's crosses. The basis for the difference between sex-linked and autosomal inheritance lies in the fact that only one X chromosome is found in males, whereas two are found in females. The Y chromosome, when present, contains essentially none of the genes found on the X chromosome. Thus, sex linkage is in reality X linkage, and males are hemizygous for practically all such genes. Autosomes, on the other hand, are present equally in both sexes as truly homologous pairs containing equivalent genes.

A demonstration of the generally nonequivalent nature of the X and Y chromosomes, as compared to autosomes, can be found in the pairing behavior of these chromosomes during meiosis in human males. The metacentric X and the small acrocentric Y each contain a short *pairing segment,* and synapsis and the formation of a synaptinemal complex occurs only between these mutually homologous regions. The rest of these chromosomes constitute *differential segments,* and pairing between these much longer regions does not occur. This phenomenon is clearly evident in Figures 6–20 and 6–25. In contrast to the behavior of the X and Y chromosomes, autosomes are consistently found paired throughout their length at meiosis. If synapsis is a valid criterion of chromosome homology, we must conclude that the X and Y chromosomes of humans are mostly nonhomologous in genetic content.

SEX-LINKED TRAITS OF DROSOPHILA AND HUMANS

Through continuing intensive study of the genetics of *Drosophila melanogaster,* numerous genes have been located on the X chromosome of this species. They comprise a wide variety of dominant and recessive mutations which exert their effects on practically every organ system of the body. A number of lethal mutations have also been identified. When such a mutant is present in the heterozygous condition in a female, only half of her sons, those inheriting the normal, nonlethal allele, will be viable. The sex ratio of males to females is altered from the normal 1 : 1 ratio to a 1 : 2 ratio. Although some of these mutations produce an abnormal morphological phenotype when heterozygous in females, whether dominant or recessive, all such mutations cause the death of hemizygous males and homozygous females. Allelic genes found on both the X and Y chromosomes are rare, and in *Drosophila* only the mutant bobbed *(bb)* has been identified as both X linked and Y linked. Bobbed and its alleles are mutations of the nucleolus organizer and the location of these mutants on both of the sex chromosomes is accounted for by the fact that a nucleolus is associated with the X as well as the Y chromosome.

In general, when one surveys the list of sex-linked genes in *Drosophila* (see Figure 8–8) it is clear that the majority of those discovered thus far exert their effects in both males and females and therefore have seemingly little to do with sex determination per se. However, all evidence points to the presence of female-determining genes in the X chromosome, and in *Drosophila* different segments of the X chromosome have been shown to contain such genes.

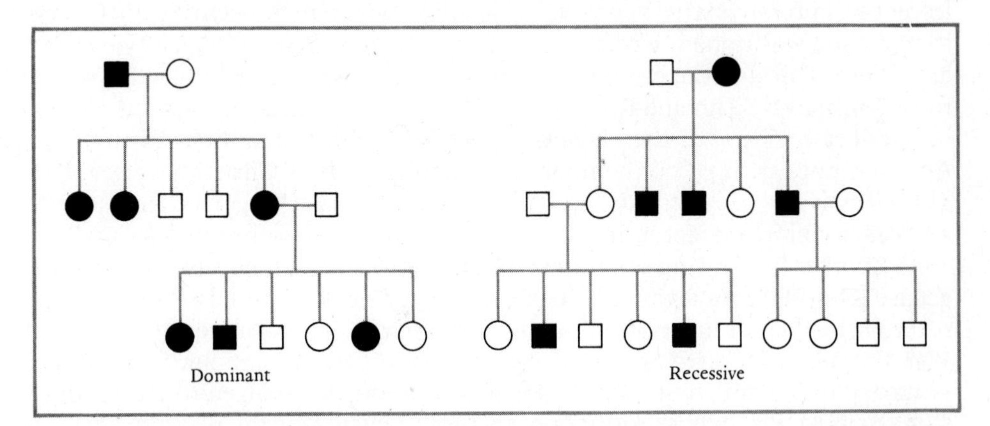

Figure 7–4. Hypothetical human pedigrees illustrating the pattern of inheritance of a sex-linked dominant trait and a sex-linked recessive trait.

In humans X linkage of a trait is identified through pedigree analysis (Figure 7–4). If the condition is dominant, the trait will be transmitted directly from fathers to daughters, but never from fathers to sons. If the trait is recessive, affected homozygous mothers will transmit the trait to all sons, while unaffected, but heterozygous mothers will contribute the condition to only half of the sons. As in the case of dominant sex-linked genes, affected fathers do not ordinarily transmit the trait to sons.

Since males are hemizygous for the X chromosome, all sex-linked genes are expressed in the phenotype, whereas full expression of a recessive sex-linked trait in females requires that the gene be homozygous. As a consequence, affected males occur with higher frequency than do affected females. In the case of lethal genes homozygous females may not be found, since the death of hemizygous males usually occurs prior to reproductive age. Over 50 sex-linked traits have been identified in humans. Some of the better known examples include the following.

Muscular dystrophy (Duchenne type) is a genetic lethal found only in males. Heterozygous females are carriers. Young children are affected by the time they begin to walk and usually do not survive their teens. The Becker type of muscular dystrophy is also a sex-linked lethal mutation, but is less common.

Colorblindness is another well known sex-linked recessive trait. Two different genes are involved, one resulting in the inability to perceive red (protan type) and the other causing the inability to perceive green (deutan type). These genes are found in higher frequency among both males and females than are most sex-linked mutants because their effects do not impair survival.

The most famous recessive sex-linked trait is hemophilia A, bleeder's disease. It is characterized by a deficiency of one of the factors necessary for normal blood clotting (factor VIII, antihemophilic globulin). As a result, hemorrhage leading to death can occur from even minor wounds, and persons suffering from this disease have a life expectancy of less than 20 years. Although this disease has been known from ancient times and is even mentioned in the Talmud, its more recent notoriety stems from the fact that a mutation from the normal to the hemophilia A allele apparently occurred in Queen Victoria and was transmitted by her to one of her sons and to at least two of her daughters. Through the carrier daughters, the disease was introduced into the royal families of Spain and Russia, and the downfall of these ruling families can be ascribed, in part, to the phenotypic effects of the hemophilia A mutation. Another, but rarer type of hemophilia, hemophilia B or Christmas disease, is also recessive and sex linked. The cause of this condition is again a deficiency of a necessary clotting factor, in this case factor IX, plasma thromboplastin.

Another interesting sex-linked mutation is glucose-6-phosphate dehydrogenase (G-6-PD) deficiency. The condition was first discovered as a hemolytic reaction to the administration of the antimalarial drug primaquine in World War II. The reaction was traced to a reduction in the amount of reduced glutathione present in red blood cells. Glutathione is changed to the reduced state by NADPH* whose formation, in turn, depends on the dehydrogenation of glucose-6-phosphate by glucose-6-phosphate dehydrogenase. A deficiency of this enzyme is thus the primary cause of the phenotype.

In addition to the allele causing enzyme deficiency, other alleles of this gene have been discovered. Each of these results in the presence of an enzyme at least partially normal in activity, but subtly altered in structure such that the net charge on the molecule differs from that of the normal enzyme or that of other variants of this enzyme. This difference in charge can be detected by the technique of electrophoresis permitting identification of the particular variant present in a cell sample. The G-6-PD deficiency occurs with high frequency in populations of Sardinians, Oriental Jews, African Negroes, and in around 15 percent of American blacks. It is interesting to note that one variant, called the Gd Mediterranean type, is relatively common in white populations of the Middle East and southern Europe. Individuals possessing this variant show no ill effects unless they happen to eat fava beans (broad beans) in which case they may experience a severe hemolytic reaction, a condition called favism. The cause of the reaction is thought to be associated with some substance produced during digestion or metabolism of the bean.

A dominant sex-linked trait in humans is the $Xg(a^+)$ phenotype. The Xg blood group system consists of a pair of alleles, Xg^a and Xg, which determine the presence or absence of a red cell antigen, the respective phenotypes being designated as $Xg(a^+)$ and $Xg(a^-)$. These blood groups can be distinguished by reactions with anti-Xg^a antibody, but are of no clinical importance. Approximately 67 percent of Caucasian males carry the dominant Xg^a allele. The Xg blood group alleles, the hemophilias, colorblindness, and the G-6-PD deficiency are extremely useful as markers in determining the linear sequence of genes on the X chromosome.

Known genes specifically linked to the Y chromosome *(holandric genes)* are extremely rare in humans. One verified example is a mutation which causes excessive growth of hair on the outer ear. The Y chromosome is not empty of genes, however, for, as discussed in the following section, the presence of the Y is responsible for the male phenotype in humans and other mammals.

* NADPH stands for the reduced state of the coenzyme NADP, nicotinamide adenine dinucleotide phosphate, also known as triphosphopyridine nucleotide, TPN.

Finally, and quite apart from sex linkage, it should be noted that gene expression in many instances may be dependent upon, or influenced by, the sex of an individual. For example, some traits are entirely *sex limited* in that they are expressed only in males or only in females, the secondary sex characters of humans being an obvious illustration. Here, although the traits of breast development, milk production, differential hair patterns, or the presence or absence of accessory reproductive glands, such as the prostate, are indeed sex limited, the numerous genes which participate in the development of these traits are not necessarily sex linked. On the contrary, many are surely autosomal and therefore present equally in both sexes, whether or not they are expressed in the phenotype. Such differential gene expression is likely conditioned or brought about not only by prior developmental events, but, more specifically, by the differing hormonal constitutions of the two sexes.

Hormonal factors are also undoubtedly involved in the expression of genes which are *sex influenced* for the penetrance and expressivity of such genes is dependent on the sex of the individual. One of the best examples in humans is pattern baldness. The gene concerned is autosomal and behaves as a dominant in males, but as a recessive in females. A similar case is the horned versus hornless condition of sheep, where the autosomal gene involved is again dominant in males and recessive in females. These conditions are obvious examples of sex-influenced gene activity. It is probable that the expression of many other genes is also modified, directly or indirectly, in response to the pervasive influence of the sex hormones.

GENETIC BASIS FOR SEX DETERMINATION

The identification of the Y chromosome and its observed restriction to the male sex in many animals initially suggested that this chromosome was the sole determiner of maleness. However, the work of Bridges indicated that, at least in *Drosophila,* the Y chromosome did not by itself determine maleness, but instead carried genes for fertility. With the continuing identification of additional mutants, both sex linked and autosomal, it soon became apparent that most phenotypic characters resulted from the concerted action of many genes. In addition, sex could no longer be considered as a simple Mendelian trait, for evidence pointing to the presence of male-determining genes in the autosomes of moths had been found by Richard Goldschmidt. In 1921 Bridges found that nondisjunction of the tiny fourth chromosome of *Drosophila,* an autosome, produced in a succeeding generation individuals with one or three representatives of this particular chromosome. Such individuals were abnormal with respect to many structures of the body, and Bridges suggested that an unbalanced chromosome complement might be responsible for the aberrant phenotype. Accordingly, he proposed that a balanced set of chromosomes and genes was required for normal development, and this hypothesis subsequently gained widespread acceptance. Since sexuality was a complex phenotype, apparently controlled by numerous genes, sex linked as well as autosomal, this theory of genic balance was extended to sex determination.

In 1922 Bridges applied this theory to the progeny produced by triploid ($3n$) females of *Drosophila*. Triploids contain three complete sets of chromosomes. In a triploid during chromosome segregation at meiosis, two of the homologues of any one kind of chromosome proceed to one pole, while the third homologue moves to the opposite pole. With the occurrence of independent assortment many of the eggs come to contain unbalanced chromosome sets. These are called *aneuploids* and they almost always lead to inviable zygotes. In addition to such abnormal aneuploid gametes, some eggs are produced that contain the following chromosome complements:

1. a single X chromosome plus one complete set of autosomes (1 X : 1 A)
2. a single X chromosome plus two complete sets of autosomes (1 X : 2 A)
3. two X chromsomes plus one complete set of autosomes (2 X : 1 A)
4. two X chromosomes plus two complete sets of autosomes (2 X : 2 A)

After fertilization of these eggs by X-bearing or Y-bearing sperm, different sexual phenotypes are produced, and Bridges was able to correlate such phenotypes with their corresponding chromosome complements (Table 7–1). He found that when the ratio of X chromosomes to complete sets of autosomes was 1 : 1 (or 1.0), the phenotype produced was that of a normal female. When this ratio was 1 : 2 (0.5), a normal male resulted. An intermediate ratio of 2 : 3 (0.67) produced an intersexual type, neither wholly male nor wholly female, but having characteristics of both. The combination of 3 X : 2 A, a ratio of 1.5, yielded a metafemale with exaggerated female characteristics, but sterile and weak, while the 1 X : 3 A complement, a ratio of 0.33, produced a metamale, or supermale, similarly weak and sterile. In addition, the Y chromosome seemingly played no role in sex determination. These data provided an elegant demonstration of genic balance as the determiner of sex in *Drosophila,* and in the absence of conflicting evidence, the hypothesis was tentatively extended to other animals, including humans.

Although it is now known that the Y chromosome of humans does in fact determine the male phenotype, this understanding was not achieved for over 35 years after the completion of Bridges's studies. It should be recalled that cytologists initially encountered great difficulties with mammalian material and that the chromosome number of humans was not known with certainty until 1956.

The foundation for establishing the association between chromosome constitution and sexual phenotype in humans was laid in 1942 when H. F. Klinefelter described an abnormal condition of male patients in which the breasts were developed and spermatogenesis was lacking, although the exter-

TABLE 7–1
Genic Balance and Sex Determination in *Drosophila**

Eggs	Sperm 1 X : 1 A	Resulting phenotype	Ratio of X : A	Sperm 1 Y : 1 A	Resulting phenotype	Ratio of X : A
1 X : 1 A	2 X : 2 A	female	1 : 1 (1.0)	1 X 1 Y : 2 A	male	1 : 2 (0.5)
1 X : 2 A	2 X : 3 A	intersex	2 : 3 (0.67)	1 X 1 Y : 3 A	metamale	1 : 3 (0.33)
2 X : 2 A	3 X : 3 A	female	1 : 1 (1.0)	2 X 1 Y : 3 A	intersex	2 : 3 (0.67)
2 X : 1 A	3 X : 2 A	metafemale	3 : 2 (1.5)	2 X 1 Y : 2 A	female	1 : 1 (1.0)

* Sexual phenotypes and corresponding X chromosome and autosome complements in the progeny of triploid (3*n*) females crossed to diploid males: 3 X : 3 A females × XY : 2 A males.

nal genitalia were male in type. This condition became known as Klinefelter's syndrome. In 1959, after the diploid chromosome number of 46 had been established for humans, P. A. Jacobs and J. A. Strong analyzed the chromosome complement of patients exhibiting Klinefelter's syndrome and found that one additional chromosome was present in the cells of such individuals. The chromosome complement consisted of two X chromosomes, one Y chromosome, and a normal double set of autosomes, a complement that can be designated as 47,XXY. This chromosome complement in conjunction with the obvious male phenotype immediately implicated the Y chromosome as the source of maleness. It was also apparent that the genic balance theory of sex determination could not be applied to humans. According to the genic balance hypothesis, a 1 : 1 ratio between sex chromosomes and sets of autosomes should result in a female, as in XXY *Drosophila,* but patients with Klinefelter's syndrome were male.

Evidence for the role of the Y chromosome as the determinant of maleness was also obtained in 1959 by C. E. Ford and his associates through the analysis of the chromosome complement of patients with a condition known as Turner's syndrome. Such individuals are sterile females with undeveloped secondary sex characters. The reproductive organs are mere streaks of ovarian tissue. Patients with Turner's syndrome were found to be X0 in sex chromosome constitution, for only a single X chromosome was present. Thus the total chromosome number of these individuals was reduced from the normal 46 to 45, and their karyotype could be designated as 45,X0. In contrast to the X0 condition in *Drosophila* which results in maleness, individuals with Turner's syndrome are always female in sexual phenotype.

These studies prompted the proposal that the determinant of sex in humans was the Y chromosome, presence of the Y being responsible for maleness and absence of the Y resulting in femaleness. Numerous additional studies have supported this hypothesis. In humans and other mammals the embryonic gonad is initially undifferentiated as to sex type. Presence of a Y chromosome induces the development of a testis through the activity of one or more genes thought to be situated in the short arm of this chromosome. In turn, cells of the seminiferous tubules of the testis produce an inhibitor substance which suppresses the development of the primitive oviduct (Müllerian duct). The interstitial cells of the testis, through their secretion of androgenic hormones, induce the differentiation of the Wolffian duct and the development of male secondary sex characters. In the absence of the Y chromosome, the primitive gonad differentiates into an ovary and the Müllerian duct into Fallopian tubes and uterus. In humans, though not in mice, it appears that both X chromosomes are required for ovarian development.

ABNORMAL SEX TYPES

Variations in Chromosome Complement

In 1949 Murray L. Barr and E. G. Bertram described a small, dark-staining body which was present in nerve cell nuclei in female, but not in male cats. Because of its association with sex, this body was named *sex chromatin*. It is also called a *Barr body* after its discoverer. Subsequent investigations in both cats and humans showed that sex chromatin could be identified in approximately a third or more of the cell nuclei of many tissues and organs in females and could also be readily detected in cells scraped from the lining of the mouth (Figure 7–5). Sex chromatin could also be identified in the polymorphonuclear leucocytes of the circulating blood of females where it appeared as a small body or "drumstick" attached to one of the lobes of the nucleus (Figure 7–6).

The association between the presence or absence of a Barr body and the sex of an individual prompted a search for these structures in the cells of sexually

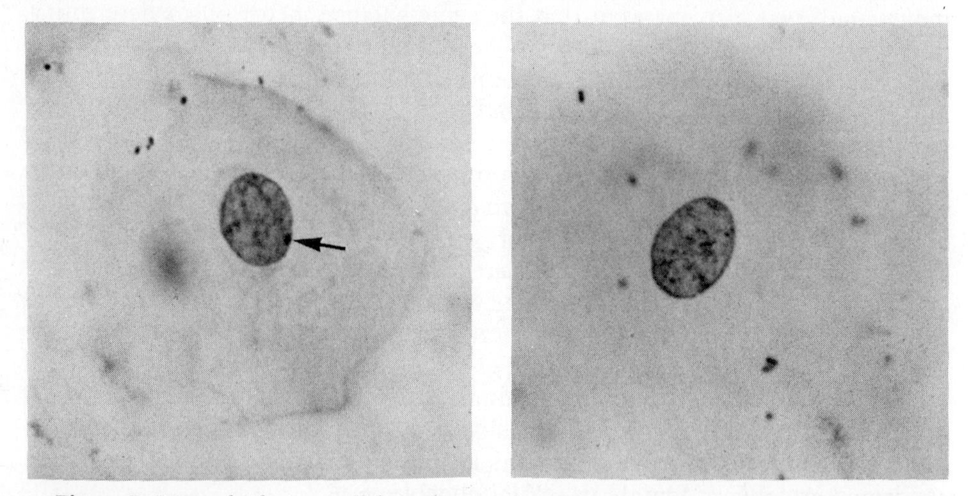

Figure 7–5. Barr body in a cell from human buccal mucosa. Left cell is from a normal female and contains a single Barr body (arrow) adjacent to the nuclear envelope. Cell at right was obtained from a male and contains no sex chromatin. (440×) [From K. L. Moore and M. L. Barr, *Lancet, 2,* 57–58, 1955.]

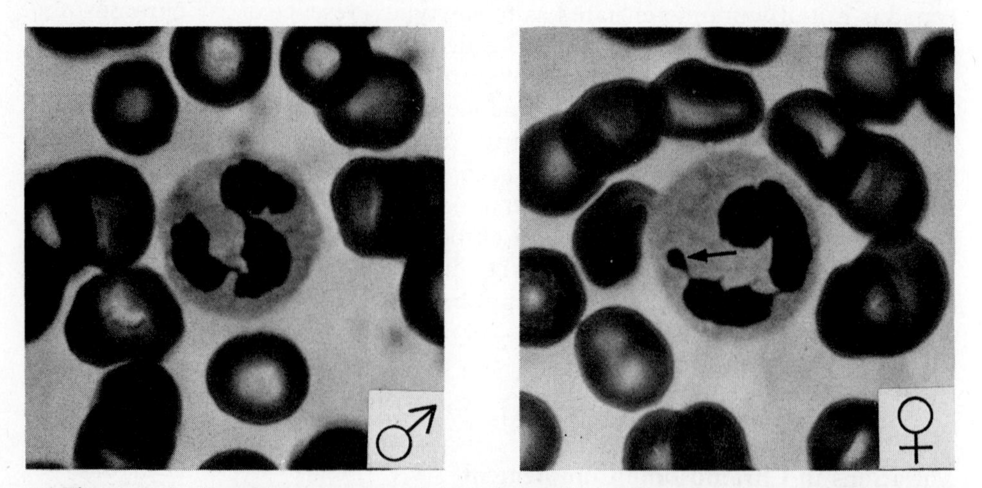

Figure 7–6. Sex chromatin in polymorphonuclear leucocyte (neutrophil) where it appears as a "drumstick" (arrow) in the female. Note that a drumstick is absent in the male. (1600×) [From M. L. Barr, *Prog. in Gynecol., 3,* 131–141 1957. By permission of Grune & Stratton, New York.]

abnormal humans, and it was soon found that X0 females with Turner's syndrome were negative, while XXY males exhibiting Klinefelter's syndrome were positive for sex chromatin. Furthermore, in a female whose chromosome complement included three X chromosomes, two Barr bodies instead of one were found. Further studies of aberrant sex phenotypes indicated the existence of a numerical relationship between the number of X chromosomes and the number of Barr bodies visible in somatic cell nuclei: there was always one less Barr body than the number of X chromosomes present.

To account for this phenomenon, Mary F. Lyon proposed that in both males and females one X chromosome was active and extended in its length and therefore essentially invisible in interphase nuclei. However, any additional X chromosomes present were inactive and condensed and therefore visible in the nucleus as sex chromatin bodies. This concept provided an explanation for the observed numerical relationship between X chromosomes and Barr bodies. Normal XY males and X0 females, having but one X chromosome, would indeed lack sex chromatin according to this hypothesis. Normal XX females and XXY males, having two X chromosomes, should evidence one Barr body, and the cells of individuals with three X chromosomes would be expected to contain two such bodies. The hypothesis that Barr bodies are in fact condensed, heterochromatic X chromosomes has since been verified (see below).

This relationship between X chromosomes and Barr bodies is of great clinical and diagnostic importance. It has also permitted the rapid screening of large numbers of individuals, thereby providing information as to the frequency with which abnormal sex chromosome complements occur in the population. In addition, the recently developed technique of quinacrine staining has made possible the identification of the Y chromosome in somatic interphase nuclei (Figure 7–7). It will be recalled that the Y chromosome, and

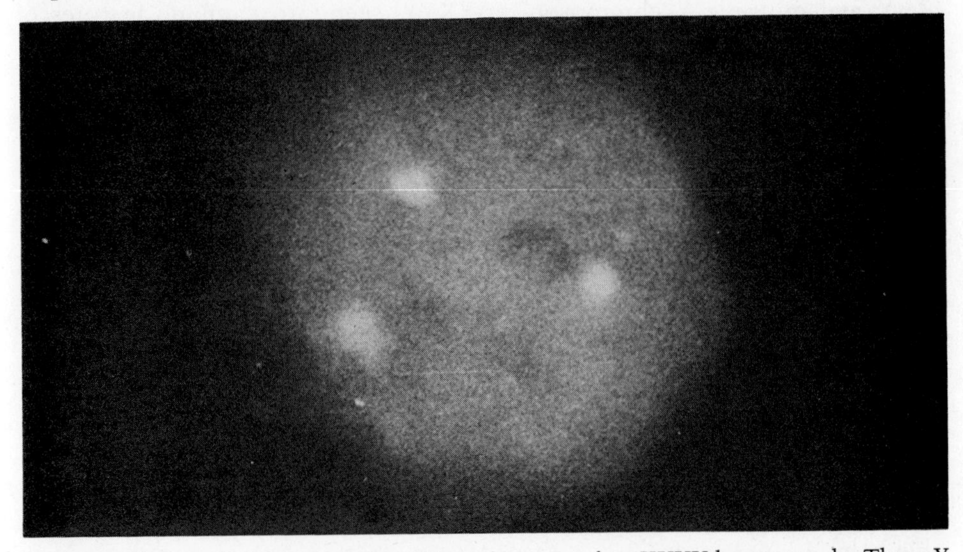

Figure 7–7. Quinacrine staining of lymphocytes of an XYYY human male. Three Y chromosomes can be seen in the nucleus as bright fluorescent bodies. [From G. S. Schoepflin and W. R. Centerwall, *J. Med. Genet.*, *9*, 356–380, 1972.]

especially its long arm, exhibits a brilliant fluorescence with this staining reagent. The use of this technique in conjunction with basic staining for sex chromatin now provides a rapid method for determining the Y as well as the X chromosome constitution of a patient.

Besides the well-known Turner's and Klinefelter's syndromes, other abnormal numbers of sex chromosomes have been identified and associated with phenotypic manifestations. One such instance is the XYY chromosome complement which has been reported to occur in significantly higher frequency in inmates of penal institutions than in the population at large. This finding has led to the notion that the Y chromosome is associated with aggression and that the presence of an extra Y results in an inherent predisposition toward antisocial behavior. This conclusion is as yet unwarranted, for the causes of aggressive behavior in institutionalized XYY individuals have not been unequivocally established. It is nonetheless true, however, that the frequency of these individuals in maximum security institutions is around 3 percent, whereas the frequency in surveyed newborn males is about 0.15 percent. XYY individuals are usually taller than average, with normal to subnormal intelligence. In those studied antisocial behavior seems to occur at a relatively early age and is directed more often at objects than at persons.

Other, rare variations in sex chromosome complement have also been reported, such as XXYY, XXXXYY, XXXX, and even XXXXX. Females with three X chromosomes occur with a frequency of one per thousand female births and are usually normal; females with four or five X chromosomes are mentally retarded, but exhibit normal sexual development. Males with multiple X chromosomes are mentally deficient and exhibit a sexual phenotype similar to that found in Klinefelter's syndrome. Males with multiple Y chromosomes are usually fertile, but mentally retarded and exhibit the tall stature and aggressive behavior of the XYY type. The phenotype of males with multiple X and multiple Y chromosomes (XXYY) seems to combine features of Klinefelter's syndrome with those associated with the XYY syndrome, in that such persons are sexually underdeveloped, mentally deficient, unusually tall, and found with higher frequency in penal institutions.

The origin of all of these variations in chromosome number can be explained most simply as a consequence of nondisjunction of the sex chromosomes at the first, the second, or both meiotic divisions. Some consequences of nondisjunction with respect to the sex chromosome constitution of the gametes are illustrated in Figure 7–8. It can be seen that if such abnormal gametes were to participate in fertilization, a variety of sex chromosome complements could be produced. That such abnormalities are rare attests to the stability of the meiotic process. It should be noted, however, that the frequency of nondisjunction of the sex chromosomes, as well as of the autosomes, rises with increasing maternal age (Figure 7–9). In addition, occasionally it has been possible to trace the source of nondisjunction to one parent or the other by establishing pedigree relationships based on the phenotype produced by sex-linked genes. In such attempts the *Xg* blood group system has been of considerable help.

Nondisjunction can also occur during mitosis in the cells of the zygote thus giving rise to *mosaic* individuals, that is, individuals whose cells differ from one another in regard to chromosome complement. Turner's syndrome in particular is thought to arise more frequently through mitotic than meiotic nondisjunction, for such individuals often have both XX and X0 type cells. In addition, no association with maternal age is evident for Turner's syndrome. Mosaicism caused by mitotic nondisjunction is also the likely origin of the rare human hermaphrodite that possesses ovarian as well as testicular tissue since, in addition to XX-containing cells, some XY-containing cells are also usually found.

The mosaics produced in *Drosophila* by mitotic nondisjunction of the X chromosome in XX females differ sharply from those found in humans. In *Drosophila* sex is an autonomous cellular phenotype determined by the number of X chromosomes versus sets of autosomes, and the loss of an X changes a cell and its descendents from female (XX) to male (X0). If this loss occurs during early development, spectacular mosaics, called *gynandromorphs,* are produced. These can be bilateral, wherein one side of the body is female and the other side male, or the disposition of male and female portions of the body can be anterior versus posterior. Gynandromorphs with smaller proportions of male tissue are also found, the amount of such tissue being dependent on the time and stage in development at which nondisjunction or chromosome loss not related to nondisjunction occurred. Gynandromorphs can be produced in *Drosophila* when female zygotes contain an abnormal X chromosome in the form of an unstable ring along with a normal

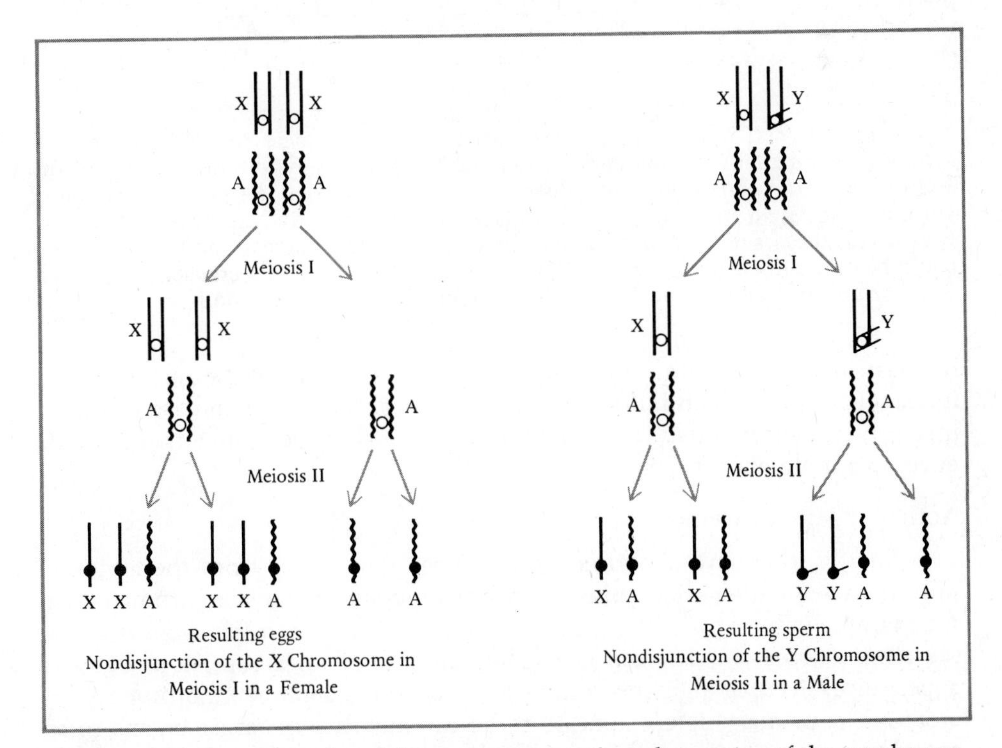

Figure 7–8. Two examples of the consequences of nondisjunction of the sex chromosomes. X and Y are the sex chromosomes; A represents the autosomes.

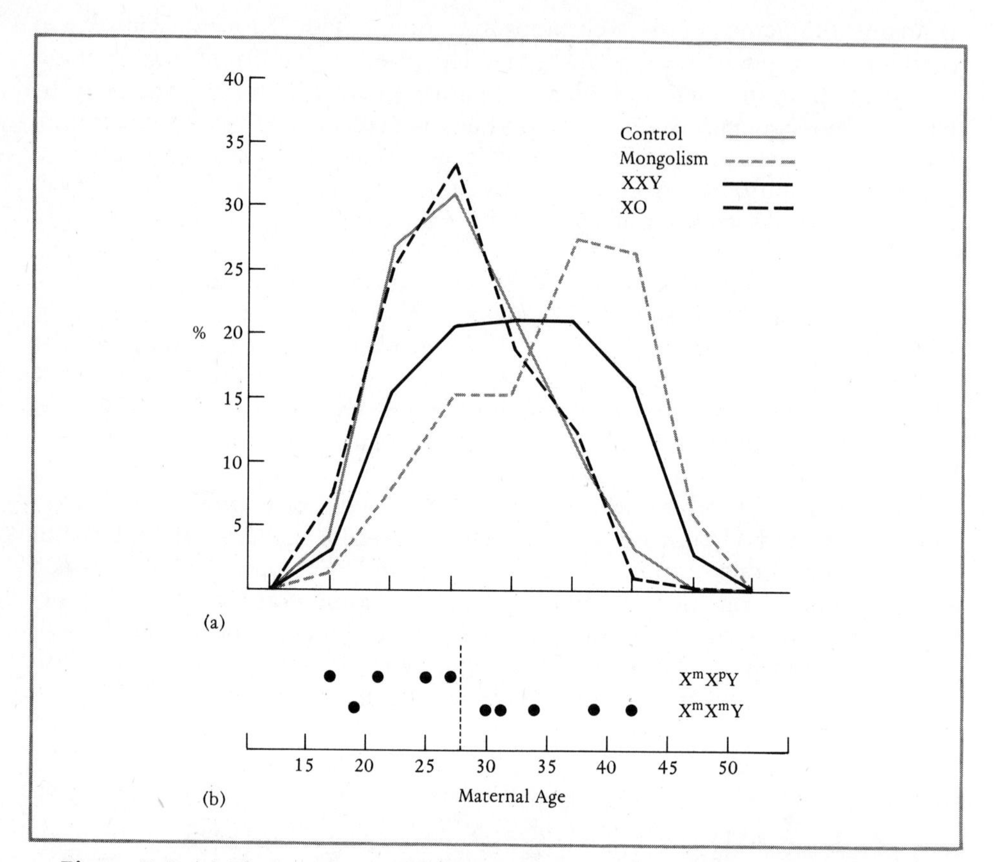

Figure 7–9. (a) The percentage distribution of maternal ages in XXY Klinefelter's syndrome compared with the distributions in XO Turner's syndrome, in mongolism, and in a control population. Mongolism is a condition traceable to nondisjunction of chromosome 21, an autosome of humans (see Chapter 9). (b) The distribution of maternal ages in 10 cases of XXY individuals where the parental source of the X chromosome could be determined as maternal (X^m) or paternal (X^p). [From M. Ferguson-Smith, *The Sex Chromatin,* K. L. Moore, Ed., W. B. Saunders, Philadelphia, 1966.]

rod-shaped X. At chromosome duplication the ring chromosome may form interlocked rings which fail to separate at mitotic telophase and are lost from nuclei of daughter cells. The daughter cells and their descendents, being X0, develop a male phenotype.

Action of Mutant Genes

Finally, there are abnormal sexual types that result from the action of mutant genes, rather than from the nondisjunction of the sex chromosomes. An example in humans is the condition known as *testicular feminization*. The sex chromosome constitution is XY, but the individual is female in outward appearance, even though the gonads are testes and male sex hormone is secreted. The primary defect appears to be the inability of target organs to respond to androgen stimulation. Such individuals are sterile and thus pedigree analysis cannot be performed. However, an equivalent mutation in the mouse is clearly sex linked, suggesting X linkage of the human mutant as well. In *Drosophila* an analogous phenomenon occurs due to the action of a recessive autosomal mutant called *transformer*. When *transformer* is present in the homozygous condition in XX females, the phenotype develops into that of a sterile male.

MECHANISMS CONTROLLING THE ACTIVITY OF WHOLE CHROMOSOMES

X Chromosome Inactivation

The proposal that a Barr body is formed by the condensation of an X chromosome was supported by cytological observations as well as by the numerical relationship between the X chromosomes and Barr bodies. It was observed that during mitotic prophase in the cells of females, one chromosome could be distinguished from all others by its greater degree of contraction. Like the Barr body, this chromosome was characteristically situated adjacent to the nuclear envelope. In XXX and in XXXX females two or three such condensed chromosomes, respectively, could be observed. However, in normal males, having only a single X, such a condensed chromosome was lacking. In addition, with the use of isotopically labeled precursors of DNA, it was shown that one of the two X chromosomes replicated later than all other chromosomes in that period of interphase during which chromosomes are synthesized, and when three X chromosomes were present, two of them were late replicating. In contrast, no such chromosomes were evident in normal males.

Evidence that a condensed X chromosome was inactive with respect to gene expression was first obtained in mice heterozygous for the dominant and recessive coat color alleles, wild type and albino. In this strain of mice the autosome containing the wild type allele had been broken and a segment which included this allele had been transferred to one of the X chromosomes, a phenomenon known as *translocation* (see Chapter 9). In its normal location, the wild type allele is dominant over albino and is always expressed in the phenotype. However, in female mice in which this allele had become transferred to one of the X chromosomes, the coat color was variegated, containing patches of albino and patches of normally pigmented hair.

This observation suggested that the wild type allele had been inactivated in those cells producing albino hair, thus permitting expression of the recessive allele. Cytological study of cells derived from these white patches revealed that the particular X chromosome which was condensed in these cells was that chromosome to which the autosomal segment was attached. It was therefore proposed that the condensation of an X chromosome inactivated or prevented the expression of genes contained within that chromosome. Since the patches of pigmented and nonpigmented fur were roughly the same in number, it was also apparent that chromosome condensation could involve either of the two X chromosomes with equal frequency (Figure 7–10).

Conclusive evidence that one of the X chromosomes of the female was not only condensed, but also for the most part inactivated was obtained by R. G. Davidson and associates in studies of cells derived from human females

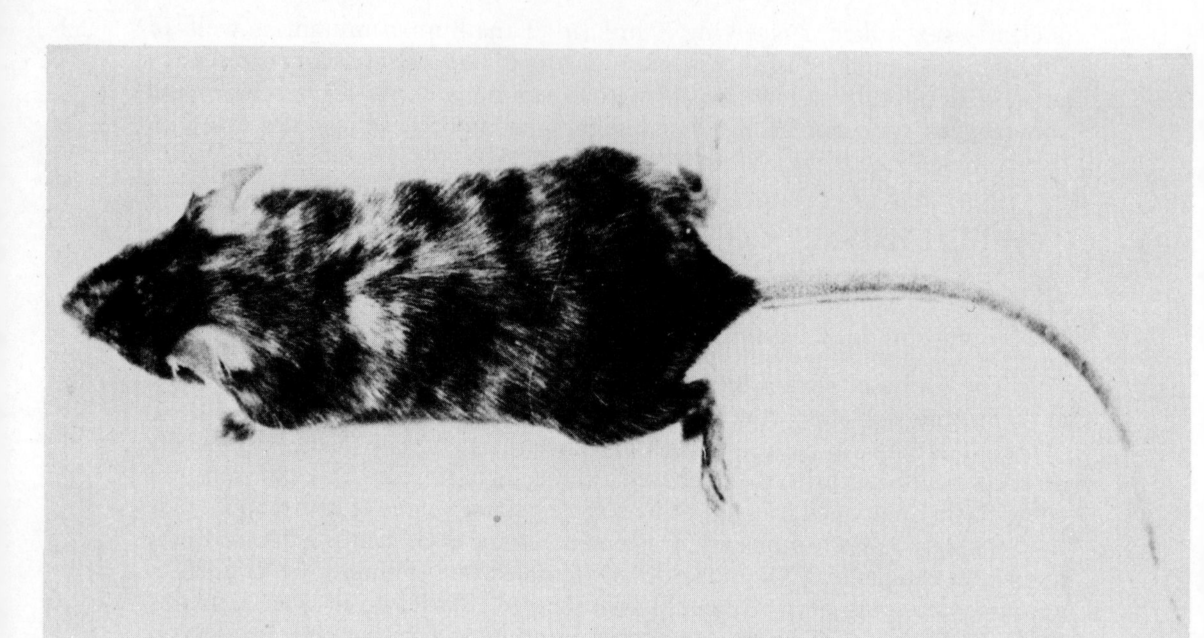

Figure 7–10. Mosaic phenotype of a female mouse heterozygous for Tortoise, a dominant sex-linked mutant affecting coat color. Note that the patches of white fur do not cross the dorsal midline as if inactivation occurred separately on each side of the body. [From M. W. Thompson, *Can. J. Genet. Cytol.*, *7*, 202–213, 1965.]

heterozygous for sex-linked genes. In a case involving a woman heterozygous for two variants of the enzyme G-6-PD, tissue taken from the woman was analyzed and initially both enzyme variants were found to be present. However, when single cells were cultured and allowed to divide and form a group of identical descendents, called a *clone,* each clone contained either one enzyme variant or the other, but never both. Since the allelic genes responsible for the production of the enzyme variants were located, one on each of the two X chromosomes, the absence of one form of the enzyme in cells from a single clone indicated nonexpression, that is, inactivation, of the allele responsible for its production. The fact that both enzyme variants were present in the initial random mixture of uncloned cells again indicated that chromosome inactivation could involve either X with equal frequency.

Similar experiments utilizing other sex-linked genes of humans as well as of mice have yielded equivalent results and indicate that many of the genes of the condensed X chromosome of the female are not expressed in the phenotype of somatic cells. Thus mammalian females are phenotypic mosaics with respect to the expression of most sex-linked traits.

Inactivation is thought to involve most, but not all genes of the condensed X chromosome, for there is evidence that at least some genes of this chromosome, for example, the Xg blood group alleles, are expressed in somatic cells. Inactivation of an X chromosome occurs in very early embryonic stages, and in any cell the choice as to which of the two X chromosomes will undergo condensation is apparently random. It has also been shown that once a given X chromosome is inactivated, it remains inactivated in that cell and in all subsequent daughter cells. These various facets of the phenomenon of X chromosome inactivation have been summarized by Lyon and are often referred to as the Lyon hypothesis.

It must also be noted that X chromosome condensation apparently does not occur in cells of the germ line in females, for Barr bodies are not found in oöcytes. Presumably, differentiation of the oöcyte requires the active participation of genes on both X chromosomes. While the single X of normal males remains extended and active in somatic cells, it is condensed in spermatocytes and may be largely inactivated throughout the process of spermatogenesis, although, as in the case of the female, it is likely that some genes of the condensed X in spermatocytes remain active.

Dosage Compensation

The phenomenon of X chromosome inactivation has much broader significance than that of clinical utility, for it provides a means whereby the activity and expression of X-linked genes can be equalized in the somatic cells of the two sexes. It must be remembered that, with some few exceptions, the Y chromosome bears essentially no genes allelic to those on the X, and consequently in XY males each X-linked gene is present in only a single dose. Females, however, possess two X chromosomes and therefore two doses of X-linked genes. If all of the genes on the X chromosome were concerned with sex, the usefulness of this arrangement might be conceded. However, if one reviews the phenotypic actions of genes present on the X chromosome, it is readily apparent that most have nothing whatever to do with sexuality, but instead are concerned with functions common to both sexes.

The X-linked gene responsible for the synthesis of glucose-6-phosphate dehydrogenase is a good example, for this enzyme is a required participant in the metabolic activity of all cells, male or female. Because females have two doses of this gene, as compared to only one dose in males, one might expect the level of activity of this enzyme to be twice as high in females as in males. Yet, the activity of this enzyme in the two sexes is exactly the same. The obvious explanation, of course, is that due to the inactivation of one set of X chromosome genes in the female, the difference in gene dosage has been equalized. This mechanism is known as *dosage compensation,* and in mammals it permits the establishment of a common cellular physiology in functions unrelated to sex.

Dosage compensation also occurs in *Drosophila* and probably, at least to some extent, in many species where the dosage of genes serving general metabolic functions is not equivalent in the two sexes. In *Drosophila* X chromosome inactivation does not occur in the somatic cells of females. Instead, there is considerable evidence that X-linked genes in males compensate by doubling their activity. Also, analysis of phenotypes produced by sex-linked genes in females suggests that a reduction in gene activity occurs. It is proposed that control over X chromosome gene expression is exerted both by autosomal and by X-linked genes.

The existence of dosage compensation, whether in mammals or *Drosophila,* points to the presence in these organisms of a regulatory system capable of controlling the activity of whole chromosomes. By whatever means this regulation is accomplished, it is clearly responsible for X chromosome inactivation and late replication in mammals and for the enhanced activity of X-linked genes in *Drosophila* males and the reduced activity of such genes in *Drosophila* females.

In addition, sustained regulation of chromosome function is evident, for in the somatic cells of female mammals X chromosome inactivation is irreversible and results in the continuing suppression of gene activity through many cell generations. The selective nature of chromosome regulation is also apparent. In female mammals X chromosome inactivation occurs only in somatic cells and not in oöcytes, while in all heterogametic males thus far studied condensation of the X regularly occurs only in spermatocytes. From all of these considerations, we can conclude that although the nature of the mechanisms which control the activity of whole chromosomes is yet obscure, the existence of such mechanisms cannot be in doubt.

Meiotic Drive

Deviations in the expected 1 : 1 sex ratio can sometimes be attributed to meiotic drive, a phenomenon in which one chromosome is recovered in the progeny with a higher frequency than that expected on the basis of chance. Meiotic drive has been analyzed extensively in *Drosophila*. For example, the presence in males of an X-linked mutation called sex ratio *(sr)* results in the almost exclusive production of female offspring. Consistent with these genetic findings are the observations, using electron microscopy, that half of the sperm degenerate in males carrying the *sr* gene; the aborted sperm presumably represent those which would have carried the Y chromosome. In another case of meiotic drive associated with the sex chromosomes (the sc^4sc^8 X chromosome), males appear with less than half the expected frequency. The cause of this abnormality has been attributed to a failure in meiotic pairing of the X and Y chromosomes, correlated with nondisjunction and the subsequent aberrant development of spermatids. Whereas both of these instances of meiotic drive in *Drosophila* are associated with structurally rearranged X chromosomes, the *sr* condition is found in natural populations of *Drosophila*.

Although specific meiotic mutants or X chromosome abnormalities directly causing deviations in the sex ratio are thus far unknown in humans, there is some evidence suggesting the presence of meiotic drive mechanisms. As an illustration, XYY human males are expected to produce normal XX and XY children along with XYY sons and XXY Klinefelter type sons. However, it has been reported that the offspring of such XYY fathers are almost always normal with respect to sex chromosome constitution and that the abnormal XYY and XXY types are rare. This observation would suggest that XYY males arise only via early mitotic events in XY males or that one Y may be lost premeiotically so that all primary spermatocytes are XY or that XY and YY sperm are nonfunctional. Whatever the cause, normal sex chromosome complements are found in the offspring in much higher frequency than that expected on the basis of chance.

SEX CHROMOSOMES IN REPRESENTATIVE DIPLOID FORMS

Many of the lower plants (mosses and liverworts) are dioecious, having separate sexes which produce either male or female spores. In some species (for example, *Sphaerocarpus*) the haploid gametophyte generation is distinguished as female or male by the presence of an X or Y chromosome. Most flowering plants are hermaphrodites in that male and female structures are found in the same flower. When found in separate flowers but on the same plant, the term monoecious is applied, corn *(Zea)* being an example of this state. Far fewer flowering plants are dioecious, having male and female floral organs on separate plants. Dioecious plants with recognized sex chromosomes include spinach, *Ginkgo, Bryonia,* a member of the cucumber family studied by Correns, the campions *(Silene),* the hop plant *(Humulus),* sheep's sorrel *(Rumex),* hemp *(Cannabis),* and the wild (not cultivated) strawberry *(Fragaria)*. The male is the heterogametic sex in most of these examples, although the female is heterogametic in *Fragaria*.

The sex-determining mechanism of a variety of insects has been investigated. Most exhibit the XY or X0 type of sex determination, the male being the heterogametic sex. However, in moths and butterflies (Lepidoptera), the female is heterogametic, as determined by breeding experiments in which the pattern of inheritance of sex-linked genes has been followed. Chromosomes of Lepidoptera are numerous and difficult to study, and specific sex chromosomes are not cytologically distinguishable. Although male-determining genes are thought to occur on the X (Z chromosome) and female determining genes on the Y (W chromosome), the final sex of an individual appears to depend on the balance between male and female factors, some of which are autosomal. Intersexes are produced in crosses of moths obtained from different geographical regions suggesting that sex-determining genes of different degrees of expression occur in different populations.

In Hymenoptera (bees, wasps, ants) haploid parthenogenesis is the usual form of sex determination. Fertilized eggs develop into females and unfertilized eggs into males. However, the cells of males do not necessarily remain haploid in constitution. Although dividing cells appear to be haploid, chromosome duplication occurs in the absence of cell division in many tissues of the body leading to the presence in such cells of multiple copies of each chromosome. A similar type of parthenogenesis also occurs in scale insects (coccids) and in mites and ticks.

In the lower vertebrates sex chromosomes are not always differentiated. The XX (♀) – XY (♂) type of sex determination is commonly found, but in certain fishes, in some amphibians such as *Xenopus,* and in birds, the female is XY and the male XX. Heteromorphic sex chromosomes also occur in females of a number of species of reptiles.

In marsupials and mammals the female is homogametic (XX) and the Y chromosome is required for the development of the male phenotype. Dosage compensation via the condensation of one of the X chromosomes is a constant feature.

Susumu Ohno has proposed that genes on the X chromosome represent a linkage group that has been conserved during the evolution of placental mammals. This proposal predicts the presence of extensive homology between X-linked genes in different mammalian species. Such homologies have indeed been found for G-6-PD, hemophilia A and B, and several other inherited traits. This hypothesis is most interesting, and the reader is referred to the publications of Ohno cited at the end of this chapter for a detailed discussion.

PROBLEMS

7–1. In *Drosophila* carnation *(c)* is a recessive eye color mutant and short wings *(s)* is a recessive wing mutant. A carnation-eyed female, otherwise wild type, was crossed to a normal-eyed, short-winged male. The male progeny all had carnation eyes, but half had short wings and half had normal wings. All female progeny had normal red eye color, but half had short wings and half had normal wings.

a. What was the genotype of the parental female?
b. What was the genotype of the parental male?
c. What are the genotypes of the male and female progeny?

7–2. A male with reduced eyes was crossed to a female with normal eyes. The F_1 consisted of a total of 67 males, all with normal eyes, and 65 females, all with reduced eyes. How would you explain the inheritance of this character?

7–3. Assume in *Drosophila* the recessive mutants purple eyes *(p)* and jagged wings *(j)*. A phenotypically wild type female was crossed to a phenotypically wild type male and the progeny were the following:

FEMALES	MALES
74 wild type	36 wild type
23 jagged wing	14 jagged wing
	37 purple eyes
	12 purple eyes, jagged wing

What were the probable genotypes of the parental male and female?

7–4. White *(w)* is a sex-linked recessive gene of *Drosophila* which blocks eye pigment formation, while scarlet *(st)* is an autosomal recessive gene which results in bright red eyes when homozygous. Wild type eye color is dull red. A white-eyed male heterozygous for scarlet was crossed to a scarlet-eyed female heterozygous for white. What phenotypes are expected in the F_1 males and in what proportions?

7–5. Vermilion *(v)* is a sex-linked recessive gene causing bright red eye color. The phenotype produced by vermilion is essentially indistinguishable from that produced by scarlet *(st)*, an autosomal recessive gene. Normal eyes are dull red. A bright-red-eyed female was crossed to a bright-red-eyed male. Of the female offspring, $\frac{3}{4}$ had bright red eyes and $\frac{1}{4}$ had dull red eyes. Of the male offspring, $\frac{3}{4}$ had bright red eyes and $\frac{1}{4}$ had dull red eyes. What were the genotypes of the parents?

7–6. Suppose the cross black-eyed females × normal-eyed males yielded $\frac{1}{2}$ black and $\frac{1}{2}$ normal males and $\frac{1}{2}$ black and $\frac{1}{2}$ normal females. What hypotheses can you construct from these data to explain the inheritance of black eyes?

7–7. In *Drosophila* normal body color is tan and normal bristles are long. A female with short bristles and tan body color was crossed to a male with long bristles and black body color. The progeny were:

FEMALES	MALES
20 short, tan	22 long, tan
19 short, black	20 long, black
21 long, tan	
18 long, black	

a. Is the gene for short bristles dominant or recessive? Could it be either?
b. Is the gene for black body color dominant or recessive? Could it be either?
c. Postulate the genotypes of the parental male and female.
d. Why are there no short-bristled males in the F_1 progeny?

7–8. The following hypothetical data concern the phenotype of brown-colored eyes in *Drosophila*.

(1) A brown-eyed female crossed to a homozygous normal male produced 40 normal females, 38 brown-eyed females, and 42 normal males.
(2) A second brown-eyed female crossed to a homozygous normal male produced 62 normal females, 62 brown-eyed females, 65 normal males, and 63 brown-eyed males. When the F_1 brown-eyed males and females were inbred, the F_2 progeny consisted of 25 normal females, 49 brown-eyed females, 23 normal males, and 47 brown-eyed males.

a. Is the brown mutant used in cross 1 the same as the brown mutant used in cross 2?
b. If it is not the same, how do the genotypes of the two females differ and how is the trait inherited in each case?

7–9. In *Drosophila* assume a dominant autosomal gene hairless *(H)*, which causes the absence of hair on the head and body. *H/H* is lethal. Another dominant, but sex-linked gene, suppressor of hairless *(Su-H)*, reverses the effects of hairless when both are present in the heterozygous condition, so that the wild type phenotype results. *Su-H* has no phenotypic effect by itself, but is lethal in the hemizygous or homozygous condition. If females heterozygous for both *H* and *Su-H* are crossed to hairless males,
a. What proportion of the male offspring will be viable?
b. What proportion of the viable male offspring will be hairless?
c. What proportion of the female offspring will be inviable?
d. What proportion of the viable female offspring will be wild type in phenotype?

7–10. In fowl barring *(B)* is sex linked and dominant, the recessive allele *(b)* producing solid black color when homozygous. Silky feathers *(s)* is a recessive autosomal gene, as opposed to nonsilky *(S)*. If black cocks heterozygous for silky are crossed to barred, silky hens, what genotypes and phenotypes will be produced and in what proportions?

7–11. In the guppy a dominant gene *(M)* results in the presence of a black spot on the dorsal fin. Spotted males transmit the trait only to sons and not to daughters, and male and female progeny of such daughters do not show the trait. Unspotted parents do not transmit the trait. How is this trait inherited?

7–12. Red-green colorblindness in humans is recessive and sex linked. If a woman heterozygous for colorblindness marries a colorblind man, what is the probability that their first child will be a colorblind daughter?

7–13. A married couple, both of whom had normal vision, produced a colorblind son. Examination of cell samples from the son showed the presence of a Barr body. What is the probable genotype of the son with respect to sex chromosomes and color blindness? What is the simplest explanation which will account for this genotype?

7–14. Classify the following sex and autosomal chromosome complements as to the phenotypic sex type which would result in *Drosophila* and in humans.

a. XXX : 2 A
b. XY : 2 A
c. XXY : 2 A
d. X : 2 A
e. XYY : 2 A
f. XXY : 3 A
g. XXXY : 3 A
h. XX : 3 A
i. XY : 3 A

In the following human pedigrees the inheritance of a single trait or two separate traits is diagramed. For single traits use *A* and *a* to designate genotypes. When two separate traits are involved, use the following key:

Trait 1: ◧ or ◐ , symbols *A* and *a*

Trait 2: ◨ or ◑ , symbols *B* and *b*

7–15. Examine the pedigree.

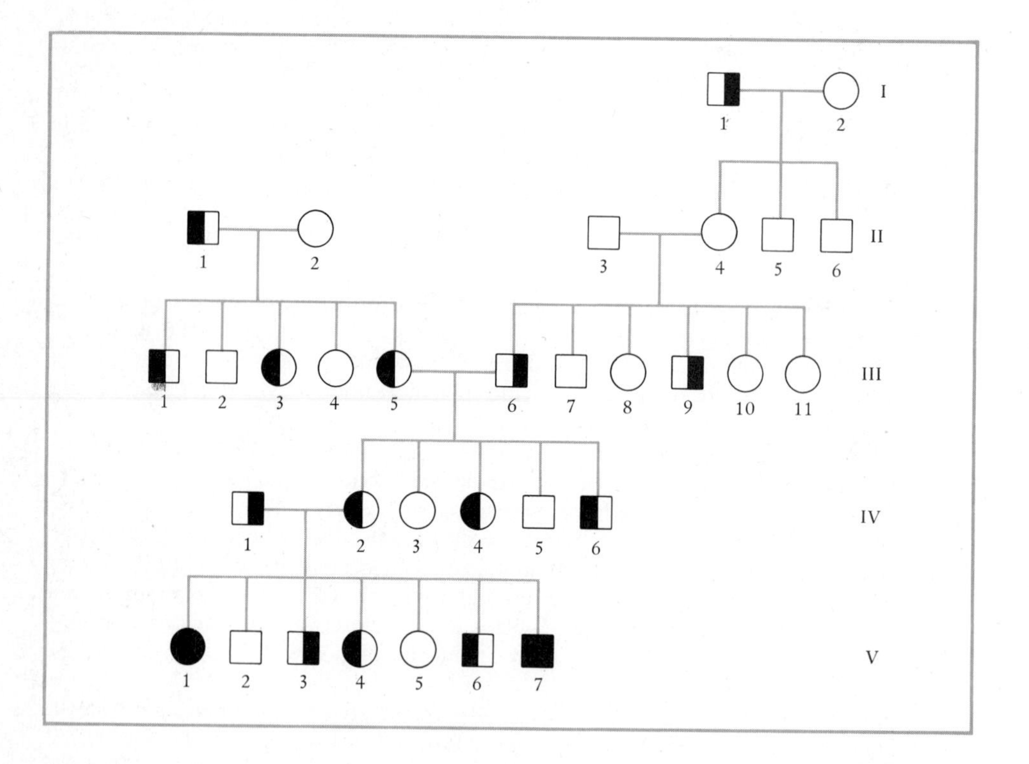

a. Characterize traits 1 and 2 as dominant or recessive and sex linked or autosomal.
b. Give the genotypes of:

individuals 1, 2, 3, and 4 of generation II
individuals 5 and 6 of generation III
individuals 1 and 2 of generation IV
individuals 1 through 7 of generation V

7–16. Examine the pedigree shown below.

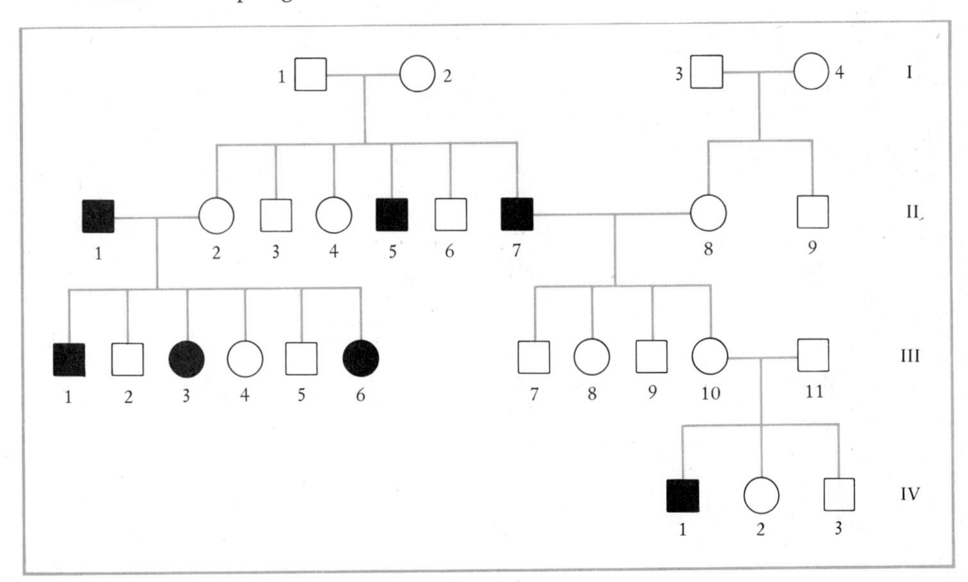

a. Is the trait illustrated in this pedigree dominant or recessive? Is it sex linked or autosomal?
b. What are the probable genotypes of:

    individuals 1 and 2 of generation I
    individuals 1, 2, and 8 of generation II
    individuals 6, 10, and 11 of generation III

7–17. Examine the pedigree.

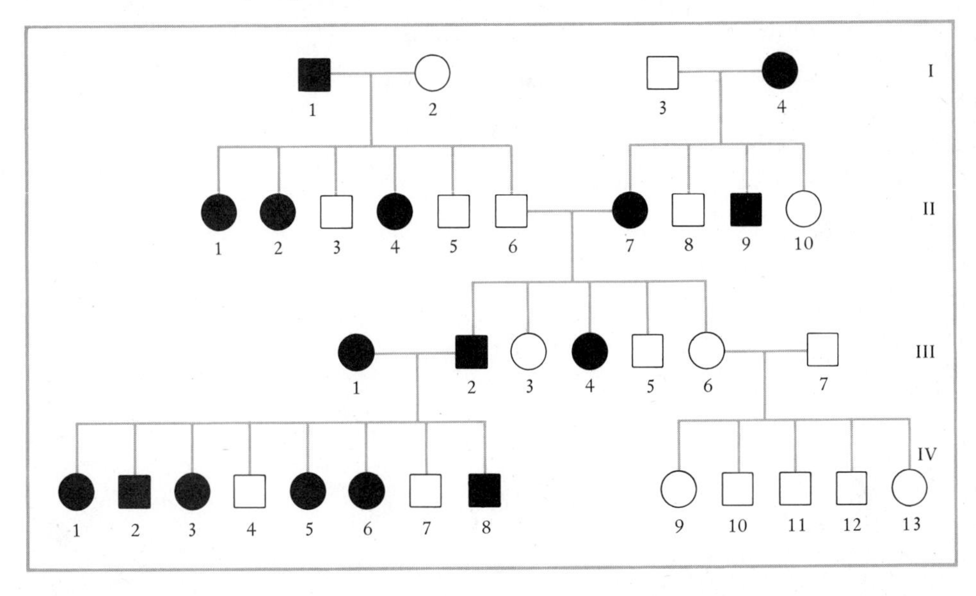

a. Is this trait dominant or recessive? Is it sex linked or autosomal?

b. What are the genotypes of:

individuals 1, 2, and 4 of generation I
individuals 1, 6, and 7 of generation II
individuals 1, 2, 6, and 7 of generation III

7–18. Examine the pedigree.

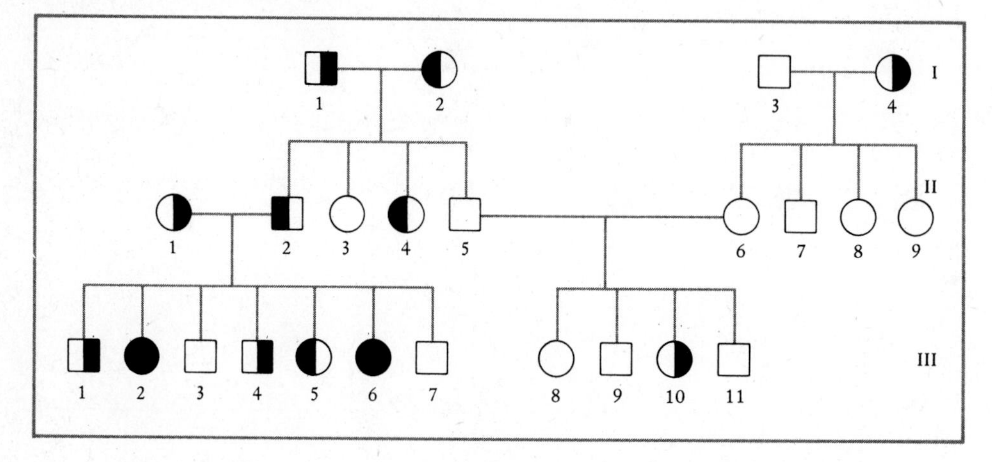

a. Characterize traits 1 and 2 as dominant or recessive and sex linked or autosomal, using the symbols *A* and *a* for trait 1 and *B* and *b* for trait 2.
b. Give the genotypes of:

individuals 1, 2, and 4 of generation I
individuals 1, 2, 5, and 6 of generation II
individuals 1, 6, and 10 of generation III

7–19. Examine the pedigree.

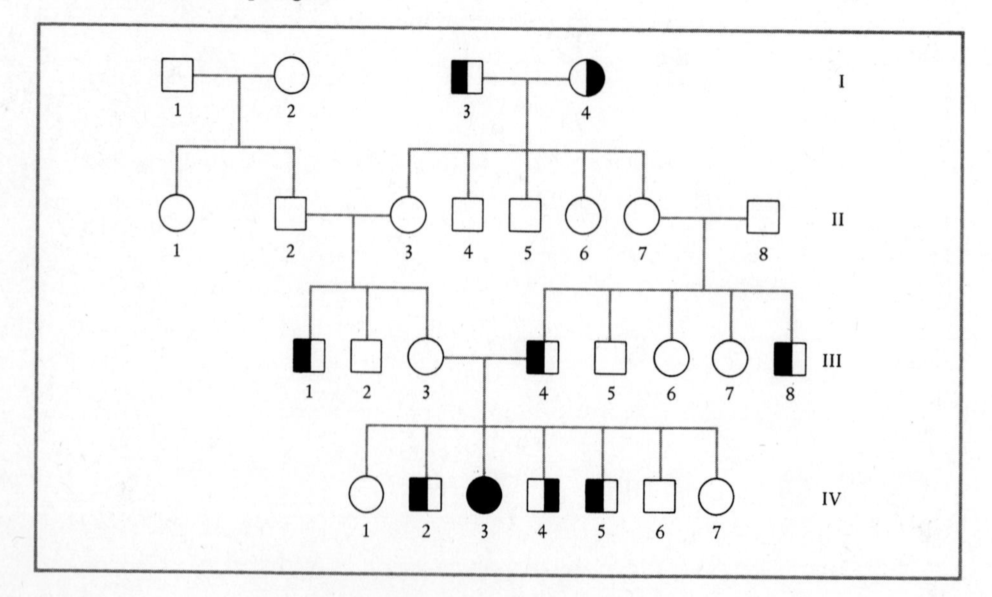

a. Characterize traits 1 and 2 as dominant or recessive and sex linked or autosomal.
b. Give the genotypes of:

individuals 3 and 4 of generation I
individuals 2, 3, 7, and 8 of generation II
individuals 3 and 4 of generation III
individuals 2, 3, and 4 of generation IV

7–20. What might be the origin of 2 X, 2 Y, 1 A sperm? Of 4 X, 1 A eggs? Of 2 X, 1 A sperm?

7–21. Give two probable means by which a zygote with an X0 chromosome complement could originate in humans via meiotic nondisjunction.

7–22. Why is the lethal phenotype of Duchenne muscular dystrophy not found in human females?

7–23. A sex-linked mutant in humans called ocular albinism results in a colorless retina in males and homozygous females. In heterozygous females the retina contains areas that are pigmented as well as areas that are albino. How can you explain this phenotype?

REFERENCES

BRIDGES, C. B., 1916. Non-disjunction as proof of the chromosome theory of heredity. *Genetics, 1,* 1.

HOOK, E. B., 1973. Behavior implications of the human XYY genotype. *Science, 179,* 139.

LEVINE, H., 1971. *Clinical Cytogenetics.* Little, Brown, Boston.

LEWIS, K. R., AND B. JOHN, 1968. The chromosomal basis of sex determination. *Int. Rev. Cytol., 23,* 277.

LYON, M. F., 1968. Chromosomal and subchromosomal inactivation. *Ann. Rev. Genet., 2,* 31.

LYON, M. F., 1972. X-chromosome inactivation and developmental patterns in mammals. *Biol. Rev., 47,* 1.

LUCCHESI, J. C., 1973. Dosage compensation in *Drosophila. Ann. Rev. Genet., 7,* 225.

MITTWOCH, U., 1967. *Sex Chromosomes.* Academic Press, New York.

MITTWOCH, U., 1973. *Genetics of Sex Differentiation.* Academic Press, New York.

MOORE, K. E. (Ed.), 1966. *The Sex Chromatin.* Saunders, Philadelphia.

MORGAN, T. H., 1910. Sex-limited inheritance in *Drosophila. Science, 32,* 120. (Reprinted in J. A. Peters (Ed.), 1959. *Classic Papers in Genetics.* Prentice-Hall, Englewood Cliffs, N.J.)

OHNO, S., 1967. *Sex Chromosomes and Sex-Linked Genes.* Springer-Verlag, Berlin.

OHNO, S., 1969. Evolution of sex chromosomes in mammals. *Ann. Rev. Genet., 3,* 495.

LINKAGE, CROSSING OVER, AND MAPPING

As we have seen, the discovery of sex chromosomes and sex-linked genes greatly strengthened Sutton's hypothesis that Mendelian factors were physically located in the chromosomes. With the continuing identification of new mutants in *Drosophila,* corn, and other species, it soon became obvious that there were many more genes inherited according to Mendelian principles than there were chromosomes, and it was therefore inferred that each chromosome must contain numerous genetic factors. A chromosome could thus be viewed as a group of genes physically associated or linked with one another to comprise a *linkage group.*

LINKAGE AND RECOMBINATION

Since chromosomes occurred as homologous pairs, with the members of any pair bearing like genes, the number of linkage groups in any species could be equated with the number of pairs of chromosomes. For example, Mendel's garden peas have seven pairs of chromosomes and consequently seven groups of linked genes, while *Drosophila,* with four pairs of chromosomes, is considered to have four linkage groups in both sexes, three autosomal and one sex linked. In a strict sense, of course, the genes of the Y chromosome represent an additional, or fifth linkage group, but this group, being normally confined to the male sex, is not found in all individuals.

The location of several mutations on one particular chromosome was first accomplished for the X chromosome of *Drosophila*. It was expected that such X-linked genes, being physically associated, would be transmitted together from parents to progeny. When the parent in question was a female homozygous for two or more X-linked mutants, this was indeed found to be true and all sons inherited such genes as a group. However, when the female was heterozygous for two or more X-linked genes, the expected two types of sons representing each maternal X chromosome were, in addition, accompanied by other sons whose phenotypes demonstrated that new associations between the maternal X-linked genes had occurred. Such new associations are called *recombinations*. To illustrate this phenomenon, two hypothetical examples employing the recessive sex-linked mutants, yellow body *(y)* and vermilion eyes *(v)* can be presented:

Let us assume a heterozygous female where one X chromosome contains both recessive mutants (*y*, *v*) and the other X chromosome contains the wild type alleles of these mutants (+), the genotype of the female thus being: $\frac{y\ v}{+\ +}$. If such a female is crossed to a wild type male $\left(\frac{y\ v}{+\ +} \times \frac{+\ +}{\rightharpoondown}\right)$, we might expect the following progeny on the basis of standard sex-linked inheritance:

1. All daughters should be wild type in phenotype $\left(\frac{y\ v}{+\ +} \text{ and } \frac{+\ +}{+\ +}\right)$ because each will receive the paternal X chromosome which carries the dominant alleles.

2. Sons should be either yellow, vermilion ($\underline{y\ v}$) or wild type ($\underline{+\ +}$), in a 1 : 1 ratio, the phenotype in each case being dependent on which of the two maternal X chromosomes has been inherited.

3. The number of daughters should be equal to the sum of the two classes of sons.

However, if this cross is performed, the actual progeny might be the following:

FEMALES	MALES	
100 all wild type	35 yellow, vermilion 36 wild type	parental types
	15 yellow 15 vermillion	nonparental types
	101	Total

Examining the data from this cross, we can see that all the daughters correspond to expectations and that the anticipated yellow, vermilion and wild type males do occur with equal frequency and appear to have inherited an unchanged maternal X chromosome. They can therefore be designated as *parental types*. However, contrary to expectations, such males comprise only 70 percent of the male progeny, the other 30 percent being found in two additional classes: yellow males, whose X chromosome necessarily contains the recessive allele *y* along with the wild type allele of *v* ($\underline{y\ +}$), and vermilion males, whose X chromosome must carry the wild type allele of *y* along with the recessive mutant *v* ($\underline{+\ v}$). Since the X chromosome constitution of these two classes of males is different from that of either of the X chromosomes of the female parent, these male progeny can be termed *nonparental* or *recombinant* types. They appear in a 1 : 1 ratio and are reciprocals of one another. The new X chromosome constitutions evident in these nonparental males are necessarily derived from the heterozygous mother. Consequently, the presence of these X chromosomes in male offspring suggests that exchanges producing new associations between linked genes must have occurred between the maternal X chromosomes either prior to or during gamete formation.

For our second example let us again assume a female heterozygous for *y* and *v*. In this case, however, one X chromosome contains the recessive mutant *y* and the wild type allele of vermilion and the other X chromosome contains the wild type allele of yellow along with the recessive mutant *v*. The genotype of this female is thus $\frac{y\ +}{+\ v}$. If such a female is crossed to a wild type male $\left(\frac{y\ +}{+\ v} \times \frac{+\ +}{}\right)$, the following progeny might result:

FMALES	MALES	
100 all wild type	36 yellow 35 vermilion	parental types
	13 yellow, vermilion 14 wild type	nonparental types
	98	Total

The data from this cross shows that all daughters are wild type as expected and that the sons are again distributed in four classes. Two of these classes, together comprising 70 percent of the males, can again be designated as parentals because an unchanged maternal X chromosome has been inherited. The parental types are yellow males $(\underline{y\ +})$ and vermilion males $(\underline{+\ v})$. The two nonparental classes consist of yellow, vermilion males $(\underline{y\ v})$ and wild type males $(\underline{+\ +})$. The new associations of X-linked genes evident in these males can again be attributed to exchanges which have occurred between the X chromosomes of the female parent.

If the progeny of the two sample experiments are compared, it can be seen that the parentals of example 1 are the nonparentals of example 2, and vice versa. The reason for this lies in the initial disposition of the mutants *y* and *v* in the X chromosome of the respective female parents and the consequences of exchange in each case. In example 1 the recessive mutants *y* and *v* were initially located on the same homologue of the mother, a condition sometimes referred to as *coupling* or the *cis* arrangement. The other homologue of the mother contained both wild type alleles. Exchanges between these homologues produced the nonparental X chromosomes $\underline{y\ +}$ and $\underline{+\ v}$. In example 2, however, the recessive mutants *y* and *v* were initially located on opposing homologues of the mother, a condition termed *repulsion* or the *trans* arrangement. Here, exchanges would produce the nonparental X chromosomes $\underline{y\ v}$ and $\underline{+\ +}$. It should also be noted that in both sample experiments parentals comprise 70 percent and nonparentals 30 percent of the male progeny. From this it can be inferred that the initial disposition of alleles on the two X chromosomes of the mother does not influence the frequency of exchange between these homologues.

That an exchange between homologous chromosomes could occur during meiosis had been proposed in 1903 by De Vries, and evidence suggesting such an exchange had been obtained in studies of plants by W. Bateson, E. R. Saunders and R. C. Punnett in 1905. Results similar to those presented above, but much more extensive, were reported by T. H. Morgan in 1910 and 1911 based on work using *Drosophila* as the experimental organism. In crosses involving the sex-linked genes white eyes and miniature wings Morgan found that nonparentals constituted 36.9 percent of the progeny. However, in crosses involving yellow body and white eyes nonparentals totaled only 1.3 percent of the progeny. In explanation Morgan suggested that the frequency of exchange between two given linked genes was directly related to their respective positions or *loci* (*sing*. locus) in the chromosome. Thus, genes situated close together would rarely be separated by an exchange and would therefore tend to be inherited together most of the time. On the other hand, exchanges between genes situated farther apart from one another might occur with greater frequency, thus accounting for the presence of larger numbers of nonparental types in this case.

Morgan called the process of genetic exchange crossing over, and his view of this event was based on a theory proposed in 1909 by F. A. Janssens. In cytological studies of the meiotic process in amphibians, Janssens had observed that at the site of any particular chiasma in the diplotene stage of meiosis I, only two of the four chromatids of any one tetrad were involved and, moreover, that these two were nonsister in origin. He proposed that chiasmata represented sites of exchange between chromatids of opposing homologues. His hypothesis, called the *chiasmatype* theory, provided an explanation for the source of the observed nonparental (crossover) classes. Janssens's theory, with some modifications, has since been shown to be essentially correct (see Chapter 14). Janssens proposed that exchanges between homologues occurred only in the four-strand (tetrad) stage of meiosis, after each homologue had undergone duplication. He visualized the exchange process as a breakage and rejoining phenomenon which involved only two of the four chromatids of the tetrad at any site. Extensive genetic and cytological studies carried out over the past 60 years have demonstrated the correctness of this interpretation.

Janssens suggested that diplotene was the stage at which exchange took place, but it was later found that crossing over occurs earlier, probably during the time when homologues are precisely paired with one another in synapsis. More recently, the time of crossing over has been associated with the presence of synaptinemal complexes, structures which are thought to provide a framework facilitating the process of exchange. In this regard, Morgan and his coworkers discovered in 1912 that crossing over did not occur in *Drosophila* males, all genes of this sex showing *complete linkage*. This finding can now be correlated with the demonstrated absence of synaptinemal complexes in the male sex of this organism. In most other animals crossing over occurs in both sexes.

If we apply Janssens's concept of crossing over to the hypothetical data presented in our first example of recombination, we can visualize the origin of the nonparental individuals. In Figure 8–1 the X chromosomes of the parental female of example 1 are shown as a tetrad of chromatids, numbered 1 through 4, with the sister strands of each homologue attached to a single centromere. Since a given crossover event can involve any two nonsister chromatids, all

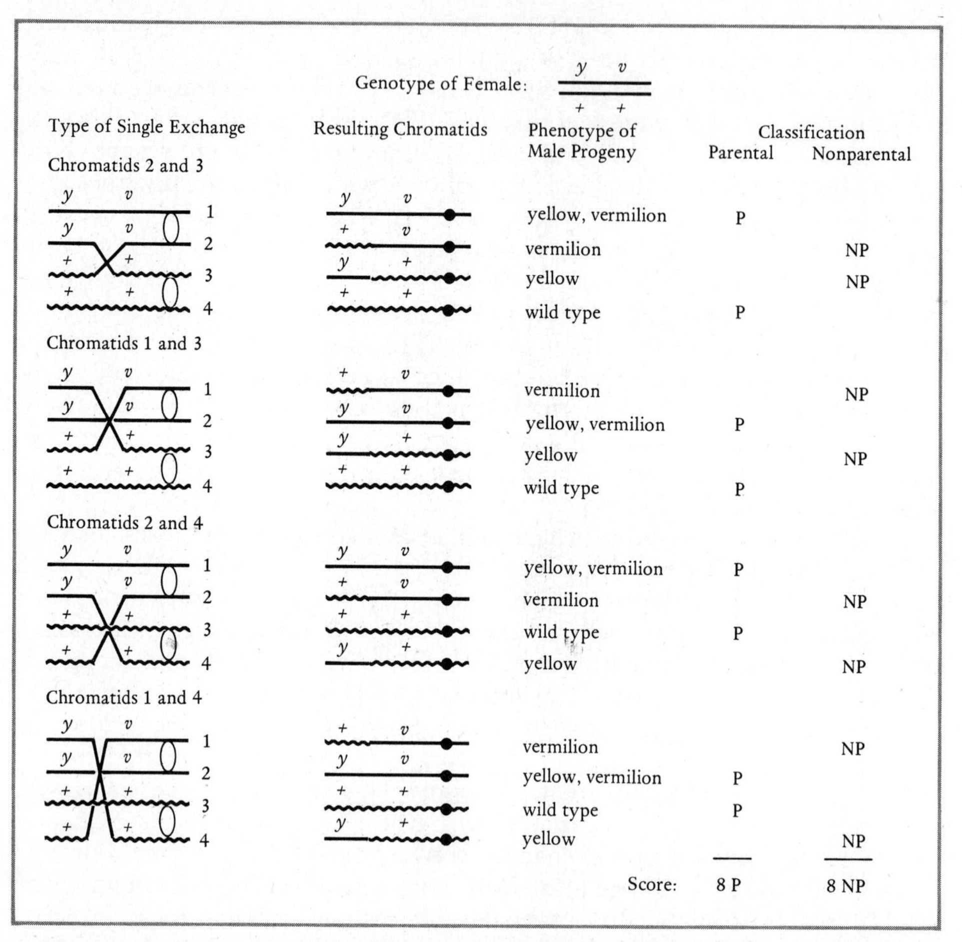

Figure 8–1. Consequences of a single exchange between linked genes. Homologous chromosomes are represented in the four-strand stage of meiosis with sister chromatids attached to a single centromere. Crossing over can involve any two nonsister chromatids. These possibilities and the consequences in each case are shown. Progeny phenotypes are scored as parental (P) or nonparental (NP). Note that parental and nonparental types occur with equal frequency.

possible single exchanges between nonsister strands are illustrated. It can be seen that in each case such an exchange occurring between the loci of the genes *y* and *v* produces two parental and two nonparental chromatids. Each of these chromatids, when inherited through the egg by a male offspring, will result in an identifiable phenotype. A similar series of diagrams could be used to illustrate crossover events for example 2.

It should also be observed that although crossing over within the tetrads diagramed in Figure 8–1 produces equal numbers of parental and nonparental *chromatids,* equal numbers of parental and nonparental *offspring* are not present in the progeny of examples 1 and 2. The reason for this is that crossing over does not occur in all oöcytes and, in addition, when crossing over does occur, a parental or a nonparental chromatid can be included in the functional egg with equal frequency. Thus, crossing over in the oöcytes of the female produces not

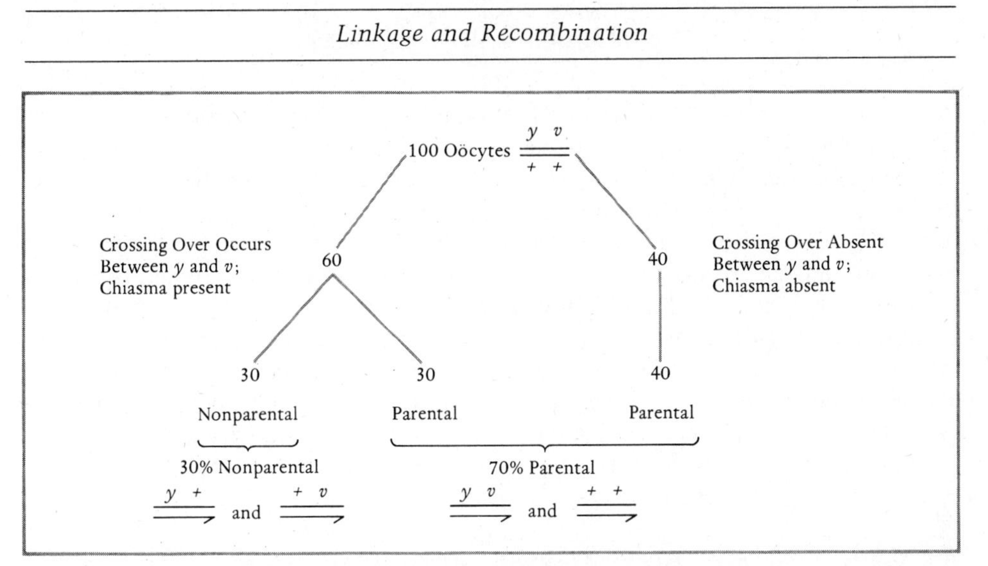

Figure 8–2. Origin of parental and nonparental type male progeny, assuming a recombination frequency of 30 percent.

only 30 percent nonparental progeny, but also 30 percent parental type progeny, and these, added to the parentals arising from oöcytes in which no exchanges occur, result in the higher frequency of parental types than nonparental types observed. The parental types of example 1 are thus derived from two sources: 30 percent originate from oöcytes in which an exchange occurred and 40 percent from oöcytes in which crossing over did not take place (Figure 8–2). In addition, we can infer that an exchange between the genes y and v must have occurred in 60 percent of the oöcytes in order to produce the 30 percent crossover progeny recovered in the experiment.

The difference in the frequency of parental versus nonparental type progeny provides a basis for the recognition of linkage, as opposed to independent assortment. As an example, let us assume two recessive autosomal mutants, a and b, along with their respective dominant alleles which we can designate as + in each case. If a and b are located on separate chromosomes and therefore not linked, a testcross between a heterozygote and an individual homozygous for both recessive alleles, i.e., $\frac{a \quad b}{+ \quad +} \times \frac{a \quad b}{a \quad b}$ will yield four phenotypic classes of progeny in a 1 : 1 : 1 : 1 ratio. These classes are + +, a +, + b, and a b, and all four classes should appear with equal frequency.

However, if a and b are linked and located close enough together on the same chromosome so that crossing over does not occur in every cell undergoing meiosis, the cross $\frac{a \quad b}{+ \quad +} \times \frac{a \quad b}{a \quad b}$ will produce more parental types (a b and + +) than nonparental types (a + and + b), and a 1 : 1 : 1 : 1 ratio between the four classes of progeny will not be observed. The frequency of parentals versus nonparentals in any instance depends entirely on how close together the genes in question are situated in the chromosome. If they are very close, nonparentals will be few; if they are farther apart, nonparentals will be more numerous. If they are so far apart that a crossover between them occurs in every cell undergoing meiosis, then equal numbers of parental and nonparental progeny will be produced and it will not be possible to distinguish linkage from independent assortment. In this case, the question of linkage could be resolved by using additional genes in testcrosses, that is, if two loci can each be shown to be linked to a third gene, then they must also be linked to each other. Obviously linkage is easiest to recognize when the loci of the genes in question are close together on the chromosome. Linkage is also easiest to analyze when the genes involved are sex linked (since any male may be used) or when a testcross is employed. In both cases recombinants are immediately evident and recombination frequencies can be obtained directly from observation of the progeny. In those instances where a testcross is not feasible, recombination frequencies can also be derived indirectly from analysis of the F_2 offspring which result from a cross of heterozygotes. Methods for obtaining linkage data from inbred populations can be found in Immer (1930) and Mather (1951).

DOUBLE CROSSING OVER AND CHROMOSOME MAPPING

The idea, proposed by Morgan, that the frequency of crossing over between two genes reflects their relative positions on the chromosome was subjected to extensive tests by A. H. Sturtevant in 1913. Sturtevant determined the frequency of exchange between six sex-linked recessive mutants of *Drosophila* and on the basis of these data prepared a linear linkage map showing the sequence and relative positions of these genes on the X chromosome. On this map the "distance" between any two genes was expressed in terms of *map units,* one map unit being equal to 1 percent recombination. (This unit is often referred to as a *centimorgan*.) As an illustration, we can apply this unit concept to the hypothetical data previously presented for the genes yellow and vermilion. In this example the frequency of nonparental or crossover individuals was 30 percent, and therefore we can say that genes *y* and *v* are 30 map units apart.

Sturtevant not only established the basic rules for mapping chromosomes, but also discovered that when two genes are positioned relatively far apart, more then one crossover can occur between them. The second crossover may restore the original parental gene association and thus go undetected in the progeny. This circumstance is diagramed in Figure 8–3 which illustrates the results of double crossing over in a female heterozygous for the genes *y* and *v* $\left(\frac{y\ v}{+\ +}\right)$. It can be seen that three general kinds of double exchanges can occur.

1. two-strand double—the same two nonsister chromatids undergo exchange twice

2. three-strand double—one homologue undergoes an exchange with each of the two nonsister chromatids of the opposing homologue

3. four-strand double—all four chromatids are involved in an exchange

In Figure 8–3 the chromatids derived from these exchanges are classified as to origin, and the phenotype which each will produce in a male offspring is listed and scored as parental or nonparental. Since crossovers are identified by analyzing the phenotypes of progeny, individuals that are yellow, vermilion and those that are wild type will be scored as parental, noncrossover types, and the fact that some of these individuals resulted from double crossing over will escape detection. Such individuals, which appear phenotypically as parentals, but are in fact derived from double crossing over, are starred (*P) in Figure 8–3.

When the possible types of progeny are classified on the basis of phenotype, it can be seen that parental and nonparental types occur in a 1 : 1 ratio. As a consequence, detectable crossing over between any two genes will not exceed 50 percent. In addition, even if triple or quadruple crossing over occurs between two given linked genes, the total number of phenotypic parentals will remain equal to the total number of nonparentals, and detectable crossing over will still not exceed 50 percent. The proof for this statement can be worked out by the reader.

In summary, an appreciable number of undetected double exchanges occurring between two genes will lead to an underestimate of their distance apart on a linkage map because not all crossovers will be included in the data on

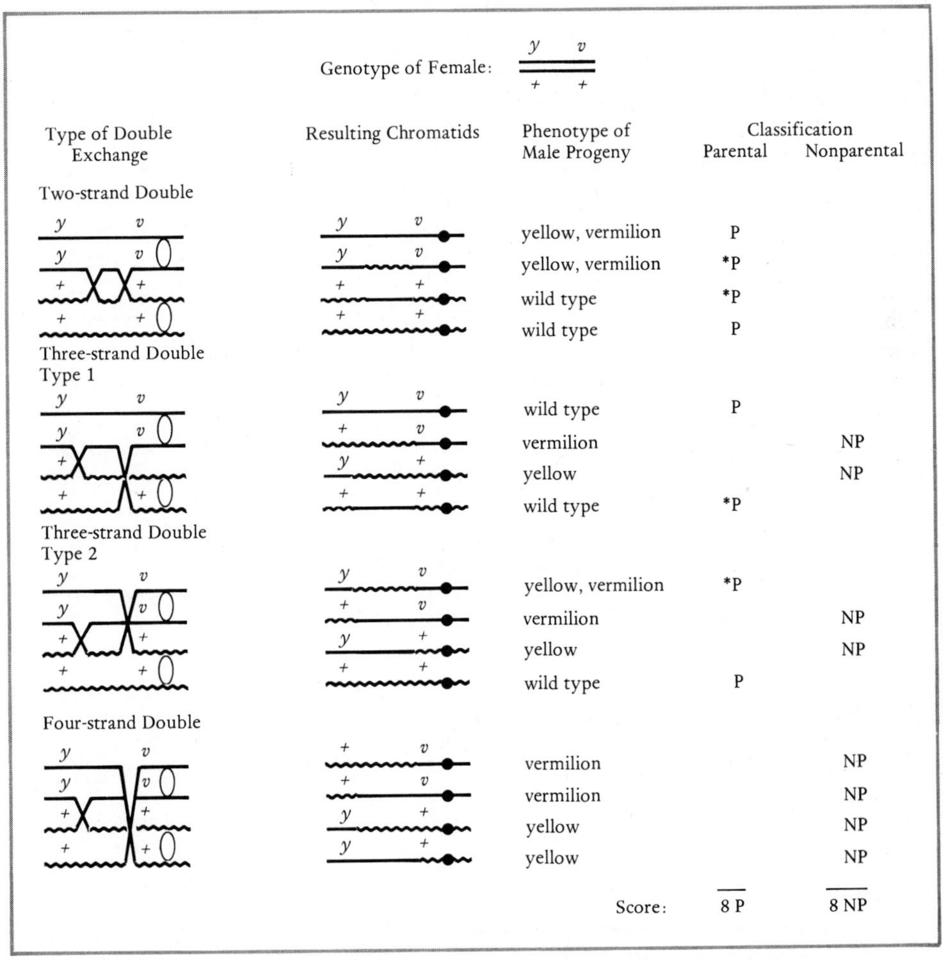

Genotype of Female: $\frac{y \quad v}{+ \quad +}$

Type of Double Exchange	Resulting Chromatids	Phenotype of Male Progeny	Classification: Parental	Classification: Nonparental
Two-strand Double	y v	yellow, vermilion	P	
	y v	yellow, vermilion	*P	
	+ +	wild type	*P	
	+ +	wild type	P	
Three-strand Double Type 1	y v	wild type	P	
	+ v	vermilion		NP
	y +	yellow		NP
	+ +	wild type	*P	
Three-strand Double Type 2	y v	yellow, vermilion	*P	
	+ v	vermilion		NP
	y +	yellow		NP
	+ +	wild type	P	
Four-strand Double	+ v	vermilion		NP
	+ v	vermilion		NP
	y +	yellow		NP
	y +	yellow		NP
		Score:	8 P	8 NP

Figure 8–3. The consequences of double crossing over between linked genes. The possible kinds of double crossovers are shown, along with the resulting chromatids. Progeny phenotypes are scored as parental (P), nonparental (NP), or phenotypically parental originating from a double exchange (*P). Note that parental and nonparental phenotypes occur with equal frequency.

which the map is based. Reliable estimates of linkage distance between genes can be made only when such genes are close enough together so that double crossing over between them occurs rarely or not at all. By this criterion our recombination frequency of 30 percent between *y* and *v* may be an unreliable estimate of the actual amount of exchange occurring between them, depending on the frequency of double crossovers.

The preceding discussion should not be interpreted to mean that double crossing over cannot be detected, for indeed it can, providing three (or more) genes are utilized in a cross. As an example, suppose that our female is heterozygous not only for *y* and *v*, but also for the recessive mutant crossveinless (*cv*). Since *cv* is situated between *y* and *v*, the genotype of the female would be $\frac{y\ cv\ v}{+\ +\ +}$.

The results of single and double crossing over in such a female are illustrated in Figure 8–4. For convenience, the region between *y* and *cv* can be designated as region I and that between *cv* and *v* as region II. Notice that a single crossover occurring in either of these regions results in new gene associations that can readily be distinguished in the phenotypes of male progeny. The simultaneous occurrence of a crossover in both regions results in a "switch" of the centrally located gene from one nonsister chromatid to the other, and this new association can also be identified in the phenotype of male offspring. Thus, the use of three genes has made possible the detection of a double crossover.

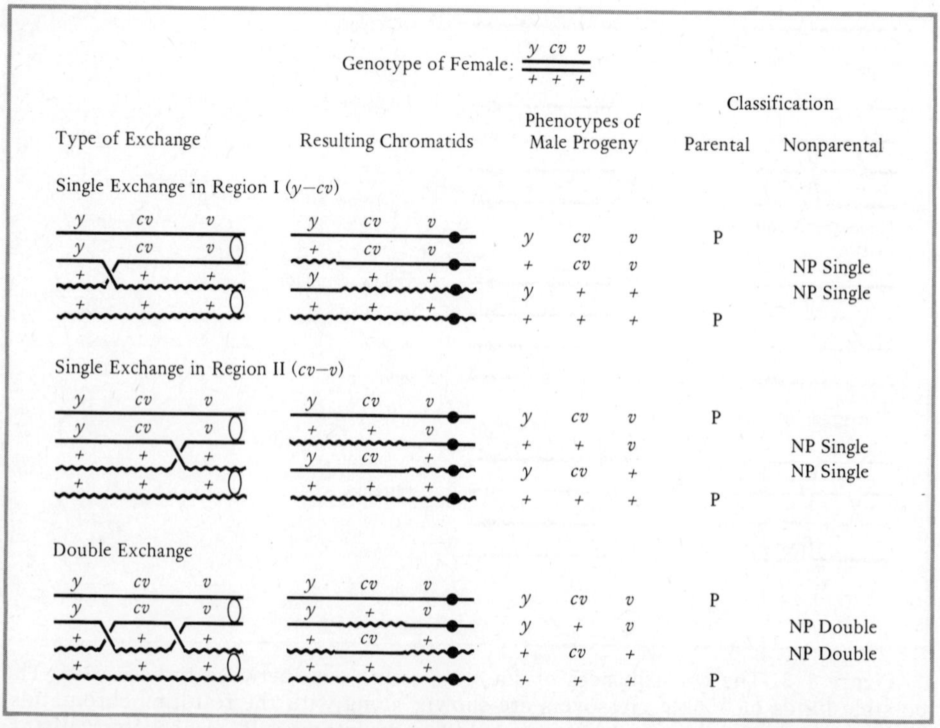

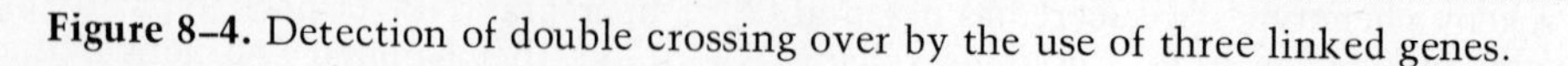

Figure 8–4. Detection of double crossing over by the use of three linked genes.

With all of these considerations in mind, we can apply Sturtevant's method of mapping genes to the hypothetical data given in Figure 8–5. Let us assume that these data were obtained from the testcross: $\frac{a\ b\ c}{+\ +\ +} \times \frac{a\ b\ c}{a\ b\ c}$. The data of Figure 8–5 are arranged in reciprocal classes according to phenotype, with + representing the wild type alternative in each case. Inspection of the data permits immediate identification of parental types (group 1), for their numbers greatly exceed those of all other classes. Further inspection reveals that group 4 contains the fewest individuals. We can therefore be quite certain that the two phenotypic classes of group 4 represent double crossovers, since doubles occur much less frequently than singles.

Having identified the doubles, they can now be used to determine the gene order on the chromosome. The data are written as if the gene order were *a—b—c*, but this is an arbitrary arrangement. If the gene order were *a—b—c*, then a double crossover would produce the nonparental progeny *a* + *c* and + *b* +, and it can be seen that these do not correspond to the actual doubles identified in group 4. The order *b—a—c* can next be tried to see whether or not the doubles produced by this arrangement will correspond to the actual data. If the order were *b—a—c*, the doubles would be *b* + *c* and + *a* +. Again, it is evident that these do not agree with the actual doubles. The only other possible

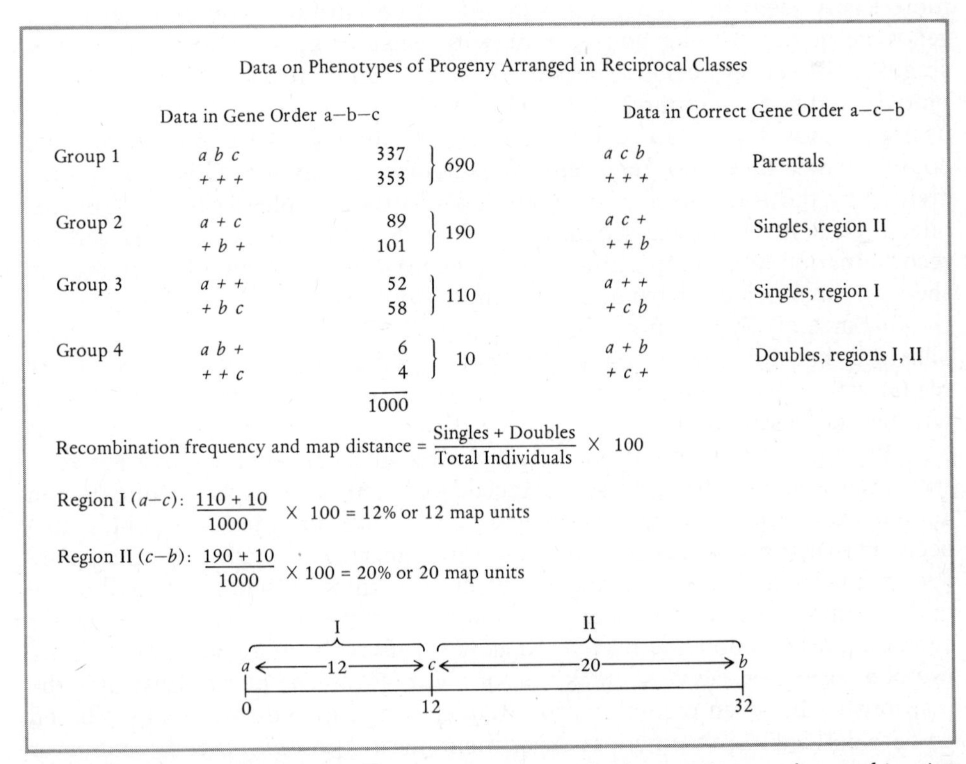

Data on Phenotypes of Progeny Arranged in Reciprocal Classes

	Data in Gene Order a—b—c			Data in Correct Gene Order a—c—b	
Group 1	*a b c*	337	690	*a c b*	Parentals
	+ + +	353		+ + +	
Group 2	*a* + *c*	89	190	*a c* +	Singles, region II
	+ *b* +	101		+ + *b*	
Group 3	*a* + +	52	110	*a* + +	Singles, region I
	+ *b c*	58		+ *c b*	
Group 4	*a b* +	6	10	*a* + *b*	Doubles, regions I, II
	+ + *c*	4		+ *c* +	
		1000			

$$\text{Recombination frequency and map distance} = \frac{\text{Singles + Doubles}}{\text{Total Individuals}} \times 100$$

$$\text{Region I } (a\text{—}c)\text{: } \frac{110 + 10}{1000} \times 100 = 12\% \text{ or 12 map units}$$

$$\text{Region II } (c\text{—}b)\text{: } \frac{190 + 10}{1000} \times 100 = 20\% \text{ or 20 map units}$$

Figure 8–5. Hypothetical data from a testcross and determination of recombination frequency and map distance.

arrangement of these three genes is the order $a—c—b$ (or $b—c—a$). A double crossover in this case will produce the nonparental types $a + b$ and $+ c +$. By comparing these with the actual doubles, it can be seen that they agree. This method of determining gene order is a trial and error technique, useful when only three genes are involved. An easier method is to examine the two types of doubles for the nonparental gene present in each; this gene will be the one in the middle.

When more than three genes are used, trial and error becomes tedious, and the gene order can be obtained by considering any two genes apart from the rest. For example, *a* and *b* entered the cross on the same chromosome (coupled) and therefore an exchange between them will be evident in all classes that are $a +$ or $+ b$ (ignoring *c*). If these are totaled (groups 2 and 3), 300 such individuals are found. When the relationships between *a* and *c* are considered (ignoring *b*), crossovers between *a* and *c* will be represented in individuals that are $a +$ and $+ c$ (groups 3 and 4). The total of these is 120. The last remaining relationship to be determined is that between *b* and *c* (ignoring *a*). Here, crossovers will be $b +$ and $+ c$. These total 200 (groups 2 and 4). On the principle that crossing over is more frequent between genes that are farthest apart, we can infer that since *a* and *b* show the highest number of exchanges (300), they are most distantly separated, and therefore gene *c* must be between them in the middle.

Having determined the order of these three genes on the chromosome, it is convenient to re-write the data in correct order so that single crossovers can be more easily identified. This has been done in Figure 8–5. By comparing the genotype of the heterozygous parent with each group of nonparental types, single crossovers in region I (between *a* and *c*) can be identified as those individuals contained in group 3 of the data, that is, $a + +$ and $+ c\, b$. Total crossing over in region I is obtained by summing all singles of this region, plus all doubles, since they, too, represent an exchange in region I. This sum is then divided by the total number of individuals in the sample. The result, multiplied by 100, is the percent recombination. Thus, for region I there is 12 percent recombination or 12 map units between genes *a* and *c*. Region II is treated in the same manner to arrive at the recombination frequency of 20 percent and the distance of 20 map units between genes *c* and *b*. It should again be emphasized that if double exchanges are not included in the calculations for any region where they occur, the frequency of recombination within such a region will be underestimated.

We have already mentioned that the recombination frequency between two linked loci will not ordinarily exceed 50 percent because of the undetected double exchanges that occur between them. Such undetected doubles also occur between genes less than 50 map units apart. Although their frequency declines with decreasing distance between loci, their presence will still cause underestimates of the true recombination frequency. It is possible to correct for such underestimation and thereby to arrive at more accurate map positions by use of a correction curve such as that of Figure 8–6. The curve illustrates the relationship between recombination frequency and map distance and is based on data derived from numerous experiments with *Drosophila,* corn, and other organisms.

Linkage maps, such as those illustrated in Figures 8–7, 8–8, and 8–9, are obtained by combining the results of many experiments. Over short distances,

50
Recombination Frequency
25
25
50
75
100
125
150
Map Distance

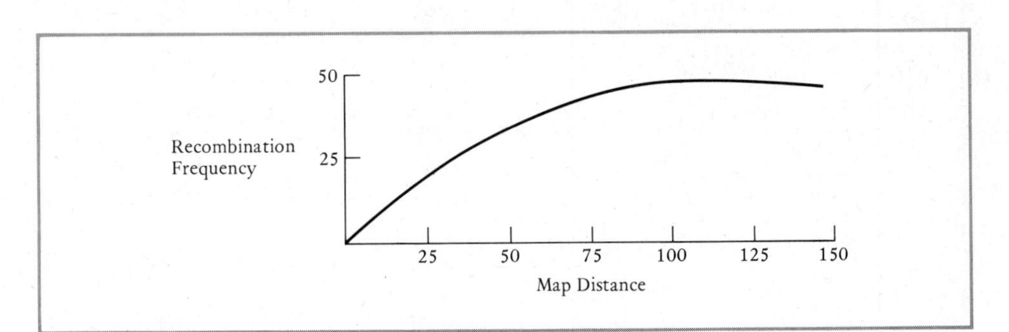

Figure 8–6. The relationship between recombination frequency and map distance. An estimate of the actual map distance between two loci is found by determining the point at which observed recombination frequency (vertical axis) intersects the curve; the distance of this point along the horizontal axis indicates the map distance. [From J. F. Crow, *Genetics Notes,* 6th ed., Burgess Publ. Co., Minneapolis, Minn., 1966.]

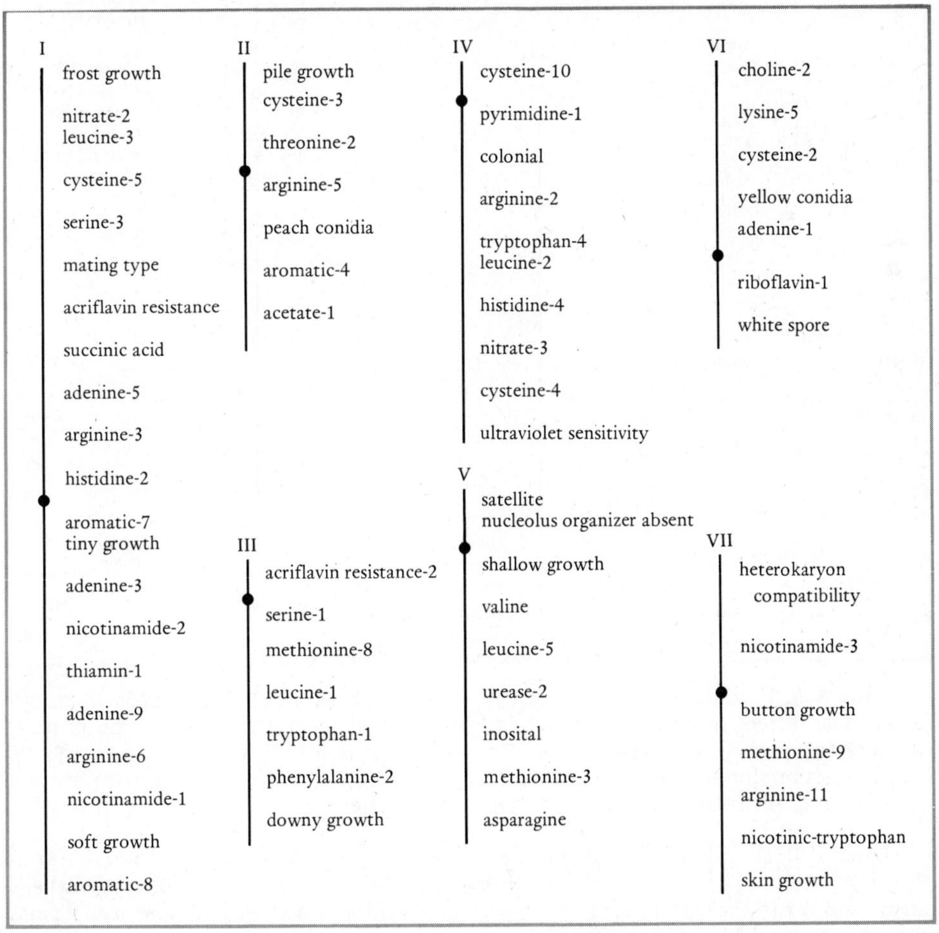

I
frost growth
nitrate-2
leucine-3
cysteine-5
serine-3
mating type
acriflavin resistance
succinic acid
adenine-5
arginine-3
histidine-2
aromatic-7
tiny growth
adenine-3
nicotinamide-2
thiamin-1
adenine-9
arginine-6
nicotinamide-1
soft growth
aromatic-8
II
pile growth
cysteine-3
threonine-2
arginine-5
peach conidia
aromatic-4
acetate-1
III
acriflavin resistance-2
serine-1
methionine-8
leucine-1
tryptophan-1
phenylalanine-2
downy growth
IV
cysteine-10
pyrimidine-1
colonial
arginine-2
tryptophan-4
leucine-2
histidine-4
nitrate-3
cysteine-4
ultraviolet sensitivity
V
satellite
nucleolus organizer absent
shallow growth
valine
leucine-5
urease-2
inosital
methionine-3
asparagine
VI
choline-2
lysine-5
cysteine-2
yellow conidia
adenine-1
riboflavin-1
white spore
VII
heterokaryon compatibility
nicotinamide-3
button growth
methionine-9
arginine-11
nicotinic-tryptophan
skin growth

Figure 8–7. Partial linkage maps of *Neurospora crassa.* Note that most mutations involve the inability to synthesize some substance required for growth. [Reprinted with permission from R. W. Barratt and A. Radford, in *Handbook of Biochemistry,* 2nd ed., H. A. Sober, Ed. Copyright CRC Press, Inc., Cleveland, Ohio, 1970.]

1 (X)

y 0 yellow (B)
sc 0+ scute (H)
Hw 0+ hairy wing (W)
w 1.5 white (E)
fa 3.0 facet (E)
ec 5.5 echinus (E)
rb 7.5 ruby (E)
cv 13.7 crossveinless (V)
ct 20.0 cut (W)
sn 21.0 singed (H)
lz 27.7 lozenge (E)
v 33.0 vermilion (E)
m 36.1 miniature (W)
s 43.0 sable (B)
g 44.4 garnet (E)
f 56.7 forked (H)
B 57.0 bar (E)
fu 59.5 fused (V)
car 62.5 carnation (E)
bb 66.0 bobbed (H)

4

bt 0 bent (W)
ci 0 cubit, interr. (V)
sv 0.1± shaven (H)
ey 0.1 ± eyeless (E)

2

al 0 aristaless (B)
S 1.3 star (E)
dp 13.0 dumpy (B)
cl 16.5 clot (E)
d 31.0 dachs (B)
J 41.0 jammed (W)
b 48.5 black (B)
rd 51.0 reduced (H)
pr 54.5 purple (E)
Bl 54.8 bristle (H)
lt 55.0 light (E)
cn 57.5 cinnabar (E)
en 62.0 engrailed (B)
vg 67.0 vestigial (W)
L 72.0 lobe (E)
c 75.5 curved (W)
hy 93.3 humpy (B)
px 100.5 plexus (W)
bw 100.7 brown (E)
sp 107.0 speck (B)

3

ru 0 roughoid (E)
ve 0.2 veinlet (V)
jv 19.2 javelin (H)
se 26.0 sepia (E)
h 26.5 hairy (H)
D 41.0 dichaete (H)
G 41.0 glued (E)
th 43.2 thread
st 44.0 scarlet (E)
Dfd 47.5 deformed (E)
p 48.0 pink (E)
cu 50.0 curled (W)
Sb 58.2 stubble (H)
ss 58.5 spineless (H)
bx 58.7 bithorax (B)
sr 62.0 stripe (B)
gl 63.0 glass (E)
Dl 66.2 delta (V)
H 69.5 hairless (H)
e 70.7 ebony (B)
cd 74.7 cardinal (E)
ro 91.1 rough (E)
ca 100.7 claret (E)
Mg 106.2 minute-g (H)

Figure 8–8. Partial linkage maps of the four chromosomes of *Drosophila melanogaster.* The letters in parentheses indicate the part of the fly affected by the mutant character concerned: B, body; E, eyes; H, hairs or bristles; V, venation of wings; W, wings. Positions of centromeres are indicated by arrows. [From A. H. Sturtevant and G. W. Beadle, *An Introduction to Genetics,* Dover, New York, 1962.]

Figure 8–9. (a) Linkage map of chromosome 9 of maize *(Zea),* a representative example of the 10 pairs of chromosomes (10 linkage groups) in corn. (b) Linkage map of the X chromosome (linkage group 20) of the house mouse *(Mus musculus).* There are 19 pairs of autosomes and 1 pair of sex chromosomes in the mouse. [(a) Reprinted with permission from M. Neuffer and E. Coe, Jr., in *Handbook of Biochemistry,* 2nd ed., H. A. Sober, Ed. Copyright CRC Press, Inc., Cleveland, Ohio, 1970; (b) reprinted with permission from M. C. Green, in *Handbook of Biochemistry,* 2nd ed., H. A. Sober, Ed. Copyright CRC Press, Inc., Cleveland, Ohio, 1970.]

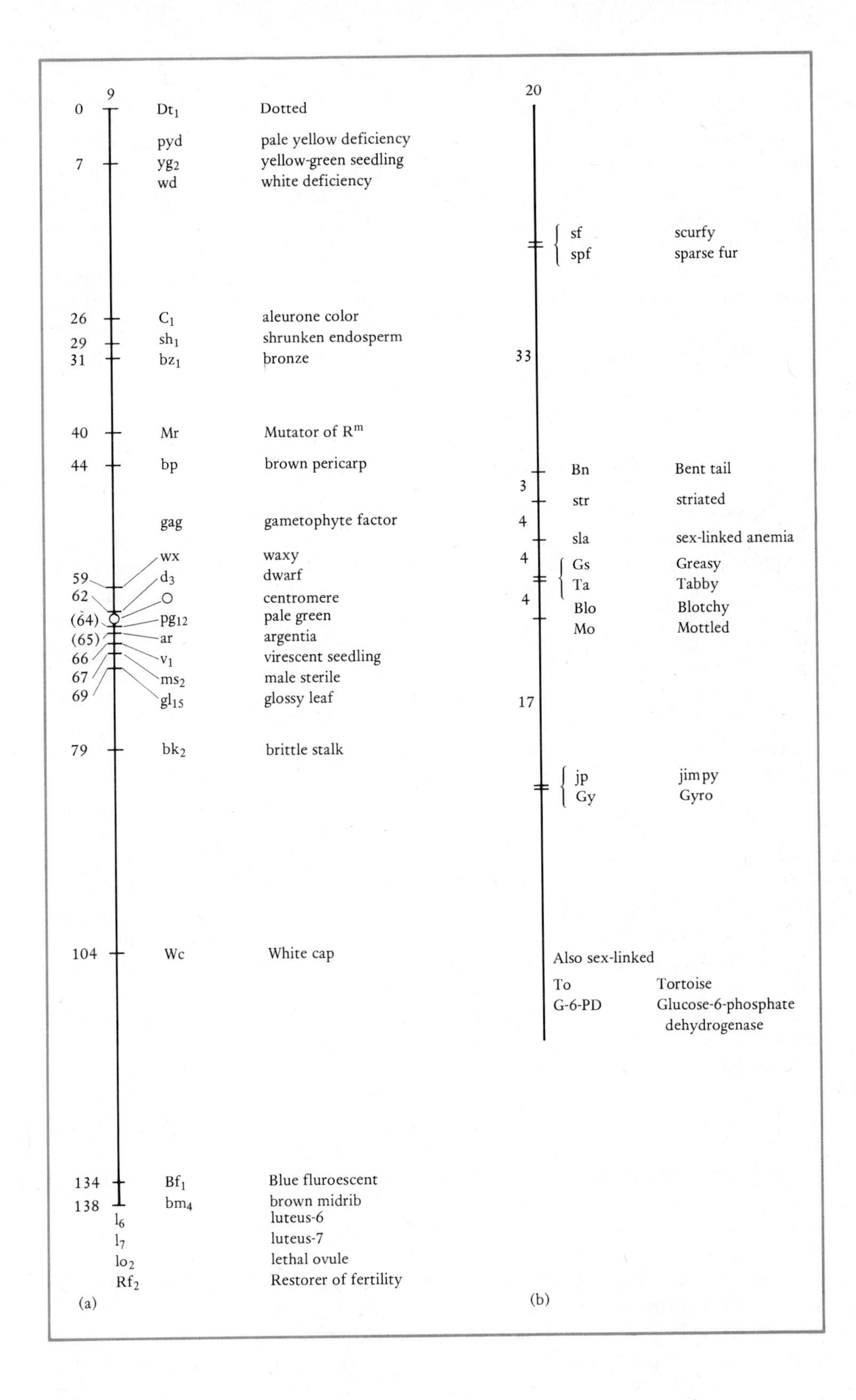

9
0 Dt1 Dotted
pyd pale yellow deficiency
7 yg2 yellow-green seedling
wd white deficiency
26 C1 aleurone color
29 sh1 shrunken endosperm
31 bz1 bronze
40 Mr Mutator of R^m
44 bp brown pericarp
gag gametophyte factor
wx waxy
59 d3 dwarf
62 O centromere
(64) pg12 pale green
(65) ar argentia
66 v1 virescent seedling
67 ms2 male sterile
69 gl15 glossy leaf
79 bk2 brittle stalk
104 Wc White cap
134 Bf1 Blue fluroescent
138 bm4 brown midrib
l6 luteus-6
l7 luteus-7
lo2 lethal ovule
Rf2 Restorer of fertility
(a)
20
sf scurfy
spf sparse fur
33
Bn Bent tail
3
str striated
4
sla sex-linked anemia
4
Gs Greasy
Ta Tabby
4
Blo Blotchy
Mo Mottled
17
jp jimpy
Gy Gyro
Also sex-linked
To Tortoise
G-6-PD Glucose-6-phosphate dehydrogenase
(b)

the frequencies of crossing over are additive and therefore the position of a gene on a map represents the sum of all recombination frequencies up to that point. As a result, linkage maps may be over 100 units long. We can apply the same principle to the linked genes *a c b*. If we consider gene *a* to be at position 0, then *c* will be at position 12, and *b* at position 32 (Figure 8–5). If additional genes are found and located between *a* and *c* or *c* and *b*, the originally determined map positions will change to reflect the more accurate estimate of the amount of recombination. Thus, with the addition of new data linkage maps are subject to expansion. The position of a gene on such a map is not immutable, but merely reflects the best estimate available at the time the map was prepared.

COINCIDENCE AND INTERFERENCE

Thus far we have assumed that crossing over is a random event and that the occurrence of one exchange will have no bearing on the occurrence of a second exchange in the same pair of homologues. This appears to be the case for genes that are located some distance apart, for example, 40 map units or more. With distances less than this, however, the closer together genes are positioned, the fewer are the double crossovers, as if the formation of one exchange somehow interferes with the nearby formation of a second exchange.

For example, in our data above, crossing over in region I occurred in 12 percent (0.12) of the progeny, while crossing over in region II occurred in 20 percent (0.20) of the progeny. If an exchange in region I is independent of an exchange in region II, then the probability of a simultaneous exchange occurring in both of these regions is the product of their separate probabilities, that is, $0.12 \times 0.20 = 0.024$ (2.4 percent) doubles expected. In the data, however, doubles actually occurred with a frequency of 10 in 1000 individuals, that is, in 0.01 or 1 percent of the progeny. The ratio of observed to expected is thus $0.01/0.024 = 0.416$, or around 0.42. This ratio is called the *coefficient of coincidence* and it expresses the degree to which expectations of the frequency of double crossing over are realized. A coefficient of coincidence of 0.42 means that only 42 percent of the expected doubles actually occur.

We can also assess the extent to which a crossover in one region inhibits a crossover in an adjacent region by determining *interference*. Interference is calculated by subtracting the coefficient of coincidence from 1.0, in this case, $1.0 - 0.42 = 0.58$. The value of 0.58 means that the presence of one exchange interfered with the formation of a second exchange 58 percent of the time. When coincidence is 1.0, observed doubles are equal in frequency to expected doubles and interference is 0. When coincidence is 0, interference will be 1.0 (100 percent) and no doubles will be observed. Thus interference measures the proportional deviation from the expectation that a crossover in one region is independent of that in another.

These relationships were first analyzed by H. J. Muller in 1916. Muller also found that under normal circumstances the amount of crossing over in one pair of homologues does not affect the frequency of exchange in other, nonhomologous pairs of chromosomes. It should be noted, however, that the presence of a heterozygous inversion in one pair of homologues causes an increase in crossover frequency in other, normal pairs (see Chapter 9).

Muller and others have also studied the relationship between the frequency of crossing over and the length of a given chromosome and have found that the two are proportional.

More recently, a phenomenon called *negative interference* has been identified in viruses, bacteria, fungi, and higher organisms. Negative interference refers to an increase in the frequency of exchanges beyond that expected with 100 percent coincidence. It is associated with nonreciprocal recombination observed within highly restricted regions of the genome (gene conversion), and is more appropriately discussed in Chapter 14.

THE USE OF LINKAGE MAPS

Linkage maps and coefficients of coincidence are useful when it is desirable to predict in advance the number of individuals of any given phenotype or genotype to be expected from a cross involving linked genes. As an illustration, suppose genes *x*, *y*, and *z* are located at positions 3, 23, and 53, respectively, on a linkage map. The frequency of recombination between *x* and *y* should be 20 percent and between *y* and *z*, 30 percent, as shown in the following: $\frac{x\ \ 20\ \ y\ \ 30\ \ z}{3\quad 23\quad 53}$. Suppose also that the coefficient of coincidence in this case is 0.4. Then in a testcross $\left(\frac{x\ y\ z}{+\ +\ +} \times \frac{x\ y\ z}{x\ y\ z}\right)$ we can predict in advance the number of individuals expected to be present in each phenotypic class when the total progeny is, for example, 500.

The first predictions made must be those for double crossover types. The frequency of expected doubles is the product of the frequency of recombination in both regions, that is, $0.2 \times 0.3 = 0.06$ (6 percent). In a population of 500, 6 percent represents 30 double crossover individuals. However, the coefficient of coincidence is 0.4, and therefore only 40 percent of the expected doubles will arise. Thus, the observed doubles will number only 12 individuals ($0.4 \times 30 = 12$). Since double crossovers occur as two phenotypic classes ($x + z$ and $+ y +$), we must assign 6 individuals to each class.

Single crossover types can next be predicted. Total recombination between genes *x* and *y* is 20 percent, and 20 percent of 500 individuals is 100 individuals. In each of these 100 individuals a crossover should have occurred between genes *x* and *y*. However, in order to predict the number of single crossovers expected, we must first subtract the 12 double crossovers, since doubles were originally included in determining the map distance. This subtraction leaves 88 individuals representing single crossovers between genes *x* and *y*, and these can be assigned equally to the phenotypic classes $x + +$ and $+ y z$. The number of individuals expected to result from a single crossover occurring between genes *y* and *z* is determined in the same way. Here, total recombination is 30 percent, and 30 percent of 500 individuals is 150 individuals. Again subtracting the 12 double crossovers, we are left with 138 individuals to be assigned equally to the phenotypes $x y +$ and $+ + z$. Finally, having predicted 12 double crossovers, 88 single crossovers between genes *x* and *y*, and 138 single crossovers between genes *y* and *z* (a total of 238), the remaining 262 individuals can be equally divided between the two noncrossover classes, *x y z* and + + +.

Similar kinds of predictions can be made for the progeny which result from a cross of individuals heterozygous for linked genes. As an illustration, the linked genes *x* and *y* are 20 map units apart. The cross $\frac{x\ y}{+\ +} \times \frac{x\ y}{+\ +}$ will not result in the typical dihybrid 9 : 3 : 3 : 1 phenotypic ratio because the four kinds of gametes, + +, *x y*, *x* +, and + *y*, are not produced in equal frequencies. Since recombination between these genes is 20 percent, we can expect that crossover type gametes, *x* + and + *y*, will occur with a frequency of 10 percent (0.1) each, while the noncrossover gametes, *x y* and + +, will occur with a frequency of 40 percent (0.4) each. Thus each parent will produce gametes in the following proportions: 0.4 + +, 0.4 *x y*, 0.1 *x* +, and 0.1 + *y*. By using these gamete frequencies, the proportion of F_2 individuals of a given genotype or phenotype can be predicted. For example, the proportion of individuals expected to be homozygous for the recessive alleles *x* and *y* is 0.16 or 16 percent ($0.4 \times 0.4 = 0.16$). In addition, if three linked genes are involved, the frequencies of the different kinds of gametes will depend on the coefficient of coincidence as well as upon the map distances between them.

CYTOLOGICAL PROOF OF CROSSING OVER

Although detailed and extensive genetic evidence that crossing over must occur in the four-strand stage of meiosis was early obtained by Bridges in 1916, by E. G. Anderson in 1925, and by others, it was not until 1931 that a direct correlation between cytologically visible exchanges in the chromosomes and crossover type progeny was obtained by Harriet Creighton and Barbara McClintock, using the corn plant, *Zea*. The particular strain of corn used was one which was heterozygous for the kernel characters, colored *(C)* versus colorless *(c)* and starchy *(Wx)* versus waxy *(wx)*. These genes are borne on chromosome 9. The two homologues of chromosome 9 in this strain of corn differed physically in that one was normal in size, while the other was characterized by the presence of a knob on one end and by the attachment of an additional chromosome segment to the opposite end. This extra segment had been transferred (translocated) to chromosome 9 from chromosome 8. Thus, the abnormal homologue could be distinguished from its partner not only by the presence of the knob, but also by its unusual length. In the cross used one parent was heterozygous for the genes *C* and *c* and *Wx* and *wx* and also heterozygous for the normal and abnormal chromosomes 9. The other parent was *c Wx/c wx* and both homologues of chromosome 9 were normal. This constitution of the parents is diagramed in Figure 8–10, together with the parental and nonparental types of offspring produced by the cross.

The genotypes of progeny with respect to the genes for colored and colorless could be ascertained immediately by examination of kernel phenotypes. Starchy individuals were either heterozygous or homozygous for the gene *Wx*, but their genotype could be determined by the kinds of pollen they produced. It will be recalled that pollen containing starch (and gene *Wx*) stains dark blue with iodine solution, but *wx* pollen grains are without starch and remain unstained by this treatment (see Figure 1–5). Homozygous *Wx/Wx* plants produce pollen all of which stains blue with iodine, whereas only half the pollen of *Wx/wx* plants will stain. Thus by visual examination for color and by testing the pollen for starch, the genotypes of the offspring could be accurately determined. The cells of genetically nonparental types could then be examined microscopically for evidence of physical exchange between the normal and the knobbed, long chromosomes. Such exchanges were found. For example, individuals genotypically homozygous for both colorless and waxy possessed a pair of ninth chromosomes, one of which was normal, the other long, but not knobbed. The reciprocal type of individual, genotypically heterozygous for colored and colorless and homozygous for starchy, contained the reciprocal types of chromosomes, one normal and the other knobbed, short. This experiment by Creighton and McClintock provided an elegant demonstration that genetic crossing over was indeed accompanied by physical exchanges between homologous chromosomes.

A similar demonstration was also carried out by Curt Stern in 1931, using *Drosophila* as the experimental organism. In this case the characters were sex linked, and the X chromosomes of the female were morphologically different, and therefore cytologically distinguishable from one another. As was the case in experiments with corn, genetic crossing over as evidenced by progeny phenotypes and genotypes could be correlated with physical exchanges between the two types of X chromosomes.

Direct evidence that crossing over occurs when the homologues are in the four-stranded condition during meiosis was obtained in *Neurospora* by Carl Lindegren in 1933. This evidence is presented later in this chapter along with the discussion of crossing over in organisms which combine briefly, undergo meiosis, and resume the haploid state.

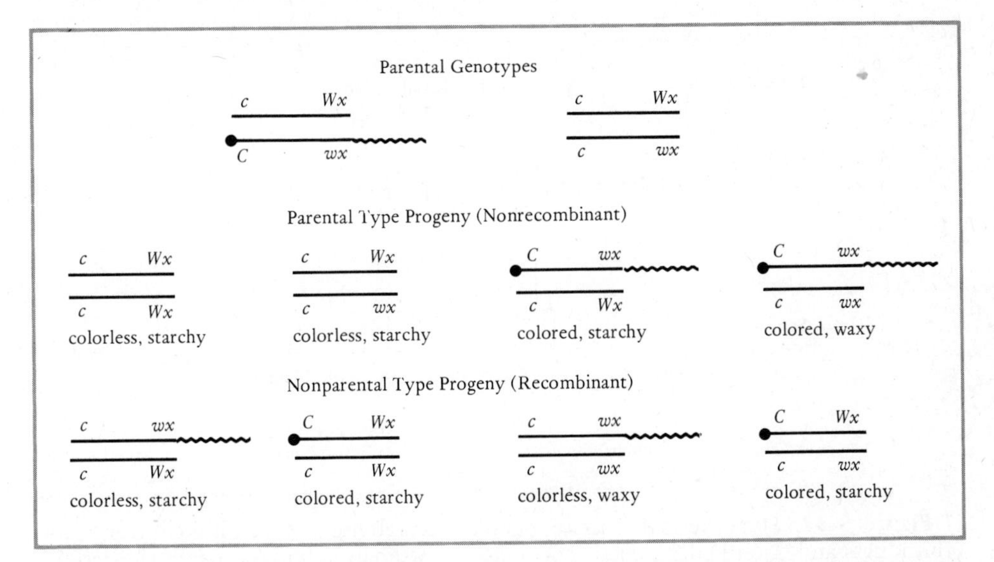

Figure 8–10. Diagram of Creighton and McClintock's experiment illustrating the correlation between genetic recombination and physical exchanges between homologous chromosomes.

LINKAGE IN HUMANS

Although the phenomenon of crossing over appears to be a normal and regularly occurring event of meiosis, the detection of crossing over requires that the genes in question be linked, that the parental genotypes be ascertained or inferred with confidence, and that the offspring be sufficiently numerous to make probable the recovery of recombinant individuals. These requirements are difficult to fulfill. In humans there are 22 autosomal linkage groups as well as those of the X and Y chromosomes. Linkage to the X chromosome can be identified through the unique pattern of sex-linked inheritance. In the case of autosomes, however, even if two or more traits evidence a linked relationship, assignment of genes to one specific pair of autosomes is extraordinarily difficult, or at least has been so in the past. The identification of parental genotypes is, in itself, often no easy matter. Compared to experimental organisms, human families are very small, and thus even when parental genotypes represent a testcross, a recombinant offspring is seldom produced. Despite these difficulties, crossover individuals have been identified by pedigree analysis, particularly in cases involving sex-linked genes.

For example, it is possible to determine that a phenotypically normal woman is heterozygous for two different sex-linked genes by analyzing the phenotypes of her father and mother. If her father is colorblind *(cb)* and her mother is deficient in the enzyme glucose-6-phosphate dehydrogenase *(G-6-PD)*, indicating homozygosity, it could be inferred that the normal daughter is heterozygous for each of these X-linked mutants, that is, $\frac{cb \quad +}{+ \quad G\text{-}6\text{-}PD}$. By examining the phenotypes of the woman's sons, a crossover individual could be recognized as one who exhibited either both traits or neither trait (Figure 8–11).

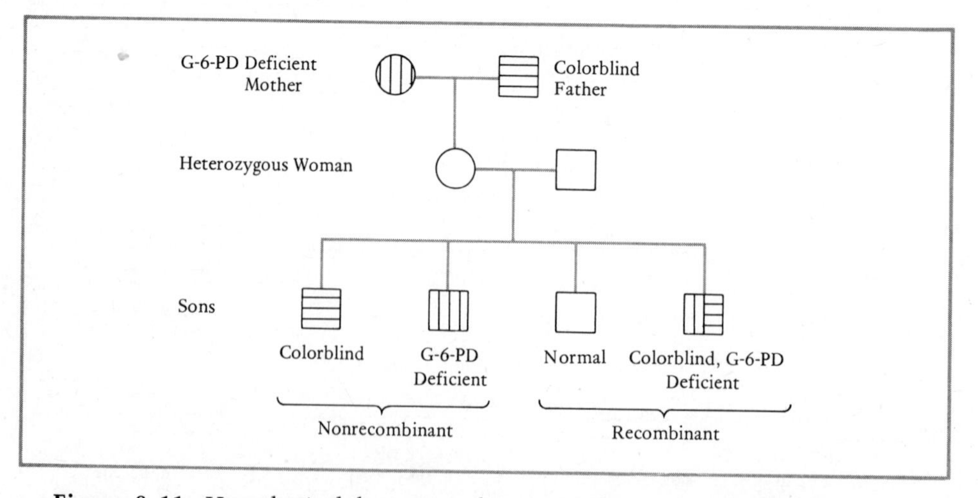

Figure 8–11. Hypothetical human pedigree involving the sex-linked traits colorblindness and G-6-PD deficiency. The sons of a woman heterozygous for these traits are ordinarily either colorblind or G-6-PD deficient, but if crossing over occurs, a son exhibiting both traits or neither trait can be produced. All four types of son are illustrated.

Although linkage can be identified in individual pedigrees, the number of recombinant individuals from one family, or several families combined, is often too few to permit a reliable estimate of recombination frequency. More recently a statistical method has been devised which utilizes computers to calculate the likelihood that data for given traits could have arisen by different recombination frequencies, with a maximum likelihood for the most probable recombination frequency. Through such determinations, probable linkage as well as estimated recombination frequencies have been found for a number of human characters. The method is called the *log-odds* or *lod* method, and the reader is referred to Renwick for a discussion of theory and application.

Except for sex-linked characters, however, neither pedigrees nor statistics, by themselves, provide the information needed to associate particular linked genes with particular chromosomes. Such information can be obtained only when genetic analysis is combined with cytological study of the chromosomes themselves. Three recent advances have made such correlated studies possible, and the assignment of linkage groups to particular chromosomes is proceeding very rapidly.

One such advance has been the application of tissue culture techniques to human cells derived from biopsies of skin or other tissues. The cells in culture can be treated as an extension of the individual from which they were taken and the biochemical phenotype of these cells can be analyzed. In many cases such studies have defined the primary cause of an inherited defect (usually an absent or defective enzyme) and by doing so have provided a phenotype recognizable at the cellular level.

A second major advance has been the development of the chromosome banding techniques which make possible the positive identification of every chromosome of the human complement.

The third advance is the use of hybrid cells grown in tissue culture. Fusion of the cytoplasm of cells from two different sources can occur spontaneously, as in the fungi, or can be induced experimentally in the case of higher organisms. If the nuclei of fused cells remain separate from one another, the condition is called *heterokaryosis,* and in higher organisms, such cells *(heterokaryons)* have been induced by Boris Ephrussi and M. C. Weiss and others. If cytoplasmic fusion is accompanied by nuclear fusion, a true *hybrid cell* is produced as first demonstrated by G. Barski and associates. Cell fusion is accomplished by treatment with a virus, called Sendai virus. The virus is first inactivated by ultraviolet radiation so that it cannot replicate in treated cells, but it retains its ability to adsorb to cell membranes, a function which causes the formation of cytoplasmic bridges between adjacent cells, followed by cytoplasmic confluence and nuclear fusion.

Both genomes contained in a hybrid cell continue to function, for the presence of biochemical traits specific to each can be identified. For example, in a mouse-human hybrid both mouse and human enzymes are produced. Mouse-human hybrids are also characterized by the preferential loss of human, not mouse, chromosomes from the cell. Chromosome loss is more or less random and different cells may come to contain different human chromosomes. Eventually, the hybrids stabilize, retaining from 2 to around 15 human chromosomes.

Such hybrids provide a powerful tool for the assignment of human genes to specific chromosomes because the presence (or disappearance) of a biochemical trait specific to the human genome can be correlated with the presence (or loss) of a given, identifiable human chromosome. As an example, the human gene for the enzyme thymidine kinase *(TK)* was assigned to chromosome 17 by such studies. Thymidine is necessary for the life of the cell and is required for the process of chromosome replication. Ordinarily, thymidine is synthesized by the cell without recourse to extracellular (exogenous) sources. However, when endogenous synthesis is blocked by the presence of an inhibitor (aminopterin), the cell survives, provided a purine source and thymidine are added to the culture medium and the enzyme thymidine kinase is present *(TK^+)*. When human cells that are TK^+ are fused with mouse cells that are mutant for the *TK* gene and thus lack this enzyme *(TK^-)*, growth will proceed

TABLE 8–1
Human Genes Assigned or Confirmed to Specific Chromosomes by Cell Hybrid Analysis

Chromosome	Abbreviated form	Full name
1	PGM_1	phosphoglucomutase-1
	PGD	6-phosphogluconate dehydrogenase
	PepC	peptidase C
2	IDH	isocitric dehydrogenase
	MOR	maleate oxidoreductase
3	—	
4–5	Adenine B^+	adenine B auxotrophy
6	MOD	NADP-malate dehydrogenase
	IPO-B	indole phenol oxidase B
7	MPI	mannose phosphate isomerase
	PK_3	pyruvate $kinase_3$
8–9	—	
10	GOT	glutamateoxaloacetate transaminase
11	LDH-A	lactate dehydrogenase A
	$Es\text{-}A_4$	esterase A_4
	KA	antigen
12	LDH-B	lactate dehydrogenase B
	Pep B	peptidase B
	Gly A^+	serinehydroxymethylase
13	—	
14	NP	nucleoside phosphorylase
15	—	
16	APRT	adenine phosphoribulosyltransferase
17	TK	thymidine kinase
18	Pep A	peptidase A
19	GPI	glucosephosphate isomerase
20	ADA	adenosine deaminase
21	IPO-A	indolephenoloxidase-A
	AVP	antiviral protein
22	—	
Chromosome X	HGPRT	hypoxanthine-guanine phosphoribosyltransferase
	PGK	phosphoglycerate kinase
	GPD	glucose-6-phosphate dehydrogenase
	α-Gal	α-galactosidase
Chromosome Y	—	

Source: F. H. Ruddle, *Nature, 242,* 165–169, 1973.

in a medium containing hypoxanthine (purine source), aminopterin, and thymidine (HAT medium). Since the utilization of thymidine for DNA synthesis depends on thymidine kinase, the hybrid cell will live only so long as the human chromosome containing the TK^+ gene is present. Thus, all cultures of fused cells that survive under these selective conditions necessarily contain the human chromosome with the TK^+ locus. Cytological studies of the human chromosomes present in such cultures showed that the one human chromosome common to all was chromosome 17. Accordingly, the gene for thymidine kinase was assigned to this chromosome. Verification of this assignment was made by use of a structural analogue of thymidine called 5-bromodeoxyuridine (5-BUdR). Cells containing thymidine kinase (TK^+) will incorporate BUdR into DNA where its presence eventually causes death, but cells that are TK^- will survive. Thus cultures that are TK^+ and survive in HAT medium always contain chromosome 17, and these are the cultures that are selectively killed in the presence of BUdR.

Other studies with hybrid cells have also successfully assigned human genes to specific chromosomes (Table 8–1). Although such experiments do not always yield unambiguous results, the method promises to provide much information on human gene locations. Map distances between genes in human linkage groups were first estimated in 1971, and since then progress has been rapid. The proposed map of the long and short arms of the X chromosome has been one result (Figure 8–12). It is likely that a more extensive map of the X

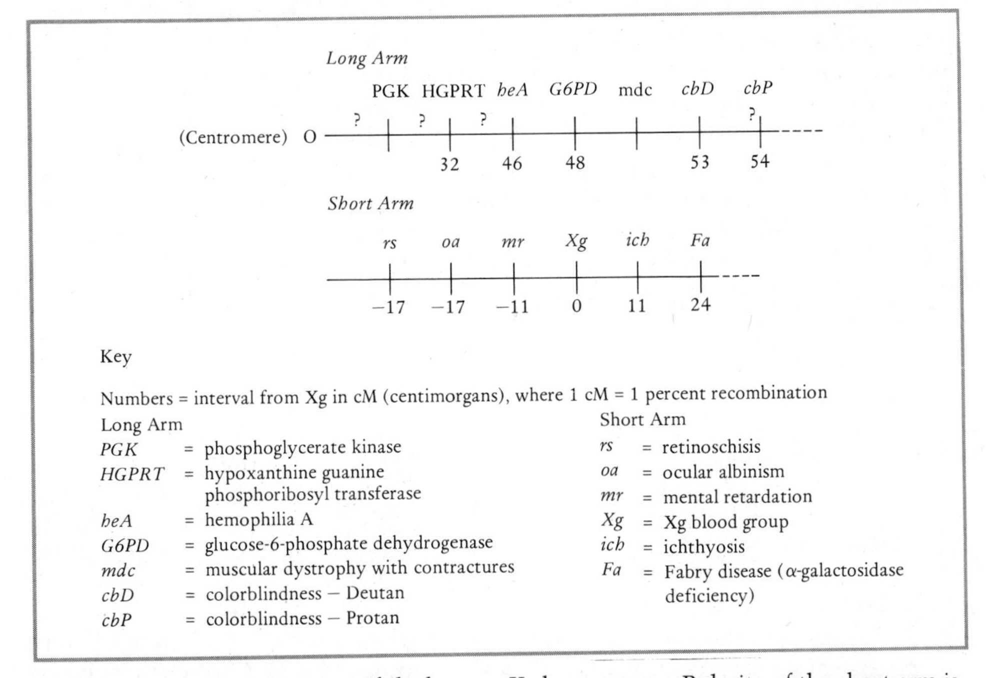

Figure 8–12. Tentative map of the human X chromosome. Polarity of the short arm is arbitrarily represented, for example, all loci may be to the right of *Xg* assuming that *Xg* is near the distal end of the short arm and the centromere is on the right end of this diagram. [Reproduced with permission from V. A. McKusick and G. Chase, *Ann. Rev. Genetics*, 7, 435–473, 1973. Copyright © by Annual Reviews Inc. All rights reserved. Linkage distances, long arm, from V. A. McKusick, Human Chromosome Mapping Newsletter, December 1972.]

chromosome as well as maps for the autosomes will be established within a relatively short time.

LINKAGE ANALYSIS IN HAPLOID EUKARYOTES

Linkage analysis in haploid eukaryotes requires that the products of meiosis (usually spores) be allowed to germinate so that a determination of phenotype can be made. Since the vegetative phase is haploid, all mutant genes are fully expressed, and thus phenotype is directly indicative of genotype.

Tetrad Analysis of Chlamydomonas

One of the first haploid organisms used in genetic studies was *Chlamydomonas* (see Figure 6–26). It will be recalled that in this alga, the four haploid cells produced by meiosis are temporarily retained together within a cystlike structure. This feature of the life cycle permits the recovery and analysis of all of the products resulting specifically from the meiotic divisions of one single diploid cell.

Each such group of four haploid cells is called a *tetrad,* and the genetic analysis of such cells which comprise all four products of meiosis is often called *tetrad analysis*. In performing an experiment, individual tetrads resulting from a cross are isolated and the cells of each grown separately to determine their genotypes. Data on tetrads with like cells are then grouped, and comparisons between the number of tetrads in each such group permit identification of independent assortment or of linkage between the genes used in the cross. The distinction between independent assortment and linkage is best illustrated by example. Let us assume that the mutant genes *a* and *b* are independent and borne on separate chromosomes, and that the following cross is made between mutant and wild type individuals: *a b* × + +. Suppose that from this cross the four cells of 100 separate tetrads are analyzed for genotype, and it is found that, on the basis of the kinds of cells present in each, the 100 tetrads can be grouped into three classes, as follows:

Group 1	Group 2	Group 3
a b	*a* +	*a b*
a b	*a* +	*a* +
+ +	+ *b*	+ *b*
+ +	+ *b*	+ +
45 parental (P)	46 nonparental (NP)	9 tetratype (T)

If we examine these data, it can be seen that in group 1 tetrads, two cells are *a b* and two cells are + + in genotype. Since these gene arrangements are the same as those found in the parents, group 1 tetrads can be designated as parental types (P). Group 2 tetrads can be termed nonparental (NP) because the two types of cells present (*a* + and + *b*) represent a gene combination different from that of either parent. Comparing the number of P and NP tetrads, it can be seen that they are approximately equal in frequency, and thus the four types of cells, *a b*, + +, *a* +, and + *b*, are present in a 1 : 1 : 1 : 1 ratio. Accordingly, we would conclude that genes *a* and *b* were assorting independently and were located on separate, nonhomologous chromosomes.

Group 3 tetrads must next be considered. All four types of gene combinations are present within each tetrad and we designate group 3 as tetratype (T). The presence of tetratypes indicates the occurrence of crossing over between gene *a* and its centromere or between *b* and its centromere. With either exchange second division segregation of the mutant gene from its wild type allele will occur, resulting in the production of a tetratype containing the two kinds of parental and the two kinds of nonparental cells (Figure 8–13). It should be noted that the occurrence of second division segregation does not alter the ratio between parental and nonparental type cells and thus does not interfere with the recognition of independent assortment.

If genes *a* and *b* were linked, the cross *a b* × + +, would produce data quite different from that given above providing, of course, that these gene loci were close enough together so that an exchange between them did not occur in every meiotic cell. In the sample data below, again for 100 tetrads, it can be seen that group 1 tetrads are parental, group 2 nonparental, and group 3 tetratype. Comparing the number of parental and nonparental tetrads (73 : 3), it is

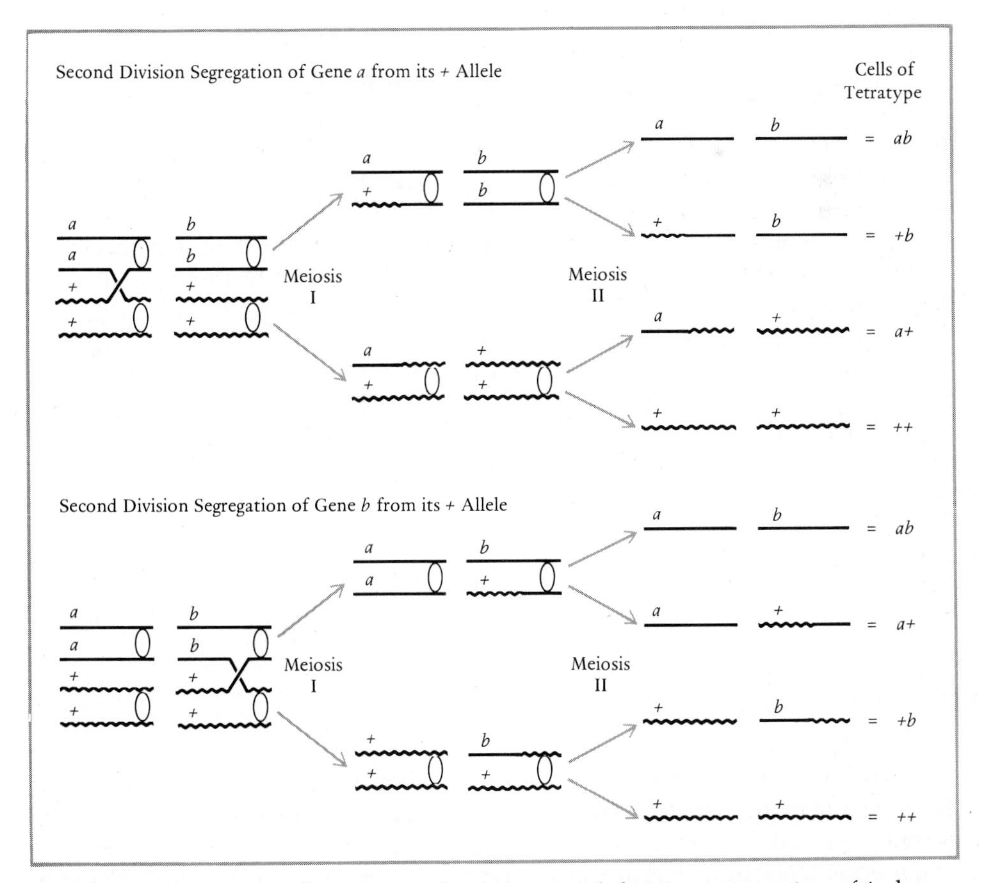

Figure 8–13. Origin of tetratypes through second division segregation of independently assorting genes *a* and *b*. Second division segregation results from crossing over between the gene locus and the centromere.

clear that they are far from equal, and on this basis we would conclude that genes *a* and *b* were linked.

Group 1	Group 2	Group 3
a b	*a* +	*a b*
a b	*a* +	*a* +
+ +	+ *b*	+ *b*
+ +	+ *b*	+ +
73 parental (P)	3 nonparental (NP)	24 tetratype (T)

With respect to origin, parental tetrads arise from meioses where crossing over between genes *a* and *b* did not occur. Nonparental tetrads, however, can arise only from four-strand double crossover exchanges (see Figure 8–3) which accounts for their rarity. The tetratypes also originate from crossing over. Each such tetrad consists of two parental and two nonparental type cells, and it will be recalled that the kinds of exchanges that give rise to this distribution are single crossovers and three-strand double crossovers. Figures 8–2 and 8–3 should be consulted for verification.

An estimate of the map distance between genes *a* and *b* can also be calculated according to the following formula:

$$\frac{\text{NP} + \frac{1}{2}(\text{T})}{\text{total tetrads}} \times 100 = \text{recombination frequency or map distance}$$

In this formula the sum of the total number of nonparentals and one-half the number of tetratypes is divided by the total tetrads scored in the experiment. Only half of the tetratypes are included because only half of the cells of this group are recombinant (a + and + b). All nonparentals are included because in these tetrads, every cell is recombinant. In this example the map distance between genes *a* and *b* would be:

$$\frac{3 + \frac{1}{2}(24)}{100} \times 100 = 15\% \text{ or } 15 \text{ map units}$$

When three linked genes are used in a cross, the gene order as well as recombination frequencies can be determined. For example, from the cross $a\ b\ c \times + + +$, assume that a total of 100 tetrads are analyzed and grouped as follows:

(1)	(2)	(3)	(4)	(5)	(6)	(7)
a b c	*a* + *c*	+ *b c*	*a* + *c*	*a b* +	*a* + *c*	*a b* +
a b c	*a b c*	+ + +	*a* + +	*a b c*	*a* + *c*	*a* + +
+ + +	+ *b* +	*a b c*	+ *b c*	+ + +	+ *b* +	+ *b c*
+ + +	+ + +	*a* + +	+ *b* +	+ + *c*	+ *b* +	+ + *c*
74	13	4	4	2	2	1

To determine recombination frequencies for these genes, it is necessary to examine the relationships between any two genes separate from the third. Each tetrad group can be scored as P, NP, or T, with respect to the arrangement of *a* and *b* only, ignoring *c*. Similarly, genes *a* and *c* can be considered apart from *b*, and genes *b* and *c* apart from *a*.

Figure 8–14. Ascospores from a cross of *Neurospora* showing segregation for an ascospore pigment mutation. [From D. R. Stadler, *Genetics, 41*, 528–543, 1956.]

TETRAD GROUP	*a—b*	*a—c*	*b—c*
1	74 P	74 P	74 P
2	13 T	13 P	13 T
3	4 T	4 T	4 P
4	4 NP	4 T	4 T
5	2 P	2 T	2 T
6	2 NP	2 P	2 NP
7	1 T	1 NP	1 T
	100	100	100
Score:	76 P 6 NP 18 T	89 P 1 NP 10 T	78 P 2 NP 20 T
Recombination frequency	15%	6%	12%

Applying the formula for recombination frequency in each case, the map distance between each pair of genes can be determined. Since those that are farthest apart will exhibit the highest frequency of exchange, we can conclude that *a* and *b* are at opposite ends, while *c* is in the middle of this region of the linkage map, the order being *a—c—b*. Note that when genes *a* and *b* are considered apart from *c*, only 15 percent recombination is evident. However, when gene *c* is included and the separate map distances are added, a total recombination frequency of 18 percent between *a* and *b* is obtained. The presence of gene *c* has permitted recognition of crossovers that would otherwise go undetected.

Ordered Tetrads in Neurospora

Although all products of a single meiosis can be recovered in *Chlamydomonas,* the arrangement of the four haploid cells of a tetrad does not permit inferences as to which chromatids participated in particular exchanges during this process. Such inferences can be made in *Neurospora* because the physical order of spores within the ascus sac directly reflects prior meiotic events. It will be recalled that in *Neurospora* meiosis occurs in a narrow sac, the ascus, where space is so confined that division spindles do not overlap. The sequence of the spores reflects the products of the first and second meiotic divisions, as well as the subsequent mitotic division, there being a total of eight spores produced in each ascus (Figure 8–14). Since the spores of an ascus can be

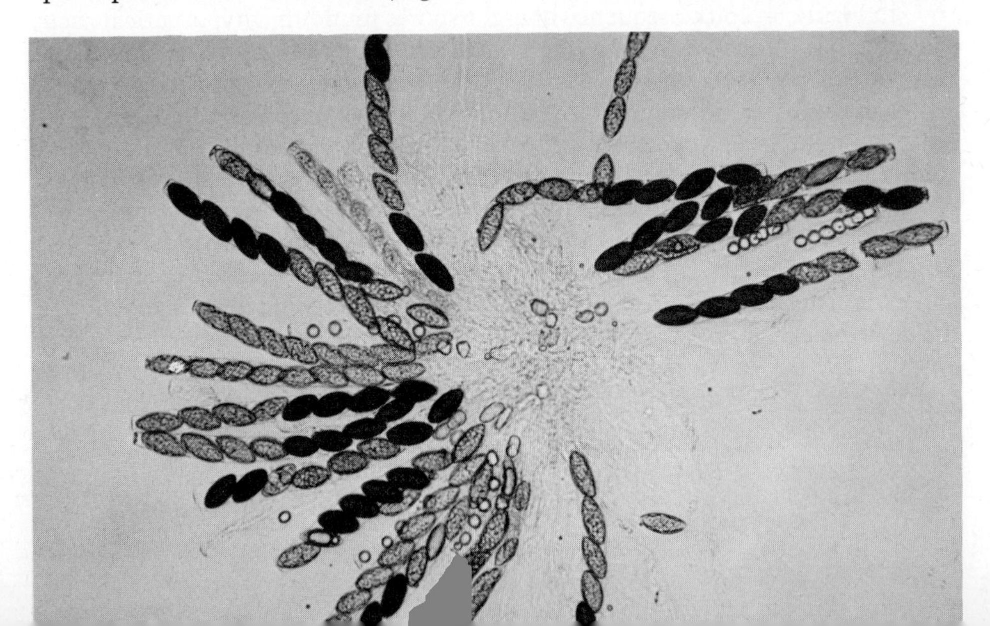

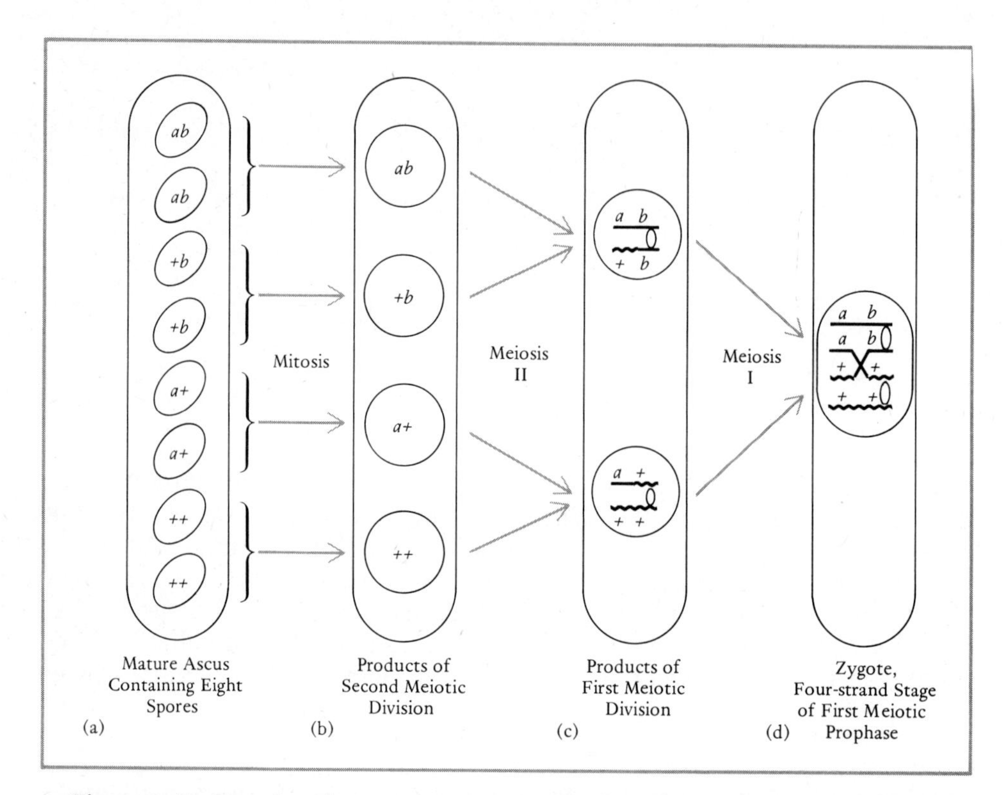

Figure 8–15. Tetrad analysis in *Neurospora.* The direction of the arrows follows the line of reasoning used in deducing preceding events. From the sequence of spores in the mature ascus, (a), the order of the preceding nuclei (b) can be deduced; and from the order of these nuclei (b), the position and orientation of nuclei (c) resulting from segregation at the first meiotic division can be inferred; finally, from the deduced constitution of nuclei (c), the crossover events occurring in the meiotic prophase (d) can be analyzed.

dissected out in order and grown separately to determine genotype, it is possible to visualize the meiotic events which must have occurred to produce any particular spore order under consideration.

In Figure 8–15 the sequence of eight spores found in a hypothetical ascus is traced to its probable origin at the four-strand stage of meiosis. Let us assume that the order of genes and the position of the centromere is known. The eight spores of this ascus are found in the following order: 2 *a b*, 2 + *b*, 2 *a* +, 2 + +. Since four kinds of spores are present, we can characterize this ascus as a tetratype. Each pair of identical spores results from the mitotic division of one of the final products of meiosis. Since the ascus is too narrow to permit movement, we can assume that the sequence of these final meiotic products is also *a b*, + *b*, *a* +, and + +, *in that order*. These nuclei, in turn, reflect the positions and constitutions of their predecessors which are derived from segregation at the first meiotic division. Finally, these predecessors reflect the occurrence of an exchange between nonsister chromatids as well as the actual orientation of homologues on the first meiotic spindle. Thus by tracing backwards from spore order, it is possible to infer much information concerning preceding meiotic events. Because eight spores are present in each ascus, their analysis can be called octad analysis. However, since only four types of spores occur, the term tetrad analysis is still applied.

In *Neurospora* and other fungi with ordered spores it is also possible to map the position of the centromere as if it were a gene. The segregation pattern of each gene involved in the cross is examined (see Figure 6–22). For example, in Figure 8–15 second division segregation is apparent for gene *a*, as opposed to first division segregation of gene *b*. All tetrads (octads) in which second division segregation of the gene in question has occurred are summed and the following formula is applied.

$$\frac{\frac{1}{2}(\text{total second division segregations})}{\text{total tetrads}} \times 100$$

= recombination frequency or map distance between gene and centromere

As an example, if second division segregation for gene *a* was found in a total of 16 tetrads out of 100, then

$$\frac{\frac{1}{2}(16)}{100} \times 100 = 8\% \text{ or } 8 \text{ map units}$$

Only half of the total tetrads showing second division segregation of the gene in question are used because only half the spores of each such tetrad are recombinant types.

Direct evidence that crossing over occurs in the four-strand, rather than the two-strand stage of meiosis was obtained by Lindegren in 1933 from the detailed analysis of ordered spores of *Neurospora*. The consequences of two-strand versus four-strand crossing over are shown in Figure 8–16. It can be seen

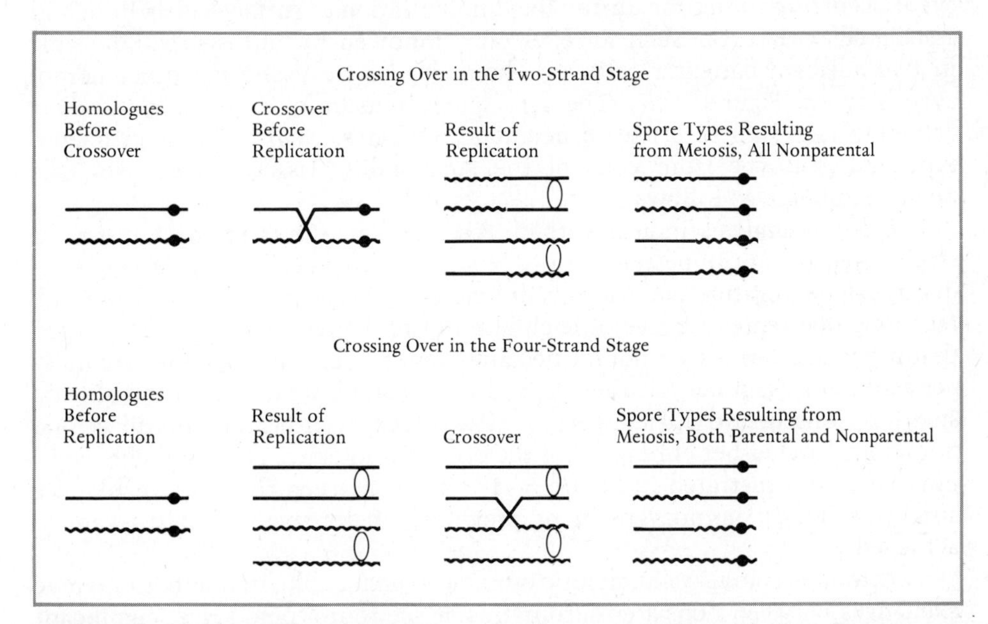

Figure 8–16. The results of crossing over in the two-strand stage before chromosome replication, compared with the results of crossing over in the four-strand stage after chromosome replication.

that crossing over between homologues at the two-strand stage prior to chromosome replication would produce spores all of which would be nonparental in type. However, a similar crossover occurring in the four-strand stage results in both parental and nonparental types. Since tetrad analysis demonstrates that both parental and nonparental spores invariably result from a single exchange, it can be concluded that crossing over occurs after chromosome replication. The analysis of ordered spores also provides clear evidence that maternally and paternally derived centromeres segregate from one another at the first meiotic division.

CROSSING OVER IN SOMATIC CELLS

Although one ordinarily associates crossing over with meiosis, exchanges between nonsister chromatids can also occur in somatic cells, albeit rarely. This phenomenon was first described by Stern in 1936 in *Drosophila* females heterozygous for the recessive sex-linked genes yellow body (*y*) and singed bristles (*sn*), both of which affect superficial and readily visible aspects of the body. Stern found that on rare occasions in females of the genotype $\frac{y\ +}{+\ sn}$, twin spots were present on the external body surface, one spot being yellow with normal bristles, the other wild type in color with singed bristles. Detailed analysis ruled out possible loss of one or the other X chromosome, as in a gynandromorph, and it was concluded that homozygosity of *y* or of *sn* was responsible for the phenotypes of these spots. Further study demonstrated that the origin of these homozygous patches could be traced to a crossover event occurring sometime during the differentiation of surface epithelium and associated structures. Such an exchange, followed by mitosis, could result in two adjacent daughter cells, one homozygous for *y* and the other homozygous for *sn* (Figure 8–17). The subsequent division of these cells and their descendents would give rise to neighboring clones, each different in phenotype and genotype from cells of the surrounding tissues. Details of this phenomenon are as follows.

Genetic analysis indicated that crossing over occurs in the four-strand stage, after chromosome replication. However, subsequent mitosis is not reductional. Segregation does not occur, and each daughter cell is fully diploid, receiving one representative of each chromosome present. It will be recalled that in mitosis tetrads are not formed and homologous chromosomes are independent of one another with regard to position and orientation on the mitotic spindle. Consequently, a daughter cell can receive either chromatid of one homologue and either chromatid of the other homologue. These possible combinations are illustrated in Figure 8–17. It can be seen that as a result of a single exchange, homozygosity of previously heterozygous alleles can be achieved.

Although somatic and meiotic crossing over are alike in that both involve exchanges between nonsister chromatids at the four-strand stage, significant differences between the two are evident. Meiotic crossing over occurs with high frequency and double exchanges are common, whereas somatic crossing over is rare and double exchanges rarer still. Meiotic crossing over most frequently involves genes situated in those areas of the chromosome deemed euchromatic by virtue of their extended and uncoiled condition and consequent invisibility during interphase. In contrast, somatic crossing over most frequently involves genes situated in heterochromatic regions near the centromere. In *Drosophila* this phenomenon may perhaps be attributable to the aggregation of centromeric regions of all chromosomes into a chromocenter, thus permitting close association between homologues in these, and perhaps other regions as well. Other obvious differences between meiotic and somatic crossing over include the absence in somatic exchange of synapsis, tetrad formation, synaptinemal complexes, and segregation.

Mitotic crossing over has also been analyzed in haploid fungi, especially *Aspergillus nidulans*. The life cycle of *Aspergillus* is similar to that of *Neurospora* in that both sexual and asexual spores are formed. Hyphal fusion between strains may also give rise to heterokaryons, and on rare occasions nuclear fusion may occur to form diploid cells. The continuing division of diploid cells forms a diploid sector in the heterokaryotic mycelium. Such sectors form diploid asexual spores which are distinguishable by their larger size and which give rise to diploid mycelia when isolated. Within the diploid mycelium reversion to the haploid state may then occur in some cells through the progressive loss of chromosomes. During this loss either member of a pair of homologues may be discarded, and as a result, when a stable haploid state is reached, the

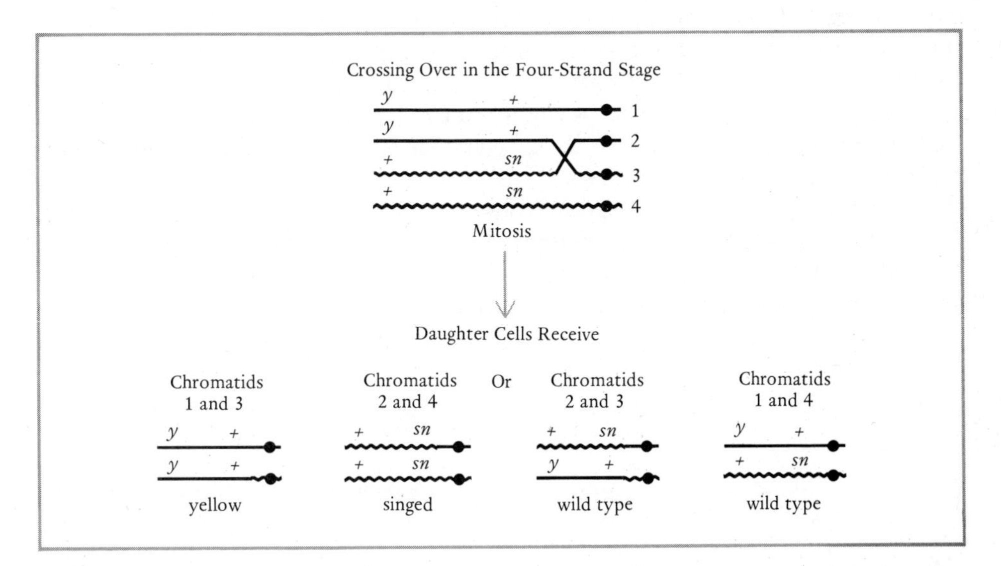

Figure 8–17. The results of crossing over in somatic cells of *Drosophila* females heterozygous for the sex-linked genes, *y* and *sn*. Crossing-over between the locus of *sn* and the centromere at the 4-strand stage, after chromosome replication; chromatids are numbered 1 to 4. Below, alternative types of daughter cells which can result from the crossover event shown above. Note that homologous chromosomes behave independently in mitosis. Each daughter cell is diploid and receives one chromatid representing each homologue.

haploid complement may contain chromosomes derived from both of the original strains whose fusion formed the initial heterokaryon.

On rare occasions mitotic crossing over takes place within a diploid nucleus and when reversion to haploidy occurs, recombinant phenotypes may be observed and their frequency used to assign genes to linkage groups. For example, suppose a heterokaryon is formed between two strains, *a* + and + *b* in genotype, and it is not known if genes *a* and *b* are linked. If nuclear fusion occurs, a heterozygote will be produced whose genotype can be symbolized as *a* +/+ *b* (linkage) or as *a*/+ *b*/+ (nonlinkage). If *a* and *b* are unlinked, reversion to haploidy will produce cells containing either homologue of a pair and therefore spores of genotypes *a b*, *a* +, + *b*, and + + will be recovered in approximately equal frequency. However, if the loci are linked, practically all spores will be parental in genotype, *a* + or + *b*, and only very rarely will the recombinant types, *a b* and + +, be found. By this means linkage between two or more genes can be determined.

The entire process of nuclear fusion, mitotic crossing over, and return to haploidy, which occurs quite apart from meiosis, has been termed the *parasexual cycle* by Pontecorvo. For some fungi it provides an alternate means by which genetic exchange can be accomplished. A detailed discussion of the phenomenon can be found in Pontecorvo and Käfer (1958).

It should be emphasized that crossing over is a normal physiological process, responsive to the internal and external environment of the organism. It is therefore not surprising that numerous factors have been found to influence the process of genetic exchange. These include intrinsic factors such as age, sex, and presence of chromosome abnormalities or of mutants whose expression affects some aspect of the meiotic process. Environmental factors such as increased temperature, use of radiation, or exposure to chemicals which cause chromosome breakage can act to increase the frequency of exchange. The experimental use of a number of these variables has provided valuable insights into the process of crossing over.

The recognition and analysis of linkage has provided major advances in our knowledge of heredity. Development of linkage maps has amply demonstrated the linear arrangement of genes on the chromosome, and the occurrence of crossing over can be seen to provide great diversity in genotype. This diversity must be added to that already made possible through independent assortment.

PROBLEMS

8–1. The following characters were identified in a haploid organism: *A, a; B, b; C, c; D, d; E, e; F, f;* and *G, g*. A series of crosses involving these characters gave the following results.

Cross	Progeny			
AE × *ae*	80 *AE*	80 *ae*	20 *Ae*	20 *aE*
AF × *af*	45 *AF*	45 *af*	5 *Af*	5 *aF*
AB × *ab*	62 *AB*	62 *ab*	63 *Ab*	62 *aB*
BC × *bc*	75 *BC*	75 *bc*	25 *Bc*	25 *bC*
BG × *bg*	95 *BG*	95 *bg*	5 *Bg*	5 *bG*

CROSS	PROGENY (cont'd)			
$CG \times cg$	40 CG	40 cg	10 Cg	10 cG
$DE \times de$	45 DE	45 de	5 De	5 dE
$DF \times df$	30 DF	30 df	20 Df	20 dF
$FG \times fg$	72 FG	73 fg	72 Fg	73 fG

a. Which genes are linked?

b. What is the order of linked genes and the recombination frequencies between them?

8–2. The trihybrid *AaBbCc* is testcrossed to the triple recessive *aabbcc,* and the following phenotypes are obtained in the progeny: 64 *abc,* 2 *abC,* 11 *aBc,* 18 *aBC,* 14 *AbC,* 17 *Abc,* 3 *ABc,* 71 *ABC.*

a. Which of these loci are linked?

b. What is the correct genotype of each parent?

c. Determine the map distances between these genes.

8–3. Assume that a crossover and resulting chiasma occurs between two gene loci 100 percent of the time.

a. What would be the percentage of recombinant chromosomes among the progeny?

b. If a chiasma occurred 50 percent of the time, what would this percentage be?

8–4. The recessive genes c^1 and c^2 produce similar phenotypic effects when homozygous. When genes c^1 and c^2 are used in testcrosses with a third recessive gene x, the following results are obtained.

CROSS	RESULT	
$+\ +/c^1\,x \times c^1\,x/c^1\,x$	$+\ +$	85
	$c^1\ x$	86
	$c^1\ +$	14
	$+\ x$	15
		200 total progeny
$c^2\ +/+\ x \times c^2\,x/c^2\,x$	$+\ +$	22
	$c^2\ x$	23
	$c^2\ +$	76
	$+\ x$	79
		200 total progeny

a. Are genes c^1 and c^2 allelic?

b. Are genes c^1 and c^2 linked?

8–5. In *Drosophila* the genes cinnabar *(cn)* and vestigial wings *(vg)* are approximately 10 map units apart. The cross $cn\ +/+\ vg \times cn\ vg/cn\ vg$ produced 1000 progeny, of which 487 were cinnabar and 513 were vestigial. Which parent was the female?

8–6. In a given plant long leaves *(S)* and green veins *(Y)* are dominant, respectively, over short leaves *(s)* and yellow veins *(y)*. The cross $SSYY \times ssyy$ produced an F_1 *SsYy*. When F_1 plants were inbred, the F_2 consisted of 570 long, green individuals and 190 short, yellow. Are the genes *S* and *Y* linked?

8–7. In *Drosophila* the recessive sex-linked genes abnormal eye facet *(fa)* and singed bristles *(sn)* show 18 percent recombination.

a. If a singed male is crossed to a $fa\ +/fa\ +$ female, what phenotypes are expected in the F_1?

b. If F_1 males and females are inbred, what phenotypic proportions would be expected to occur in F_2 males and females?

8–8. In *Drosophila* females heterozygous for the recessive genes *p*, *q*, and *r* were crossed to males homozygous for these three mutant genes. The phenotypes of the resulting progeny were:

+ + +	71	*p* + +	4
p q r	62	+ *q r*	3
p q +	27	*p* + *r*	405
+ + *r*	36	+ *q* +	392
			1000 total progeny

Determine the order of the genes on the chromosome, the map units between these genes, and the coefficient of coincidence.

8–9. Assume the four genes *a*, *b*, *c*, and *d* are all recessive to the wild type allele. A female heterozygous for these genes was crossed to a male homozygous for these mutants. From the following progeny data, determine the order of these genes on the chromosome and the distances between them in map units.

+ *b c* +	261	*a* + + *d*	304
a b c d	4	*a b* + *d*	64
a b + +	90	+ + *c d*	102
a + *c d*	52	*a* + + +	1
+ *b c d*	2	+ + + *d*	2
a b c +	3	+ + *c* +	76
+ + + +	3	+ *b* + +	36
			1000 total progeny

8–10. The genes *w*, *s*, and *t* are recessive, autosomal, and linked, with *w* lying at position 10, *s* at 20, and *t* at 40 of the linkage map. The coefficient of coincidence is 0.6. A testcross *w* + *t*/+ *s* + × *w s t*/*w s t* results in a total progeny of 500. Predict the number of individuals expected to be + + *t*, + *s t*, and + *s* + in phenotype.

8–11. The genes *x*, *y*, and *z* are linked and have the following positions:

x	*y*	*z*
10	26	36

a. A testcross *X Y Z*/*x y z* × *x y z*/*x y z* results in a progeny of 1000 individuals. On the basis of the linkage map how many double crossovers theoretically would you expect from the cross?

b. If the coefficient of coincidence were 0.25, how many doubles would actually arise?

8–12. Assume that in *Drosophila* the recessive autosomal mutants black *(b)*, net wing *(n)*, and light eyes *(lt)* are located at positions 48, 50, and 55, respectively, on the linkage map and that the coefficient of coincidence is 0.1. If you were to perform a testcross involving these three genes, theoretically how many flies would you have to raise, assuming that all eggs laid developed to maturity, to obtain one individual that showed a double crossover?

8–13. Assume the presence in corn of the recessive genes *ds* and *mp*. The genes are linked and 20 map units apart. From the cross *ds mp*/ + + × *ds* + / + *mp*, what percentage of the progeny would be expected to be both *ds* and *mp* in phenotype?

8–14. In sweet peas two recessive linked genes *a* and *b* are 8 map units apart. When plants of the genotype *a* +/+ *b* are allowed to self-fertilize, what percent of the resulting progeny would be expected to be + *b* in phenotype?

8–15. What is the percent recombination between two loci in a species where crossing over occurs with equal frequency in both sexes, if a mating between identical dihybrids *(Ab/aB)* produces four equally viable classes of offspring, the smallest class comprising 0.8 percent of all offspring?

8–16. Assume a diploid somatic cell of the genotype

a + (o = centromere)
——o—
——o—
+ *b*

a. If a crossover occurred between the loci of *a* and *b*, what genotypes could occur in the daughter cells after mitosis?
b. If the crossover occurred between gene *b* and the centromere, what daughter cells could result?
c. If a two-strand double crossover occurred, one exchange between genes *a* and *b*, the other between *b* and the centromere, what daughter cells could result?

8–17. The following data were obtained in *Neurospora* by analysis of ordered spores from the cross *a* + + × + *b c*. The total number of tetrads was 100.

(1)	(2)	(3)	(4)	(5)	(6)
a + +	*a b c*	*a b c*	*a* + +	*a* + +	+ + +
a + +	*a b c*	+ *b c*	+ + +	*a* + *c*	+ + *c*
+ *b c*	+ + +	*a* + +	*a b c*	+ *b* +	*a b* +
+ *b c*	+ + +	+ + +	+ *b c*	+ *b c*	*a b c*
40	40	5	5	5	5

a. Which of these genes are linked?
b. What is the map distance between them?
c. What is the map distance between gene *a* and its centromere?

8–18. In *Neurospora* a cross between strains *x y* and + + yielded 100 tetrads in the following groups.

(1)	(2)	(3)	(4)	(5)	(6)	(7)
x y	*x* +	*x y*	*x y*	*x y*	*x* +	*x* +
x y	*x* +	+ +	*x* +	*x* +	*x y*	*x y*
+ +	+ *y*	+ +	+ *y*	+ +	+ *y*	+ +
+ +	+ *y*	*x y*	+ +	+ *y*	+ +	+ *y*
92	2	2	1	1	1	1

a. Are the genes *x* and *y* linked?
b. If they are not linked, determine the crossover distance between each gene and its centromere. If linked, construct a map showing crossover distances between genes *x*, *y*, and the centromere.

8–19. In *Neurospora* the genes *a*, *b*, and *c* are linked in the order *a b c*—o (centromere). Tetrad analysis of ordered spores resulting from the cross + *b c* × *a* + + gave the following spore orders.

(1)	(2)	(3)	(4)
a b c	+ *b c*	*a* + *c*	+ *b c*
+ + *c*	+ *b* +	*a b c*	*a b c*
+ *b* +	*a* + *c*	+ + +	*a* + +
a + +	*a* + +	+ *b* +	+ + +

For each group sketch the four chromatids at meiosis, indicating the points where exchanges must have occurred to result in the spore order given.

8–20. Assume that the crossover events diagramed below occur during meiosis in a *Neurospora* zygote of genotype $a\ b\ +/+\ +\ c$, where the centromere is to the left of gene a. List the spore order that would result from the exchanges illustrated; assume that spore types occur in the order in which the chromatids are numbered.

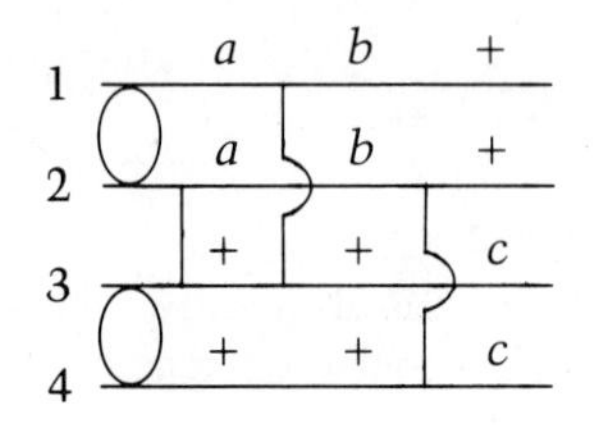

REFERENCES

ANDERSON, E. G., 1925. Crossing over in a case of attached X chromosomes in *Drosophila melanogaster*. *Genetics, 10,* 403.

DAVIDSON, R. L., 1973. *Somatic Cell Hybridization: Studies on Genetics and Development*. Addison-Wesley, Reading, Mass.

DAVIDSON, R. L., 1974. Gene expression in somatic cell hybrids. *Ann. Rev. Genet., 8,* 195.

IMMER, F. R., 1930. Formulae and tables for calculating linkage intensities. *Genetics, 15,* 81.

KROOTH, R. S., G. A. DAVIDSON, AND A. A. VELAZQUEZ, 1968. The genetics of cultured mammalian cells. *Ann. Rev. Genet., 2,* 141.

MATHER, K., 1951. *The Measurement of Linkage in Heredity,* 2nd ed. Wiley, New York.

PONTECORVO, G., AND E. KAFER, 1958. Genetic analysis based on mitotic recombination. *Adv. Genet., 9,* 71.

PONTECORVO, G., 1958. *Trends in Genetic Analysis.* Columbia University Press, New York.

RENWICK, J. H., 1971. The mapping of human chromosomes. *Ann. Rev. Genet., 5,* 81.

RUDDLE, F. H., 1972. Linkage analysis using somatic cell hybrids. *Adv. Human Genet., 3,* 173.

STERN, C., 1936. Somatic crossing over and segregation in *Drosophila melanogaster*. *Genetics, 21,* 625.

Chapter 9

CHANGES IN CHROMOSOME NUMBER AND STRUCTURE

CONTINUING application of the methods of cytogenetic analysis initiated by Bridges soon revealed the occurrence of several types of chromosomal variation in a number of plants and animals. Such variations include (1) the presence of multiple sets of chromosomes, (2) the gain or loss of individual chromosomes from a set, (3) the gain or loss of individual genes or chromosome segments, and (4) modifications in chromosome structure leading to new linkage relationships between genes.

It was also discovered that some of these variations could be induced in experimental organisms by exposure to mutagenic agents such as X rays. Structurally modified chromosomes produced by such treatment were then exploited in efforts to locate genes in cytological, as opposed to linkage, maps. These efforts met with resounding success in *Drosophila* because of the presence of giant chromosomes in larval tissues. The study of these chromosomes has added greatly to our understanding of the organization and activities of the genome. In this chapter the nature, consequences, and cytological analysis of chromosomal variation are reviewed.

MULTIPLE SETS OF CHROMOSOMES: POLYPLOIDY

The life cycle of eukaryotic organisms is typically characterized by an alternation between the haploid (n) and the diploid $(2n)$ states. However, chromosome complements other than haploid and diploid are known, and the general term *euploidy* is used to refer to the presence of any multiple of whole chromosome sets beyond the haploid condition. The condition of the presence of two sets is called diploidy, three sets triploidy ($3n$), four sets tetraploidy ($4n$), five sets pentaploidy ($5n$), and so on. Individuals with more than two sets are usually called *polyploids*. Although such a condition is exceedingly rare in bisexual animals, a large proportion of the higher plants (almost half the angiosperms) are characterized by the presence of more than two sets of chromosomes. It is clear that polyploidy has played a significant role in their evolution. For this reason, the origin of some types of polyploids will be considered.

The triploid or $3n$ state can arise from the union of an egg with two sperm or from the union of a haploid with a diploid gamete which has resulted from the failure of segregation at meiosis. Diploid gametes are also produced by tetraploids. Once formed, triploid plants are usually robust and viable although highly sterile, due to the low frequency with which gametes with balanced chromosome sets are produced. It will be recalled that when three homologues of any chromosome are present, segregation at meiosis I results in the movement of two of these homologues into one daughter cell and the remaining homologue into the other daughter cell. With the occurrence of independent assortment the resulting gametes come to contain varying numbers of different homologues and it is rare indeed that a true haploid or diploid gamete is formed. Although such unbalanced sets of chromosomes lead to sterile pollen or inviable zygotes, triploid plants overcome these disadvantages to some extent through asexual reproduction, and plant breeders propagate such strains with cuttings or by grafting. Triploids are economically useful because they produce few seeds and are therefore more palatable to the consumer.

A tetraploid, containing four sets of chromosomes, is usually also partially sterile. Although one might expect that segregation should routinely produce diploid gametes, this is not always the case. Synapsis can occur at any given point between any two homologues and consequently very complex associations are formed when four homologues are present. These associations frequently lead to segregation abnormalities.

Presumably because of these segregation abnormalities, polyploids whose chromosomes are derived all from the same species *(autopolyploids)* are uncommon. By far the more frequent case in higher plants is a polyploid whose chromosome sets are derived not from the same species, but from two or more closely related species *(allopolyploids)*. Many cultivated varieties of plants (for example, cotton, tobacco) have originated as allopolyploids.

As an example, suppose species A is diploid and has the chromosome complement PP QQ RR SS, and species B, also diploid, has the chromosome complement WW XX YY ZZ. Natural or artificial hybridization between the two will produce offspring P Q R S W X Y Z in constitution. Such progeny, though perhaps viable, will be sterile. No chromosome present has a pairing partner, and each can pass at random to either pole during meiosis I. As a result, gametes containing exclusively the full set of chromosomes of species A or of species B or all chromosomes of both species will be exceedingly rare. However, if the chromosome complement of the hybrid were to undergo doubling, either spontaneous or induced during early development, an individual diploid for the chromosomes of both species would result: PP QQ RR SS WW XX YY ZZ. Such an allopolyploid individual is called an *amphidiploid*. Amphidiploids are vigorous and fertile. Each chromosome has a single pairing partner, thus avoiding the difficulties found with autotetraploids, and meiosis results in gametes containing a haploid set from each original parental species. Such amphidiploids can resemble one parent more closely than the other and often combine the desirable features of both. For this reason they have been induced between closely related species of many ornamental and economically important plants. Induction of amphidiploidy has been achieved by treatment of buds or seeds with colchicine, a cell poison which causes dissolution of the mitotic spindle. With the dissolution of the spindle replicated chromosomes are not distributed to daughter cells, and a doubling of the chromosome number in treated cells is thereby obtained.

Polyploidy is possible in plants because of the general absence of sex chromosomes, most plants being hermaphrodites. In addition, even when partial sterility is present, plants are able to establish themselves through asexual reproduction. In contrast, most animals have well-defined separate sexes characterized by the presence of specific sex chromosomes. Multiple chromosome sets lead to variations in the numbers of the sex chromosomes (for example, XXXY or XYYY), a situation incompatible with the normal sex-determining mechanism. As a result animal polyploids are restricted to groups that reproduce asexually or are hermaphroditic or parthenogenetic. Examples are earthworms, some types of shrimp, and parthenogenetic species among the insects and lizards.

As opposed to an increase in the number of sets of chromosomes, a decrease is also possible, as for example, in the case of a *monoploid,* that is, an individual that is haploid at a stage in the life cycle when diploidy should be present. Monoploids have been produced in animals by artificial parthenogenesis. Development is usually abnormal and survivors few. Monoploids in plants are somewhat stronger, though sterile. In both groups monoploids are essentially curiosities, rather than viable alternatives to diploidy or polyploidy.

GAIN OR LOSS OF INDIVIDUAL CHROMOSOMES: ANEUPLOIDY

Aneuploids are individuals with unbalanced chromosome sets. Their chromosome complement contains at least one more or one less chromosome than the normal diploid number. If one homologue of the normally diploid complement is present three times, the condition is called *trisomy,* an individual of this type is called a *trisomic,* and the condition can be designated by the symbol $2n + 1$. *Tetrasomy* ($2n + 2$) indicates that one chromosome of the complement is present four times. A $2n + 1 + 1$ individual would be trisomic for two different homologues. The reciprocal situation, involving loss of a chromosome from a pair, is called *monosomy* ($2n - 1$) and the loss of a whole pair of chromosomes is known as *nullisomy* ($2n - 2$).

The usual origin of these aneuploids is nondisjunction of homologues during meiosis. If one pair of homologues fails to disjoin, $n + 1$ and $n - 1$ gametes will result. If these gametes participate in fertilization with a normal gamete of the opposite sex, $2n + 1$ and $2n - 1$ individuals will be formed. The consequences of aneuploidy involving the sex chromosomes have already been reviewed. It will be recalled that nondisjunction of the sex chromosomes can give rise to X0 ($2n - 1$), XXX ($2n + 1$), XXY ($2n + 1$), and even XXYY ($2n + 2$), and other types. Such individuals, though abnormal, are frequently viable.

Nondisjunction of autosomes is more serious for it leads to the loss (or gain) of components essential to the genome of both males and females. Loss of one autosome of a pair (monosomy) is almost invariably lethal. One exception, found in *Drosophila,* is the loss of one of the two tiny, dotlike fourth chromosomes. This condition does permit the survival of some few, but abnormal offspring.

The addition of an extra autosome to the diploid complement is often better tolerated, especially in plants, and usually gives rise to an identifiable phenotype. For example, A. F. Blakeslee was able to distinguish twelve specific phenotypes different from the normal in the seed pod of the Jimson weed, *Datura*. Each such phenotype was the result of trisomy for one of the twelve chromosomes of the diploid complement. In *Drosophila* trisomy for the tiny fourth chromosome is a viable condition accompanied by few phenotypic manifestations. However, trisomy for other and larger autosomes is lethal.

In humans the most frequently encountered aneuploid condition is trisomy for chromosome 21 which gives rise to a phenotype known as Down's syndrome or mongolism (Figure 9–1). Among other phenotypic manifestations

Figure 9–1. 21 Trysome (formerly called Down's Syndrome) in a 1-year-old girl. [Courtesy Sanford Schneider, M.D., Loma Linda School of Medicine.]

TABLE 9–1
Risk of Down's Syndrome in Child

Age of mother	Frequency at any pregnancy
Under 29	1 in 1500
30–34	1 in 700
35–39	1 in 300
40–44	1 in 100
45–49	1 in 50
All mothers	1 in 910

Source: J. S. Thompson and M. W. Thompson, *Genetics in Medicine,* 2nd ed., W. B. Saunders, Philadelphia, 1973.

infants with Down's syndrome exhibit flat noses, protruding tongues, epicanthal folds giving a somewhat slanted appearance to the eyes, and severe mental retardation. The risk of trisomy 21 increases with increasing maternal age (Table 9–1; see Figure 7–9). Other and rarer human trisomic conditions have also been identified. Trisomy for chromosome 18 results in malformations and is lethal during the first six months of life, while trisomy 13 produces even more severe abnormalities.

Besides nondisjunction, partial monosomy and trisomy can also result from the presence of a chromosome aberration (translocation) in the parents (see below). Chromosome heteroploidy is known to account for some 40 percent of early spontaneous abortions. These variations in chromosome number include monosomy for the X chromosome (X0) or for various autosomes, autosomal trisomy, and also, more rarely, triploidy and tetraploidy. Chromosome mosaicism, or *mixoploidy,* also found in humans, is caused by nondisjunction occurring during an early cleavage division. The individual is composed of at least two cell lines, and the tissues of the body are composites of cells containing different chromosome numbers. Usually the sex chromosomes are involved, an early chromosome loss producing more serious effects than one occurring later in development.

GAIN OR LOSS OF GENES OR CHROMOSOME SEGMENTS: DEFICIENCY AND DUPLICATION

Deficiency

A deficiency is a loss or deletion of one or more gene loci from a chromosome. Deficiencies arise through chromosome breakage caused by radiation, chemicals, drugs, viruses, or other agents. Although a broken chromosome can undergo restitution, a larger or smaller fragment may fail to be incorporated once more into a chromosome during repair processes. If the fragment lacks a centromere, it will be unable to form spindle fiber attachments at the next cell division and consequently will not be included in the nucleus of either daughter cell. Such fragments are lost from the genome and ultimately degraded.

Gametes carrying deficiencies and duplications are also produced through unequal crossing over and by individuals heterozygous for inversions and translocations (to be discussed later in this chapter).

As a general rule (but with some exceptions) deficiencies are lethal when present in the homozygous condition. In the heterozygous state small deficiencies are usually viable. However, if the deletion is large, the condition will approach monosomy and will be inviable. Homozygous deficiencies are often lethal even in cells surrounded by normal tissue. For example, in *Drosophila* somatic crossing over can produce homozygosity of alleles in daughter cells, these giving rise to twin spots or patches of tissue. If through somatic crossing over, a heterozygous deficiency is rendered homozygous, the condition is almost always cell-lethal, no spot representing this type being found.

Deficiencies occur in humans as well as in other organisms. For example, partial deletion of the short arm of chromosome 5 results in a condition called the *cri du chat* syndrome, because the cry of infants with this abnormality resembles the mewing of a cat. Such infants are microcephalic and exhibit severe mental regardation (Figure 9–2). Deletions involving one or the other arm of chromosome 18 have also been reported. In either case affected individ-

Figure 9–2. An infant with *cri du chat* syndrome. [From J. S. Thompson and M. W. Thompson, *Genetics in Medicine,* 2nd ed., W. B. Saunders, Philadelphia, 1973.]

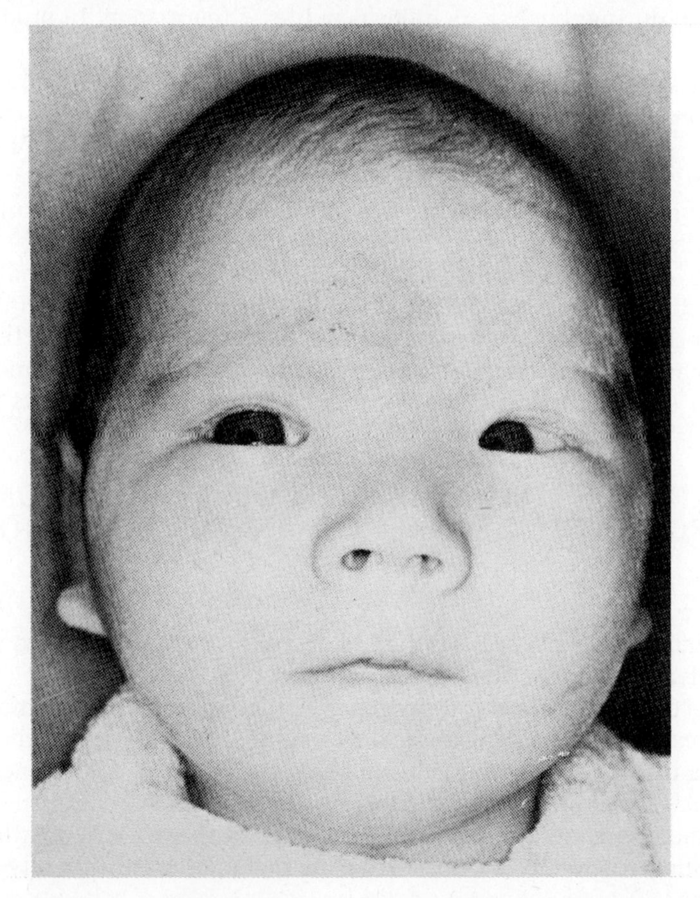

uals are retarded in both mental and physical development. One type of leukemia (chronic myelocytic leukemia) has been consistently associated with the presence of a deleted chromosome 22, called the Philadelphia chromosome because of its original discovery in a resident of that city.

The presence of a deficiency for a given gene can be recognized by the phenotypic expression of a recessive allele when the deficiency and the recessive are present together in the same individual. For example, suppose wild type *Drosophila* males are irradiated and a deficiency for the wild type allele of the recessive mutant white eyes is produced. If the males are crossed to homozygous white-eyed females, fertilization of an egg by a deficient X-bearing sperm will produce a daughter heterozygous for white and the deficiency for white. In the absence of the wild type allele the recessive gene will be expressed in the phenotype and the daughter will have white eyes. This unexpected expression of a recessive in the presence of a deficiency for its wild type allele is called *pseudodominance*.

Pseudodominance presents a means whereby the physical position of a gene in a chromosome can be assigned. The deficient chromosome is compared cytologically with its normal counterpart and the gene location is assigned to the region present in the normal and absent in the deficient chromosome. This is a most difficult task where chromosomes are small and lacking in topographical features such as knobs, bands, or constrictions, and often the most that can be done is assignment to one chromosome arm or the other or to major regions within an arm. In *Drosophila,* however, cytological mapping, while not easy, can be carried out with great precision because of the presence of giant chromosomes.

In a number of groups of insects, particularly the Diptera to which *Drosophila* belongs, tissues specific to the larval stage grow by increase in cell size, rather than by increase in cell number. As a result cells of the salivary glands, the midgut, and other tissues become relatively enormous by late larval stages. The chromosomes of these cells appear to be in a permanent interphase to early prophase condition and exhibit somatic synapsis, in that homologues are precisely paired with one another. Although mitosis has been suspended, these chromosomes undergo repeated replication until, in the largest cells, a pair of homologues comes to contain around 1024 separate strands (chromomonemata), all aligned exactly parallel to one another. Such chromosomes, called *polytene* chromosomes, give the appearance under the microscope of giant cables composed of a multitude of separate fibers. Their great usefulness in cytological analysis stems not only from their size, but also from their longitudinal differentiation by crossbands of varying sizes (Figure 9–3). A crossband can be interpreted as a region where each chromonema of the bundle is coiled into a knotlike arrangement or chromomere, and the parallel alignment of numerous such identical, condensed chromomeres presents the appearance of a vertical crossband (Figure 9–4). The size, shape, and linear arrangement of these bands is the basis for salivary chromosome maps in which the crossbands present in each pair of homologues are figured, enumerated, and catalogued by division and subdivision (Figure 9–5).

It is easy to see how a deficient region can be recognized in such chromosomes. If the cell is heterozygous for a deficient and a normal chromosome, somatic synapsis will insure the precise pairing, band for band, of all homolo-

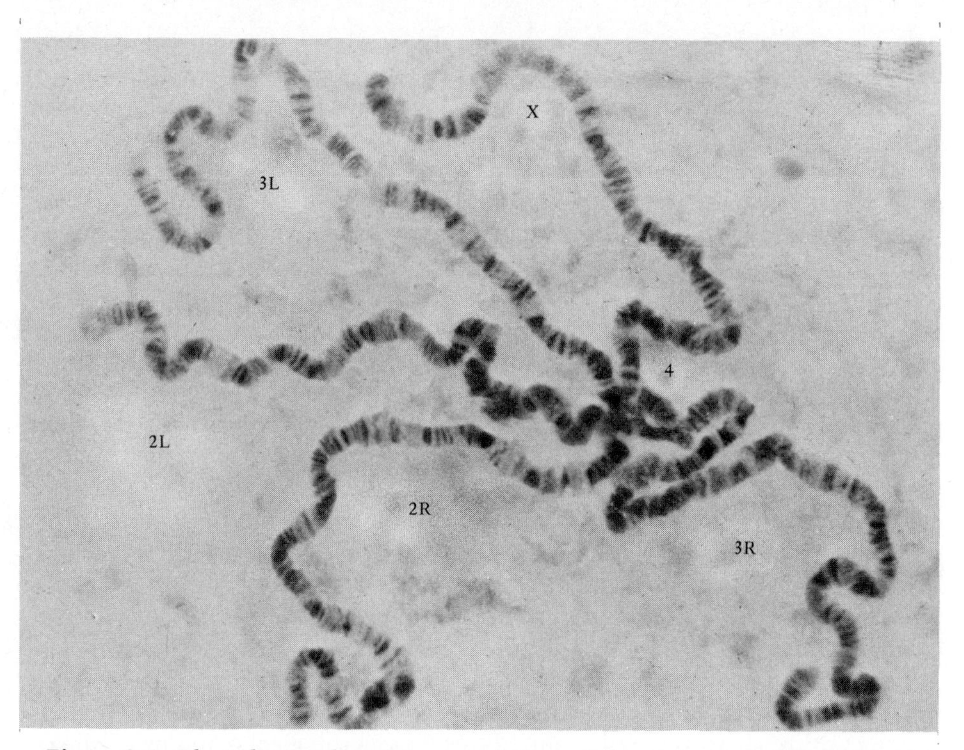

Figure 9–3. The salivary chromosomes of *Drosophila melanogaster* (female). Centromeric regions of all chromosomes are aggregated into a chromocenter from which extend the synapsed arms of the homologues: X chromosome (X), right and left arms of chromosome 2 (2R, 2L), right and left arms of chromosome 3 (3R, 3L), tiny fourth chromosome protruding from chromocenter (4). [B. P. Kaufman, Induced chromosomal rearrangements in *Drosophila melanogaster*, *J. Hered. 30,* 179–190, 1939.]

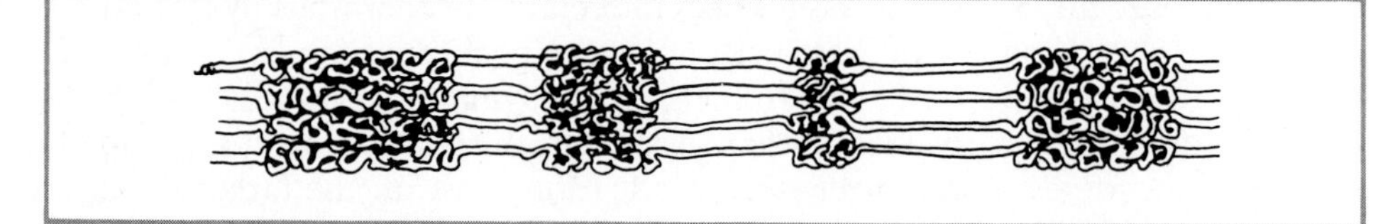

Figure 9–4. A model of band-interband structure in a giant polytene chromosome. [From E. J. Dupraw and P. M. M. Rae, *Nature, 212,* 598–600, 1966.]

Figure 9–5. Tip end of the left arm of chromosome 2 of *Drosophila melanogaster* illustrating classification of bands in salivary chromosomes. [From C. B. Bridges, *J. Heredity, 26,* 60–64, 1935.]

gous regions. In the deficient region, however, the normal chromosome will have nothing with which to pair and will therefore form a small unpaired buckle-out to the side. Genes showing pseudodominance in such a heterozygote can then be assigned to the group of bands forming the buckle-out (Figure 9–6).

Precise localization of a gene to a single band is very difficult. The usual procedure employs overlapping deletions. As a simple example, suppose recessive genes *a* and *b* are very close together on the linkage map. Gene *a* shows pseudodominance with deficiency x, but not with deficiency y; gene *b* is pseudodominant when heterozygous with deficiency y, but not with x; and both genes *a* and *b* are expressed in the presence of deficiency z (Figure 9–7).

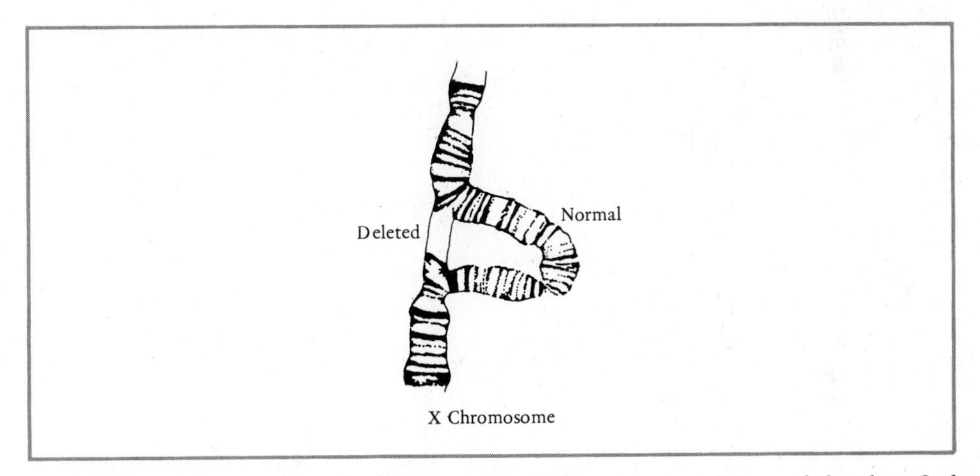

Figure 9–6. The presence of a deletion in the X chromosome of *Drosophila,* identified by a buckling out of the normal chromosome. [From T. S. Painter, *J. Heredity, 25,* 464–476, 1934.]

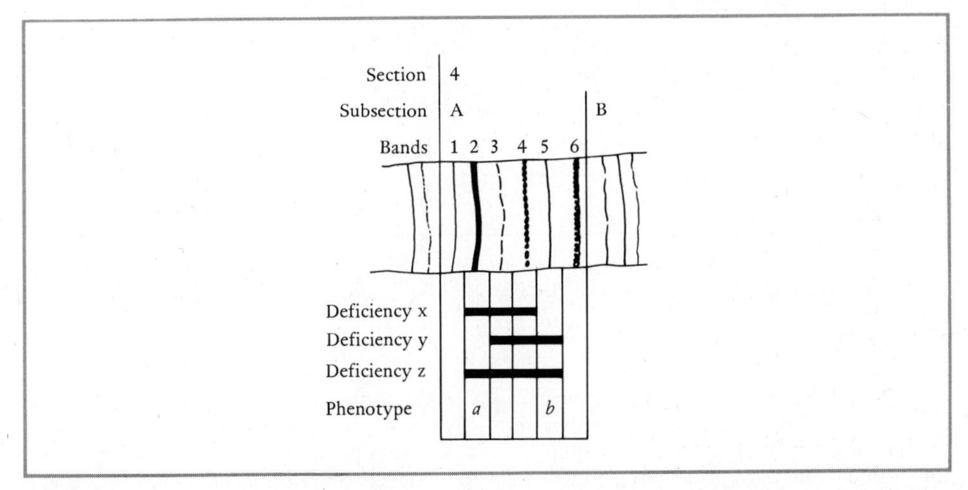

Figure 9–7. Hypothetical example of deletion mapping which localizes gene *a* to band 4A2 and gene *b* to band 4A5.

Salivary chromosome analysis in this case might show that deficiency x included bands 4A2 through 4A4, deficiency y included bands 4A3 through 4A5, and deficiency z included bands 4A2 through 4A5. By correlating phenotype with the extent of the cytological deficiency, we could conclude that gene *a* was associated with band 4A2 and gene *b* with band 4A5.

In actual practice the possibility of obtaining three deletions that would exactly bracket a given locus is remote, and thus a number of overlapping deletions must ordinarily be analyzed if any success is to be achieved. Despite these difficulties a number of genes of *Drosophila* have been localized to specific bands of the salivary chromosomes by the use of deletions (as well as translocations). A particularly elegant example of this method of deficiency mapping is the localization of genes in the white-facet region of the X chromosome (Figure 9–8).

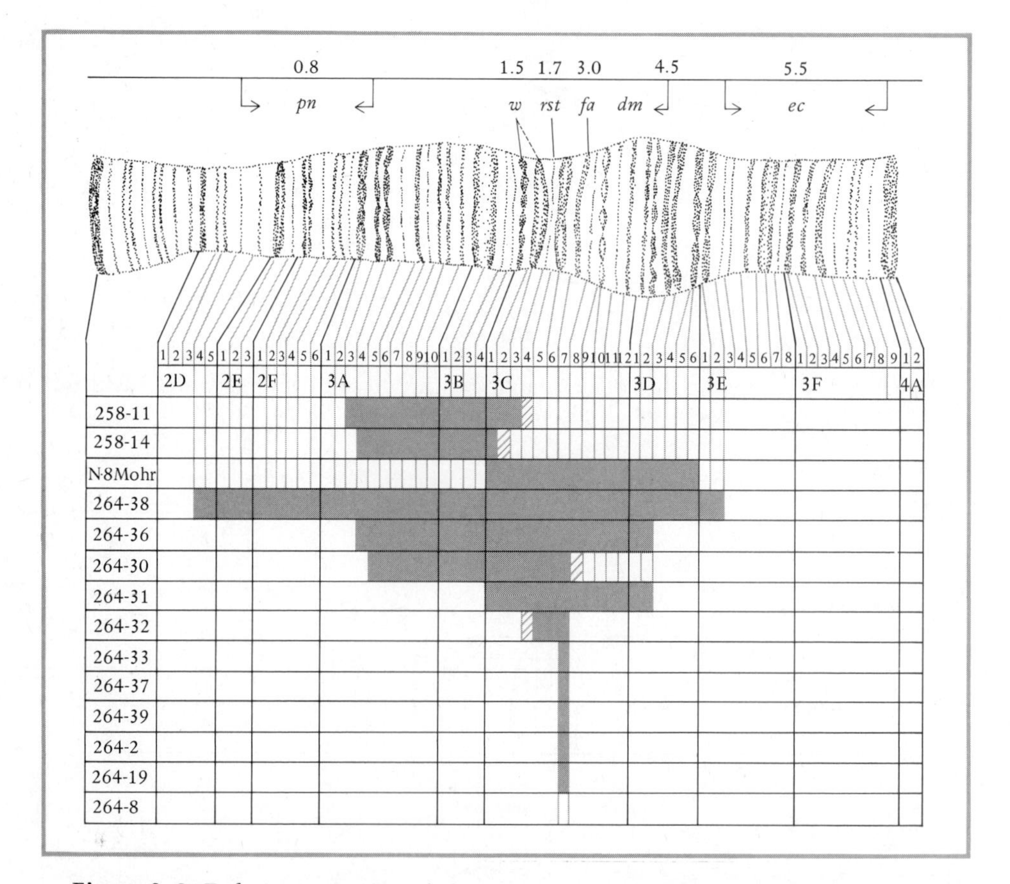

Figure 9–8. Deletion mapping of the white-facet region of the X chromosome of *Drosophila.* The linkage map is shown above the cytological map, with arrows pointing to the band or bands with which specific mutants can be associated. Below are shown the deficiencies used and their extent in the salivary chromosome. White *(w)* is associated here with either band 3C1 or 3C2, roughest *(rst)* with band 3C4, and facet *(fa)* with band 3C7. [From H. Slizynska, *Genetics, 23,* 291–299, 1938.]

The development of cytological maps has permitted comparisons between the chromosomal locations of genes and their positions in linkage maps. Although the same linear sequence is present in both kinds of maps, some discrepancies are evident with respect to the distance between linked genes. One would expect that genes showing frequent recombination would be physically located a considerable distance apart, but in cytological studies such genes are often found to be situated fairly close together. On the other hand, genes showing infrequent recombination have often been shown to be located closer together on the cytological map than linkage data would suggest. Since "distance" in linkage maps is based on frequency of recombination and not on physical distance, the source of these discrepancies appears to reside in the differential participation in crossing over of genes located in different regions of the chromosome. For example, genes located in heterochromatic regions near the centromere engage in meiotic exchanges much less frequently than do those more distally situated in euchromatic areas. Such genes therefore appear to be closely linked whether or not their loci are immediately adjacent. These differences in susceptibility to exchange between euchromatic and heterochromatic regions of the chromosome likely reflect differences in condensation.

Duplications

Duplications are repeats of chromosome regions, and their effect on the phenotype can range from negligible to severe depending on size, on location, or on the particular genes multiply represented in the genome. In general, the effects are not lethal unless the duplication is large enough to approach trisomy.

Careful analysis of salivary chromosome cytology has revealed the presence of a number of such repeats within the chromosome complement of *Drosophila* species. These repeats can be arranged in tandem (ABCDABCD) or in reverse order (ABCDDCBA). Small duplicated regions are usually present within the same chromosome, but in some cases duplications of bands characteristic of one chromosome can be found in a nonhomologue. In the latter case the origin of the repeat is most likely due to a previous instance of chromosome breakage followed by the incorporation of a fragment into the nonhomologue. When small repeated regions are present in the same homologue, origin can be attributed to prior breakage and the insertion of the repeated region into one of the homologues or, in some cases, to unequal crossing over.

Ordinarily, crossing over between nonsister chromatids appears to occur at precisely homologous points, resulting in an equal exchange between the two chromatids. Exchanges do sometimes occur not between identical points, but between adjacent points on the two chromatids, and as a result one chromatid will emerge from the exchange with a duplication and the other chromatid with a deficiency (see Figure 9–11). This type of exchange is called unequal crossing-over. Its cause is not known, but has been tentatively attributed to inexact synapsis between homologues.

The first case of unequal crossing over was discovered and analyzed by Sturtevant in 1925 in *Drosophila* females exhibiting the dominant phenotype called Bar eyes, a condition characterized by a reduction in the number of facets of the compound eye (Figure 9–9). Subsequent salivary chromosome

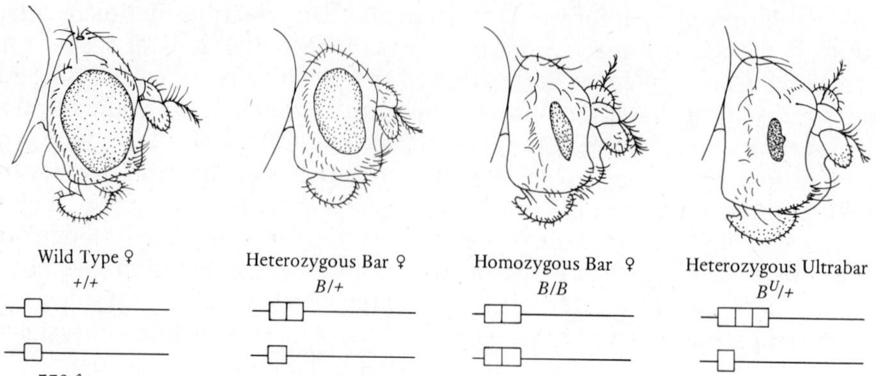

Wild Type ♀ +/+ 779 facets

Heterozygous Bar ♀ B/+ 358 facets

Homozygous Bar ♀ B/B 68 facets

Heterozygous Ultrabar $B^U/+$ 45 facets

Figure 9–9. The Bar eye phenotype and gene duplication in *Drosophila* females. Note that a total of four units is present in both *B/B* and $B^U/+$ females. The difference in phenotype between these two can be attributed to position effect. [From left to right, first three from A. H. Sturtevant, *Genetics, 10,* 117–147, 1925; fourth from *General Genetics,* Second Edition, by Adrian M. Srb, Ray D. Owen and Robert S. Edgar. W. H. Freeman and Company. Copyright © 1965.]

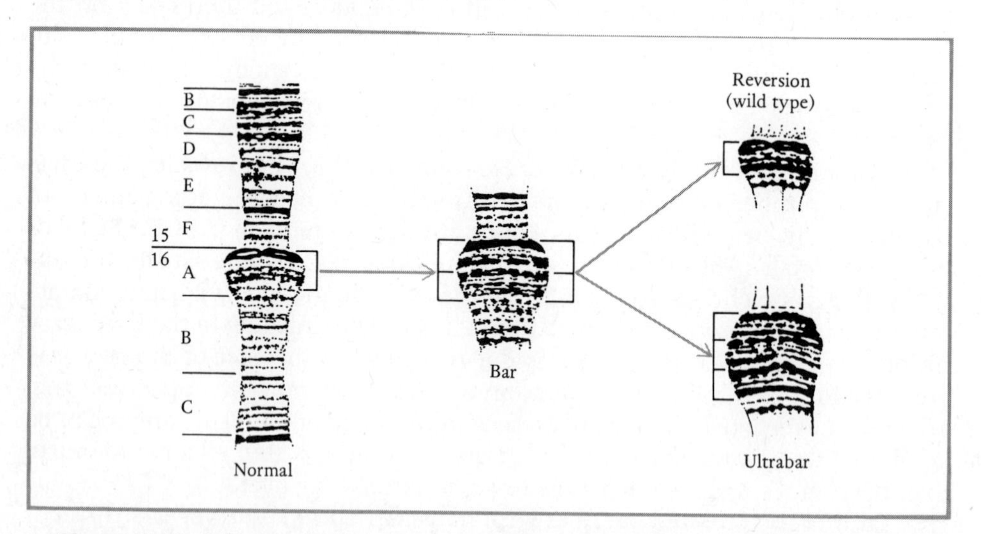

Figure 9–10. Salivary chromosome cytology of Bar. With the exception of the last crossband, Bar is a duplication of the 16A region of the X chromosome. Compare the origin of the triplication (Ultrabar) and reverted types with Figure 9–11. [From C. B. Bridges, *Science, 83,* 210–211, 28 February 1936.]

analysis by Bridges demonstrated that the various manifestations of Bar were associated with the presence of one or more duplications of the 16A region of the X chromosome (Figure 9– 10). The wild type non-Bar chromosome contains one such set of bands. However, when these bands are present in duplicate in one X chromosome of a female, the phenotype of Bar eyes and a reduced facet number results (358 facets as compared to the normal number of around 779). If

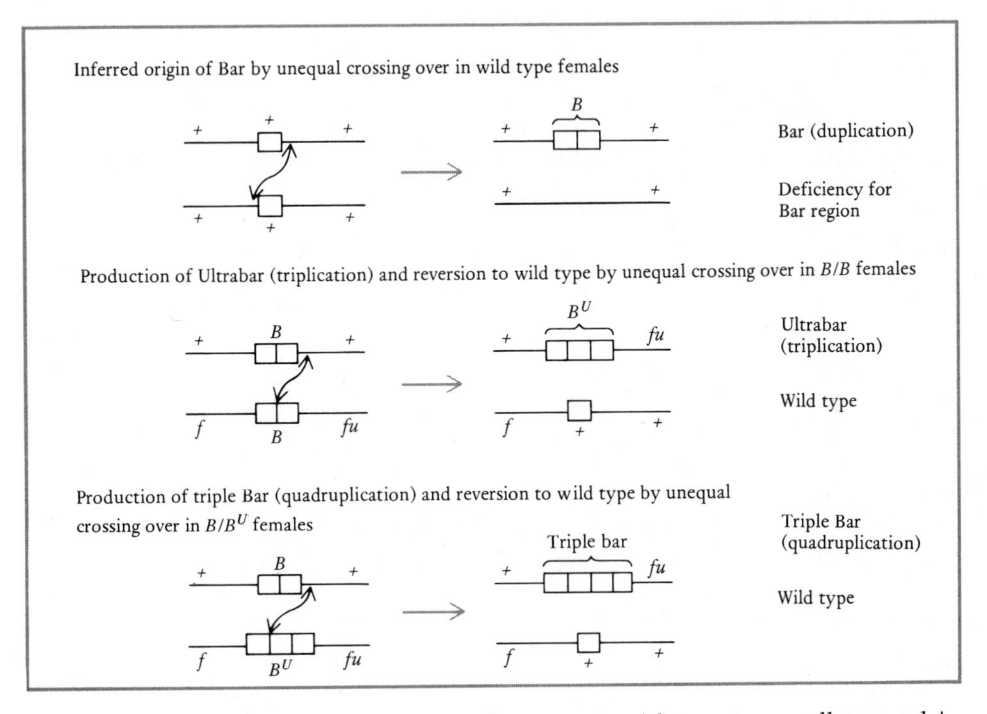

Figure 9–11. Unequal crossing over in the Bar region (shown as a small rectangle). One such region is normally present in the wild type chromosome. Note that the production of Ultrabar, triple Bar, and the reversion to wild type is accompanied by the exchange of the outside marker genes, forked *(f)* and fused *(fu)*.

this duplication is homozygous in a female, a further reduction in facet number to 68 is observed. When the 16A region is present in triplicate on one X chromosome of a female, the phenotype Ultrabar (B^U) with only 45 facets results. Finally, if the triplication (Ultrabar) is homozygous, the eye is reduced to only 25 facets.

That these changes in the Bar phenotype are derived from crossing over and not some other phenomenon is indicated by the fact that the production of Ultrabar and triple Bar (quadruplication) as well as the reversion to wild type, is accompanied by the exchange of outside marker genes (Figure 9–11). Unequal crossing over has also been proposed as the source of a rare type of human hemoglobin (see Chapter 10). The Bar phenotype can also be viewed as an example of *position effect,* that is, the phenomenon whereby the degree of expression of a given gene or genes is modified by changes in their physical location within the genome (see below).

The small duplications discernable in the salivary chromosomes of a number of Dipteran species constitute an increase in the amount of genetic material present. Since the original set of genes may be presumed to be fully operative, mutations in such duplicated regions can increase variability without necessarily influencing viability. In this respect duplications are thought to play a role in the evolution of new species. It should be additionally noted that the genomes of all eukaryotic organisms are characterized by the presence of reiterated sequences (see Chapter 19). Although the origin and activities of such sequences are unknown, their presence has undoubtedly played a major role in evolution.

CHANGES IN CHROMOSOME STRUCTURE: INVERSION AND TRANSLOCATION

Inversion

An inversion is a chromosome aberration in which the gene order within a segment of the chromosome becomes reversed. Inversions require chromosome breakage. If a chromosome arm is broken in two places, the fragment may be reincorporated during restitution, but in the reverse order. For example, the chromosome $\underline{A\ B\ C\ D\ E\ F\ G\ H\ I\ J}$ is broken between genes *C* and *D* and between *H* and *I*, the central fragment $\underline{D\ E\ F\ G\ H}$ may be reunited in reversed sequence to produce the chromosome $\underline{A\ B\ C\ H\ G\ F\ E\ D\ I\ J}$. An inversion thus produces a change in sequence, but usually no gain or loss of genetic material. The new gene sequence can be detected by testing for linkage relationships. In the above illustration map distances between genes *C* and *D* and *C* and *H* in the inverted chromosome would be quite different from those in the normal chromosome.

When an inversion is present in the homozygous condition, no difficulties in chromosome pairing or crossing over at meiosis are encountered. However, when a heterozygous inversion is present, synapsis results in the formation of a characteristic loop by means of which the pairing of homologous inverted and noninverted regions is achieved. Such loops can be identified in meiotic chromosomes of many species and are clearly seen in the salivary chromosomes of *Drosophila* (Figure 9–12).

Two types of inversions can be distinguished: (1) *paracentric* is an inversion which is situated in a chromosome arm to one side of the centromere and

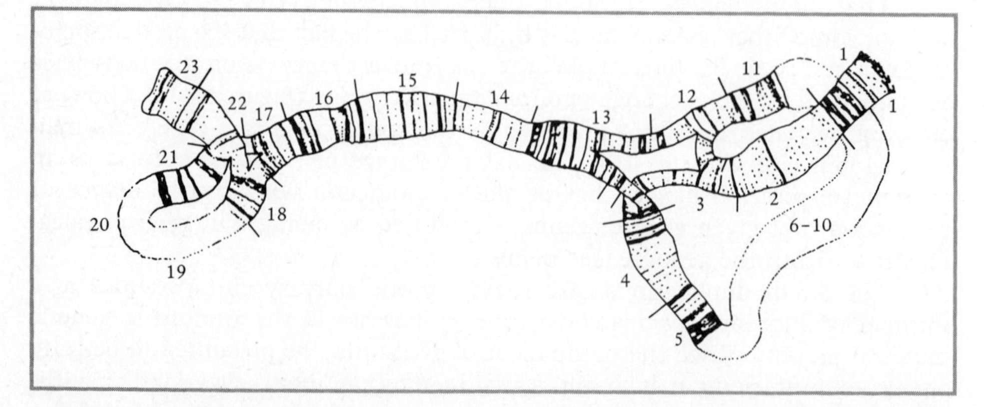

Figure 9–12. Inversion loops formed in the X chromosome in heterozygotes of two strains of *Drosophila azteca.* (From Th. Dobzhansky and J. Socolov, *J. Heredity, 30,* 3–19, 1939.)

(2) *pericentric* is an inversion which includes the centromere. In the heterozygous condition crossing over within either of these kinds of inversions produces aberrant chromatids. If a single crossover occurs within a heterozygous paracentric inversion, the products of the first meiotic division include a dicentric (two centromeres) chromatid, an acentric (no centromere) fragment, and two noncrossover chromatids (Figure 9–13).

The fate of the abnormal products (dicentric and acentric) differs in different organisms. In *Drosophila* they are not included in the functional egg nucleus because of the particular arrangement of the four nuclei produced by meiosis. In *Drosophila* polar bodies are not extruded as separate cells; the four nuclei arising from meiosis remain in the egg cytoplasm where they are arranged in linear order. Only the first or the last of the four participates in fertilization, the middle two being consistently relegated to polar body status

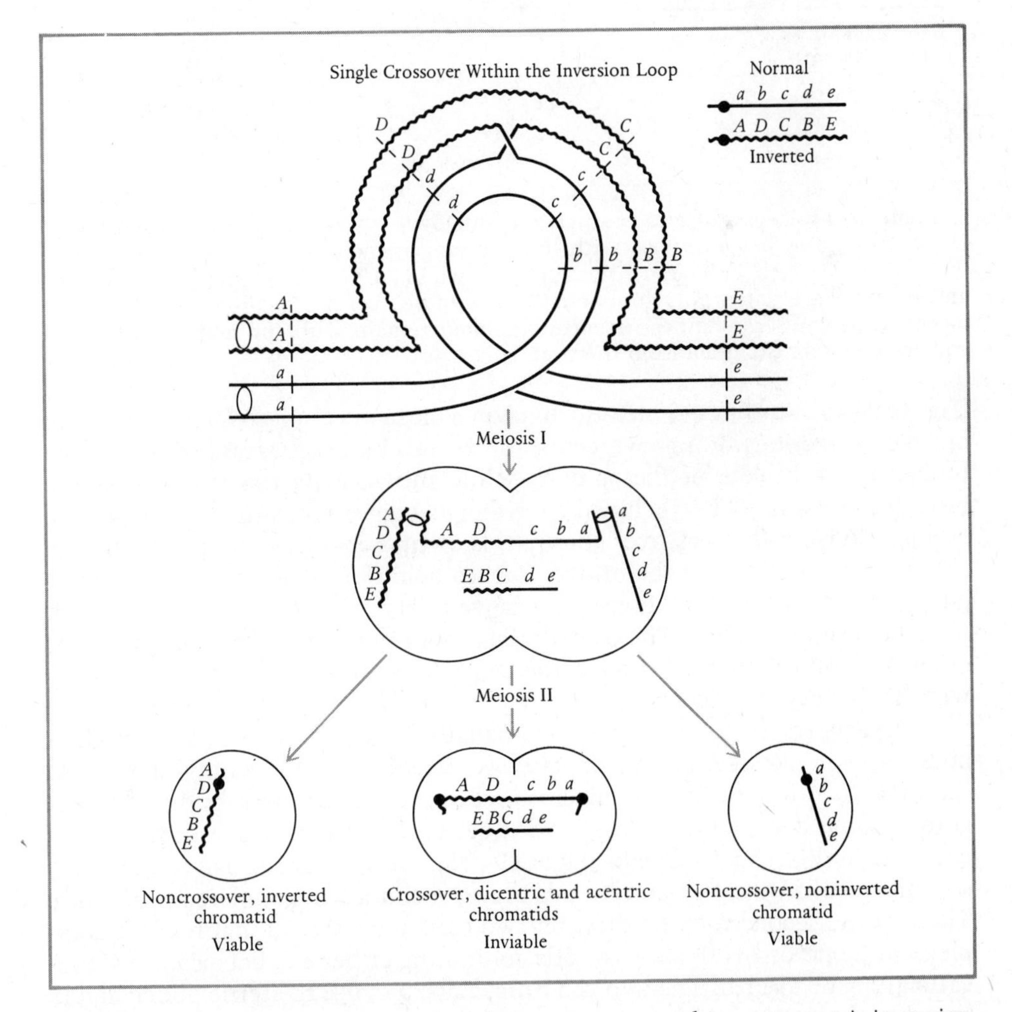

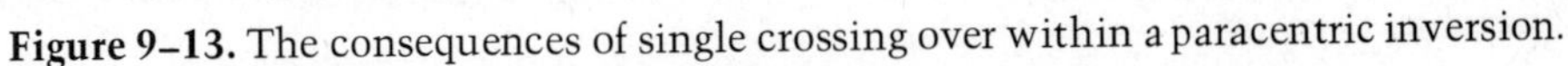
Figure 9–13. The consequences of single crossing over within a paracentric inversion.

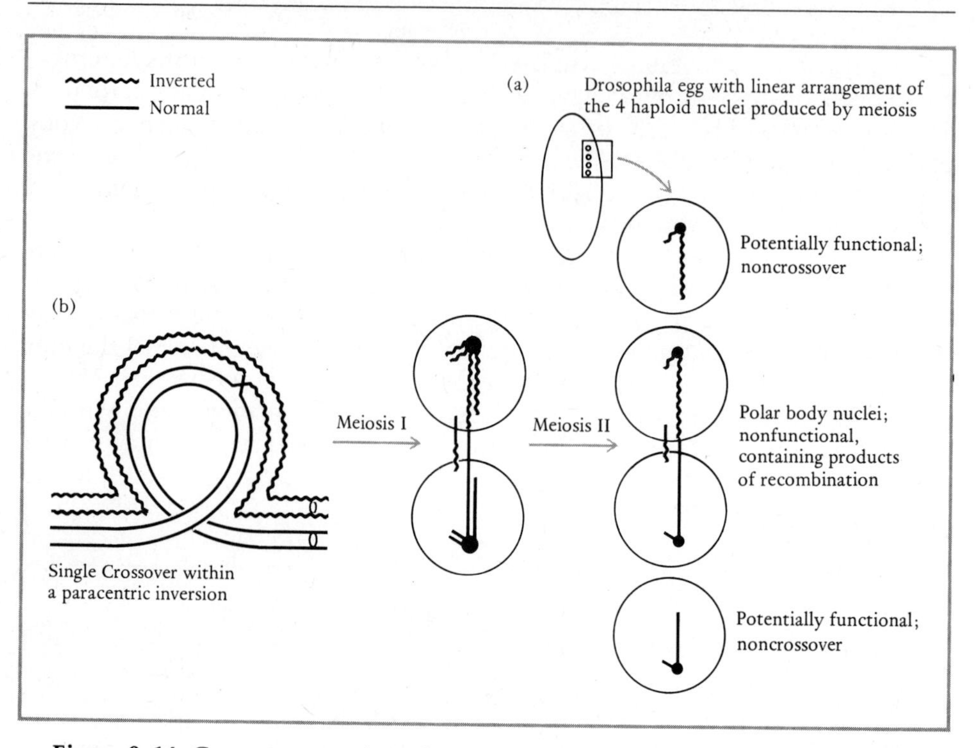

Figure 9–14. Consequences of single crossing over within a paracentric inversion in *Drosophila.* (a) *Drosophila* egg, with linear arrangement of the 4 haploid nuclei produced by meiosis. (b) Enlargement of nuclei shown in (a). Only the outer two nuclei (top and bottom) are potentially functional; the middle two are polar body nuclei. The dicentric chromosome and the acentric fragment remain with the middle two nuclei and are therefore excluded from the zygote.

(Figure 9–14). At the first meiotic division a dicentric chromosome is pulled in opposite directions by its two centromeres and becomes stranded, forming a bridge in the middle of the spindle, while the acentric fragment, lacking a centromere, fails to be included in either daughter nucleus. At the second meiotic division the crossover chromatids comprising the dicentric, being unable to move because of the bridge, remain behind in the middle two nuclei, but the noncrossover chromatids can move freely and are therefore segregated into the terminal nuclei. Since fertilization will involve only either one or the other of the outer nuclei, and not the centrally positioned nuclei, all progeny will be noncrossover in type.

An acentric fragment in corn will usually fail to be included in a daughter nucleus, as in *Drosophila,* but the fate of a dicentric is somewhat different. At the first meiotic division the dicentric chromosome forms a bridge, but the chromosome does not remain stranded between the two daughter nuclei because the bridge usually breaks (Figure 9–15). Since breakage can occur at any site, the resulting broken chromatids may not contain a complete set of genes. Thus, at the second meiotic division two cells with the normal haploid complement (noncrossover) and two cells containing either a deficiency or a duplication may be formed. The two abnormal cells give rise to sterile pollen and to inviable zygotes.

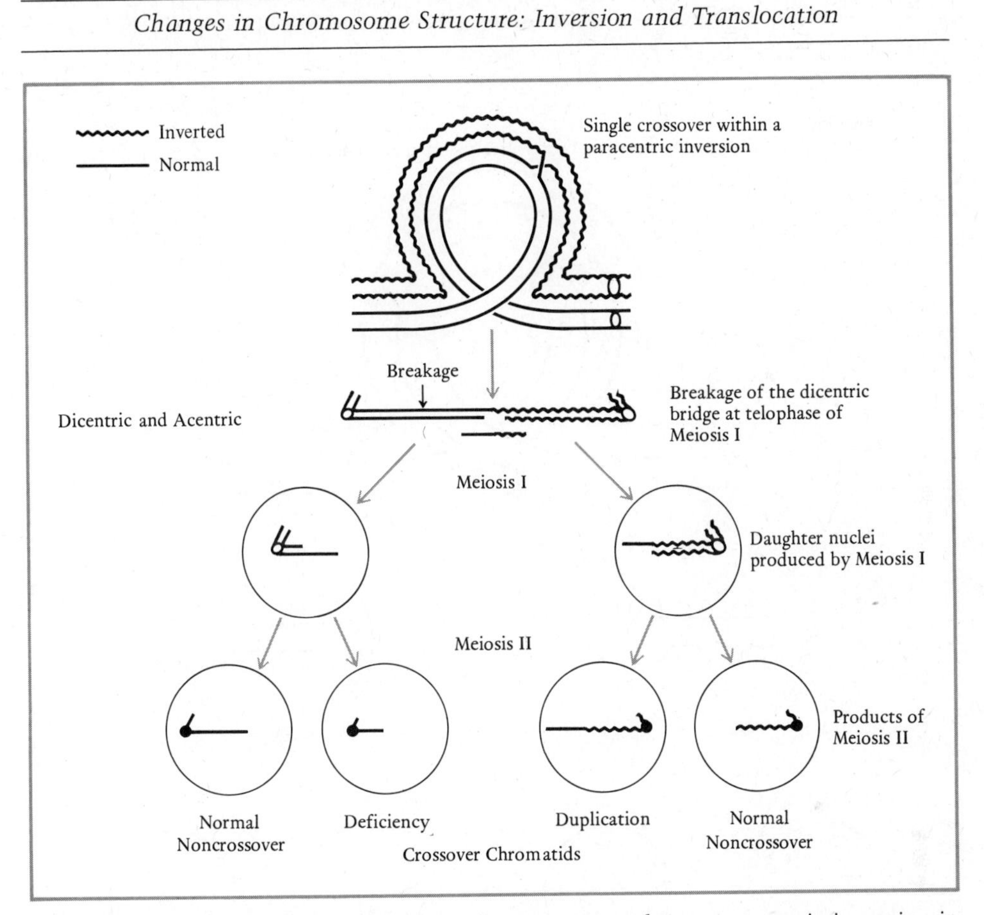

Figure 9–15. Consequences of single crossing over within a paracentric inversion in corn.

The results of single crossing over within a pericentric inversion are diagramed in Figure 9–16. Although in this case dicentrics and acentrics are not formed, the crossover chromatids contain duplications for some genes and are deficient for others. These conditions lead to unbalanced chromosome complements in both gametes and zygotes and are usually inviable with the result that only noncrossover progeny are generally recovered. Since chromosome bridges are not formed by crossing over within a pericentric inversion, unbalanced chromosome complements may occur within the functional egg nucleus of *Drosophila*. Although inversions are sometimes called *crossover suppressors,* their presence does not eliminate crossing over and thus the term is appropriate only in the sense that while crossover chromatids are produced, they are rarely recovered in progeny.

In *Drosophila* special stocks have been developed which contain inversions in the X chromosome or in each arm of the large autosomes or in both the X chromosome and autosomes. These special chromosomes also contain dominant genes associated with the inversions and these serve to mark the presence of that chromosome in any individual. Particularly useful are dominant genes which are lethal when homozygous. When two dominant mutants of this type, along with the associated inversions, are present, one on each homologue, the

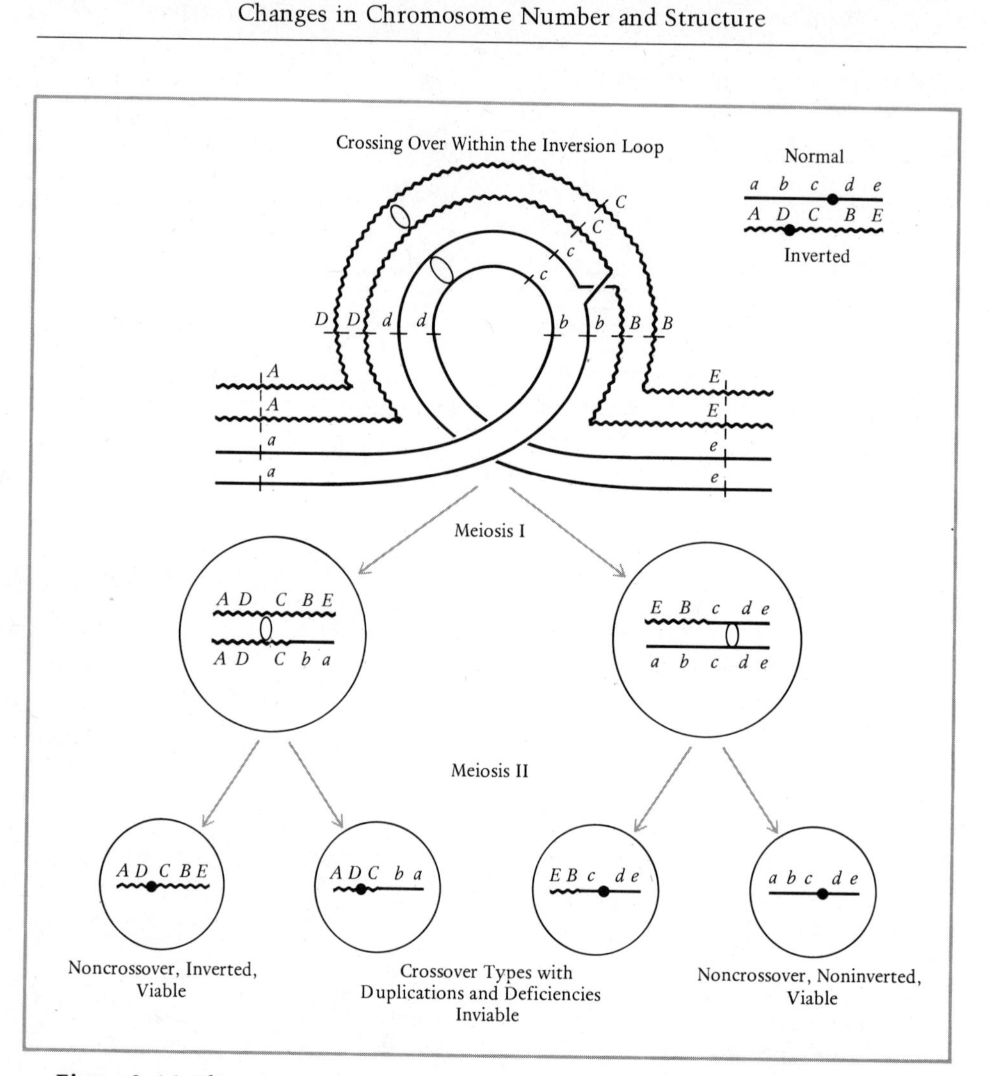

Figure 9–16. The consequences of single crossing over within a pericentric inversion.

stock becomes a true breeding heterozygote (Figure 9–17). Such stocks, called balancers or balanced lethals, provide shortcuts in assigning genes to linkage groups. In addition, because of the absence of viable recombinants, new mutants can be maintained without selection when the chromosome on which they are carried is made heterozygous to a balancer chromosome. For this reason, balancer stocks are employed extensively in studies of mutation.

Translocation

A translocation is a transferral of a region, segment, or arm from one chromosome to a nonhomologous chromosome. Translocations originate through chromosome breakage and reunion and can result in complex rearrangements. An interstitial region from one chromosome may be inserted into one of the arms of a nonhomologous chromosome. In other and more frequent instances an exchange of arms *(reciprocal translocation)* may occur. As a result

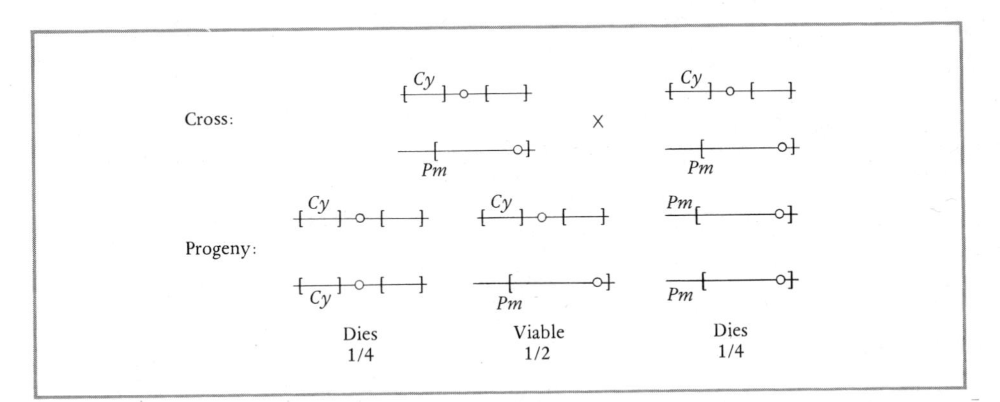

Figure 9–17. A heterozygous genotype that breeds true. *Cy,* dominant curly wings; *Pm,* dominant plum-colored eyes; both *Cy* and *Pm* are lethal when homozygous. Brackets indicate the presence of inversions.

of such exchanges genes formerly belonging to one linkage group will show new and different linkage group relationships. Homozygous translocations give rise to viable individuals whose homologues undergo normal pairing, crossing over, and segregation at meiosis. Individuals bearing heterozygous translocations are viable as long as all constituents of the genome are present, regardless of position.

Because more than one pair of homologues is involved, synapsis at meiosis becomes complicated. With a heterozygous reciprocal translocation, meiotic pairing between homologous regions will draw together two separate pairs of chromosomes into a complex association composed of eight, instead of four, chromatids. This association typically results in a cross-shaped figure, easily distinguishable under the microscope. Crossing over and chiasma formation occur normally in synapsed regions.

With the cessation of pairing at diplotene, the cross-shaped figure opens to form a ring composed of the four chromosomes now associated only at their tips. Such rings can be either open or twisted and, depending on the positions and associations of the chromosomes forming the rings, different patterns of segregation can result (Figure 9–18).

Ordinarily, homologous centromeres segregate from one another and move to opposite poles at the first meiotic division. This pattern is found in both open and twisted rings. In a twisted ring orientation of chromosomes is such that both normal chromosomes proceed to one pole and both translocated chromosomes proceed to the opposite pole *(alternate segregation)*. As a result, both daughter cells, as well as the gametes formed from them, will contain a complete set of genes and can give rise to viable individuals.

If the ring is open, two patterns of segregation are possible. When homologous centromeres move to opposite poles as expected *(adjacent 1 segregation),* each daughter cell will come to contain one normal and one translocated chromosome and will thus lack a complete set of genes. Such cells give rise to gametes with duplications and deficiencies and these chromosome complements are inviable in the pollen of plants and the zygotes of animals.

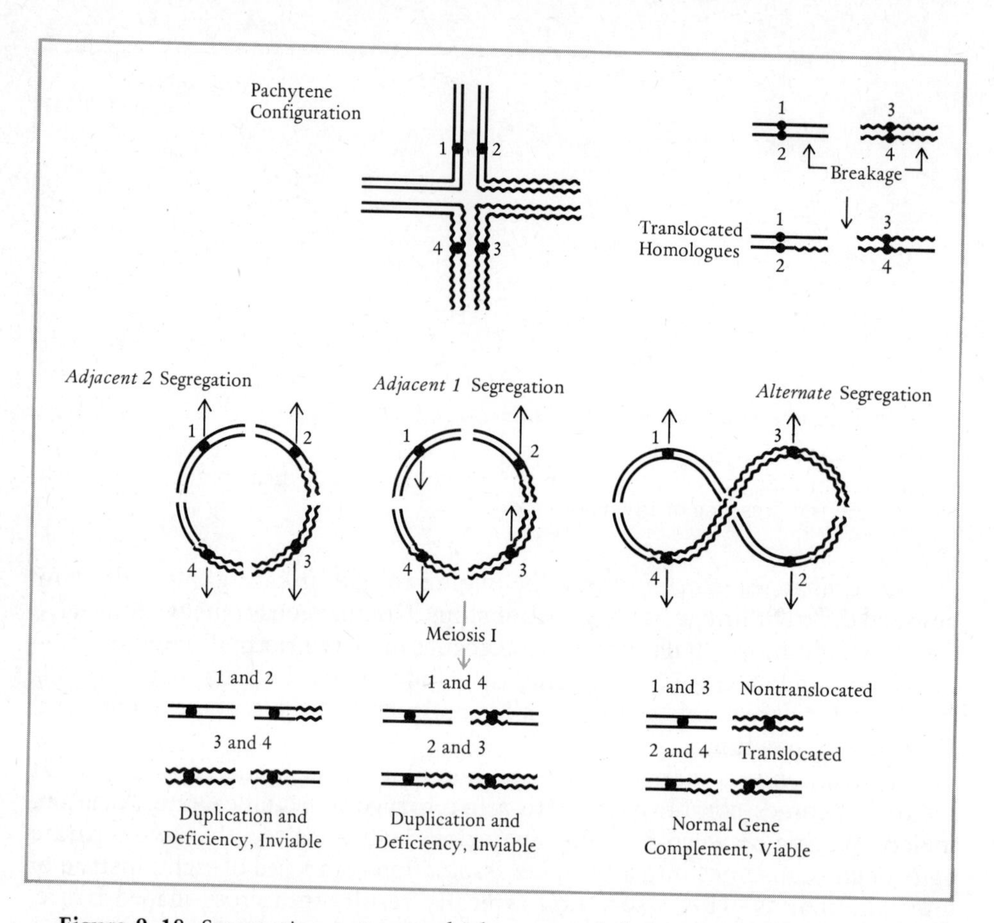

Figure 9–18. Segregation patterns which can result from a heterozygous translocation. Centromeres are numbered 1 through 4, 1 and 2, and 3 and 4 being homologous. Arrows indicate the direction of segregation.

Homologous centromeres may proceed to the same pole *(adjacent 2 segregation)* and, again, duplication and deficiency gametes will arise. As a result of these segregation patterns, gametes capable of functioning in fertilization or giving rise to viable offspring are limited to those that contain either both normal or both translocated chromosomes. Consequently, individuals carrying heterozygous translocations of the type described produce fewer offspring than normal and appear semisterile.

The results of these patterns of segregation can be used to advantage in identifying individuals heterozygous for a translocation. In the testcross, *A/a B/b* × *a/a b/b,* where alleles *A* and *a* and *B* and *b* are known to be located on nonhomologous chromosomes, independent assortment is expected to produce four types of progeny in equal frequency: *AB, ab, Ab,* and *aB*. However, if the parent heterozygous for these genes is also heterozygous for a translocation involving the chromosomes on which genes *A* and *B* are carried, the only viable offspring which will result from the cross will be those *AB* and *ab* in phenotype (Figure 9–19). Since exclusive recovery of parental types from a testcross is

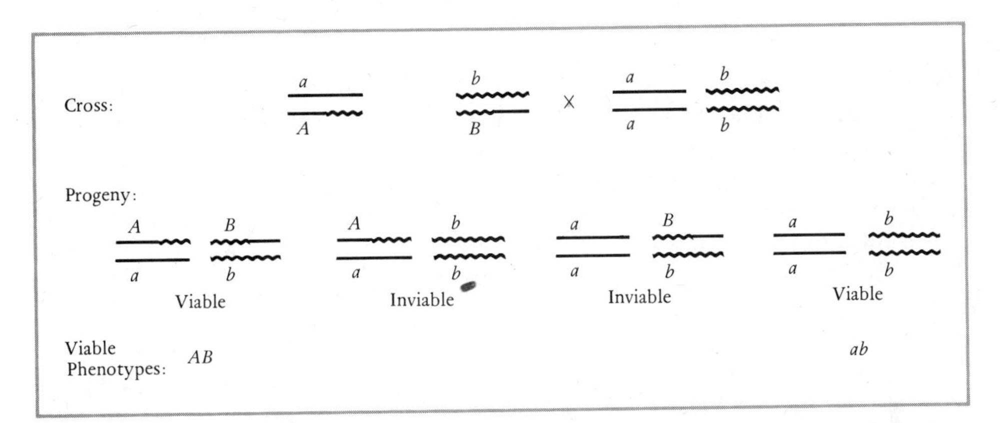

Figure 9–19. A testcross involving a heterozygous translocation. Viable progeny are limited to those carrying either both normal or both translocated chromosomes. The results of such a cross resemble those obtained with linked genes and can be used to identify the chromosomes involved in a translocation.

characteristic of closely linked genes, *A* and *B* will appear to be linked, when in fact, they are located on different, nonhomologous chromosomes. This kind of genetic test can be applied to individuals suspected of carrying a translocation. Usually such an individual is made heterozygous for marker genes known to be located on nonhomologous chromosomes. A testcross is then performed and the offspring analyzed for independent assortment of the genes involved in the cross. Since the linkage groups of marker genes are known, any two or more such genes showing spurious linkage will identify the specific chromosomes involved in the translocation. In some plants, such as *Oenothera,* multiple translocations between some or all chromosomes of the complement are normally present. As a consequence, the genes associated with such translocated chromosomes are transmitted as if they were completely linked.

Translocations are also known to occur in humans. Approximately 2 percent of individuals with Down's syndrome arise not from nondisjunction, but from partial trisomy of chromosome 21 caused by a translocation between chromosome 21 and some other autosome (usually chromosome 14). In such cases one parent is heterozygous for the translocation, and partial trisomy 21 in an offspring can be attributed to fertilization involving a duplication-bearing gamete. The *cri du chat* syndrome can arise through partial monosomy of chromosome 5 caused by a deficiency-bearing gamete produced by a parent heterozygous for a translocation involving this chromosome. Other instances of translocation in humans have also been observed and infants bearing duplications or deficiencies are in most cases grossly abnormal, if not aborted spontaneously.

In addition, evidence for interspecific translocation in hybrid mouse-human cells has been reported. In one case involving the fusion of thymidine kinase deficient *(TK⁻)* mouse cells with *TK⁺* human cells, the production of human enzyme continued after apparent loss of all human chromosomes from the hybrid, suggesting that human genes had been inserted into the mouse genome.

Although inversions and translocations would appear to have rather drastic effects on meiosis, they have been employed in many experiments which attempt to probe the meiotic pairing and segregational activities of chromosomes. For example, it has been found that when these kinds of aberrations are used to impede synapsis and crossing over in one pair of homologues, an increased amount of crossing over occurs in other, normal homologues of the complement. Since biochemical studies of meiosis have indicated that the events leading to pairing and exchange are temporally ordered, the difficulties in achieving synapsis encountered by aberrant chromosomes are thought to increase the time span allotted for crossing over, thus providing an extended opportunity for exchange to occur between unimpeded, normal chromosomes. It has also been found that chromosomes unable to enter a pairing relationship with a homologous partner, due to the presence of complex aberrations, may subsequently pass at random to either pole in meiosis I or may segregate from a similarly abnormal, unpaired nonhomologue. The results of many such studies indicate that normal pairing between homologous chromosomes at meiosis is a prerequisite for normal disjunction and segregation.

POSITION EFFECT

The term *position effect* refers to a change in the activity of a gene when its position or location within the genome is altered. Position effects can arise from inversion, translocation, and, more rarely, unequal crossing over. The first position effect discovered was that of Bar eye in *Drosophila,* in which the duplication and triplication of the 16A region of the salivary chromosome causes a progressive decrease in the number of facets of the eye. Most simply, one could attribute the phenotype to an increase in the number of genes regulating facet number. However, this explanation does not account for the difference in phenotype exhibited by females homozygous for Bar (with 68 eye facets) versus those heterozygous for Ultrabar (with 45 eye facets). Homozygous Bar females contain two 16A regions on each X chromosome, a total of four in all. Heterozygous Ultrabar females also contain four 16A regions, but three are present on one X chromosome and one on the opposing X chromosome (see Figure 9–9). Evidently the position or arrangement of these regions produces a phenotypic effect apart from that resulting from the number of 16A regions present. This effect appears as a stable and more severe suppression of the normal phenotype.

Many other position effects have been discovered and analyzed in *Drosophila* and in mice. In *Drosophila* the movement by inversion or translocation of a wild type gene into centromeric or other heterochromatin results in suppression of that gene's activity in many cells, resulting in a variegated or mosaic phenotype. Suppose the X chromosome of *Drosophila* is broken in two places, adjacent to the wild type allele of white eyes and in the centromeric heterochromatin. Suppose also that reunion of broken ends occurs in reverse sequence and that the inversion so formed brings the wild type allele for white into the heterochromatin adjacent to the centromere. If a male carrying this rearranged chromosome is crossed to a homozygous white-eyed female, daughters heterozygous for the inversion, as well as heterozygous for the dominant red and recessive white eye color alleles, are produced. One would expect such daughters to have uniformly red eyes because of the presence of the dominant allele, but instead the eye color is variegated, with patches of white as well as patches of red tissue. If chromosome loss and somatic crossing over are ruled out as causative factors, one can account for the phenotype by assuming that the normal activity and expression of the dominant allele has somehow been suppressed in the cells of white tissue, but not in red tissue. It can also be demonstrated that mutation of the wild type allele is not involved, for if this allele is restored to its original euchromatic position by a second inversion, its activity returns to normal.

A position effect of this type is called a *variegated* or *V-type* position effect. It appears that almost any euchromatic gene of *Drosophila* will exhibit this phenomenon when moved into broken heterochromatin. The suppression of gene activity can extend beyond the locus newly placed adjacent to heterochromatin, for other genes farther away from the breakpoint can be similarly affected. Thus, inactivation or suppression of gene function can exhibit an appreciable spreading effect along the chromosome. In addition, the presence of a Y chromosome, as in an XXY female, tends to restore normal gene activity, for variegation is reduced. Position effects are also produced when genes normally located in heterochromatic regions are moved by translocation or inversion into distant euchromatic regions or into broken heterochromatin. In this case, presence of an extra Y chromosome will enhance variegation. Why the Y chromosome suppresses position effect variegation involving euchromatic genes and enhances variegation involving heterochromatic genes is unknown.

Variegated position effects are not limited to *Drosophila,* for they have also been observed in the mouse and are thought to constitute a phenomenon common to all eukaryotic organisms. In the mouse position effects are associated with translocations of autosomal genes to the X chromosome. The variegation observed is not found in XY males or in X0 females, but occurs exclusively in XX females and is therefore associated with the X chromosome itself and not with the sex of the individual. Analysis of a number of different translocations of autosomal genes to the X chromosome has indicated the presence of a site on the X chromosome which suppresses the expression of inserted autosomal genes. This site may be associated with the inactivation which regularly occurs in one of the X chromosomes of female mammals.

The phenomenon of position effect is of considerable significance in regard to chromosome organization. From the existence of position effect we must infer that the genes present in any chromosome do not represent a miscellaneous assortment or random collection of hereditary units. Instead, we must visualize the chromosome as a highly organized entity in which the locations of genes have meaning in terms of activity and expression. We can also infer that regulatory mechanisms exist within this organized entity and that when this organization is disrupted and the arrangement of contained genes changed, such genes may no longer be able to respond to their original controls. Instead they may be placed under controlling mechanisms not normally encountered, and their activity modified accordingly.

PROBLEMS

9–1. Both autopolyploidy and allopolyploidy can result in a species with double the original chromosome number. If you had a plant with a $4n$ complement, how would you determine cytologically if it were an autopolyploid or an allopolyploid?

9–2. The Chinese hamster has a diploid number of 22 and the Golden hamster is a fertile tetraploid, females having 44 chromosomes. Suggest a possible (abnormal) sex chromosome constitution which could be maintained indefinitely in a Golden hamster male.

9–3. A human female has an abnormal chromosome constitution consisting of chromosome 21, chromosome 14, and a chromosome 14.21. Chromosome 14.21 consists of chromosome 14 and its centromere plus most of the genetic material of chromosome 21. Other chromosomes are normal.

a. What distributions of these chromosomes are possible as a result of segregation?

b. If the woman marries a normal male, what is the chance that the first child will have Down's syndrome?

9–4. If a plant were trisomic for one of its chromosomes and these chromosomes carried the alleles A, A^1, and A^2, respectively, what would be the genotypes of the gametes produced with respect to these genes?

9–5. Assume a viable trisomic plant where two of the chromosomes involved bear the dominant gene *B* and the other chromosome the recessive gene *b*. This plant is crossed to a diploid of the genotype *b/b*.

a. What proportion of the progeny will be diploid?

b. What proportion of the progeny will be trisomic and *Bbb* in genotype?

9–6. Assume a trisomic plant where the chromosomes involved in trisomy are marked as follows:

(1) $\underline{A\ b\ c}$
(2) $\underline{a\ B\ c}$
(3) $\underline{a\ b\ C}$

Assuming no crossing-over, suppose you crossed the trisomic to a normal diploid of the genotype *a b c/a b c* and the phenotypes of the progeny from this cross were:

A b c and *a B C*	45%
A B c and *a b C*	45%
A b C and *a B c*	10%

Which of the three homologous chromosomes of the trisomic were most frequently involved in pairing and segregating from each other?

9–7. Assume the heterozygous paracentric inversion: $\underline{D\ A\ B\ C\qquad}$o— (inverted)/ $\underline{D\ C\ B\ A\qquad}$o— (normal). Diagram the products of the first and second meiotic divisions if the following exchanges were to occur within this inversion: two-strand double, three-strand double, four-strand double. Which of these events yields the largest proportion of normal (viable) gametes?

9–8. Diagram the gametes which would be produced by the above types of exchanges occurring within the following heterozygous pericentric inversion: $\underline{A\ C\quad}$o$\underline{\quad B\ D}$ (inverted)/$\underline{A\ B\quad}$o$\underline{\quad C\ D}$ (normal).

9–9. Suppose a plant heterozygous for an inversion, P Q R S T/P Q S R T, is crossed to one homozygous for the same inversion, P Q S R T/P Q S R T. What percentage of all gametes produced by both parents would carry a complete haploid set of genes (euploid) in the following cases.

a. The inversion between *R* and *S* is so small that synapsis and crossing over between these genes does not occur.

b. The inversion is sufficiently large so that synapsis and one chiasma occurs between *R* and *S* in every meiotic cell.

9–10. The mutants white eyes (*w*), miniature wings (*m*), and forked bristles (*f*) are recessive and sex-linked in *Drosophila.* Wild type males are irradiated with X rays and then crossed to homozygous *w m f* females, and the F_1 daughters testcrossed individually to *w m f* males. In most cultures of this testcross standard recombination frequencies between *w* and *m* (35 percent) and *m* and *f* (20 percent) are obtained. However, two exceptional cultures give the following results.

CULTURE 1		CULTURE 2	
w m f and + + +	67%	*w m f* and + + +	80%
w + + and + *m f*	32%	*w* + + and + *m f*	2%
w m + and + + *f*	1%	*w m* + and + + *f*	18%

How would you explain these results?

9–11. A given plant is heterozygous for a reciprocal translocation between two nonhomologous chromosomes. The normal chromosomes are *a b c d* and *w x yz;* the translocated chromosomes are *A B C Z* and *W X Y D*.

a. Diagram the figure which would result at synapsis.

b. Diagram the kinds of gametes that would result if a single crossover occurred between genes *W* and *X*.

c. Diagram the kinds of gametes resulting from a single crossover between genes *B* and *C*.

9–12. In *Drosophila* brown eyes (*bw*) is a recessive mutant on chromosome 2, ebony color (*e*) is a recessive on chromosome 3, and eyeless (*ey*) is a recessive mutant on chromosome 4. Wild type males were treated with X rays and crossed to homozygous *bw, e, ey* females. The F_1 were wild type in phenotype and were individually testcrossed to *bw, e, ey*. In most cultures independent assortment of these genes was observed, but in the following cultures only the phenotypes listed were obtained in approximately equal frequency.

(1) + + +, + + *ey*, *bw e* +, *bw e ey*
(2) + + +, *bw* + *ey*, + *e* +, *bw e ey*
(3) + + +, *bw* + +, *bw e ey*, + *e ey*
(4) + + +, *bw e ey*

a. What kinds of chromosome aberrations are indicated?

b. Which chromosomes are involved in the aberration in each culture?

9–13. Assume that closely linked genes in the order *p, rt, m,* and *q* are located somewhere within the series of polytene chromosome bands A1-2-3-4-5-B1-2-3-4-5-6-C1-2-3. A wild type strain is irradiated and testcrossed individually to members of a strain homozygous for *p, rt, m,* and *q*. In the following eight cultures the indicated phenotypes were obtained and could be associated with deficiencies for the chromosome bands given (inclusively).

(1) *p* + + +	Bands A1–2
(2) + *rt m* +	Bands A3–B2
(3) + + *m* +	Bands B1–4

(4) + + + *q*	Bands B5–C3
(5) + + + +	Bands B6–C3
(6) *p rt* + +	Bands A1–5
(7) + + *m* +	Bands A4–B3
(8) + + + *q*	Bands B3–C3

Diagram the chromosome bands, indicating the extent of each deletion and determine with which bands the genes *p, rt, m* and *q* are associated.

REFERENCES

Astaurov, B. L., 1969. Experimental polyploidy in animals. *Ann. Rev. Genet., 3,* 99.

Baker, W. K., 1968. Position-effect variegation. *Adv. Genet., 14,* 133.

Blakeslee, A. F., 1934. New Jimson weeds from old chromosomes. *J. Hered., 25,* 80.

Bridges, C. B., 1935. Salivary chromosome maps. With a key to the banding of the chromosomes of *Drosophila melanogaster. J. Hered., 26,* 60.

Carr, D. H., 1971. Genetic basis of abortion. *Ann. Rev. Genet., 5,* 65.

Cleland, R. E., 1962. The cytogenetics of *Oenothera. Adv. Genet., 11,* 147.

Grell, R. F., 1967. Pairing at the chromosomal level. *J. Cell. Physiol., 70* (Suppl. 1), 119.

Lucchesi, J. C., and D. T. Suzuki, 1968. The interchromosomal control of recombination. *Ann. Rev. Genet., 2,* 53.

Smith, D. W., 1970. *Recognizable Patterns of Human Malformation*. Saunders, Philadelphia.

Stebbins, G. L., 1971. *Chromosomal Evolution in Higher Plants*. Addison-Wesley, Reading, Mass.

Thompson, J. S., and M. W. Thompson, 1973. *Genetics in Medicine,* 2nd ed. Saunders, Philadelphia.

White, M. J. D., 1969. Chromosomal rearrangements and speciation in animals. *Ann. Rev. Genet., 3,* 75.

Chapter 10

GENES AND THE ORIGIN OF THE PHENOTYPE

By the early 1940s the principles of transmission genetics were firmly established, the linear arrangement of genes on the chromosome was universally accepted, and the analysis of recombination was providing linkage maps for a number of species. From these studies there emerged a concept of the gene as a hereditary particle, having a fixed chromosomal locus, subject to alteration by mutation and to exchange via crossing over, but essentially indivisible in nature. How such a genetic entity could initiate the development of a given phenotypic character remained speculative, however, and provoked the search for a primary product of gene expression.

MUTATIONS IN METABOLIC PATHWAYS

One early clue was provided by A. E. Garrod, who in 1909 proposed that inherited metabolic diseases of humans were due to the absence or deficiency of a required enzyme. He called such diseases "inborn errors of metabolism" and associated the primary cause of such conditions with mutant genes.

Garrod's hypothesis was based primarily on his analysis of the rare hereditary trait, alkaptonuria. This condition is characterized by the excretion of large amounts of homogentisic acid (alkapton), a substance not present in the urine of normal individuals. Due to oxidation of this compound, the urine of alkaptonurics turns black on standing in air, and by this means the presence of the disease is readily detected. Garrod fed homogentisic acid to patients with alkaptonuria and found that it could be recovered in equivalent quantities in the urine. Normal individuals fed homogentisic acid successfully utilized this substance, for none was present in urine. Garrod also found that diets containing proteins rich in the amino acids phenylalanine and tyrosine, or the feeding of these amino acids alone, increased the excretion of homogentisic acid in alkaptonurics.

On the basis of these findings Garrod proposed that homogentisic acid occupied an intermediate position in a metabolic pathway through which phenylalanine and tyrosine were utilized. In normal persons homogentisic acid would be metabolized as rapidly as it was formed, but in alkaptonurics the absence of a specific enzyme would block the metabolic pathway at this step, resulting in the accumulation and subsequent excretion of this substance. Since Garrod observed that alkaptonuria was inherited as a simple recessive, the association between an abnormal gene and a missing or abnormal enzyme was postulated as early as 1909.

Proof of his hypothesis was not obtained until 1958, however, when the complete metabolic pathway for the breakdown of tyrosine and phenylalanine in liver was determined. It was found that in alkaptonurics the only enzyme missing from the entire pathway was homogentisic acid oxidase, the enzyme catalyzing the oxidation of homogentisic acid to the next succeeding compound of the chain (see Figure 10–7).

The association between genes and sequential biochemical reactions was also postulated from studies of flower pigments, coat colors of mice, and eye pigments of *Drosophila,* where modified dihybrid ratios (for example, 9 : 3 : 4, 9 : 7, and others) were obtained. One of the early experimental studies was that of G. W. Beadle and B. Ephrussi who in 1936 investigated the synthesis of eye pigments in *Drosophila* by the technique of tissue transplantation.

It will be recalled from Chapter 4 that normal red eye color is actually a mixture of red and brown pigments (pterins and ommachromes, respectively) which are synthesized by separate pathways controlled by different sets of genes. In mutants characterized by brilliant orange-red eyes, such as vermilion *(v)* and cinnabar *(cn),* the production of brown pigment is blocked, while in mutants such as brown *(bw)* and sepia *(se)* the synthesis of red pigment is affected. The compound eye of the adult develops from clusters of cells called imaginal discs which are formed in the embryo and which grow rapidly in later larval stages. Such discs can be removed from one larva and transplanted into the body of another larva of the same or different genotype. When the host undergoes metamorphosis to form adult structures, the transplanted eye disc also metamorphoses into an adult compound eye whose color can be observed after dissection from the host.

In their experiments Beadle and Ephrussi found that in most instances a genotypically mutant eye disc placed in a wild type host retained the mutant eye color after metamorphosis. Since the mutant tissue was not influenced by the host, it was termed *autonomous*. Wild type discs transplanted to mutant hosts were also autonomous. However, of those genotypes studied, two eye color mutations, vermilion and cinnabar, were found to be nonautonomous, for when eye discs of either of these genotypes were transplanted to wild type hosts, the eye color of the transplant became wild type (Figure 10–1). This finding implied that some substance or substances not synthesized by these mutant genotypes was being supplied via the body fluids of the host. To determine whether the same or different substances were involved, *v* discs were transplanted to *cn* hosts with the result that the transplanted tissue became wild type, although the host remained mutant. This observation suggested that the *cn* host provided some requirement that enabled *v*, but not *cn*, to synthesize brown eye pigment. In the reverse experiment in which *cn* discs were transplanted to *v* hosts, the discs remained mutant, suggesting that the *v* host could not supply the substance needed by *cn*.

From the results of these experiments, Beadle and Ephrussi proposed that two separate substances were sequentially involved in the production of brown pigment. The first acting substance was evidently that produced in the presence of the wild type allele of *v*, and therefore named the *v*+ substance. It was visualized that absence of this substance effectively blocked an early step in the pathway, thus preventing the formation of a subsequent intermediate required by *cn*. This hypothesis explained the inability of the *v* host to supply the requirements of a *cn* transplant. The second substance, absent in *cn*, but presumably produced in the presence of its wild type allele, was called the *cn*+ substance. It was inferred that the *cn*+ substance was required at a later step in pigment formation because a *cn* host (containing the wild type allele for *v*) was able to supply the *v*+ substance to a *v* transplant. Since the *v* transplant contained the wild type allele of *cn*, the transplant was able to form brown pigment even though the host could not. Subsequent studies have verified these

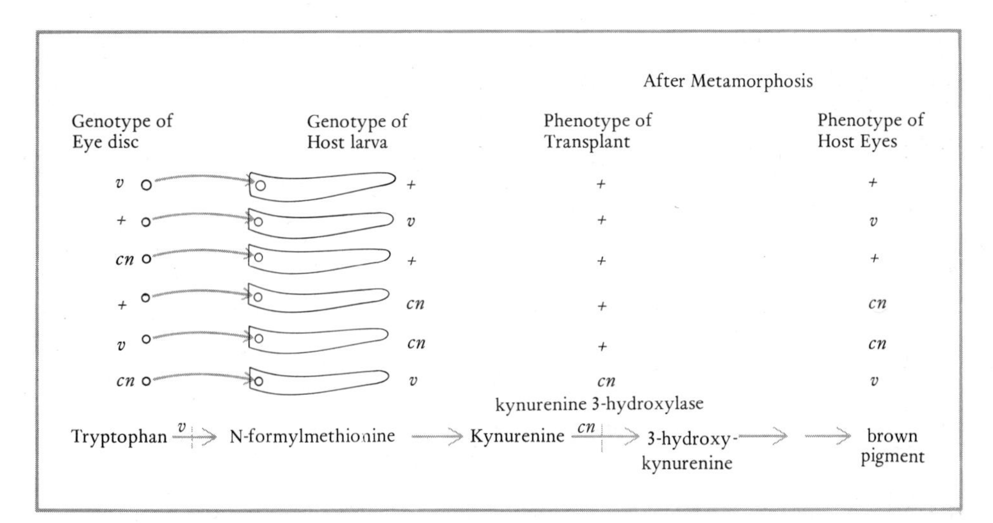

Figure 10–1. Results of Beadle and Ephrussi's studies of reciprocal transplants of eye discs between + and *v*, + and *cn*, and *v* and *cn*. Eye discs are transplanted to a region of the host other than that where the host eye discs are located. Vertical lines through arrows indicate the reactions blocked in the presence of *v* and *cn*, respectively. Double arrows (→→) indicate the presence of additional steps in the pathway beyond 3-hydroxykineurenine.

proposals and the source of the mutant phenotype in each case has been traced to the absence of a specific enzyme. A condensed outline of the pathway is shown in Figure 10–1, and the interested reader is referred to Phillips et al (1973) for a detailed discussion.

Although these studies of Beadle, Ephrussi, and others strengthened the hypothesis that genes were expressed through cellular chemical reactions, a more precise demonstration of this fact was first obtained with nutritional mutants of *Neurospora. Neurospora* offers numerous advantages for studies of the genetic control of metabolism. Since it is haploid, all genes are immediately expressed, so that the phenotype directly reflects the genotype and the complexities caused by dominance are absent. Also, since *Neurospora* offers little in the way of morphological differentiation, the plant body being merely a mass of filamentous hyphae, the primary mode of gene expression is not obscured by the complex developmental events characteristic of higher eukaryotes. Added to these advantages is the fact that the ordered ascospores permit precise genetic analysis.

Wild type *(prototrophic) Neurospora* can be grown in large quantities on a defined minimal medium containing only inorganic salts, biotin, and a carbon source such as sucrose, all other required substances being synthesized by the organism itself. Although a few mutations do cause changes in pigmentation or in the general morphology of the mycelium, most result in the inability of the organism to grow on minimal medium. Such mutants are called *auxotrophs* and they will grow only when the substance that they can no longer synthesize for themselves is added to the medium. The specific compound required in any case is determined by adding individual sugars, amino acids, and other substances, one at a time, to minimal medium until the material needed for growth is identified.

When large numbers of auxotrophic mutants of independent origin are tested in this way, more than one of them may be found to exhibit the same nutritional requirement. If so, the question arises as to whether such mutants are alleles and therefore deficient for the same product or function, or whether they are nonalleles whose individual functions contribute to the same end result. To distinguish between these alternatives, a test for allelism, called a *complementation test,* is used.

COMPLEMENTATION AND FUNCTIONAL ALLELISM

It will be recalled that allelism can be recognized without knowing either the exact gene product or the linkage position. The criterion applied is whether or not the phenotype is mutant when two different mutated genes are present together in the same cell or organism. This criterion for allelism for diploids has already been discussed (see Chapter 4), but can be briefly reviewed here. When a diploid is heterozygous for two nonallelic recessive mutations (for example, *a* and *b*), the phenotype is wild type and the genotype can be designated as *a*/+ *b*/+ or, if linked, *a b* / + +. Although the observed normal phenotype is ordinarily attributed to dominance, in terms of gene action dominance means that the respective products of the wild type alleles compensate for the lack of product or the abnormal product of the mutant genes. This compensating action of the wild type alleles is called *complementation*. On the other hand, if genes *a* and *b* are allelic, the phenotype of a heterozygote *(a/b)* will be mutant because no complementing wild type allele is present for either gene.

In haploid fungi such as *Neurospora* complementation tests involve the use of *heterokaryons*. When different mutant strains of the same mating type are grown together, hyphal fusion may occur to form cells containing the separate haploid nuclei of both strains within the united cytoplasm. Hyphae of this nature are called heterokaryons. If the mutations present in each strain are alleles, the heterokaryon, like each original strain, will fail to grow on minimal medium, thus retaining the mutant phenotype. However, if the strains which form the heterokaryon are mutant for different genes, the wild type phenotype will result and growth will occur on minimal medium (Figure 10–2).

Complementation is observed in this case because the genomes of the two strains are able to supply or compensate for one another's deficiencies. For example, if strain 1 is mutant for gene *a,* but wild type for gene *b,* and strain 2 is wild type for *a,* but mutant for *b,* the separate nuclei of these two strains will contain between them a wild type allele of each gene. Since these nuclei reside within the same cytoplasm, the activity of each normal allele permits the heterokaryon to assume the wild type phenotype. That the separate nuclei are unchanged and retain their individual deficiencies despite their joint presence in the same cell can be demonstrated by testing the genotypes of the asexual spores produced by the heterokaryon. These spores each contain a single hap-

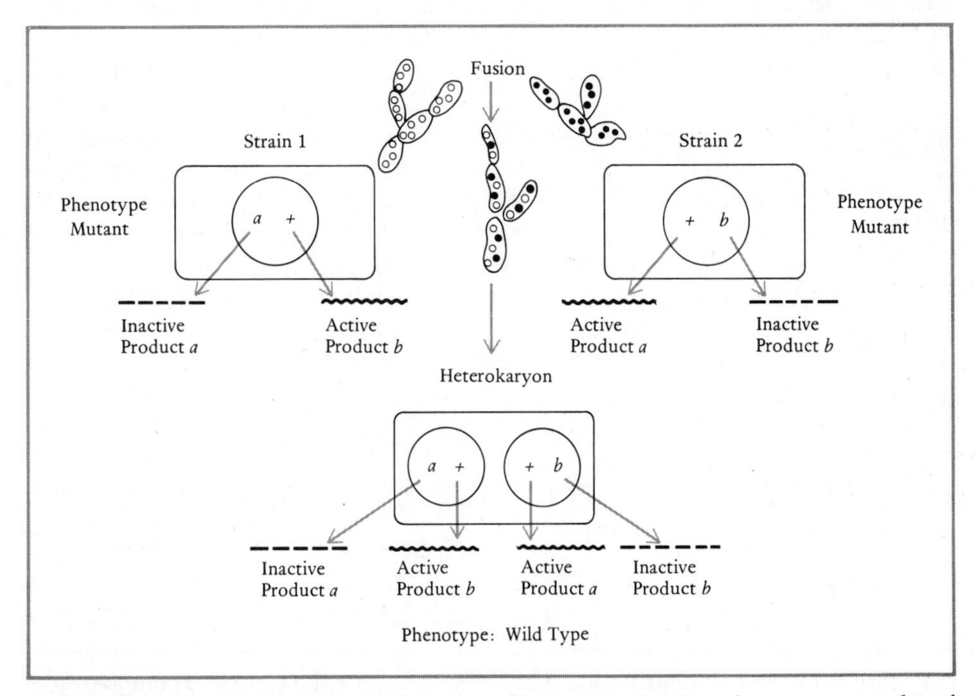

Figure 10–2. Complementation in a heterokaryon containing the separate nuclei of mutant strains 1 and 2.

loid nucleus and give rise to hyphae of one or the other of the two mutant strains which underwent fusion to form the heterokaryon.

Complementation tests for allelism are ordinarily applied to mutants that exhibit the same general phenotype, such as a similar nutritional requirement. The results of such tests are recorded in the form of a matrix or grid where the symbols + and 0 denote the presence or absence of the wild type phenotype for each paired combination of mutants tested (Figure 10–3). The grid permits easy identification of noncomplementing mutants (alleles) versus complementing mutants (nonalleles).

The data can then be summarized in the form of a complementation map in which separate, nonoverlapping lines are used to represent different, nonallelic functions. In the example in Figure 10–3 complementation tests reveal that mutants 1 through 6 can be assigned to four different functional units, represented by four separate lines in the complementation map. Mutants 1 and 6 are alleles, as are mutants 2 and 5, while mutants 3 and 4 each belong to different functional units. Because mutant 7 fails to complement with either 3 or 4, it is placed on a separate line which overlaps those of mutants 3 and 4. Mutant 7 can be interpreted either as a double mutant or as a deletion which includes some or all of both genes 3 and 4. The latter explanation is acceptable if no reverse mutations back to wild type occur when large numbers of individuals are observed, the basis for this assumption being that if a gene is missing, it can no longer undergo mutation.

It must be emphasized that complementation maps provide information only with respect to the functional identity or nonidentity of a series of mutants and give no indication as to the linkage group or the position of these genes

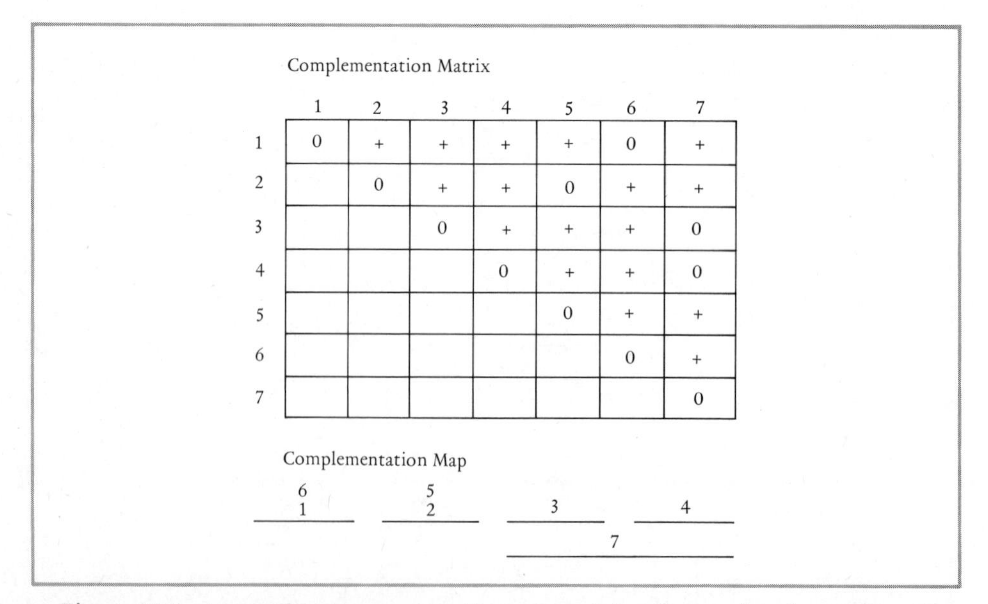
Complementation Matrix

	1	2	3	4	5	6	7
1	0	+	+	+	+	0	+
2		0	+	+	0	+	+
3			0	+	+	+	0
4				0	+	+	0
5					0	+	+
6						0	+
7							0

Figure 10–3. A complementation matrix summarizing the results of complementation tests performed on all paired combinations of mutants 1 through 7 and the complementation map derived from the matrix.

on the chromosome. Linkage group and chromosomal locus must be determined separately through the results of appropriate crosses. It should also be noted that on rare occasions complementation may occur between alleles. A discussion of this type of complementation (intragenic) will be deferred to Chapter 14.

BIOCHEMICAL GENETICS

Gene-Controlled Metabolic Pathways in Neurospora

Complementation tests are particularly useful in studies which combine genetics and biochemistry for they permit the resolution of seemingly similar mutations into separate nonallelic genes, each of which can be inferred to control a different step within the same metabolic pathway. Such tests were first employed in studies of *Neurospora* by Beadle and V. L. Coonradt in 1944 to demonstrate that two different auxotrophic mutants, each requiring nicotinic acid, were in fact nonallelic, since complementation occurred in a heterokaryon containing nuclei from each strain.

A. S. Srb and N. H. Horowitz used the same technique to separate a series of arginine-requiring mutants into functional classes and found that seven different nonallelic genes were involved in the biosynthesis of arginine. Three steps within this metabolic pathway were then identified by these workers. It was found that one group of alleles would grow only with the addition of arginine to the medium, while representatives of two other allelic groups would grow when either arginine or citrulline was added. The other four allelic groups would multiply in the presence of arginine, citrulline, or ornithine (Figure 10–4). From these results it appeared that strains that would grow only with the addition of arginine represented a terminal step in the pathway and that those able to grow with either arginine or citrulline were blocked at preceding steps. It was also reasoned that the groups able to grow with arginine, citrulline, or ornithine must be involved in earlier reactions which converted a precursor to ornithine. This proposed stepwise biosynthesis of argnine in *Neurospora* agreed with similar findings based on studies with mammalian liver. It was thus evident that the investigation of gene-controlled metabolic pathways in fungi might provide models for the analysis of equivalent pathways in humans.

This prediction was strengthened when subsequent investigations of the tryptophan-requiring mutants of *Neurospora* showed for the first time how this amino acid is synthesized by living organisms. In addition, these mutants

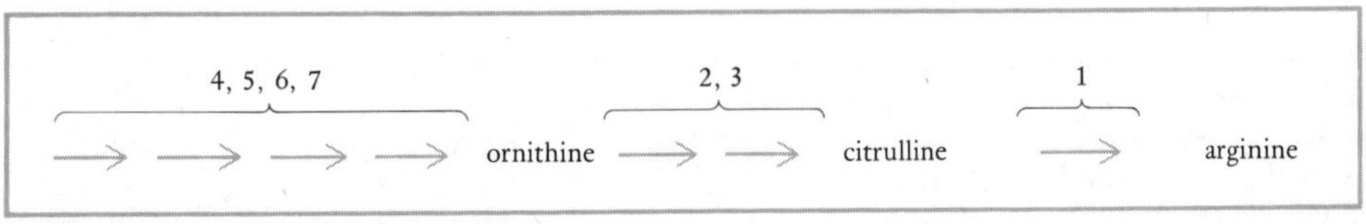

4, 5, 6, 7 — 2, 3 — 1

→ → → → ornithine → → citrulline → arginine

Figure 10–4. Gene controlled steps in the pathway for arginine synthesis as proposed by Srb and Horowitz for *Neurospora*. Numbers 1 through 7 represent different groups of alleles as defined by complementation tests.

provided the first clear-cut demonstration of the direct relationship between genes and enzymes, for it was found that mutants blocked in the final step of the tryptophan pathway lacked the specific identifiable enzyme (tryptophan synthetase) which catalyzes this reaction in wild type cells.

Gene-Controlled Metabolic Pathways in Bacteria

While complementation tests in *Neurospora* were being developed and explored, the genetic mechanisms of bacteria were under intensive study (see Chapter 13), and when complementation tests were devised for these organisms, further analysis of gene-controlled biosynthetic pathways became possible. Two examples of such studies with bacteria can be cited.

In 1955 S. Brenner and others were able to separate tryptophan-requiring mutants of *Salmonella typhimurium* into functional groups each of which

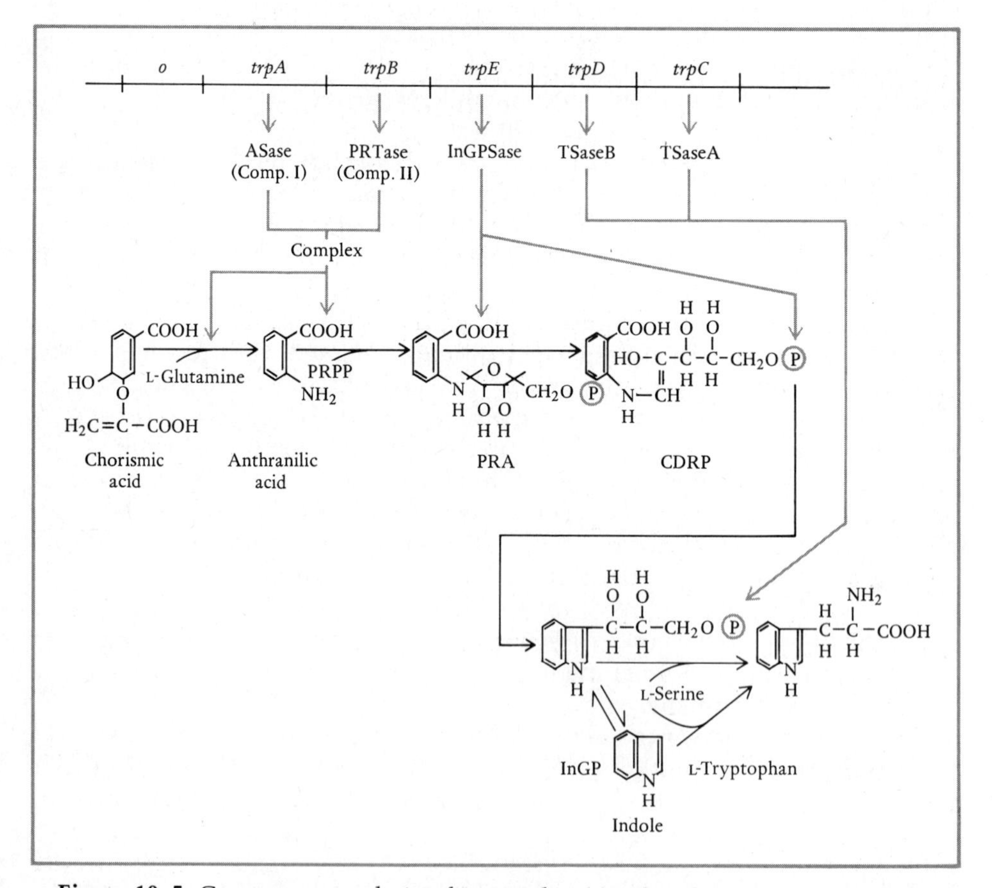

Figure 10–5. Gene-enzyme relationships in the trytophan biosynthetic pathway of *Salmonella typhimurium*. The order of *trp* genes is the same as the steps in the pathway. Locus *o* (operator) is a control element for the pathway. Enzyme abbreviations: ASase (anthranilate synthetase); PRTase (phosphoribosyl transferase); InGPSase (indole glycerol phosphate synthetase); TSaseA, TSaseB (trytophan synthetase A and B). Intermediates of the pathway: PRA (phosphoribosyl-anthranilic acid); CDRP (carboxyphenylamino-1-deoxyribulose-5-phosphate) InGP (indole-3-glycerol phosphate). [From G. Wuesthoff and R. H. Baurle, *J. Mol. Biol., 49,* 171–196, 1970.]

could be presumed to block a different step in the pathway for tryptophan synthesis. These steps were then characterized biochemically and arranged in a functional sequence according to the precursor substances accumulated and the compounds which, when added to the medium, would permit growth to occur in each case (Figure 10–5). When the pathway for tryptophan biosynthesis discovered in *Salmonella* was compared with that in *Neurospora* and subsequently with that in *Escherichia coli* and yeast, it was found that the steps of this pathway were essentially similar for all, thus further demonstrating a fundamental unity of living cells with respect to biochemical mechanisms. The tryptophan mutants of *Salmonella* also provided an intriguing insight into the arrangement of genes in bacteria. Genetic mapping experiments carried out by M. Demerec and Z. Hartman and others revealed that these mutations occupied one or another of five contiguous, but nonoverlapping loci. In addition, it was found that the order of these loci was the same as the order of the sequential steps in the tryptophan pathway.

A second and classic example of combined genetic and biochemical analysis in bacteria is that of loci mediating histidine synthesis in *Salmonella*. Over 900 independent mutations exhibiting a requirement for histidine were resolved by P. E. Hartman, B. N. Ames, and others into 9 separate functional genes by complementation tests. Biochemical studies then enabled the association of each such gene with a particular enzyme-catalyzed step in the pathway for histidine synthesis (Figure 10–6). Over 500 histidine mutants were then mapped to 9 distinct and separate loci which were clustered together within a small region of the genome. Although in this case the order of the genes was not the same as the order in which their products (enzymes) participated in the pathway, the study once again illuminated a characteristic feature of gene arrangement in prokaryotes, that is, the grouped association of functionally related loci.

Gene-Controlled Metabolic Pathways in Humans

Although alkaptonuria was the first inherited human disease to be associated with an enzyme deficiency, the enzyme implicated in the expression of this trait is only one of several belonging to the metabolic pathway through which the amino acids phenylalanine and tyrosine are utilized or degraded. Inspection of Figure 10–7 shows that the blocks A, B, C, and D, each result in a different inherited trait. Block A involves a deficiency of the liver enzyme phenylalanine hydroxylase which results in the failure of the conversion of phenylalanine to tyrosine. As a consequence high levels of phenylalanine are found in the blood and, by ordinarily unimportant reactions, phenylalanine is converted to phenylpyruvic acid and other by-products which are excreted in the urine. The condition is called *phenylketonuria* (PKU).

PKU is inherited as an autosomal recessive, the frequency of homozygotes being around 1 in 15,000 births. If untreated in infancy, it results in such severe mental retardation that lifelong institutional care is required. The brain undergoes exceedingly rapid growth in the postnatal period and the excessive amounts of phenylalanine present in the blood of homozygous individuals inhibit this development. The mechanism of inhibition is not well understood, but it has been postulated that high levels of phenylalanine interfere with

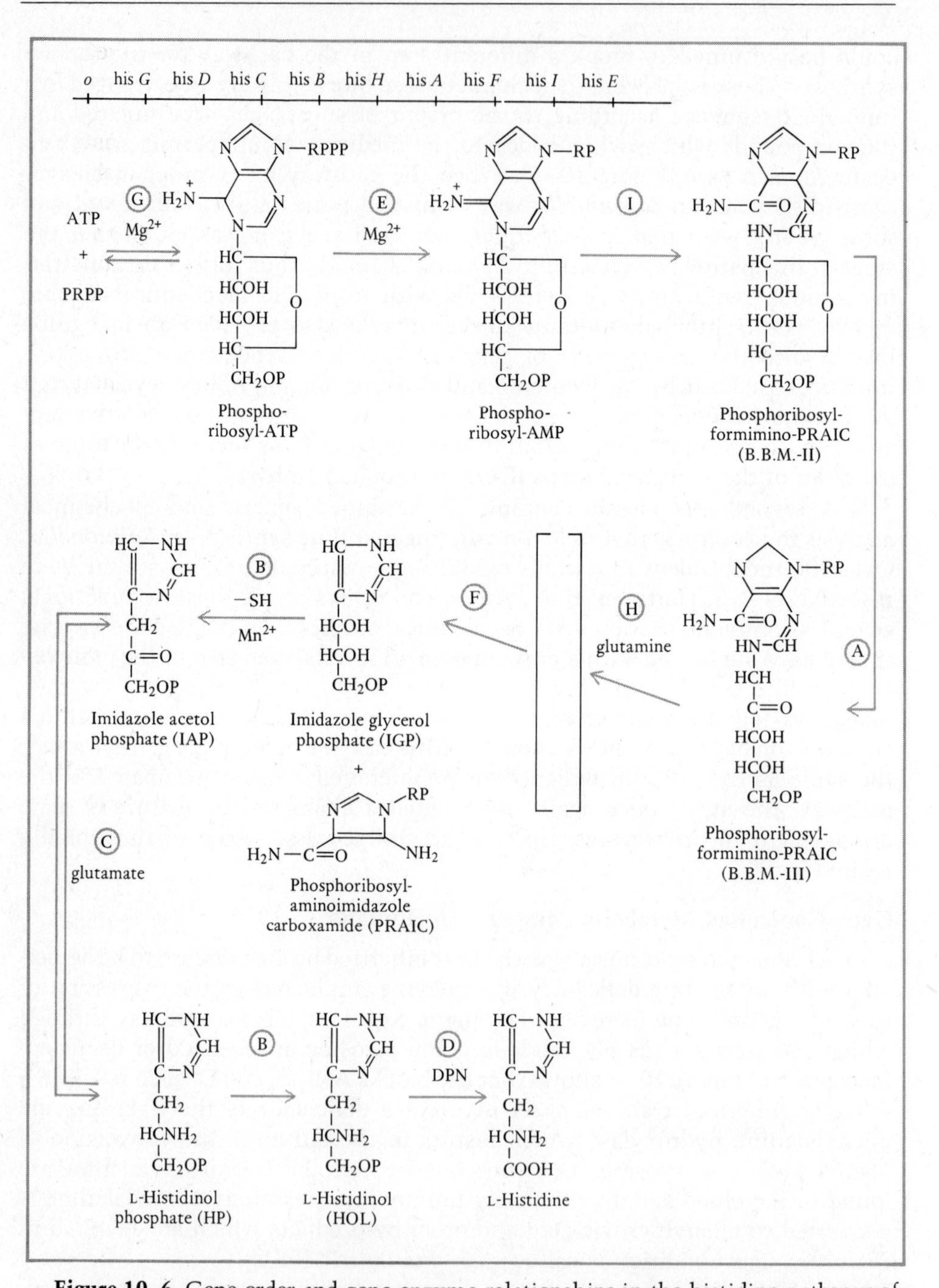

Figure 10–6. Gene order and gene-enzyme relationships in the histidine pathway of *Salmonella typhimurium*. *His* genes *G* through *E* specify enzymes involved in the biosynthesis of histidine. Locus *o* (operator) is a controlling element. In the biosynthetic pathway circled letters indicate the site of action of enzymes specified by the corresponding *his* genes. The structure of the compound indicated by brackets is uncertain. Note that the order of genes is not the same as that in which their products participate in the reaction sequence. Abbreviations: P (phosphate); R (ribose); PRAIC (phosphoribosynaminoimidazole carboxamine). [From M. Brenner and B. N. Ames, in *Metabolic Pathways,* 3rd ed., vol. V, *Metabolic Regulation,* H. Vogel, Ed., Academic Press, New York, 1971.]

cellular transport mechanisms or inhibit other enzyme systems. Treatment consists of dietary control of phenylalanine, so that while a sufficient supply of this essential amino acid is provided for growth, the level of phenylalanine in the blood is brought within the normal range. Such treatment is best instituted within the first month of postnatal life, for a significant delay in dietary management causes irreversible brain damage. Long term studies of homozygotes indicate that with dietary control, mental ability can be brought into the normal range. Because of the possibility of treatment, widespread screening of the newborn for PKU is practiced in many areas of the United States as well as in other regions of the world.

The metabolic block labeled B in Figure 10–7 results in an extremely rare recessive trait called *tyrosinosis* (tyrosineamia). This condition is characterized by the inability to convert *p*-hydroxyphenylpyruvic acid to homogentisic acid due to the deficiency of the enzyme which normally catalyzes this reaction. The block at this point in the pathway causes an accumulation of both tyrosine and *p*-hydroxyphenylpyruvic acid, both of which are found in high levels in blood and urine. The condition leads to liver damage and is frequently fatal in infants and young children.

An enzyme deficiency in the next step in this pathway produces the inherited trait *alkaptonuria* (block C). As previously discussed, this relatively mild disorder is characterized by the accumulation and excretion of large amounts of homogentisic acid. The defective enzyme in this case is homogentisic acid oxidase.

Finally, disorders in tyrosine metabolism can lead to disturbances in melanin production resulting in albinism (block D). At least three types of albinism are known. Two are inherited as autosomal recessive conditions and affect both skin and eye pigmentation, while one affects only the eyes (ocular albinism) and is inherited as a sex-linked recessive. One type of autosomal albinism has been attributed to the absence or defective function of tyrosinase, the enzyme which catalyzes two initial steps in the formation of melanin. In the second type of autosomal albinism, tyrosinase is present and functional when tested *in vitro,* but is apparently unable to act *in vivo*. The basis for ocular albinism is at present unknown.

It can be seen from the preceding survey that an interruption of a biosynthetic or degradative pathway can occur at any one of a variety of sites as the result of a missing or dysfunctional enzyme. An additional example of this phenomenon is provided by Figure 10–8 which illustrates a pathway concerned with the metabolism of glucose in red blood cells. Since deficiencies of individual enzymes of this, as well as the preceding pathway, arise from separate, identifiable inherited defects, the direct association of specific genes with specific enzymes in humans is clearly evident.

One Gene–One Enzyme Hypothesis

The experiments carried out with *Neurospora* and bacteria as well as the studies of human metabolic diseases have provided abundant evidence that genes control the overall biochemical phenotype by controlling the sequential steps of cellular reactions. Since each such step is ordinarily catalyzed by a specific enzyme, it can be inferred that an enzyme represents the primary product of gene action. This hypothesis, known as the *one gene–one enzyme* hypothesis, was enunciated by Beadle as early as 1945. The main tenets of the hypothesis can be summarized as follows.

The synthesis (or degradation) of any substance proceeds as a stepwise sequence of reactions catalyzed at each point by a specific enzyme whose structure and consequent activity is determined by a corresponding gene. Accordingly, the genotype determines the particular complement of enzymes

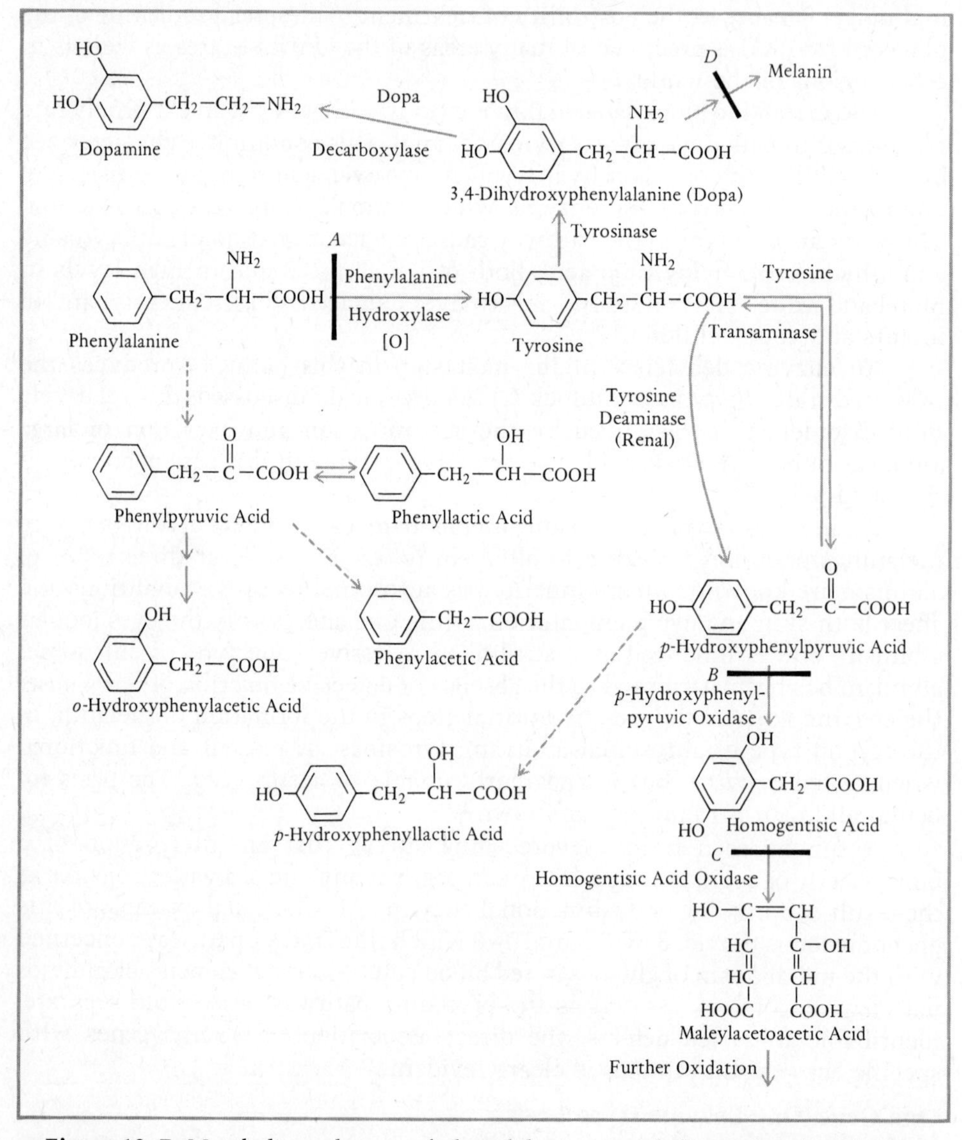

Figure 10–7. Metabolic pathways of phenylalanine metabolism, showing blocks in phenylketonuria (A), tyrosinosis, (B), alkaptonuria (C), and albinism (D). Reactions shown by broken arrows are quantitatively unimportant in normal persons, but become important in phenylketonuria. [From *An Introduction to Human Genetics,* Second Edition, by H. Eldon Sutton. Copyright © 1975, 1965 by Holt, Rinehart and Winston. Reprinted by permission of Holt, Rinehart and Winston.]

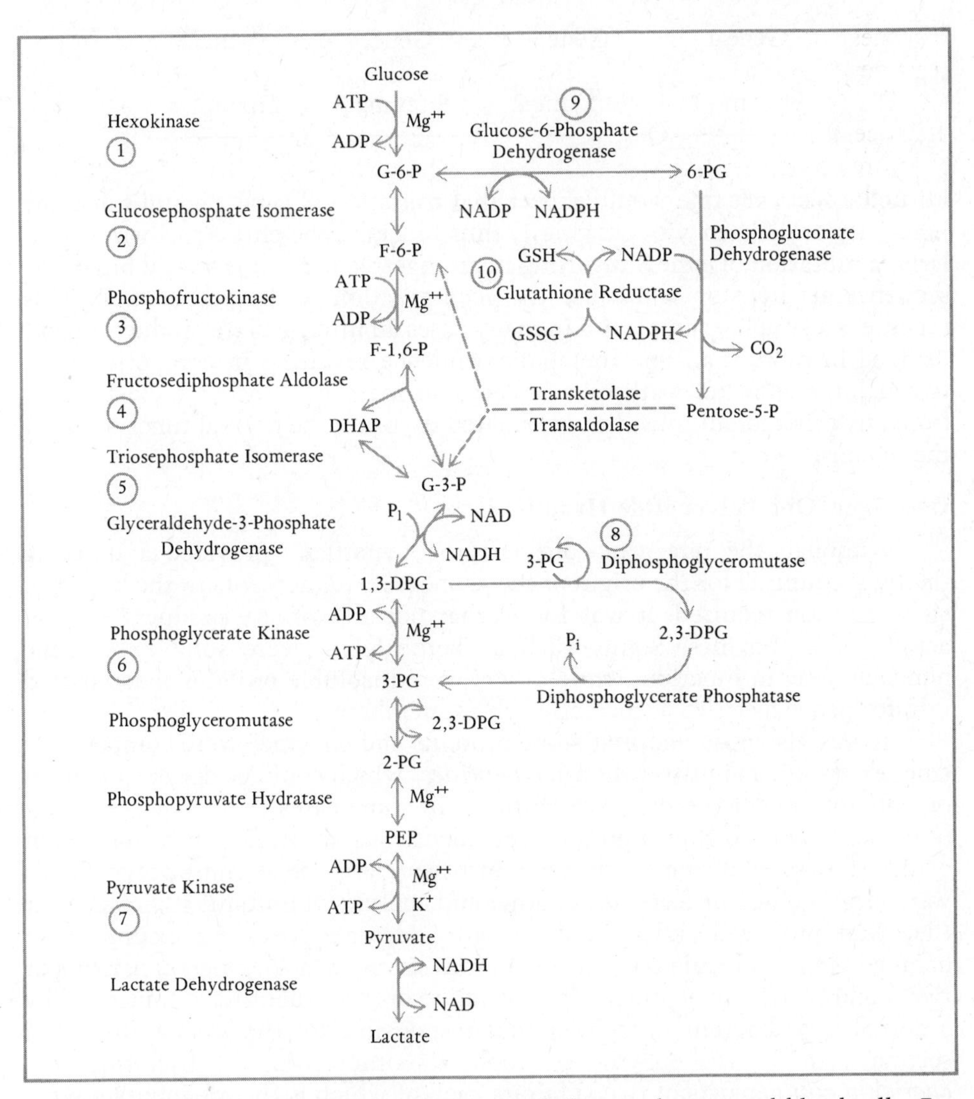

Figure 10–8. The pathways of glucose metabolism in human red blood cells. Enzymes numbered 1 to 10 are those whose dysfunction or deficiency has been associated with a specific inherited disease. Most such traits appear phenotypically as hemolytic anemias of varying degrees of severity. A deficiency of aldolase (4) in the liver causes hereditary fructose intolerance. K^+ and Mg^+ indicate ions whose presence is required for enzyme activity. [From W. N. Valentine, *The Western Journal of Medicine* (formerly *California Medicine*), *108*, 280–294, 1968.]

available to an organism, and this complement determines the biochemical phenotype and, through it, the morphological phenotype. We can visualize a gene-controlled biosynthetic sequence as follows, where precursor (P) enters an organism through the diet and is metabolized to the final product (T) through a series of intermediate steps, each catalyzed by a specific enzyme (enzymes 1, 2, 3, 4), the structure of the enzymes being determined by the corresponding genes (genes 1, 2, 3, 4).

Gene 1	Gene 2	Gene 3	Gene 4
↓	↓	↓	↓
Enzyme 1	Enzyme 2	Enzyme 3	Enzyme 4

$$P \xrightarrow{\text{Enzyme 1}} Q \xrightarrow{\text{Enzyme 2}} R \xrightarrow{\text{Enzyme 3}} S \xrightarrow{\text{Enzyme 4}} T$$

In such a sequence one would expect that mutation of gene 1 could cause the inactivity or deficiency of enzyme 1, thus blocking the entire pathway. Similarly, a mutation of gene 3 resulting in an inactive enzyme 3 would block the sequence at this step and cause the accumulation of the precursor R. This scheme is a simple example, for in many cases an intermediate product may be utilized in more than one metabolic sequence resulting in competition between interconnected pathways. Indeed, no metabolic pathway exists apart from any other for all must be coordinated to insure the normal functioning of the organism.

One Gene–One Polypeptide Hypothesis

Although the one gene–one enzyme hypothesis provided a unifying theory accounting for the origin of the phenotype, refinements in the hypothesis were soon required. It was found that not all primary products of gene action were enzymes. Some, such as hemoglobin, were soluble proteins, nonenzymatic in function, while others were insoluble proteins incorporated into formed structures of the cell such as membranes.

It was also observed that some proteins and enzymes were composed of smaller protein subunits, called *polypeptides,* which could be dissociated from one another under the proper conditions. In some cases the polypeptide subunits of a single enzyme or protein were identical and therefore their formation could be attributed to a single gene. In other cases the subunit polypeptides were clearly different from one another and analysis of mutants indicated that they were produced under the direction of separate genes. For example, the hemoglobin A molecule is composed of four subunits (a *tetramer*) consisting of two α and two β polypeptides. The formation of the α subunits is controlled by a completely different gene from that responsible for the β subunits. As a second example, the enzyme tryptophan synthetase of the bacterium *Escherichia coli* consists of two subunits each of which is the product of a separate gene.

In view of these and other examples, the one gene–one enzyme hypothesis was modified to a one gene–one polypeptide hypothesis. Insofar as proteins of the cell are concerned, this theory has been supported by all subsequent evidence. Since proteins and polypeptides are the primary gene products which establish the metabolic and structural phenotype, it is pertinent to examine their nature in order better to understand and appreciate the activities of the genome.

THE NATURE AND STRUCTURE OF PROTEINS

Proteins form an integral part of almost every cellular structure. All cell products, all synthetic and degradative reactions, all steps in the entrapment of energy, in short, almost every activity of the cell is mediated by proteins, either in their role as structural elements or as catalytic agents (enzymes). To a large extent, we are what our proteins are since almost all else is formed through their activity. In addition, the complexity of their structure endows proteins with specificity so that each kind of protein characteristically performs only one, or at the most, a very few functions in the cell.

The enormous variety of proteins is a function of their structure. A protein molecule is a linear, unbranched polymer composed of individual units called amino acids which are connected by covalent bonds to form a long chain. Since the bond involved is referred to as a *peptide bond*, a single chain of amino acid residues so linked is called a polypeptide or, if long enough, a protein. (The point at which a polypeptide becomes a protein is arbitrary, so the terms polypeptide and protein will be used here interchangeably.)

Variety is built into the molecule by the incorporation of different kinds of amino acids at different positions in the polypeptide chain. There are approximately 20 different, naturally occurring amino acids commonly found in proteins and the number of possible arrangements of these in an average sized polypeptide of around 300 amino acids would be 20^{300}. This number seems too large even to imagine, but nevertheless provides a vivid illustration of the possibilities available for protein structure. Since the positioning of the different amino acids in the chain determines the subsequent biological activity of the molecule, it can readily be seen that an enormous variety of activities is also available to the cell.

Amino acids, regardless of type, have a common basic structure. One end of the molecule consists of an amino group (H_2N—), while the other end is a carboxyl group (—COOH). A central or alpha (α) carbon is bonded to these end groups and also to hydrogen and a side chain or R group, the symbol R standing for any one of a number of possibilities.

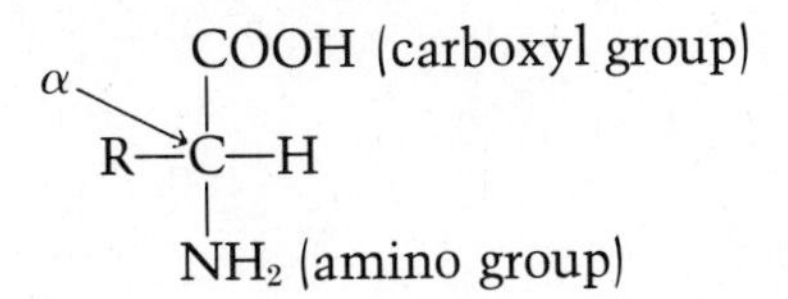

Amino acids are linked together by the covalent peptide bond which joins the carboxyl group of one with the amino group of another amino acid, with the elimination of water.

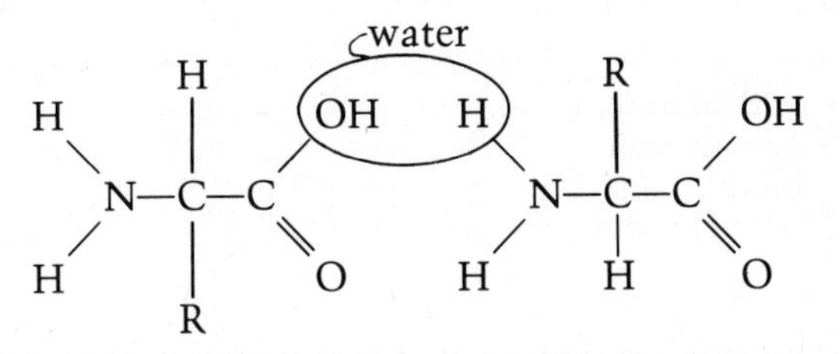

A series of amino acids so joined is thus a long chain with an amino group at one end and a carboxyl group at the other end.

Since the continuous peptide linkages are identical, the character and individuality of any protein depends on the side chains (R groups) of the different amino acid residues (Figure 10–9). The R groups of amino acids may be

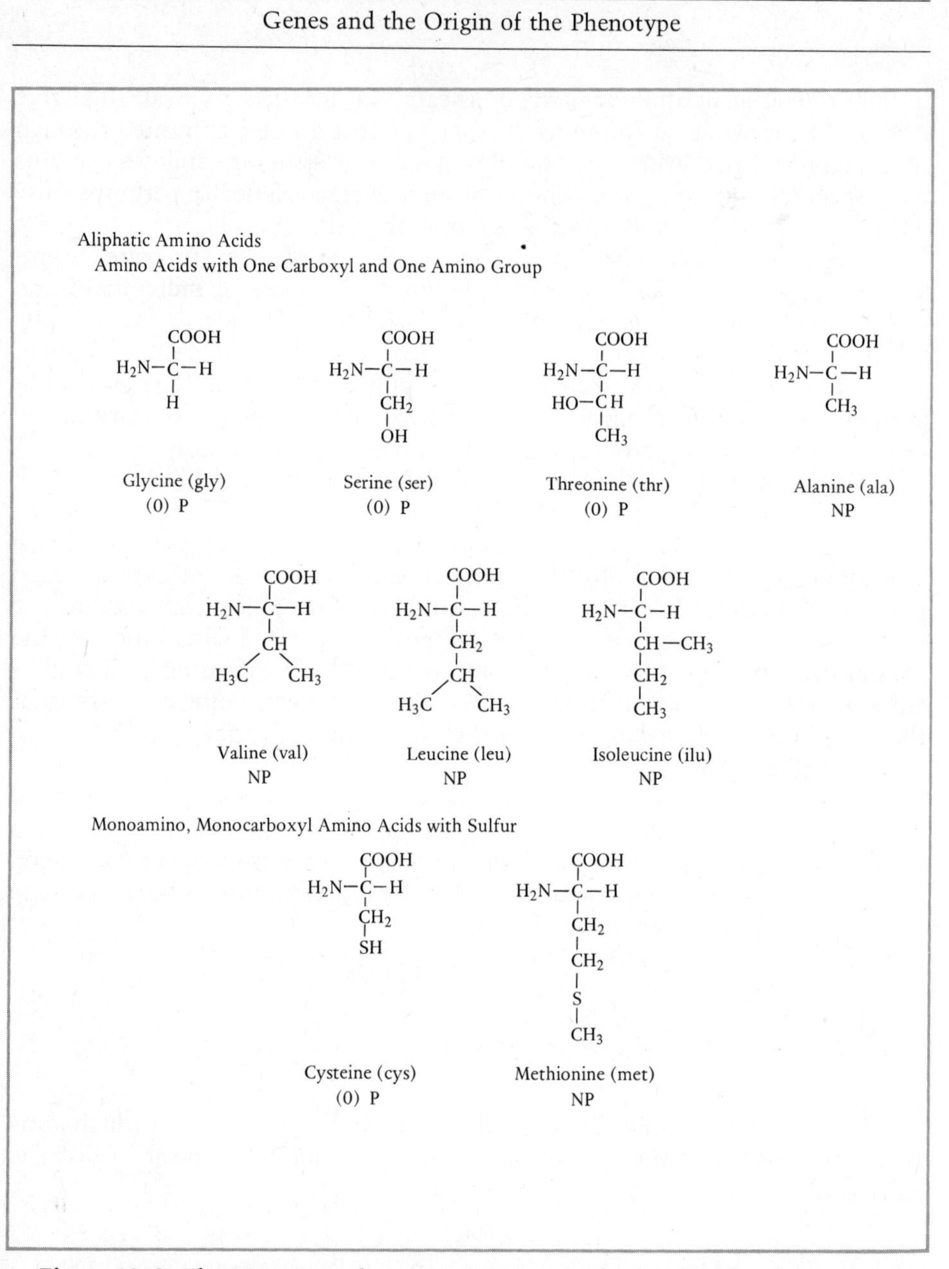

Figure 10–9. The 20 commonly occuring amino acids classified by chemical structure. The notation beneath each group indicates polarity (P) or nonpolarity (NP) of the R groups. The charge on the polar R groups at the pH of the cell is shown: uncharged (0), positively charged (+), negatively charged (−).

nonpolar, that is, hydrophobic (not soluble in water) or they may be polar, that is, hydrophilic (soluble in water). Of the polar amino acids, some are neutral, some contain basic groups, and others contain acidic groups. In an aqueous environment at the pH of the cell, the basic and acid groups ionize, resulting in local positive and negative charges, respectively. A polypeptide containing

Dicarboxylic Amino Acids and Their Amides

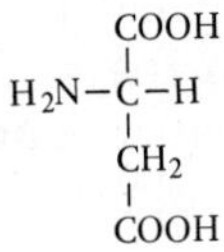

$COOH$
H_2N-C-H
CH_2
$COOH$

Aspartic Acid (asp) (−) P

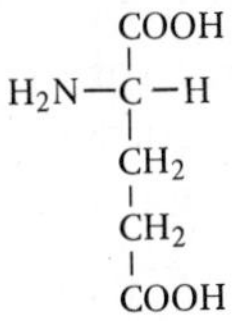

$COOH$
H_2N-C-H
CH_2
CH_2
$COOH$

Glutamic Acid (glu) (−) P

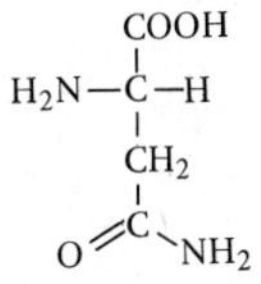

$COOH$
H_2N-C-H
CH_2
$O{=}C-NH_2$

Asparagine (asn) (0) P

$COOH$
H_2N-C-H
CH_2
CH_2
$O{=}C-NH_2$

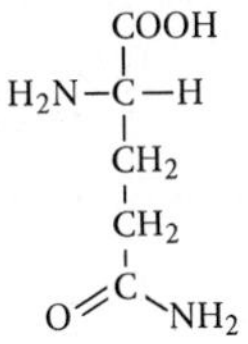

Glutamine (gln) (0) P

Basic Amino Acids

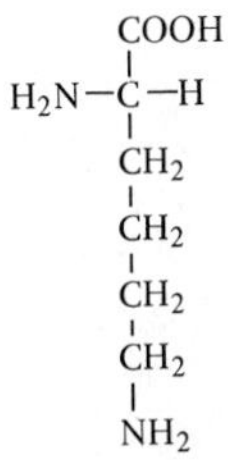

$COOH$
H_2N-C-H
CH_2
CH_2
CH_2
CH_2
NH_2

Lysine (lys) (+) P

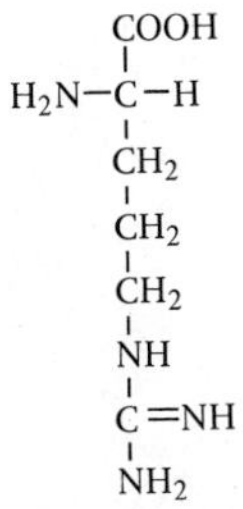

$COOH$
H_2N-C-H
CH_2
CH_2
CH_2
NH
$C{=}NH$
NH_2

Arginine (arg) (+) P

$COOH$
H_2N-C-H
CH_2
$C-NH$
$HC-N{=}CH$

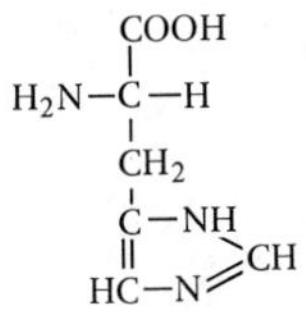

Histidine (his) (+) P

Aromatic Amino Acids (Includes Histidine, Given Above)

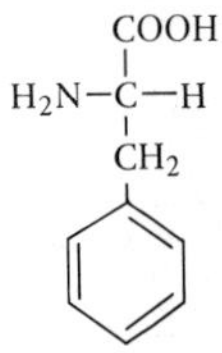

$COOH$
H_2N-C-H
CH_2

Phenylalanine (phe) NP

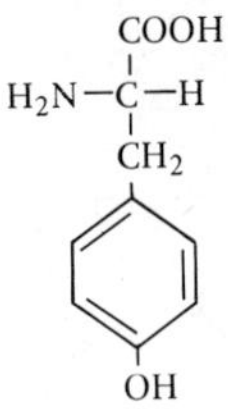

$COOH$
H_2N-C-H
CH_2
OH

Tyrosine (tyr) (0) P

$COOH$
H_2N-C-H
CH_2
$C{=}CH$
NH

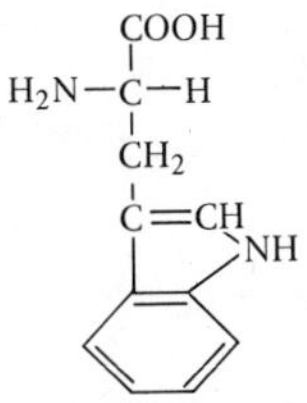

Tryptophan (trp) NP

Cyclic Amino Acids

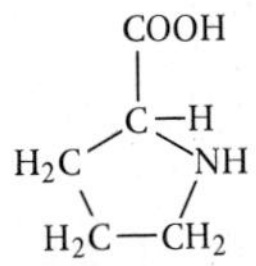

$COOH$
$C-H$
H_2C NH
H_2C-CH_2

Proline (pro) NP

many basic amino acids will have a net positive charge and is referred to as a *basic protein,* while one with an excess of acidic groups will have a net negative charge and is called an *acidic protein*. A protein with equal frequencies of both will be close to neutrality.

Polypeptides also exhibit different levels of structure. Primary structure refers to the sequence of amino acids present in the chain, while secondary structure refers to a regular helical shape (α-helix) caused by the formation of weak bonds, such as hydrogen bonds, between the carboxyl and amino groups of adjacent residues. Tertiary structure (Figure 10–10) involves the folding of the molecule into a complex globular shape which is held in place not only by weak bonds, but also by covalent bonds, particularly by disulfide bridges (—S—S—) which are readily formed between the sulfhydryl (—SH) groups of cysteine residues. Through the formation of such bonds, strategically situated cysteine residues can bring and hold together various portions of the polypeptide chain (Figure 10–11).

The net result of all of these interactions is the spontaneous assumption by the polypeptide of a specific three dimensional configuration in which charged and uncharged polar residues are on the outside of the molecule adjacent to the aqueous environment, while hydrophobic residues are tucked away inside. This final shape confers upon the protein the all important attribute of

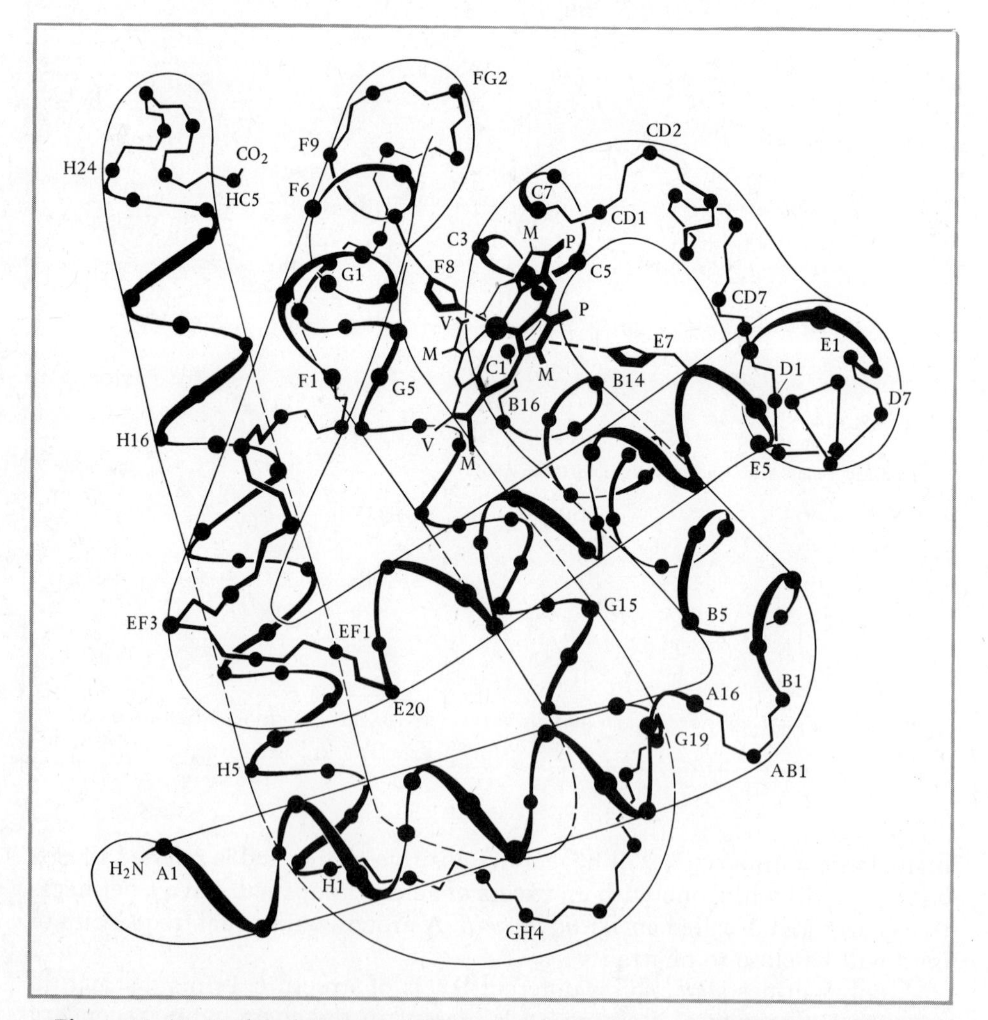

Figure 10–10. Model of the myoglobin molecule. Amino terminal (H_2N), bottom left at A1; carboxyl terminal (CO_2), top left at HC5. Letters and numbers designate amino acids in different segments of the chain. Large dots are alpha carbons. [From R. E. Dickerson, in *The Proteins*, 2nd ed., vol. II, 1964, H. Neurath, Ed., Academic Press, New York.]

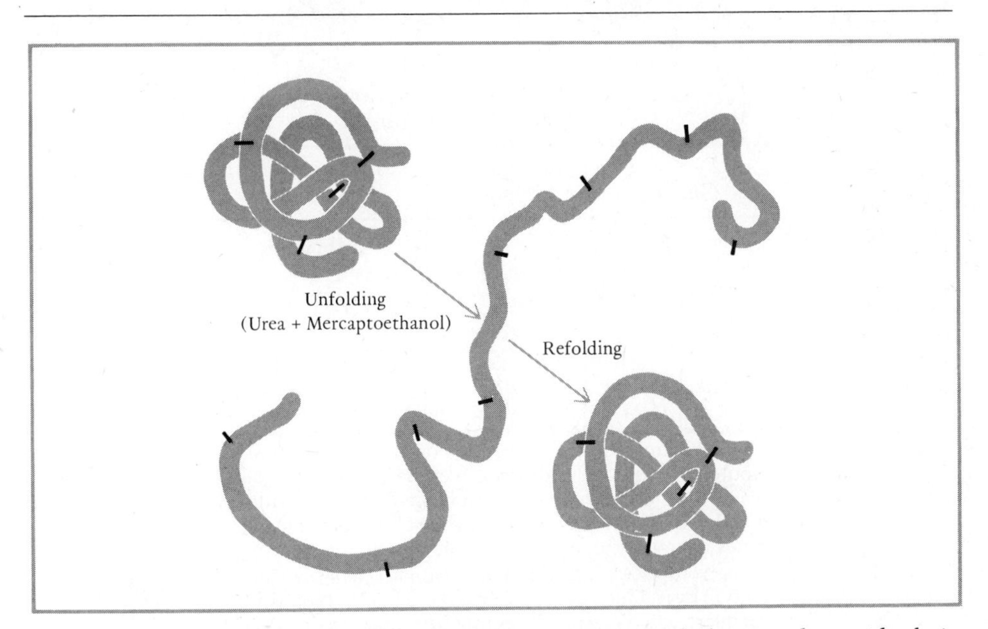

Figure 10–11. Model system for studying the conversion of a linear polypeptide chain to a folded protein. A native protein, with intramolecular disulfide bonds, is reduced and unfolded with B-mercaptoethanol and 8 M urea. After removal of these reagents, the reduced protein is allowed to undergo spontaneous refolding and reoxidation. [From C. J. Epstein, R. F. Goldberger, and C. B. Anfinson, *Cold Spring Harbor Sympos. Quant. Biol.*, *28*, 439–449, 1963.]

specificity and is a consequence of the particular sequence of amino acids that were originally built into the molecule, the designation of this sequence being a function of the genotype. The genotype dictates the selection and incorporation of appropriate amino acids in an order which will automatically result in the formation of bonds at specific points along the molecule, thus assuring the proper, biologically active conformation.

The Hemoglobins

More is known about the human hemoglobin molecule than about almost any other animal, plant, or bacterial protein. It was the first multiunit protein in which the entire amino acid sequence of all component polypeptides was determined from the amino end to the carboxyl end. Hemoglobin also provided the first instance in which the effects of gene mutation could be clearly defined in terms of a specific change in the molecular structure of a polypeptide. For an understanding of mutant types, it is helpful to consider first the structure of normal hemoglobins.

Several kinds of hemoglobin are synthesized during the course of life (Table 10–1). All types consist of four polypeptide subunits and are hybrid in the sense that they consist of two kinds of subunits. Four iron-containing haem residues are also present, providing oxygen carrying capacity. The predominant type of hemoglobin present depends on the stage of life: embryonic, fetal, or adult. Human embryos contain a hemoglobin which consists of two identical *alpha* (α) chains and two identical *epsilon* (ϵ) chains. Accordingly, embryonic hemoglobin can be represented symbolically as $\alpha_2\epsilon_2$. Synthesis of this hemoglo-

TABLE 10–1
Genetic and Molecular Relationships of Human Hemoglobins*

Genes		Hemoglobin types	Principal period of synthesis
	Hb_{ϵ}	$\alpha_2\epsilon_2$	embryonic
Hb_{α}	Hb_{γ^G} Hb_{γ^A}	$\alpha_2\gamma_2^G$ $\alpha_2\gamma_2^A$	fetal
	Hb_{β} Hb_{δ}	$\alpha_2\beta_2$ $\alpha_2\delta_2$	postnatal

* Hemoglobin Portland 1 is omitted but presumably would represent an additional locus, Hbζ, whose product reacts with the γ chains. It is probable that some synthesis of all Hb chains occurs throughout life, but the non-α genes operate primarily during one period of ontogeny.

Source: *An Introduction to Human Genetics*, Second Edition by H. Eldon Sutton. Copyright © 1975, 1965 by Holt, Rinehart and Winston. Reprinted by permission of Holt, Rinehart and Winston.

bin decreases sharply by the third month of gestation. It is replaced by fetal hemoglobin (HbF) which consists of two α chains plus two *gamma* (γ) chains. Apparently two different types of γ chains occur, type A and type G, so that fetal hemoglobin can be designated as $\alpha_2\gamma_2^A$ and $\alpha_2\gamma_2^G$. Synthesis of HbF essentially ceases at birth and very little can be detected by the age of 6 months (Figure 10–12). During the latter part of fetal life and throughout adulthood, hemoglobin A (HbA) is synthesized, and it comprises over 90 percent of the hemoglobin found in red cells from age 6 months onward. Hemoglobin A consists of two α chains and two *beta* (β) chains and can be symbolized as $\alpha_2\beta_2$. An additional, but very small component of adult hemoglobin is hemoglobin A^2 (HbA^2) which contains two α chains plus two *delta* (δ) chains, $\alpha_2\delta_2$.

More recently, another embryonic hemoglobin, called Portland 1, has been detected. It contains no α chains, but instead is a combination of a different polypeptide, *zeta* (ζ) with the fetal polypeptide chain gamma (γ), and is symbolized as $\zeta_2\gamma_2$.

Thus, disregarding Portland 1, about which very little is known, the major and well-studied kinds of normal hemoglobin are the embryonic ($\alpha_2\epsilon_2$), the fetal ($\alpha_2\gamma_2$), and the adult hemoglobins A and A^2 ($\alpha_2\beta_2$ and $\alpha_2\delta_2$, respectively). All of these contain the α polypeptide chain which consists of 141 amino acids. The complete structure of the γ, β, and δ chains is also known. Although they differ somewhat from one another in amino acid sequence, each is composed of 146 amino acids. The utility of these many hemoglobins appears to reside in their differing capacity to carry oxygen. Fetal hemoglobins have a greater affinity for oxygen than adult hemoglobins, this affinity being correlated no doubt with the fact that the fetus must derive its oxygen secondarily through the placenta, rather than directly from the air. The evolutionary origin of these multiple hemoglobin genes has been attributed to gene duplication (see Chapter 24).

From the preceding review we can see that at least seven different types of

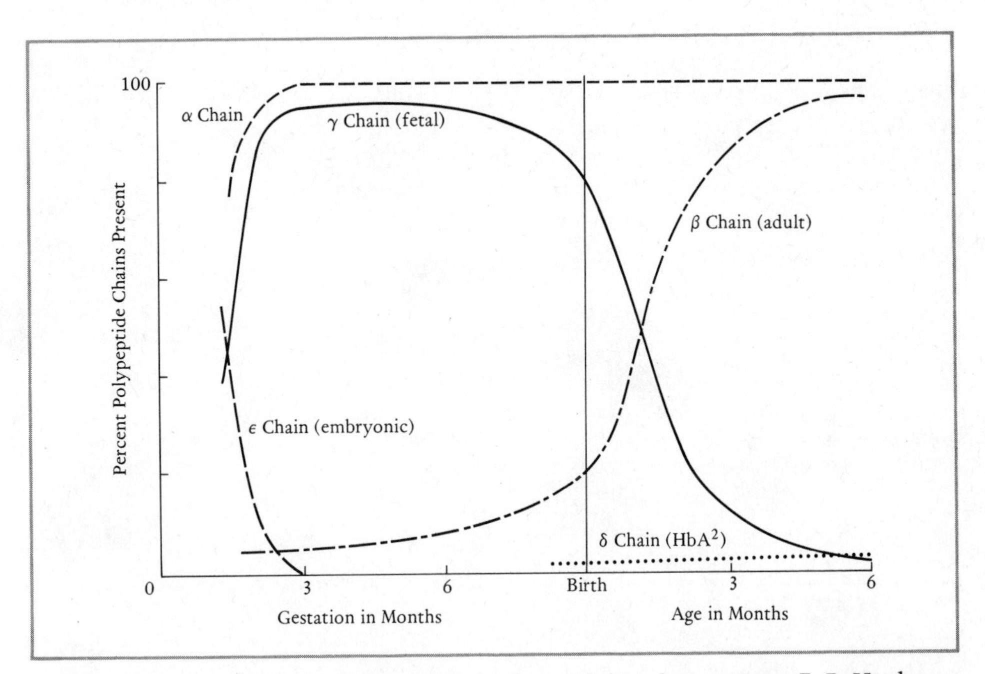

Figure 10–12. The development of human hemoglobin chains. [From E. R. Huehns et al., *Cold Spring Harbor Sympos. Quant. Biol.*, *29*, 327–331, 1964.]

polypeptides occur in normal hemoglobin molecules over the course of life (ζ, ϵ, γ^A, γ^G, α, β, and δ). Applying the hypothesis that a single gene dictates the amino acid sequence of but one kind of polypeptide, we must postulate the presence in humans of seven different hemoglobin genes. It also follows that a mutation occurring in one of these genes should result in a changed amino acid sequence in only one kind of polypeptide. Such a changed amino acid sequence should in addition affect only the particular type of hemoglobin in which that polypeptide occurs. Accordingly, a change in either kind of γ chain should result in an aberrant fetal hemoglobin, a change in the β chain should affect adult hemoglobin A only, while a change in the δ chain should produce an abnormal hemoglobin A^2. Since the α chain is found in embryonic, fetal, and adult types of hemoglobin, an alteration in the α chain should affect all of these. Numerous inherited hemoglobin variants have been shown to conform to these expectations in that only one type of polypeptide is affected by a given mutation, and the abnormal hemoglobin molecules present are restricted to those in which that polypeptide occurs.

In addition to causing clinical symptoms of anemia, alterations in the primary structure of a subunit often result in a change in the charge on the hemoglobin molecule, detectable by electrophoresis (see the appendix to this chapter). After identification by this method of a variant of hemoglobin A, for example, the constituent polypeptides of the variant hemoglobin can be analyzed for amino acid composition and sequence. The technique is complex, but usually involves partial digestion by proteolytic enzymes followed by separation of the resulting peptides by chromatography and electrophoresis (fin-

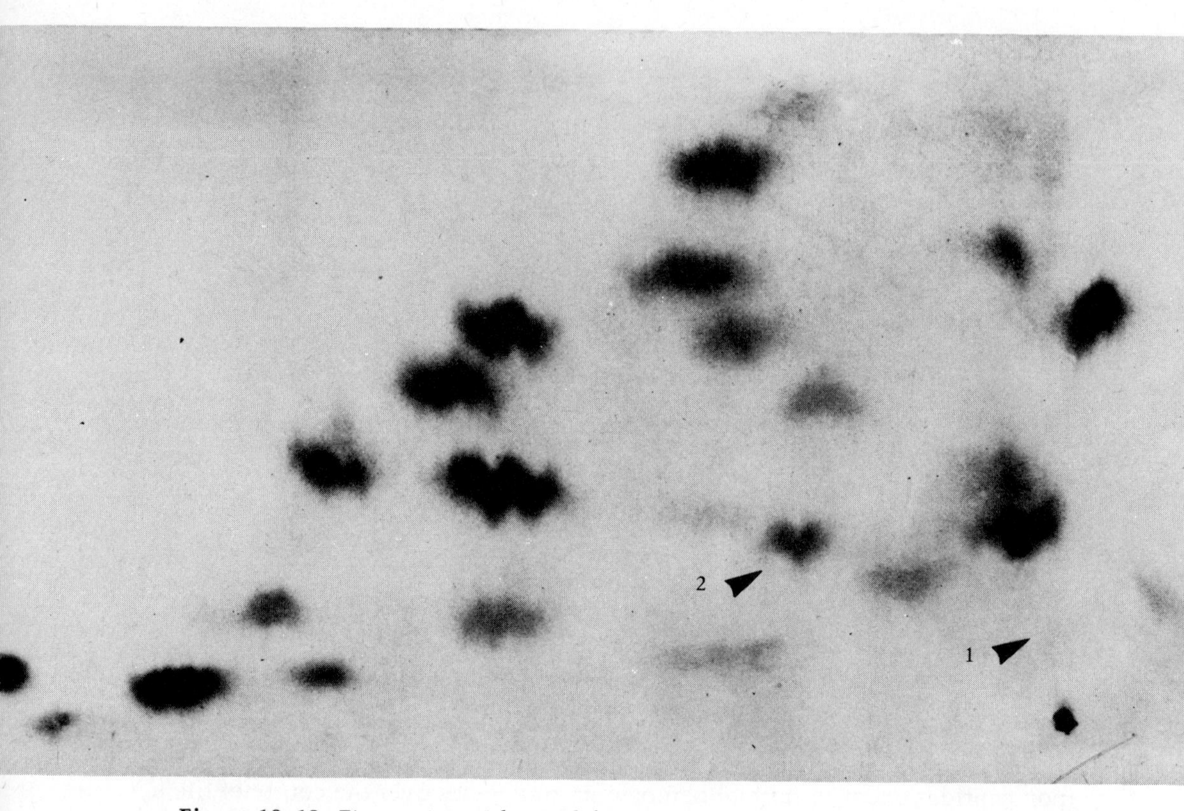

Figure 10–13. Fingerprint of hemoglobin S. Arrow 1 indicates the position of a peptide normally present in HbA, but absent in HbS. Arrow 2 indicates a new peptide found in HbS, but absent in HbA. [From H. Lehmann and R. G. Huntsman, *Man's Haemoglobins*, North-Holland, Amsterdam, 1966.]

gerprinting, Figure 10–13). Since the pattern of separation of these peptides is a consequence of their amino acid composition, comparison with the normal pattern serves to identify the presence of a peptide containing a "wrong" amino acid. The specific amino acid sequence of the aberrant peptide can then be determined by further analysis.

Sickle cell hemoglobin was the first variant to be studied in this manner. It will be recalled (Chapter 4) that persons homozygous for this mutant gene suffer red blood cell destruction and severe anemia. Instead of the normal hemoglobin A, the blood cells of sickle cell homozygotes contain an aberrant hemoglobin called hemoglobin S (HbS) which can be distinguished by its electrophoretic mobility (Figure 10–14). Since HbF and HbA^2 are normal in such individuals, indicating the presence of unchanged α, γ, and δ chains, a lesion in the β chain can be inferred. This prediction is verified by studies of HbS.

The molecule of HbS is composed of the usual two α and two β chains. The amino acid sequence of the α chain is identical to that found in the α chain of normal HbA. However, the β chain of HbS differs from the normal β subunit with respect to one particular amino acid residue. In the normal β chain the sixth amino acid, counting from the free amino terminus, is glutamic acid,

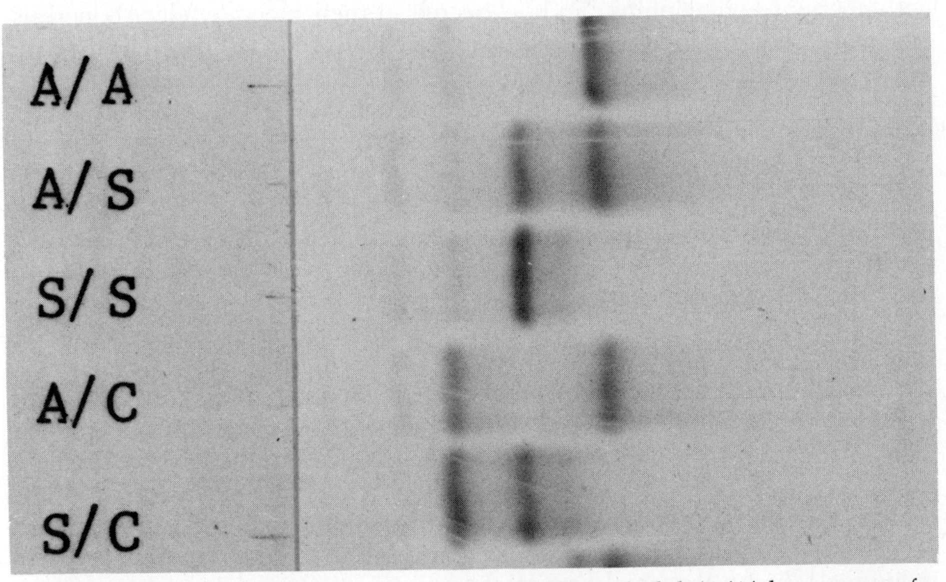

Figure 10–14. Electrophoresis of hemoglobins in normal adult (A/A), heterozygote for the sickle cell trait (A/S), sickle cell homozygote (S/S), heterozygote for the hemoglobin C trait (A/C), and heterozygote for sickle cell and hemoglobin C traits (S/C). Electrophoresis carried out at pH 8.6 in cellulose acetate. [Courtesy of Dr. Donald L. Rucknagel, Department of Human Genetics, University of Michigan, Ann Arbor.]

whereas in the β chain of HbS this position is occupied by valine. Other than this one amino acid substitution, the two polypeptides are alike. The presence of valine instead of glutamic acid has profound effects on the properties of the β chain. It is known from studies of the three dimensional configuration of the molecule that the sixth amino acid residue lies on the outside in contact with the aqueous environment of the cell. Glutamic acid, being polar, is well adapted for this position, but valine is nonpolar or hydrophobic, and its presence decreases the solubility of the molecule such that hemoglobin containing the altered β chains precipitates from solution, particularly under conditions of low oxygen tension. As a result, the characteristic sickling distortion of the red cells occurs and this, in turn, gives rise to the clinical symptoms of the disease. The blood cells of heterozygotes contain HbA as well as HbS. Such persons are clinically normal under ordinary circumstances.

The second inherited variant of hemoglobin to be discovered was called hemoglobin C (HbC). As in the case of sickle cell disease the abnormality was traced to a change in the amino acid sequence of the β chain. In HbC the α chains are normal, but the β chains contain an amino acid substitution, again in the sixth position, whereby lysine is present instead of glutamic acid. That this substitution occurs in the same position as in sickle cell hemoglobin is considered merely coincidental. The presence of lysine does not cause sickling of red blood cells. Homozygotes exhibit only a mild anemia. Heterozygotes are clinically normal. Marriages between carriers of HbS and HbC produce individuals heterozygous for these two hemoglobin variants (HbS/HbC). These heterozygotes suffer an anemia of intermediate severity, and the presence of both aberrant hemoglobins can be detected in their blood by electrophoresis. Since 8 percent of American blacks are heterozygous for sickle cell disease and 3 percent are heterozygous for hemoglobin C disease, the expected marriage frequency between such persons is 0.24 percent (0.08×0.03). One-fourth of the progeny of such marriages would be expected to be HbS/HbC and thus the expected frequency of this heterozygote is 0.06 percent. Most hemoglobin variants are extremely rare and the unusually high frequency of HbS and HbC in persons of African descent is thought to be associated with a greater resistance of their red cells to infection by malarial parasites, as compared to normal individuals. Malaria is endemic in many parts of Africa, and heterozygotes thus enjoy a selective advantage in these regions, although this advantage disappears in nonmalarial regions of the world.

Continuing studies of hemoglobin in human populations throughout the world have resulted in the identification of over 100 different mutations affecting this molecule. Over 80 of these have been traced to the substitution of a single "wrong" amino acid for its normal counterpart in one or another of the subunit polypeptide types. These variant hemoglobins are usually rare, their occurrence being limited to a single family or to a small group of related individuals. Some abnormal hemoglobins give rise to hemolytic disease or to unstable molecules, while others produce no detectable phenotypic effect on the health of the individual. Table 10–2 presents a recent summary of some of the abnormal hemoglobins identified in humans along with specific amino acid substitution involved. The variants are named according to the locality in which they were found.

Both somatic cell hybrid analysis and pedigree analysis of families exhibiting variants of the α and β chains indicate that the α and β hemoglobin genes are probably located on different autosomes. However, the γ, β, and δ

TABLE 10–2
List of Known Hemoglobin Substitutions and Deletions

Residue number	Amino acid substitution	Name	Residue number	Amino acid substitution	Name
		α Chain Variants			
5	ala → asp	J Toronto	78	asn → lys	Stanleyville II
12	ala → asp	J Paris	80	leu → arg	Ann Arbor
15	gly → asp	J Oxford	84	ser → arg	Etobicoke
16	lys → glu	I	85	asp → asn	G Norfolk
22	gly → asp	J Medellin	87	his → tyr	M Iwate
23	glu → gln	Memphis	90	lys → asn	Broussais
23	glu → val	G Audhali	92	arg → gln	J Cape Town
23	glu → lys	Chad	92	arg → leu	Chesapeake
30	glu → gln	G Chinese	102	ser → arg	Manitoba
43	phe → val	Torino	112	his → gln	Dakar
47	asp → gly	L Ferrara	114	pro → arg	Chiapas
47	asp → his	Hasharon	115	ala → asp	J Tongariki
50	his → asp	J Sardegna	116	glu → lys	O Indonesia
51	gly → arg	Russ	136	leu → pro	Bibba
54	gln → arg	Shimonoseki	141	arg → pro	Singapore

TABLE 10–2 *(Continued)*

Residue number	Amino acid substitution	Name	Residue number	Amino acid substitution	Name
		α Chain Variants			
54	gln → glu	Mexico	141	Arg (split off on hemolysis in plasma)	Koellicker
57	gly → asp	Norfolk			
58	his → tyr	M Boston			
68	asn → lys	G Philadelphia			
68	asn → asp	Ube II			
		β Chain Variants			
2	his → tyr	Tokuchi	63	his → tyr	M Saskatoon
6	glu → val	S	63	his → arg	Zürich
6	glu → lys	C	67	val → glu	M Milwaukee
6 or 7	glu deleted	Leiden	67	val → ala	Sydney
7	glu → gly	San José	69	gly → asp	J Cambridge
7	glu → lys	Siriraj	73	asp → asn	Korle Bu
9	ser → cys	Pôrto Alegre	6	glu → val	C Harlem
14	leu → arg	Sogn	73	asp → asn	
16	gly → asp	J Baltimore	76	ala → glu	Seattle
16	gly → arg	D Bushman	77	his → asp	J Iran
22	glu → ala	G Coushatta	79	asp → asn	G Accra
22	glu → lys	E Saskatoon	87	thr → lys	D Ibadan
23	val deleted	M Freiburg	88	leu → pro	Santa Ana
25	gly → arg	G Taiwan Ami	90	glu → lys	Agenogi
26	glu → lys	E	91	leu → pro	Sabine
28	leu → pro	Genova	92	his → tyr	M Hyde Park
30	arg → ser	Tacoma	94	asp → asn	Oak Ridge
35	tyr → phe	Philly	95	lys → glu	N
42	phe → ser	Hammersmith	91–95		
43	glu → ala	G Galveston	or	leu, his, cys,	
46	gly → glu	K Ibadan	92–96	asp, lys	Gun Hill
47	asp → asn	G Copenhagen	or	deleted	
56	gly → asp	J Bangkok	93–97		
58	pro → arg	Dhofar	98	val → met	Köln
61	lys → glu	N Seattle	99	asp → his	Yakima
61	lys → asn	Hikari	99	asp → asn	Kempsey
102	asn → thr	Kansas	130	tyr → asp	Wien
113	val → glu	New York	132	lys → gln	K Woolwich
120	lys → glu	Hijiyama	136	gly → asp	Hope
121	glu → lys	O Arab	143	his → asp	Hiroshima
121	glu → gln	D Punjab	145	tyr → his	Rainier
126	val → glu	Hofu			
		δ Chain Variants			
2	his → arg	A_2 Sphakia	22	ala → glu	A_2 Flatbush
12	asn → lys	A_2 N.Y.U.	136	gly → asp	A_2 Babinga
16	gly → arg	A_2 (or B_2)			
		γ Chain Variants			
5	glu → lys	F Texas I	12	thr → lys	F Alexandra
6	glu → lys	F Texas II	121	glu → lys	F Hull

Source: H. Lehmann and R. W. Carrell, *Brit. Med. Bull.*, *25*, 14–23, 1969.

hemoglobin genes appear to be closely linked as in the mouse, a suggested order being $\gamma^G-\gamma^A-\delta-\beta$. Although this sequence has not yet been established with certainty, analysis of rare Lepore type hemoglobins has strengthened the hypothesis that the δ and β loci are adjacent and in the order given.

Lepore hemoglobins are named for the family in which they were first identified. Individuals heterozygous for the condition are clinically normal, but their red cells contain not only HbA and HbA^2, but also a new and unusual hemoglobin. This variant contains a normal α chain, but the β chain is represented by a polypeptide which consists of amino acid sequences derived from the δ as well as from the β chain. The source of these unusual molecules has been attributed to the occurrence of unequal crossing over resulting from displaced pairing during meiosis. The consequences of such an event are illustrated in Figure 10–15 and are similar to those observed with unequal crossing over at the Bar locus in *Drosophila* (see Chapter 9). It can be seen that one product of such a crossover event would be a chromatid containing both the δ and β loci, plus a new "hybrid gene," part δ and part β in constitution. The other chromatid would contain only the reciprocal hybrid condition and would lack the δ as well as the β gene. The occurrence of unequal crossing over between these loci suggests their close linkage since a slight displacement of normal synaptic pairing might be expected to occur on rare occasions only between loci situated adjacent to one another.

The order of these loci and the general site of unequal exchange can also be inferred from the structure of Lepore hemoglobin. Normal δ and β chains each contain 146 amino acids and differ from one another only with respect to 10 residues, those at positions 9, 12, 22, 50, 86, 87, 116, 117, 125, and 126, counting from the amino terminus. Lepore hemoglobin also contains 146 amino acid residues. Analysis has shown that one Lepore type, called Hollandia, contains the δ sequence from positions 1 through 22 and the β sequence from position 50 to the carboxyl terminus. The middle region, residues 23 to 49, is the same in both the δ and β chains and it is postulated that the exchange must have occurred within that portion of the δ and β genes that corresponds to this middle region. In another Lepore type, called Washington, residues 1 through 87 are δ in origin and 116 to the carboxyl terminus are β in origin. Since the intervening residues 88 to 115 are the same in the δ and β polypeptides, the site of exchange can be localized to the respective portions of the two loci that specify this region. It is known from other evidence (see Chapters 14,

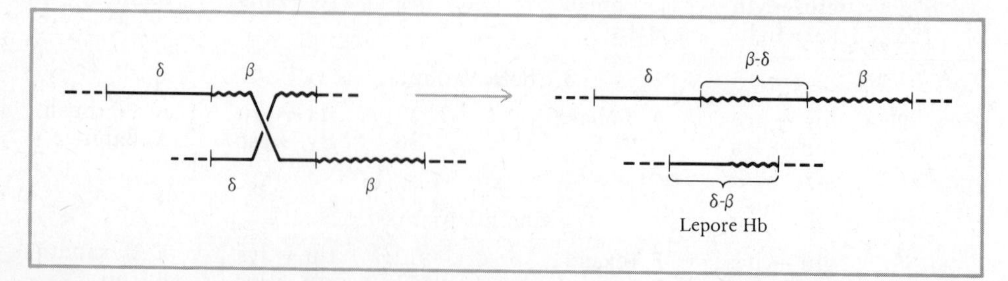

Figure 10–15. Origin of Lepore hemoglobin variant by unequal crossing over between the δ and β hemoglobin loci.

15) that the amino acid sequence of a polypeptide is specified starting from the amino terminus. Since Lepore hemoglobin begins with the δ sequence at this terminus and ends with the latter portion of the β sequence at the carboxyl terminus, it can be inferred that the order of these loci is δ—β, and not β—δ. The occurrence of Lepore hemoglobins also indicates that recombination, even if abnormal, can occur within the limits of a single functional gene.

SOME INFERENCES REGARDING THE NATURE OF THE GENE

A number of inferences can be drawn concerning the nature of the gene from the study of gene-specified polypeptides alone and without reference to the structure of DNA or to studies of recombination within the limits of a single gene. Analysis of such polypeptides, particularly the hemoglobin variants, indicates that the gene cannot be visualized as an indivisible particle that mutates in an all-or-none fashion. If the gene invariably mutated "as a whole," we would expect the resultant polypeptide to be totally different from the normal. However, in the majority of hemoglobin variants only one amino acid out of an entire polypeptide sequence is changed as the result of mutation, the rest of the sequence being normal. If the nature of the mutant polypeptide can be considered to reflect the nature of its corresponding specifying gene, it follows that in these hemoglobin variants most of the gene is unchanged and only a small region within the gene has been altered. Since an amino acid substitution can occur at any position within the polypeptide as the result of mutation, it would also appear that the gene must be subject to change at many different sites. Accordingly, the concept of the gene which emerges from the study of the hemoglobin variants is no longer that of an indivisible particle. Instead, one is forced to conclude that the gene must be composed of numerous subunits which are independently subject to mutation.

This concept necessitates consideration of the meaning of the term allele. A restricted definition of this term would require that two mutants be altered at the identical site within the gene and result in an amino acid change at an identical position in the polypeptide product. By this definition few mutants could be considered allelic, an example being the variants β^{S} and β^{C} which contain amino acid substitutions at the same position. On the other hand, the term can be employed in the functional sense to refer to genes that, in addition to occupying the same chromosomal locus, specify the same final product. By this definition all mutations in the β hemoglobin gene, regardless of site, would be considered alleles. Since relatively few proteins have been completely sequenced and changes therein correlated with specific mutations, this latter, functional definition of allelism is more useful.

It should be emphasized that the inferences drawn above concerning the nature of the gene are based entirely upon the analysis of mutant polypeptides. Although detailed studies of gene-controlled metabolic pathways began in the early 1940s, the complete sequence of a mutant polypeptide, the β chain of HbS, was not known until 1957. Meanwhile other investigations led to similar concepts of the gene. Among the foremost was the discovery of the structure of DNA whose nature and properties are the subject of the next chapter.

APPENDIX

SEPARATION AND IDENTIFICATION OF AMINO ACIDS AND PROTEINS

A mixture of proteins or amino acids can be separated into individual components by techniques that take advantage of charge and size differences. *Electrophoresis* is one such widely used method. Zone electrophoresis is a procedure in which mixtures of proteins, applied to paper or to starch or polyacrylamide gels, are placed in a suitable buffer and subjected to an electric current. Depending on the net charge on the molecule, a given protein will then move to the positive or negative pole at a characteristic rate, and by this means different proteins can be separated from one another and obtained in pure form. Subsequent staining of the gel with a dye reveals the positions of the different components which can then be sliced from the gel and studied further. Abnormal types of hemoglobin, such as sickle cell hemoglobin, have been identified by this technique (see Figure 10–14). Enzyme variants and isozymes (see Chapter 19) are also distinguished by their electrophoretic mobility.

Another technique appropriate for the separation of mixtures of amino acids, peptides, or polypeptides is *chromatography*. Although numerous chromatographic methods have been designed, separations achieved with all of these depend on differences in the rates of movement of the components of the starting mixture when applied to some supporting substance such as a gel, an ion-exchange resin, or to filter paper. With column chromatography, the supporting material is placed in a glass tube and the mixture to be separated is poured on top. One or more solvents or buffers are then allowed to flow through the tube and the effluent is collected in a series of test tubes. When a gel is used as the supporting medium, separation of aqueous and nonaqueous phases depends on molecular size, either the smaller or the larger molecules appearing first in the effluent fractions depending on the type of gel employed. When an ion-exchange resin is used, the separation depends on the net charge of each constituent of the mixture. Ion-exchange chromatography is thus most useful for the separation of amino acids. A sample of a mixture can also be applied to filter paper and an appropriate solvent system allowed to creep upward or downward by capillary action. In this case separation of components depends on differential solubility in the particular solvent system used, and the various constituents present will be carried different distances from the starting point. Their respective positions are then visualized by spraying the paper with a dye that stains amino acids. More complete separation can be achieved with two-dimensional paper chromatography. With this technique, separation in one direction is obtained first, after which the paper is dried, rotated 90 degrees, and then subjected to a second chromatographic separation, usually with a different solvent system.

Complex mixtures of peptides obtained by digestion of proteins can also be analyzed on filter paper by the use of electrophoretic separation in one direction followed by chromatographic separation in a second direction. The method is called fingerprinting (see Figure 10–13) and it has permitted the rapid detection of the presence of abnormal amino acid sequences in peptides obtained from proteins such as hemoglobin.

PROBLEMS

10–1. If a gene mutation occurred such that a cysteine residue was replaced by an amino acid such as serine, what effect might this have on the final configuration of a polypeptide?

10–2. Individuals homozygous for phenylketonuria frequently exhibit lighter pigmentation of skin and hair than normal persons of the same family. Refer to Figure 10–7 and suggest a reason for this phenomenon.

10–3. Two pure breeding white-flowered plants were crossed and the F_1 had purple flowers. An inbreed of the F_1 produced an F_2 generation in a ratio of 9 purple : 7 white plants. Propose a hypothetical metabolic pathway to account for this ratio.

10–4. In *Neurospora* two different single gene mutants have been found, each of which requires choline for normal growth. Mutant A will grow with the addition of dimethylaminoethanol or monomethylaminomethanol instead of choline. Mutant B will grow with dimethyl- but not with monomethylamino-methanol. Furthermore, mutant B accumulates the monomethyl compound. Propose a pathway for the synthesis of choline.

10–5. A strain of mutant *Neurospora* will grow only if two separate compounds are added to the culture medium. Does this finding require that two separate genes be mutant?

10–6. The independent genes *a, b, c,* and *d* control the synthesis of compounds A, B, C, and D. Mutants of each of these genes will grow if the substances listed below are added to the medium. From these data, propose a metabolic pathway for the synthesis of compounds A, B, C, and D.

Gene Mutated	Substances Producing Growth
a	A
b	B or C or A
c	C or A
d	D or B or C or A

10–7. Independent genes *w, x, y,* and *z* control the synthesis of compounds W, X, Y, and Z. From the data given below, propose a metabolic pathway for the synthesis of these compounds.

Genes Mutated	Substances Producing Growth
w	W
x	X or Y and W
y	Y
z	Z or X or Y and W

10–8. The independent genes listed control the synthesis of compounds A, B, X, Y, P, Q, R, and S. From the data given, propose a metabolic pathway for the synthesis of these compounds.

Genes Mutated	Substances Producing Growth
a	A or B
b	B
x	X or A or B
y	Y or R or S
p	P or Q or X and Y or B and S
q	Q or X and Y or B and S
r	R or S
s	S

10–9. The independent mutants listed control the synthesis of compounds A, B, C, D, E, F, and G. From the data given, propose a metabolic pathway for the synthesis of these compounds.

GENES MUTATED	SUBSTANCES PRODUCING GROWTH
a	A or F
b	B or D or A and G or E and C and F
c	C
d	D or A and G or E and C and F
e	E
f	F
g	G or E and C

10–10. A given enzyme consists of two subunits (a dimer) which are specified by a single gene locus.

a. If a = normal subunit and a′ = mutated subunit, what subunit combinations might be found in a heterozygote, assuming random association between these subunits?

b. If the mutation resulted in a change in the net charge on the molecule, how many bands would be detectable by electrophoresis?

c. If this enzyme was a tetramer composed of four randomly associated subunits, what different subunit combinations might be found in a heterozygote?

d. How many bands would be detectable by electrophoresis?

10–11. A polypeptide specified by the wild type gene contains 100 amino acids, while the polypeptide found in the presence of the mutated gene contains 90 amino acids. Comparison of sequence between the two indicates that residues 67–77 are missing in the abnormal polypeptide. What kind of genetic change would account for this?

10–12. Five genotypes and the symbols used to express them are given below.

(1)	α/α	β^S/β	δ/δ
(2)	α/α	β^S/β^C	δ/δ
(3)	α/α^E	β/β^S	δ/δ
(4)	α/α^E	β/β	δ/δ^x
(5)	α/α^E	β^S/β^C	δ/δ^x

α = normal gene for α chain
α^E = mutant gene for α chain
β = normal gene for β chain
β^C = mutant gene for β^C chain of HbC
β^S = mutant gene for β^S chain of HbS
δ = normal gene for the δ chain
δ^x = mutant gene for the δ chain

a. For each genotype list the types of hemoglobin molecules that could be identified in the red cells of an adult.

b. Would any of these genotypes result in a variant of fetal hemoglobin (HbF)? If so, what types of HbF would occur in each case?

10–13. Take a piece of string around 8 inches long and imagine it to be a polypeptide. Suppose that 6 cysteine residues are located 1 inch apart along the string, that is, at points 1, 2, 3, 4, 5, and 6 inches from one end of the string. Form a three dimensional figure by assuming disulfide bridges between 1 and 4, 2 and 5, and 3 and 6. By rearranging these bridges, can you achieve other configurations?

10–14. Complementation tests between all paired combinations of 10 different mutants of independent origin resulted in the data presented in the complementation matrix below. Prepare a complementation map based on these data. How many functional groups are represented? Which mutants are alleles? Which mutants are multisite, that is, contain mutations or deletions in more than one gene?

	1	2	3	4	5	6	7	8	9	10
1	0	0	+	+	+	+	0	0	+	+
2		0	0	+	+	0	0	0	+	0
3			0	+	+	+	+	+	+	+
4				0	+	+	+	+	0	+
5					0	+	+	+	0	+
6						0	+	+	+	0
7							0	0	+	+
8								0	+	+
9									0	+
10										0

10–15. The following complementation map is based on complementation tests between 15 mutants of independent origin. Prepare a complementation matrix that represents the relationships contained in the map.

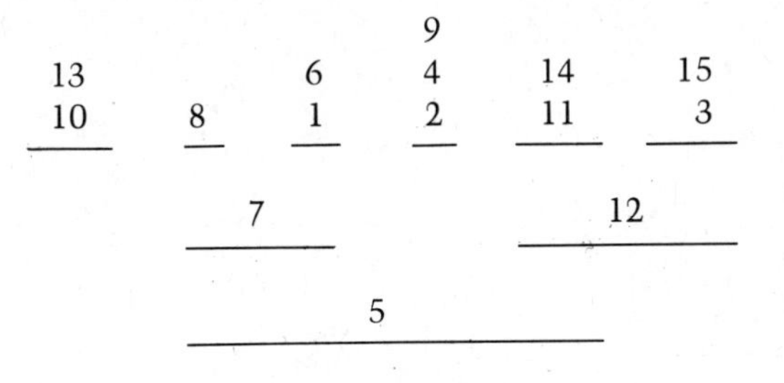

REFERENCES

AEBI, H., 1967. Inborn errors of metabolism. *Ann. Rev. Biochem.*, *36*, 271.

BEADLE, G. W., 1945. Genetics and metabolism in *Neurospora*. *Physiol. Rev.*, *25*, 643.

BEADLE, G. W., AND E. L. TATUM, 1941. Genetic control of biochemical reactions in *Neurospora*. *Proc. Nat. Acad. Sci.*, *27*, 499.

DICKERSON, R. E., AND I. GEIS, 1969. *The Structure and Action of Proteins*. Harper & Row, New York.

GARROD, A. E., 1909. *Inborn Errors of Metabolism*. Oxford University Press, London.

HARRIS, H., 1970. *The Principles of Human Biochemical Genetics*. Elsevier, New York.

HAYES, W., 1968. *The Genetics of Bacteria and Their Viruses*, 2nd ed. Wiley, New York.

LEHMANN, H., AND R. G. HUNTSMAN, 1966. *Man's Haemoglobins*. North-Holland, Amsterdam.

MOTULSKY, A. G., 1973. Frequency of sickling disorders in U.S. Blacks. *New Eng. J. Med.*, *288*, 31.

PHILLIPS, J. P., H. S. FORREST, AND A. D. KULKARNI, 1973. Terminal synthesis of xanthommatin in *Drosophila melanogaster:* Mutational pleiotropy and pigment granule association of phenoxazinone synthetase. *Genetics*, *73*, 45.

PORTER, I. H., 1968. *Heredity and Disease*. McGraw-Hill, New York.

RAIVIO, K. O., AND J. E. SEEGMILLER, 1972. Genetic diseases of metabolism. *Ann. Rev. Biochem.*, *41*, 543.

STAMATOYANNOPOULOS, G., 1972. The molecular basis of hemoglobin disease. *Ann. Rev. Genet.*, *6*, 47.

STANBURY, J. B., J. B. WYNGAARDEN, AND D. S. FREDRICKSON (Eds.), 1972. *The Metabolic Basis of Inherited Disease*, 3rd ed. McGraw-Hill, New York.

SUTTON, H. E., 1975. *An Introduction to Human Genetics*, 2nd ed. Holt, Rinehart and Winston, New York.

WAGNER, R. P., AND H. K. MITCHELL, 1964. *Genetics and Metabolism*, 2nd ed. John Wiley, New York.

WITKOP, C. J., JR., 1971. Albinism. *Adv. Human Genet.*, *2*, 61.

Chapter 11

NUCLEIC ACID: THE MOLECULAR BASIS OF THE GENOME

Two types of nucleic acid occur within living cells: *deoxyribonucleic acid (DNA)* and *ribonucleic acid (RNA)*. DNA is known to constitute the genome of all organisms, all cells, all cell organelles, and most viruses. Although RNA acts as an intermediary in the process of protein synthesis in cellular organisms, it too can function as a hereditary molecule, for it forms the genome of the RNA viruses.

The recognition and conclusive proof that nucleic acids constitute the hereditary material required many years and the contributions of many investigators. Friedrich Miescher, in 1871, had demonstrated that the primary constituent of nuclei was a material called nucleoprotein which consisted of a nucleic acid (DNA) bound to basic protein. Further analysis suggested complexity of the protein component, but degradation of the nucleic acid revealed simplicity, in that a sugar, phosphate, and only four organic bases were present. This apparent simplicity seemed inconsistent with the high degree of specificity expected of a hereditary molecule, and therefore attention for many years was focused on the protein moiety as the probable candidate for the role of gene, the nucleic acids being considered merely structural elements.

In 1928 F. Griffith reported an unusual result from an experiment involving two strains of pneumococcus bacteria. One strain possessed a polysaccharide capsule conferring a smooth (S) appearance to the colony. Although living S-type cells are virulent, causing pneumonia in experimental animals, heat-killed S cells do not cause disease. The other bacterial strain had no capsule, a colony of such cells presenting a rough (R) appearance. R cells are not pathogenic and do not produce pneumonia in experimental animals. The presence (S) or absence (R) of the polysaccharide capsule is an inherited character of these bacteria. In his experiment Griffith mixed heat-killed S–type cells with living, nonvirulent R-type cells and injected the mixture into mice. The mice developed pneumonia and died, and from them, living, encapsulated, virulent S-type bacteria were recovered (Figure 11–1). It appeared that somehow within the mice, the nonvirulent strain had become endowed with the genetic characters of the virulent strain. The phenomenon was called *transformation*.

Griffith's experiments were refined by M. H. Dawson and R. H. P. Sia in 1931. These workers cultured heat-killed S cells with living R cells in a test tube and found that living S cells could be recovered from the culture. It was thus evident that transformation could occur *in vitro,* that is, outside the body of an experimental animal.

J. L. Alloway then extended these findings and demonstrated that transformation occurred when R-type cells were exposed to cell-free extracts of the S-type strain. Alloway also subjected these extracts to chemical tests which indicated that the "transforming substance" could not be the polysaccharide coat present in smooth-type cells.

An exhaustive analysis of extracts of S-type cells was then undertaken by O. T. Avery, C. M. MacLeod, and M. McCarty in an effort to define the nature of the transforming substance. The extracts were fractionated and every class of compounds present was tested for activity by adding each, separately, to cultures of R-type cells. An antiserum whose antibodies reacted only with R cells was also added, so that R cells were precipitated to the bottom of the culture tube. Upon incubation absence of transformation was indicated by the presence of a clear supernatant containing no cells, while the occurrence of

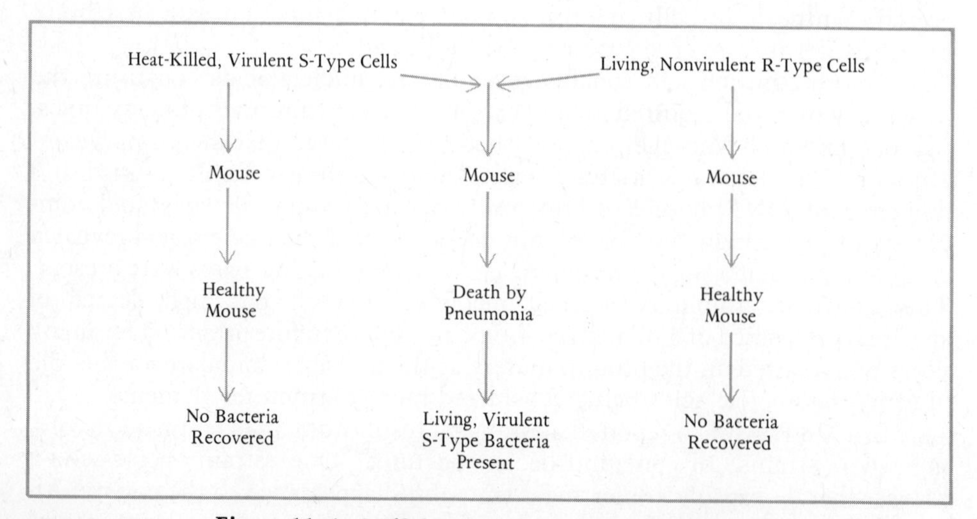

Figure 11–1. Griffith's transformation experiment.

Components of Extracts of Heat-Killed S-Type Cells:	Carbohydrates, Polysaccharides + R Cells + Anti-R Serum	Lipids + R Cells + Anti-R Serum	Protein + R Cells + Anti-R Serum	RNA + R Cells + Anti-R Serum	DNA + R Cells + Anti-R Serum	DNA + DNAse + R Cells + Anti-R Serum
Transformed Cells	0	0	0	0	+	0

Figure 11–2. The results of Avery, MacLeod, and McCarty's experiment with transformation in which different components of extracts of heat-killed S-type cells were added to cultures of living R-type cells. The addition of anti-R serum caused precipitation of R cells. The presence of living S-type transformants was observed only when the fraction containing DNA was used.

transformation could be readily detected by a cloudy appearance of the supernatant which contained multiplying, transformed S-type cells (Figure 11–2). As a result of these assays, it was found that transformation occurred only with the fraction containing DNA. In addition, the transforming activity of this fraction could be obliterated by the addition of deoxyribonuclease (DNAse), an enzyme that degrades DNA, whereas treatment with other enzymes such as proteases or ribonuclease did not affect or inhibit transforming activity. To eliminate the possibility that some impurity present in the DNA-containing fraction was the causative agent, an exacting purification procedure was employed and it was demonstrated that transforming activity increased in parallel with increasing purity of DNA. These studies, reported in 1944, provided strong evidence that DNA functioned as the transforming principle in bacteria.

Other evidence pointing to nucleic acids, and not protein, as the hereditary material came from studies of mutations induced by ultraviolet light where the most effective wavelength (260 nm) proved to be that most strongly absorbed by DNA. In addition, the amount of DNA in somatic tissue was found to be twice that of sperm, a condition which agreed with the known diploid versus haploid constitution of these cells. Studies of gametes, particularly sperm, also implicated DNA as the hereditary substance. The sperm nuclei of fish, such as salmon and herring, contain DNA bound to an unusually small protein, molecular weight (MW) 5000, called protamine, 70 to 80 percent of which consists of but a single amino acid, arginine. Sperm carry a haploid complement of genes and function as a link between generations and such a simple protein could not conceivably possess sufficient specificity to perform this role. Therefore of necessity it could be inferred that DNA was the gene-bearing material of these gametes.

More direct evidence came from experiments carried out in 1952 by A. D. Hershey and M. Chase, using a virus called T2 which attacks the bacterium *Escherichia coli*. In essence, this virus is composed of DNA contained within a protein capsule. Hershey and Chase devised an experiment which took advantage of the differences in the chemical constitution of these two components.

The protein capsule contains sulfur, but no phosphorus, while DNA contains phosphorus, but no sulfur. By allowing the virus to multiply in *E. coli* cells supplied with either radioactive ^{32}P or ^{35}S, progeny virus particles containing, respectively, ^{32}P-labeled DNA or ^{35}S-labeled protein were obtained. When these labeled viruses were used as infective agents, it was found that 30 percent of the ^{32}P was transmitted to a new generation of virus particles. However, when ^{35}S-labeled virus was used, the progeny contained practically no ^{35}S. Furthermore, it was found that upon the infection of a bacterium, the protein capsule of the virus remained on the surface of the host cell and only the DNA entered to multiply within.

The results of these experiments demonstrated that DNA, not protein, was involved in the transmission of hereditary characters from parent to progeny. The Hershey and Chase experiments in conjunction with the studies of Avery, MacLeod, and McCarty, and the contributions of many others provided conclusive proof that DNA was the genetic material of at least these viruses and bacteria and, presumably, of eukaryotic organisms as well.

The role of RNA as a hereditary molecule was demonstrated somewhat later, in 1957, by H. Fraenkel-Conrat and B. Singer using tobacco mosaic virus (TMV). This virus is composed of RNA contained in a protective protein sheath. A. Gierer and G. Schramm had earlier shown that these two components could be separated from one another and that only the RNA was infective. To obtain additional proof for the hereditary role of RNA in this virus, Fraenkel-Conrat and Singer isolated the RNA and protein from two strains of TMV which differed from one another with respect to the amino acid composition of the protein capsule, as well as the type of lesion produced on tobacco leaves. By recombining these components two types of active virus particles were produced. One type consisted of RNA from strain A plus protein from strain B (RNA^A-$protein^B$), the other contained RNA from strain B plus protein from strain A (RNA^B-$protein^A$). These parental particles were then used to infect tobacco plants (Figure 11–3). The type of lesion produced was observed and the progeny virus particles were recovered and analyzed for the type of protein sheath present. In all cases both the lesion and the protein found in the progeny corresponded to the RNA, not the protein, constitution of the parent virus used to produce the infection. In other words, parent RNA^A-$protein^B$ particles produced only RNA^A-$protein^A$ progeny, and parent RNA^B-$protein^A$ virus yielded only RNA^B-$protein^B$ progeny. These experiments demonstrated that the RNA moiety alone specified the constitution of the progeny virus and, therefore, that RNA could function as a hereditary molecule.

In the meantime, studies of the molecular configuration, or geometry of DNA were being pursued by R. Franklin and M. H. F. Wilkins, and these findings enabled J. D. Watson and F. H. C. Crick, in 1953, to define the three dimensional structure of DNA. The molecule was shown to possess three fundamental properties with which any candidate for the role of genome must be endowed: specificity, chemical stability, and, most important, the potential for self-reproduction. Since these characteristics provide the basis for the transmission of hereditary traits and the origin of the phenotype, it is pertinent to inquire into the molecular structure that makes such properties possible.

STRUCTURE OF NUCLEIC ACIDS

The two nucleic acids, DNA and RNA, are similar in that both are unbranched polymers constructed of individual units called *nucleotides*. Each nucleotide is a compound consisting of three essential parts: a five-carbon sugar (pentose), an organic nitrogen-containing base, and a phosphate group. Let us first consider the kinds of sugars and bases present in the nucleotides of DNA and RNA.

The names of the two nucleic acids are derived from the type of sugar present in each. The nucleotides of DNA contain only the pentose, 2 deoxy-D-ribose, while those of RNA contain only D-ribose. The structure of these two sugars is almost identical. The only difference between them lies in the R group

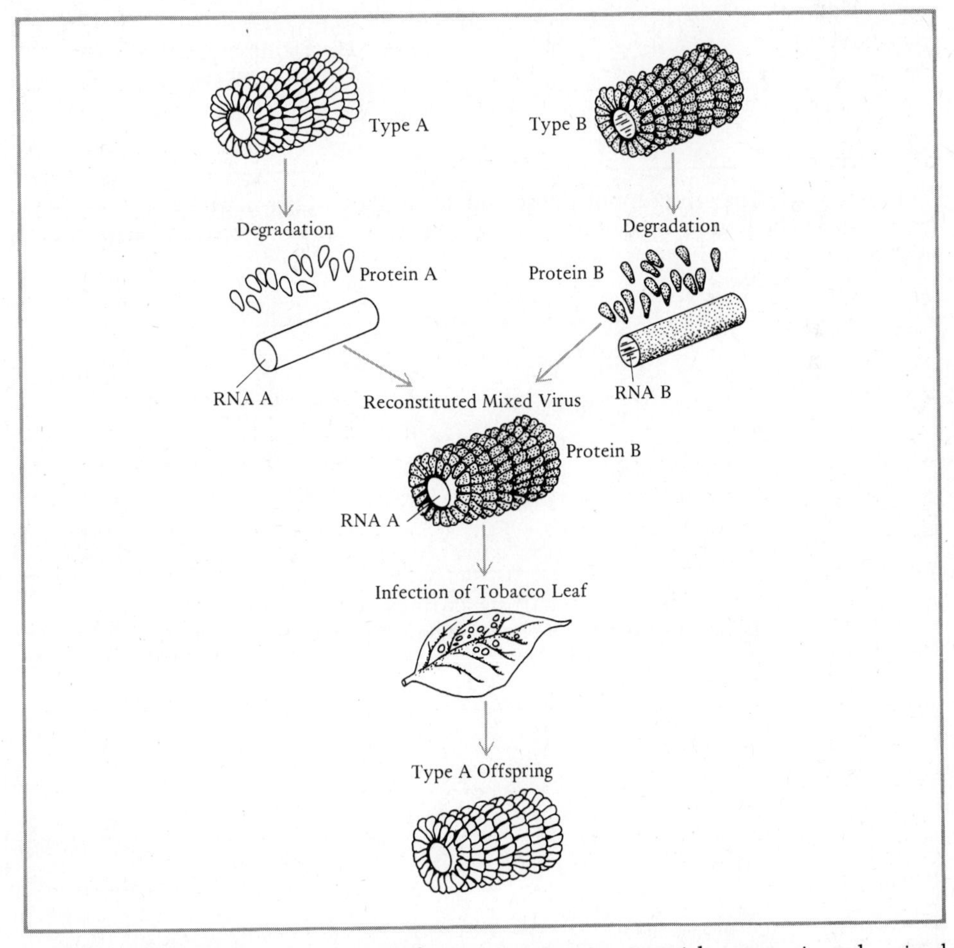

Figure 11–3. Fraenkel-Conrat and Singer's experiment with reconstituted, mixed tobacco mosaic virus which demonstrated that the RNA determined the nature of the progeny virus particles. [Reprinted with permission of Macmillan Publishing Co., Inc. from *Genetics,* 2nd ed., by Monroe Strickberger. Copyright © 1976 Monroe W. Strickberger.]

of carbon 2, where the hydroxyl group (—OH) of ribose is replaced by a hydrogen atom (—H) in deoxyribose (Figure 11–4).

The organic bases found in the nucleotides of DNA and RNA are nitrogen-containing ring compounds of two general types, *purines* and *pyrimidines*. Two kinds of purines are present in both DNA and RNA. These purines are *adenine* (A) and *guanine* (G). Their structure is shown in Figure 11–5.

Although alike in regard to the purines, DNA and RNA differ in their

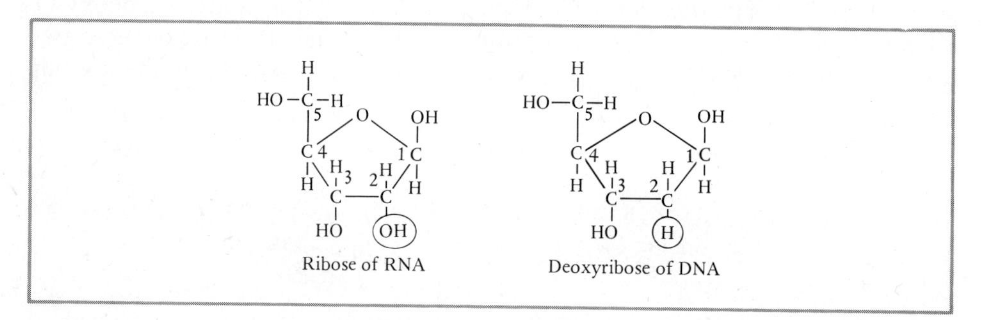

Figure 11–4. The structure of ribose and deoxyribose. Carbon atoms are numbered from 1 to 5. The R group of carbon 2 is circled to indicate the difference between these sugars.

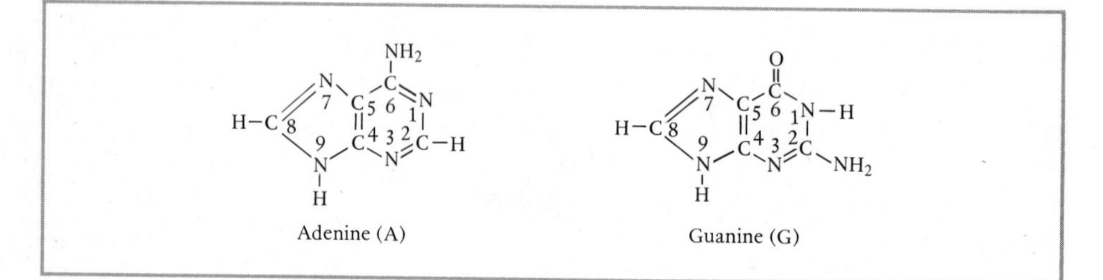

Figure 11–5. The commonly occurring purines of DNA and RNA, adenine and guanine. Conventional numbering of the atoms of the rings is indicated.

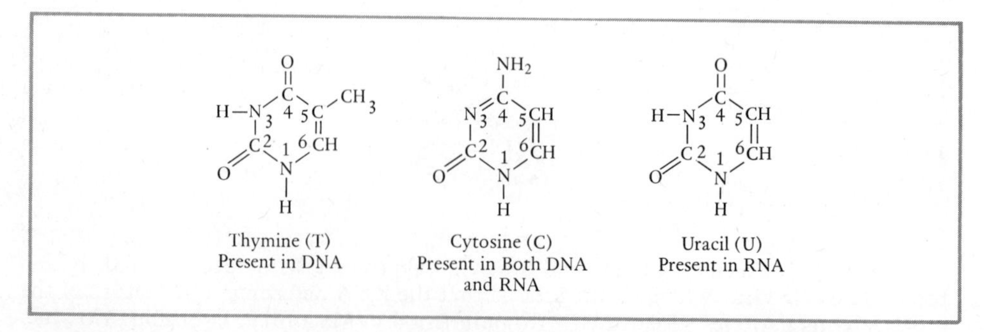

Figure 11–6. Commonly occurring pyrimidines of nucleic acids: cytosine and thymine in DNA and cytosine and uracil in RNA. Conventional numbering of the atoms is indicated.

respective pyrimidines. DNA nucleotides contain the pyrimidines *cytosine* (C) and *thymine* (T), while those of RNA are characterized by the presence of *cytosine* and *uracil* (U). It can be seen in Figure 11–6 that thymine differs from uracil only with respect to the presence of a methyl group (—CH_3) at position 5 in thymine. Because of this group, thymine is also named 5-methyluracil. A simple mnemonic device is helpful in remembering which bases are purines and which pyrimidines: PAG, for the *p*urines *a*denine and *g*uanine and PyCUT for the *py*rimidines *c*ytosine, *u*racil, and *t*hymine. It should be noted that a few different and unusual bases have been identified as components of the DNA of viruses and bacteria. In addition, various kinds of RNA molecules have also been found to contain small proportions of bases other than those described above.

The combination of the sugar moiety with any one of the organic bases constitutes a compound known as a *nucleoside*. Linkage occurs between carbon 1 of the sugar and the nitrogen in the first position of a pyrimidine or between carbon 1 of the sugar and the nitrogen in the ninth position of a purine (Figure 11–7). Notice that when the sugar is bonded to the nitrogen base, the carbon atoms of the sugar are primed, 1′, 2′, 3′, and so on, to distinguish them from the positions in the base.

When the third component of a nucleotide, the phosphate group, is added to carbon 5′ of the sugar, the nucleoside becomes a nucleoside monophosphate, or a nucleotide. The compound shown in Figure 11–8 would be given the name adenosine 5′-monophosphate, abbreviated 5′-AMP, if the sugar were ribose (R = OH). If the sugar were deoxyribose (R = H), the name given to the compound would indicate this fact by the title deoxyadenosine 5′-monophosphate or 5′-dAMP. As a general rule, nucleotides are named to indicate the presence of either ribose or deoxyribose. In an abbreviation the presence of the lowercase letter d indicates that deoxyribose is the sugar involved; the absence of the d automatically implies that ribose is involved. One exception to this rule is compounds containing thymine, a base that only very rarely occurs in RNA in association with ribose. Since for all practical purposes thymine occurs only with deoxyribose, the small d is dropped from the title for convenience.

A nucleotide containing one phosphate group can be further phosphorylated to a diphosphate or a triphosphate, the additional phosphates being added on to the one already present (Figure 11–9). If the sugar moiety of compounds A and B of Figure 11–9 were ribose, the appropriate names would be adenosine

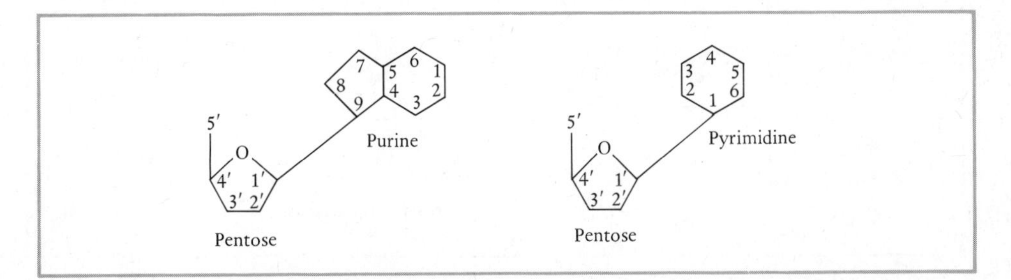

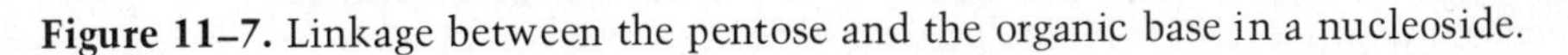
Figure 11–7. Linkage between the pentose and the organic base in a nucleoside.

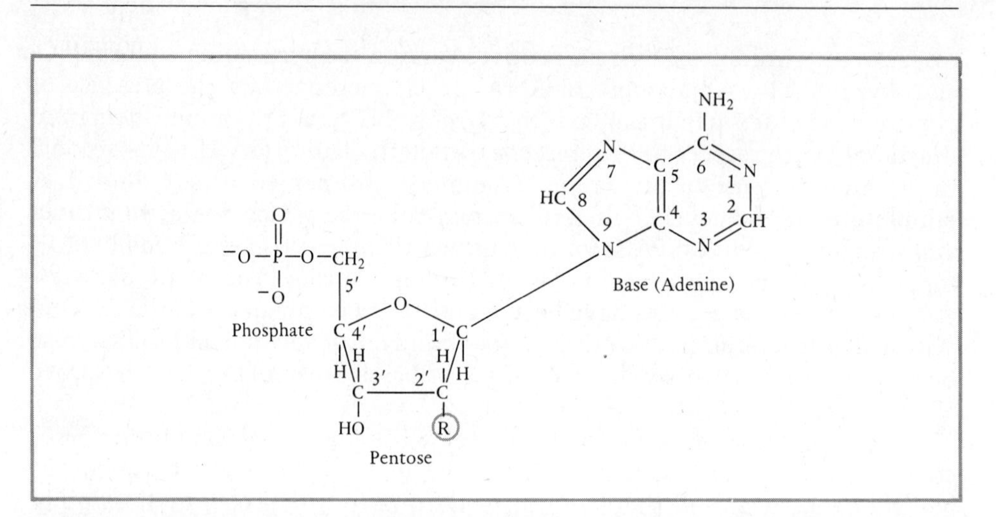

Figure 11–8. A nucleotide composed of adenine, ribose (or deoxyribose), and a phosphate group attached to carbon 5′ of the sugar. If R = OH, the sugar is ribose; if R = H, the sugar is deoxyribose.

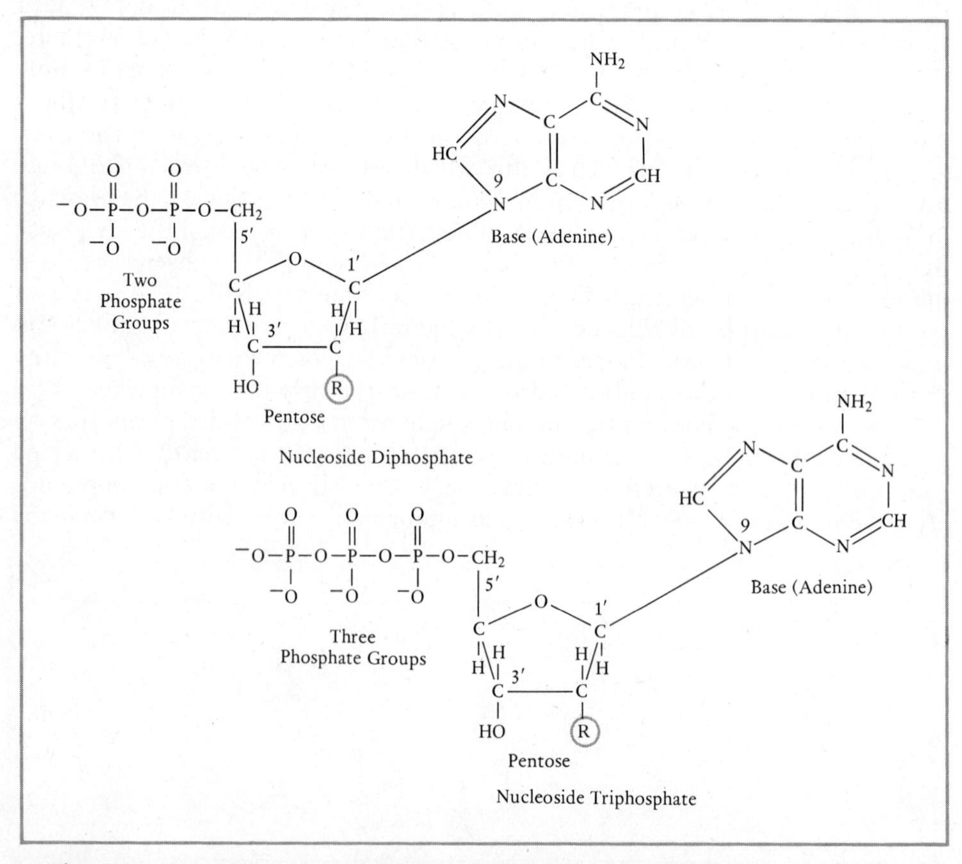

Figure 11–9. A nucleoside diphosphate and a nucleoside triphosphate. If R = OH, the pentose is ribose; if R = H, the pentose is deoxyribose.

5′-diphosphate (5′-ADP) and adenosine 5′-triphosphate (5′-ATP). If the sugar were deoxyribose, the names would change accordingly to deoxyadenosine 5′-diphosphate (5′-dADP) and deoxyadenosine 5′-triphosphate (5′-dATP). Names of other nucleoside di- and triphosphates are given similarly. A list of such names, synonyms, and abbreviations can be found in any biochemistry text.

It must be remembered that nucleosides and nucleotides are ubiquitous in the cell and perform other functions besides that of being integrated into the structure of nucleic acids. In nucleoside triphosphates, for example, the bonds involved in the linkage of the second and third phosphate groups are known as high energy bonds, and a nucleoside triphosphate so endowed becomes a high energy compound. Such triphosphates, ATP for example, carry and provide chemical energy in usable form to drive synthetic processes and other activities of the cell and organism. Indeed, one of these energy-requiring reactions is the polymerization of nucleotides to form nucleic acids, a process which occurs during the replication of DNA and the synthesis of RNA. The starting materials for these reactions are the triphosphates of the respective nucleosides whose high energy phosphate bonds provide the chemical energy required for the linkage of each nucleotide into the nucleic acid chain. Before considering this reaction, however, we must first inquire into the nature of the polynucleotide itself.

In overall structure, DNA and RNA each consist of an unbranched polymer of repeating units, each such unit being a nucleotide composed of phosphate-sugar-base. The individual nucleotides of the polymer are joined together by phosphodiester linkages which connect the 3′ carbon of one sugar with the 5′ carbon of the next sugar via the phosphate group (Figure 11–10). The sugar-phosphate linkages form the repetitive backbone of the polymer. The bases attached to carbon 1′ of the sugar moieties project outward and impose no restriction on bond formation. If we visualize the polynucleotide as a linear structure, the backbone has two ends: one is a 3′ end with a free hydroxyl group and the other is a 5′ end with a free phosphoryl group. Circular DNA molecules are also found, notably in bacteria and viruses. We can imagine the formation of a circle by the addition of another phosphodiester bond which would link together the 5′ phosphate group of one terminal nucleotide with the 3′ hydroxyl group of the other terminal nucleotide.

Nucleic Acids and the Potential for Specificity

Having examined the primary structure of a polynucleotide, we must now ask how such a polymer fulfills the requirements of a hereditary molecule. Superficially, both DNA and RNA might seem to be fairly uncomplicated molecules with respect to primary structure. The sugar-phosphate backbone is unvarying and therefore the only source of specificity must lie with the organic bases and their pattern of occurrence within the polymer. That there are only four bases might seem to impose limitations on possible specificity, but the length of the molecule provides an enormous potential in this regard. The RNA polymers which serve as the genome of some viruses are extremely long by any standards, while DNA polymers are the largest biological molecules known, in higher organisms having lengths of several centimeters and comprising millions of nucleotides. The possibilities for pattern and specificity with such a molecule are truly astronomical. In a polymer of only one million units, the four different nucleotides can be arranged in $4^{1,000,000}$ possible ways—clearly providing an almost inconceivable variety in sequence and, consequently, in specificity. This specificity, inherent in the primary structure of the DNA polymer, is ultimately expressed in the order of the amino acids which constitute the thousands of different proteins found in the cell and which, in turn, provide the basis for the phenotype.

Besides specificity, the other properties required of nucleic acids are stability and the potential for self-reproduction. These are more easily understood in terms of the secondary structure of the polymer. Since RNA is usually found as a single stranded molecule, the following discussion will be focused primarily upon DNA.

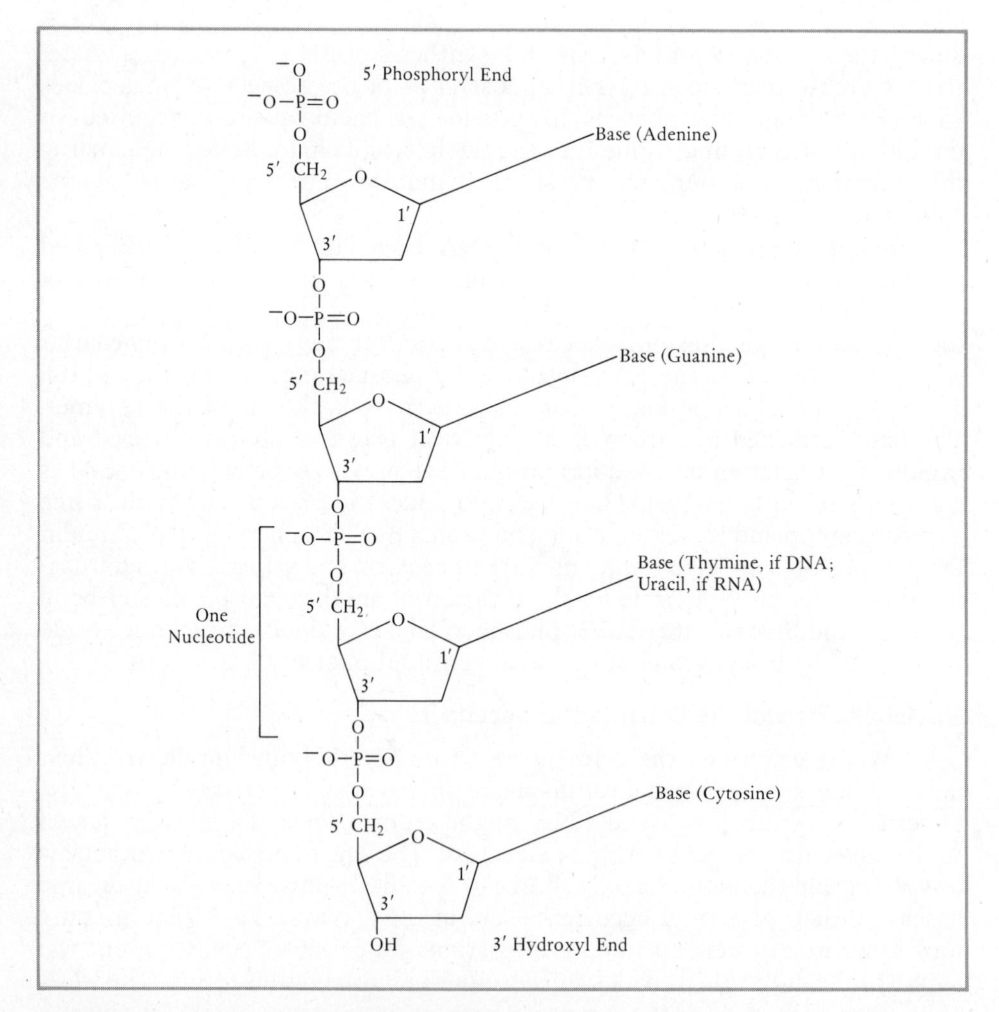

Figure 11–10. Backbone structure of DNA or RNA. Carbons 1′, 3′, and 5′ of the pentose are numbered. An individual nucleotide of the chain is indicated by a bracket. Note that the two ends of the polymer are different.

THE DOUBLE HELIX

DNA, whether circular or long and filamentous, has been described thus far as a single polymer, whereas in fact in only a few viruses has it been found in the single-stranded state. In all other forms possessing DNA as the genome the molecule exists as a double polymer composed of two complementary strands wound about one another in a helical arrangement, the so called double helix. This helical structure was first proposed by Watson and Crick in 1953 based on evidence obtained from X-ray diffraction studies of DNA polynucleotides. The double helix consists of two DNA polymers running in opposite directions and held together crosswise by hydrogen bonds between the bases. The sugar-phosphate linkages of the respective strands are located at the periphery, while the bases are placed centrally, stacked on top of one another. In addition, the entire two part structure is twisted gently into a helical form which results in a sequence of regularly arranged coils throughout its length.

Of the two chains of sugar-phosphate linkages, one begins with a 5′ phosphoryl group and ends with a 3′ hydroxyl, thus giving a 5′ to 3′ direction to the strand. The other backbone begins with a 3′ hydroxyl and ends with the 5′ phosphoryl, giving a 3′ to 5′ direction to this strand. This arrangement of the two polynucleotides, whereby they run in opposite directions, is called *antiparallel*.

The Watson-Crick-Wilkins model of the double helix also clarified a relationship between the bases which had not previously been understood. Studies by E. Chargaff of the proportions of the different bases in DNA obtained from a variety of organisms revealed obvious differences in overall composition. Despite such differences, the amounts of adenine (A) and thymine (T) present in any sample were always found to be approximately equal. The same relationship held between guanine (G) and cytosine (C). In other words, regardless of the source of the DNA, [A] equalled [T], and [G] equalled [C]. This finding aided the interpretation of the X-ray diffraction data with the result that a double helix model was proposed in which the respective bases projecting from the two opposing polynucleotides were placed adjacent to one another in a specific spatial arrangement. Since the double helix was shown to have a uniform diameter of 20 Å, the only spatial arrangement that would satisfy this dimensional requirement was the consistent pairing of a thymine residue of one strand with an adenine residue of the opposite strand, along with comparable pairing of guanine with cytosine, in each case, the pairing occurring between a pyrimidine and a purine. As a consequence of such base pairing, the amount of adenine would be expected to equal thymine, and cytosine to equal guanine. As mentioned above, this is indeed the case (see appendix to this chapter).

The base pairs form the core of the double helix and lie flat, or in a planar arrangement, approximately perpendicular to the longitudinal axis of the helix. In addition, the members of each such base pair are linked together by hydrogen bonds, there being two such bonds between adenine and thymine and three between cytosine and guanine (Figure 11–11).

A model of the three-dimensional structure of the double helix is shown in Figure 11–12. A careful examination of this figure will reveal the following details. In side view, the double helix is visualized as a three-dimensional cylindrical structure whose outer limits are indicated by dash lines. The imaginary axis of the structure is shown by the central dash line. The diameter of the double helix is 20 Å. Although the phosphate-sugar linkages form the outer boundaries, as well as the backbones of the double helix, for the sake of clarity, their paths about the circumference of the model are shown as ribbons I and II. Because of the spatial arrangement of the bases, these ribbons are only 120 degrees, rather than 180 degrees, apart. Consequently, lengthwise grooves that are unequal in width appear between the ribbons as they twist about, the larger being called the major groove and the smaller, the minor groove. Although the flat ribbons represent the surface paths of the backbones, the actual structure is also indicated. Notice that the sugar-phosphate moieties do not lie flat on the surface, but extend about a quarter of the way into the central part of the helix. Each backbone is composed of a deoxyribose sugar, shown as a pentagon (S), linked to a phosphate (P), both above and below. The oxygen atom of the sugar

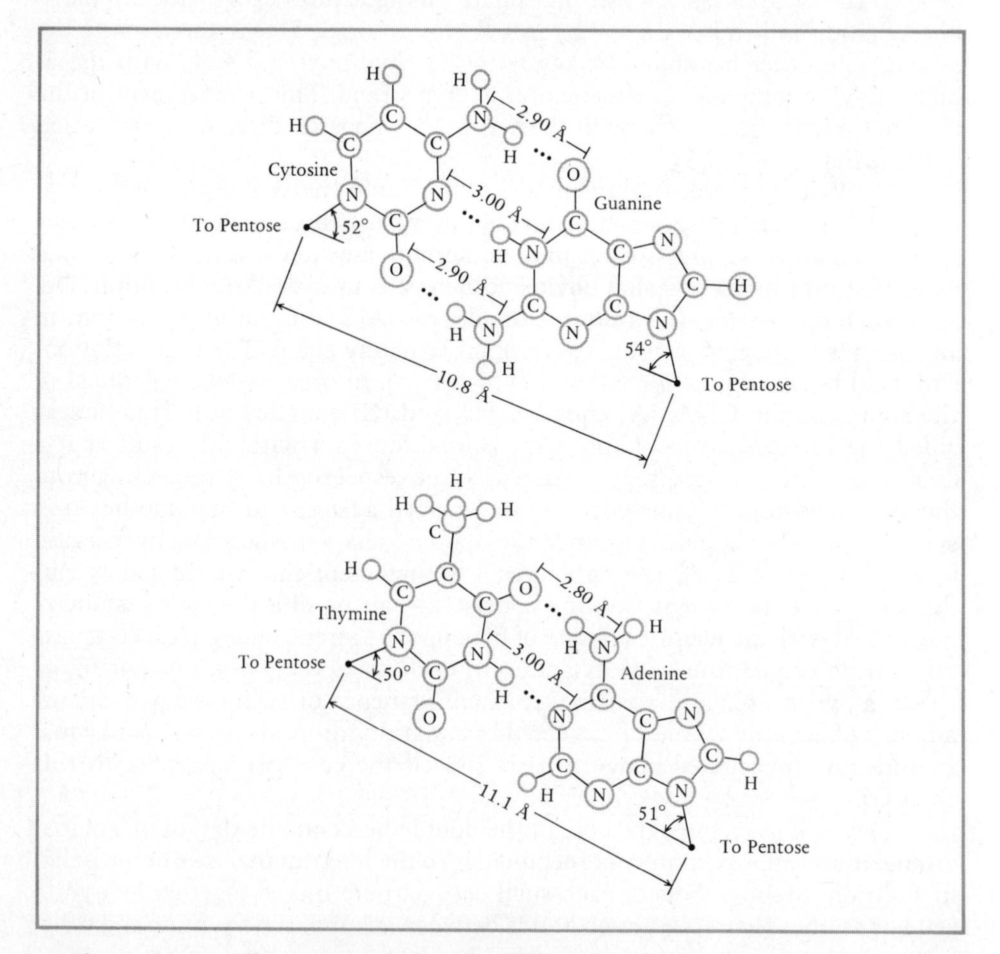

Figure 11–11. Illustration of the three hydrogen bonds between guanine and cytosine, and the two hydrogen bonds between thymine and adenine. [From L. Pauling and R. B. Corey, *Arch. Biochem. Biophys.*, *65*, 164–181, 1956.]

appears as a small circle at the apex of the pentagon. The positions of carbons 3′ and 5′ of deoxyribose are numbered, and carbon 5′ is also indicated by a small dot. As shown by the orientation of the sugar molecules, the two backbones are antiparallel, running in opposite directions.

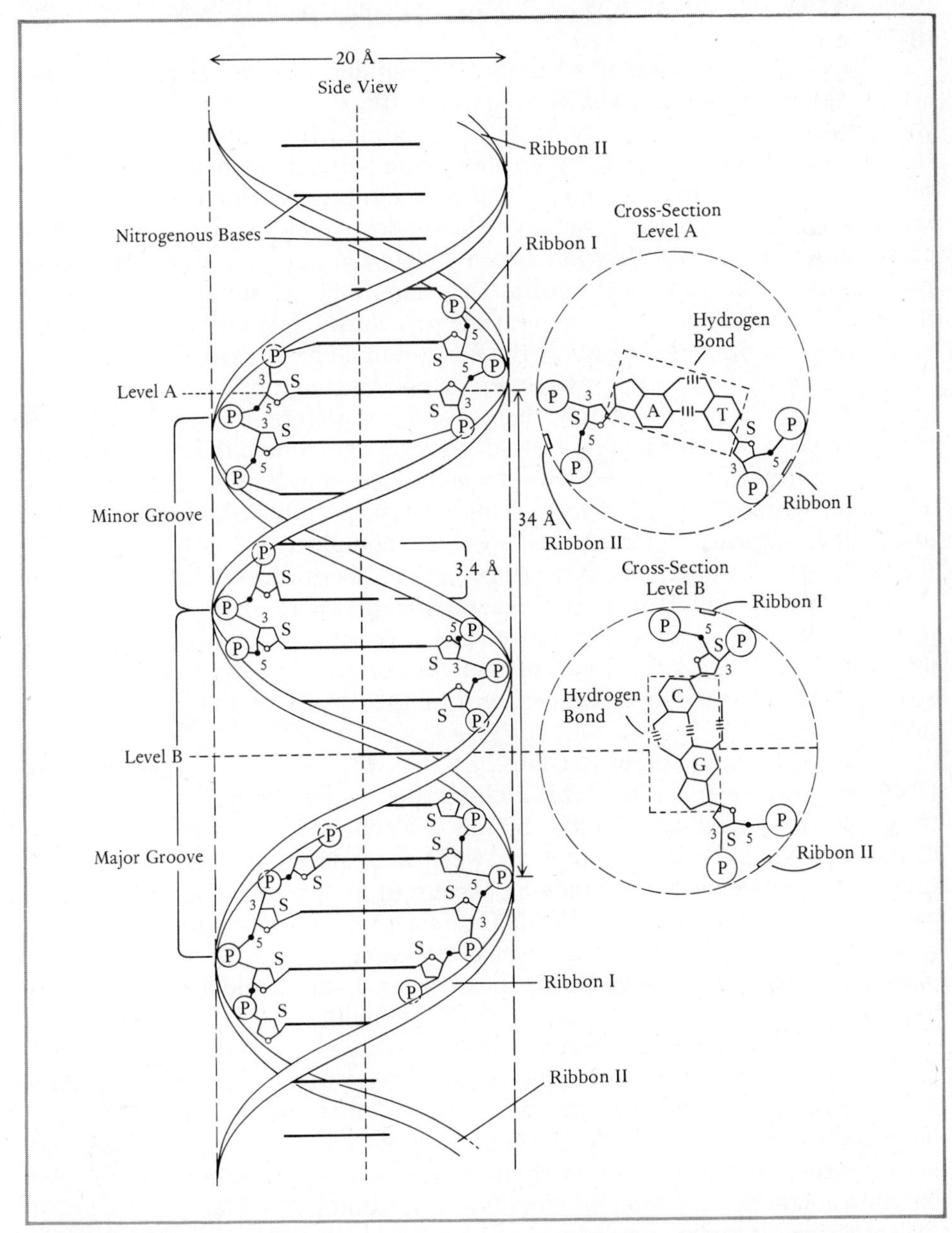

Figure 11–12. A model of the double helix in the hydrated (B form) as found in the cell. [Reprinted from "A Representation of the Structure of DNA" by William Etkin in *BioScience* 23 (November 1973) p. 653 with permission of the author and the American Institute of Biological Sciences.]

The bases, extending inward from the deoxyribose molecules, occupy the central area of the structure. They are flat molecules, stacked on top of one another. The heavy lines denote the base pairs and are longer or shorter depending on whether the bases are seen in side or end view. The distance between successive base pairs is 3.4 Å. The helix rotates 36 degrees with each set of base pairs, so that one full turn is 34 Å long and consists of 10 base pairs (count them).

To clarify the internal structure, the double helix is shown in cross-section at two levels. Level A illustrates the orientation of adenine and thymine and Level B that of cytosine and guanine within the central region of the helix. Hydrogen bonds between these base pairs are shown in each cross-section. The bases are illustrated as planar molecules which is their actual orientation. Although the respective backbones should be seen "end on" in cross-section, for clarity they have been shown flat and projected in the same plane as that of the base pairs. In summary, Figure 11–12 should be considered thoughtfully and in conjunction with the preceding explanation so that a correct three-dimensional concept of the DNA double helix may be grasped.

Because of the presence of numerous phosphate groups, the double helix carries a high density of negative charges and is therefore strongly acidic at the pH of the cell. The mutual repulsion of such charges would ordinarily create a stiff and unbending structure. This is not the case, however, for DNA is found associated with cations and also with highly basic proteins which in eukaryotic (nucleated) organisms form nucleoprotein complexes. Such proteins are thought to occupy positions in the major and minor grooves of the helix. Since the length of the double helix is characteristically many times that of the cell in which it occurs, neutralization of the negative charges by cations and basic proteins permits the coiling and packing of the double helix into a compact structure which can be contained within the limits of the virus, cell, or nucleus.

Having become somewhat familiar with the molecular architecture of DNA, we must now ask how this architecture confers upon the double helix the property of chemical *stability*. Stability is conferred partly by the presence of the millions of hydrogen bonds between the paired bases. Of even greater importance, however, is the stacking pattern of the parallel arrays of the base pairs which results in Van der Waals interactions between them. Individually these interatomic forces are quite weak. However, in an array of stacked base pairs of the length found in DNA, these forces assume sufficient strength to assure the integrity and stability of the double helix.

As previously noted, RNA forms the genome of certain viruses and is found in either the single- or double-stranded state within the infectious virus particle (virion). As in DNA, the two antiparallel strands of an RNA double helix are held together by hydrogen bonding between guanine and cytosine and between adenine and uracil (not thymine). As a result, double-stranded RNA exhibits a base composition wherein [G] = [C], and [A] = [U].

A survey of nucleic acid structure is incomplete without mention of some of the ways by which a double helix can be analyzed. These methods are reviewed in the appendix to the chapter.

REPLICATION OF NUCLEIC ACIDS

The third and most important requirement of a hereditary molecule is the potential for self-reproduction or self-replication. The reproduction of the genome is a prerequisite for the reproduction of the organism itself, thereby providing the continuity that exists between generations.

The potential for self-replication is inherent in the structure of DNA. As described above, the double helix consists of two single, antiparallel strands, each such strand containing a sequence of bases which is precisely complementary to the sequence found in the opposing strand. Thus, if a polymer, starting at the 5′ phosphoryl end, possesses the starting base sequence of A-T-C-C-T-A-G-, then the opposite polymer, beginning at the 3′ hydroxyl end, will contain the complementary starting sequence of T-A-G-G-A-T-C-.

The theoretical scheme for the duplication of such a polymer requires, first of all, the presence of the starting raw materials, the four deoxyribonucleoside triphosphates. In addition, a proper pH and a proper ionic environment are needed, along with an enzyme capable of joining the separate nucleotides into a new sugar-phosphate backbone. The final requirement is the presence of the original double helix itself, which is needed to serve as the template upon which the new strands must be modeled. The enzyme which participates in this reaction is a replicase, called DNA polymerase. Since it requires DNA as a guide or template for its activity, it can more precisely be called a DNA-dependent DNA polymerase.

Given these starting requirements, we can visualize that replication of a region of the double helix occurs as follows (Figures 11–13 and 11–14). The two strands of the original double helix separate progressively from one another by the dissolution of the hydrogen bonds between the paired bases. Upon strand separation, the bases of each single polymer become available and free to form new hydrogen bonds with appropriate bases of the nucleoside triphosphates present as raw materials. The position taken by any triphosphate is dictated by hydrogen bonding properties, these properties functioning as discriminating agents to insure the correct complementarity of the new base pairs. Once these base pairs are properly aligned on each of the original single strands, the polymerizing enzyme catalyzes the formation of phosphodiester bonds between successive nucleotides. This reaction results in the formation of two new backbones, each of which is hydrogen bonded to one of the original strands of the starting double helix. Because of the complementary nature of the base pairs, each new double helix is an accurate copy of the parent double molecule.

The polymerizing enzyme is thought to move along the major groove of the original double helix, progressively catalyzing the formation of new phosphodiester bonds as separation of the original strands and alignment of bases proceeds. For its activity, DNA polymerase requires magnesium ions as well as the presence of template or primer DNA. The energy for phosphodiester bond formation is derived from the hydrolysis of a high energy phosphate bond of each nucleoside triphosphate, pyrophosphate ($P\text{-}P_i$) being split out in this reaction as illustrated in Figures 11–15 and 11–16. The overall reaction can be summarized as:

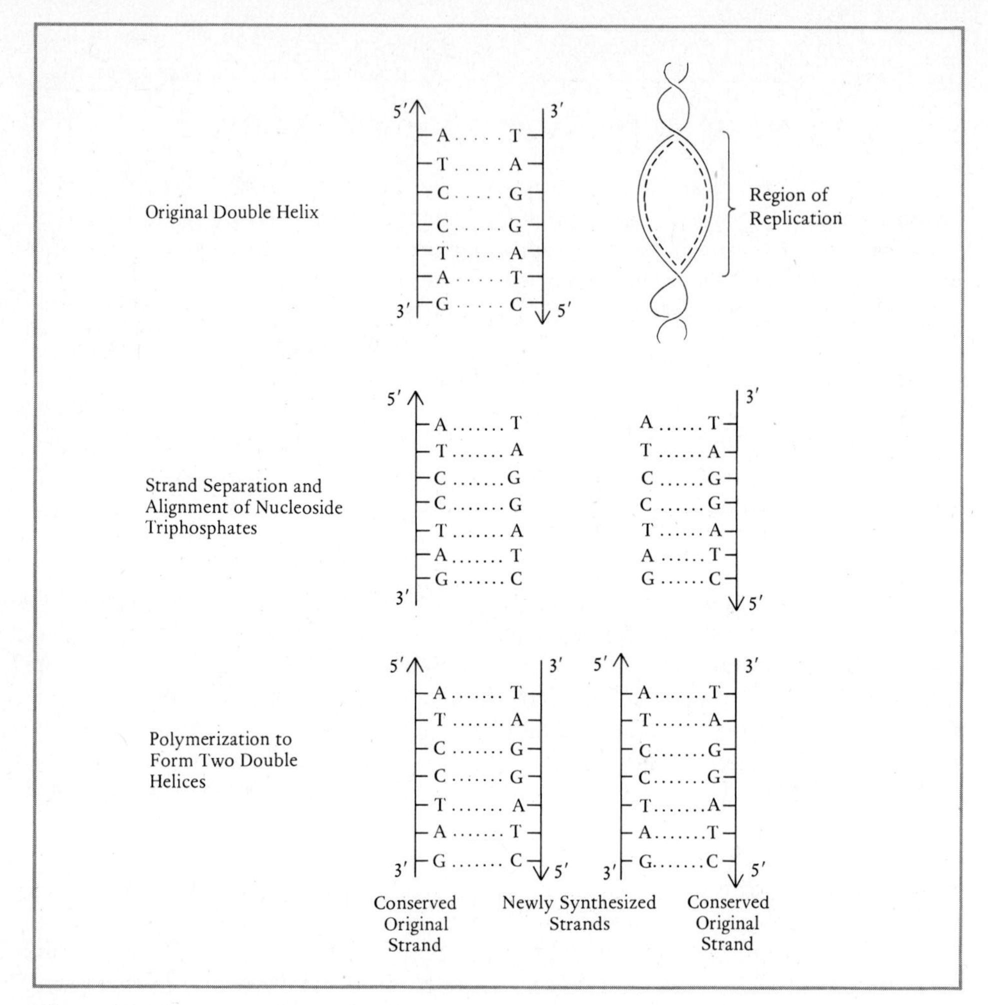

Figure 11–13

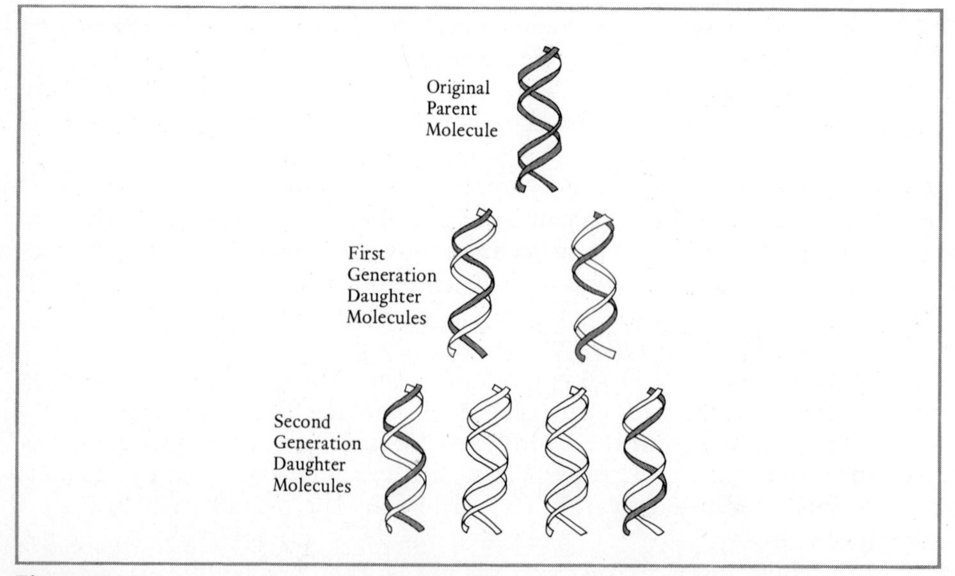

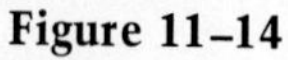

Figure 11–14

Figure 11–13. Overall scheme of DNA replication.

Figure 11–14. Illustration of the mechanism of DNA replication proposed by Watson and Crick. Each daughter molecule contains one of the parental chains (black) paired with one new chain (white). Upon continued duplication, the two original parent chains remain intact, so that there will always be found two molecules each with one parental chain. [From M. Meselson and F. Stahl, *Proc. Nat. Acad. Sci.*, *44*, 671–682, 1958.]

Figure 11–15. Addition of cytosine nucleotide to the 3′—OH end of a growing DNA polymer, with the release of pyrophosphate.

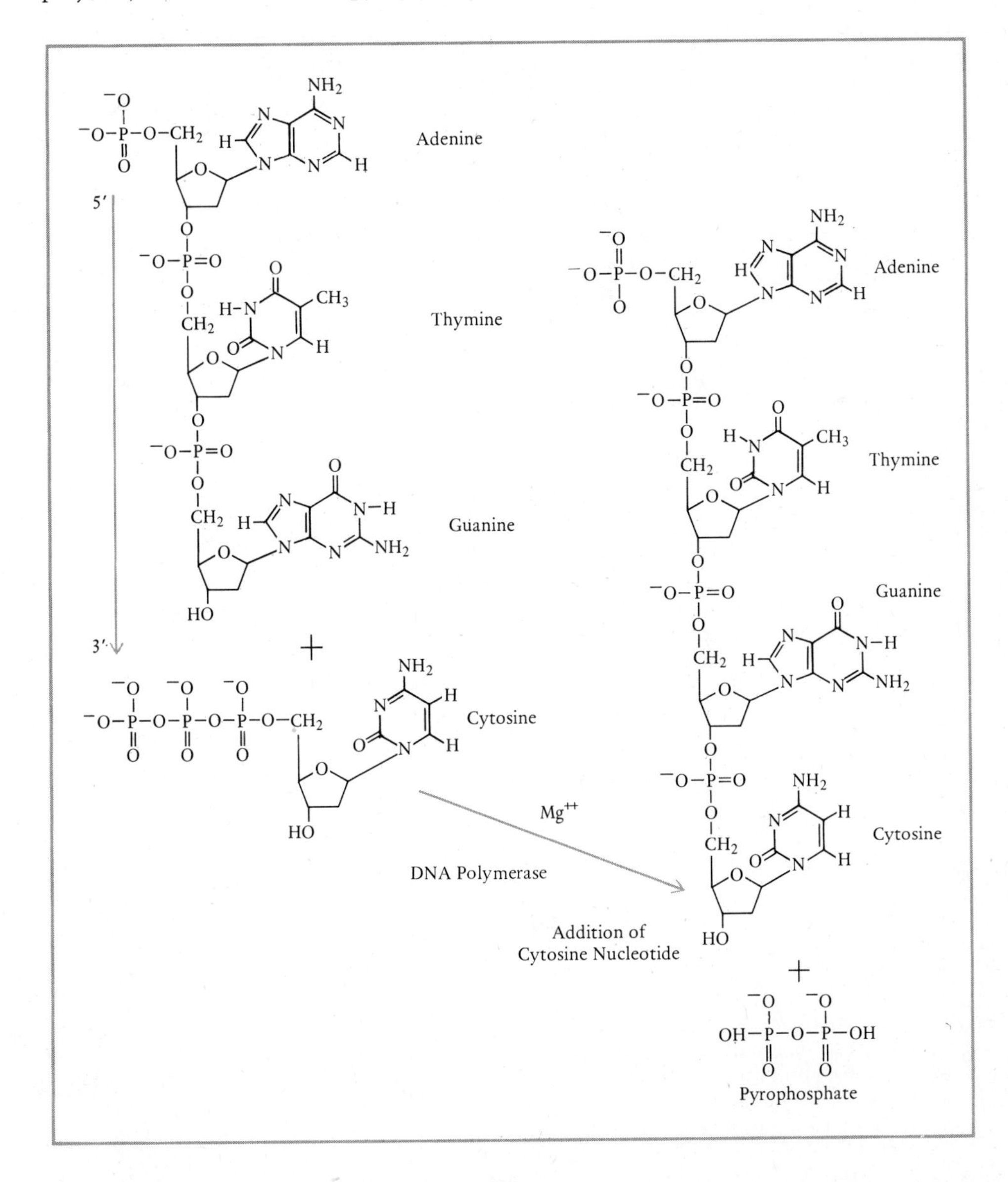

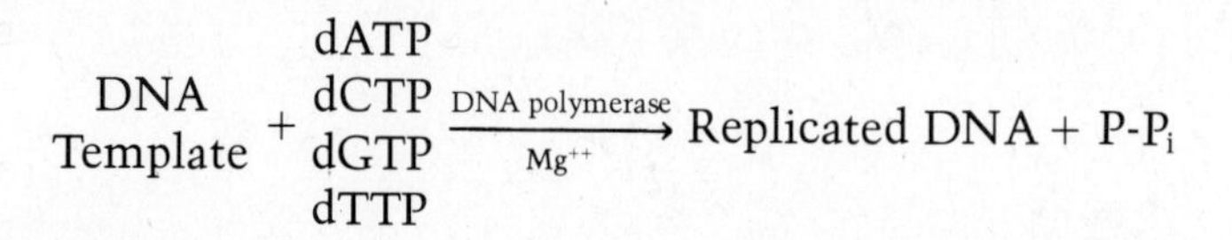

where dATP, dCTP, dGTP, and dTTP represent the deoxynucleoside triphosphates of adenine, cytosine, guanine, and thymine, and P-P_i represents pyrophosphate.

This mechanism for the replication of DNA is termed *semiconservative* because each strand of the original two-stranded structure is "conserved" to become half of a daughter double helix. In turn, each half of such a daughter

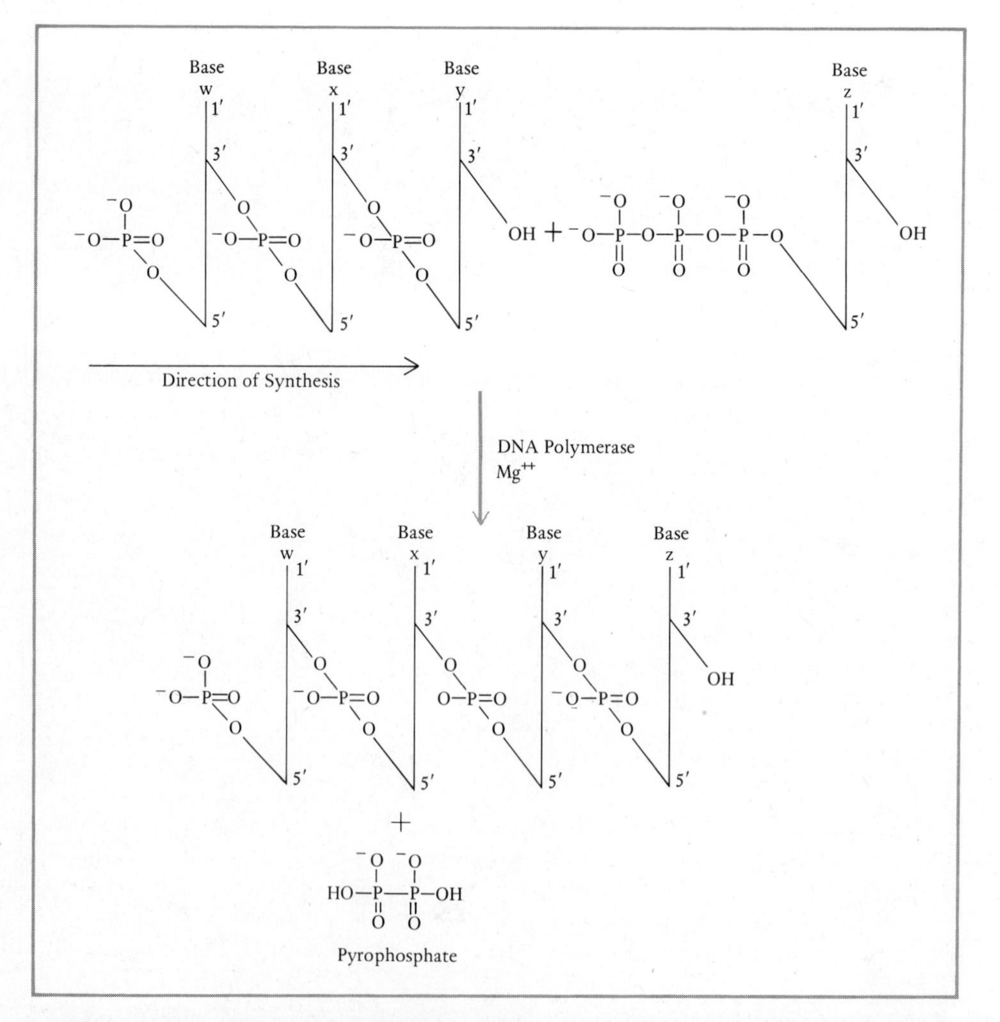

Figure 11–16. Diagram of addition of a nucleotide with base z to the growing 3′—OH end of a DNA polymer. Vertical line represents the sugar with attached base; linkage by phosphodiester bonds is shown diagonally. Overall direction of synthesis indicated by arrow. This type of stick diagram is used conventionally by chemists.

molecule becomes the template or model for the synthesis of a second generation of double helices, and so on and on. It is easy to understand how by means of this mechanism DNA can be reproduced so faithfully through countless generations of molecules and through countless generations of individual organisms. Before the process of semiconservative replication had been unequivocally demonstrated, other possible methods of replication had been proposed, such as the conservation of the original double helix intact accompanied by the synthesis of a new one *de novo* or the breaking up of the original helix, its parts being dispersed into new helices. Such proposals have long been abandoned in favor of the semiconservative mode of reproduction.

Evidence supporting the semiconservative mode of replication was obtained by M. Meselson and F. W. Stahl in 1958. In their experiment bacterial cells were initially cultured in a medium containing a heavy isotope of nitrogen, ^{15}N, so that this atom became incorporated into all nitrogen-containing compounds of the cells, including the DNA. After being so labeled, the cells were transferred to fresh medium containing the usual and lighter ^{14}N and allowed to grow and divide. Some cells were removed from the culture at successive intervals after transfer, and their DNA was isolated and sedimented in the ultracentrifuge. The method employed was a technique called density gradient centrifugation. The suspending solution used was 8M cesium chloride (CsCl), a substance that by itself exhibits a known range or gradation in density when subjected to the forces of centrifugation. DNA mixed with this solution sediments down from the top of the centrifuge tube and also exhibits flotation up from the bottom of the tube until, after an extended centrifugation, all the DNA is concentrated in a narrow band where its density exactly matches that of the CsCl at that position in the gradient. If DNA's of different densities are used, each will move to that region of the tube where its density equals that of the CsCl at that point. In the Meselson-Stahl experiment initial samples of DNA from cells labeled with ^{15}N exhibited a single band in the centrifuge tube at a point where DNA containing this heavy isotope was expected. However, with DNA obtained from cells after one cycle of division in unlabeled medium, this single band was replaced by a new band whose density was intermediate between that of ^{15}N and that expected for ^{14}N-containing DNA. This new band represented hybrid daughter double helices, each such daughter helix containing one ^{15}N-labeled original strand and one newly synthesized ^{14}N strand. After a second cycle of cell division in unlabeled medium, the isolated DNA showed two density bands, one characteristic of ^{14}N-^{15}N hybrid molecules and the other of ^{14}N-^{14}N molecules.

These results are easily interpreted. The first generation hybrid ^{14}N-^{15}N molecules underwent strand separation and subsequent replication, the ^{15}N strand becoming hydrogen bonded to a new ^{14}N polymer, thus yielding a hybrid molecule once more. The opposing ^{14}N strand of the first generation hybrid was replicated to a new ^{14}N-^{14}N double helix. Subsequent sampling of the culture showed a gradual increase in the amount of ^{14}N DNA as the cells continued to replicate their DNA using the ^{14}N-containing compounds of the medium (Figure 11–17).

This experiment clearly showed that the pattern of replication involves strand separation and the synthesis of new complementary polymers. If the

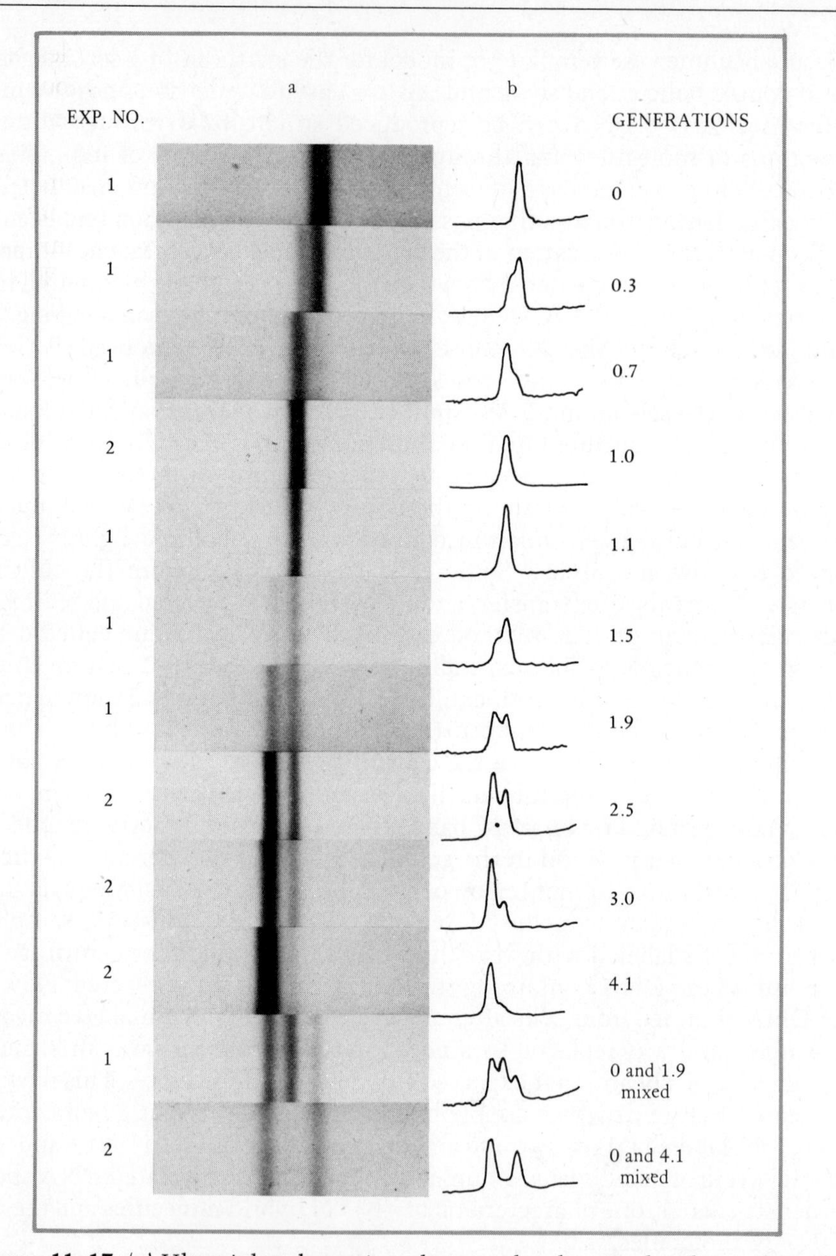

Figure 11–17. (a) Ultraviolet absorption photographs showing bands resulting from density gradient centrifugation of bacterial DNA and (b) densitometer tracings indicating amounts of different density DNA's present. Higher density ^{15}N is represented by the right band, lower density ^{14}N by the left, and hybrid ^{15}N-^{14}N DNA by the middle. DNA at 0 generation exhibits one peak and one band at the density of ^{15}N. At one generation the DNA band has an intermediate density between that of ^{15}N DNA and that of ^{14}N DNA and consists of only one peak; it is therefore composed of only one molecular type, a hybrid ^{15}N-^{14}N molecule. At approximately two generations two bands and two peaks appear, one for ^{15}N-^{14}N hybrid DNA and one for ^{14}N-^{14}N DNA. Bands for the third and fourth generation DNA show a progressive increase in ^{14}N DNA as the cells continue to replicate their DNA using ^{14}N precursors. [From M. Meselson and F. Stahl, *Proc. Nat. Acad. Sci., 44*, 671–682, 1958.]

original double helix had been conserved intact, no hybrid molecules would have been formed in the first generation, and two bands, not one, would have resulted. On the other hand, if the original helix had become fragmented and the resulting pieces incorporated into new DNA, a series of bands of different densities would have been found and such was not the case. In summary, the Meselson-Stahl experiment is an elegant demonstration of the molecular events that underlie the reproduction of the genome. This same technique has also been used to demonstrate semiconservative replication of the DNA of eukaryotic organisms. The results of one such experiment, for cells of the tobacco plant *Nicotiana,* again utilizing differences in density between ^{15}N and ^{14}N, are illustrated in Figure 11–18. The banding pattern obtained is equivalent to that found for bacteria by Meselson and Stahl.

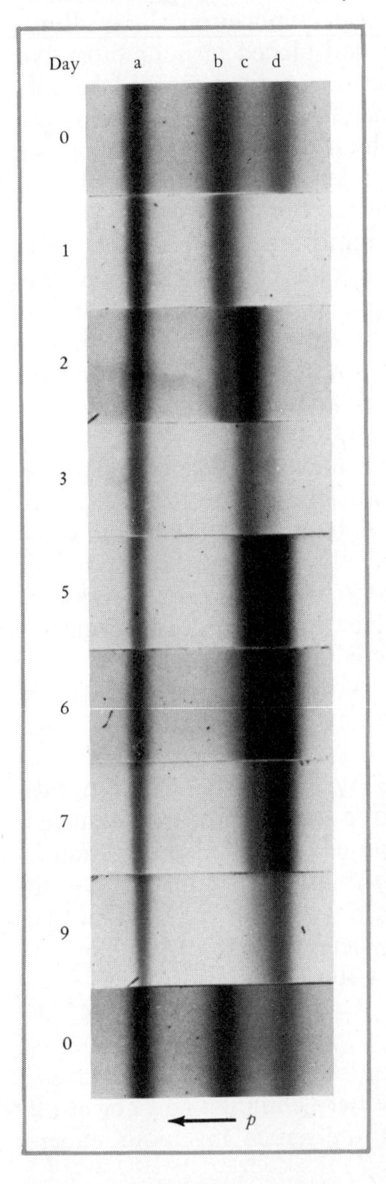

Figure 11–18. Cesium chloride density gradient photographs of DNA isolated from tobacco plant cells. Density increases to the left. (←). Top and bottom frames show the resolution of a mixture of reference ^{15}N and ^{14}N DNA. Band a is bacterial DNA of known density used as a reference marker, band b is ^{15}N-^{15}N DNA, band c is hybrid ^{15}N-^{14}N DNA, and band d is ^{14}N-^{14}N DNA. Generation time is two days. Starting (day 1) DNA is ^{15}N labeled. At 3 days (generation 1) DNA is present as a single band of intermediate ^{15}N-^{14}N density. At 5 days (generation 2) hybrid DNA and ^{14}N-^{14}N DNA are both present. A third generation (day 7) shows a continuing increase in the amount of ^{14}N DNA. [From P. Filner, *Exp. Cell Res., 39,* 33–39, 1965.]

Further evidence supporting the semiconservative mode of replication was obtained by J. H. Taylor, P. S. Woods, and W. L. Hughes in 1957, in mitotic chromosomes of bean *(Vicia)* root tips. These workers employed the technique of autoradiography which relies on the fact that radioactivity exposes photographic film in the same manner as X rays. As briefly mentioned with Figure 6–11, if a cell is synthesizing DNA, RNA, or other constituent, precursor molecules containing a radioactive atom will be incorporated in the product synthesized. Such unstable atoms produce emissions which expose a photographic emulsion applied to the cells. After development, the presence of silver grains in the emulsion will mark the location of the synthesized product.

In their experiment Taylor and associates placed the root tips of seedlings in a medium containing tritium-labeled (^3H) thymidine (thymine nucleoside). After a timed period sufficient to permit chromosome replication, the seedlings were rinsed to remove the tritiated thymidine and placed in nonradioactive medium containing colchicine. The colchicine not only prevented spindle formation and mitotic division, but also had the effect of spreading apart the arms of daughter chromosomes (chromatids) which remained attached to one another only at the centromere. The root tips were either treated with colchicine for a short time and then prepared immediately or they were allowed to complete one or more additional cycles of chromosome duplication in the nonradioactive-colchicine medium before preparation for cytological examination. In either case, after staining and squashing on a microscope slide, these root tip chromosome preparations were covered with a thin film of photographic emulsion and placed in the dark. After a suitable period of exposure, the emulsion, still fixed to the slide, was developed and the chromosomes were examined under the microscope.

The results of these studies provided strong evidence for semiconservative replication and for the theory that the chromosome (chromatid) consists of but one single long DNA double helix (Figure 11–19). In general, Taylor and associates found the following. (1) After one cycle of chromosome duplication in the ^3H-thymidine medium followed by exposure to colchicine, each metaphase chromosome consisted of two chromatids, both labeled. (2) After two cycles of duplication, one in labeled and one in unlabeled medium (plus colchicine), each chromosome was represented by two pairs of chromatids (tetraploid), each such pair consisting of one labeled and one unlabeled chromatid. (3) After three cycles of duplication, one in labeled and two in unlabeled medium (plus colchicine), each chromosome was represented by four pairs of chromatids (octoploid). Of these four pairs, two consisted of a labeled plus a nonlabeled chromatid, while the other two pairs were composed of unlabeled chromatids only. It can be seen that these cytological results are fully consonant with semiconservative replication of the double helix (Figure 11–19). It should be noted that Taylor also found some instances where the pattern of labeling indicated that exchanges had occurred between sister chromatids.

Mechanisms of Replication

Although the overall process of DNA replication is easily comprehended, the underlying molecular mechanisms are extremely complex and not at all well understood. To clarify these mechanisms, numerous ingenious exper-

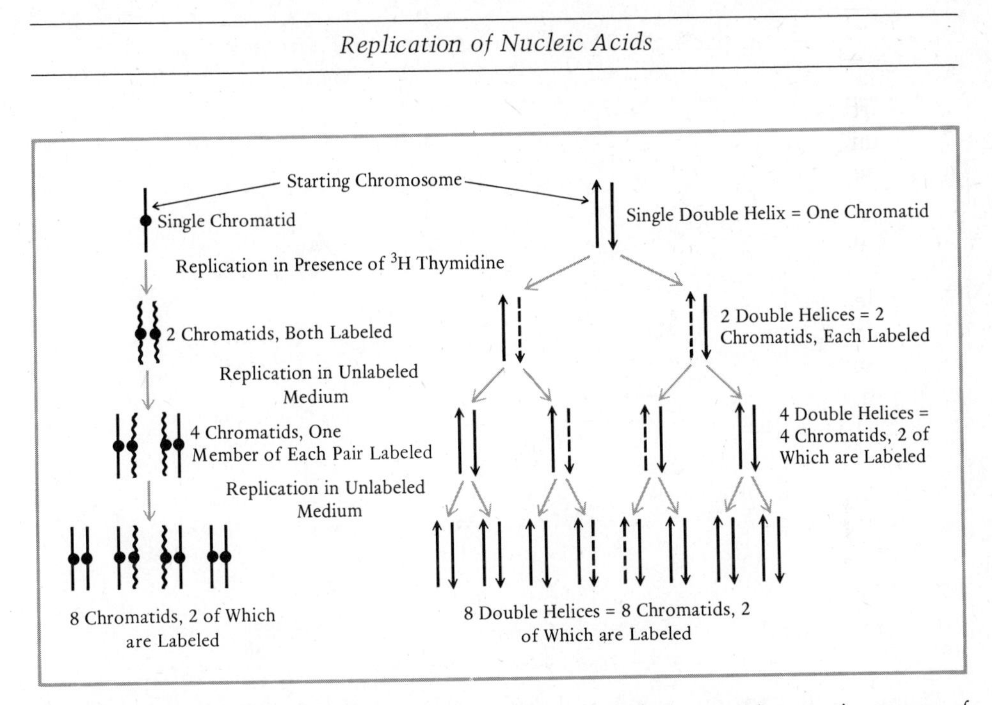

Figure 11–19. Interpretation of the results of Taylor's experiments in terms of semiconservative replication of the DNA double helix. ^{3}H labeled chromatids are represented by wavy or dash lines.

iments have been devised in attempts to explore the process *in vivo* and to duplicate the process *in vitro*. Some of the insights gained are of interest.

It will be recalled that the semiconservative mode of replication requires that strand separation occur to permit the synthesis of new complementary polymers. This requirement imposed upon DNA presents a problem: DNA is constructed of two antiparallel strands wound about one another in the form of a helix, and the helix must be unwound in order to achieve strand separation. In addition, after replication the daughter double strands must be rewound into the helical form. At first glance, this unwinding and rewinding requirement might seem to impose an insuperable obstacle to the replication of the enormously long DNA molecule. However, the obstacle seems less insuperable if we consider that strand separation and replication need not occur simultaneously everywhere along the length of the molecule, but instead can be brought about progressively, so that unwinding and rewinding is restricted to the actual site of synthesis.

The most common pattern of DNA replication is one in which synthesis is initiated in a localized region and at a specific initiation site. The separation of strands begins at this site and produces a small bubble or "eye" within the double helix. Synthesis of complementary polymers then occurs within the bubble, proceeding in opposite directions from the initiation site via two growing points (Figure 11–20). C. Levinthal and H. R. Crane have compared each such growing point to a Y-shaped fork, the two arms representing daughter helices and the stem representing the parent, as yet unreplicated helix. They have postulated that if the arms and the stem were to revolve simultaneously, the parent would be unwinding at the same time that the daughters were

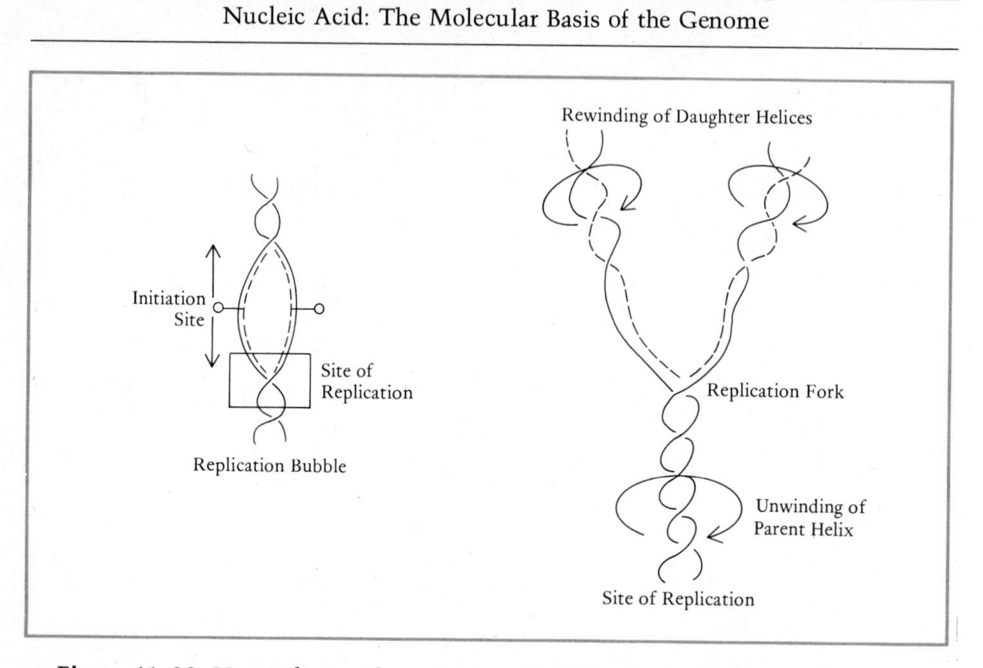

Figure 11–20. Unwinding and rewinding of DNA during replication. The sites of replication (replication forks) progress in opposite directions (arrows) from the initiation site.

rewinding. According to this concept, such a replication fork would travel progressively along the parent molecule, with unwinding occurring ahead of it, and rewinding taking place behind.

Levinthal and Crane have calculated the energy required for this rotation and found it to be insignificant compared to that needed for the formation of the requisite number of phosphodiester bonds per turn. The initiation of the replication fork, however, requires some sort of swivel mechanism which will permit the unwinding to begin. A break or nick produced by an endonuclease in one of the two strands in advance of the fork would serve this purpose by allowing one strand to separate from and revolve around its partner. After completion of synthesis, any such break could be repaired enzymatically. Enzymes, called polynucleotide ligases, possessing the properties and specificities required for this repair reaction are known to be present and active during DNA synthesis in both prokaryotes and eukaryotes. The presence of specific "unwinding proteins" which perform the function of a swivel for strand rotation has been proposed. Efforts to isolate and delineate the actions of such proteins are currently in progress and two such proteins have been tentatively identified, one in the DNA virus T4 and another in the mouse. Although actual rotation of the strands has never been observed, replication forks have been identified in the DNA molecules of bacteria and the chromosomes of higher organisms.

The polymerizing enzymes involved in DNA replication are also of interest. In 1960 such an enzyme, DNA polymerase I, was isolated by Arthur Kornberg from *E. coli* cells and used in efforts to duplicate the process of replication *in vitro*. Kornberg found that all four deoxyribonucleoside triphosphates were required as substrates and that Mg^{++} was necessary for enzyme activity. In addition, a primer, preferably single-stranded or partially single-stranded DNA, was needed to serve as a template for synthesis. When the primer used was prepared from DNA isolated from a variety of sources, from viruses to mammals, the product formed by the reaction was a high molecular weight double-stranded molecule whose base composition closely resembled that of the template used in the experiment (see the appendix to this chapter). To test whether or not such a product was biologically active, M. Goulian, Kornberg, and R. L. Sinsheimer in 1967 utilized the virus ϕX174, whose DNA genome is naturally single stranded. When DNA from this virus was used as the template, the product ultimately recovered was capable of infecting host cells and multiplying within them to produce progeny virus particles—convincing evidence that the new molecules generated *in vitro* matched the template employed.

However, although this experiment demonstrated that a biologically active DNA molecule could be produced outside the cell through the activity of polymerase I, this enzyme is not thought to be the replicase primarily involved in genome reproduction *in vivo*. Polymerase I usually functions most efficiently when the primer DNA is denatured and single stranded and is less effective and sometimes inactive when presented with the natural double-stranded template. Also, in extended reactions abnormal branched DNA molecules may be formed. These activities are opposite to what one might expect of a true replicase. Polymerase I also has degradative capabilities in that, given a break in a DNA strand, it can remove nucleotides in either the 3′ to 5′ or in the 5′ to 3′ direction. In addition, P. DeLucia and J. Cairns have shown that *E. coli* cells possessing a mutation causing lack of this enzyme are nevertheless able to carry out the replication of their genome. In view of all of these findings, it has been postulated that in the intact cell polymerase I may function as a repair enzyme, replacing and correcting defects and breaks in the double helix caused by radiation, chemical mutagens, or other adverse environmental circumstances.

Two other polymerases, polymerase II and III, have been successfully isolated from *E. coli*. These appear to possess more of the properties expected of a genome replicase in that they, and particularly polymerase III, seem capable of initiating DNA replication in intact cells. Since the molecular mechanisms involved in the synthesis of DNA are obviously extremely complex, it would not be surprising if additional enzymes are found to participate in various steps of the overall process.

A further problem in replication is that posed by the antiparallel nature of the two strands of the double helix. Replication in the 5′ to 3′ direction would seem to require the addition of nucleotides to the 3′ hydroxyl end of a growing polymer, while duplication of the opposing strand would require addition to the 5′ phosphoryl terminus. Since the two ends are different, one would anticipate the involvement of two enzyme systems, one for synthesis in the 5′ to 3′ direction, the other for synthesis in the 3′ to 5′ direction. However, all of the polymerizing enzymes thus far discovered can add nucleotides only to the 3′—OH end (5′ to 3′ synthesis), and none have been found to catalyze synthesis in the opposite direction. Yet, in the intact cell the replication of both strands appears to proceed simultaneously and in parallel.

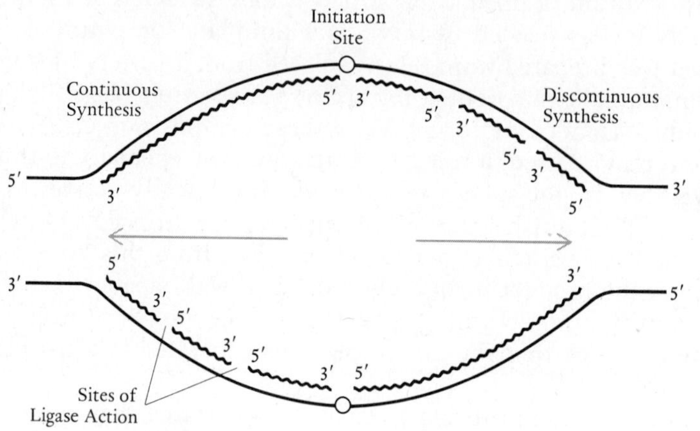

Figure 11–21. Model of DNA replication illustrating continuous and discontinuous synthesis at the two replication forks of a replication bubble. The replication forks are shown proceeding in opposite directions from the initiation sites by the arrows. At each replication fork one daughter strand is synthesized continuously, the other discontinuously, but both are synthesized in the 5′ to 3′ direction. Note that each daughter strand is antiparallel to its parent strand.

As a solution to this seeming paradox, it has been postulated that at a replication fork one strand may be synthesized continuously in the 5′ to 3′ direction, while replication of the other strand may occur through the formation of separate small pieces, each piece being synthesized in the usual 5′ to 3′ direction as before and utilizing the same enzyme system (Figure 11–21). It is proposed that such short segments are then joined together by a polynucleotide ligase to form the required antiparallel strand.

Experimental evidence in support of the discontinuous synthesis of at least one strand has been obtained from studies of *E. coli*. When cells of *E. coli* are grown for very short periods of time in the presence of ^{3}H-labeled thymidine, newly synthesized DNA containing the label is found in the form of short fragments, called *Okazaki fragments* after their discoverer. However, if a longer exposure to labeled precursors is permitted, the label is found in much longer DNA segments, suggesting that the longer molecules are formed through union of previously synthesized short pieces. Similar Okazaki fragments are also found as a transitory phenomenon during the replication of eukaryotic chromosomes. It has also been discovered that RNA is covalently bonded to newly synthesized Okazaki fragments in both *E. coli* and in a eukaryote (Chinese hamster). This association is fleeting, the RNA being cleaved from the DNA fragment before union of fragments occurs. This finding has prompted the proposal that RNA serves as a primer during DNA synthesis.

In summary, it is clear that an understanding of the molecular mechanism of DNA synthesis has not yet been achieved. Certainly, the present interpretations of this process will require revision as new information gained from current intensive investigation in this area becomes available.

ORGANIZATION OF REPLICATION IN REPRESENTATIVE FORMS

While the underlying mechanisms of DNA replication, such as the formation of replication forks and the utilization of polymerases in synthesis, appear to be common to most forms, the way in which replication is organized depends on the nature of the genome and on the kind of organism and its life cycle. The genome found in bacteria and viruses consists of a single molecule of nucleic acid which is essentially naked. In viruses this molecule can be either single- or double-stranded DNA or RNA. The genome of eukaryotes is composed of a number of separate, lengthy molecules in which the double-stranded DNA is tightly bound to protein to form the nucleoprotein complexes called chromosomes.

The amount of nucleic acid comprising the genome is also startlingly different in eukaryotes as opposed to prokaryotes and viruses. In Figure 11–22, it can be seen that the average number of nucleotide pairs present in the genome of viruses is only around 10,000 and that the bacterial genome contains the equivalent of only 1 percent of the DNA content of a haploid mammalian

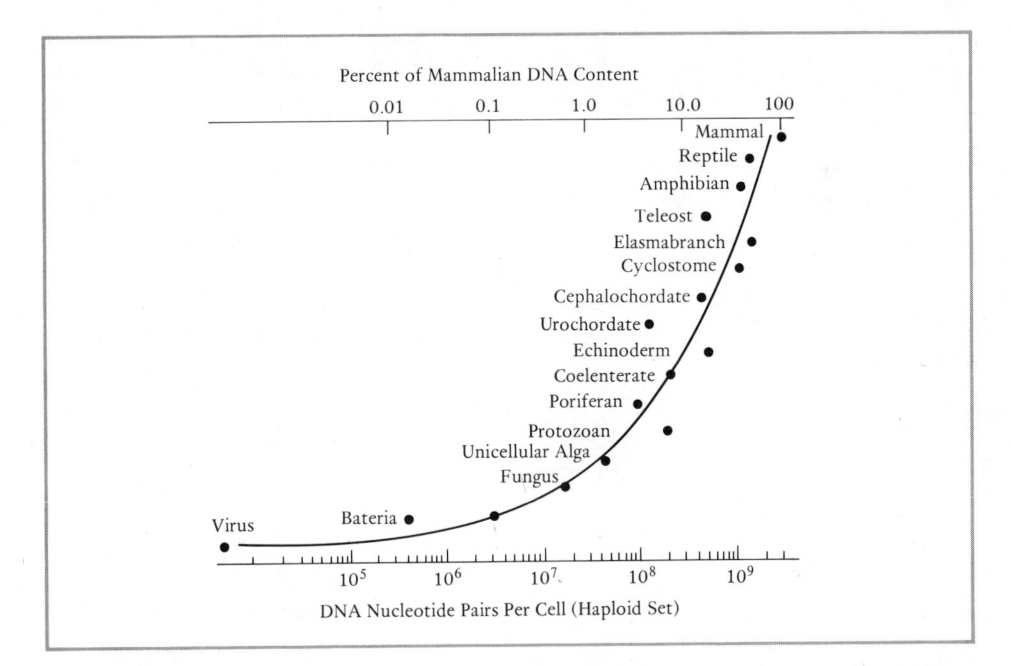

Figure 11–22. The minimum amount of DNA observed for species of various grades of organization. Each point represents the measured DNA content per cell for a haploid set of chromosomes. In the cases of mammals, amphibians, teleosts, bacteria, and viruses, enough measurements exist to give the minimum value meaning. However, for the intermediate grades few measurements are available, and the values shown may not be truly minimal. The ordinate is not a numerical scale, and the exact shape of the curve has little significance. The figure shows that a great increase in DNA content is a necessary concomitant to increased complexity of organization. [From R. J. Britten and E. H. Davidson, *Science, 165,* 349–357, 25 July 1969. Copyright © 1969 by the American Association for the Advancement of Science.]

cell nucleus. As an example in terms of length, the genome of the virus T4 is estimated to measure approximately 0.055 mm (55 μm) and that of the bacterium *E. coli* around 1.1 mm (1100 μm). In contrast, the combined length of all of the chromosomes comprising a haploid human genome is close to a meter.

An additional and striking difference between the respective genomes of eukaryotes on the one hand and prokaryotes and viruses on the other is the presence in eukaryotes of reiterated nucleotide sequences within the chromosomes (see Chapter 19). Such sequences are essentially absent in both prokaryotes and viruses.

Because of the significant differences in structure as well as in size between the chromosomes of eukaryotes and the tiny nucleic acid molecules of prokaryotes and viruses, the term chromosome is properly restricted to the former, while chromoneme or genophore is applied to the latter. Although chromosome is often loosely applied to both, the more precise designation will be observed here. How replication is organized in genophoral and chromosomal forms will next be considered.

Viruses

Viruses are minute parasites of prokaryotic and eukaryotic cells and consist of an outer protein capsule *(capsid)* within which is packed the genome. The viruses that attack bacteria are given the special name of *bacteriophage* or

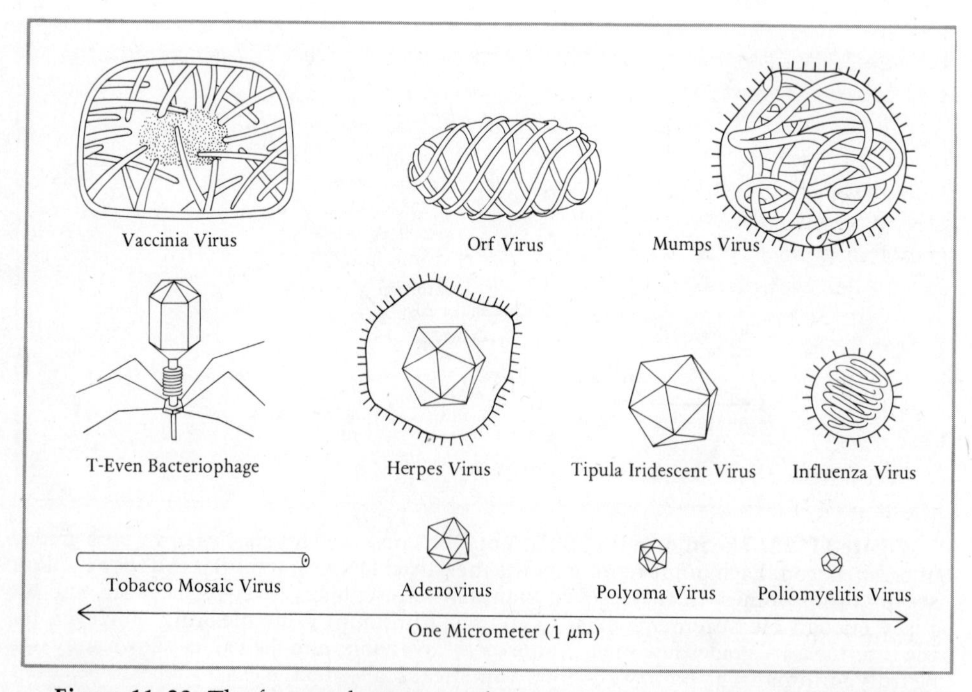

Figure 11–23. The form and structure of some representative viruses. Relative size can be judged by comparison with the 1 μm arrow. [From R. W. Horne, "The Structure of Viruses," *Sci. Amer., 208,* 48–56, January 1963. Copyright © 1963 by Scientific American, Inc. All rights reserved.]

phage for short. Regardless of host, the various types of viruses are designated either by symbol, such as T4, R17, λ, and so on, or by the name of the specific pathological condition which they cause, for example, the herpes, poliomyelitis, smallpox, influenza, or tobacco mosaic viruses. Some typical forms are illustrated in Figure 11–23. Although viruses exhibit great diversity in size and complexity, two major groups can be distinguished on the basis of whether the genome consists of DNA or RNA. These two groups can be considered separately.

DNA Viruses. These viruses and bacteriophages include forms in which the genome may be single- or double-stranded DNA, either circular or linear in form. Although replication is semiconservative and thought to proceed in the usual 5′ to 3′ direction, the ways by which replication is achieved are various and not well understood.

One novel mechanism, called the *rolling circle,* occurs in the *E. coli* phage ϕX174. The stages of the mechanism are shown in Figure 11–24. (1) The genome of this phage consists of a single-stranded closed ring. (2) Upon entrance into the host, a complementary strand is synthesized via host polymerases and ligases, so that a double-stranded closed ring results. This duplex ring thus consists of the original infectious DNA polymer (plus strand) and its newly synthesized complementary polymer (minus strand). (3) It is postulated that an endonuclease produces a nick or break at a specific site in the plus strand to produce free 5′ and 3′ ends. The 5′ end then begins to

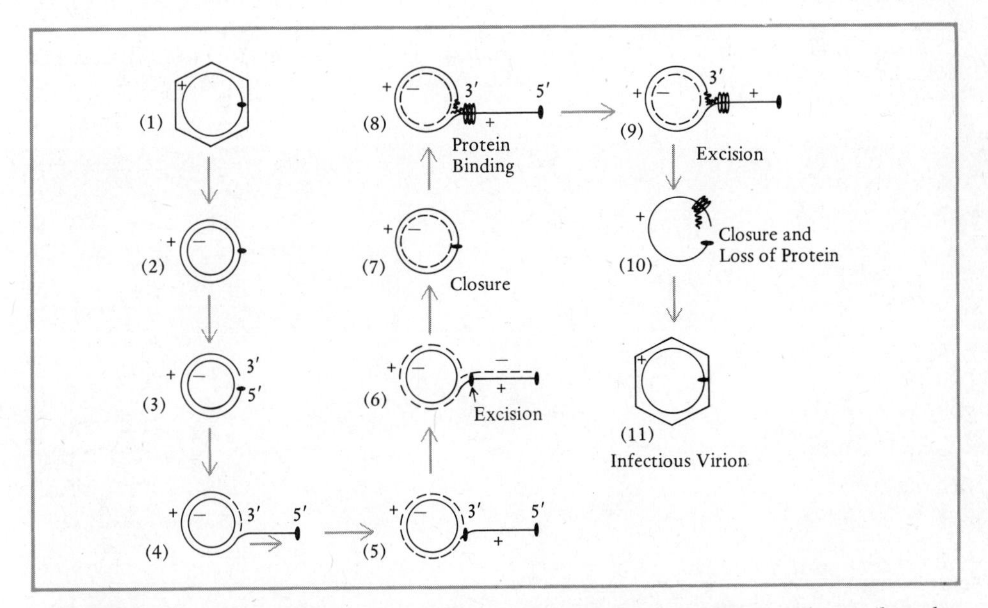

Figure 11–24. The rolling circle in the *E. coli* bacteriophage ϕX174. After nicking by endonuclease the 5′ end of the positive strand is extended as a long linear tail, while the negative closed circular strand is used as an endless template. Replication and circularization of the tail produces double-stranded circles. Single-strand positive circles are formed when protein binding prevents tail replication. [Adapted from D. Dressler, *Proc. Nat. Acad. Sci., 67,* 1934–1942, 1970.]

separate progressively from its complementary minus partner. (4) As this occurs, DNA synthesis is initiated at the free 3′ end, using the minus circle as a template, thereby maintaining the duplex condition of the ring. (5) The overall result is the generation of a long single-stranded tail whose 5′ end lies in the cytoplasm attached to host membrane and whose 3′ end lies on the duplex ring. (6) A strand complementary to the plus tail arises via DNA synthesis to render the tail double-stranded except at its ends. The tail is then excised from the ring. (7) The single-stranded complementary ends overlap, and ligase action produces a closed daughter double-stranded ring called a replicating form. Meanwhile, using the minus circle as a continuous template, the parental duplex ring unrolls to produce additional daughter replicating forms. (8) Formation of the infectious single-stranded rings occurs in much the same way. A plus strand is unreeled from a replicating form, but in this case synthesis of a minus strand complementary to the plus tail is prevented by the binding to the tail of a protein specified by the virus genome. (9) The tail is then cleaved from the replicating form, (10) circularization follows, and (11) the single-stranded plus circle becomes enclosed within the protein capsule to form an infectious particle. This unique method of replication has been postulated to occur in other viruses and has been applied to the transfer of the bacterial genophore to recipient cells (see Chapter 13). It may also be involved in the amplification of the ribosomal genes in the oöcytes of animals (see Chapter 19).

Linear genophores are also found in DNA bacteriophages and other viruses. In some cases, such as phage λ, the double-stranded genome becomes circularized after entrance into the host (Figure 11–25). The newly circularized

Figure 11–25. Electron micrograph of circular λ genophore extracted from infected *E. coli* cells. (35,466×). [From V. C. Bode and L. A. MacHattie, *J. Mol. Biol.*, *32*, 673–679, 1968.]

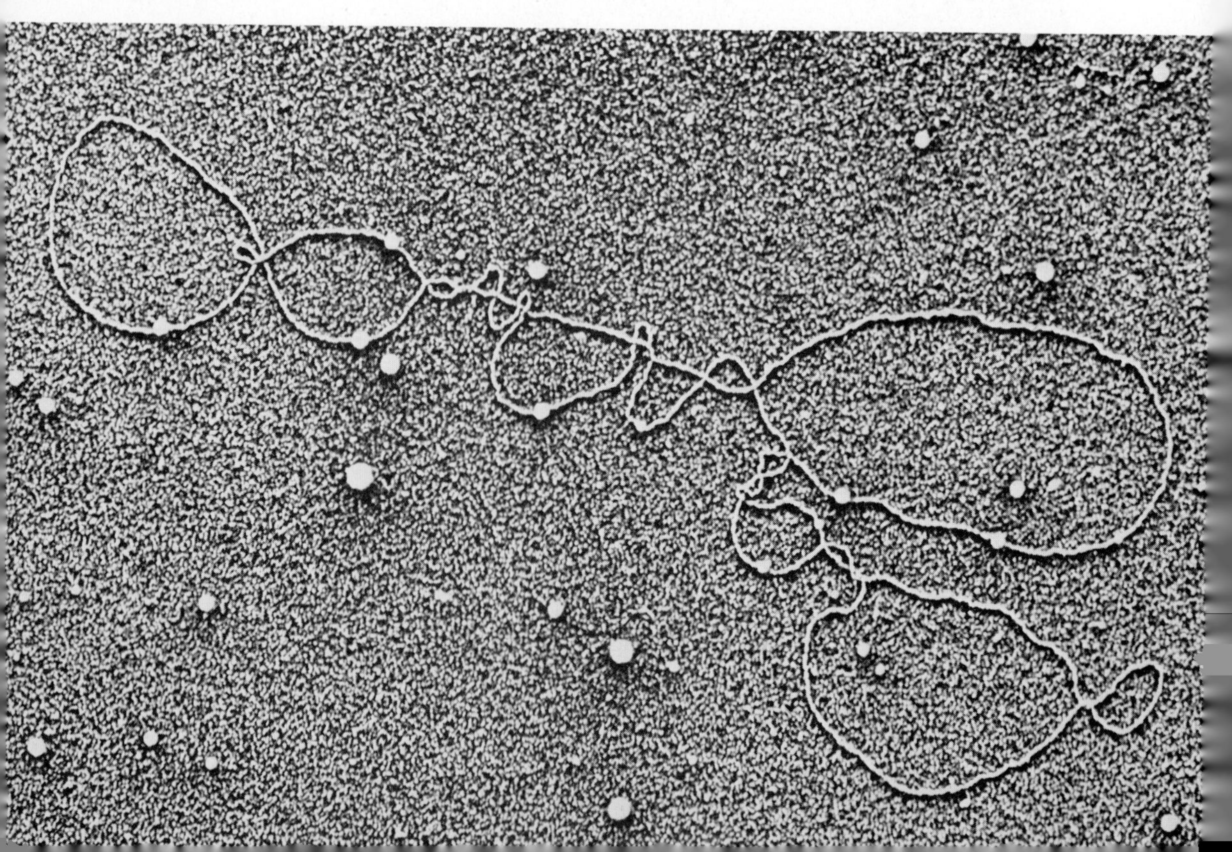

genome may then replicate autonomously in the cytoplasm or it may become inserted directly into the host genome, where it replicates in synchrony with the host genophore. The circular genome of mammalian viruses, such as simian virus 40 (SV40) and polyoma virus, are also thought to integrate within the cellular genome. In contrast to such forms, the linear genome of phage T7 remains linear within the host where it replicates semiconservatively in a bidirectional pattern. A single initiation site located not far from one end gives rise to growing points which travel to opposite ends of the rod-shaped genophore. Continuous cycles of bidirectional replication produce numerous daughter genophores which are then packaged within newly synthesized capsules.

From these limited examples it can be seen that the DNA viruses utilize a variety of mechanisms for the duplication of the genome. Some of these mechanisms will be explored more fully in connection with the genetics of viruses.

RNA Viruses. The genome of most RNA viruses is single stranded and is found within the virion as a single molecule. Examples of such types are tobacco mosaic virus, the RNA bacteriophages of *E. coli,* and polio virus. In other single-stranded RNA viruses the genome consists of separate pieces of different lengths, examples being influenza virus, Rous sarcoma, and murine leukemia viruses. In still other forms, such as the reoviruses of mammals, the genome is double-stranded and divided into separate fragments. However, whether single- or double-stranded, the genophore is linear, no circular forms having as yet been found.

Genophore reproduction is accomplished by a replicating enzyme, called an *RNA-dependent RNA polymerase,* which is often contained within the virion itself and enters the host along with the viral RNA. The single-stranded RNA bacteriophages that infect *E. coli,* such as MS2 or R17, do not carry their own enzyme, but instead direct the synthesis of a replicase immediately upon entrance into the host cell. This enzyme appears to be "hybrid," in that one subunit is contributed by the parasite while the other subunits are appropriated from an enzyme of the host. By means of this enzyme, the single, infectious plus strand of the virus generates a double-stranded RNA intermediate from which new plus strands arise. Thse are then encapsulated into new virus particles.

The RNA tumor viruses, such as the mouse leukemia virus, generate a DNA intermediate. Such viruses contain an *RNA-dependent DNA polymerase,* also called a *reverse transcriptase*. This enzyme catalyzes the formation of RNA-DNA hybrid molecules which in turn give rise to DNA-DNA molecules. The double-stranded DNA molecules can then be inserted into the genome of the host where they replicate in synchrony with the host chromosome and are passed on to daughter cells by mitosis. Such viruses may cause the malignant transformation of cells and they have been associated with the origin of a number of different types of cancer in domestic animals, rats, mice, and humans (see Chapter 20).

Prokaryotes

The prokaryotes include the blue-green algae and the bacteria; the best known representative of these is the bacterium *Escherichia coli*. The genophore of *E. coli* is a single, closed, double-stranded ring of DNA which, with the electron microscope, appears as a fibrous mass lying in direct contact with the cytoplasm of the cell, no nuclear envelope being present (Figure 11–26). The entire genophore constitutes a single replicon. Semiconservative replication begins at a specific initiation site which is bound to an extension of the cell membrane called a *mesosome*. Cairns originally described the replication pattern as *unidirectional synthesis,* that is, a single growing point proceeds in a one-way direction around the circle (Figures 11–27 and 11–28). However, more recent studies indicate that bidirectional synthesis is probably the more common pattern. In bidirectional synthesis two growing points arise at the initiation site and proceed in opposite directions, eventually to fuse with one another (Figure 11–28). The rate of movement of a replication fork in *E. coli* is estimated to be around 20–30 μm per minute and the duplication of the entire genophore requires about 40 minutes.

Synthesis of proteins needed for cell division proceeds concurrently. The separate daughter genophores become attached to either side of the mesosome which grows and divides to form new cell membranes and a septum which partitions the original cell into two daughter cells. Each daughter cell so formed contains a complete genophore. The regular process of mitosis with accompanying spindle formation is absent in prokaryotes.

In *E. coli* a second round of DNA synthesis can be initiated before the first is completed and, as a result, additional copies of some genes may be present, depending on the state and progress of replication.

Figure 11–26. A prokaryotic cell, the bacterium *Escherichia coli.* The genophore (G) occupies the center of the cell and contains the hereditary material, DNA. Plasma membrane (PM) lies beneath the cell wall (WC). Note the absence of nuclear membrane, cell organelles, and endoplasmic reticulum. (84,000×). [Dr. G. Cohen-Bazaire]

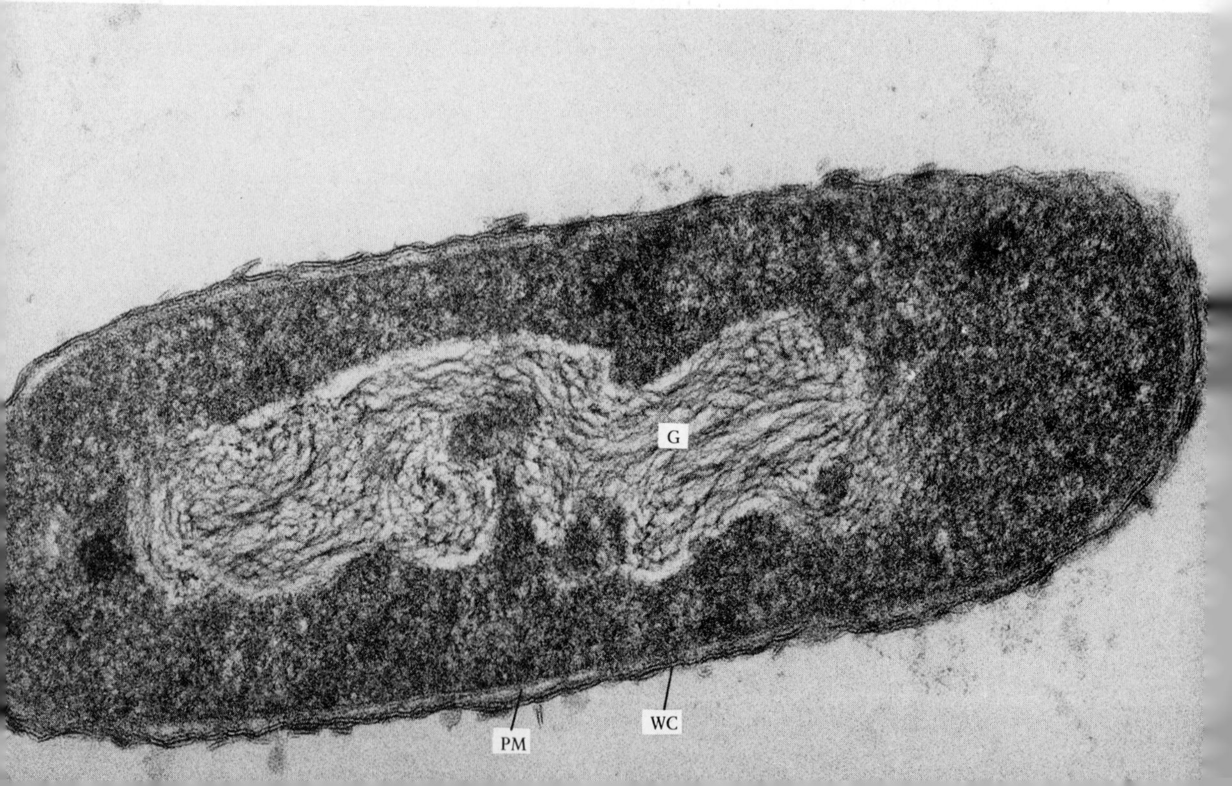

G
WC
PM

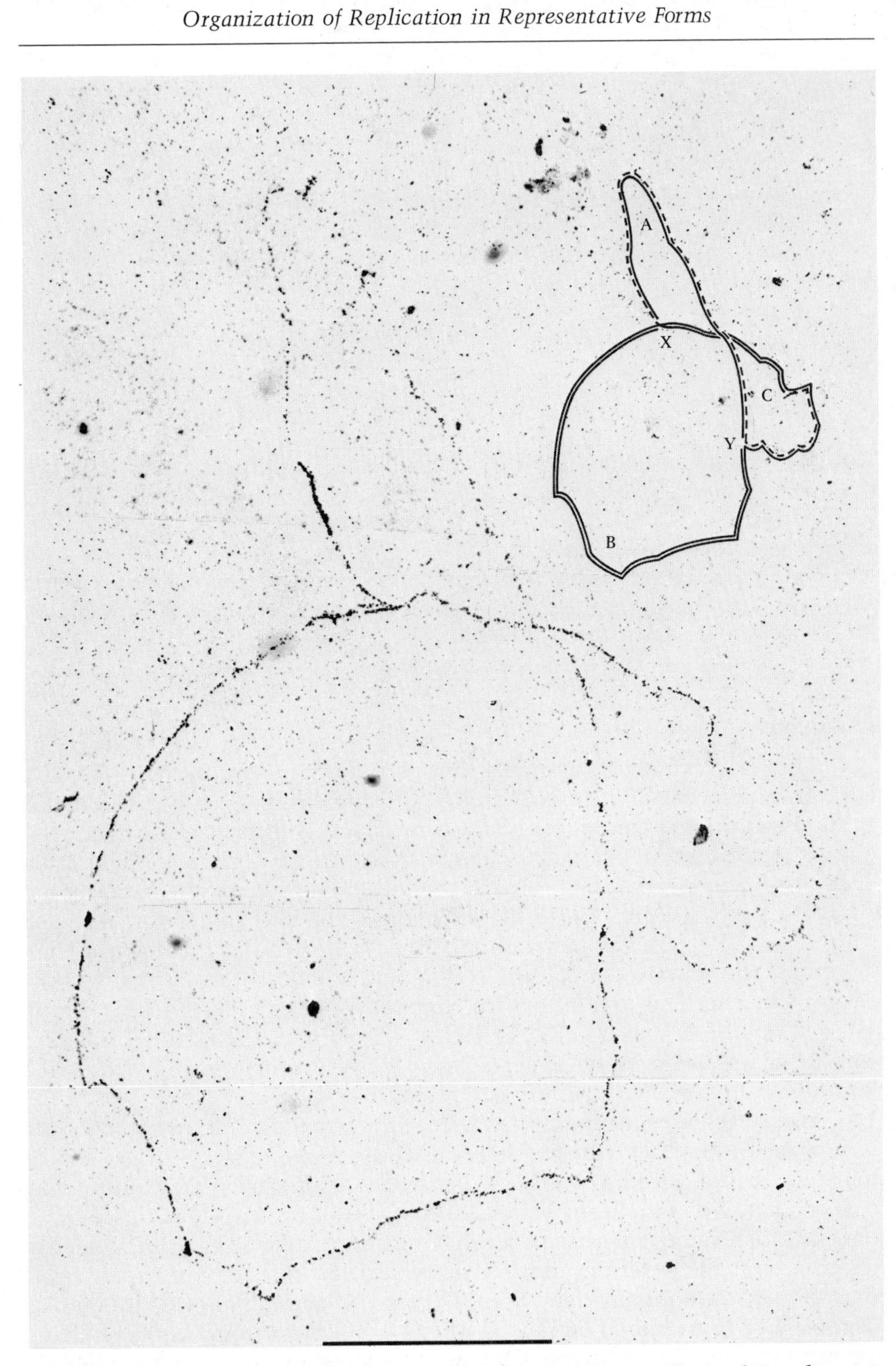

Figure 11–27. Autoradiograph of a replicating circular genophore of *E. coli.* strain K-12 Hfr labeled with tritiated thymidine. Inset is a diagram of the same structure divided into sections A, B, and C, and forks X and Y. The count of silver grains indicates unidirectional replication is occurring in this genophore, with X as the initiation site and Y the replication fork. [From J. Cairns, *Cold Spring Harbor Sympos. Quant. Biol., 28,* 43–46, 1963.]

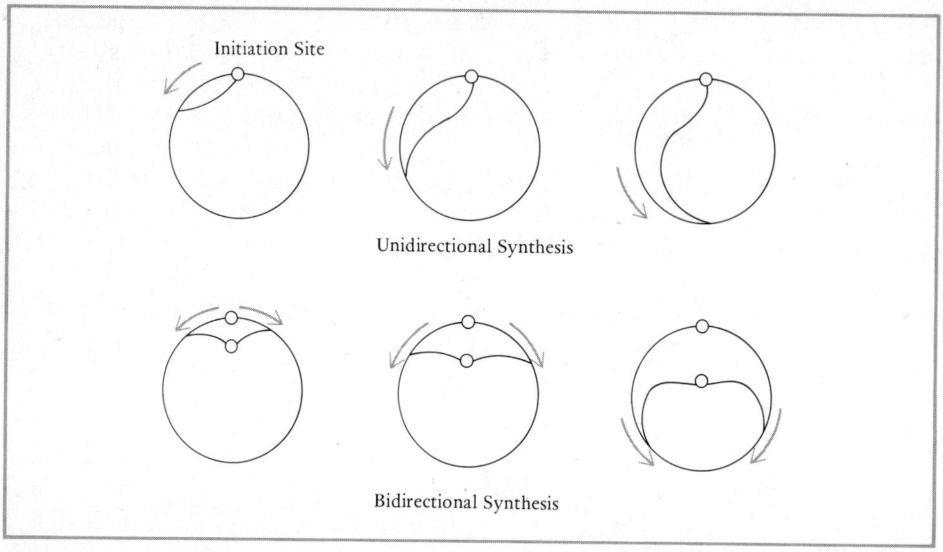

Figure 11–28. Unidirectional and bidirectional synthesis of the genophore of *E. coli.* Arrows indicate the direction of synthesis.

Eukaryotes

The early cytologists assumed that chromosome duplication took place during the prophase of mitosis because the chromosomes first become visible as double entities at this stage. This assumption was incorrect, however, for a number of experiments have since demonstrated that chromosome replication occurs during interphase. For example, by measuring ultraviolet absorbance at a wavelength of 260 nm, the DNA content of a cell nucleus can be quantitated. The technique is called microspectrophotometry. When this method is used to determine the amount of nuclear DNA present at different times in the life of an individual cell, it can be shown that the DNA content doubles during interphase prior to the onset of prophase. In addition, the results of studies employing autoradiography indicate that labeled precursors are incorporated into DNA only during the interphase period.

Precise analyses carried out by A. Howard and S. R. Pelc and others have shown that interphase can be subdivided into three stages. At the end of telophase the cell enters a phase called G_1 (gap 1). This stage is then followed by a *synthetic* or *S phase,* during which DNA synthesis occurs. Upon the completion of chromosome replication the cell enters a phase called G_2 (gap 2) which is terminated at the onset of mitosis. The typical life cycle of a eukaryotic cell thus consists of interphase and the G_1, S, and G_2 stages, followed by mitosis (Figure 11–29). Although protein synthesis can occur throughout interphase, chromosome replication is restricted to the S phase preceding mitosis.

J. A. Huberman and A. D. Riggs, H. G. Callan, and others have analyzed the pattern of chromosome replication in a variety of higher organisms with the following results. The lengthy eukaryotic chromosome appears to be linearly organized as a series of replicating units, called *replicons,* each containing at or near its midpoint an initiation site for DNA synthesis. It is postulated that attachment of the polymerizing enzyme occurs at the initiation site and is followed by strand separation and the formation of two replication forks. These travel in opposite directions from the initiation site so that the overall pattern of DNA synthesis within the replicon is *bidirectional* (Figure 11–30).

Presumably, a single-strand break brought about by endonuclease action at the initiation site or at the terminal ends of the replicon or both permits the rotation required for separation of the two strands of the double helix. The rate at which the replication forks proceed has been variously estimated at anywhere from 0.5 μm per minute to a maximum of around 2.5 μm per minute. Assuming that all replication forks maintain a constant rate, the duplication of the entire genome within the time span of an S phase requires the presence of

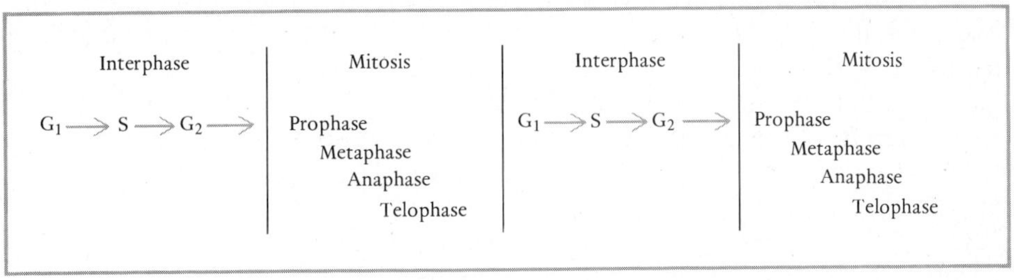

Interphase	Mitosis	Interphase	Mitosis
$G_1 \rightarrow S \rightarrow G_2 \rightarrow$	Prophase Metaphase Anaphase Telophase	$G_1 \rightarrow S \rightarrow G_2 \rightarrow$	Prophase Metaphase Anaphase Telophase

Figure 11–29. The eukaryotic cell cycle.

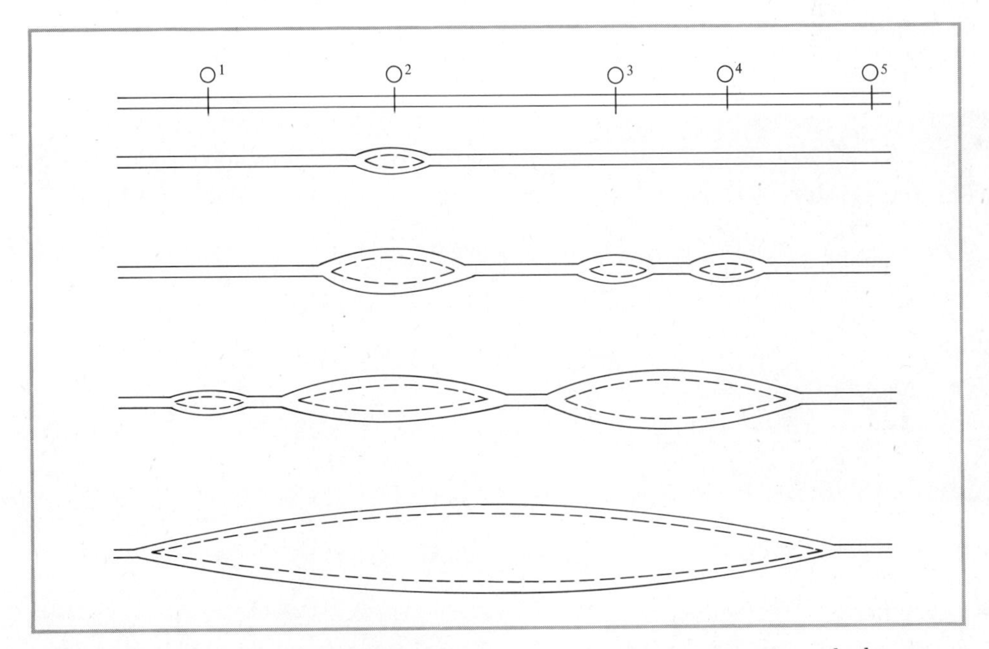

Figure 11–30. Replication of DNA in a eukaryotic chromosome through the operation of tandemly arranged units of replication, with initiation points at o^1 to o^5. Progressive stages in replication are shown. [From H. G. Callan, *Brit. Med. Bull., 29,* 192–195, 1973.]

numerous replicons and initiation sites within each chromosome, or a total of several thousand for a whole complement.

The duration of the S phase, and therefore the time allotted for chromosome replication, varies. In differentiated cells DNA synthesis may extend over a period of many hours, while in early embryos the entire cell cycle from interphase through mitosis may occupy only one hour or less. As an illustration, the nuclei of *Drosophila* embryos typically undergo a mitotic division every 9 to 10 minutes, a rate much faster than that observed in cells of later developmental stages or in those maintained in tissue culture. In view of such differences in timing, it has been suggested that the speed with which the chromosome complement is replicated is dependent upon the number of initiation sites active at any given time, rather than on an increase in the rate of movement of replication forks. Evidence supporting this proposal has recently been obtained for *Drosophila* embryos by A. B. Blumenthal and associates (Figure 11–31).

In differentiated cells it has also been observed through autoradiography that different chromosomes or regions of chromosomes engage in DNA synthesis according to a characteristic temporal pattern such that the replication of one region may be completed before the replication of an adjacent region begins. Such temporal patterns have been used as criteria to identify specific chromosomes. For example, the condensed X chromosome of mammals is

Figure 11–31. Fragment of a replicating chromosome isolated from cleavage nuclei of *Drosophila* embryos. This fragment is about 119,000 base pairs in length and contains 23 replication bubbles or "eye forms." Insert is a diagrammatic representation of the molecule (kb = kilobase or 1000 base pairs). [From H. J. Kriegstein and D. S. Hogness, *Proc. Nat. Acad. Sci., 71*, 135–139, 1974.]

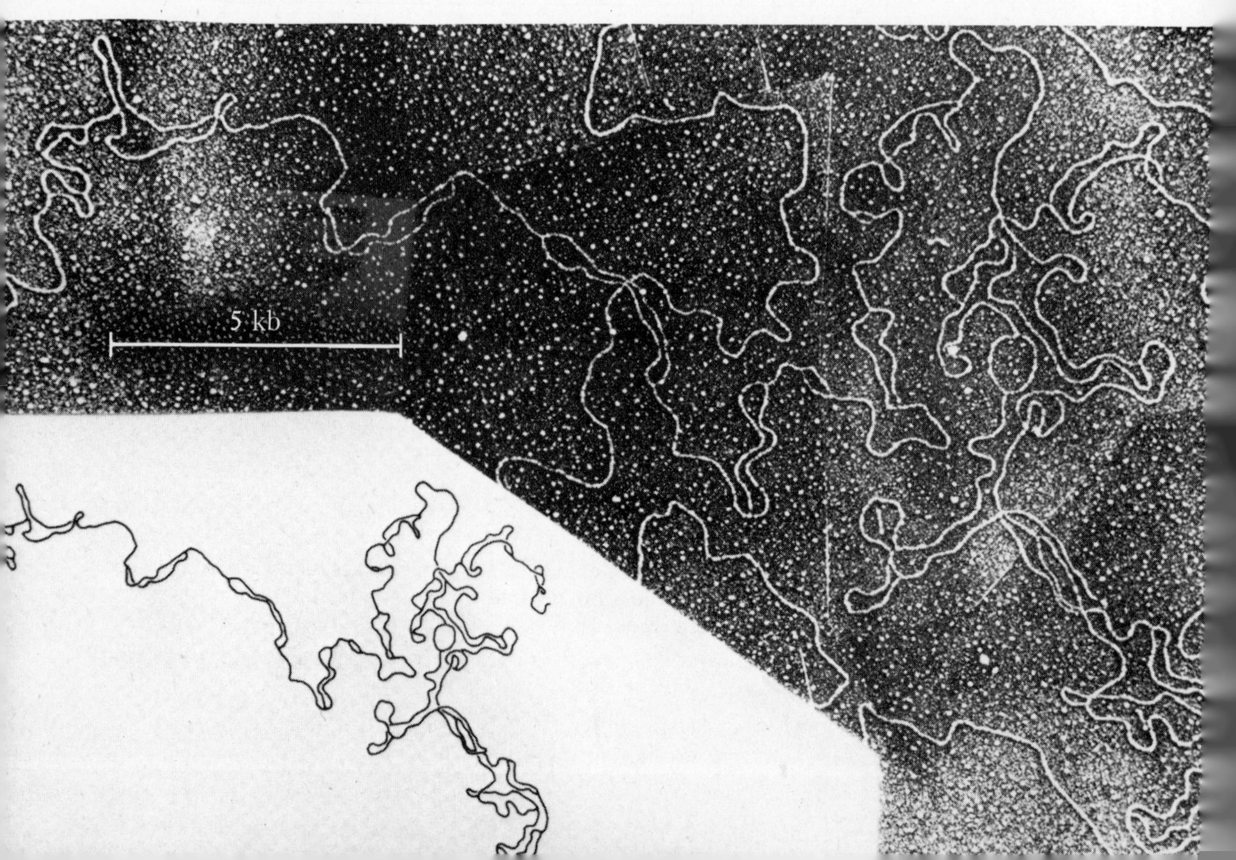

typically late replicating and can be recognized by this characteristic. In *Drosophila* salivary gland cells, mitosis is absent and cyclic DNA synthesis serves to double the DNA content with each round of replication so that polytene chromosomes containing over 1000 chromatids may be found in the largest of these cells (see Figure 9–3). Careful quantitative studies by G. Rudkin have demonstrated that this geometric increase in DNA content does not apply to centromeric regions which remain underreplicated, as compared to the rest of a chromosome. Presumably, these regions perform no function in salivary gland cells and are not duplicated for reasons of economy. In general, the varying patterns of DNA synthesis characteristically evident in eukaryotic chromosomes suggest the regular imposition of precise controls upon the process of chromosome duplication.

As a final consideration, the DNA of eukaryotic chromosomes is found intimately complexed with proteins, both basic and acidic. Of these, the basic proteins, or *histones,* have been implicated as structural elements which may influence or control the degree of coiling or condensation of the DNA fiber. In this regard, it is significant that the synthesis of histones is coordinated with DNA replication, for it also occurs during the S phase. Thus, as chromosome duplication proceeds, the newly synthesized DNA can be immediately complexed with concurrently available histone proteins.

DNA AND THE GENE

Our review of nucleic acids and the various modes of genome duplication should not be allowed to obscure the overall significance of these molecules to heredity. The structure of DNA, along with its evident potential for self-reproduction, permitted a concept of the nature of the gene that was entirely consonant with that derived from the study of mutant polypeptides. It will be recalled that by the mid-1950s the role of the gene in initiating the phenotype had been defined as the specification of the amino acid sequence of a polypeptide. Recognition that mutations were expressed as amino acid substitutions or changes at different sites within a protein had also led to the realization that the gene itself was mutable at many different internal sites. The structure of DNA, with its numerous linearly arranged subunit nucleotides, provided a physical basis for this concept, for it permitted visualization of the gene as a sequence of nucleotides whose order was directly related to the order of the amino acids of a corresponding polypeptide. Mutation could be viewed as a gain or loss of nucleotides or as a change in the sequence of the particular nucleotides that comprised a gene. Proposals relating the base sequence of DNA with amino acid sequence were brought forth by A. L. Dounce in 1952 and by G. Gamow in 1954, and it was suggested that the four bases found in DNA might constitute a code that could be translated by the cell into protein. That such a code exists, that it is essentially universal, and that it does indeed specify amino acid sequence has since been demonstrated. Proof for these postulates has been obtained in large part through the study of viruses and microorganisms whose genetic mechanisms must therefore be our next consideration.

APPENDIX
ANALYSIS OF DNA

Besides X-ray diffraction which yields information on the geometry of the double helix, a number of other methods for analyzing DNA have proved extremely valuable. Some of these are outlined below.

Nearest Neighbor Sequence Analysis

This technique has been utilized to demonstrate (1) that the DNA synthesized *in vitro* has the same base sequence as the template used for its synthesis and (2) that the two strands of the double helix are of opposite polarity, that is, antiparallel.

The method involves analysis of products derived from the enzymatic degradation of DNA. The degradative enzymes employed are of wide occurrence and can be broadly categorized as phosphatases or phosphodiesterases because they attack the phosphodiester bond. Some begin destruction at a free end of the helix and are thus called *exonucleases*. Others produce single and eventually double strand cuts in the middle of a DNA chain and are thus termed *endonucleases*. If their substrate is DNA, they are *deoxyribonucleases* or *DNAses;* if their substrate is RNA, they are *ribonucleases* or *RNAses*.

The action of these enzymes is very specific. Some, such as snake venom phosphodiesterase, cleave the phosphodiester bond between the phosphate group and the 3′ carbon to yield 5′ monophosphate products, that is, nucleotides with the phosphate group still attached to the 5′ carbon of the sugar. Other enzymes, such as *Micrococcus* or cattle spleen DNAse, cleave the phosphodiester bond between the phosphate group and the 5′ carbon to yield 3′ monophosphate products, that is, nucleotides with a phosphate group attached to the 3′ carbon instead of the 5′ carbon (see Figure 11–32). It is the latter type of enzyme that is used in nearest neighbor sequence analysis.

Note that as a result of these enzymes' action, the products now contain a phosphate group that was originally part of an immediate neighboring nucleotide. This transfer of a phosphate group from one neighbor to the next is utilized as follows. The 5′ phosphorus atom of one of the four nucleotides required for DNA synthesis is labeled with ^{32}P and used along with the other three unlabeled precursors, DNA polymerase, and a suitable template for the *in vitro* synthesis of DNA. The DNA produced in this reaction is then degraded with DNAse and the 3′ monophosphates are recovered and analyzed for the association of the ^{32}P label with each kind of base. If the label was originally incorporated into adenine nucleotide (5′-dATP), the frequency with which the label is associated with cytosine, or thymine, or guanine in the monophosphate products, indicates the frequency with which A and C, A and T, and A and G were immediate neighbors within the original linear polymer. The experiment is then repeated with the ^{32}P label incorporated, in turn, into each of the other precursors and the frequency of all 16 possible paired associations determined.

This technique was employed by Kornberg and associates to demonstrate that the product of the *in vitro* DNA synthesis catalyzed by polymerase I matched the template used. In these experiments DNA extracted from various sources was used as the template for *in vitro* synthesis, and the product of this reaction was then employed as a template to produce a second generation of DNA molecules. Nearest neighbor sequence analysis performed on both the first and second generation DNA indicated that the frequency with which the different bases were neighbors to one another was the same in each case and therefore that the product of the second generation was identical to the original template of the first generation.

The antiparallel orientation of the two strands of the double helix can also be demonstrated by this method. Recall that base pairing between the two strands occurs between A and T and between G and C. Assuming a nonrandom sequence of bases, if the two strands are of the same polarity, the frequency with which ^{32}P will be transferred by the action of DNAse from A to C will be the same as the frequency with which ^{32}P is transferred from T to G, because transfer occurs in the same direction in each strand. However, if the two strands are antiparallel, the ^{32}P will be transferred in opposite directions in the

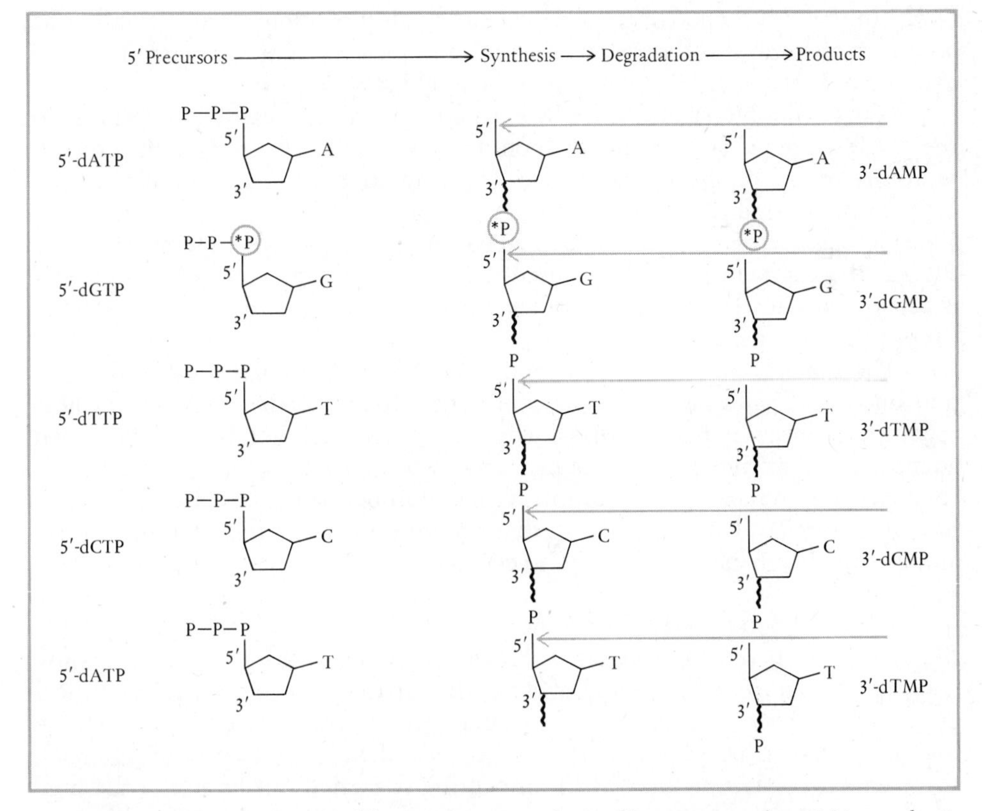

Figure 11-32. Nearest neighbor sequence analysis. 5′ precursors for DNA synthesis are given in column to left. The 5′ phosphorus atom of 5′-dGTP is labeled with ^{32}P, as indicated by the circled *P. Phosphodiester bonds between successive nucleotides formed during 5′ to 3′ synthesis are shown by wavy lines within the DNA chain. Hydrolysis by *Micrococcus* DNAse (large arrows) yields the 3′ monophosphate products shown in the column to the right. Note that the ^{32}P label, originally incorporated in the guanine-containing (G) precursor, is now present within the adenine-containing product (3′-deoxyadenosine monophosphate, 3′-dAMP).

two strands, and therefore the frequency of A and C as neighbors will be different from the frequency with which T and G occur as neighbors. By applying these criteria to the frequencies with which all paired combinations of bases were present as neighbors, J. Josse, A. D. Kaiser, and Kornberg were able to demonstrate that the two strands of the DNA derived from a variety of sources, from bacteria to mammals, were indeed antiparallel.

Denaturation, Renaturation, and Hybridization

Purines and pyrimidines absorb light in the near ultraviolet range of the spectrum and this property is used in the analysis of DNA samples. The wavelength usually chosen is 260 nm. When nucleotides are free in solution, they absorb strongly. However, when they are incorporated into the organized structure of the double helix, they become *hypochromic,* that is, their absorbance of light is sharply decreased, because they are held firmly in position by hydrogen bonding and Van der Waals forces. Disruption of the double helix so that it becomes single stranded, either regionally or totally, results in an increase in the absorbance of light which can readily be monitored. The amount of this increased absorbance, or *hyperchromicity,* is directly related to the amount of DNA that has been rendered single stranded.

Native, double-stranded DNA extracted from cells can be *denatured* to the single-stranded state by heat. Upon separation of the base pairs, the ordered structure of the double helix is lost. The transition from the double- to the single-strand condition occurs quite suddenly over a temperature range of about 5° (approximately 85° to 90°C). The midpoint of this temperature range, the melting temperature or T_m, is used to characterize a given DNA sample and is correlated with the sharp rise in absorbance which occurs upon the separation of the strands.

The exact melting temperature of any DNA sample depends on the proportion of G-C and A-T base pairs present. Guanine and cytosine are held together by three hydrogen bonds, while only two such bonds occur between adenine and thymine. Since more heat energy is required to disrupt three bonds than two, a DNA sample containing a higher proportion of G-C base pairs will have a higher T_m. Consequently, a determination of the melting temperature along with absorbance can be used to estimate the G-C versus the A-T content of a DNA sample.

Denaturation is reversible. If the sample is heated and then cooled rapidly, the strands will remain separated. However, under conditions of very slow cooling over an extended period of time, the separated strands of DNA anneal or reunite to form once more the original two-stranded structure. Thus, the double helix can be taken apart, the process of denaturation, and put back together again, *renaturation*. If the annealing is carried out properly, the final double-stranded product will be almost perfectly matched with respect to base pairs and will exhibit a melting temperature and ultraviolet absorbance almost identical to that of the original sample (Figure 11–33). Any significant mismatching of base pairs or the presence of single-stranded regions will be reflected in changes in absorbance and a lower melting temperature. Thus, the degree of precision of base pairing and the consequent extent of double-strandedness of any annealed sample can be assessed quite rigorously.

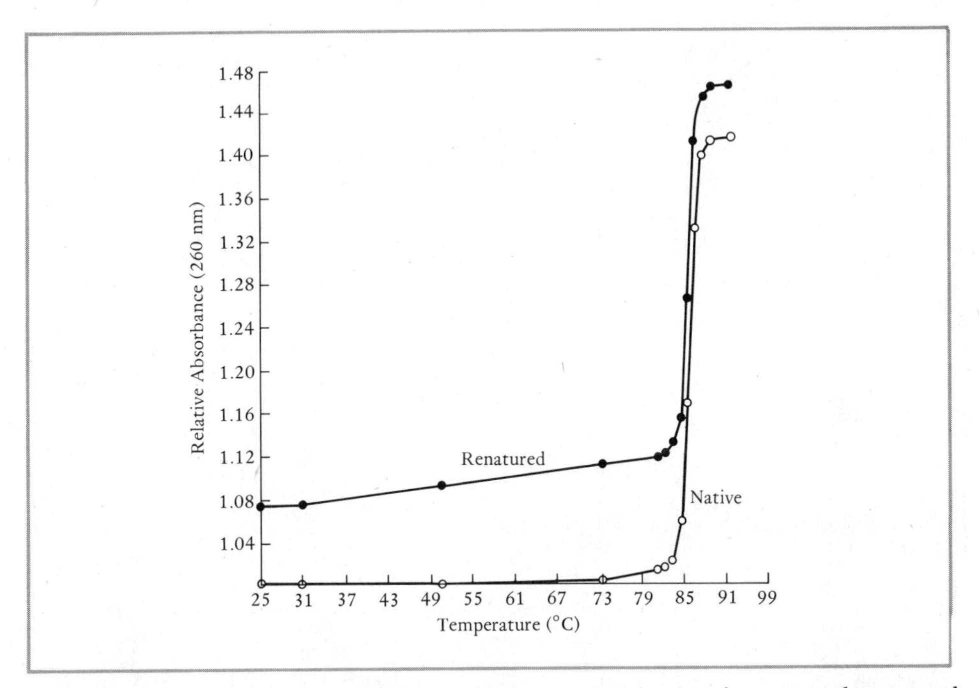

Figure 11–33. Effect of temperature on relative absorbance of native and renatured DNA. The transition from the double to the single-strand condition occurs over a narrow temperature range. Note the similarity in the thermal profiles of native DNA and annealed (renatured) DNA. [From J. Marmur and P. Doty, *J. Mol. Biol.*, 3, 585–594, 1961.]

The technique of annealing can also be used in attempts to unite into hybrid molecules nucleic acids derived from different sources or species. This procedure, called *hybridization,* allows estimates of the degree of similarity or complementarity existing between the sequence of base pairs in the DNA strands involved. As might be expected, DNA's from closely related species hybridize to a greater extent than do those from distantly related species. In any hybridization experiment both kinds of DNA are first denatured to the single-stranded state. Usually one or the other has previously been labeled either with a radioactive isotope or with a heavy isotope whose presence confers a different and identifiable density. Labeling is achieved by growing cells for a short time in a medium containing precursors for DNA synthesis, such as thymine nucleotide, which contains a radioactive isotope, for example, ^{3}H, ^{14}C, or ^{32}P. Heavy nitrogen, ^{15}N, is also used. The presence of these unusual atoms in one of the DNA's used for annealing makes possible the identification of hybrid molecules, since radioactivity can be assayed by appropriate detectors and differences in density can be recognized by use of the ultracentrifuge.

The extent of annealing can also be visualized with the electron microscope. Figure 11–34 shows the results of hybridization between the genomes of two *E. coli* phages, T3 and T7. The DNA genomes of each phage were first rendered single stranded to permit annealing to occur between the complementary strands of the two types. The result is a T3/T7 *heteroduplex* which

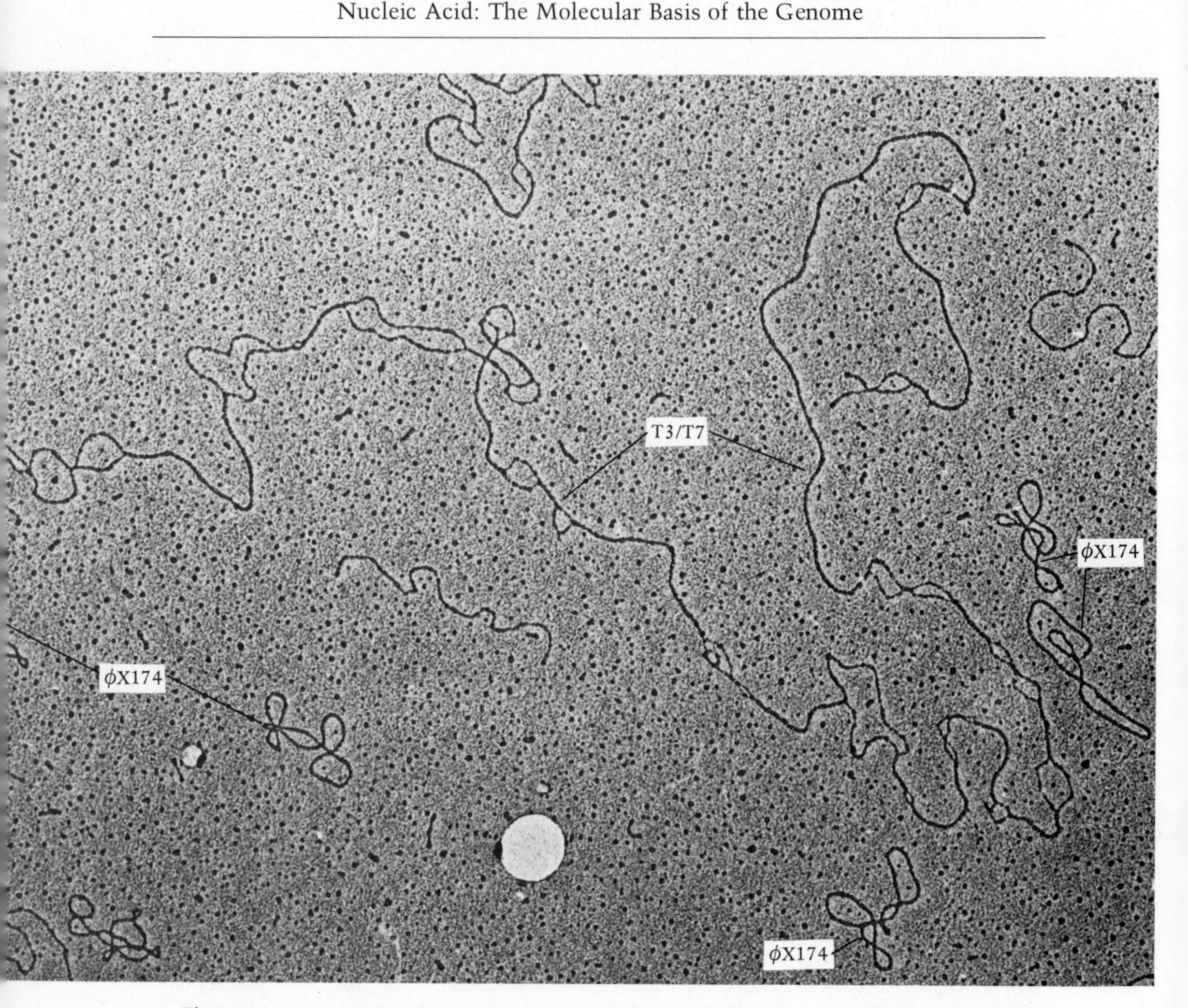

Figure 11–34. T3/T7 heteroduplex formed by annealing complementary single strands derived from each phage genome. The numerous single-stranded bubbles are regions of nonhomology where the base sequences of the two genomes do not match. The small circles are genomes of phage ϕX174, whose known length (1.7 mμ) provides a reference standard against which the length of the heteroduplex can be measured. [From R. W. Davis and R. W. Hyman, *J. Mol. Biol.*, *62*, 287–301, 1971.]

exhibits numerous bubblelike areas where the two strands do not match and therefore remain separate. This technique is most usefully applied to viruses whose small genomes can be visualized *in toto*. The method is now being used for the physical mapping of phage genophores (see Chapter 12).

It should also be noted that a variety of RNA molecules are synthesized through the action of RNA polymerase with DNA being used as the template for their formation (see Chapter 15). Since such RNA molecules contain a base sequence complementary to the region of the DNA from which they were copied, they too can be hybridized with single-stranded DNA. Such RNA-DNA hybridization has been successfully used to localize sites within the chromosomes where particular kinds of RNA are synthesized (see Chapter 19).

PROBLEMS

11–1. Suppose the following base sequence was found in a 20 base DNA polymer: 3′ A-T-T-C-G-A-C-C-T-T-A-T-T-A-C-T-G-C-A-C 5′.

a. What would be the first 5 bases of the 3′ end of the complementary strand?
b. What would be the first 10 bases of the 5′ end of the complementary strand?
c. Assuming the presence of complementary strands, what is the percent composition of the polymer with respect to A-T base pairs and with respect to G-C base pairs?
d. Using this double stranded polymer, diagram the process of semiconservative replication.

11–2. Analysis of four double-stranded DNA samples yielded the following information.

(1) 15% cytosine
(2) 12% guanine
(3) 35% thymine
(4) 28% adenine

a. What would be the percentage of the other bases in each sample?
b. Could any of these samples have been obtained from the same organism? If so, which ones?

11–3. Four samples of nucleic acid were analyzed for the proportion of the different bases present with the following results.

(1) A = 30%, C = 30%, G = 20%, T = 20%
(2) A = 27.5%, C = 22.5%, G = 22.5%, T = 27.5%
(3) A = 18%, C = 32%, U = 32%, G = 18%
(4) A = 18%, C = 32%, U = 18%, G = 32%

Which of these samples were DNA and which were RNA? Which were single and which were double stranded?

11–4. Ratios of the bases present in different samples of nucleic acid yielded the following results.

(1) $(A + C)/(T + G) = 1$
(2) $(A + C)/(U + G) = 0.8$
(3) $(A + G)/(T + C) = 1.5$

Which were RNA and which were DNA? Which were single and which double stranded?

11–5. If one DNA sample had a melting temperature of 85.5°C and another showed a melting temperature of 88°C, what might you conclude concerning the base composition of the two samples?

11–6. Assume an average sized gene consisted of a linear sequence of 1000 bases and there were 1000 genes in a chromosome.

a. How many bases would such a chromosome contain in each strand of the double helix?
b. If 10 nucleotides = 34 Å, how long would this chromosome be in millimeters?

11–7. The average length of the double helix in a human chromosome is 3.8 cm.

a. Approximately how many base pairs would be present in such a chromosome?
b. If all chromosomes of the human complement were 3.8 cm in length, approximately how many base pairs would be present in a sperm nucleus?

11–8. A species of insect has a diploid chromosome number of 2. A somatic cell from this insect was allowed to replicate its chromosomes in the presence of ^{3}H-thymidine and was then removed to unlabeled medium.

a. Upon mitotic division, how many chromosomes of a daughter cell would be labeled?

b. If one of these cells was allowed to undergo a second mitotic division, with respect to the presence of labeled or unlabeled chromosomes, what types of daughter cells could be formed?

11–9. Suppose human cells in tissue culture were grown in the presence of ^{3}H-thymidine during interphase (one replication cycle). Suppose also that the cells were washed to remove ^{3}H-thymidine and allowed to undergo mitosis and further growth and division in unlabeled medium.

a. As a result of the first mitotic division, what proportion of the chromosomes of the daughter cells would be labeled?

b. If one of these daughter cells underwent a second mitotic division, what is the probability that all of the chromosomes of a granddaughter cell would be labeled?

11–10. Assume a single stranded DNA polymer of the following sequence: 5′ C-A-T-A-C-G-T-G-C-A-A-A-C-T-T-G-T-C 3′ Suppose this polymer was replicated *in vitro* and among the precursors, only guanine nucleotides were labeled with ^{32}P at the 5′ position. If nearest neighbor sequence analysis were performed on the daughter polymer, what labeled monophosphates would be recovered? If cytosine, thymine, or adenine precursor nucleotides had been labeled, what labeled monophosphates would be recovered in each case?

11–11. A small, closed, single-stranded DNA ring contains the sequence illustrated. If, in separate experiments, each kind of precursor nucleotide is labeled with ^{32}P at the 5′ position, what nearest neighbor base pairs will be obtained in each experiment?

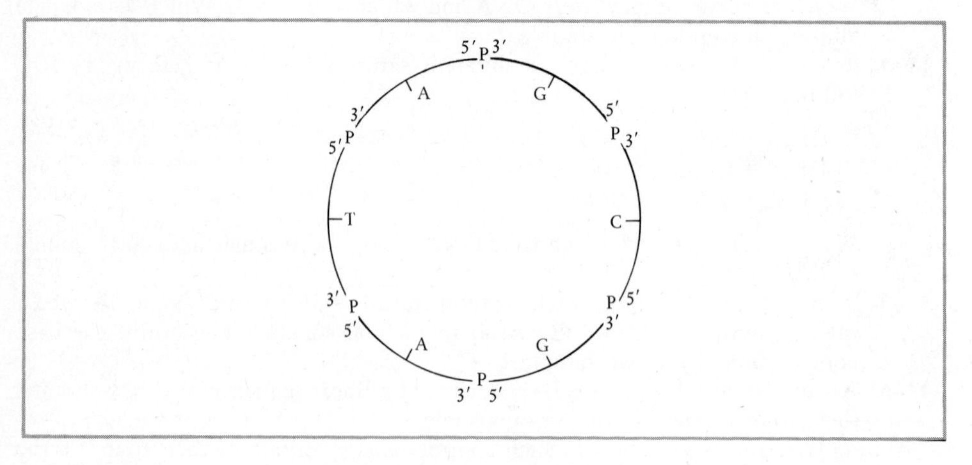

REFERENCES

AVERY, O. T., C. M. MACLEOD, AND M. MCCARTY, 1944. Studies on the chemical nature of the substance inducing transformation of pneumococcal types. Induction of transformation by a deoxyribonucleic acid fraction isolated from *Pneumococcus* Type III. *J. Exp. Med., 79,* 137.

CAIRNS, J., 1963. The chromosome of *Escherichia coli. Cold Spring Harbor Sympos. Quant. Biol., 28,* 43.

EDENBERG, H. J., AND J. A. HUBERMAN, 1975. Eukaryotic chromosome replication. *Ann. Rev. Genet., 9,* 245.

KLEINSCHMIDT, A. K., 1969. Chromosomes of viruses. In A. Lima-de-Faria (Ed.). *Handbook of Molecular Cytology*. North-Holland, Amsterdam.

KORNBERG, A., 1974. *DNA Synthesis.* Freeman, San Francisco.

WATSON, J. D., 1968. *The Double Helix*. Atheneum, New York.

WATSON, J. D., AND F. H. C. CRICK, 1953. A structure for deoxyribose nucleic acid. *Nature, 171,* 737.

WATSON, J. D., AND F. H. C. CRICK, 1953. Genetical implication of the structure of deoxyribonucleic acid. *Nature, 171,* 964.

Chapter 12

GENETIC MECHANISMS OF BACTERIOPHAGES

ALTHOUGH the field of medical virology was early established, studies on the genetics of viruses were not initiated until the 1930s. The report by M. J. Schlesinger in 1936 that bacteriophages contained only nucleic acid and protein, along with the discovery of a mutant phage by F. M. Burnet in the same year prompted the idea that these most simple forms might be useful tools for genetic research. This possibility was pursued by M. Delbrück, and subsequently by A. D. Hershey, R. Rotman, S. Luria, M. Chase, A. H. Doermann, and many others.

For genetic studies, some of the most useful viruses are the DNA-containing bacteriophages of *E. coli* (RNA phages have not been found to undergo recombination). These bacteriophages can be divided into two groups on the basis of their growth cycle: *virulent* phages that invariably kill the host and *temperate* phages that can either kill or enter into a stable relationship with the host. Temperate bacteriophages possess the ability to insert their genophore into that of a bacterial cell and also may transmit bacterial genes from one cell to another. Because of their differing capabilities, these two kinds of phages will be described separately.

LIFE CYCLE OF A VIRULENT BACTERIOPHAGE

Among the best known virulent phages are members of the T series which consists of seven different viruses all of which infect strain B of *E. coli* and are numbered as types 1 to 7. The T-even phages (T2, T4, T6) are identical in morphology, but the odd numbered types (T1, T3, T5, T7) differ somewhat in appearance among themselves as well as from those of the T-even group. Since the genetics of the T-even phages has been intensively studied, the following discussion will be restricted to these forms.

The morphology of a T-even phage is illustrated in Figures 11–23 and 12–1. The hexagonal head of the virus contains the double-stranded DNA genophore and is attached to a straight tail. The tail is a complex structure composed of a hollow protein core surrounded by a contractile protein sheath. At the distal end of the tail is a base plate to which six short spikes and six long tail fibers are attached.

Since the T-even phages are virulent, infection initiates a *lytic cycle* resulting in the destruction of the host. The time span, measured from the entrance of the parasite's DNA into a sensitive cell until the release or *burst* of phage progeny, occupies about 25 to 30 minutes and is known as the *latent period*.

The process of infection requires only a minute and is initiated by chance contact between the phage particle and a host cell, the ability of the parasite to

Figure 12–1. (a) Diagram of phage T2 showing morphological components and their arrangement. (b) Electron micrograph of an isolated T2 phage particle showing relations of filled head, contracted sheath, core and tail fibers. (570,000×). [(a) William Hayes, *The Genetics of Bacteria and Their Viruses,* 2nd ed., Blackwell Scientific Publications Ltd., Oxford, 1968). (b) courtesy of R. W. Horne. From S. Brenner et al., *J. Mol. Biol., 1,* 282–292, 1959.]

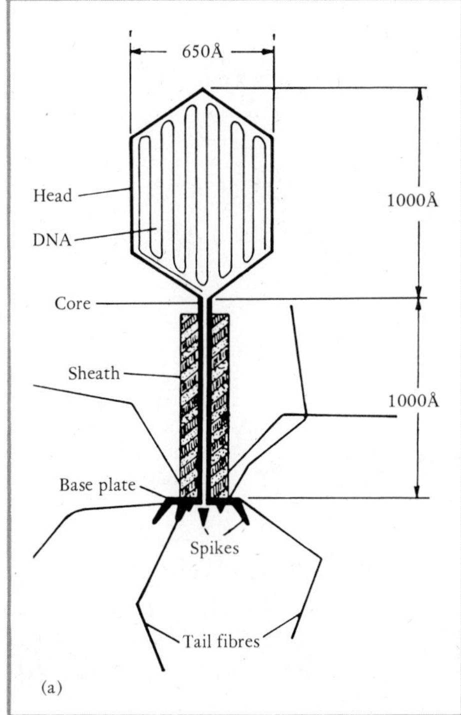

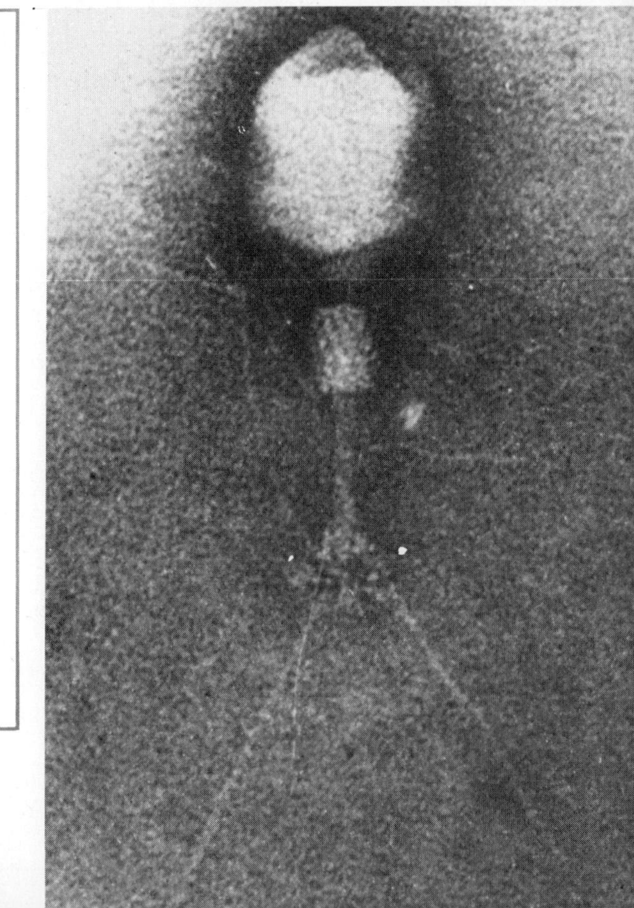

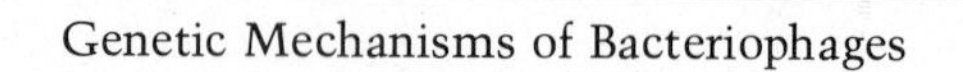

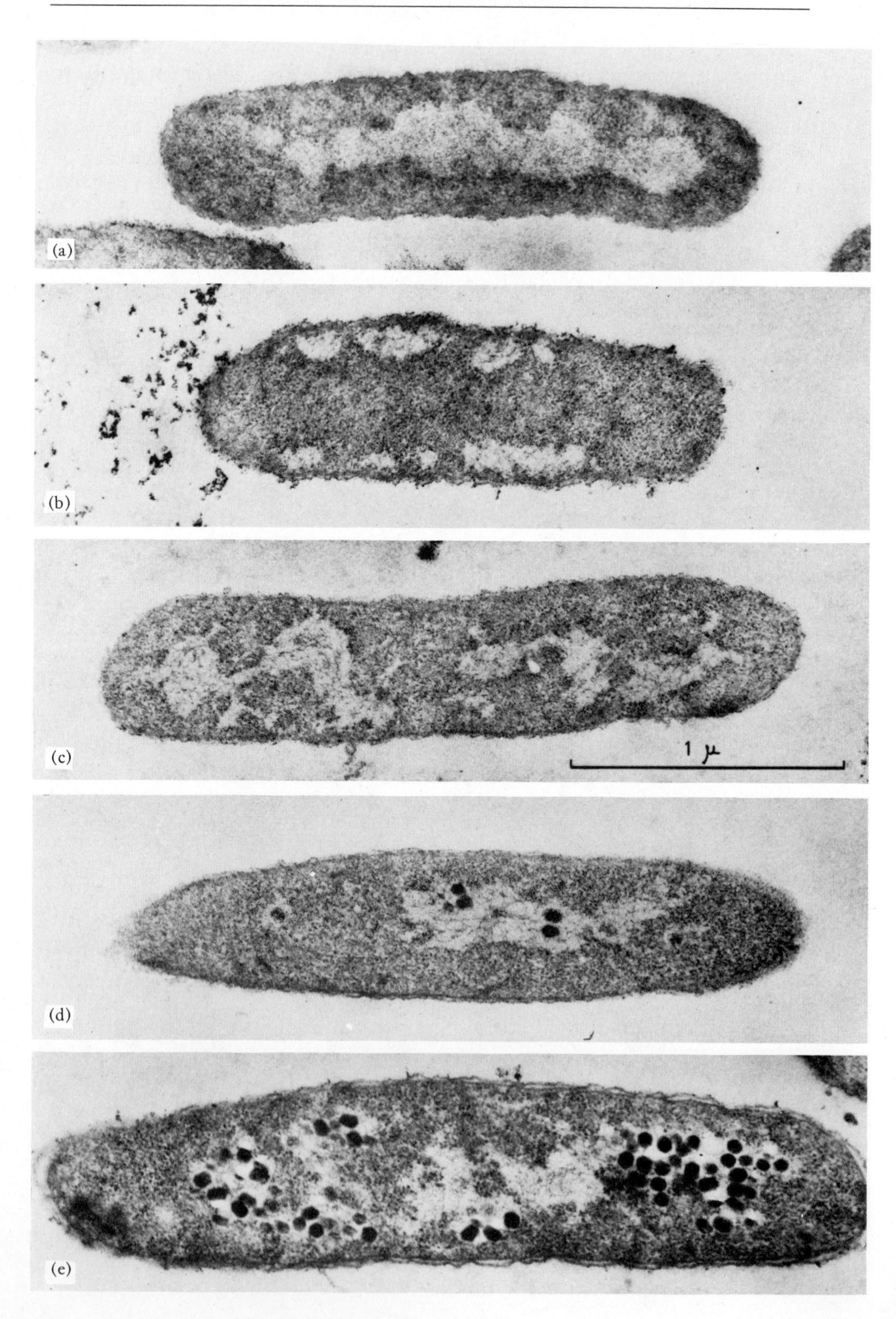
(a)
(b)
(c)
1 μ
(d)
(e)

adsorb to its host being governed by the antigenic properties of the bacterial cell wall. The adsorption of the infectious particle is immediately followed by the unwinding of the previously coiled tail fibers and their specific attachment to the cell wall. The hollow protein core then penetrates the cell wall and, aided by the contractile sheath, the DNA genome is injected into the host.

Within the cell the T-DNA utilizes the synthetic machinery of the host for the formation of a number of early appearing phage enzymes required for the synthesis of its own DNA as well as for the dissolution of the host genophore. Rapid exponential replication of the T-genome follows, and a pool of vegetative phage DNA molecules is thereby generated. Genome replication is accompanied by the synthesis of those proteins which form the head capsule and other structures of the mature virion. About half way through the cycle (around 15 minutes) the assembly of mature phage particles begins, and DNA molecules are withdrawn from the vegetative pool and packaged within the newly formed protein coats (Figure 12–2).

Within approximately 30 minutes from the start of the infection the phage DNA specifies the synthesis of a lysozymal enzyme which digests the cell wall permitting the escape of 100 or more mature phage progeny from their dead host. The DNA which fails to be encapsulated into a protein coat remains behind in the vegetative pool. These DNA molecules are noninfectious and therefore cannot be analyzed for their genetic constitution. The packaging of DNA into its protein coat appears to be a random process. If a bacterial cell is doubly infected with two different phages, for example, with both T2 and T4, the genophores of each type can be included in their own protein coat or in the coat of the other phage with equal probability. The phenomenon is called *phenotypic mixing*. It will be recalled that Fraenkel-Conrat and Singer demonstrated the hereditary role of RNA in tobacco mosaic virus with similarly "mixed" virions.

BACTERIOPHAGE PHENOTYPES

As in other organisms, the genetic analysis of bacteriophages depends on the identification of mutant phenotypes. Such phenotypes can be morphological, such as the presence of heads without tails or vice versa, the absence of tail fibers, or some other structural aberration. Many such mutants are *temperature-sensitive conditional lethals* in that normal progeny will be produced at a permissive temperature, but not at a different and restrictive temperature. Such mutants are particularly interesting from the standpoint of phage morphogenesis (see Chapter 18), but their structural anomalies can be

Figure 12–2. Development of T-even virus inside a bacterial host. (50,000×). (a) Uninfected *E. coli* cell; clear areas are genophores; more than one is present. (b) Dissolution of bacterial genophores is underway. (c) Bacterial genophores have disappeared; clear areas are pools of vegetative phage DNA. (d) Progeny virus particles are forming. (e) More than 50 mature phage particles are evident; cell lysis is imminent. (50,000×) [Original photographs by Drs. E. Boy de la Tour and E. Kellenberger.]

discerned only with the electron microscope. It is apparent, however, that at least 50 genes are involved in morphogenesis.

Other characters are those that define *plaque morphology* and *host range*. When *E. coli* is grown on nutrient agar in a petri plate, the cells multiply to form a uniform, opaque layer called a *lawn* which covers the surface of the agar. If a dilute suspension of phage is added to a lawn of sensitive bacteria, each virion attacks and lyses a host cell, releasing progeny which continue the lytic cycle by adsorption to immediately adjacent cells. After several such cycles, sufficient bacteria are destroyed to create a clear zone in the opaque lawn. Such a zone is called a *plaque,* and the size, shape, clarity or turbidity, as well as the nature of the border, whether sharply defined or irregular, constitute phenotypic characters by which different phages can be identified. The r mutants of T-even phages are good examples. The wild type phages, r^+ in genotype, form small plaques with irregular borders on *E. coli* strain B, as opposed to the large plaques with sharp borders produced by r mutants (Figure 12–3).

Figure 12–3. Types of plaques produced on a mixed lawn of *E. coli* strains B and B/2 cells as a result of infection by phage T2 particles $h^+ r$ and $h\ r^+$ in genotype. Parental type plaques are turbid, large (h^+r) and clear, small $(h\ r^+)$. The reciprocal recombinant plaques are turbid, small $(h^+ r^+)$ and clear, large $(h\ r)$. [Photographs by Maureen de Saxe and Janet Mitchell; From W. Hayes, *The Genetics of Bacteria and Their Viruses,* 2nd ed., Blackwell Scientific Publications Ltd., Oxford, 1968.]

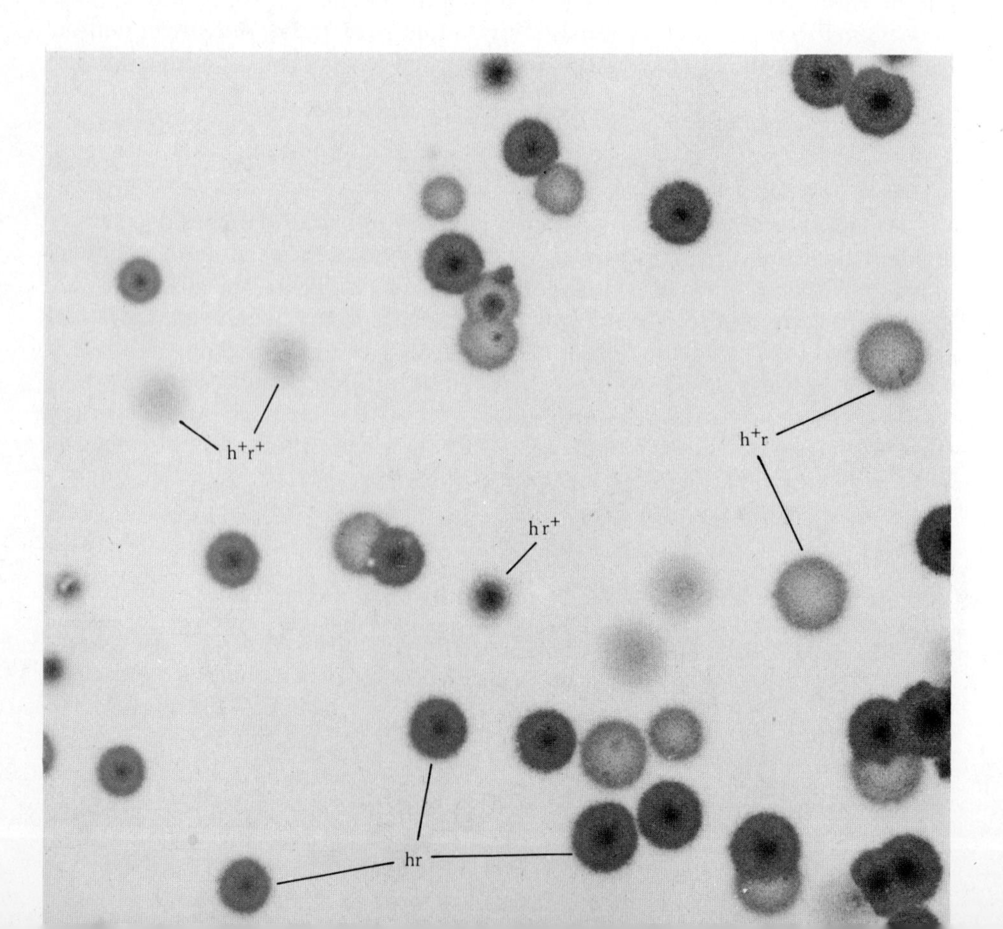

Host range mutants are defined by their ability to adsorb to different bacterial strains. As an illustration, the T2 mutant *h* can attack and form plaques on both strains B and B/2 of *E. coli,* but h^+ phage can produce plaques only on strain B. In a suspension of phage containing particles of both genotypes, the presence of each can be identified by plating on a lawn composed of a mixture of B and B/2 cells. The *h* mutant will lyse cells of both strains to form a clear plaque. The h^+ phage can lyse only B cells, and since the lawn is a mixture, the plaques produced by h^+ phage are turbid due to the presence of living B/2 cells within the plaque. Thus the two kinds of plaques so formed can be used to distinguish the two phage genotypes present.

COMPLEMENTATION TESTS IN BACTERIOPHAGES

Complementation tests are carried out with mixed infections using two different mutated phages, both of which can enter the host, but neither of which alone can produce progeny particles or plaques on a bacterial lawn. If the two phages are mutant for different functions, one will complement the other and plaques will be formed. On the other hand, if both phages are mutant for the same function, no plaques will be observed. An exception to this rule is the rare instance where recombination occurs between mutated sites present within allelic genes. As more fully explained in Chapter 14, such an event can produce a wild type recombinant capable of plaque formation. However, such recombinants arise so infrequently that they pose no difficulty in distinguishing complementing from noncomplementing genes.

RECOMBINATION IN BACTERIOPHAGES

Recombination in bacteriophages was first reported by Delbrück and W. T. Bailey and by Hershey in 1947, and a more detailed analysis employing the T2 plaque-type *r* mutants and the host range *h* mutants was carried out by Hershey and R. Rotman in 1948 and 1949. Their studies employed the single burst experiment devised by E. L. Ellis and Delbrück. In such an experiment a "cross" involves two (or more) types of parental phages which are used to infect a liquid broth culture of a bacterial strain sensitive to both, the concentration of phage particles being sufficient to assure a mixed double infection of the host cells. Any unadsorbed phage is then inactivated by the addition of specific antisera, and the culture is appropriately diluted to a series of tubes such that on the average, no more than one infected cell is present per tube. After lysis has occurred, the entire contents of each tube is plated on a bacterial lawn and scored for plaque type. By this means, the phenotypes of all progeny produced in a single host cell can be ascertained. Because of phenotypic mixing, it is necessary to test the progeny to make sure that the phenotype corresponds with the genotype. This is done by allowing progeny particles to infect a second host under conditions where the host cell receives no more than one phage. The host cells are then plated before lysis can occur. The phenotype of the second generation plaque then accurately reflects the genotype of the single infecting particle.

Hershey and Rotman employed parental phages of genotypes $h\ r^+$ and $h^+\ r$ to produce a mixed infection on strain B cells. Lysates were then plated on

TABLE 12–1
Types of Progeny Resulting from the T2 Phage Cross $h^+ r \times h\, r^+$

Plaque phenotype	Genotype	Classification
Clear, small	$h\, r^+$	parental
Turbid, large	$h^+ r$	parental
Turbid, small	$h^+ r^+$	recombinant
Clear, large	$h\, r$	recombinant

lawns of mixed B and B/2 bacteria, and the resulting plaques were of four different types (Figure 12–3). Two of these appeared with relatively high frequency and could be scored as parentals in that one type was clear, but small ($h\, r^+$), while the other was turbid, but large ($h^+ r$). The other two kinds of plaques appeared with lower frequency and exhibited recombinant phenotypes, one class being turbid, but small ($h^+ r^+$), the other class being clear, but large ($h\, r$) (Table 12–1).

The experiments performed by Hershey and Rotman thus provided evidence that recombination could occur in bacteriophages. Although the two reciprocal recombinant classes rarely, if ever, appeared with equal frequency in the progeny of any one single burst, the combined data from many infected cells indicated that approximately equal numbers of the two kinds of recombinants could be recovered. In addition, it was found that the number of recombinant progeny obtained from the cross $h^+ r^+ \times h\, r$ was the same as that recovered from the cross $h^+ r \times h\, r^+$. The frequency of exchange between the two loci could be calculated in the following manner: (Total recombinant plaques/total plaques scored) $\times$ 100 = percent recombination. On the assumption that the frequency of recombination between any two genes is proportional to the distance between them, it then became possible to establish a linear relationship between these and other phage mutations and thus to map the genophore. For example, Hershey and Rotman were able to assign rapid lysis (r) mutations to three different loci on the basis of the recombination frequencies observed with h, h^+, and other mutants.

Continuing studies of recombination have demonstrated that the T-even bacteriophages contain a single linkage group which is circular, rather than linear, in nature (Figure 12–4). Although it might be deduced that the genophore is structurally circular as well, such a conclusion does not necessarily follow. It is now known that the T-even viruses, such as T4, contain a linear, double-stranded DNA genophore approximately 55 μm in length. The vegetative pool produced by replication of the viral DNA contains numerous, large double-stranded molecules, each of which is much longer than the original genophore. These long molecules are called *concatamers* because they consist of the viral gene sequence repeated over and over again, the number of repeats being correlated with the length of the molecule.

When the proteins required for the head and other structures of the mature phage particle have been synthesized, the DNA of these long molecules is cleaved into smaller segments which are then packaged within the head capsule, each mature phage receiving a headful of DNA. The amount of DNA so included is longer than that which contains a complete set of genes and therefore some genes of the series are present more than once. For example, if the long molecule contained five genes repeated serially, such a molecule could be represented as 12345123451234512345123451234512345. If one headful of DNA were an amount equivalent to seven genes, then (starting from the left) the first headful would contain genes 1234512, the second headful the genes 3451234, the third headful the genes 5123451, and so on. In this way all mature phage particles contain the same genes, but the DNA molecules of the population begin and end at different points in different molecules, that is, they are

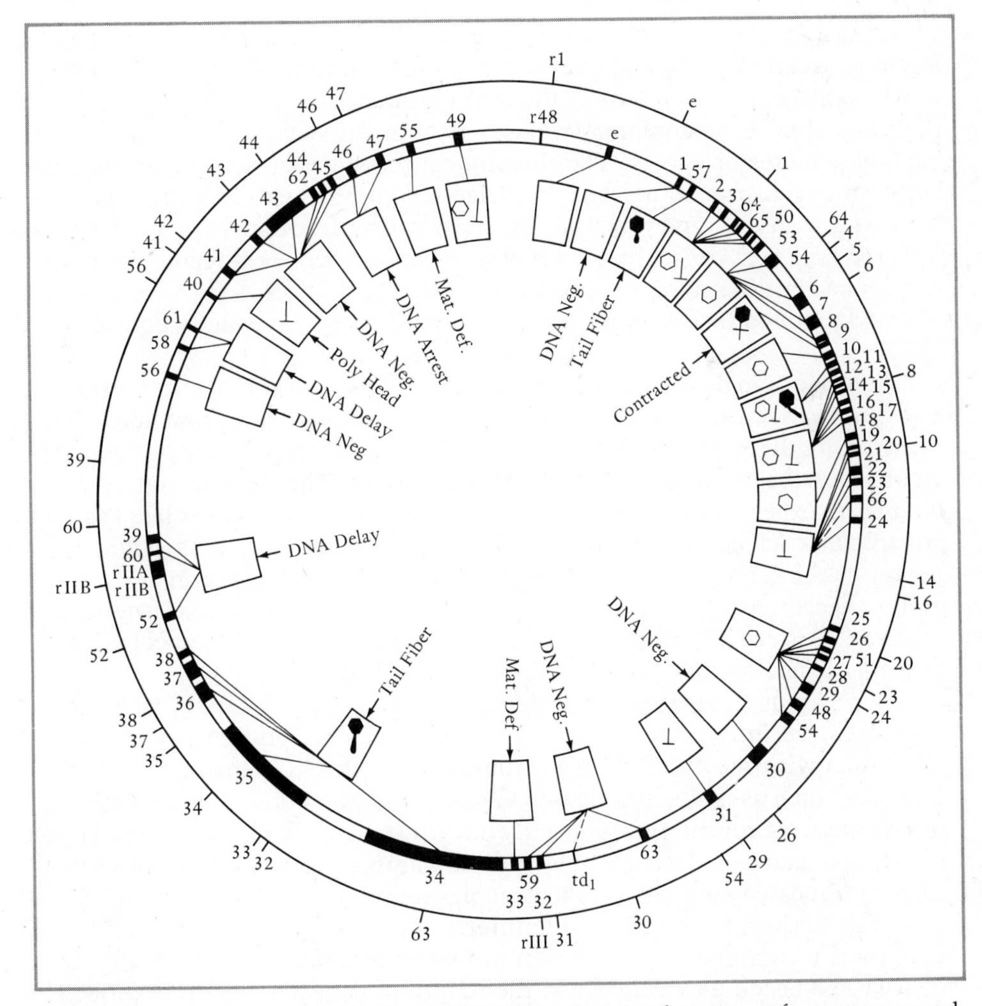

Figure 12–4. Genetic map of the T4 genophore showing functions of some genes and their linkage distances. Around 70 separate genes have been identified. The inner circle illustrates the functions of genes and indicates linkage distances between them; the outer circle represents an estimate of physical distance between markers. [Reprinted with permission from G. Mosig, in *Handbook of Biochemistry*, 2nd ed., H. A. Sober, Ed. Copyright CRC Press, Inc., Cleveland, Ohio, 1970.]

circularly permuted, and the circular permutations give rise to the circular linkage map. In addition, the DNA molecules are terminally repetitious for different genes of the series. The origin of the long DNA molecules found in the vegetative pool has been attributed to recombination occurring between the terminal genes of two different molecules, for such an exchange would serve to link the two molecules together, as shown below:

```
1  2  3  4  5  1  2
___________________
                  \ /
                  / \
            5     1  2  3  4  5  1 → 1  2  3  4  5  1  2  3  4  5  1
```

The Character of Bacteriophage Recombination

The events surrounding recombination in phage differ greatly from those found in eukaryotes. In eukaryotes recombination typically occurs as a reciprocal exchange between two of the four chromatids of a meiotic tetrad, with the result that two recombinant classes are found in equal numbers among the haploid gametes or spores. The maximum frequency of detectable exchange between two loci is 50 percent. In phage sexual processes, mating types, and meiosis are absent, and when all progeny arising from an individual doubly infected host cell are analyzed, equal numbers of reciprocal recombinants are not found. Furthermore, the maximum frequency of crossing over between widely separated loci does not exceed 40 percent and is usually considerably less.

An additional distinctive phenomenon, called *negative interference,* is also found in crosses with phage. It will be recalled that coincidence is a measure of the extent to which the number of observed double exchanges approaches the number expected on the basis of chance (ratio of observed doubles to expected doubles, see Chapter 8). When coincidence is less than 1.0, *positive interference* has occurred, that is, the formation of a chiasma in one region has interfered with the formation of a second chiasma in an immediately adjacent region of the chromosome. In contrast, negative interference refers to the recovery of more double exchanges than can be expected on the basis of chance, that is, coincidence is greater than 1.0.

Two kinds of negative interference can be recognized. One type, *localized high negative interference,* occurs in phage, bacteria, and eukaryotes and is associated with recombination within extremely small segments of the chromosome, such as a single gene (see Chapter 14). A second type, *low negative interference,* is routinely found only in viruses and is characterized by the recovery of additional doubles beyond the number expected in crosses involving genes located far apart on the genophore.

The reasons for low negative interference, absence of equal numbers of reciprocal recombinants, and maximum recombination frequencies of 40 percent or less become evident when the nature of the recombination process in T-even phages is examined. In a host cell doubly infected with phage of opposing genotypes, such as *p q r* and + + +, the genophores of both parental types are replicated and form a mixed pool of vegetative DNA. Within this pool these DNA molecules undergo successive, random exchanges with one another without discrimination as to genotype or source of partner. Genophores that have engaged in one exchange are free to seek another, and five or more rounds of such haphazard molecular "matings" may take place during the latent period. The occurrence of more than one round of mating can be observed when bacterial cells are triply infected with three kinds of phage, for example, with $p + +$, $+ q +$, and $+ + r$. The recovery of progeny containing genes from all three $(p\ q\ r)$ suggests either that some sort of group mating has taken place or that exchange events have successively involved the three parental types. The successive involvement of parental types has been shown to be the correct explanation.

Under these conditions of repeated matings, recombinants arising from an initial exchange are likely to be altered by later associations so that the recovery of equal numbers of reciprocal recombinants from any one infected cell is improbable. In addition, phenotypic double recombinants are derived not only from true double exchanges between two genophores, but also from successive single exchanges during separate matings. As a result, coincidence values greater than 1.0 are obtained, and a generalized low negative interference is observed.

The failure to achieve 50 percent recombination between widely separated marker genes can also be traced to the random nature of the molecular mating process. Suppose two genophores, $p +$ and $+ r$, are present in equal numbers after one cycle of replication. Each such genophore has an equal probability of associating with another of the same type or with a genophore of the opposing parental type. If the frequency of $p +$ is ½ and that of $+ r$ is also ½, then the following random matings could result.

MATING	PROBABILITY OF MATING
$p +$ with $p +$	$\frac{1}{2} \times \frac{1}{2} = \frac{1}{4}$
$p +$ with $+ r$	$\frac{1}{2} \times \frac{1}{2} = \frac{1}{4}$ (recombinants possible)
$+ r$ with $+ r$	$\frac{1}{2} \times \frac{1}{2} = \frac{1}{4}$
$+ r$ with $p +$	$\frac{1}{2} \times \frac{1}{2} = \frac{1}{4}$ (recombinants possible)

Only two $(\frac{1}{2})$ of these four kinds of matings can give rise to recombinants. If genes p and r are so widely separated as to appear unlinked, then the genophores emerging from a mixed mating will consist of $\frac{1}{4}\ p +$, $\frac{1}{4} + +$, $\frac{1}{4}\ p\ r$, and $\frac{1}{4} + r$, a distribution characteristic of independent assortment. Since recombinants arise from only half of the matings and constitute only half of the products of such matings, the maximum recombination frequency that can be obtained in the initial round of mating is 25 percent $(\frac{1}{2} \times \frac{1}{2} = \frac{1}{4})$.

Other factors also influence the recombination process, such as the physiological state of the host and the input of phage types, that is, the number that gain entrance into a bacterial cell. For example, if the infection is unequal such that the input of one parental type greatly exceeds that of the other, it is possible to recover more recombinants than parentals of the low input strain. Also, if the two parental phages enter at opposite ends of a host cell, the DNA pools produced by the replication of each parental genophore may not immediately become confluent, thereby decreasing the opportunities for exchange. Since recombination in phage is a complex phenomenon which occurs through the random association of genophores within a population of replicating molecules, it is best analyzed as a problem in population genetics. Although a review of the relevant mathematical procedures is outside the scope of this book, an interested reader will find them fully detailed in Visconti and Delbrück (1953), and in Visconti (1966) and Mosig (1970).

HETEROZYGOSITY IN BACTERIOPHAGES

Ordinarily, heterozygosity would not be expected to occur in a haploid form such as a virus. However, in 1951 Hershey and Chase described an experiment in which bacteria doubly infected with r and r^+ phage were plated on bacterial lawns before lysis of the host had occurred. A number of mottled, turbid plaques were observed and could be distinguished from those caused by either r or r^+ phage particles alone (Figure 12–5). Their origin was attributed to the presence at the same site of both the r and r^+ types of progeny.

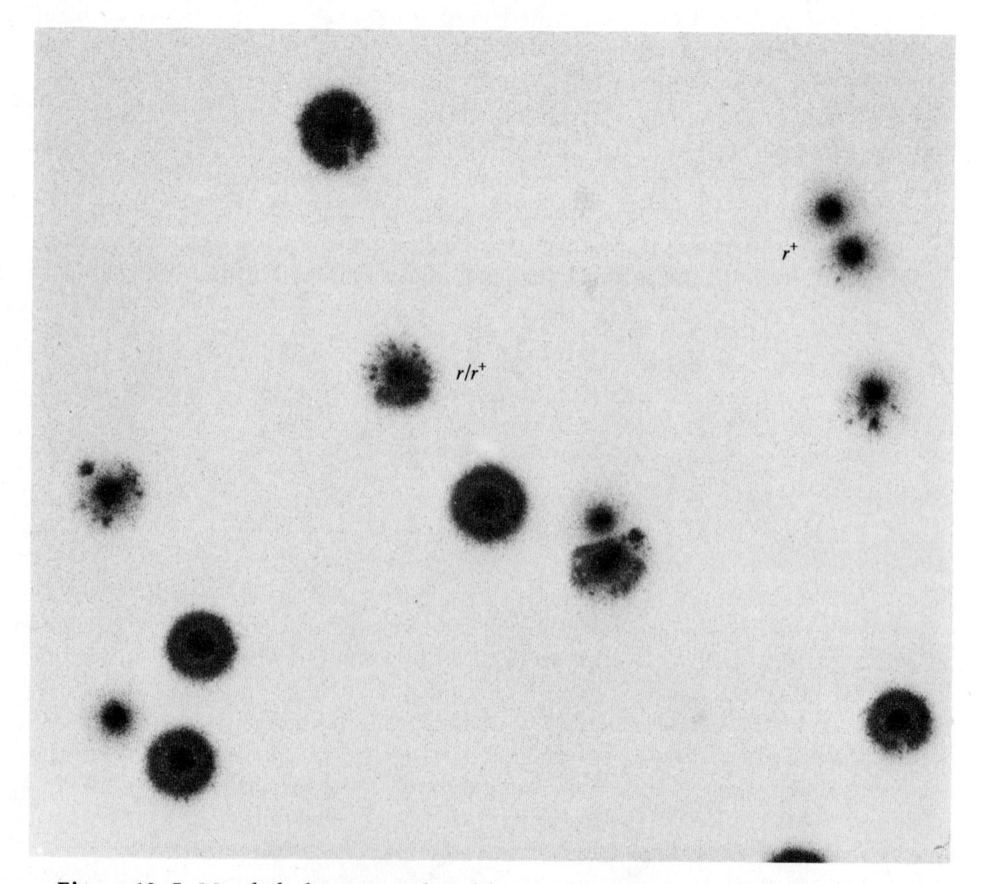

Figure 12–5. Mottled plaques produced by r/r^+ heterozygotes arising from a mixed infection of *E. coli* strain B cells by r and r^+ phage T2 particles. [Photographs by Maureen de Saxe and Janet Mitchell; From W. Hayes, *The Genetics of Bacteria and Their Viruses,* 2nd ed., Blackwell Scientific Publications Ltd., Oxford, 1968.]

In a later experiment Hershey and Chase permitted lysis to occur before plating, so that each plaque subsequently formed was derived from an individual infectious particle. Mottled plaques were again observed, appearing with a frequency of 2 percent. When the particles from such plaques were themselves plated, they gave rise to approximately equal numbers of pure *r* and r^+ plaques, along with 2 percent mottled plaques once more. Since the mottled plaques in each case arose from a single particle which produced both *r* and r^+ progeny, it was concluded that the original particle must have contained both genes, thereby being heterozygous in this region of the genophore. It has since been learned that such partial heterozygosity is not confined to the *r* region, but can occur at any locus.

It has been postulated that such heterozygotes are of two distinct types. One type can be attributed to the terminal redundancy present in the phage genophore. For example, through recombination a genophore terminally redundant for the *r* region can come to contain both the *r* and r^+ genes. By the same process such a particle would be expected to produce both *r* and r^+ progeny, as well as the original r/r^+ parental type.

The other kind of heterozygote appears to reflect the process of recombination itself and is thought to originate from the presence of a heteroduplex region within the genophore. In such a region one DNA strand would contain the nucleotide sequence of one allele, while the other strand would contain the sequence of the alternate allele. It is proposed that during recombination a heteroduplex region of hybrid DNA may be formed at the site of an exchange (see Chapter 14). The presence of a heterozygous locus in a phage particle is attributed to encapsulation of the genophore before replication has occurred following recombination. Upon subsequent entrance of such a genophore into a new host, semiconservative replication would generate equal frequencies of genophores carrying the alternate alleles in question. The rapid and random withdrawal of genophores from the DNA pool into new phage coats would again be likely to include some in which the process of recombination was still in progress.

LIFE CYCLE OF A TEMPERATE BACTERIOPHAGE

The best known temperate bacteriophage is phage λ which parasitizes *E. coli*. The linear genome of λ is 17.3 μm in length and is composed of DNA which is double stranded except at the ends, where the 3′ terminus of each strand extends 12 nucleotides beyond the 5′ terminus of the opposing strand. The base sequences of these extended single-stranded regions are complementary to one another and therefore *cohesive*.

When the viral DNA is injected into a sensitive host cell, the linear genophore assumes a circular form which is held together by overlapping and hydrogen bonding between the base sequences of the extended 3′ ends (Figure 12–6). Covalent closure of each strand is then accomplished via a DNA ligase.

At closure, either one of two alternative events can occur: lysis or lysogeny. The viral DNA may enter a lytic cycle during which the circular genophore undergoes rapid multiplication, possibly involving a rolling circle

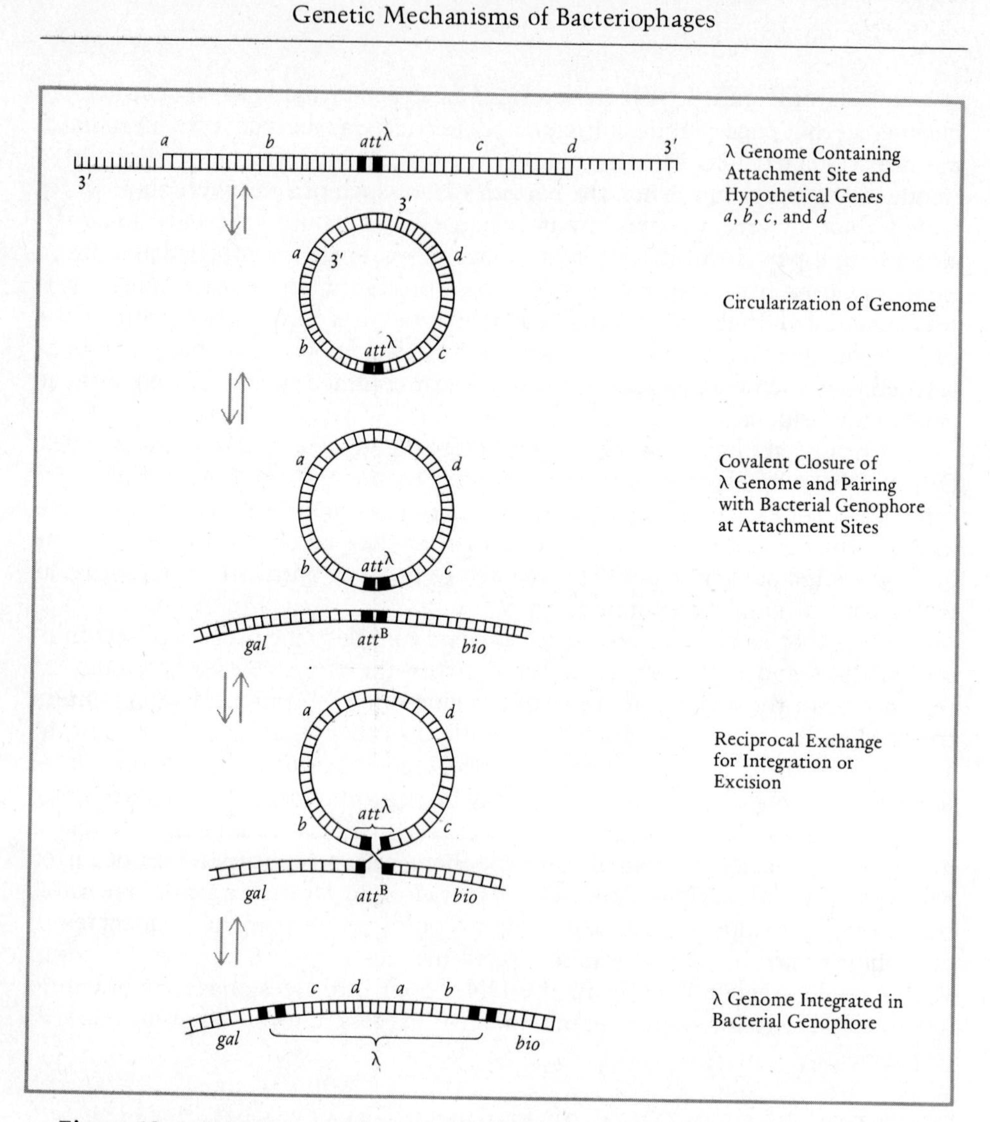

Figure 12–6. Integration and excision of λ genome. Integration leads to lysogeny; excision leads to lysis.

mechanism. At maturation nicks are enzymatically produced in daughter circles to generate linear genophores which are packaged into newly synthesized head capsules. Subsequent lysis of the host cell releases from 50 to 100 progeny phage particles.

Alternatively, the circular viral genome may be covalently inserted into the genome of the host *(lysogeny)*. Once integrated, the viral DNA, now called a *prophage,* replicates with the host genophore and is transmitted to daughter cells as a stable genetic entity. Bacteria which carry such a latent virus are termed *lysogenic,* that is, subject to lysis in the event that the prophage is excised from the host genophore. If such excision occurs, the viral DNA replicates rapidly, progeny phage particles are formed, and the host cell is destroyed.

A genetic entity such as λ DNA which can exist either integrated within the host genome or as an autonomous cytoplasmic element is called an *episome*.

Whether entrance of phage DNA into a susceptible cell is followed by lysis or by lysogeny in any given instance depends on the production of a repressor protein which is specified by a λ gene called c_I (Figure 12–7). If present in sufficient quantity, the repressor blocks the synthesis of those viral proteins which initiate the lytic cycle. The continuing presence of repressor is also essential for the maintenance of the lysogenic state.

Integration of the λ DNA occurs at preferred attachment sites present within both the λ and the bacterial genophores. The λ attachment site (att^{λ}) is a relatively short sequence estimated to consist of less than 50 nucleotide pairs. The corresponding attachment site on the bacterial genophore (att^{B}) is situated between the genes for galactose fermentation *(gal)* and biotin synthesis *(bio)*. Since the two attachment sites, in themselves, have no affinity for one another, it is thought that a λ-specified protein involved in the integration process is capable of recognizing both sites. Insertion of the λ genome into the bacterial genophore takes place via a reciprocal exchange occurring somewhere in the middle of the two sites; both sites are divided and come to lie at either end of the inserted phage genome (Figures 12–6 and 12–7).

Excision of the phage DNA from the bacterial genophore is visualized as the reverse sequence, that is, the prophage forms a circular loop, thus bringing together the two halves of the attachment sites, and a reciprocal exchange (crossover) results in excision of the prophage. Excision may occur spontaneously, or it can be induced by exposure of lysogenized cells to low doses of ultraviolet radiation. This treatment interferes with host DNA synthesis and, for reasons not well understood, causes λ to cease production of the repressor protein. In the absence of this protein, the lysogenic state cannot be maintained.

Induction of the prophage may also occur as the result of sexual union (conjugation) between a lysogenic and a nonlysogenic cell. During this process one cell donates a portion of its genome to its mate (see Chapter 13). If the donor cell is lysogenic for λ, the portion of the genome transferred to the recipient may include the integrated prophage; if so, absence of repressor in the recipient will cause excision of the prophage and lysis of the recipient cell. This phenomenon is called *zygotic induction*.

MAPPING OF VIRAL GENOMES

Genetic Analysis

Since λ directs a lytic cycle in sensitive strains of *E. coli*, the frequency of recombinants arising from mixed infections can be used to determine gene order. Progeny particles plated on lawns of susceptible bacteria produce plaque phenotypes which differ with respect to size, clarity, and host range. Additional phenotypes involve such characters as the number of progeny produced per host cell or the ability to lysogenize the host. It should be noted that recombination in λ does not involve continuous exchanges between genophores as in

T2 or T4. In λ many genophores undergo no exchanges at all, some participate in one exchange, and very few are involved in two separate exchanges.

The mapping of the λ genophore has been facilitated by the use of mutations called *suppressor-sensitives (sus)* which can develop and form progeny only within certain mutant strains of *E. coli*. When used to infect nonmutant (nonpermissive) strains, these defective phages cannot form progeny for a variety of reasons, such as failure to replicate upon induction or to synthesize head or tail proteins or to lyse the host cell.

Using the *sus* mutants, Campbell performed complementation tests in which nonpermissive host cells were infected with two different defective phages to determine whether or not the two genotypes acting in concert could promote the formation of progeny and therefore of plaques. If the two mutants failed to complement one another, the reason for the lack of normal progeny in

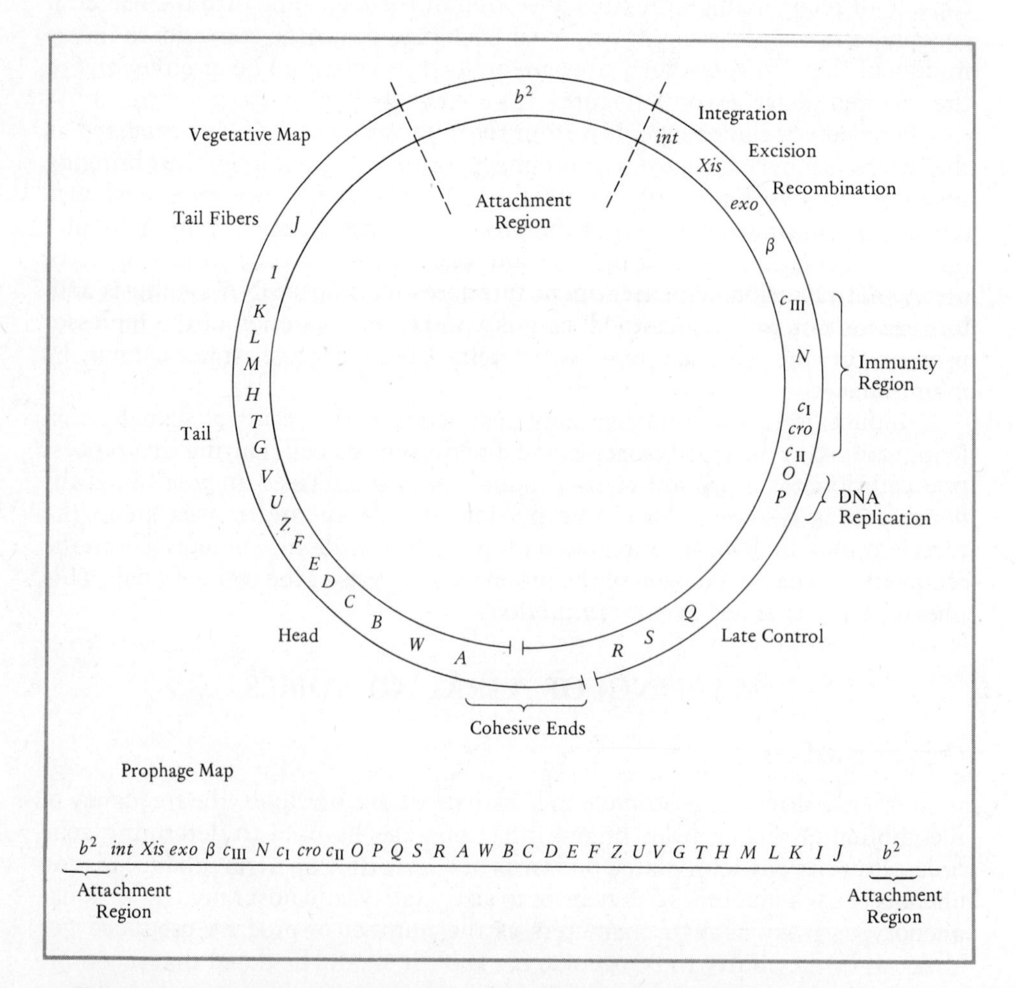

Figure 12–7. Circular vegetative map and linear prophage map of phage λ. [Adapted from W. Szybalski et al., *J. Cell. Physiol., 74, Suppl. 1,* 33–70, 1969.]

any case could then be determined by chemical tests or other means. For example, absence of head capsules or tails can be ascertained by electron microscopy, absence of antigens by testing with antisera, absence of lysozyme by testing for the dissolution of cell walls, and so on. By analyzing the results obtained with many such pairs of *sus* mutants, Campbell identified a series of distinct functions reflecting the presence of separate and distinct genes and the respective linkage positions of these genes were established through the analysis of recombinant phages produced in lytic cycles. Additional genes have since been discovered, and a current map of the λ genophore is presented in Figure 12–7 and Figure 18–12.

Physical Mapping

Methods for the physical mapping of viral genomes have recently been developed. These involve the isolation of genophores from wild type and mutant viruses, the denaturation of these DNA molecules to the single-stranded state, and the annealing of complementary strands from the two different genophores into a heteroduplex. These heteroduplex molecules are then examined with the electron microscope. The method permits visualization of the entire genome, and the presence of insertions, deletions, or mutated regions where the bases of the two strands are not matched can be identified and their positions and extent measured with great precision. It will be recalled that a similar heteroduplex between the complementary strands of T3 and T7 permits an evaluation of the degree of homology present between their respective genomes. The value of heteroduplex molecules for physical mapping can be illustrated with phage λ. Figure 12–8 is an electron micrograph of a heteroduplex formed by renaturation between complementary single strands of wild type (λ^+) and a defective λ (λ*b2b5*). The genotype of the heteroduplex is thus + +/ *b2b5*. Two single-stranded regions are evident in the micrograph. One ($b2^+$) is a region where the wild type strand forms a loop at the side, a configuration which indicates that the *b2* mutation present in the opposing strand is in fact a deletion of this region. The other single-stranded bubble marks an area of nonhomology between the two genomes, where the *b5* region of the defective genophore does not match the i^λ region of the wild type genophore. Through other physical mapping experiments, the source of this nonhomology has been traced to the replacement of λ genes normally present in this region with genes from an entirely different phage, called phage 21.

This example demonstrates that physical mapping techniques provide an extremely powerful tool for the precise localization of genes, particularly those of viruses. Viral genophores readily lend themselves to this kind of combined genetic and physical analysis because of their small size. For example, the λ genophore contains only enough DNA (approximately 46,500 nucleotide pairs) for about 50 genes. Bacterial plasmids (see Chapter 13) fall within the same range and can also be isolated and manipulated with relative ease. In contrast, the genophores of bacteria are immensely more complex and they, in turn, are dwarfed by the enormous size and genetic capacity of the chromosomes of higher organisms. Nevertheless the extension of this technique to other forms besides viruses holds great promise.

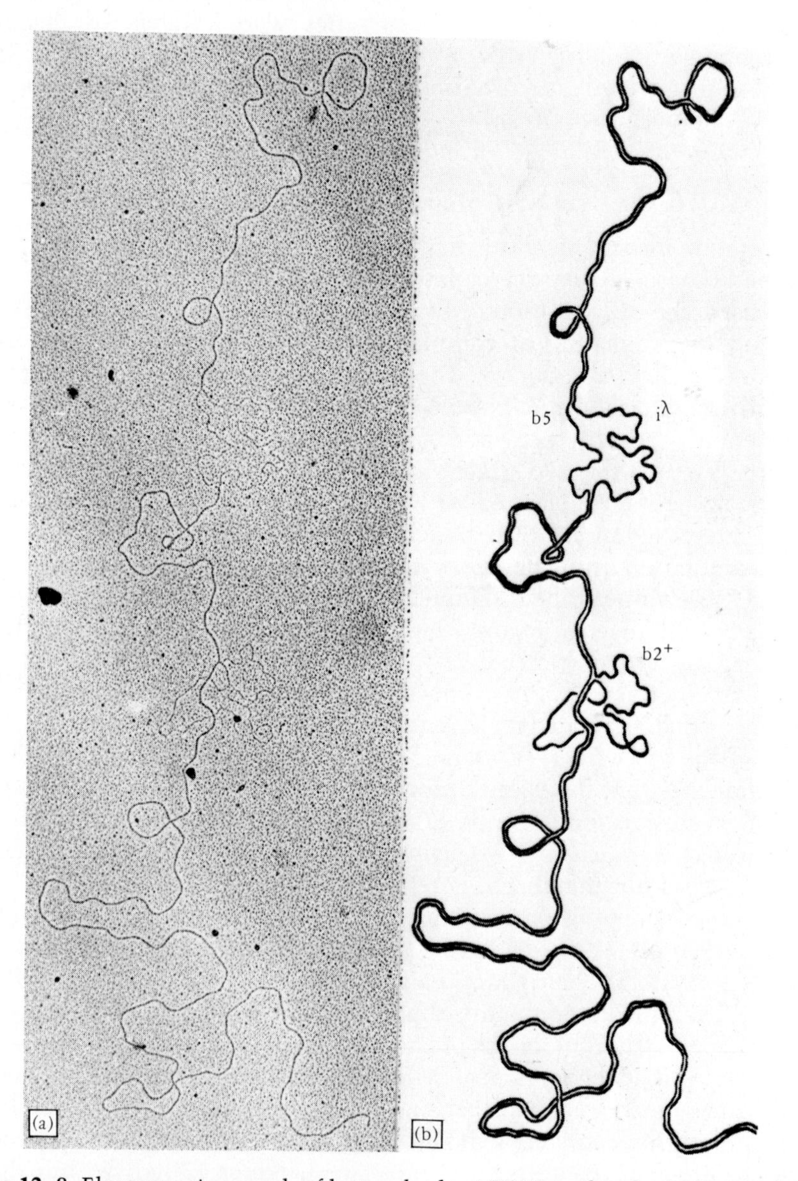

Figure 12–8. Electron micrograph of heteroduplex λDNA molecule. (a) Heteroduplex between complementary single strands of λ^+ *and* λ*b2b5* (++/*b2b5*). (b) An interpretive drawing of the ++/b2b5 heteroduplex showing a single-stranded $b2^+$ loop and the unpaired i^λ/b5 (+/b5) region. [From B. Westmoreland, W. Szybalski, and H. Ris, *Science, 163,* 1343–1348, 21 March 1969. Copyright © 1969 by the American Association for the Advancement of Science.]

PROBLEMS

12–1. Explain what is meant by a "cross" in bacteriophage. What factors influence the results of such a cross?

12–2. Two independent temperature sensitive mutations were recovered in a bacteriophage. Both mutants formed plaques at 25°, but not at 37°. If one mutant was defective in DNA replication and the other was defective in lysozyme formation, how would you determine that these phages were mutant for different functions?

12–3. How can a haploid bacteriophage be heterozygous? Explain how such a phage, heterozygous for genes *a* and *a'*, can give rise to equal numbers of *a* and *a'* plaques.

12–4. Explain the meaning and origin of the negative interference observed in phage crosses.

12–5. Five independent mutations were induced in a virulent bacteriophage of *E. coli*. When tested as pairs in mixed infections, the following results were obtained (+ and 0 indicate presence or absence, respectively, of plaques).

	1	2	3	4	5
1	0	+	0	0	+
2		0	+	+	+
3			0	0	+
4				0	+
5					0

How many different functions can be identified? Which phages are mutant for the same function?

12–6. What are the characteristics of a temperate vs a virulent phage? Contrast the developmental cycles of T2 and λ.

12–7. On a bacterial lawn a wild type phage produces large (s^+), clear (t^+) plaques with a smooth border (f^+). Three mutants of this phage produce the following plaque phenotypes:

Mutant 1 produces small plaques (s), as opposed to large (s^+).
Mutant 2 produces fuzzy-bordered plaques (f), as opposed to smooth-bordered plaques (f^+)
Mutant 3 produces turbid plaques (t), as opposed to clear plaques (t^+)

From the results of the following crosses (mixed infections), determine the order of these mutants and the linkage distance between them.

Cross 1: large smooth $(s^+ f^+)$ × small, fuzzy $(s\ f)$

Progeny:	
large, smooth $(s^+ f^+)$	447
small, fuzzy $(s\ f)$	473
large, fuzzy $(s^+ f)$	32
small, smooth $(s\ f^+)$	48
	1000 Total

Cross 2: clear, smooth $(t^+ f^+)$ × turbid, fuzzy $(t\ f)$

Progeny:	
clear, smooth $(t^+ f^+)$	1150
turbid, fuzzy $(t\ f)$	1202
clear, fuzzy $(t^+ f)$	18
turbid, smooth $(t\ f^+)$	30
	2400 Total

Cross 3: large, clear $(s^+ t^+)$ × small, turbid $(s\ t)$

Progeny:		
	large, clear $(s^+ t^+)$	775
	small, turbid $(s\ t)$	845
	large, turbid $(s^+ t)$	95
	small, clear $(s\ t^+)$	85
		1800 Total

REFERENCES

BROKER, T. R., AND A. H. DOERMANN, 1975. Molecular and genetic recombination of bacteriophage T4. *Ann. Rev. Genet., 9,* 213.

CAMPBELL, A. D., 1969. *Episomes.* Harper & Row, New York.

DOERMANN, A. H., 1973. T4 and the rolling circle model of replication. *Ann. Rev. Genet., 7,* 325.

DOVE, W. F., 1968. The genetics of the lambdoid phages. *Ann. Rev. Genet., 2,* 305.

HAUSMANN, R., 1973. The genetics of T-odd phages. *Ann. Rev. Microbiol., 27,* 51.

HAYES, W., 1968. *The Genetics of Bacteria and Their Viruses,* 2nd ed. Wiley, New York.

HERSHEY, A. D., 1947. Spontaneous mutations in bacterial viruses. *Cold Spring Harbor Sympos. Quant. Biol., 11,* 67.

HERSHEY, A. D. (Ed.), 1971. *The Bacteriophage Lambda.* Cold Spring Harbor Laboratory, Cold Spring Harbor, N.Y.

HERSHEY, A. D., AND M. CHASE, 1951. Genetic recombination and heterozygosis in bacteriophage. *Cold Spring Harbor Sympos. Quant. Biol., 16,* 471.

HERSHEY, A. D., AND R. ROTMAN, 1949. Genetic recombination between host range and plaque-type mutants of bacteriophage in single cells. *Genetics, 34,* 44.

MOSIG, G., 1970. Recombination in bacteriophage T4. *Adv. Genet., 15,* 1.

STENT, G. S., 1971. *Molecular Genetics.* Freeman, San Francisco.

STREISINGER, G., R. S. EDGAR, AND G. H. DENHARDT, 1964. Chromosome structure in phage T4. I. Circularity of the linkage map. *Proc. Nat. Acad. Sci., 51,* 775.

VISCONTI, N., 1966. Mating theory. In J. Cairns, G. S. Stent, and J. D. Watson (Eds.). *Phage and the Origins of Molecular Biology*. Cold Spring Harbor Laboratory, Cold Spring Harbor, N.Y.

VISCONTI, N., AND M. DELBRÜCK, 1953. The mechanism of recombination in phage. *Genetics, 38,* 5.

Chapter 13

GENETIC MECHANISMS OF BACTERIA

A bacterial cell is a minute haploid organism that reproduces by simple fission; mitosis, meiosis, and all structures associated with these processes are absent. The mutations which occur in bacteria are expressed through a variety of phenotypes, such as resistance to drugs, antibiotics, or various types of bacteriophages or the inability to utilize a given sugar or to synthesize some essential substance required for growth. These nutritional mutants are called *auxotrophs,* as opposed to wild type *prototrophs*.

The various methods by which genetic exchange is achieved in bacteria are unique. All involve the one-way transfer of DNA from a donor cell to a recipient cell. As a result, the recipient becomes partially diploid, but only for the genes that have been received, and recombination is necessarily restricted to these genes. A bacterial cell can acquire donor genes by three general mechanisms: (1) *transformation,* whereby a donor DNA fragment present in the environment enters a recipient cell and is incorporated into the genophore; (2) *conjugation,* a form of sexuality in which gene transfer is accomplished through union of a donor cell with a recipient cell; and (3) *transduction,* whereby a bacteriophage acts as a vector and carries bacterial genes from one host cell to another.

In bacteria recombinants arise from double or an even number of exchanges. Because the genophore is physically circular, single exchanges open the circle, and an odd number of exchanges leads to the loss of one or more segments, both conditions being inviable. The recovery of recombinants from among the millions of cells in a bacterial culture often necessitates the use of a selective medium on which only recombinants can form colonies.

TRANSFORMATION

Transformation involves the passage of a DNA molecule from the environment into a recipient cell, whereupon it replaces a homologous sequence originally present in the recipient's genophore. If the entering DNA fragment carries an allele different from that already present, replacement of one allele by the other will give rise to a new phenotype detectable in daughter cells. Such cells are called *transformants*. Although early investigations utilized strains of *Pneumococcus (Diplococcus),* transformation has been found to occur in a variety of bacteria such as *Escherichia coli, Bacillus subtilis, Hemophilus influenzae,* and others.

Transformation is observed when recipient cells are grown in the presence of killed donor cells or in filtrates from donor cell cultures. The requirements of the process are better analyzed, however, through the use of DNA artificially extracted from donor cells. As a result of the extraction procedure, the entire bacterial genome is broken and sheared into several hundred pieces, and it has been found that only those fragments above a certain minimum size are effective. This size varies depending on species. In *Bacillus* a fragment of at least 16,000 nucleotide pairs and a molecular weight above 10 million is necessary, while in *Pneumococcus* transformation is successful with much smaller molecules. As an additional requirement, the fragment must be double stranded, for denatured single-stranded DNA is not taken up by recipient cells.

In any experiment the amount of DNA added per recipient cell is also an important factor. Although uptake of a single fragment of high molecular weight DNA can give rise to a transformant, the overall frequency of transformation is directly proportional to the concentration of DNA up to the point where approximately 10 molecules of DNA per cell are present. Increasing the concentration of DNA beyond this level has no further effect, suggesting that only a limited number of receptor sites are present.

Successful transformation also requires that the bacterial cell be in a physiologically competent state. Competence is a transitory condition exhibited during a limited period in the growth of a culture. Experimental evidence suggests that the onset of competence is associated with changes in the cell wall and that the ability to achieve this state is an inherited character. In *Streptococcus, Pneumococcus,* and *Bacillus* competence can be induced by a protein factor released by cells that have already reached the competent state. Genophore replication is not involved in competence, for transformation is not accompanied by significant DNA synthesis. It is interesting to note that the adsorption of DNA is not species specific, for bacterial cells may take up DNA derived from other species and even from mammals. However, although "foreign" DNA may gain entrance into the cell, only DNA of the same or closely related species exhibits transforming activity.

Although initially the DNA fragment is reversibly attached to the cell wall, it shortly becomes permanently bound and is then actively transported to the interior of the cell. Either during entry or inside the cell, the DNA is rendered single stranded; one strand is degraded, the other conserved, choice in the matter being random. Although the details of the process are not well understood, it is thought that this single strand then pairs with the homologous region of the genophore and that pairing is followed by the replacement of one or more genes in the recipient's genophore by those present in the transforming DNA (Figure 13–1). The segment of recipient DNA which is removed during this process is lost through degradation. Since replacement involves only one of the two DNA strands of the recipient, the genophore becomes a partial heteroduplex at the site of integration. This condition is temporary, for the next round of semiconservative replication produces daughter genophores each containing DNA strands entirely complementary to one another. The two daughters would necessarily differ from one another, however, with respect to any allele introduced by transformation.

Evidence that transformation proceeds via replacement of recipient genes rather than by addition was obtained through a demonstration of the reversible nature of the process. As an illustration, if donor a^- DNA is used to transform

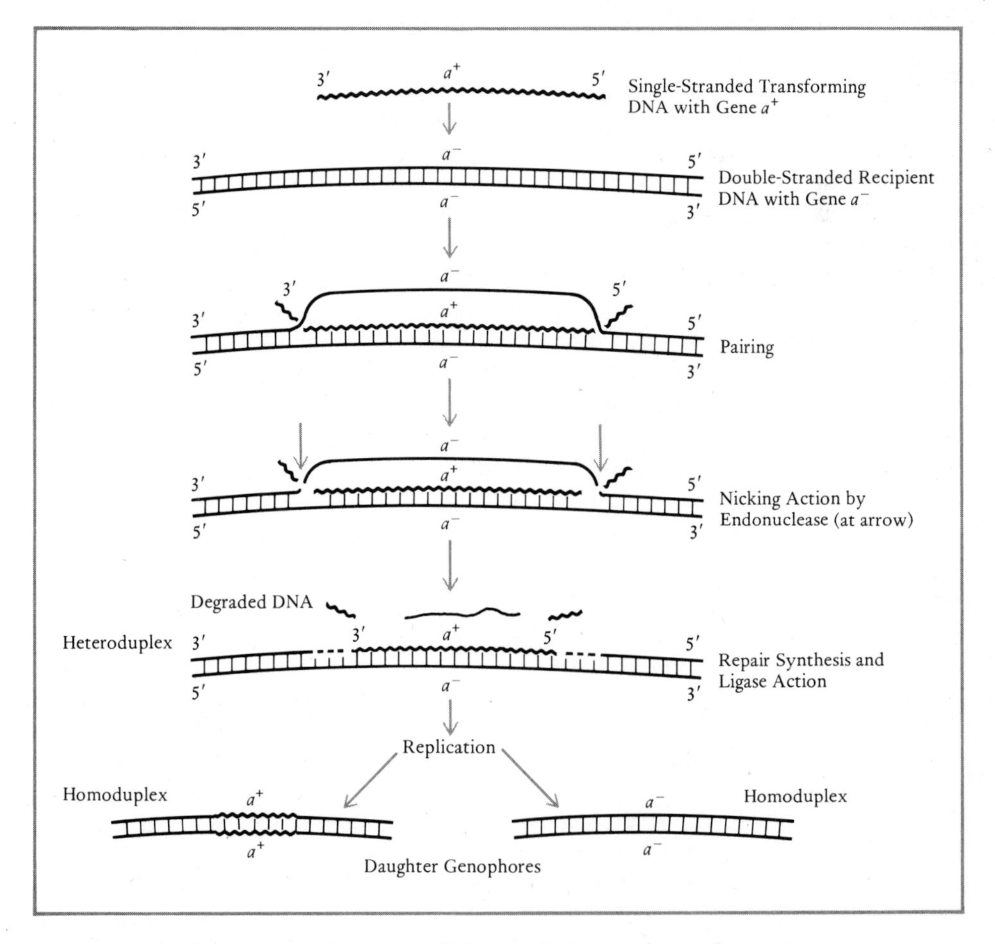

Figure 13–1. Diagram of the mechanism of transformation

a^+ recipients, transformants are a^- in phenotype. When these a^- transformants are then used as recipients for donor a^+ DNA, the resulting phenotype is once more a^+. These results would be difficult to interpret on the basis of an addition which would result in heterozygosity (a^-/a^+) in the first experiment and in a genotype of $a^-/a^+/a^+$ in the second. Since there is no evidence for the presence of such genotypes among transformants, the substitution of donor for recipient genes is a more logical explanation.

That transformation results in the incorporation of only one strand of the donor DNA can be demonstrated, for when donor DNA is isotopically labeled, the genophore of transformants contains a labeled segment in only one of its two DNA chains.

Determination of Linkage by Transformation

Transformation can be used to determine gene order in bacteria. Since donor DNA is randomly broken into fragments by the extraction process, genes located far apart in the genophore will be found in different fragments. Therefore two separate and independent fragments and two transformation events will be required to produce a doubly transformed cell. Closely linked or adjacent genes are likely to be present on the same fragment, and in this case only a single transformation event will be required to produce a doubly transformed cell. Since the occurrence of a single event is more probable than the simultaneous occurrence of two separate events, more double transformants will arise when the genes in question are linked and present on the same fragment than when they are unlinked and present on different fragments.

Suppose that in separate experiments it is found that locus p is transformed to p^+ with a frequency of 5 percent and that locus q is transformed to q^+ also with a frequency of 5 percent. To determine if these genes are linked, donor DNA of genotype $p^+ q^+$ is used to transform $p\ q$ cells. If the two loci are distant from one another, the extracted donor DNA will contain p^+ fragments as well as q^+ fragments. Since the probability of a p^+ transformant is 0.05 and the probability of a q^+ transformant is 0.05, the expectation for the occurrence of double transformation is 0.0025 (the product of the separate probabilities), that is, an expectation of 25 double transformants per 10,000 cells.

However, if p and q are closely linked and on the same fragment, double transformants should arise with much higher frequency approaching or exceeding that observed for either gene alone. In this event linkage could be inferred, but unequivocal proof would require additional experiments in which the frequency of single and double transformation was correlated with DNA concentration. The reason for this precaution is that at saturation levels, a recipient cell can take up several DNA molecules, thus increasing the chance for double transformation to occur via two separate recombination events. Below saturation levels, however, the frequency of transformation is directly dependent on the concentration of DNA. When decreasing concentrations of DNA are used, the number of single transformants will decrease accordingly and to about the same extent for each gene. If both genes are carried on the same fragment, as the DNA concentration is lowered, the decrease in frequency of double transformants should parallel the decrease in the frequency of single transformants, and if a graph is constructed to illustrate transformation fre-

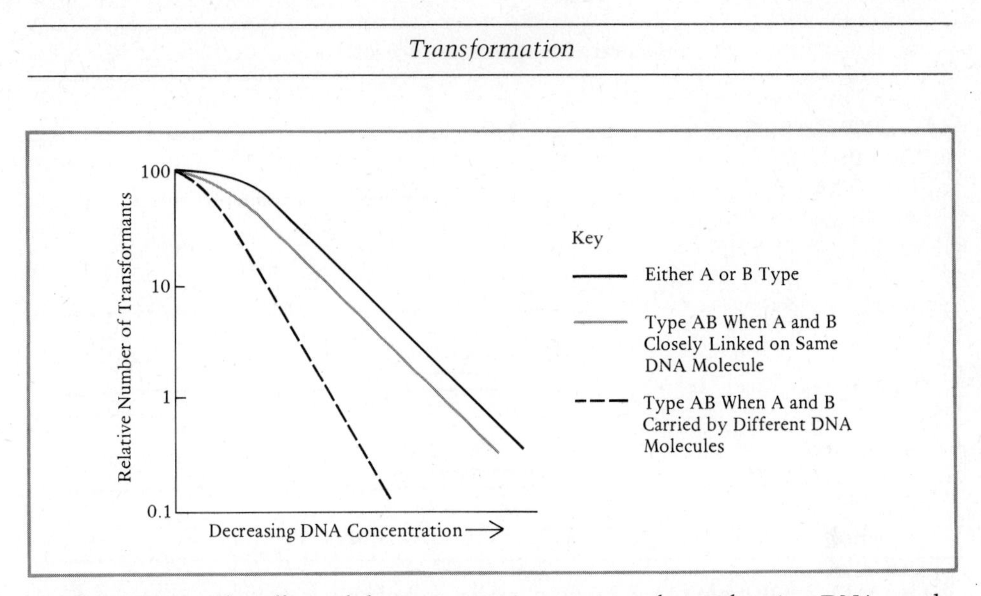

Figure 13-2. The effect of decreasing concentration of transforming DNA on the relative number of single and double transformants for the two markers, A and B. [From W. Hayes, *The Genetics of Bacteria and Their Viruses,* 2nd ed., Blackwell Scientific Publications Ltd., Oxford, 1968. After S. Goodgal, *J. Gen. Physiol., 45,* 205–228, 1961.]

quency versus DNA concentration, the slope of the curves for single and double transformants will be the same (Figure 13–2). In contrast, if the two genes are on separate fragments, a decrease in the concentration of DNA will result in a much lower frequency of double transformation, and when represented graphically, the slope of the curve will be steeper and therefore different from that which represents single transformants.

Mapping by Transformation

Assuming that the transforming fragments are approximately the same size and that precautions with regard to DNA concentration are taken, it is possible to estimate the frequency of exchange between two linked genes. As an example, suppose recipient cells, $a^- b^-$ in genotype, are exposed to transforming DNA extracted from $a^+ b^+$ donors and that the resulting transformants are distributed as follows.

Class 1	$a^+ b^-$	127
Class 2	$a^- b^+$	98
Class 3	$a^+ b^+$	275
		500 Total transformants

It can be seen that classes 1 and 2 are single transformants for genes a^+ and b^+, respectively, and that class 3 consists of double transformants. Since the number of doubles is as high as, and in this case higher than singles, linkage is assumed. Reference to Figure 13–3 shows that each class arises from a double exchange between transforming DNA and the genophore of recipients. Note that double transformants arise from exchanges occurring *to either side* of genes *a* and *b,* while single transformants arise from exchanges *between* genes *a* and *b*. Only the latter are recombinants for these genes. The proportion of

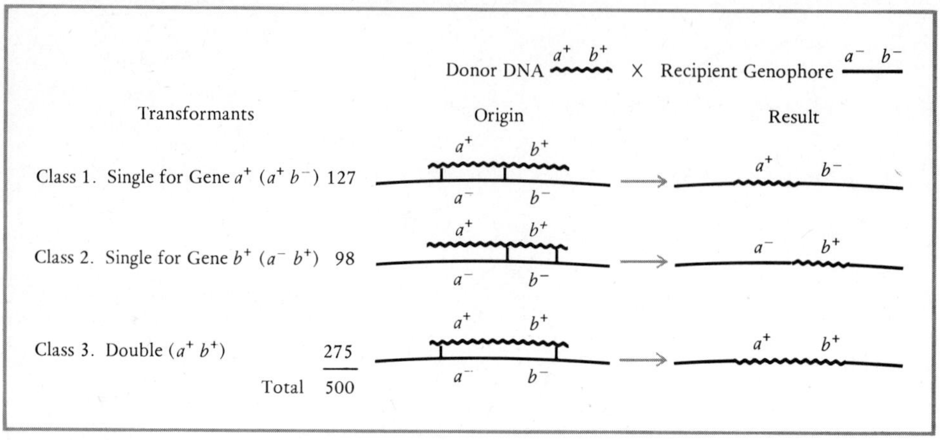

Donor DNA $a^+ \; b^+$ × Recipient Genophore $a^- \; b^-$

Transformants		Origin	Result
Class 1. Single for Gene a^+ ($a^+ b^-$)	127	$a^+ \; b^+$ / $a^- \; b^-$	→ $a^+ \; b^-$
Class 2. Single for Gene b^+ ($a^- b^+$)	98	$a^+ \; b^+$ / $a^- \; b^-$	→ $a^- \; b^+$
Class 3. Double ($a^+ b^+$)	275	$a^+ \; b^+$ / $a^- \; b^-$	→ $a^+ \; b^+$
Total	500		

Figure 13–3. Diagram of the origin of single and double transformants arising from the cross donor $a^+ \, b^+ \times$ recipient $a^- \, b^-$.

recombinants among total transformants can be used as a measure of the linkage distance between these loci. The total number of transformants scored was 500; of these, recombinants numbered 225 in all (127 + 98 = 225). Applying the formula for calculating linkage distance, we have 225/500 = 0.45, that is, a map distance between genes *a* and *b* of 0.45 or 45 percent recombination.

By using triply marked donors and recipients, it is also possible to establish the order of genes known to be linked. As an illustration, suppose donor DNA of genotype $a^+ b^+ c^+$ is used to transform recipient cells carrying the linked genes $a^- b^- c^-$. In such an experiment the distribution of transformants might be:

Single Transformants		Double Transformants		Triple Transformants	
$a^+ b^- c^-$	600	$a^+ b^+ c^-$	300	$a^+ b^+ c^+$	2800
$a^- b^+ c^-$	125	$a^- b^+ c^+$	575		
$a^- b^- c^+$	280	$a^+ b^- c^+$	10		

To determine gene order, it is first necessary to calculate the linkage distance between any two genes, that is, the distance between *a* and *b* (disregarding *c*), the distance between *a* and *c* (disregarding *b*), and the distance between *b* and *c* (disregarding *a*). Linkage distance is calculated as the proportion of recombinants among the total transformants for the pair of genes specifically under consideration. The calculations for this example are given in Figure 13–4. It can be seen that the linkage distance between genes *a* and *c* is greater than that between either *a* and *b* or *b* and *c*, and, therefore, the order of genes is *a—b—c*.

The origin of each class of transformants is shown in Figure 13–5. Note that all arise from a double exchange with the exception of one class ($a^+ b^- c^+$) whose formation requires a quadruple exchange. By reference to the data in Figure 13–4 it can be seen that this particular class appears with the lowest frequency (10 transformants), a consequence to be expected in view of the rarity of such an event. Since a quadruple exchange must occur to transform both outer genes, but not the centrally located gene, the class with the lowest frequency of transformants can be used to determine gene order, essentially by

Donor $a^+b^+c^+$ × Recipient $a^-b^-c^-$

	Transformants for Genes *a* and *b* (Disregarding *c*)			Transformants for Genes *a* and *c* (Disregarding *b*)			Transformants for Genes *b* and *c* (Disregarding *a*)		
Single Transformants:	$a^+b^-(c^-)$	600	1310	$a^+(b^-)c^-$	600	1755	$(a^-)b^+c^-$	125	715
	$a^-b^+(c^-)$	125		$a^-(b^-)c^+$	280		$(a^-)b^-c^+$	280	
	$a^-b^+(c^+)$	575		$a^+(b^+)c^-$	300		$(a^+)b^+c^-$	300	
	$a^+b^-(c^+)$	10		$a^-(b^+)c^+$	575		$(a^+)b^-c^+$	10	
Double Transformants:	$a^+b^+(c^-)$	300	3100	$a^+(b^-)c^+$	10	2810	$(a^-)b^+c^+$	575	3375
	$a^+b^+(c^+)$	2800		$a^+(b^+)c^+$	2800		$(a^+)b^+c^+$	2800	
	Total		4410			4565			4090
Linkage Distance:	Gene *a* to Gene *b* 1310/4410 = 0.297			Gene *a* to Gene *c* 1755/4565 = 0.384			Gene *b* to Gene *c* 715/4090 = 0.174		

Order of Genes: *a* ←— .29 —→ *b* ←— .17 —→ *c*; *a* to *c* |←— .38 —→|

Figure 13–4. Determination of gene order from transformants arising from the cross donor DNA a^+ b^+ c^+ × recipient a^- b^- c^-.

Donor $a^+b^+c^+$ × Recipient $a^-b^-c^-$

Transformants		Origin	Result	Number of Exchanges Required
Single:	$a^+b^-c^-$	a^+ b^+ c^+ / a^- b^- c^-	a^+ b^- c^-	2
	$a^-b^+c^-$	a^+ b^+ c^+ / a^- b^- c^-	a^- b^+ c^-	2
	$a^-b^-c^+$	a^+ b^+ c^+ / a^- b^- c^-	a^- b^- c^+	2
Double:	$a^+b^+c^-$	a^+ b^+ c^+ / a^- b^- c^-	a^+ b^+ c^-	2
	$a^-b^+c^+$	a^+ b^+ c^+ / a^- b^- c^-	a^- b^+ c^+	2
	$a^+b^-c^+$	a^+ b^+ c^+ / a^- b^- c^-	a^+ b^- c^+	4
Triple:	$a^+b^+c^+$	a^+ b^+ c^+ / a^- b^- c^-	a^+ b^+ c^+	2

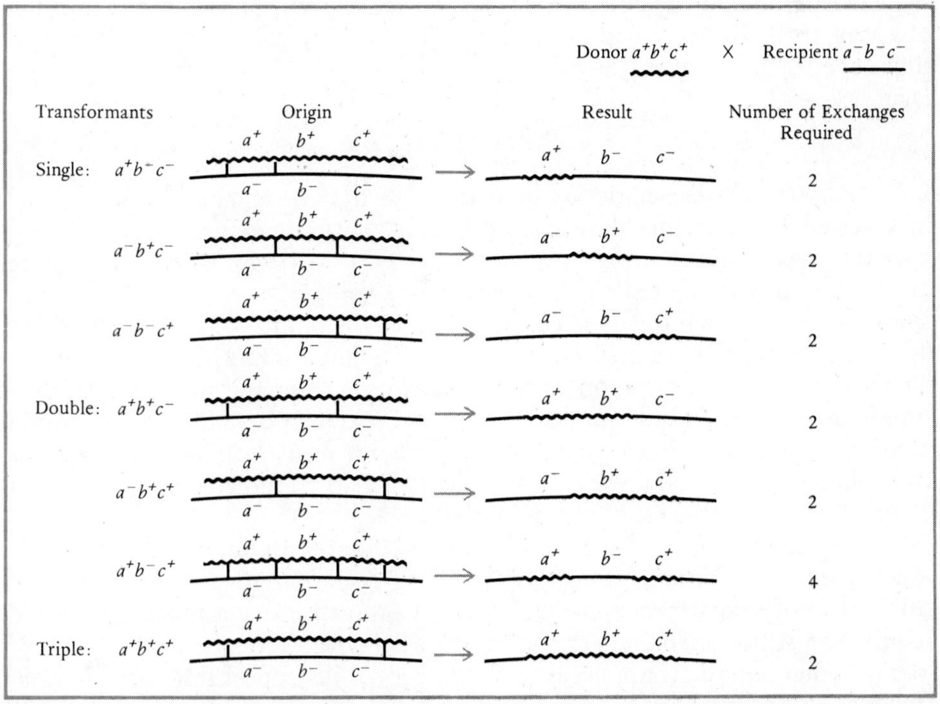

Figure 13–5. Origin of single, double, and triple transformants arising from the cross donor DNA a^+ b^+ c^+ × recipient a^- b^- c^-.

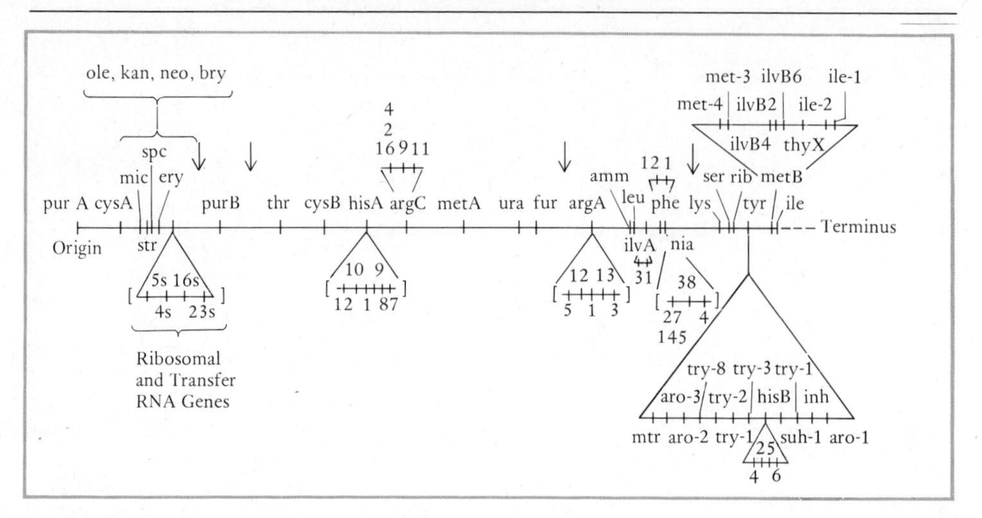

Figure 13–6. Linkage map of *Bacillus subtilus*, strain 168. The locations of some of the marker genes on this map were established by means of transformation. [From D. Dubnau et al., *J. Mol. Biol.*, *27*, 163–185, 1967.]

inspection. It should be recalled that similar reasoning applied to the double crossover classes in diploid eukaryotes, can be used to infer gene order.

Linkage distances obtained by transformation provide useful estimates for mapping, particularly when correlated with the frequency of transformants for an unlinked marker gene. By employing many different mutants, the general sequence of loci may be determined. As an example, the positions of a number of genes in the linkage map of *Bacillus subtilis* were assigned by transformation experiments (Figure 13–6).

CONJUGATION

Although the association of bacterial cells in pairs had been observed and interpreted as signifying some sort of sexuality, genetic evidence for recombination by sexual mechanisms was not obtained until 1946, when J. Lederberg and E. L. Tatum reported the occurrence of conjugation in *Escherichia coli*. These workers mixed together two different auxotrophic strains which can be designated simply as strain 1, $a^- b^- c^- d^+ e^+ f^+$, and strain 2, $a^+ b^+ c^+ d^- e^- f^-$. Neither strain could grow on minimal medium. After the mixture was cultured in complete medium, samples were taken and plated on minimal medium. Upon incubation prototrophic colonies in a frequency of 1 per 10^6 cells plated were found (Figure 13–7). This phenotype indicated a genotype of $a^+ b^+ c^+ d^+ e^+ f^+$, that is, wild type for all loci.

The origin of these colonies could be attributed to one of three possible mechanisms: mutation, transformation, or recombination resulting from some type of sexual process. The origin of prototrophs via mutation would require the simultaneous reversion to the wild type allele of all three mutant loci of either parent strain, an event so rare as to be improbable. Transformation was also eliminated as an explanation when the use of filtrates or extracts of either parent strain failed to produce recombinants. Accordingly it appeared that the phenomenon depended on the presence of living cells of both strains,

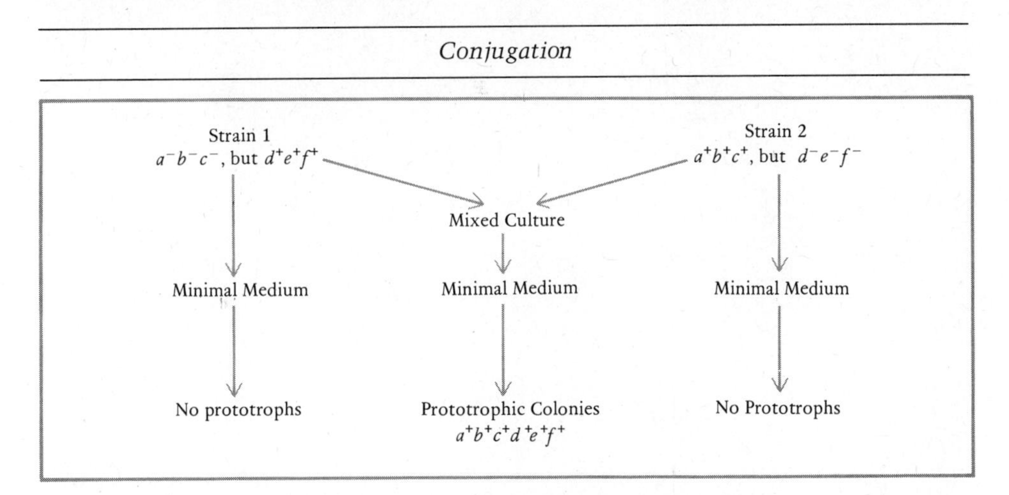

Figure 13–7. Diagram of Lederberg and Tatum's experiment with *E. coli* demonstrating the production of prototrophs by sexual mechanisms.

suggesting some kind of mating process. Further experiments with *E. coli* carried out by W. Hayes, Lederberg, F. Jacob, E. L. Wollman, and others have defined this process which is called *conjugation*.

Conjugation involves the union of sexually differentiated cells and is accompanied by the transfer of DNA across a cytoplasmic bridge from one partner (donor) to the other (recipient). The ability to act as a donor is conferred by the presence of a *sex* or *fertility factor* called *F*, cells containing this factor being F^+ and those lacking the factor being F^-. The *F* factor is one of several cytoplasmic elements of *E. coli* which are known as *plasmids*. These genetic entities are capable of autonomous replication in the cytoplasm and can also promote their own transfer to other cells by conjugation. Transfer of the sex factor occurs with great efficiency, for when F^+ and F^- cells are mixed, the F^- cells are converted to the F^+ state within the space of an hour.

The sex factor is a circular, double-stranded DNA molecule, approximately $\frac{1}{50}$ the length of the *E. coli* genophore. It contains genes which control the appearance on the cell surface of specific hairlike extensions called *sex pili* or *F pili* which function in conjugation (Figure 13–8). Contact between an F^+ and an F^- cell is established by the F pilus; retraction of the pilus draws the mating pair into close association to form a cytoplasmic bridge across which the sex plasmid can pass from the F^+ donor to the F^- recipient. Transfer appears to involve a rolling circle mechanism analogous to that found in phage (see Figure 11–24). One strand of the double-stranded plasmid DNA is nicked by an endonuclease to produce free 5′ and 3′ ends, and the 5′ end, called the origin, then proceeds across the cytoplasmic bridge into the F^- cell. The unbroken circular strand of the donor plasmid serves as a template for DNA synthesis, so that after transfer is completed, the donor *F* factor is once more a double-stranded, covalently closed circle. Synthesis of a strand complementary to that transferred is also carried out in the recipient cell, and after completion of this synthesis and covalent closure of the plasmid DNA by ligase action, the recipient is no longer F^-, but has become F^+ in phenotype (Figure 13-9).

This transfer of the sex factor rapidly changes F^- cells into F^+ donors, but it does not account for the transfer and recombination of bacterial genes present in the genophore. An explanation for this event has been provided by the

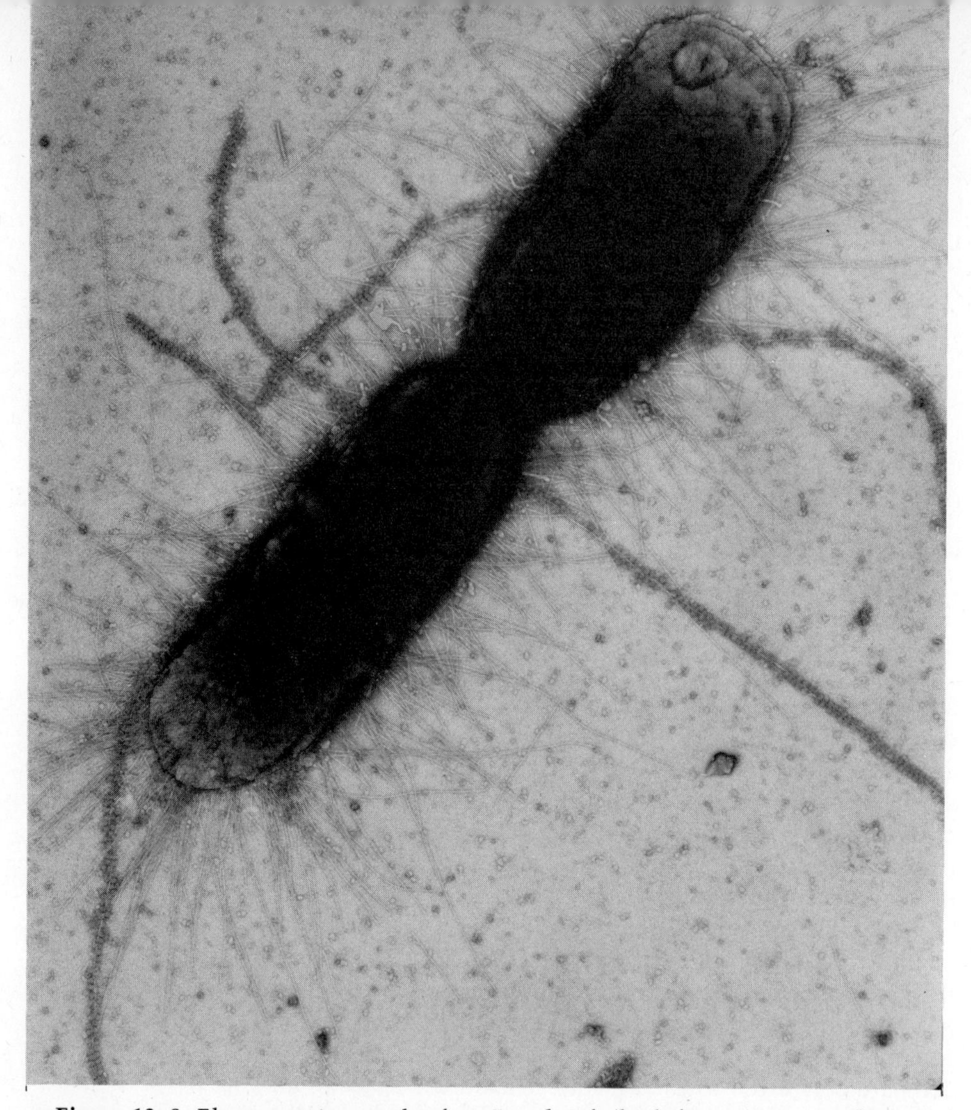

Figure 13–8. Electron micrograph of an *E. coli* Hfr (high frequency recombination) cell covered with numerous hairlike appendages (pili). The sex pili appear larger and thicker because of their specific adsorption of particles of an RNA bacteriophage. Other extensions, devoid of adsorbed phage, are common pili. (34,750×) [Photograph by Dr. Lucien G. Caro; from W. Hayes, *The Genetics of Bacteria and Their Viruses,* 2nd ed., Blackwell Scientific Publications Ltd., Oxford, 1968.]

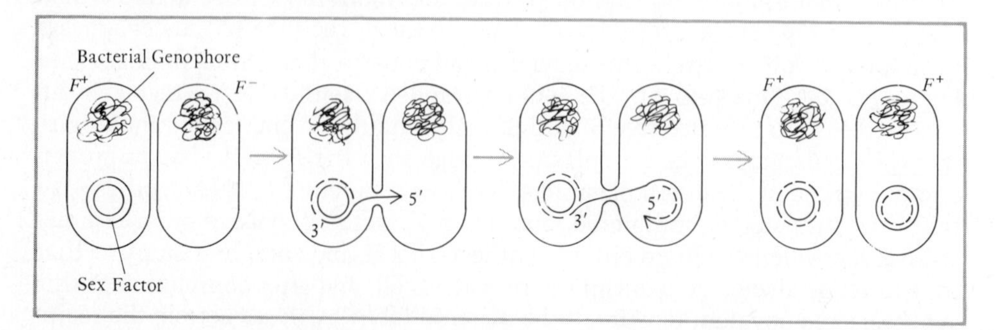

Figure 13–9. Transfer of the sex factor from an F^+ donor to an F^- recipient by conjugation. Upon receiving the sex factor, the recipient becomes F^+ in phenotype. Dotted lines indicate replication of the sex factor.

discovery that the sex factor can exist in two different states: as an autonomous cytoplasmic entity or as an integral part of the genophore. When free in the cytoplasm, the sex factor alone is transmitted to F^- cells during mating, but when integrated into the genophore, it promotes the transfer of the genophore itself. *E. coli* strains containing an integrated sex factor form F pili and behave as donors in conjugation. They are called *high frequency recombination (Hfr) strains* because a portion (or sometimes all) of the genophore is transferred to a recipient cell, with recombination occurring thereafter.

Transfer of the genophore is thought to occur in the following manner (Figure 13–10). As conjugation between an *Hfr* and an F^- cell is initiated, a

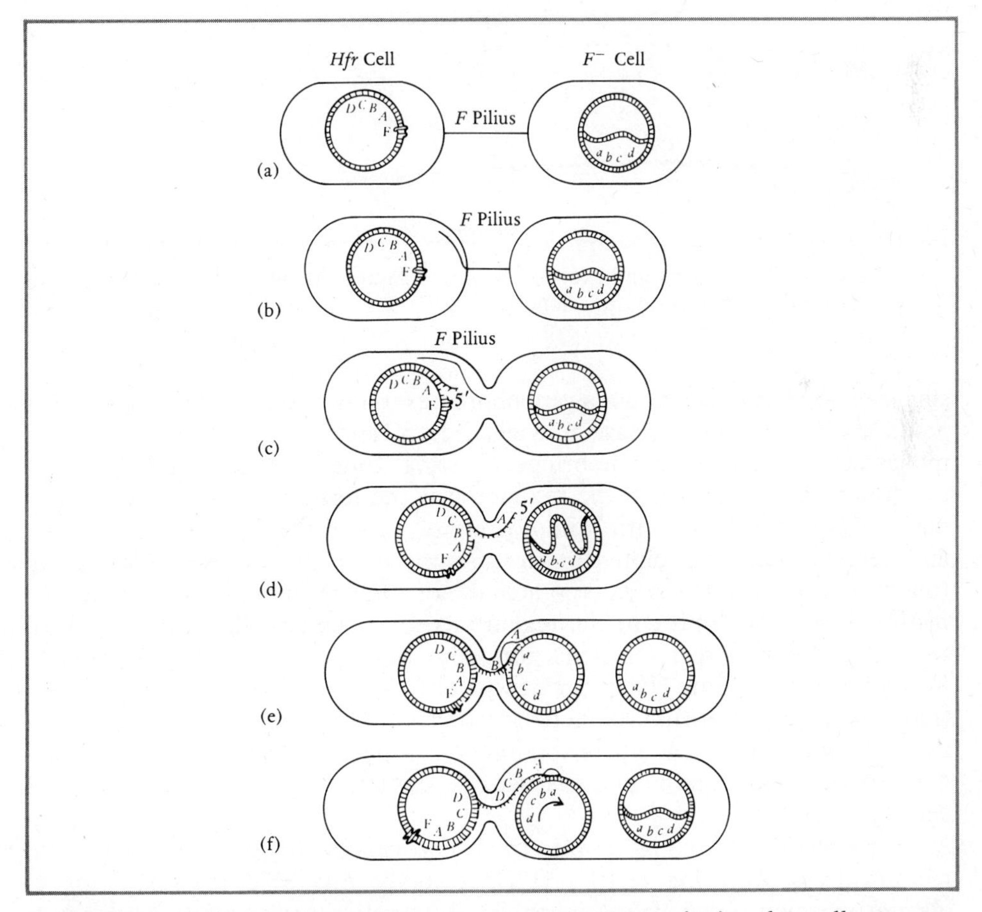

Figure 13–10. Stages of bacterial conjugation. (a) Joining of *Hfr* and F^- cells via a sex pilus; replication of F^- genome is in progress. (b, c) Retraction of the pilus bringing conjugants into contact; nicking of the donor genome next to the locus of the sex factor *F* occurs to produce a 5′ end. (d) The 5′ end of a single-stranded copy of the donor genome proceeds across the cytoplasmic bridge to the recipient; replication of *Hfr* genome is initiated (dotted line). (e, f) Pairing between homologous regions of donor and recipient genomes followed by replacement of F^- genes with those of the donor. [Reproduced, with permission, from R. Curtiss, *Ann. Rev. Microbiol., 23,* 69–127, 1969. Copyright © 1969 by Annual Reviews Inc. All rights reserved.]

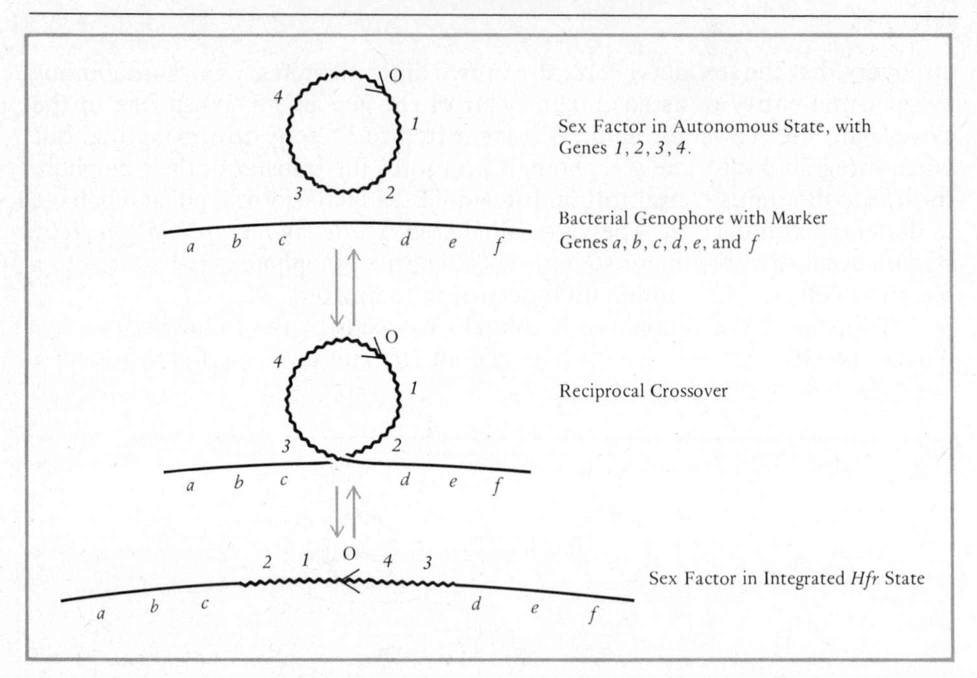

Figure 13–11. Integration and excision of the sex factor. Arrow at O (Origin) indicates the direction of transfer at conjugation.

single-strand cut in the donor's genophore next to the locus of the sex factor is produced by an endonuclease to yield free 5′ and 3′ ends. The origin then moves across the cytoplasmic bridge, drawing along a single-stranded copy of the donor's genophore with the sex factor at the terminus. This newly introduced DNA then pairs with homologous regions of the recipient genophore, and replacement of recipient genes by those of the donor occurs as in transformation. If a newly incorporated donor allele is different from that of the recipient, a heteroduplex in the region of the exchange will be present until genome replication restores strict complementarity of the two DNA chains. Recipient cells become *Hfr* in phenotype only if the entire donor genophore is transferred and the entire sex factor introduced. This rarely occurs since mating cells ordinarily break apart spontaneously before complete transfer is accomplished. Thus, recipient cells usually remain F^- in sex phenotype after conjugation.

Insertion of the sex factor into the *E. coli* genophore occurs by means of a reciprocal crossover (Figure 13–11). Once stably integrated, the sex factor replicates in synchrony with the genophore. Integration can occur at a number of different sites and with either a forward or reverse polarity, so that the order in which given genes are transferred to the recipient cell differs in different *Hfr* strains. For example, in *Hfr* strain 1 specific genes might enter recipients in the order Origin—*a*—*b*—*c*—*d*—*e*—*f*—*g*—*F*. In strain 2 the order might be Origin—*e*—*f*—*g*—*a*—*b*—*c*—*d*—*F,* while in Strain 3 an order of Origin—*d*—*c*—*b*—*a*—*g*—*f*—*e*—*F* might be observed (Figure 13–12). Excision of the sex factor *(F)* from the *Hfr* genophore may also occur, and if so, the sex factor

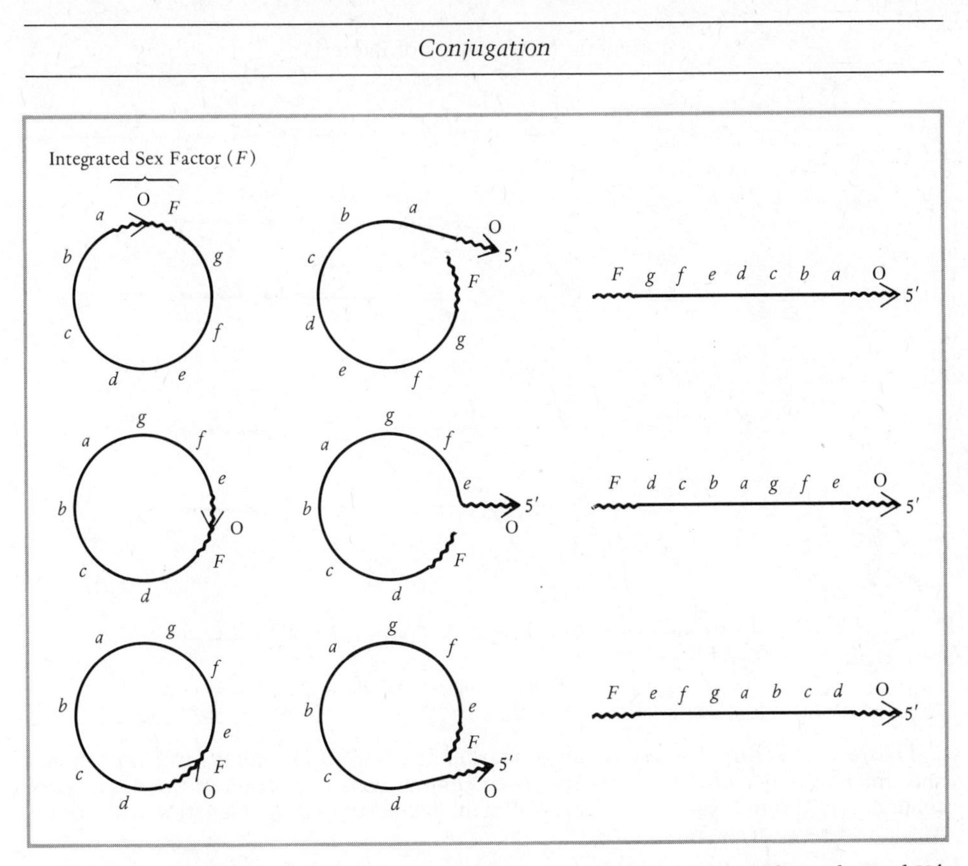

Figure 13–12. The order of gene transfer during conjugation in three hypothetical *Hfr* strains of *E. coli*; *F*, sex factor; *O*, origin.

resumes its autonomous existence in the cytoplasm and the phenotype of the cell reverts to the F^+ state. The sex factor thus exhibits a behavior which in many respects parallels that of the bacteriophage λ, and because of its two modes of existence, it can be considered an *episome* as well as a plasmid. The two terms are not synonymous, since some plasmids do not appear to integrate within the bacterial genome.

Mapping by Conjugation

The interrupted mating experiments of Wollman and Jacob have demonstrated that the order in which particular loci are transferred from one cell to another can be used for gene mapping. These workers utilized as donors an *Hfr* strain prototrophic for threonine (thr^+) and leucine (leu^+), sensitive to the cell poison sodium azide *(azi-s)*, susceptible to infection by phage T1 *(T1-s)*, able to ferment lactose (lac^+) and galactose (gal^+), and sensitive to streptomycin *(str-s)*. The genotype was thus: thr^+ leu^+ *azi-s T1-s* lac^+ gal^+ *str-s*. The F^- recipients were of the opposite genotype, that is, auxotrophic for threonine and leucine, resistant to azide and phage T1, auxotrophic for lactose and galactose, and resistant to streptomycin (thr^- leu^- *azi-r T1-r* lac^- gal^- *str-r*). Donor and recipient cells of these genotypes were mixed and plated on minimal medium containing streptomycin. This medium was selective for recombinants prototrophic for both threonine and leucine and growth of donors was prevented by

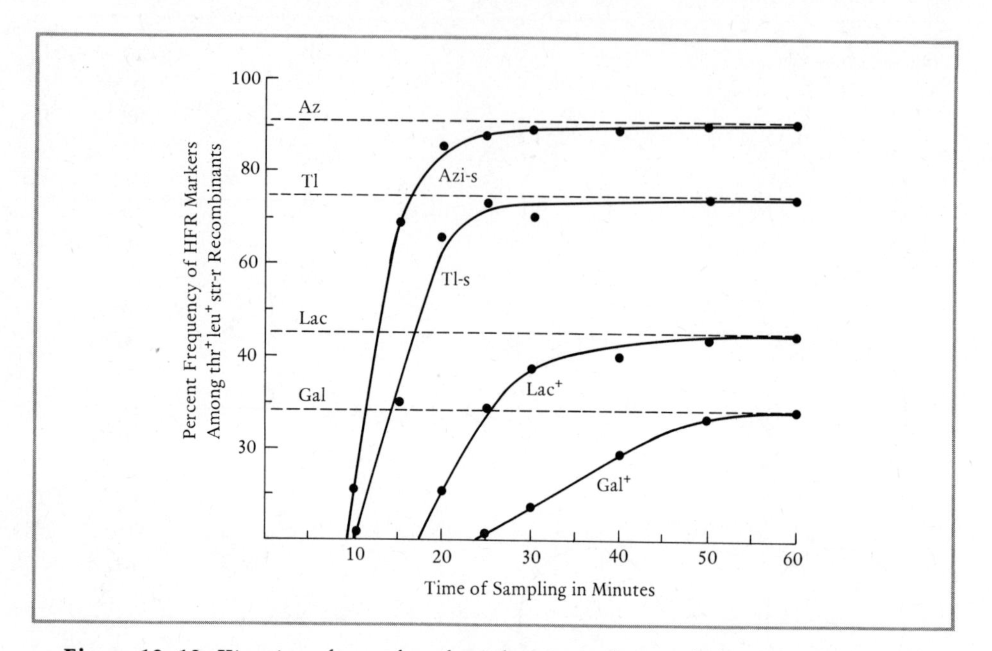

Figure 13–13. Kinetics of transfer of marker genes from *Hfr* donors to F^- recipients and the appearance of these markers in recombinants as a function of the duration of conjugation. [From F. Jacob and E. L. Wollman, *Sexuality and the Genetics of Bacteria*, Academic Press, New York, 1961.]

streptomycin. Any prototrophic colonies recovered were then appropriately tested for the presence of the other marker genes.

This experiment was unique, not for the strains employed nor for the use of a selective medium, but because the production of recombinant types was correlated with the timed duration of conjugation. After mixing donor and recipient cells, samples of these conjugants were removed at timed intervals after the initiation of mating, agitated in a blender to break apart donor and recipient cells, and plated and tested for the presence of recombinants. It was found that thr^+ and leu^+ recombinants appeared after around 8 minutes and were accompanied by azide recombinants by 9 minutes. In 10 minutes recombinants for T1 phage sensitivity were recovered. By 18 minutes lactose recombinants began to appear, followed by galactose recombinants in about 25 minutes after the start of mating (Figure 13–13).

In each case the number of recombinants for any marker gene rose to a plateau which represented the maximum yield for that gene. In addition, more recombinants were found for "early" genes than for "later" genes. Wollman and Jacob proposed that the sequence of genes on the *E. coli* genophore could be determined by the order in which recombinants for each locus appeared, and their studies demonstrated that map distance between genes could be accurately expressed in terms of the time required for the transfer of each gene. Since the map based on time intervals agreed with the map based on recombination frequencies, a time map was developed for *E. coli* in which 1 minute = 20 recombinational map units. Because the time required for transfer of the entire genophore was 90 minutes, the *E. coli* map in use until very recently was 90 minutes long (Figure 13–14). However, because of the discovery of new loci and the reevaluation of time of entry data, the 90-minute map was recalibrated in 1976 to a 100-unit map. The known length and number of nucleotides in the *E. coli* genophore has permitted the estimate that around 35,000 to 36,000 nucleotide pairs, corresponding to a DNA length of approximately 12 μm, are transferred per minute during conjugation. If it is assumed that around 800 to 1000 nucleotide pairs constitute a gene, then it can be inferred that around 35 genes are transferred per minute.

Gene mapping as a function of time is most useful for genes two and more minutes apart. For those closer together, recombination frequencies are more reliable and are determined by methods similar to those used in the analysis of transformants. For example, suppose the order of genes is known to be Origin—a—b and that donor cells that are $a^+ b^+$ (and streptomycin sensitive) are allowed to conjugate with recipient cells that are $a^- b^-$ (but streptomycin resistant). Conjugants would be plated on a medium containing streptomycin plus the growth requirement for a^-, but not that for b^-. As a result, only streptomycin-resistant daughter cells containing the b^+ allele could grow, and the presence of such colonies would provide proof that the b^+ allele had actually been transferred to recipients. These colonies could then be tested on minimal medium to determine whether a^- or a^+ was present (Figure 13–15). Colonies found to be $a^- b^+$ would be recombinant because this genotype could arise only through an exchange occurring between genes a and b. Colonies found to be $a^+ b^+$, however, arise by an exchange occurring to either side of these loci. Accordingly, the recombination frequency between a and b is calculated as the number of $a^- b^+$ colonies divided by the sum of the number of $a^- b^+$ colonies plus the number of $a^+ b^+$ colonies, that is,

$$\frac{(a^- b^+)}{[(a^- b^+) + (a^+ b^+)]} \times 100 = \text{linkage distance}$$

Sometimes mapping as a function of time or by recombination frequency fails to resolve gene order, particularly if two of the genes being mapped are very close together. In this case, as with transformation, advantage is taken of the fact that a rare quadruple exchange is required when replacement of the two flanking genes, but not the centrally located gene occurs. As an illustration, suppose the auxotrophic mutants x^-, y^-, and z^- are linked, but it is not known whether the order is x—y—z or x—z—y. To resolve this question, the following reciprocal crosses can be used:

Cross 1: donor $x^- y^- z^+ \times$ recipient $x^+ y^+ z^-$
Cross 2: donor $x^+ y^+ z^- \times$ recipient $x^- y^- z^+$

When plated on minimal medium, only prototrophic cells ($x^+ y^+ z^+$) will be able to form colonies. If the actual gene order is x—y—z—, such prototrophs arise from double exchanges in both cross 1 and cross 2, and the number of prototrophic colonies recovered will not differ greatly in the two crosses. However, if the actual gene order is x—z—y, a prototroph is formed by a double exchange in cross 1, but can arise only from a quadruple exchange in cross 2

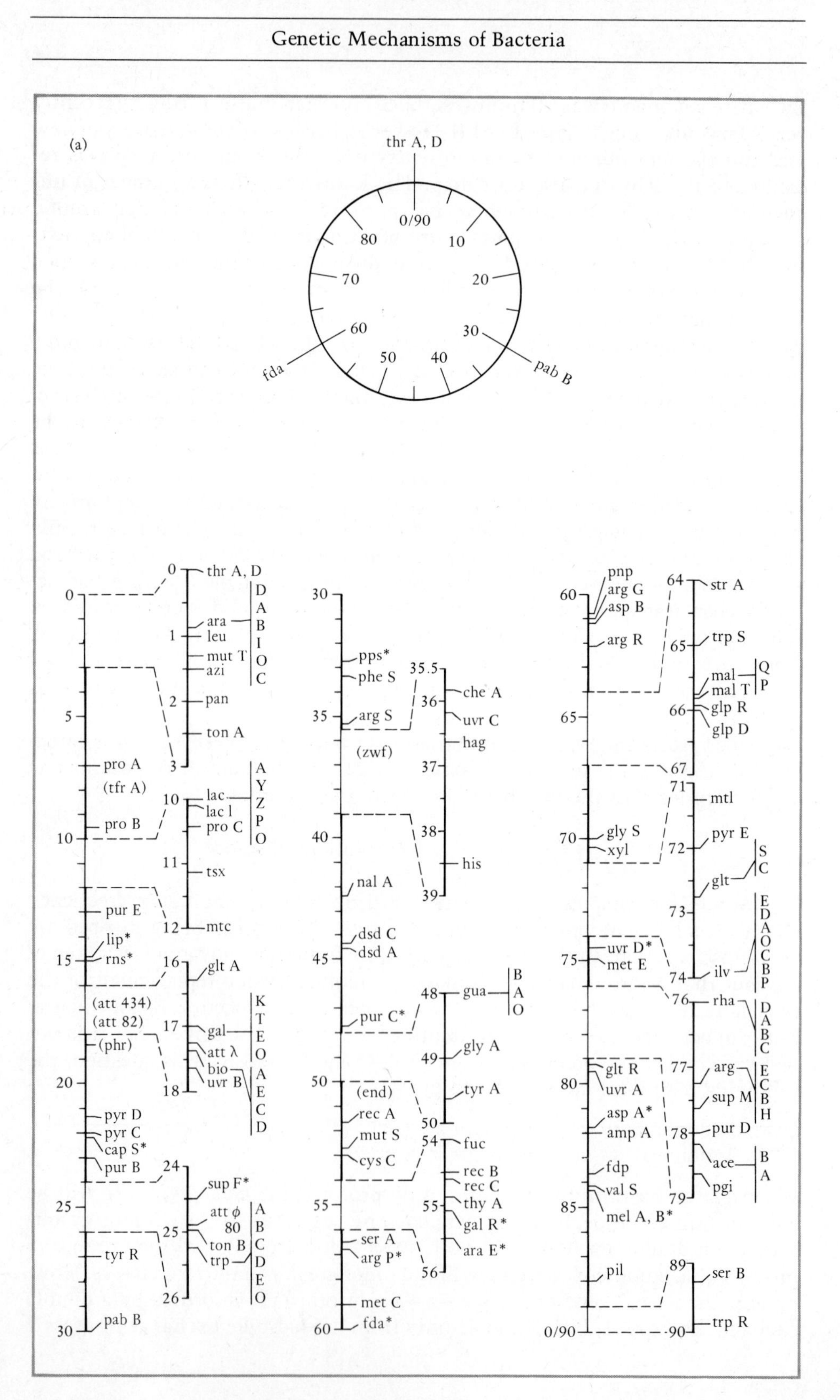
(a)
thr A, D
0/90
10
20
30
40
50
60
70
80
pab B
fda
0
thr A, D
D A B I O C
ara
1
leu
mut T
azi
2
pan
ton A
3
5
pro A
(tfr A)
pro B
10
lac
lac I
pro C
A Y Z P O
11
tsx
12
mtc
pur E
lip*
rns*
15
16
glt A
(att 434)
(att 82)
(phr)
17
gal
att λ
bio
uvr B
K T E O
A E C D
18
20
pyr D
pyr C
cap S*
pur B
24
sup F*
25
att φ 80
ton B
trp
A B C D E O
26
tyr R
pab B
30
30
pps*
phe S
35.5
36
che A
uvr C
hag
35
arg S
(zwf)
37
38
40
his
nal A
39
dsd C
dsd A
45
48
gua
B A O
pur C*
49
gly A
50
(end)
rec A
mut S
cys C
tyr A
50
54
fuc
rec B
rec C
thy A
gal R*
ara E*
55
55
ser A
arg P*
56
met C
fda*
60
60
pnp
arg G
asp B
arg R
64
str A
65
trp S
mal
mal T
Q P
66
glp R
glp D
65
67
71
mtl
70
gly S
xyl
72
pyr E
glt
S C
73
uvr D*
met E
75
74
ilv
E D A O C B P
76
rha
D A B C
glt R
uvr A
80
77
arg
E C B H
sup M
asp A*
amp A
78
pur D
ace
B A
pgi
fdp
val S
mel A, B*
85
79
pil
89
ser B
0/90
90
trp R

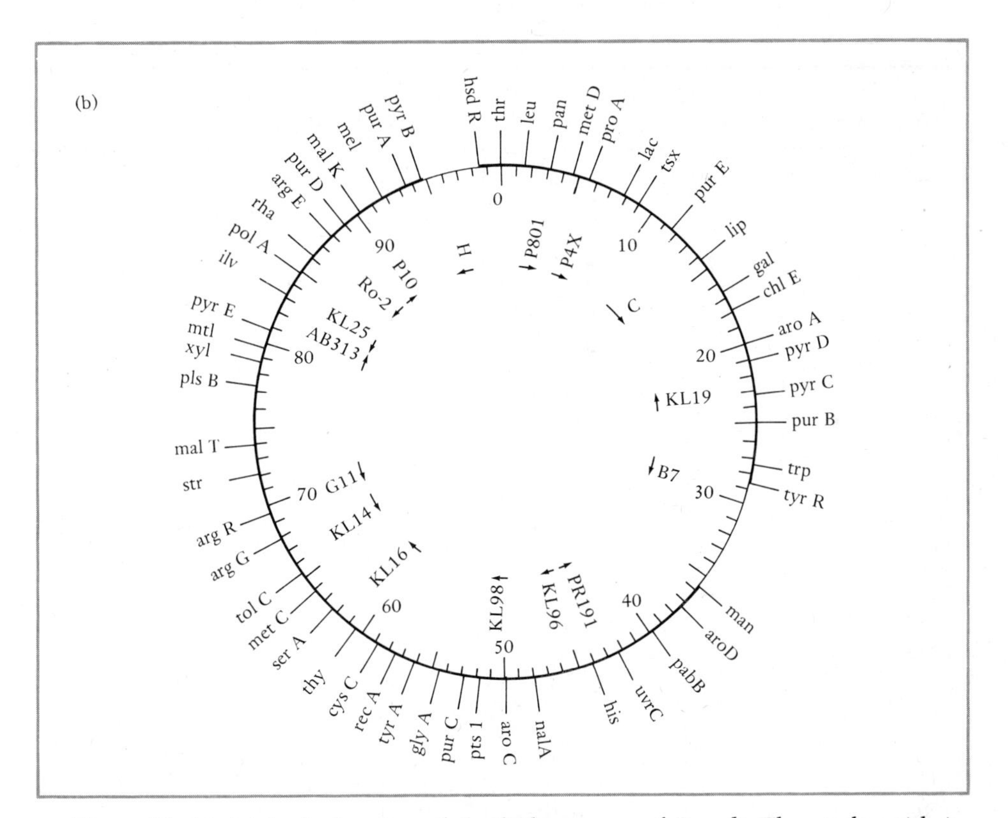

Figure 13–14. (a) Scale drawing of the linkage map of *E. coli.* The circle, with its associated time scale, depicts the intact closed linkage group which has a total length of 90 minutes. The time scale begins arbitrarily with minute zero at the *thr* locus. The three linear drawings depict successively the 0–30, 30–60, and 60–90 minute segments of the circular map. Selected portions of the map are drawn on a separate expanded scale to accomodate markers in crowded regions. Gene symbols are explained in accompanying Table 13–1. Markers in parentheses are approximately mapped at positions shown. Genes with asterisk (*) have been mapped more precisely than those in parentheses, but their order relative to adjacent genes marked by asterisks is unknown. For a complete listing of the genetic markers of *E. coli* see Taylor (1970), Taylor and Trotter (1972), and Bachmann, Low, and Taylor (1976). (b) 100-minute circular reference map of *E. coli* strain K-12. Large numbers refer to map positions in minutes relative to the *thr* locus. The two thin portions of the circle represent the only two map intervals not spanned by a continuous series of phage P1 cotransduction linkages. Inside the circle, the leading transfer regions of a number of Hfr strains are indicated. [(a) Abridged and reprinted with permission from A. L. Taylor, in *Handbook of Biochemistry,* 2nd ed., H. A. Sober, Ed. Copyright CRC Press, Inc., Cleveland, Ohio, 1970. (b) From B. J. Bachman, K. B. Low, and A. L. Taylor, *Bact. Revs., 40,* 116–167, 1976.]

TABLE 13-1
Some Genetic Markers of *E. coli*

Gene Symbol	Map Position (min.)	Trait or Function Affected
aceA, B	78	acetate utilization
ampA	82	penicillin sensitivity
araA, B, C, D	1	arabinose
araI	1	initiator locus
araO	1	operator locus
araE	56	L-arabinose permease
argB, C, E, H	77	arginine
argG	61	arginine
argP	57	arginine permease
argR	62	regulatory gene
argS	35	arginyl tRNA synthetase
aspA	82	aspartase
aspB	62	aspartate requirement
*att*λ, *434, 82*	17	integration site for prophages λ, 434, 82
*att*φ*80*	25	integration site for prophage φ80
azi	2	Na azide sensitivity
bioA, C, D, E	17	biotin
capS	22	regulatory gene for capsule polysaccharide synthesis
cheA	36	chemotactic motility
cysC	53	cysteine
dsdA	45	D-serine deaminase
dsdC	45	regulatory gene
end	50	endonuclease I
fda	60	fructose-1,6 diphosphate aldolase
fdp	84	fructose diphosphate
fuc	54	fucose utilization
galE, K, T	17	galactose utilization
galO	17	operator locus
galR	55	regulatory gene
glpD	66	α-glycerophosphate dehydrogenase
glpR	66	regulatory gene
gltA	16	glutamate requirement
gltS	73	glutamate permease
gltC	73	operator locus
gltR	79	regulatory gene for gene *gltS*
glyA	49	glycine
glyS	70	glycyl tRNA synthetase
guaA, B	48	guanine
guaO	48	operator locus
hag	37	flagellar antigens
his	39	histidine requirement
ilvA, B, C, D, E	74	isoleucine-valine
ilvO	74	operator locus for genes *ilvA, D, E*
ilvP	74	operator locus for gene *ilvB*
lacA, Y, Z	10	lactose utilization
lacI	10	regulatory gene
lacO	10	operator locus
lacP	10	promotor locus

TABLE 13-1
Some Genetic Markers of *E. coli*

Gene Symbol	Map Position (min.)	Trait or Function Affected
leu	1	leucine
lip	15	lipoic acid requirement
malP, Q	66	maltose
malT	66	probably a positive regulatory gene
melA, B	84	melibiose
metC	59	cystathionase
metE	75	methionine
mtc	12	sensitivity to mitomycin C
mtl	71	D-mannitol utilization
mutS	53	high mutability
mutT	1	high mutability, specifically AT → CG transversions
nalA	42	sensitivity to nalidixic acid
pabB	30	*p*-aminobenzoate requirement
pan	2	pantothenic acid requirement
pdxC	20	pyridoxine requirement
pgi	79	phosphoglucoisomerase
pheS	33	phenylalanyl tRNA synthetase
phr	17	photoreactivation of UV-damaged DNA
pil	88	presence of pili
pnp	61	polynucleotide phosphorylase
pps	33	pyruvate or lactate utilization
proA, B, C	7, 9, 10	proline
purB, C, D, E	23, 48, 78, 13	purine
pyrC, D, E	22, 21, 72	pyrimidine
recA, B, C	52, 55	UV sensitivity; competence for genetic recombination
rhaA, B, D	76	rhamnose utilization
rhaC	76	regulatory gene
rns	15	ribonuclease I
serA, B	57, 89	serine
strA	64	streptomycin sensitivity
supF	25	suppressor of *amber* (UAG) mutations
supM	78	suppressor of *ochre* (UAA) mutations
tfrA	8	resistance to phages T4, T3, T7, λ
thrA, D	0	threonine
thyA	55	thymidylate synthetase
tonA	2	resistance to phages T1, T5
tonB	25	resistance to phages T1, φ80
trpA, B, C, D, E	25	tryptophan
trpO	25	operator locus
trpR	90	regulatory gene
trpS	65	tryptophanyl tRNA synthetase
tsx	11	resistance to phage T6
tyrA	50	tyrosine
tyrR	27	regulatory gene for gene *tyrA*
uvrA, B, C, D	80, 18, 37, 74	repair of UV-damaged DNA
valS	84	valyl tRNA synthetase
xyl	70	D-xylose utilization
zwf	35	glucose-6-phosphate dehydrogenase

Source: Abridged and reprinted with permission from R. W. Barratt and A. Radford, in *Handbook of Biochemistry*, 2nd ed., H. A. Sober, Ed. Copyright CRC Press, Inc., Cleveland, Ohio, 1970.

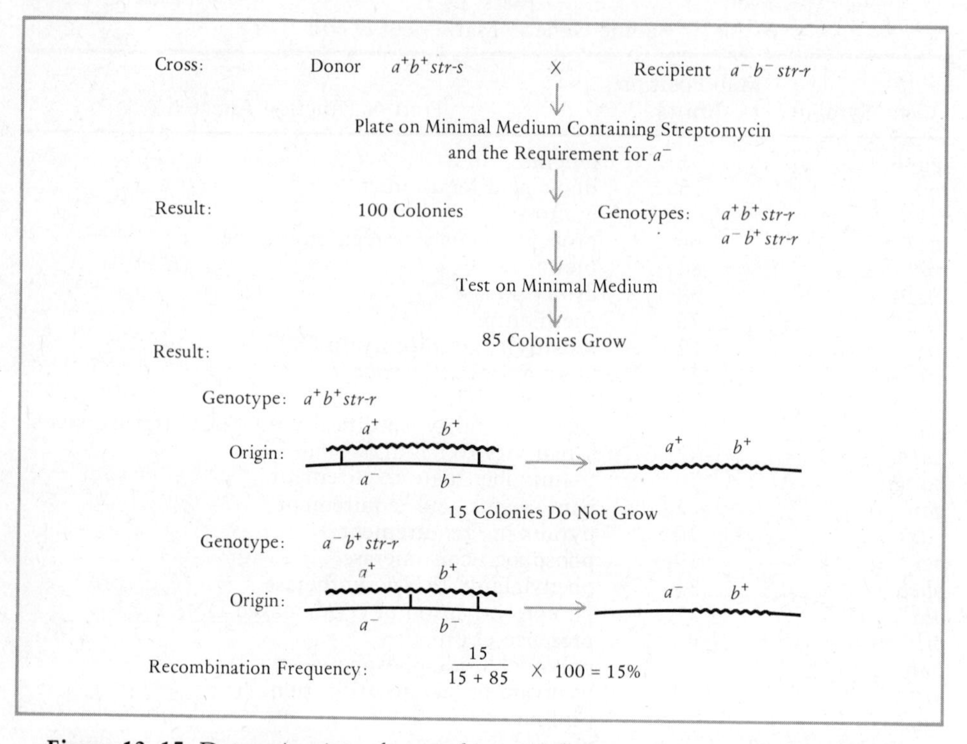

Figure 13–15. Determination of recombination frequency as a result of conjugation in the hypothetical cross between an a^+ b^+ *str-s* (streptomycin-sensitive) donor and an a^- b^- *str-r* (streptomycin-resistant) recipient.

(Figure 13–16). Thus, the number of prototrophic colonies produced by cross 1 will greatly exceed the number produced by cross 2. By scoring the protrophic colonies resulting from the two crosses, the gene order can be inferred.

Sexduction

As described above, the sex factor is capable of integration within the genome and also of excision from the genome. Ordinarily, excision occurs without loss or gain of genetic material for either the genophore or the sex factor. However, on rare occasions excision is imperfect, and the crossover which permits release of the sex factor occurs at the wrong site. As a result, the sex factor retains a small segment of the bacterial genophore. Such a sex factor is called an *F′ element.* In contrast to the usual F^+ plasmid, F' elements integrate into the genome with high frequency, usually at the same site within a recipient genophore that it occupied in the *Hfr* strain from which it was derived. This behavior is explained on the basis of pairing between the genes present in the genophore and their homologous alleles present in the F' element. Since the F' element is acting as a vector in gene transmission, the phenomenon is a form of *transduction* called *sexduction*. Numerous instances of sexduction have been analyzed and it has been found that from one to many genes can be transferred to recipient cells by this mechanism.

When an F' element carrying bacterial genes enters a recipient cell and is integrated into the genophore by crossing over, the recipient becomes diploid for those genes. Such partial diploids are called *merozygotes* or *heterogenotes*, and they provide a means by which complementation tests for allelism can be carried out. For example, suppose two different strains of bacteria exhibit a requirement for the same amino acid and the question arises as to whether these strains possess mutations of the same functional gene or mutations of different nonallelic genes whose products contribute to the same metabolic pathway. To distinguish between these alternatives, genes derived from one strain are introduced by an F' element into cells of the other strain. If the mutation present in the two strains are allelic, the mutant phenotype, characterized by auxotrophy for the required amino acid, will be maintained. However, if the mutations are not allelic, complementation and prototrophy will result. Complementation tests in bacteria can also be carried out with the use of abortive transduction and specialized transduction (see below).

TRANSDUCTION

A third mode of gene transfer in bacteria is *transduction*, a process in which a bacteriophage carries genes from donor to recipient cells. Transduction appears to be a widespread phenomenon and has been described in *E. coli*, *Salmonella*, *Shigella*, *Pseudomonas*, *Staphylococcus*, *Proteus*, and *Bacillus subtilis*. Two types of transduction are recognized: *generalized transduction* and *specialized transduction*. A bacteriophage serves as a vector of donor genetic material in each case, but in other respects the two types of transduction differ significantly from one another.

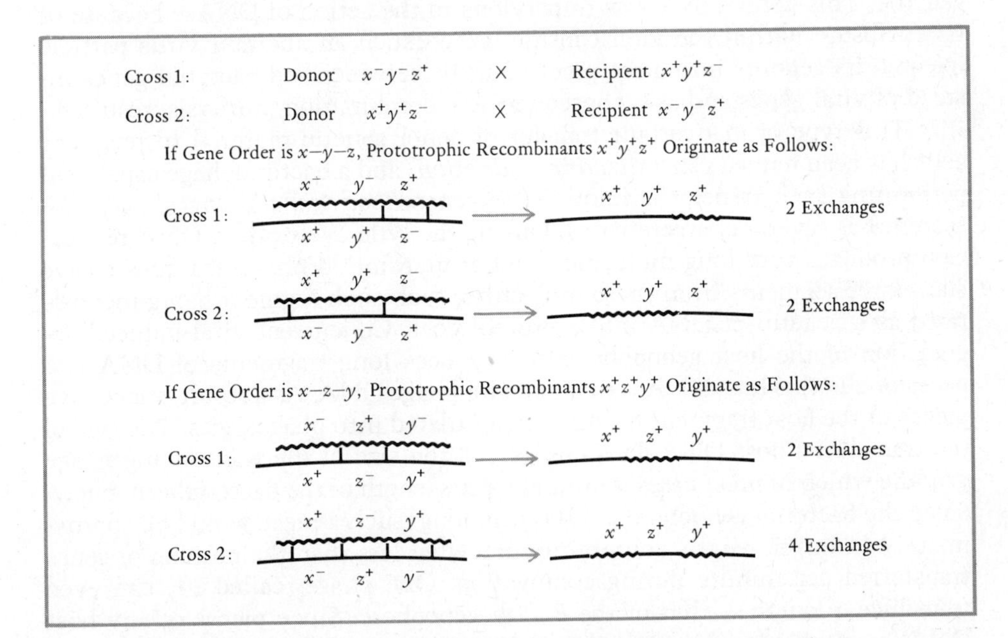

Figure 13–16. The use of reciprocal crosses to determine gene order.

Generalized Transduction

Transduction was first discovered by Lederberg, Lederberg, Zinder, and Lively in 1952 in *Salmonella typhimurium*. The initial purpose of their study was to determine if gene transfer and recombination via conjugation occurred in this species as well as in *E. coli*. Cells of two auxotrophic strains of *Salmonella* were utilized: one strain required phenylalanine, tryptophan and tyrosine, the other methionine and histidine. These strains can be symbolized as $a^- b^- c^- d^+ e^+$ and $a^+ b^+ c^+ d^- e^-$, respectively. When cultures of these two strains were mixed, prototrophic recombinants ($a^+ b^+ c^+ d^+ e^+$) were recovered with low frequency. Such prototrophs were also found when cells of one strain were exposed to cell-free filtrates of the opposite strain, indicating that cell contact and conjugation were not involved in the observed production of recombinants. A test for the presence of transforming DNA in the cell-free filtrate was then carried out with the use of deoxyribonuclease (DNAse), an enzyme that degrades DNA, and it was found that treatment with DNAse did not affect the frequency of recombination. This result effectively ruled out transformation as a possible transfer mechanism and suggested that donor genetic material was being transmitted to recipient cells in some kind of protected, but noncellular form.

Additional studies identified the agent of transfer as a bacteriophage called P22 whose genome was harbored as a prophage in the donor strain. Lysis of a donor cell occasionally released infectious virus particles which were small enough to pass through filters designed to check the far larger bacterial cell. The recipient strain was sensitive to P22, and therefore such virus particles ordinarily infected and lysed recipient cells. However, approximately 1 in 10^4 particles carried a fragment of the donor cell genophore rather than the P22 genome. This donor DNA was impervious to the action of DNAse because of its enclosure within the viral capsule. When such an aberrant virus particle injected its genome into a recipient cell, the cell received bacterial genes instead of viral genes and as a consequence, recombination, not lysis, resulted.

This type of inadvertant transfer of donor genetic material to recipient cells has been named *generalized transduction,* and a bacteriophage capable of performing such transfer is known as a *generalized transducing phage*. The system has several characteristics. During the lytic cycle phage DNA replication produces very long molecules, and at maturation phage enzymes cleave successive segments from these molecules, each such segment being incorporated as a headful of DNA into a protein coat. Concurrent viral-induced degradation of the host genophore also produces long fragments of DNA and occasionally this host DNA is mistaken for phage DNA. As a result, successive pieces of the host fragment become encapsulated into phage coats. The size of the transduced host DNA does not exceed the size of the transducing phage genome which in most cases is around $\frac{1}{100}$ the length of the bacterial genophore. Since the bacterial genophore is 1100 μm, long, such a piece would be approximately 11 μm in length, containing somewhat less than the number of genes transferred per minute during conjugation. One phage, called P1, can even transduce a length $\frac{1}{50}$ that of the *E. coli* genophore. Since pieces of any host DNA fragment can be packaged in this manner, it follows that any bacterial gene can be transduced. Although these mistakes in packaging are fairly rare, they are nevertheless characteristic of generalized transducing phages, as opposed to nontransducing phages such as T4.

Generalized transduction provides a useful method for mapping bacterial genes. Recombination between a transduced donor DNA fragment and a recipient genophore occurs via replacement of recipient genes as in transformation. However, in contrast to transformation, it is unnecessary to determine if a double recombinant arose from the incorporation of more than one separate DNA fragment since the relative rarity of phage particles containing host DNA makes such a possibility as remote as the frequency of mutation. Therefore, the recovery of double or triple transductants is immediate evidence for linkage between the genes involved. If two marker genes are almost always transduced together, close linkage is assumed, and the recombination frequency between such genes can be estimated by the methods applied in transformation. For example, if donor cells are $a^+ b^+$ and recipients are $a^- b^-$, any $a^+ b^-$ or $a^- b^+$ transductants must arise from an exchange occurring between the loci of genes a and b. Such transductants are therefore recombinants. In contrast, a double transductant $(a^+ b^+)$ must arise from exchanges to either side of genes a and b. As with transformation, linkage distance is calculated as the proportion of total recombinants to total transductants. When three genes are cotransduced, the order of genes can also be determined. This is usually done through reciprocal crosses in which one of the three genes is an outside marker. The method is essentially similar to that previously described (see Figure 13–16).

An interesting phenomenon associated with generalized transduction is that of *abortive transduction*. In abortive transduction the bacterial DNA transported to a recipient cell by a transducing phage particle fails to undergo recombination with the recipient's genophore and instead remains in the cytoplasm of the recipient cell. Although unable to replicate, these newly introduced genes are functional, and their activity may bring about a change in the phenotype of the recipient cell. When this cell divides, the unreplicated transduced DNA is passed to only one of the two daughter cells, this unilinear inheritance continuing as long as the transduced genes are present. When the transduced DNA contains an allele rendering a recipient cell prototrophic, abortive transduction is recognized by the presence of tiny colonies which can readily be distinguished from the large colonies formed as the result of recombination (complete transduction). The tiny colonies arise because of the functioning of the transduced genes. Even though these genes are present in only one cell at any given time, their products may persist for a generation or two in all daughter cells, rendering them temporarily prototrophic and able to divide. Since abortive transduction results in temporary diploidy for the particular genes that have been introduced, it is used in complementation tests for functional allelism.

Specialized Transduction

Specialized transduction involves the transfer of certain specific host genes (and no others) by a temperate bacteriophage. The particular genes transferred are those that lie on either side of the integrated prophage (*gal* or *bio* in the case of λ). It is visualized that when induction occurs and the prophage is

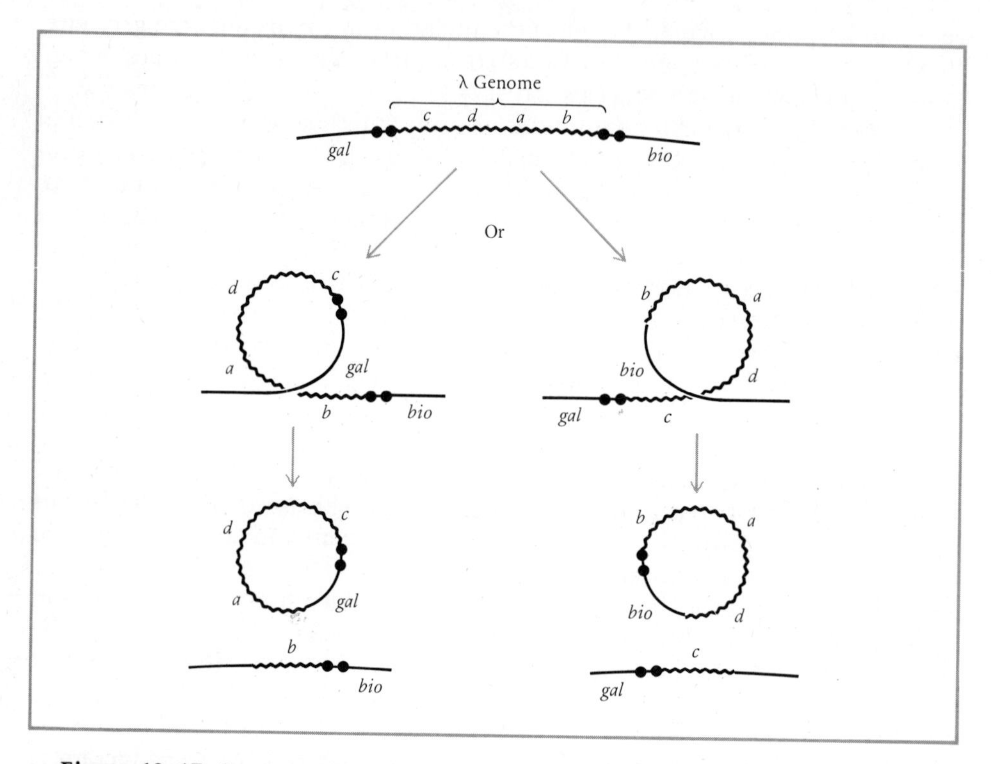

Figure 13–17. Improper excision of λ from the bacterial genophore to produce a specialized transducing phage carrying either the *gal* or the *bio* bacterial locus. Attachment sites are shown as dots; *a, b, c,* and *d* are hypothetical λ genes. (See also Figures 12–6 and 12–7.)

excised from the host genome, the crossover which causes excision may, on rare occasions, occur at the wrong site. If so, the released prophage will come to contain a small segment of the host genophore, but will leave behind an equivalent segment of its own genome.

In the case of λ, it can be seen (Figure 13–17) that if the abnormal crossover involves the region adjacent to *gal,* the resulting phage particle will include the host *gal* gene, but will be deficient for phage genes located at the opposite end of the genome, next to the *bio* locus. Conversely, if the crossover includes the locus of *bio,* the resulting phage particle will include the host gene *bio,* but will be deficient for phage genes situated next to *gal*. Such particles are termed λ*dgal* (deficient, but carrying *gal*) and λ*dbio* (deficient, but carrying *bio*). They contain an amount of DNA approximately equivalent to that present in a normal λ genome, and upon induction of lysogenic cells, they arise with a frequency of around 1 in 10^6 virus particles.

λ*dgal* and λ*dbio* are *specialized transducing particles* because they transfer these particular genes to new hosts. They differ from generalized transducing particles in the following respects. Generalized transducing particles carry bacterial genes *instead of* phage genes. Upon entrance into a recipient cell, the transduced genes may *replace* those of the recipient cell. Specialized transducing particles carry bacterial genes *as well as* phage genes. If the phage integrates within the host genophore, the transduced genes are *added to* the host genome. Since this addition results in partial diploidy, specialized transduction can be used in tests for functional allelism.

Besides λ, other temperate bacteriophages of *E. coli* mediate specialized transduction. One of these, phage $\phi 80$, specifically transduces genes for tryptophan synthesis and in doing so becomes defective as in the case of λ. Other inducible prophages of *E. coli* include phages 82 and 434. Each such phage appears to integrate at a particular attachment site on the *E. coli* genophore.

PROBLEMS

13–1. Two concentrations of donor DNA were used to transform recipient cells of genotype $a^- b^- c^-$. Judging from the frequencies of transformants, which genes were carried on the same DNA fragment?

Donor DNA	Transformant	0.05 μg DNA/ml	0.005 μg DNA/ml
$a^+ b^- c^-$	a^+	13.0%	1.3%
$a^- b^+ c^-$	b^+	12.8%	1.28%
$a^- b^- c^+$	c^+	12.9%	1.29%
$a^+ b^+ c^-$	$a^+ b^+$	11.2%	0.017%
$a^+ b^- c^+$	$a^+ c^+$	11.7%	0.017%
$a^- b^+ c^+$	$b^+ c^+$	12.6%	1.27%

13–2. In a species of bacteria the genes arginine *(arg)* and tryptophan *(tryp)* are linked. DNA isolated from a prototrophic $arg^+ \ tryp^+$ strain was used to transform an auxotrophic $arg^- \ tryp^-$ strain. From the data below, what is the linkage distance between these genes?

Cross: donor $arg^+ \ tryp^+ \times$ recipient $arg^- \ tryp^-$

Transformants:	
$arg^+ \ tryp^-$	142
$arg^- \ tryp^+$	125
$arg^+ \ tryp^+$	325

13–3. DNA from a strain $thr^+ \ his^+ \ tyr^+$ was used to transform $thr^- \ his^- \ tyr^-$ recipients. The transformants recovered are listed below. What is the order of these genes and the linkage distance between them?

$thr^+ \ his^- \ tyr^-$	135
$thr^- \ his^+ \ tyr^-$	313
$thr^- \ his^- \ tyr^+$	90
$thr^+ \ his^+ \ tyr^-$	8
$thr^+ \ his^- \ tyr^+$	327
$thr^- \ his^+ \ tyr^+$	185
$thr^+ \ his^+ \ tyr^+$	2232

13–4. Below is given the order in which genes *p, q, r, s, t, x,* and *y* enter a recipient cell during conjugation with four different *Hfr* donor strains.

(1) O—*x*—*y*—*p*—*q*—*r*—*s*—*t*
(2) O—*q*—*p*—*y*—*x*—*t*—*s*—*r*
(3) O—*t*—*x*—*y*—*p*—*q*—*r*—*s*
(4) O—*t*—*s*—*r*—*q*—*p*—*y*—*x*

For each of the above, diagram a circular genophore containing these genes. Indicate the site of integration of the sex factor and the polarity (orientation) of the origin (O).

13–5. Following conjugation between a penicillin-sensitive, *arg*$^+$ *pro*$^+$ *Hfr* strain with an *arg*$^-$ *pro*$^-$ penicillin-resistant F^- strain, exconjugants were plated on a medium containing arginine and penicillin, but not proline. After incubation 650 *pro*$^+$ colonies were recovered; of these, 638 were not only *pro*$^+$, but also *arg*$^+$, while 12 were *pro*$^+$ and *arg*$^-$. If the order of gene transfer is Origin—*arg*—*pro*, what is the frequency of recombination between *arg* and *pro*?

13–6. An interrupted mating experiment was performed with an *Hfr* strain carrying markers $a^+ b^+ c^+ d^+ e^+ f^+$ and an F^- strain of genotype $a^- b^- c^- d^- e^- f^-$. Recombinants recovered at timed intervals were the following: after 10 mins, $a^+ b^- c^- d^- e^- f^-$; after 13 mins, $a^+ b^+ c^- d^- e^- f^-$; after 18 mins, $a^+ b^+ c^+ d^- e^- f^-$. When the duration of conjugation was 20 mins or more, no recombinants for genes *d, e,* or *f* were recovered, but instead, recipient cells underwent lysis. How would you explain these results?

13–7. In *E. coli,* assume that mapping experiments have shown that genes *b* and *c* are both approximately 2.5 linkage units from gene *a*, but the data are insufficient to resolve the exact order. If the actual order was *a*—*c*—*b*, which of the following sets of reciprocal crosses would prove this order? What type of medium and therefore what kind of recombinants are scored in the determination?

	Donor		Recipient
(1)	$a^- b^- c^+$	×	$a^+ b^+ c^-$
	$a^+ b^+ c^-$	×	$a^- b^- c^+$
(2)	$a^- b^+ c^+$	×	$a^+ b^- c^-$
	$a^+ b^- c^-$	×	$a^- b^+ c^+$
(3)	$a^- b^+ c^-$	×	$a^+ b^- c^+$
	$a^+ b^- c^+$	×	$a^- b^+ c^-$

13–8. Assume that genes *cys, leu, tyr* are almost invariably cotransduced. A generalized transducing phage infects and lyses *cys*$^+$ *leu*$^-$ *tyr*$^+$ host cells. Progeny phage particle are then used to infect *cys*$^-$ *leu*$^+$ *tyr*$^-$ cells, and the following recombinants are recovered.

cys$^+$ *leu*$^+$ *tyr*$^-$	291
cys$^+$ *leu*$^-$ *tyr*$^+$	782
cys$^+$ *leu*$^+$ *tyr*$^+$	601
cys$^-$ *leu*$^-$ *tyr*$^+$	20
cys$^+$ *leu*$^-$ *tyr*$^-$	510

What is the order of these genes?

13–9. Cotransduction of genes *his*$^+$ and *thr*$^+$ to recipient *his*$^-$ *thr*$^-$ cells produced the following transductants: 474 *his*$^+$ *thr*$^+$; 42 *his*$^-$ *thr*$^+$; and 36 *his*$^+$ *thr*$^-$. What is the linkage distance between these genes?

13–10. A streptomycin-sensitive *Hfr* strain, prototrophic for threonine (thr), leucine (leu), arginine (arg), and histidine (his), was allowed to conjugate with a streptomycin-resistant F^- strain whose genotype was *thr*$^-$ *leu*$^-$ *arg*$^-$ *his*$^-$. When plated on complete medium containing streptomycin, a number of colonies arose. Cells from five of these colonies were then tested on minimal medium supplemented with added nutrients as indicated below. What were the genotypes of the five colonies?

Colonies	Minimal Medium Plus THR LEU	THR HIS	LEU ARG	LEU HIS	ARG HIS
1	0	+	0	+	+
2	+	0	0	0	0
3	0	0	+	0	+
4	0	+	0	0	0
5	+	+	0	0	0

+ = growth; 0 = no growth

REFERENCES

BACHMAN, B. J., K. B. LOW, AND A. L. TAYLOR, 1976. Recalibrated linkage map of *Escherichia coli* K-12. *Bact. Rev., 40,* 116.

CAMPBELL, A. D., 1969. *Episomes*. Harper & Row, New York.

CURTISS, R., 1969. Bacterial conjugation. *Ann. Rev. Microbiol., 2,* 69.

HAYES, W., 1968. *The Genetics of Bacteria and Their Viruses,* 2nd ed. Wiley, New York.

HERSHEY, A. D. (Ed.), 1971. *The Bacteriophage Lambda*. Cold Spring Harbor Laboratory, Cold Spring Harbor, N.Y.

HOTCHKISS, R. D., AND M. GABOR, 1970. Bacterial transformation with special reference to recombination process. *Ann. Rev. Genet., 4,* 193.

JACOB, F., AND E. L. WOLLMAN, 1961. *Sexuality and the Genetics of Bacteria*. Academic Press, New York.

STENT, G. S., 1971. *Molecular Genetics*. Freeman, San Francisco.

SUSMAN, M., 1970. General bacterial genetics. *Ann. Rev. Genet., 4,* 135.

TAYLOR, A. L., 1970. Current linkage map of *E. coli. Bact. Rev., 34,* 155.

TAYLOR, A. L., AND C. D. TROTTER, 1972. Linkage map of *E. coli* K-12. *Bact. Rev., 36,* 504.

Chapter
14

DEFINING THE GENE

WE have previously defined the gene in terms of function and have classified mutations as nonallelic or allelic on the basis of whether or not complementation occurs when two such mutations are present together in the heterozygous condition. From the analysis of gene products in the form of faulty polypeptides, we have also inferred that the gene is composed of numerous subunits independently subject to mutation, and the structure of DNA with its linearly arranged nucleotides provides a physical basis for this concept. Since recombination is a fundamental property of DNA, we can assume that on occasion it will occur between two different mutated sites of the same locus, thereby providing the means to analyze the gene in genetic terms.

INTRAGENIC RECOMBINATION AND FINE STRUCTURE MAPPING

That genetic recombination could occur within the limits of a single functional gene was first suggested by C. P. Oliver in 1942, but because the concept of the gene current at the time was that of an indivisible particle, Oliver's suggestion was not widely accepted. Some years later, recombination between noncomplementing alleles was again described in *Drosophila,* this time by M. M. Green and K. C. Green in 1949 and by E. B. Lewis in 1951. In both instances it was found that flies heterozygous for a pair of allelic mutations (a^1/a^2) produced recombinant-type progeny whose origin could be traced to crossing over between the two mutated sites of the locus.

We can examine this finding more closely with the use of a hypothetical example. Suppose a^1 and a^2 are independent, allelic mutations that produce a mutant phenotype when heterozygous, that is, they are noncomplementing. Let us further suppose that the mutated site in a^1 is at the left end, and the mutated site in a^2 is at the right end of locus a, and that in each case, except for these sites, the rest of the gene is normal (indicated by + in Figure 14–1). When these mutations are present on opposite homologues in a heterozygote, the configuration is called the *trans* arrangement. If, during meiosis, crossing over between nonsister chromatids occurs somewhere in the interval between a^1 and a^2, two kinds of recombinant chromatids will be produced: one containing both mutated sites on the same chromatid (called the *cis* arrangement) and one containing neither mutation and therefore wild type for the a locus.

In testcross progeny the recombinant class with the doubly mutated locus will usually be difficult to distinguish phenotypically, but the wild type progeny can be easily identified, and their appearance indicates that recombination between sites a^1 and a^2 has occurred. Since the interval between two mutated sites within the same functional gene is very small, recombinants for such sites are expected in very low frequency, and their recovery requires that testcrosses be done on a large scale.

Although Figure 14–1 illustrates the principle of intragenic crossing over and the chromatids which result, recombination within the gene is better demonstrated with the use of outside marker genes located to the left and right of the locus under study. An example is shown in Figure 14–2, where the flanking markers, x and y, are positioned to either side of the a locus. From the diagram of the testcross, it can be seen that the production of recombinants for the mutated sites a^1 and a^2 is accompanied by the exchange of these outside

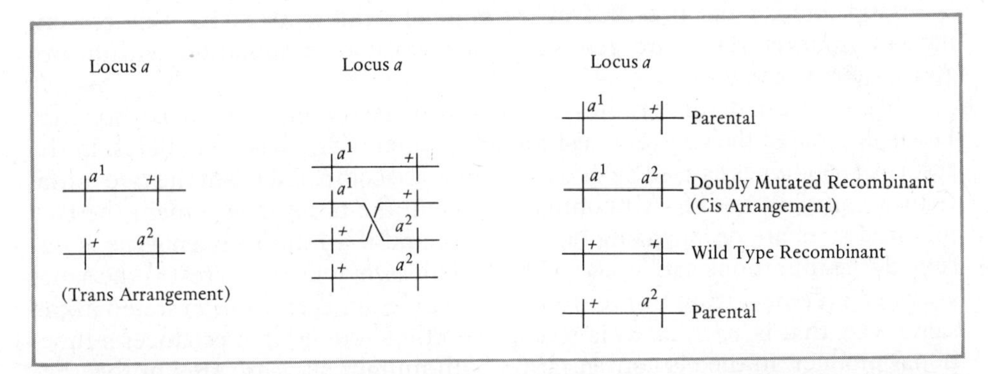

Figure 14–1. Example of crossing over between two different mutated sites, a^1 and a^2, present in the trans arrangement in a heterozygote. Sites a^1 and a^2 are different mutations of the same functional gene, locus a. Recombination between a^1 and a^2 produces chromatids containing a doubly mutant and a wild type locus, respectively. When the two mutations are present on opposing homologues, the arrangement is called *trans*; when they are both present on the same chromosome, the arrangement is called *cis*.

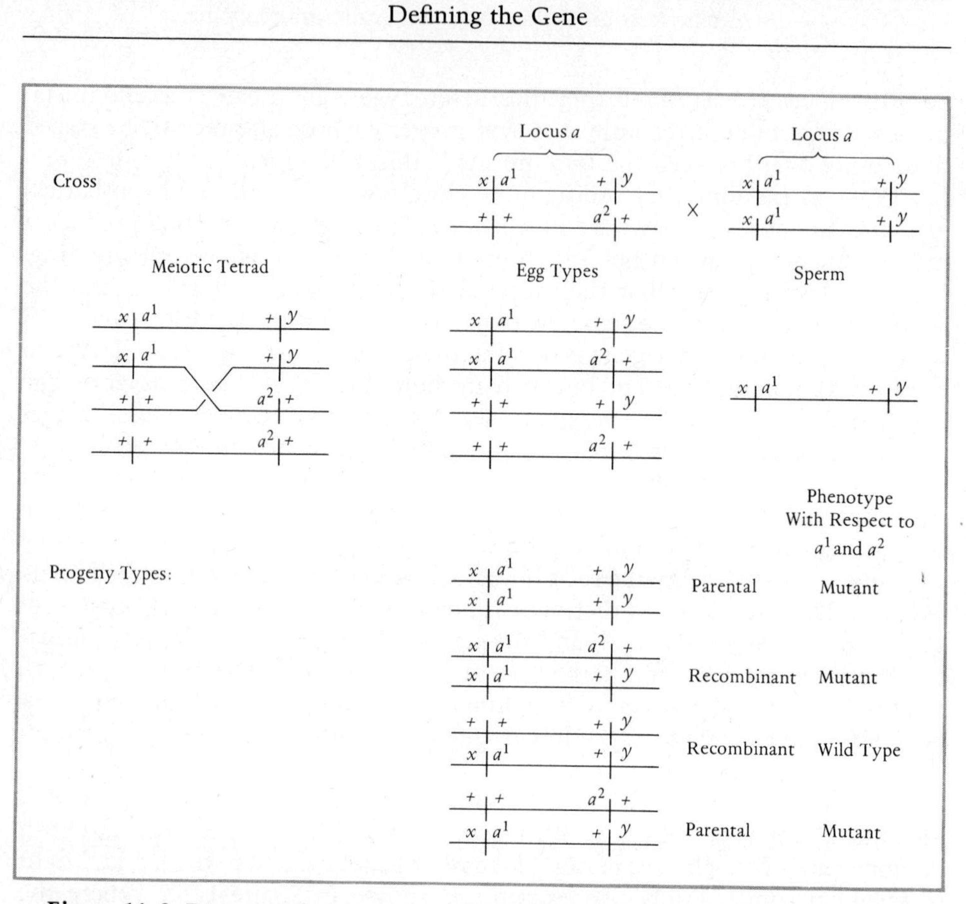

Figure 14–2. Diagram of a testcross of females heterozygous for allelic mutations a^1 and a^2 of locus a and for outside marker genes x and y. Recombination between sites a^1 and a^2 is accompanied by the exchange of the outside marker genes.

marker genes, indicating that crossing over, and not some other phenomenon, is responsible for the presence of recombinant individuals. For this reason, flanking marker genes are generally used when recombination within any given locus is analyzed.

As a result of his studies of intragenic recombination in *Drosophila,* Lewis originated the *cis-trans* test for complementarity between alleles. In the test the phenotypes caused by two mutations are compared when the two mutations are present in different configurations. In the trans arrangement the two mutated sites are on opposing homologues, and if a mutant phenotype is observed, the mutations can be considered allelic. Obviously this test is the same as the complementation test for functional alleles and can be interpreted in the same way, that is, each locus is mutant and therefore neither produces a functional product. In the cis configuration both mutant sites are present together on the same homologue, while the opposing homologue contains the wild type locus. If the phenotype is mutant in the trans arrangement, but wild type, or nearly so, in the cis arrangement, the mutations can be considered allelic. The wild type phenotype seen in the cis configuration arises because although one homologue contains a doubly mutated gene, the other homologue contains a compensating normal allele. Since a trans arrangement which produces a mutant phenotype is sufficient to indicate functional allelism, the cis portion of the test is merely a formal control. However, its significance lies in the fact that it can be achieved only by crossing over between mutated sites within the same locus.

When many different allelic mutations of a locus are available, it is possible to determine the sequence of these mutated sites by the frequency with which recombination occurs between them, and by this means a linear map of a single functional gene may be derived. For example, Green and Green prepared such a map of the lozenge locus (*lz*, abnormal eye) of *Drosophila* through the use of recombination frequencies obtained from testcrosses of females heterozygous for different *lz* alleles. Choosing a representative allele for each mutated site, the map distances found were the following.

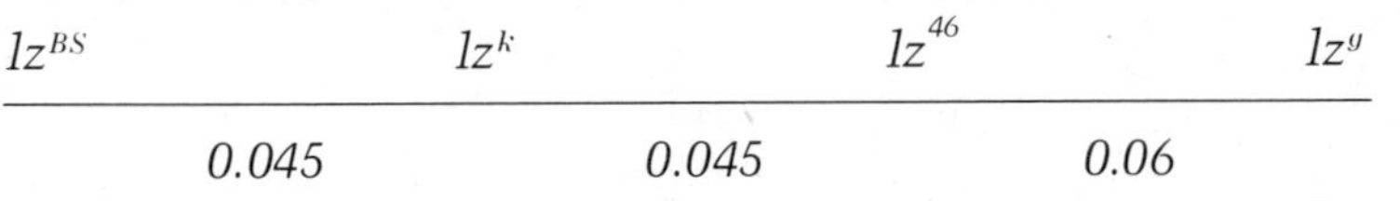

When these map distances are converted to numbers of individuals, it can be seen that recombination between these sites is rare. The map distance between lz^{BS} and lz^{k} is only 0.045 units, and therefore total recombinants of both reciprocal types represent only 0.045 percent of the progeny, or 4 to 5 per 10,000. As mentioned above, recovery of such recombinants requires large populations.

The studies of intragenic crossing over in *Drosophila* were followed by similar studies in *Neurospora, Aspergillus,* and bacteria, where again it was found that mutated sites within the same functional gene underwent recombination with one another and, in view of the general nature of the phenomenon, the idea of the gene as an indivisible particle was abandoned. Although the recombination frequencies obtained in these investigations permitted the construction of linear maps of the gene, the complexity of the genomes of bacteria and eukaryotes precluded estimates relating the minimum size of a mutational unit or the minimal distance required for exchange to a given length or number of nucleotides in DNA. Such estimates were made possible, however, through the analysis of intragenic recombination in bacteriophages whose total DNA content and gene number were known. Such high resolution studies of recombination are called *fine structure mapping,* and the ultimate in such studies was carried out by Seymour Benzer with phage T4.

For his experiments Benzer utilized mutations of the rapid lysis *(r)* character. The wild type phage (r^{+}) is capable of forming plaques on *E. coli* strain B as well as on another *E. coli* strain which is lysogenic for phage λ, and is called K-12λ. Mutations of r^{+}, called *rII* mutants, can lyse only strain B, and not strain K-12λ. The ability to form plaques on strain K-12λ was used as a selective procedure in the recovery of recombinant wild type particles. By the use of double infections, Benzer performed complementation tests between pairs of *rII* mutants and found all could be assigned to one or the other of two functional groups, indicating the presence of two separate genes, called *A* and *B*.

Benzer then proceeded to map the mutated sites within each of these genes by the following method (Figure 14–3). Mutants belonging to the same functional group (r^{x} and r^{y}) were used in a mixed infection on strain B cells. During the ensuing vegetative reproduction, recombination occurs between the respective genomes of the two phages present in the host, and if each

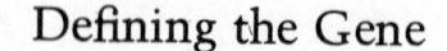

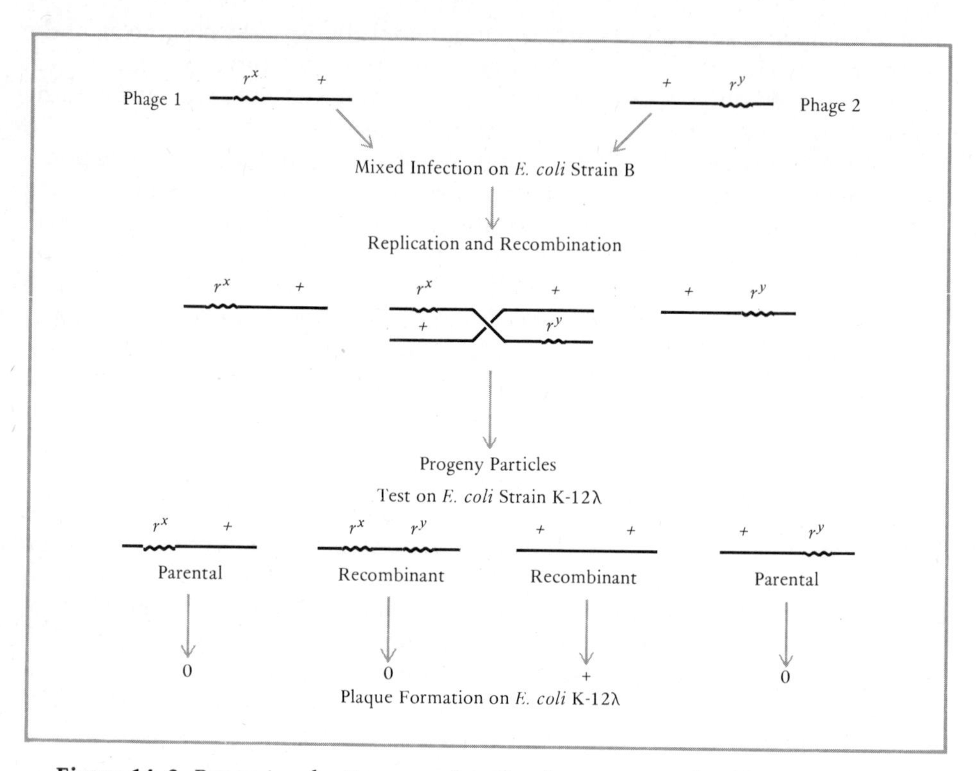

Figure 14–3. Benzer's selective procedure for the recovery of wild type (r^+) recombinants in phage T4. Two phages, r^x and r^y each mutant at different sites within the same gene, are used to produce a mixed infection on *E. coli,* strain B. Progeny phage are then used to infect *E. coli* strain K-12λ. Since only wild type (r^+) phage can form plaques on K-12λ, any plaques arising from this infection originate from wild type recombinants.

genome is mutated at a different site within the same gene, a rare exchange may occur between these sites. As a result, two types of recombinants will be formed: one will be doubly mutant, while the other will be wild type. When progeny particles from the infection on strain B cells are plated on strain K-12λ, neither the parental genotypes nor the doubly mutant recombinants can form plaques, and therefore any plaques that do arise can be attributed to the presence of wild type recombinants. On the assumption that both kinds of recombinants are produced with equal frequency, the number of wild type plaques multiplied by two provides an estimate of total recombinants which can be used for mapping. This method of selection is so sensitive that it permits the detection of wild type particles when their frequency is only 1 per 1 million. However, the rate at which *rII* mutations revert back to wild type is such that the lowest frequency of recombination between two *rII* mutants that is unambiguously detectable is 0.02 percent.

Benzer had available over 2000 *rII* mutations and performing the above procedures with all paired combinations of these was an impossible task. Therefore he devised a method whereby a mutated site could first be localized to a small segment of the *A* or *B* gene. His method made use of deletions and is called *deletion mapping.* In phage, deletions are recognized by their stability and their failure to revert by back mutation to the normal condition. The extent of a deletion can be determined by performing crosses with previously mapped, single site mutations. For example, a deleted mutant and a mutant of known position can be used to infect *E. coli* strain B, and their progeny plated on strain K-12λ. If the deletion overlaps the position of the known mutation, no wild type recombinants are possible, and therefore no plaques on K-12λ will be formed. However, if the two are not overlapping, a wild type recombinant can arise (Figure 14–4). By testing a deletion against a series of mutations whose intragenic map positions are known, the extent of the deletion can be estimated.

Once a series of deletions of graded extent is available, they can be used to locate unknowns. Such a series of hypothetical deletions is shown in Figure 14–5. It can be seen that unknown mutant *x* produces no wild type recombi-

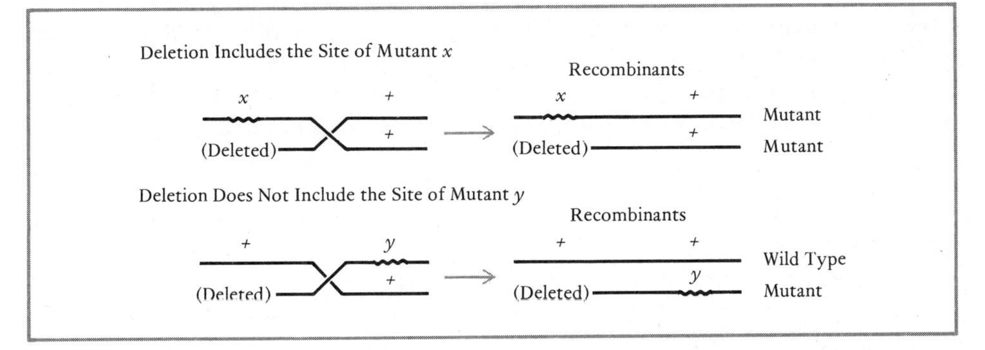

Figure 14–4. Principle of deletion mapping. If the deletion includes the site of a mutation, no wild type recombinants are possible. If the deletion does not include the site of a mutation, wild type recombinants can be formed.

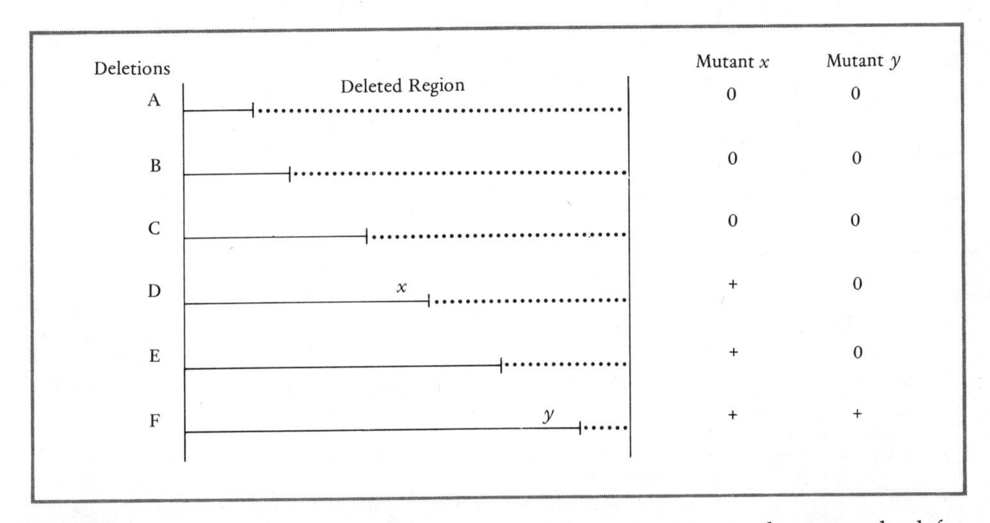

Figure 14–5. A series of graded deletions of known extent is shown to the left. Presence (+) or absence (0) of wild type recombinants resulting from crosses of deleted mutants with unknown mutants *x* and *y* is given on the right. The data indicate that mutation *x* lies in the region missing in deletion C, but present in deletion D, while unknown mutation *y* lies in the region present in deletion F, but absent in all other deletions of the series.

nants in crosses with deletions A, B, or C, but does yield such recombinants with deletion D, as well as with E and F. Accordingly, unknown *x* can be localized to the region present in deletion D, but absent in deletion C. Similarly, if wild type recombinants are produced only in crosses between unknown mutant *y* and deletion F, unknown *y* can be localized to the region present in deletion F, but absent in all other deletions of the series.

Benzer applied such a system to the *rII* mutants (Figure 14–6). To provide rough localization as to region, seven large deletions were used. A second series of deletions specific to each region was then employed to narrow further the mutated site present in an unknown, and finally, crosses between mutants localized to a given restricted segment of the *A* or *B* gene were carried out to determine recombination frequencies and map positions. By these methods, Benzer mapped more than 2400 mutations to a linear sequence of over 300 sites, 200 within the *A* gene, and 108 within the *B* gene (Figure 14–7). It was found that all mutations assigned to functional group A by complementation tests mapped within gene *A* and, similarly, all mutants assigned to function B mapped within gene *B*. In addition, it could be inferred that genes *A* and *B* were

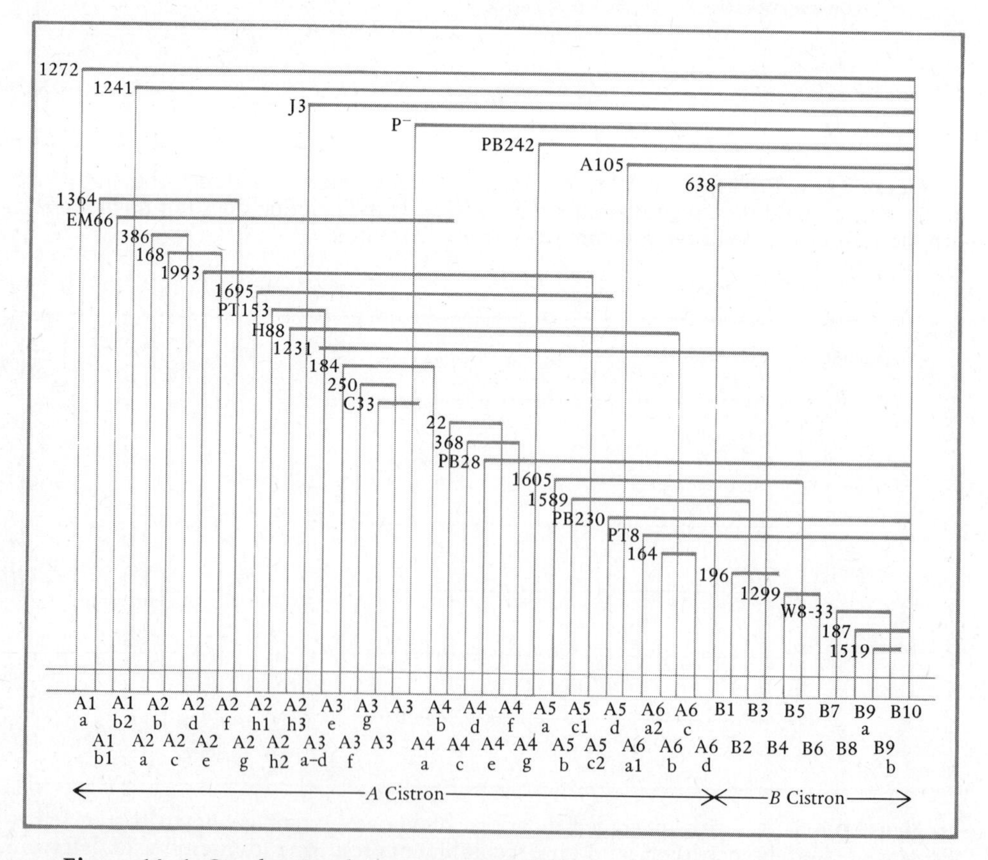

Figure 14–6. Overlapping large and small deletions (solid lines) used by Benzer to localize *rII* mutations within cistrons *A* and *B*. [From S. Benzer, *Proc. Nat. Acad. Sci.*, *47*, 403–415, 1961.]

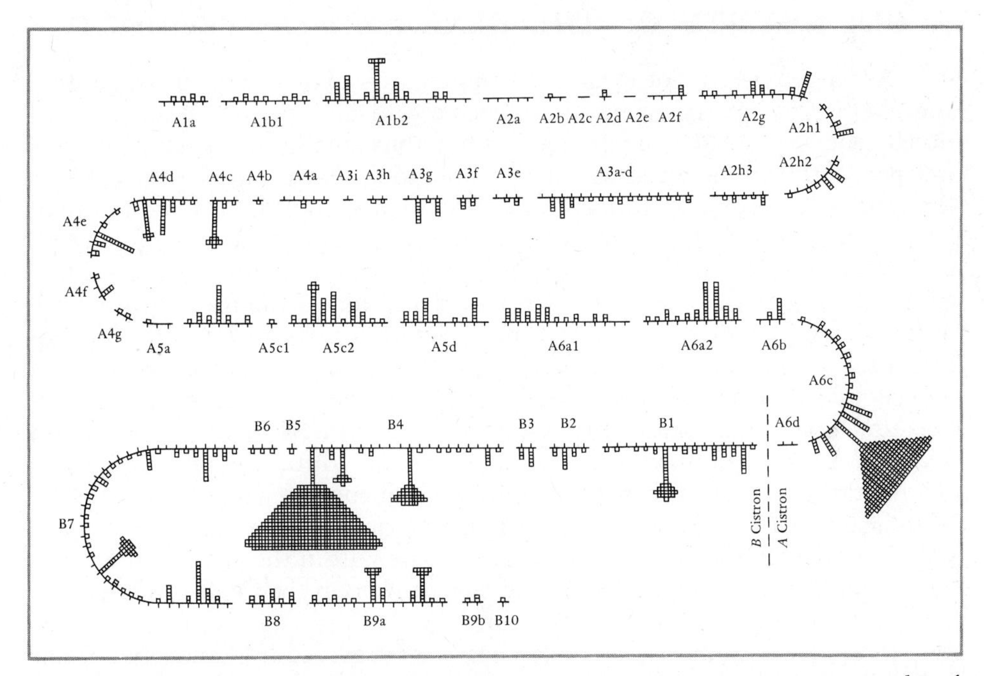

Figure 14–7. Topographic map of spontaneous mutations within cistrons *A* and *B* of the *rII* region of phage T4. Each square represents one occurrence of mutation. Note the high frequency of mutation at particular sites within these genes. [From S. Benzer, *Proc. Nat. Acad. Sci.*, *47*, 403–415, 1961.]

contiguous since the frequency of recombination between the closest *A* and *B* mutations was approximately the same as that which occurred between adjacent *A* sites or between adjacent *B* sites. It was also evident that some sites within these genes are far more prone to spontaneous mutation than are others (Figure 14–7). Why these sites mutate with significantly greater frequency than do their neighbors is unknown.

Benzer coined the term *cistron* to denote a genetic unit that has a single function as demonstrated by failure to complement in the trans configuration of the cis-trans test. Thus defined, a cistron is the equivalent of a functional gene. Benzer also applied the term *muton* to the smallest unit of a gene whose mutation can produce a mutant phenotype, and the term *recon* to the smallest distance within which recombination can occur. The lowest frequency of recombination unambiguously observed in these experiments was 0.02 percent, and when this frequency was applied to the known length of the T4 map and the known number of nucleotides present in the T4 genome, it could be estimated that the minimum distance within which recombination had occurred was 2 to 3 nucleotides. We should note here, however, that subsequent studies by C. Yanofsky (see Chapter 16) have shown that the muton is equivalent to a single nucleotide and the recon to the distance separating two adjacent nucleotides. Benzer's work was important because it strengthened the concept of the gene as a unit of function and in addition permitted visualization of the gene in physical terms of DNA.

COLINEARITY OF THE GENE AND ITS PRODUCT

As we have seen, fine structure mapping carried out with bacteriophages, and also with bacteria, fungi, and higher organisms, revealed that the functional gene, or cistron, is composed of numerous smaller units which mutate independently and between which recombination may occur. This concept of the gene is consistent with that previously inferred from the study of mutant polypeptides and is also in accord with the idea of the gene as a discrete sequence of base pairs within DNA. Thus, evidence derived from entirely different modes of analysis lead to the same model of gene structure: The gene can be visualized as a polarized sequence of nucleotide pairs which together constitute a set of symbols specifying a polarized sequence of amino acids within a polypeptide. It is evident that recombination can occur within this set of symbols and that mutations can alter any one of them with the result that a "wrong" amino acid may be present at some site within the final product. It can be further reasoned that if the polarized sequence of symbols in the DNA bears a linear relationship to the polarized sequence of amino acids in the polypeptide, then the position of a mutated site within the gene should correspond to the position of an amino acid substitution in the product, that is, the gene and its product should be colinear.

A demonstration of colinearity has been provided by Yanofsky and his colleagues in studies which combined fine structure mapping with sequence analysis of faulty polypeptides. These workers utilized the tryptophan synthetase A protein of *E. coli.* Tryptophan synthetase is an enzyme which catalyzes the terminal step in tryptophan biosynthesis (see Figure 10-5), and in *E. coli* it is composed of two subunits, A and B, each specified by different, but adjacent cistrons.

The genetic positions within the *A* cistron of a series of different mutations of this gene were first established by fine structure mapping using transduction. This study was accompanied by chemical analysis of the A polypeptide, from the amino to the carboxyl terminal, and the position and nature of the amino acid substitution present in the case of each mutation was determined. The genetic map of the mutated sites within the *A* cistron was then compared with the sequence of amino acid substitutions, and it was found that the two were the same (Figure 14–8). In addition, the linear distances between the amino acid substitutions were generally proportional to the recombination frequencies observed between the different mutated sites within the cistron. Yanofsky's experiments thus demonstrated colinearity between the gene and its product. A similar demonstration of colinearity has since been carried out by Sherman and his coworkers with the yeast gene, cy_1, and its product, cytochrome c.

Application of the principle of colinearity often permits inferences as to the region of a polypeptide most sensitive to the effects of amino acid substitution via mutation. For example, the phenotype produced by different allelic mutations can be examined and related to the position of those mutations within the cistron, as determined by fine structure mapping. When this is done, it can usually be observed that those mutations that produce the greatest departure from the normal condition occupy nonrandom positions within the cistron. Since gene and polypeptide are colinear, it can be inferred that such positions may dictate the amino acids concerned with the active site of an enzyme or those involved in maintaining the folded configuration of the amino acid chain. Derangements in either of these functions would be expected to alter drastically the activity of the polypeptide product and, through it, the phenotype.

INTRAGENIC COMPLEMENTATION

Insights into the origin of a mutant phenotype have also been provided by examination of phenotypes produced when different alleles of a series are present together in the heterozygous condition. Although as a rule mutants deficient for the same function fail to complement one another, some exceptions exist and a partial restoration of the normal phenotype is occasionally observed. This phenomenon is called *intragenic* or *interallelic complementation* because it occurs between different mutant alleles of the same gene. Intragenic complementation has been most thoroughly examined in fungi and bacteria where gene products can be more readily isolated and identified. In such studies any complementation that occurs between functional alleles has been traced to enzymes composed of two or more subunits which may be specified by the same or different genes. Intragenic complementation is not observed when an enzyme performs its function as a single polypeptide (monomer).

It will be recalled that the ability of polypeptides to form such subunit associations is conferred by conformation, which in turn is a consequence of primary structure, that is, the particular sequence of amino acids specified by the gene. When two mutant alleles of the same gene are present, each will specify an abnormal polypeptide and these may be so altered in amino acid sequence that functional associations with one another or with other

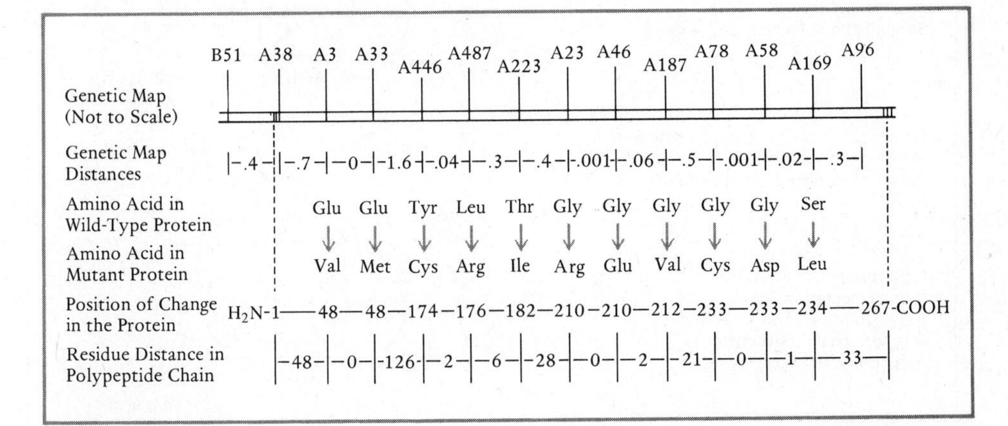

Figure 14–8. Genetic map of the tryptophan synthetase A gene and the corresponding amino acid changes in the A protein. The order of mutated sites within the gene is the same as the order of the amino acid changes in the protein, and the distance between mutational sites on the genetic map is representative of the distance between amino acid changes in the polypeptide chain. [From C. Yanofsky et al, *Proc. Nat. Acad. Sci.*, *57*, 296–298, 1967.]

polypeptides cannot be achieved. In this case enzyme activity will be absent and complementation will not occur. On the other hand, the separate derangements of each mutant polypeptide may be so situated that association of subunits becomes possible and thus some catalytic activity will be retained (Figure 14–9). The level of enzyme activity in such cases is usually 25 percent or less than that exhibited by either the normal organism or through complementation between different nonallelic genes. It must be stressed that the compensating activity observed in intragenic complementation arises at a secondary level of product interaction, and not at the primary level of gene expression.

As an example, the enzyme alkaline phosphatase of *E. coli* has been shown by Schlesinger and Levinthal to consist of two subunits which can be reversibly dissociated and characterized. Although the separate subunits (monomers) exhibit no enzyme activity by themselves, their reassociation *in vitro* restores full catalytic function. When the enzyme is isolated from strains that exhibit intragenic complementation, the separate monomers are each defective, but their union *in vitro* permits some resumption of normal activity. Similar studies have been performed with the enzyme glutamic dehydrogenase of *Neurospora* where intragenic complementation between mutants for this

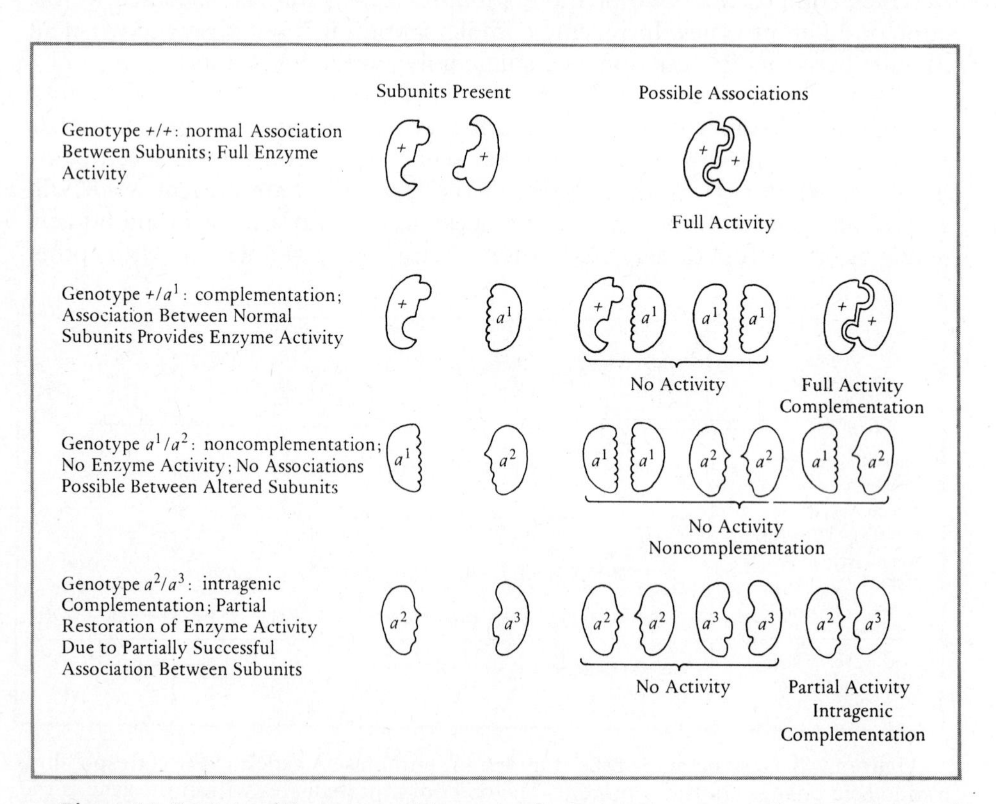

Figure 14–9. Normal Enzyme activity, complementation, noncomplementation, and intragenic complementation resulting from subunit association. In this example the two subunits of the enzyme are specified by the same locus.

enzyme has been traced to successful association between individually defective subunits.

Complementation maps can be constructed to represent the pattern of compensating activities exhibited by a series of alleles. In such maps groups of alleles that do not complement one another are placed together on a line. If the members of one such group fail to complement members of another such group, the two lines are drawn so as to overlap. However, when complementation between two groups does occur, their respective lines do not overlap. As an illustration, 45 allelic mutants of the rudimentary locus *(r)* of *Drosophila* have been described and a fine structure map of this locus has been developed by P. S. Carlson (Figure 14–10). While all mutations of this locus produce the phenotype of abnormal wings and female sterility when homozygous, certain heterozygous combinations exhibit intragenic complementation. In the absence of knowledge concerning the product specified by the rudimentary gene or the role of this product in the development of wings and reproductive organs,

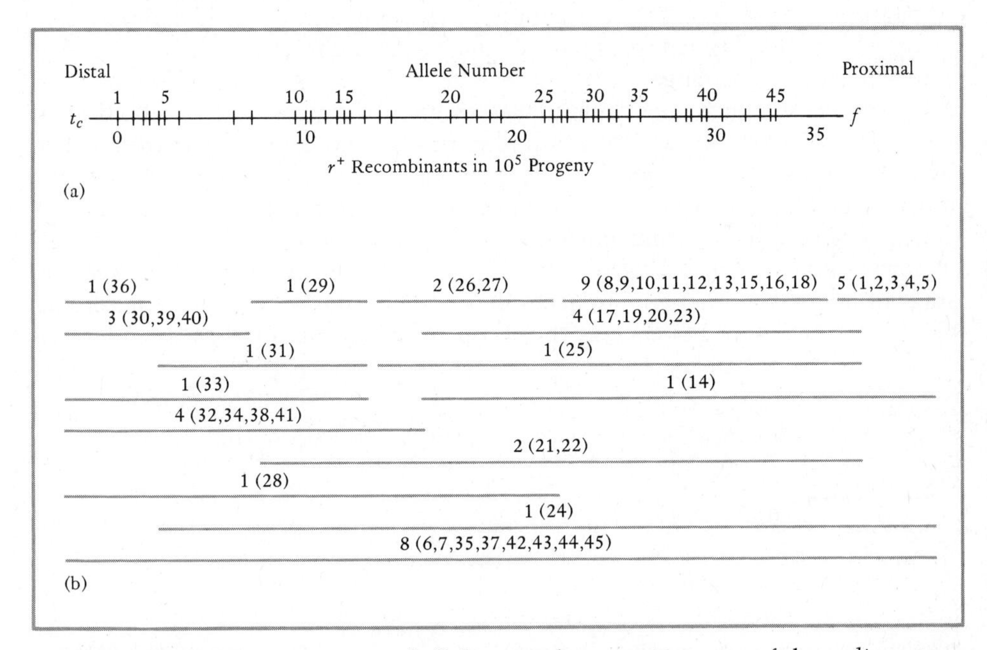

Figure 14–10. Fine structure and allelic complementation maps of the *rudimentary* *(r)* locus of *Drosophila*. (a) Genetic fine structure map. Numbers above the line are designations of each allele from 1 to 45. Numbers below the line represent the genetic distance along the fine structure map in terms of wild type recombinants in 10^5 progeny. All 45 alleles reside within a genetic distance of approximately 35 wild type recombinants in 10^5 progeny or a distance of 0.07 map units; t_c (tiny chaetae) and *f* (forked bristles) are linked flanking marker loci. (b) Complementation map. The locus includes sixteen complementation groups (solid lines), each represented by one or more alleles. The number of alleles in each complementation group is stated above the line and is followed, in parentheses, by the allele designation of the *r* mutants which belong to that group. [From Peter S. Carlson, "A Genetic Analysis of the *rudimentary* Locus of *Drosophila melanogaster.*" *Genet. Res.*, *17*, 53–81, 1971. Cambridge University Press.]

such a complementation map can tentatively be interpreted as indicating successful or unsuccessful interactions between mutant subunits that are part of a multimeric enzyme.

MECHANISMS OF RECOMBINATION

The process of recombination is central to genetic analysis, and two entirely different mechanisms have been postulated to explain this phenomenon (Figure 14–11). The *copy-choice* hypothesis proposes that recombination occurs during chromosome replication by the alternate copying of first one and then the other of the two homologous parental chromosomes and is restricted to the time of chromosome duplication. A recombinant chromosome or genophore would be composed entirely of newly synthesized DNA and would not contain any portion physically derived from either of the two original parental DNA molecules.

The *breakage-reunion* hypothesis, first proposed by C. D. Darlington in 1935, postulates that recombination occurs through the breakage and reunion of parental chromosomes. Recombinant chromosomes would contain segments physically derived from both of the parental chromosomes that participated in genetic exchange.

The copy-choice mechanism was originally suggested by J. Belling in 1931, over 20 years before DNA was identified as the hereditary material. The hypothesis has several major flaws. It is inconsistent with the semiconservative mode of DNA replication, and it cannot be reconciled with experimental evidence which indicates that chromosome duplication occurs during the premeiotic S phase, well before the time of chromosome pairing at the zygotene stage of meiosis. It also fails to account for the formation of hybrid (heteroduplex) DNA whose occurrence as the result of recombination has been demon-

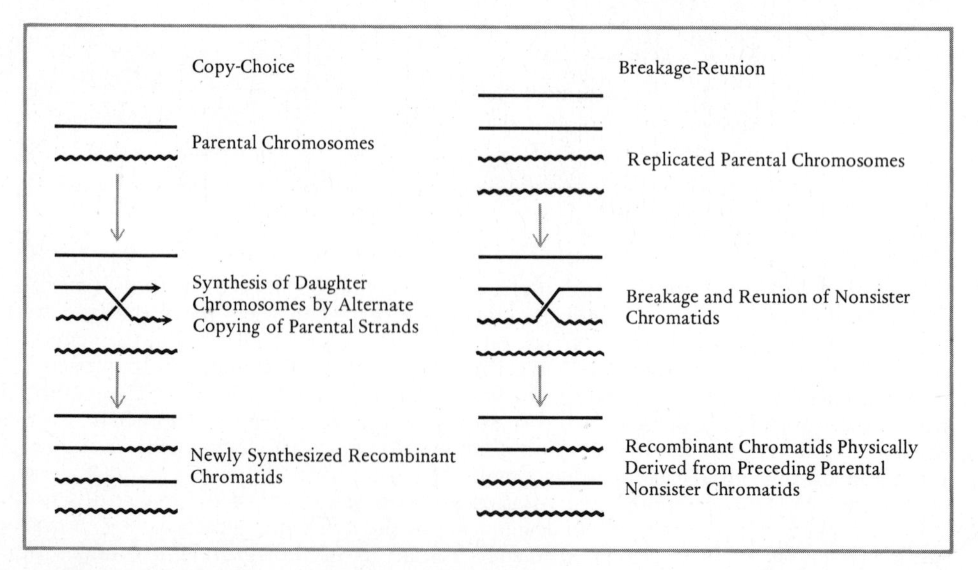

Figure 14–11. The copy-choice and breakage-reunion hypotheses to explain recombination.

strated in viruses and bacteria and can be inferred in fungi and higher eukaryotes. In contrast, the breakage-reunion hypothesis has been substantiated, for recombination accompanied by the incorporation of parental DNA into recombinant chromosomes or genophores has been demonstrated for all forms.

The advent of fine structure analysis and the development of methods for studying DNA and the enzymes associated with its replication and repair have required revision of some of the more traditional assumptions concerning recombination. It is clear that genetic exchange must be preceded by the pairing of homologous regions. According to the classical concept based on studies with the light microscope, homologous chromosomes at meiosis are paired precisely and continuously, gene for gene, throughout their length, and recombination occurs randomly between any two nonsister chromatids. In addition, the presence of one exchange appears to inhibit the nearby formation of a second exchange, a phenomenon called positive interference (see Chapter 8). It will be recalled that positive interference decreases with increasing distance between loci, so that no interference at all is detectable when the loci concerned are positioned some distance apart.

However, when recombination between immediately adjacent loci or between alleles of the cistron is analyzed, the phenomenon of localized high negative interference is observed (see Chapter 12). In other words, coincidence values higher than 1.0 are frequently observed, and the number of double exchanges within such a restricted region may be many times greater than that expected on the basis of random chance. This phenomenon should be distinguished from the low negative interference generally evident in phage recombination and attributable to multiple rounds of mating between DNA molecules of the vegetative pool. Localized high negative interference is found in phage, bacteria, and eukaryotes and has been interpreted to indicate the occurrence of multiple exchange events arising within a restricted, small region of *effective pairing* no longer than one or two cistrons. Such effectively paired regions are postulated to occur discontinuously and at random between two parental chromatids or DNA molecules. Thus the traditional concept of continuous intimate pairing has been replaced by the concept of discontinuous sites of effective pairing. If three or more mutations involved in a testcross are located far enough apart to prevent their being included in the same site of effective pairing, the usual and expected recombination frequencies will be obtained. However, if their loci are close enough together to be included in the same region of effective pairing, multiple exchanges and localized high negative interference can be detected.

The molecular events which occur within effective pairing sites and which result in recombination are not understood, and numerous interpretations and models which attempt to account for the process have been proposed. In general, it is thought that chromosome or genophore breakage is brought about by endonucleases. New arrangements between parental chromosomes are then postulated to occur through the overlapping and annealing of single-stranded DNA regions derived from each parent duplex molecule. If the region of overlap occurs at a site wherein the parent molecules are heterozygous for one or more mutations, the bases constituting the overlap will be partially or wholly mismatched and the region will consist of hybrid, or heteroduplex, DNA. Although this hybrid region may at first consist of fragments or segments held together only by hydrogen bonding, the subsequent action of DNA ligase insures the covalent joining of each strand of the recombinant DNA duplex molecule.

It is also postulated that before covalent union, the hybrid region may, in some forms, be corrected by removal of the mismatched bases of one of the two strands, followed by repair synthesis to fill any gaps and restore strict complementarity. The apparent multiple exchanges which give rise to the phenomenon of localized high negative interference are thought to be a consequence of molecular events occurring within the region of overlap and/or as the result of enzymatic repair within this region. Some of the evidence on which these proposals is based is reviewed below.

Viruses

It will be recalled that recombination in phage occurs within a pool of replicating DNA molecules and, with low frequency, gives rise to heterozygotes containing a short region of hybrid DNA. When outside marker genes are used along with the locus being followed, heterozygosity of that locus is often observed to be accompanied by the exchange of these outside marker genes, suggesting that the presence of hybrid DNA is related to the mechanism of recombination. These findings, plus the fact that recombination in phage is mostly nonreciprocal, initially prompted a revival of the copy-choice mechanism or a variation thereof. However, subsequent extensive studies with the T-even phages and with λ have provided clear evidence for the breakage and reunion hypothesis.

In 1961 M. Meselson and J. J. Weigle, using phage λ, demonstrated that recombination could occur in the absence of replication, a finding that effectively ruled out copy-choice as a mechanism. These workers utilized λ particles which had been labeled with the heavy isotopes ^{13}C and ^{15}N by previous infection of *E. coli* cells growing in a medium containing these isotopes. The heavy particles also contained linked marker genes affecting plaque clarity and morphology. The heavy particles, along with wild type, light ^{12}C–^{14}N particles were used to produce a mixed infection on light *E. coli*. Progeny virus particles were then collected and subjected to cesium chloride density gradient centrifugation, a procedure that does not affect the viability of the virus.

It will be recalled that such centrifugation causes the separation of particles (or molecules) into distinct bands according to their density. After centrifugation the particles within each band can be recovered by puncturing the bottom of the centrifuge tube and collecting the contents sequentially, drop by drop. Each drop can then be assayed on a bacterial lawn for the presence of virus particles and their genotypes, as indicated by the kind of plaques formed. It was expected that if recombination occurred by breakage and reunion, recombinant genotypes would be characterized by an intermediate density between the respective light and heavy densities of the parental molecules, and this result was, in fact, observed.

Although these findings showed that recombinant DNA molecules contained segments physically derived from a parental genome, they did not exclude the possibility that a portion of the recombinant molecule might be copied from the alternate parental genome. This possibility was eliminated through the results of further experiments in which the two parental phages of opposing genotypes were both labeled with heavy isotopes and used to infect light *E. coli* cells. Recombinants were recovered from these mixed infections which were composed entirely, or almost entirely, of heavy density DNA, indicating that the genome was derived from the breakage and reunion of unreplicated parental molecules. These results established breakage and reunion as the essential mechanism by which recombination was accomplished in these phages. Meselson also reported evidence which suggested that a very small amount of DNA was degraded and resynthesized during the recombination process.

Analogous results have been reported by J. Tomizawa and N. Anraku with phage T4. When phage DNA synthesis is inhibited by treatment of doubly infected cells with potassium cyanide, phage recombination is not affected, indicating that a copy-choice mechanism is not involved in recombination. In addition, when T4 particles of opposing genotypes are differentially labeled with heavy versus radioactive isotopes and used in a mixed infection on *E. coli*, recombinant phage particles whose DNA contains both labels can be recovered. Studies by Anraku and Tomizawa have revealed that recombinant molecules at first consist of a heteroduplex held together only by hydrogen bonding and containing single-strand gaps to either side of the hydrogen bonded region. These gaps are subsequently filled in by repair synthesis, and covalent bonding is accomplished by a DNA ligase specified by gene 30 of T4 (Figure 14–12). Initial steps leading to the formation of the joint molecule are not known, but it is postulated that pairing of homologous DNA molecules must occur first, probably followed by the production of single-strand cuts in parental molecules with digestion of the free ends or perhaps by the partial digestion of one end of each molecule so as to expose single-stranded regions that can join by base pairing. Whatever the means by which the single-stranded regions that contribute to joint molecules are formed, the process appears to be highly specific and enzymatically induced, and there is evidence that the single-strand binding protein specified by gene 32 of T4 is somehow involved.

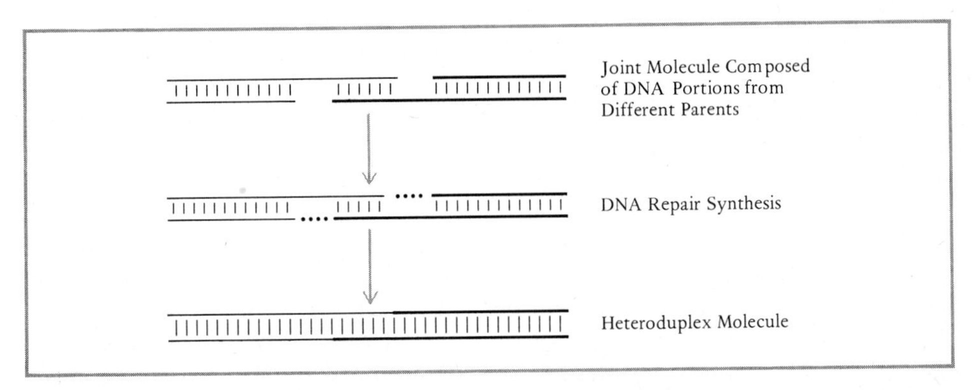

Figure 14–12. Model for the formation of recombinant molecules in phage T4. [From J. Tomizawa, *J. Cell. Physiol., 70, Suppl. 1,* 201–214, 1967.]

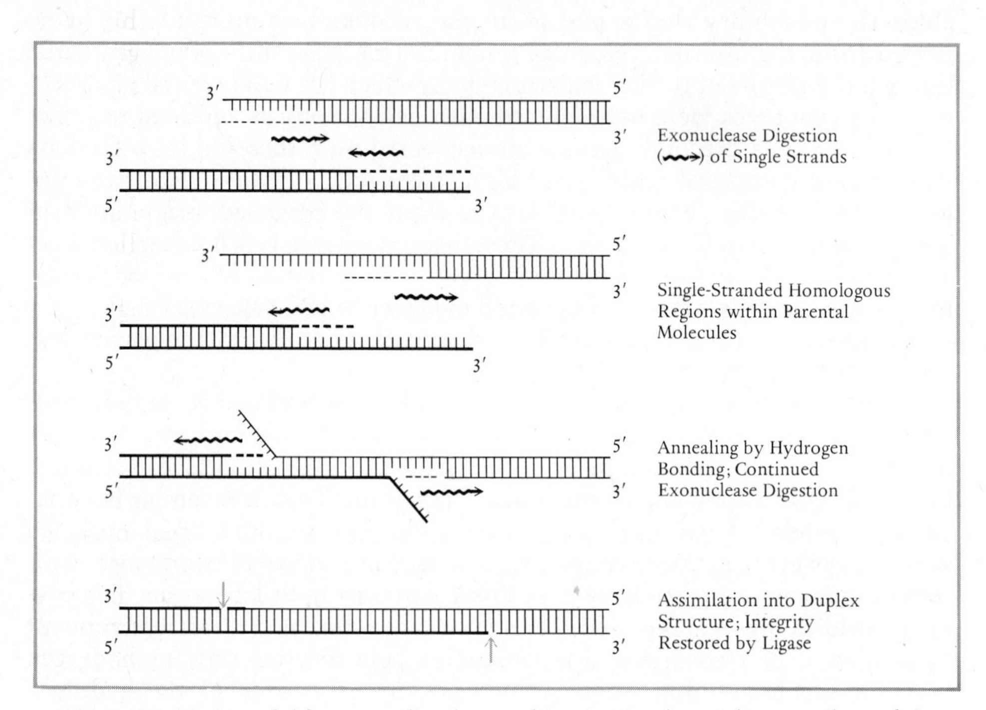

Figure 14–13. A model for generalized recombination in phage λ between homologous regions of two parental molecules (dark and light lines). Exonuclease digests one strand of the duplex DNA of both molecules, starting at a free 5′ end, either terminally located or produced internally by prior endonuclease action. Digestion produces single-stranded homologous regions within the parental molecules, which are annealed by hydrogen bonding. Exonuclease digestion of parental strands continues to permit complete annealing and terminates when all single-stranded regions are assimilated into a duplex structure. Integrity of strands is restored by ligase (at arrows). [Adapted from E. Cassuto and C. M. Radding, *Nature New Biol.*, *229*, 13–16, 1971.]

Casuto and Radding have proposed a scheme for phage λ similar to the partial digestion process postulated for T4. They analyzed recombination-deficient *(red)* mutants involving the *exo* gene which specifies an endonuclease active along with a protein termed β in the molecular process of recombination. On the basis of their studies they have proposed a model for recombination in λ in which neither gaps nor repair synthesis are required (Figure 14–13). Other models for generalized recombination in different phages have also been designed, such as that of T. Boon and N. Zinder for phage *fl* of *E. coli*. Many of these proposals are quite complex, and the interested reader is referred to Radding (1973) and Hotchkiss (1974) for a review of such hypotheses.

It must be noted once more that generalized recombination occurring in a lytic cycle in temperate phages such as λ differs significantly from the integration mechanism by which these phages become inserted into the host genome. In integration crossing over is site specific; in λ the gene *int* is required for integration, and both the genes *int* and *Xis* are needed for excision (see Figure 12–7). The respective attachment sites present in the λ and the *E. coli* genophores have no homology as indicated by the mismatching of bases in these regions in experimentally annealed heteroduplex molecules viewed by electron microscopy. In the absence of significant base homology, it has been proposed that the products of the *int* and *Xis* genes can somehow recognize the base sequence present within the attachment region of both the host and the phage genomes and facilitate integration and excision at this site.

Prokaryotes

The mechanism of recombination in prokaryotes such as *E. coli* appears to conform, in general, to that proposed for viruses. It will be recalled that transformation as well as transduction occurs through the physical incorporation of single-stranded donor DNA into the genophore of recipient cells with the formation of a heteroduplex (see Chapter 13). Similar findings have been obtained in studies of conjugation. Thus, recombination by breakage and reunion accompanied by the presence of hybrid DNA has been demonstrated in these bacteria. It has also been shown by A. B. Oppenheim and M. Riley that at the site of recombination following conjugation hydrogen bonding between donor and recipient regions precedes the formation of phosphodiester bonds.

The activities of enzyme systems which function to repair damage to DNA have also been intensively studied since it has long been known that radiation as well as some chemical mutagens not only cause mutation, but also produce an increase in recombination in practically all organisms. Ultraviolet radiation has been the principal tool in studies with prokaryotes. Three different repair mechanisms, mediated by separate genes and enzyme systems, have been reported for *E. coli*. Although these are best described along with the subject of mutation in Chapter 17, it should be noted here that one such repair system (postreplication recombination repair) is associated with the activity of recombination-deficient *(rec)* loci. Of these, mutants of the *recA* locus, in particular, are extremely sensitive to the killing effects of ultraviolet radiation and are unable to effect repairs to damaged DNA. In addition, recipient cells containing the *recA* mutation are unable to incorporate donor DNA into the genome. The *recB* and *recC* mutations cause a similar, but somewhat less extreme phenotype.

The kind of repair accomplished by the normal products of these loci is thought to take place after the replication of a previously damaged genome has occurred. Repair appears to involve the transfer of segments from one DNA chain to another to fill in gaps, excision of damaged regions, and some DNA synthesis (see Figure 17–16). As a result of these activities, recombinant-type cells are produced.

Mutations of the *recB* and *recC* loci also appear to lack a specific ATP-dependent DNAse which can act either as an exonuclease on double-stranded DNA or as an endonuclease on single-stranded DNA. This activity together with the evident effects of these loci on the incorporation of donor DNA into a recipient genome suggest their participation in normal recombination events. Only further characterization of these and similar mutations will clarify their respective roles in recombination and repair.

Eukaryotes

Although the early studies of Anderson with *Drosophila* and Lindegren with *Neurospora* which established that crossing over took place at the four-strand stage of meiosis provided strong circumstantial evidence for breakage and exchange as the most probable mechanism for recombination in eukaryotes, direct visual evidence for this hypothesis was not obtained until 1965, when Taylor applied his technique for autoradiography of chromosomes to meiotic cells.

In his studies Taylor supplied ^{3}H-thymidine to cells of the grasshopper testis during the next to last replication cycle before meiosis. The semiconservative mode of replication assured that each daughter chromosome would be composed of one original (and unlabeled) plus one newly synthesized (and labeled) DNA chain (Figure 14–14). Thus all of the chromosomes became visually labeled by this procedure, and these were distributed to daughter cells by the subsequent mitotic division. The daughter cells so formed were primary spermatocytes. The ^{3}H-thymidine medium was then removed, and the primary spermatocytes allowed to undergo the final chromosome duplication prior to meiosis in unlabeled medium. Due to the semiconservative mechanism of chromosome replication, one of the two daughter chromatids of any chromosome would thus contain a labeled and a newly synthesized unlabeled DNA chain, while the other daughter chromatid would contain two unlabeled DNA chains. As an overall result, one sister chromatid was visibly labeled, the other unlabeled. At the onset of meiosis, synapsis between homologous chromosomes thus produced a tetrad composed of two labeled and two unlabeled chromatids. Although exchanges between the two labeled or between the two

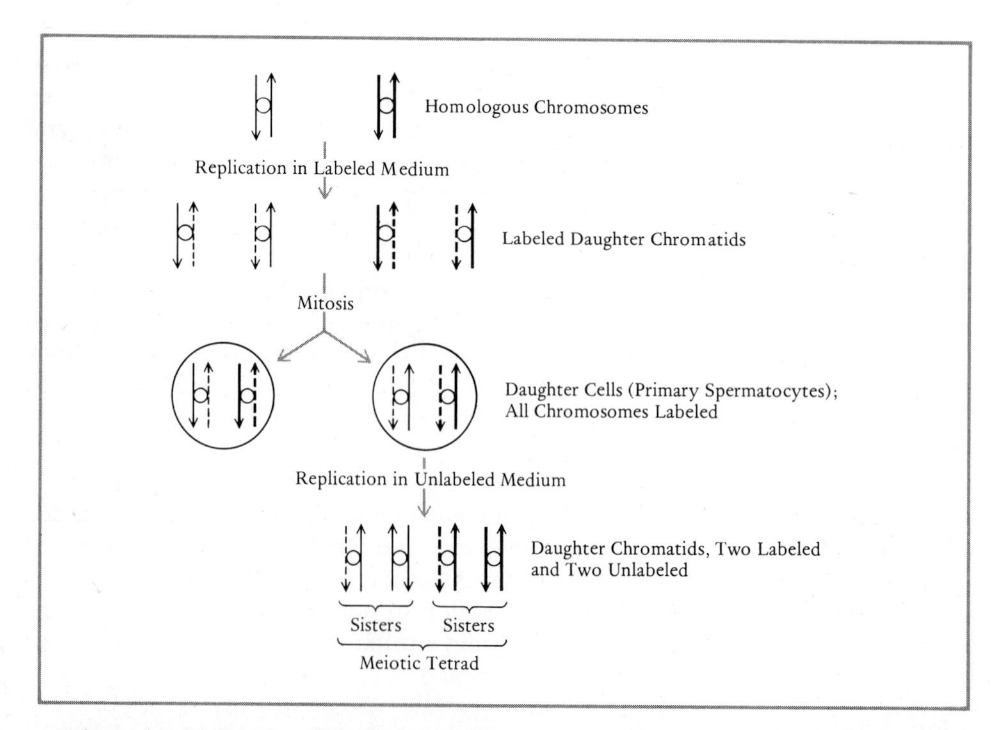

Figure 14–14. Design of Taylor's experiment which demonstrated that crossing over in eukaryotes takes place by breakage and reunion. Labeled DNA is represented by dash line.

unlabeled strands could not be detected, those which involved a labeled plus an unlabeled chromatid were clearly visible under the microscope through autoradiography.

By observing alterations in the labeling pattern, Taylor was able to correlate chiasma frequency and distribution, as seen at diplotene, with the physical exchanges between chromatids evident at later stages of the meiotic process. Since he found excellent agreement between the two, his findings provided direct visual evidence for Janssen's original chiasmatype theory as well as for Darlington's hypothesis that crossing over in eukaryotes takes place by breakage and exchange. In addition, Taylor's results conclusively demonstrated that homologous centromeres separate from one another at the first meiotic division.

These results were subsequently confirmed and extended by W. J. Peacock and by G. H. Jones, again with the technique of autoradiography. For example, by means of a temperature shock applied after premeiotic chromosome replication, Peacock was able to alter the chiasma frequency in grasshopper primary spermatocytes, and this treatment produced a corresponding alteration in the frequency of label exchange between nonsister chromatids. These findings further substantiated the hypothesis of breakage and reunion and, in addition, eliminated copy-choice as a mechanism for recombination.

Inferences as to the events underlying recombination have been drawn principally from studies of fungi where the presence of ordered spores permits detailed genetic analysis (see Chapter 8). It will be recalled that in fungi such as *Neurospora* the four haploid products of meiosis undergo a mitotic division to produce eight spores which are contained within the ascus, and that the walls of this structure are so narrow that spindle overlap or the movement of nuclei past one another is prevented. Such spores can be dissected from the ascus, in order, and their genotypes determined. When this is done and the genotypes correlated with the respective positions of the spores in the ascus, it is possible to infer prior orientation of chromatids at the first and second meiotic divisions and to determine the probable sites of crossing over between linked genes.

Ordinarily, Mendelian segregation of alleles is the pattern of segregation observed (Figure 14-15), but on rare occasions anomalous segregations are found, and these have provided insights into underlying molecular processes of recombination. We will consider three such unusual patterns of segregation. In the second pattern diagramed in Figure 14-15, the eight spores of an ascus arising from mitosis and meiosis of a cell heterozygous for the alleles *a* and + are shown. In contrast to the expected recovery of equal numbers of spores of each type (4 *a* : 4 +), a nonreciprocal segregation of 2 *a* : 6 + is observed. It is evident that the eight spores of the 2 : 6 segregation were formed by the mitotic divisions of four preceding nuclei of genotypes *a*, +, +, + (in order). These four nuclei were produced as the result of the second meiotic division occurring in two predecessor nuclei whose genotypes must have been *a*/+ and +/+. Going back one step further, these nuclei arose as the result of the first meiotic division and were derived from the tetrad chromatids formed from the cross *a* × +. Such a tetrad is expected to yield a 4 : 4 segregation, but instead, has produced a 2 : 6 segregation. Having traced this 2 : 6 segregation all the way back to the original tetrad, we must conclude that, prior to the first meiotic

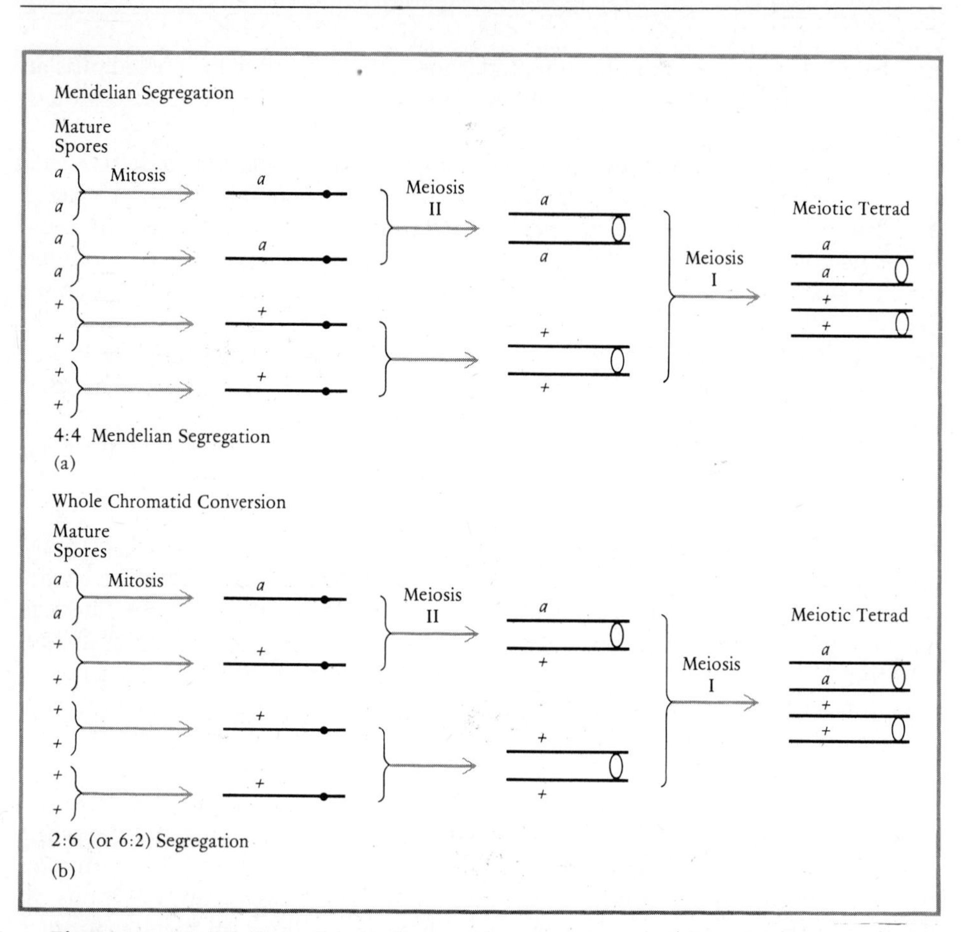

Figure 14–15. Whole and half-chromatid conversion as deduced from segregation patterns in ordered spores of fungi arising from the hypothetical cross $a \times +$. The direction of the arrows follows the line of reasoning used in deducing preceding events. See also Figure 8–15.

division, the *a* allele present in one parental chromatid was somehow transformed to the wild type allele.

This phenomenon, called *gene conversion,* is thought to reflect the underlying processes of recombination. It is not a mutation event, for it can be detected only in heterozygotes and, as opposed to mutation, a given allele is always converted to the alternate allele already present in the tetrad. Conversion can be in either direction, that is, either *a* to + or + to *a*. That conversion is directly associated with recombination can also be inferred from crosses where the segregation of three closely linked genes is followed. It has been found that conversion of the centrally positioned locus is usually accompanied by reciprocal exchange between the flanking marker genes, and analysis of spores in such cases has indicated that both conversion and exchange occurred within the same chromatid.

The 2 : 6 segregation also permits an additional inference with respect to the DNA of a converted chromatid. In fungi the chromosomes of the four

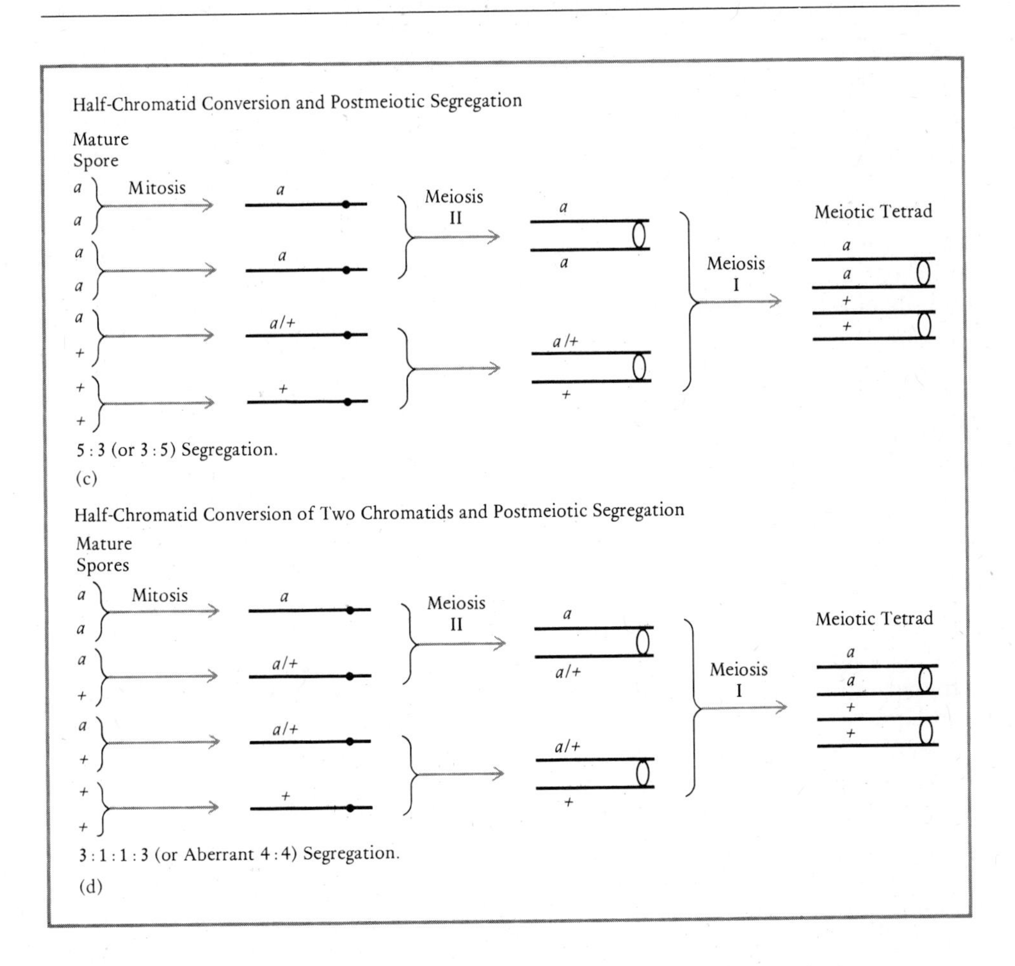

haploid nuclei produced by meiosis undergo replication and mitosis to yield the eight spores. In the present example these spores occur by two's: 2*a*, 2 +, 2 +, 2 +. This fact, together with the known semiconservative mode of replication, indicates that both strands of the duplex DNA comprising a converted chromatid were changed to the nucleotide sequence of the alternate allele prior to the cycle of replication which precedes mitosis. For this reason, the segregation pattern indicates *whole chromatid conversion*.

Figure 14–15 illustrates still another pattern—a different and rare 5:3 nonreciprocal segregation, that is, 2 *a*, 2 *a*, 1 *a*, 1+, 2 +. When the reasoning used above is applied to this pattern, it can be seen that the first pair of spores, both *a*, were derived from a preceding nucleus of genotype *a*. The same is evident for the second pair of spores. The third pair of spores is unlike, one being *a*, the other +. Since both are derived from the mitotic division of a single preceding nucleus, we must conclude that this nucleus was heterozygous, *a*/+. This nucleus was also haploid, and therefore the chromatid containing these alleles must have been composed of hybrid DNA. This phenomenon is called *postmeiotic segregation*, for here, the two parental alleles are segregated to separate nuclei after the completion of meiosis. It can also be considered an example of *half-chromatid conversion,* since only one of the two strands of duplex DNA of the chromatid has been converted to the alternate allele. The origin of this half-converted chromatid can again be traced back to the meiotic tetrad, suggesting that the formation of hybrid DNA occurs during recombination.

As a further example, Figure 14-15 shows the half-conversion of two different chromatids, a phenomenon which results in an aberrant 4 : 4 segregation pattern (3 : 1 : 1 : 3). The relative rarity of half-chromatid, as compared to whole chromatid conversions suggests that regions of hybrid DNA containing mismatched bases are ordinarily recognized and corrected to the sequence of one or the other allele by enzymes active during recombination. Within the restricted region of exchange, correction of a mismatched joint molecule could account for localized high negative interference.

An additional observation associated with recombination has also been reported. When different mutated sites within a single cistron are subjected to fine structure analysis, the frequency of conversion is often found to be high at one end of the cistron, with a gradual decrease toward the opposite end, a phenomenon called *polarity*. It has been postulated that polarity may reflect the presence of fixed sites at which recombination is initiated.

Evidence that all of the above phenomena observed in fungi also occur in higher eukaryotes has recently been obtained for *Drosophila* by A. Chovnick and coworkers, as well as by P. S. Carlson. It therefore appears that the mechanism of recombination may be the same for all nucleated organisms. In general, this mechanism is postulated to involve single-strand nicking, perhaps at a specific site recognized by an endonuclease, followed by strand separation and the overlapping of single-stranded regions derived from the two original parental chromatids. If the region is heterozygous, correction of mismatching by excision and repair synthesis is postulated to occur, with subsequent restoration of strand integrity by ligase action. All of these events involve the activities of endo- and exonucleases, DNA polymerases, and DNA ligases, enzymes known to be present and functional during recombination. A number of models which attempt to account for the mechanism of recombination have been proposed. One such model is illustrated in Figure 14–16.

Biochemical evidence in support of the above postulates has been obtained primarily in lily meiocytes by H. Stern, Y. Hotta, and coworkers. Whereas during the S phase which precedes mitosis, replication of the genome is carried to completion, during the premeiotic S phase, only 99.7 percent of the genome is replicated, and synthesis of the remaining 0.3 percent is postponed to the zygotene stage. Autoradiographs show that this zygotene-DNA is generally distributed throughout the chromosomes, and biochemical studies indicate that it becomes covalently integrated within a chromatid only after synapsis and the pachytene stage. If zygotene-DNA synthesis is inhibited, chromosome pairing does not occur.

Of great interest is the identification of a DNA-binding protein found not only in the lily, but also in mammalian spermatocytes. This protein is similar to the gene 32-binding protein of phage T4 in that it facilitates hydrogen bonding between single strands of DNA. The protein appears to be specific to meiotic cells where it exhibits a cyclical activity, rising in leptotene and zygotene, and gradually decreasing during pachytene. In addition, a distinctive endonuclease that produces single-strand nicks in native DNA has been found in lily meiotic nuclei. The activity of this particular endonuclease is also cyclical, rising to a peak during pachytene and decreasing and disappearing thereafter. Also associated with the pachytene stage is a small amount of DNA synthesis, evidently of the repair type, since no net increase in DNA occurs. Pachytene DNA is synthesized at scattered sites distributed throughout the synapsed chromatids. It has been observed by autoradiography in mouse, salamander, and wheat chromatids and has been isolated from lily meiocytes.

Although all of these findings, in conjunction with the data derived from genetic studies, are suggestive of underlying mechanisms, it should be evident that the molecular processes of recombination in eukaryotes are far from being resolved. Therefore, any proposal or model which attempts to account for all facets of the process merely represents an informed, but unsubstantiated guess.

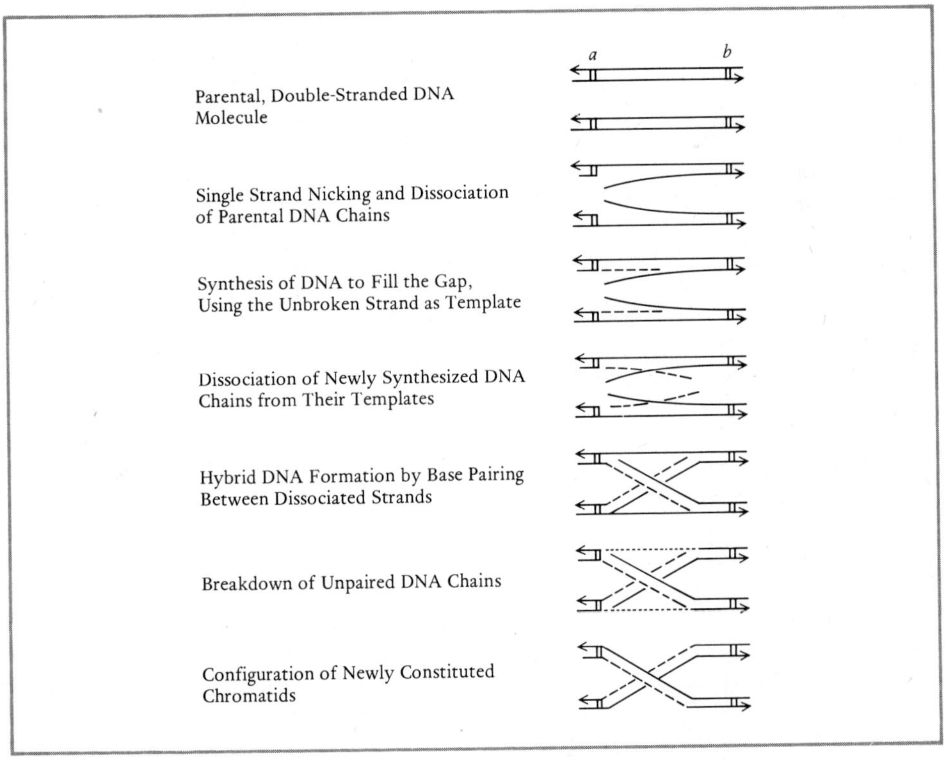

Figure 14–16. Diagram of hybrid DNA model of genetic recombination. Solid lines represent the nucleotide chains of two homologous DNA molecules. Orientation of the DNA chains is shown by arrows. Sites *a* and *b* indicate fixed points where strand nicking by an endonuclease initiates chain separation. Dash lines represent newly synthesized DNA; dotted lines represent breakdown of DNA. [From H. L. K. Whitehouse and P. J. Hastings, "The Analysis of Genetic Recombination on the Polaron Hybrid DNA Model". *Genet. Res.*, *6*, 27–92, 1965. Cambridge University Press.]

PROBLEMS

14–1. Suppose you were given two recessive mutants of *Drosophila,* short wing *(s)* and reduced wing *(r),* both autosomal and both belonging to the same linkage group.
a. What one cross could you use to determine if *s* and *r* are alleles?
b. What F_1 phenotype would indicate allelism?
c. Assuming allelism, what cross could you use to determine whether *s* and *r* occupied the same or different sites within the cistron?
d. What data would indicate that *s* and *r* occupied different sites?
e. If the sites of *s* and *r* were 0.02 map units apart, theoretically how many flies must you raise to recover one wild type recombinant?

14–2. The recessive genes *a* and *b* are allelic, each producing a narrow eye phenotype when homozygous. Females heterozygous for these genes were crossed to *b/b* males. Of a total of 50,000 progeny, 10 wild type individuals were recovered, the rest of the progeny showing a narrow eye phenotype.
a. Assuming reciprocal classes are equal, how many map units apart are these mutants?
b. If one map unit corresponds to 10,000 nucleotides of DNA, how many nucleotides separate these mutant sites?

14–3. In *Neurospora* enzyme A is inactive when either of the allelic genes, mutant 1 or mutant 2, is present. Enzyme A is also inactive in heterokaryons formed between these strains. Enzyme B is inactive when either of the allelic genes, mutant 3 or mutant 4, is present, but heterokaryons between these strains exhibit a partial restoration of enzyme B activity. Do enzymes A and B function as single polypeptide chains or is either multimeric? If multimeric, what would be the minimum number of subunits present in the active enzyme? The partial function of enzyme B in the heterokaryon is an example of what phenomenon?

14–4. Strains A through E are deletions of cistron *x*. Their extent is indicated by solid lines in the diagram below. Crosses of mutants 1 through 4 with each of strains A through E produced the data shown below (+ = appearance of and 0 (zero) = absence of wild type recombinants in the progeny). From left to right, what is the order of mutations 1 to 4 within cistron *x*?

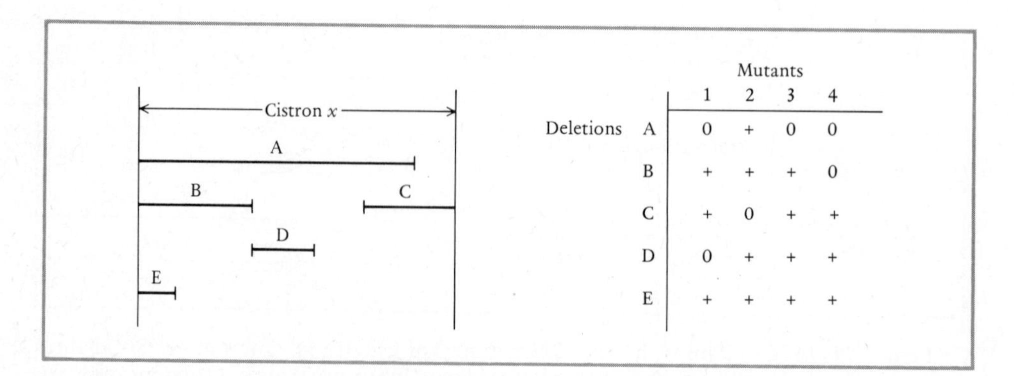

Deletions	1	2	3	4
A	0	+	0	0
B	+	+	+	0
C	+	0	+	+
D	0	+	+	+
E	+	+	+	+

14–5. Strains A through J are deletions of cistron *y*. Their extent is indicated by solid lines in the diagram below. Crosses of mutants 1 through 10 with each of strains A through J produced the data given below (+ = appearance of and 0 = absence of wild type recombinants in the progeny). From left to right, what is the order of mutations 1 to 10 within cistron *y*?

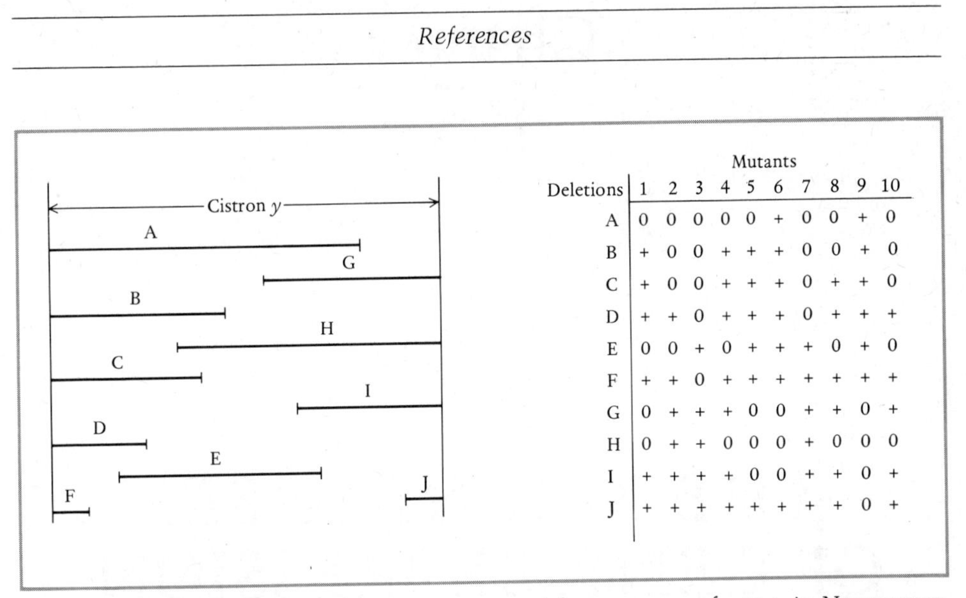

Deletions	1	2	3	4	5	6	7	8	9	10
A	0	0	0	0	0	+	0	0	+	0
B	+	0	0	+	+	+	0	0	+	0
C	+	0	0	+	+	+	0	+	+	0
D	+	+	0	+	+	+	0	+	+	+
E	0	0	+	0	+	+	+	0	+	0
F	+	+	0	+	+	+	+	+	+	+
G	0	+	+	+	0	0	+	+	0	+
H	0	+	+	0	0	0	+	0	0	0
I	+	+	+	+	0	0	+	+	0	+
J	+	+	+	+	+	+	+	+	0	+

(Column headers 1–10: Mutants)

14–6. The following spore orders were recovered from crosses of $a \times +$ in *Neurospora*. In each case what were the genotypes of the nuclei produced as a result of the first meiotic division?

(1)	(2)	(3)	(4)	(5)
+	*a*	+	*a*	+
+	*a*	+	+	+
a	*a*	+	+	*a*
a	*a*	+	+	+
+	+	*a*	+	*a*
+	+	+	+	*a*
+	*a*	*a*	*a*	+
+	*a*	*a*	*a*	*a*

REFERENCES

Broker, T. R., and A. H. Doermann, 1975. Molecular and genetic recombination of bacteriophage T4. *Ann. Rev. Genet., 9,* 213.

Clark, A. J., 1973. Recombination deficient mutants of *E. coli* and other bacteria. *Ann. Rev. Genet., 7,* 87.

Hastings, P. J., 1975. Some aspects of recombination in eukaryotic organisms. *Ann. Rev. Genet., 9,* 129.

Hotchkiss, R. D., 1974. Models of genetic recombination. *Ann. Rev. Microbiol., 28,* 445.

Miller, R. C., Jr., 1975. Replication and molecular recombination of T-phage. *Ann. Rev. Microbiol., 29,* 355.

Radding, C. M., 1973. Molecular mechanisms in genetic recombination. *Ann. Rev. Genet., 7,* 87.

Schlesinger, M. J., and C. Levinthal, 1965. Complementation at the molecular level of enzyme interaction. *Ann. Rev. Microbiol., 19,* 267.

Stadler, D. R., 1973. The mechanism of intragenic recombination. *Ann. Rev. Genet., 7,* 113.

Stern, H., and Y. Hotta, 1973. Biochemical controls of meiosis. *Ann. Rev. Genet., 7,* 37.

Whitehouse, H. L. K., 1970. The mechanism of genetic recombination. *Biol. Rev., 45,* 265.

Yanofsky, C., 1967. Structural relationships between gene and protein. *Ann. Rev. Genet., 1,* 117.

Chapter 15

THE MECHANISM OF GENE EXPRESSION

IN describing the mechanism of gene expression, we will adopt the premise that DNA constitutes a triplet code for protein in which a set of three nucleotides, called a *codon*, stands for a single amino acid. We will further assume that the DNA code contains no punctuation and that it is translated into protein from a fixed point. Although these assumptions have been validated, their proof requires an initial understanding of the mechanism of protein synthesis.

TRANSCRIPTION OF RNA

Although the primary information for the construction of proteins resides within the genome, DNA itself does not participate directly in protein synthesis. Instead, the information contained within the nucleotide sequences, or cistrons, of DNA is transferred to a variety of more or less expendable RNA intermediaries which can be synthesized whenever required. These RNA molecules can be considered gene copies in that a given DNA sequence has served as the template for their synthesis.

The mechanism for the transfer of information from DNA to RNA is called *transcription*, and an RNA molecule so produced is termed a *transcript*. Transcription requires the presence of the ribose nucleoside triphosphates of adenine, guanine, cytosine, and uracil, the DNA template, a suitable ionic environment, and a specific polymerizing enzyme, a DNA-directed RNA polymerase which catalyzes the formation of phosphodiester bonds. The process is similar to that previously described for the polymerization of DNA, in that successive nucleotides are added at the 3′ hydroxyl end of a growing RNA polymer. RNA is thus synthesized in the 5′ to 3′ direction exactly as in DNA synthesis. The overall reaction can be summarized as:

$$\begin{matrix} \text{ATP} \\ \text{GTP} \\ \text{CTP} \\ \text{UTP} \end{matrix} \xrightarrow[\substack{\text{Mg}^{++} \\ \text{DNA template}}]{\text{RNA polymerase}} \text{RNA} + P - P_i$$

where ATP, GTP, CTP, and UTP represent the ribose nucleoside triphosphates of adenine, guanine, cytosine, and uracil, respectively, and P-P_i is pyrophosphate.

A simple representation of the process is diagrammed in Figure 15–1. At a localized site of transcription the DNA double helix unwinds and undergoes strand separation. The four ribose nucleotides present as raw materials become aligned with complementary bases present in the single-stranded DNA template, proper alignment being dictated by hydrogen bonding properties. Through the action of RNA polymerase, the ordered RNA nucleotides are then sequentially joined together to form a single-stranded polymer of RNA. Upon detachment of the transcript the DNA chain which served as the template forms new hydrogen bonds with its partner, and rewinding restores the original double helical structure. By means of this overall mechanism the nucleotide sequence of the RNA product is the exact complement of the DNA region from which it was transcribed and contains information equivalent to that present within the DNA template. As an illustration of complementarity, if the DNA template began C C A T G T G C T . . . , the corresponding RNA copy would be 5′ G G U A C A C G A . . . 3′.

Studies of transcription of RNA from viral DNA have indicated that only the growing end of the RNA polymer is hydrogen bonded to the template, previously synthesized regions having already been freed from such attachment. The remarkable photograph of transcription occurring in a newt chromosome (Figure 15–2) provides visual support for this hypothesis. That the RNA synthesized via transcription is indeed the exact complement of its DNA template has been shown in viruses through analysis and comparison of the base compositions of template and product.

An intriguing question related to transcription is whether one or both strands of a DNA region are copied. Thus far, experimental evidence indicates that one and the same DNA chain is consistently utilized as the "sense" strand for RNA synthesis *in vivo.* An exception to this general rule is known in the case of some viruses such as T4 and λ (see Chapter 18). In viruses and bacteria the molecular basis for template selection may be associated with the presence of runs of pyrimidines, particularly cytosine, within a DNA chain, and it has been proposed that such regions can perhaps be recognized by one of the subunits of the complex enzyme, RNA polymerase, as sites for the initiation of transcription.

The RNA polymerase of bacteria is composed of five subunits which together form a *core enzyme.* An additional subunit, called *sigma* (σ), combines with the core enzyme to cause attachment of the enzyme at the proper initiation site for transcription, known as the *promotor*. Promotors have been mapped genetically and are known to occur in association with gene clusters (see Chapter 18). The presence of a number of different sigma factors has also been

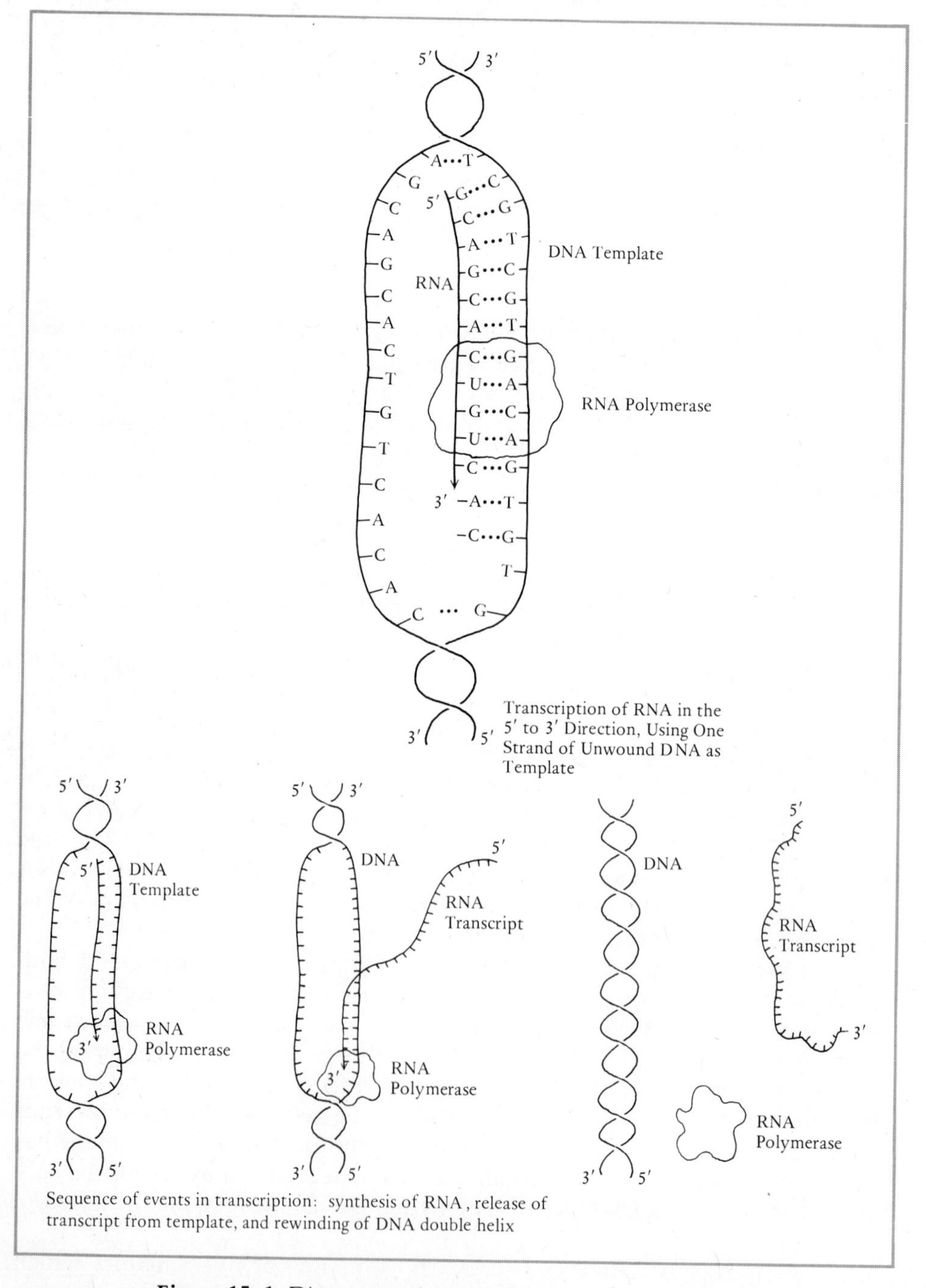

Figure 15–1. Diagrammatic representation of transcription.

postulated and evidence supporting this suggestion has been obtained for phage T4. The presence of different sigma factors implies a corresponding difference in the nature of the initiation sites within the T4 genome and suggests that the promotor sites of other organisms may also be diverse.

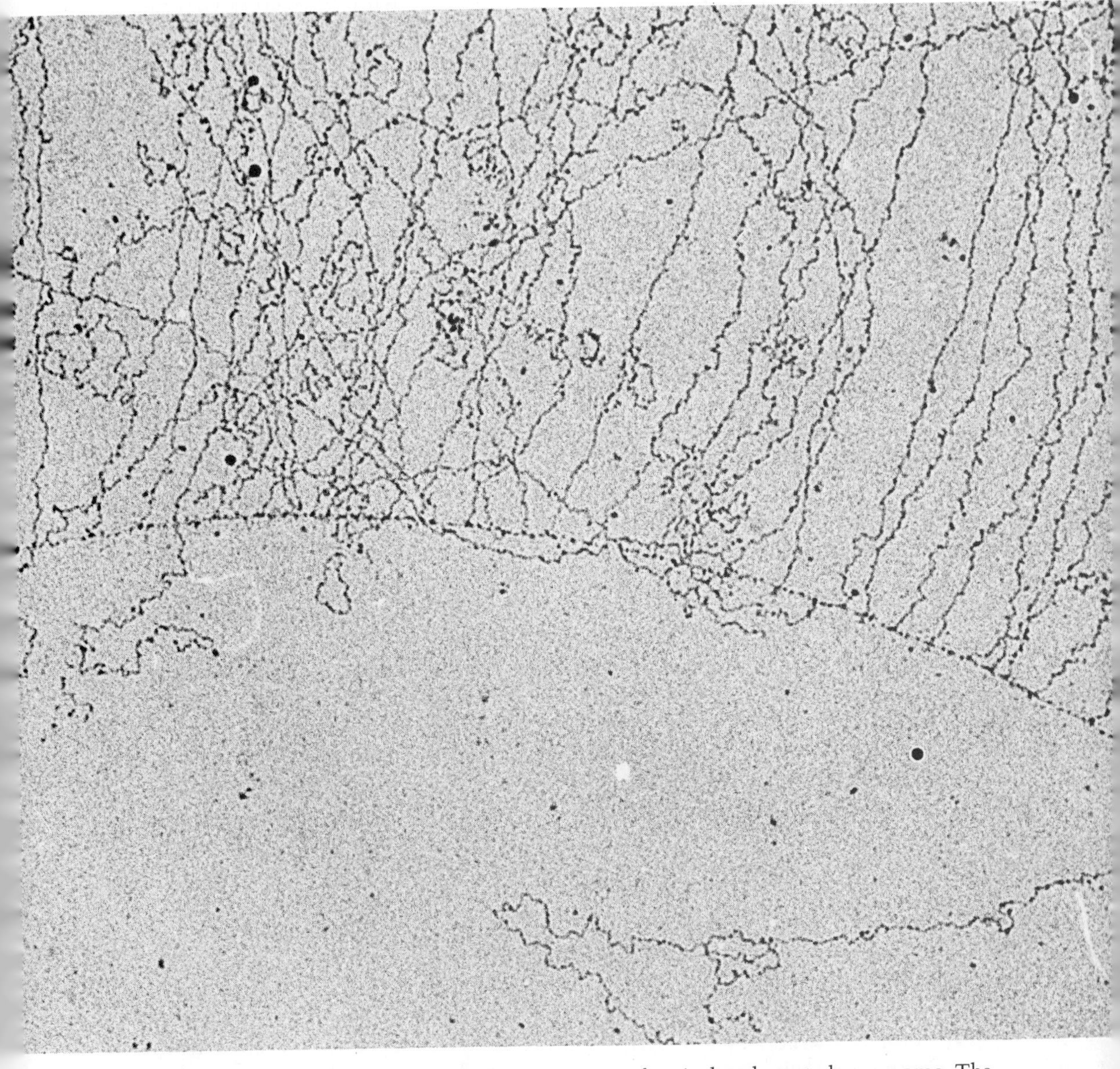

Figure 15–2. Electron micrograph of a portion of an isolated newt chromosome. The horizontal axis is DNA being transcribed into RNA which extends outward as numerous fibrils. The chromosome portion being transcribed is a single loop from a lampbrush chromosome (see Chapter 19). The RNA is already combined with protein. RNA polymerase molecules can be seen along the DNA axis and appear as granules at the base of each RNA fibril. (46,100×) [From O. L. Miller, Jr. and B. R. Beatty, *J. Cell Physiol., 74, Suppl. 1,* 225–232, 1969.]

In addition to sigma factors a termination factor called *rho* (ρ) has been identified in bacteria. It has been characterized as a dissociable subunit which attaches to the core enzyme to cause the cessation of RNA synthesis. Presumably, it is capable of recognizing "stop" signals in the form of specific nucleotide sequences within the DNA chain being transcribed.

The RNA polymerases of eukaryotes are less well characterized than those of viruses and bacteria, but they appear to be extremely large aggregates. In contrast to the soluble enzyme of bacteria, RNA polymerases of plants and animals are components of a DNA-nucleoprotein complex from which they are difficult to separate. Two such enzymes differing somewhat in properties have been identified. One, RNA polymerase I, is thought to restrict its activity to the nucleolus, the site of ribosomal RNA synthesis, while RNA polymerase II appears to be a more general transcribing enzyme. The complexity of the eukaryotic genome undoubtedly also requires the presence of initiating as well as terminating factors.

It must be emphasized that not all nucleotide sequences of the genome are transcribed at any one time. Instead, specific cistrons are selected as templates according to the needs of the cell. Some sequences of DNA may be utilized for only a short period during the entire life of an organism, for example, only during embryonic development, and thereafter not transcribed again, while others may be copied intermittently or even continuously to meet metabolic demands. An appropriate response to such demands may require the simultaneous transcription of several template regions so that a number of needed enzymes may be synthesized. If such regions are adjacent to one another, as is frequently the case in bacteria, the problem is somewhat simplified, for they can be transcribed together as a long continuous segment. However, if the required template regions are widely separated spatially, coordinated transcription must somehow be achieved. This situation is characteristic of eukaryotes and requires mechanisms leading to a precise, specific, and fail-safe coordination. We can assume that the directions for assembling such a mechanism are present in the genome, but what they are is as yet a mystery. In summary, the process of transcription cannot be viewed merely as a method for synthesizing RNA. Instead, it must be considered in relation to the totality of the cell and as an integrated and controlled response to environmental change.

TYPES OF TRANSCRIBED RNA

In both prokaryotes and eukaryotes three kinds of RNA required for protein synthesis are transcribed from the genome: *ribosomal RNA (rRNA), messenger RNA (mRNA),* and *transfer RNA (tRNA).* Ribosomal RNA is an intrinsic constituent of the ribosomal particles which provide the structural framework on which protein synthesis proceeds. Messenger RNA serves as the template on which a polypeptide is constructed and contains within its base sequence the codons that dictate the order of amino acids. Transfer RNA carries amino acids to the ribosome and provides for their orientation on the mRNA template. Although both mRNA and in some cases tRNA are transcribed from viral DNA, genes for ribosomal RNA are not present, and therefore during the synthesis of viral proteins, host ribosomes are utilized. Before reviewing the overall process of protein synthesis, the three kinds of RNA should be examined in more detail.

Ribosomal RNA

Ribosomes are small cytoplasmic particles composed of two subunits, each of which contains RNA and protein. Dissociation and reassociation of these subunits can be induced experimentally and also appears to occur naturally in the intact cell depending on whether or not the ribosome is engaged in protein synthesis, the union of subunits being required for this process.

Ribosomes and their contained RNA are classified by size as reflected by sedimentation constants (S values) obtained through ultracentrifugation. In bacteria whole ribosomes exhibit a sedimentation constant of 70 S, which corresponds to a molecular weight of around 2.6×10^6. The two subunits consist of a larger, dome-shaped 50 S particle and a smaller cap-shaped 30 S particle (Figure 15–3). The 50 S subunit contains two types of RNA: a large 23 S macromolecule and a relatively small 5 S species. These occur in association with 34 different proteins. The smaller 30 S subunit contains only one 16 S RNA molecule along with 15 distinct proteins.

Ribosomal RNA exhibits extensive intrastrand base pairing, and a ribosomal subunit can be visualized as a ribonucleoprotein strand which, by complex folding, assumes its characteristic shape. The sequence of bases within 5 S RNA has been determined, but the molecular function of 5 S, 16 S, or 23 S RNA and most of their associated proteins remains unknown.

In bacteria the three kinds of rRNA are transcribed as a single, long macromolecule which is subsequently cleaved and fashioned into the three types of RNA found within the ribosome. Since one long transcript contains all

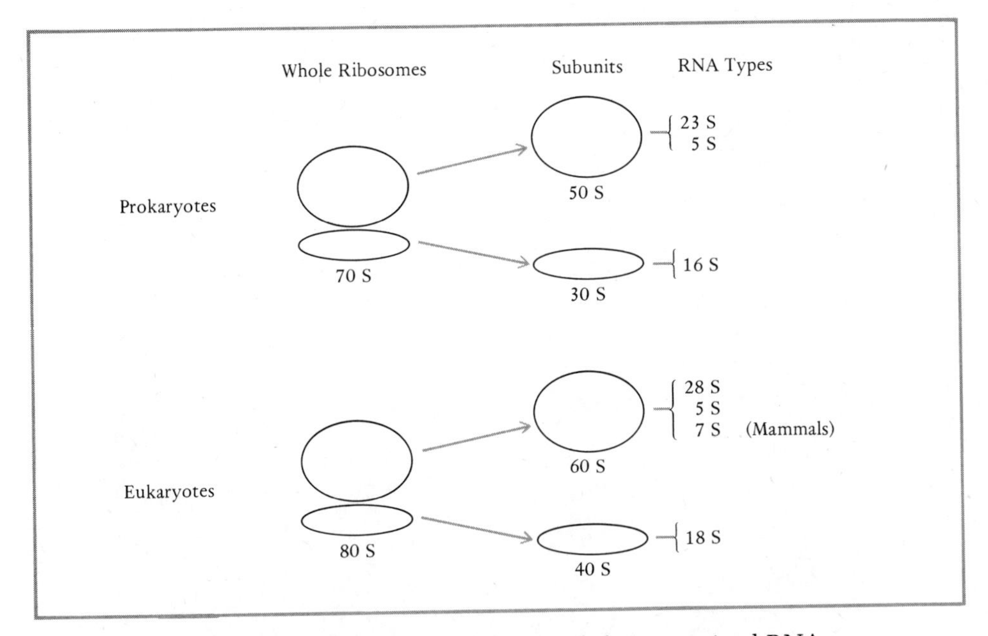

Figure 15–3. Ribosomal subunits and their contained RNA.

three RNA's, it is evident that their corresponding templates are located immediately adjacent to one another. In addition, it has been found that these tripartite transcripts are produced at around six different sites, indicating that the loci for rRNA are reiterated and present as repeats within the genome. The almost continuous transcription occurring at these multiple sites accounts for the fact that ribosomes constitute over a third of the dry mass of bacterial cells and that rRNA represents about 80 percent of the total RNA present.

The ribosomes of eukaryotes are larger than those of prokaryotes and exhibit a sedimentation constant of around 80 S. The larger subunit is a 60 S particle, containing a 28 S and a 5 S RNA molecule (Figure 15–3). A 7 S RNA is also present in the larger subunit of mammals. The smaller 40 S subunit contains an 18 S RNA molecule. The RNA of each subunit occurs in association with numerous proteins, there being around 20 in the smaller and about 35 in the larger subunit. The number of ribosomes present within a cell varies dramatically with the intensity of protein synthesis. Although the ribosomes may occur free and unattached, they are most often found in association with the outer surfaces of the rough endoplasmic reticulum (Figure 15–4). In bacteria ribosomes are always free in the cytoplasm, no endoplasmic reticulum being present.

In eukaryotes the nucleolus organizer has been shown to be the site of synthesis of a large transcript, 40 S in amphibia and 45 S in mammals, which contains the base sequences of both 18 S and 28 S rRNA, indicating that templates for these molecules are adjacent to one another. Some of the bases of this large molecule are modified after transcription by the addition of methyl groups (as in bacteria), and the initial transcript is also cleaved into fractions

Figure 15–4. Endoplasmic reticulum (ER) and ribosomes (R) in an exocrine cell of rat pancreas. The ribosomes are both free and associated with the ER. The amount of ER and ribosomes is typical of cells rapidly synthesizing protein. A poorly fixed mitochondrion (M) is also present. (66,000×) [Courtesy of Dr. George Palade]

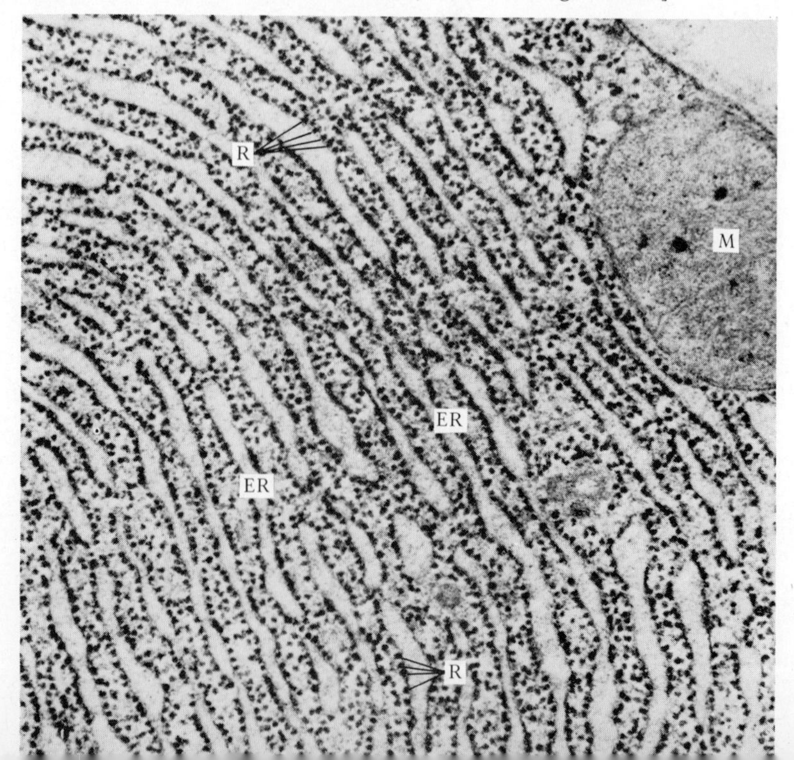

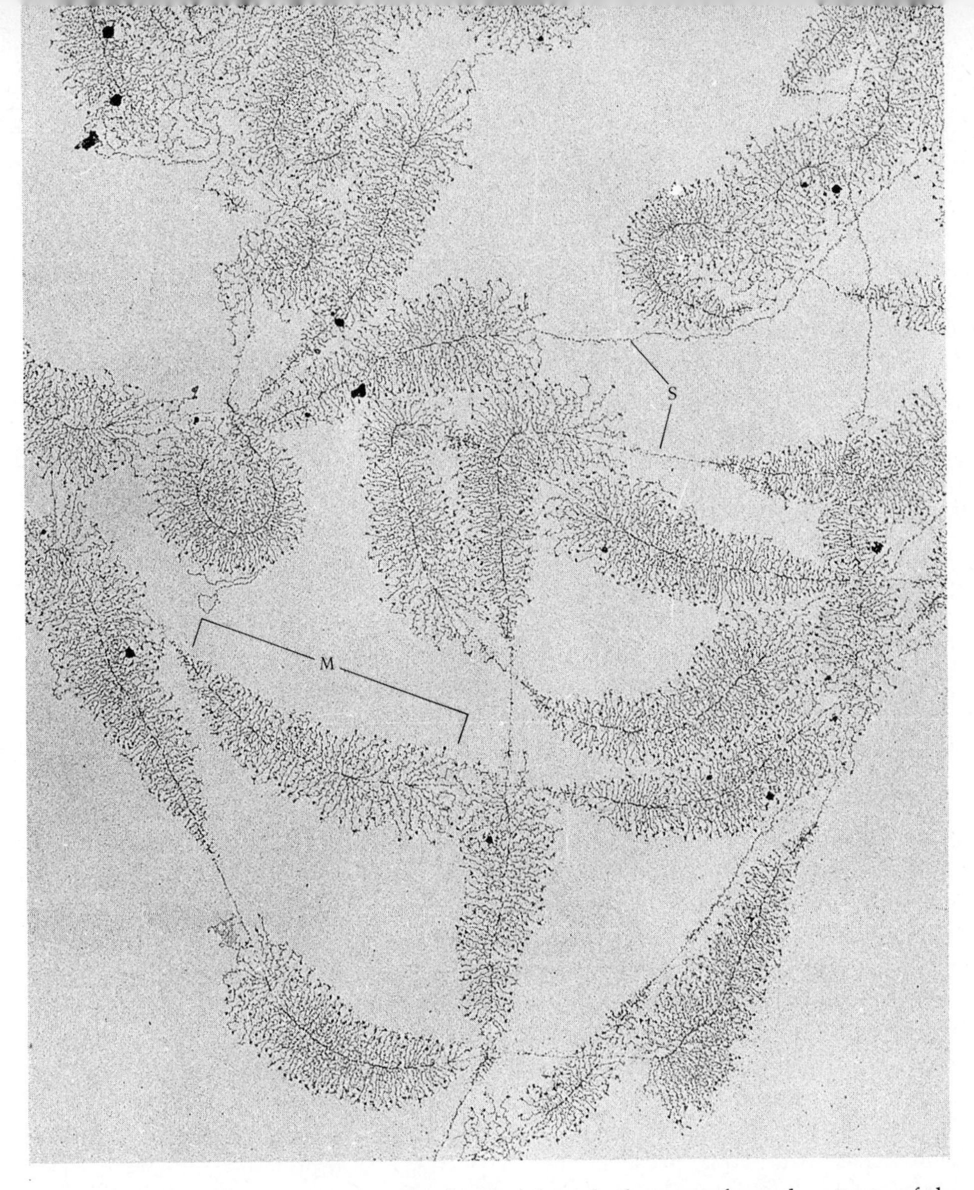

Figure 15–5. Electron micrograph of isolated nucleolar cores from the oöcyte of the newt, *Triturus.* Elongate axis filaments are nucleolar DNA from which is being transcribed fibrils of rRNA. The rRNA is already combined with protein to form ribonucleoprotein complexes. Transcription is occurring in units (M), about 2.5 um long, and from each such unit about 100 fibrils of rRNA are being produced. These transcribing units alternate with transcription free spacer sequences (S). (19,000×) [From O. L. Miller Jr. and B. R. Beatty, "Visualization of Nucleolar Genes," *Science, 164,* 955–957, 23 May 1969. Copyright © by the American Association for the Advancement of Science.]

from which additional sequences are deleted, the final result being the separate 18 S and 28 S RNA molecules.

In amphibia it has been demonstrated that within the nucleolus organizer, the genes for 18 S and 28 S RNA occur in a series of tandemly arranged repetitive sequences, any one such sequence being transcribed as the 40 S macromolecule. Separating such a sequence from its neighbor is a region of DNA, called *spacer DNA,* which does not appear to be transcribed (Figure 15–5). The overall arrangement can be visualized as: 40 S template . . . spacer . . . 40 S template . . . spacer . . . and so on. The number of such repetitive sequences varies, there being around 230 in *Drosophila*, around 300 in the rat, and up to 400 or more in the nucleolus of the toad *Xenopus*. These sequences undergo amplification in amphibian oöcytes with the result that transcriptional capacity is greatly increased in these cells (see Chapter 19).

The 5 S rRNA of eukaryotes is transcribed separately and from an entirely different region of the genome than the nucleolus. The templates for 5 S rRNA also occur in multiple copies of several hundred to over a thousand, depending on species. In *Xenopus* the genes for 5 S rRNA have been located at the tips (telomeres) of the chromosomes and have been shown to be separated by spacer sequences. The purpose of such spacers is obscure and their presence raises many questions as to the organization and evolution of the eukaryotic genome.

It should be noted that mitochondria and chloroplasts, present as organelles within eukaryotic cells, possess their own ribosomes which resemble those of prokaryotes (see Chapter 20). This observation has been used to support the hypothesis that these organelles represent hereditary, prokaryotic-type symbionts whose association with other cells occurred early in the course of evolution.

The great complexity of ribosomes reveals a need for precise coordination. In order to produce a ribosome, a cell must transcribe the required types of RNA. These may occur within a single transcript, as in bacteria, or as separate transcripts from different regions of the genome, as in eukaryotes. In eukaryotes the synthesis of the two transcripts (45 S and 5 S) must be temporally coordinated, and in both eukaryotes and prokaryotes enzyme systems for cleavage and methylation of bases must be present. Ribosomes also contain numerous different kinds of proteins and these must be available at the precise time when the ribosome is assembled. Before such assembly can occur, each of these proteins must be synthesized via a messenger RNA transcript on preexisting ribosomes, utilizing adapter transfer RNA molecules and enzyme systems which, if not already present, must themselves be synthesized. The evident simultaneous presence of all of these components reflects a very sophisticated regulatory mechanism whose nature is yet unknown. The fact that in many cells hundreds of ribosomes can be constructed within minutes is even more amazing.

Messenger RNA

Messenger RNA specifies through its base sequence the particular amino acids and their order to be built into a polypeptide. Messenger RNA is thus a copy of a structural gene transcribed in response to the particular needs of the cell. With respect to base composition mRNA is similar to the DNA of the genome, and it is therefore often characterized as DNA-like RNA. In contrast, both rRNA and tRNA have a higher proportion of G-C base pairs than does mRNA, a phenomenon thought to be associated with the extensive intrastrand hydrogen bonding exhibited by the rRNA and tRNA molecules.

In eukaryotic species the message initially transcribed within the nucleus appears to be considerably larger than that which reaches the cytoplasm. Indeed, giant 80 S and larger RNA molecules may be found within the nucleus, and it has been postulated that these and other large nuclear RNA's are precursors to the functional mRNA which participates in protein synthesis. Such large molecules have been called *heterogeneous nuclear RNA (hnRNA)* to distinguish them from other types of RNA whose activities have been defined.

From studies of cells which produce large amounts of a single protein, such as globin, it is known that the initial transcript is modified within the nucleus by the deletion of certain sequences during a maturation process. In addition, an RNA sequence containing only adenine nucleotides, poly(A), is added to the 3′ end (see Chapter 19). It is postulated that after such modifications in structure, the functional mRNA is then combined with protein for transport to the cytoplasm through the pores in the nuclear envelope.

The mRNA of bacteria and viruses does not appear to undergo major modification after transcription. However, it may contain information for the construction of several different proteins, that is, it may be *polycistronic.* It will be recalled that in bacteria genes specifying enzymes belonging to the same metabolic pathway are usually clustered together within the genophore. The members of such clusters are transcribed as a single long RNA molecule which is translated into protein even as transcription occurs (Figure 15–6).

Figure 15–6. Polyribosomes of *E. coli* translating a message into protein. The thin horizontal line is DNA from which mRNA is being transcribed by RNA polymerase. A polymerase molecule must be present at the base of each cluster of ribosomes and one enzyme molecule is clearly shown (arrow) at a probable initiation site for transcription. Messenger RNA appears as a delicate fiber connecting the ribosomes and is being translated even while transcription is occurring. Notice that the message is longer and is being read by more ribosomes as transcription proceeds along the DNA template from left to right. [From O. L. Miller Jr., Barbara A. Hamkalo, and C. A. Thomas Jr., "Visualization of Bacterial Genes in Action," *Science, 169,* 392–395, 24 July 1970. Copyright © 1970 by the American Association for the Advancement of Science.]

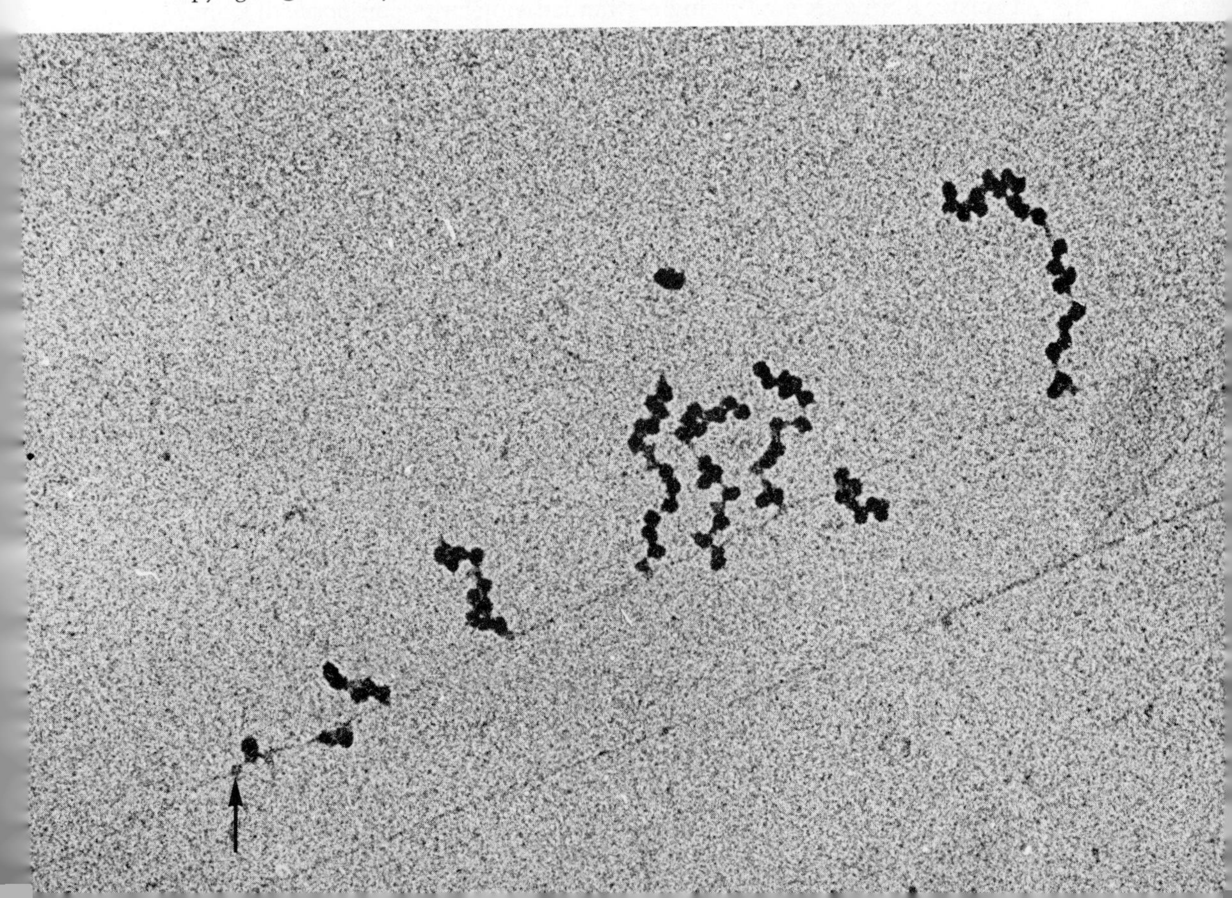

The stability of mRNA varies. In bacteria it is utilized and degraded within a few minutes. In higher organisms the lifetime of a messenger generally varies from three to twenty-four hours, and in some cells, such as those synthesizing hemoglobin, the mRNA for this protein may be utilized over and over again for several days. Amphibian eggs represent an extreme example of longevity, since mRNA is stored in inactive or masked form during the entire hibernating season and becomes functional only after the egg has been fertilized the following spring.

Transfer RNA

The existence of transfer RNA was early predicted by Crick on theoretical grounds, there being no obvious structural or chemical correspondence between polynucleotides and polypeptides. Crick postulated that proteins might be assembled on a nucleic acid template, appropriate amino acids being brought to such a template by an RNA carrier or adapter. By base pairing with the template, such a carrier RNA molecule could orient amino acids in the proper sequence necessary for the construction of an active polypeptide. This proposal has been substantiated, for tRNA has been shown to function in the manner predicted.

Cells produce at least 40 different kinds of tRNA, each of which is a copy of a specific tRNA gene. In bacteria only one gene per tRNA type is thought to be present, but in nucleated cells multiple copies of each tRNA gene are known to occur. Since transcription at these gene sites produces an RNA which is somewhat larger than the final tRNA, enzymatic modification of the molecule is apparently required for function.

Transfer RNA's are small, single-stranded molecules composed of from 73 to 93 nucleotides. The characteristic sedimentation constant is 4 S, which corresponds to a molecular weight of around 25,000. Their small size has made possible the determination of the entire sequence of bases in over 40 tRNA's from bacteria, yeast, and animal cells. Within transfer RNA are a number of unusual bases, many of them being methylated derivatives of the more common types. Such odd bases include inosine, 1-methylguanine, 5-methylcytosine, 6-methylaminopurine, and 2-methyladenine. Pseudouridine (Ψ), dihydrouridylic acid, and even thymine also occur, among others. Present, also, is a considerable amount of hydrogen bonding between bases located in different regions of the same molecule. As a result, the final shape of each tRNA is complex, marked by double-stranded sequences alternating with loops where no pairing occurs. The positions of the loops within the molecule, as well as the particular bases which form these loops, represent specific configurations required for interaction with amino acid activating enzymes, messenger RNA, and the ribosomal subunits.

The base sequences of all tRNA's form a similar hydrogen-bonded secondary structure which is illustrated in the generalized cloverleaf diagram of Figure 15–7. It is postulated that the lengths of stems a, c, and e, and the loops II and IV of Figure 15–7 are the same for all tRNA's. Differences between tRNA's in the number of nucleotides present are accommodated by variations in the lengths of stems b and d and in the size of loops I and III. Two sites within the structure whose role is certain are the 3′ hydroxyl end of the

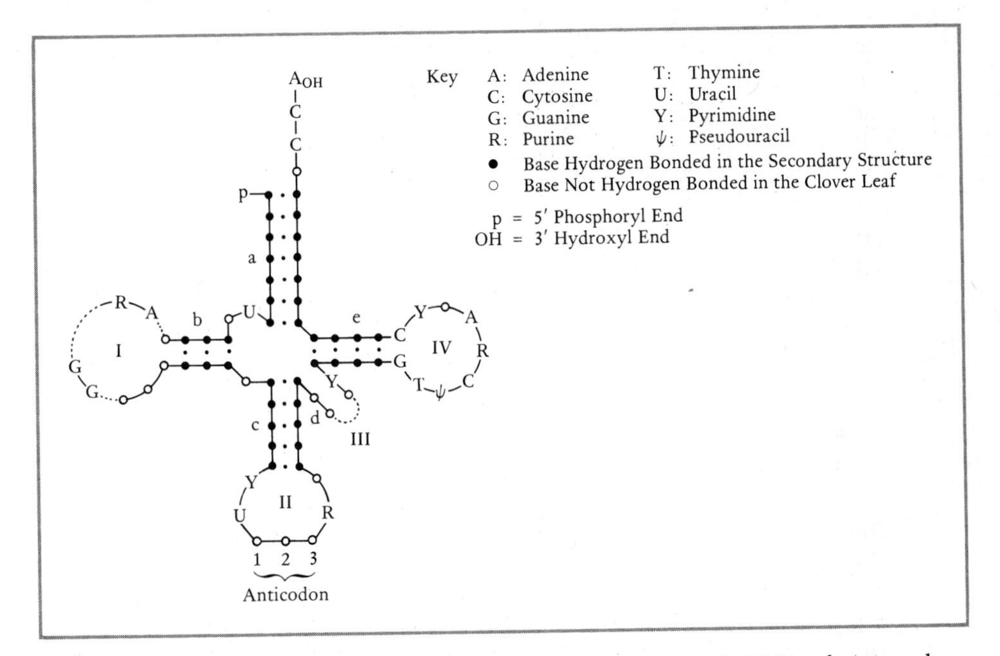

Figure 15–7. Generalized cloverleaf secondary structure of tRNA, showing those bases constant in tRNA sequences. (See text for meaning of label regions.) [From J. D. Smith, *Brit. Med. Bull., 29,* 220–225, 1973.]

molecule, to which an amino acid becomes attached, and the anticodon loop which functions in concert with mRNA during protein synthesis. Although the cloverleaf diagram permits easier analysis of the structure of tRNA, the molecule does not actually look like a cloverleaf. Figure 15–8 represents the actual folded structure of yeast phenylalanine tRNA as compared to a cloverleaf diagram of the same molecule.

In Figure 15–8 it can be seen that the folding of the molecule is such that the 5′ phosphoryl and the 3′ hydroxyl termini are adjacent and that the 5′ end, which begins with guanine, is hydrogen-bonded to the subterminal bases of the 3′ end. Beyond this region of pairing extend the last three nucleotides of the 3′ terminus. These nucleotides consist of the sequence cytosine-cytosine-adenine (C-C-A). When the function of amino acid transport is assumed by any tRNA, the amino acid is temporarily bonded via its carboxyl group (—COOH) to the terminal nucleotide (A) of this C-C-A sequence. Although all tRNA's exhibit this same end configuration, the bases preceding the C-C-A sequence differ in different tRNA's, thus providing specificity.

The attachment of an amino acid to a tRNA occurs in two steps, both of which are mediated by the same enzyme, called an *amino-acyl tRNA synthetase* or *activating enzyme.* At least 20 such enzymes exist in all cells, each recognizing only 1 of the 20 different kinds of amino acids and only 1 particular tRNA type. The two reactions catalyzed by these enzymes are:

1. An amino acid via its carboxyl group (—COOH) is combined with the AMP moiety of ATP to yield an amino-acyl-AMP enzyme complex, with the

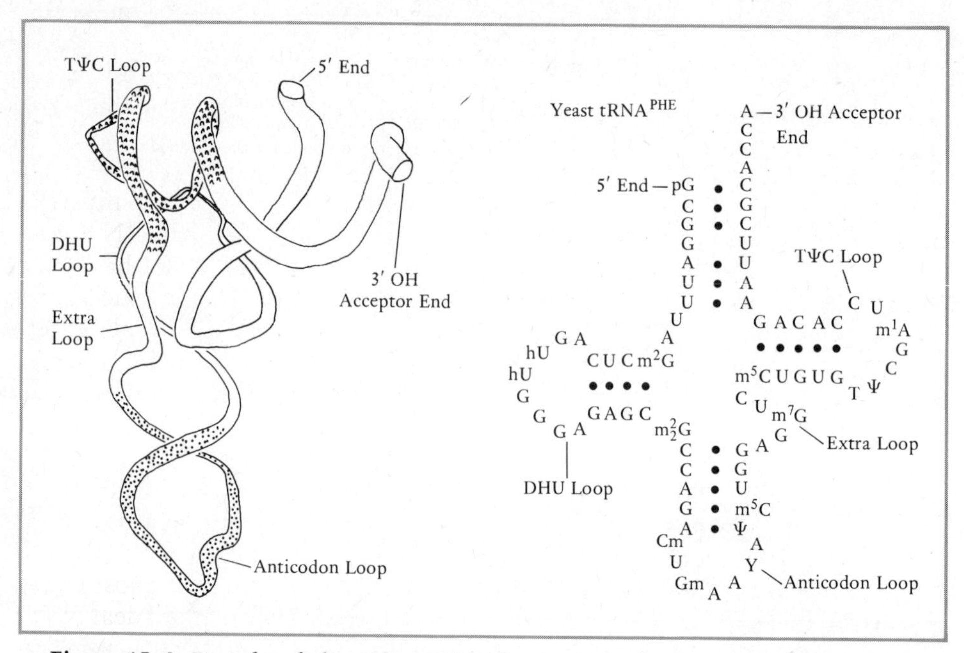

Figure 15–8. Postulated three-dimensional structure of yeast phenylalanine tRNA, in which the molecule is represented as a continuous coiled tube. The positions of the various loops of the molecule should be compared to the respective equivalents shown in the cloverleaf to the right. (From S. H. Kim et al, *Science, 179,* 285–288, 19 January 1973. Copyright © 1973 by the American Association for the Advancement of Science.)

release of pyrophosphate ($P\text{-}P_i$). Using the amino acid glycine (gly) to indicate the specificities of the reactants, this reaction would be:

$$\underset{\text{(gly)}}{\text{Amino acid}} + \text{ATP} + \underset{\text{(gly)}}{\text{Enzyme}} \rightleftarrows \underset{\text{(gly)}}{\text{Amino-acyl-AMP}} \cdot \underset{\text{(gly)}}{\text{Enzyme}} + \text{P-P}_i$$

2. The amino-acyl-AMP enzyme complex reacts with an appropriate tRNA with the result that the amino acid is transferred to the terminal adenine nucleotide of the C-C-A sequence of the tRNA. This reaction is accompanied by the release of AMP from the amino acid.

$$\underset{\text{(gly)}}{\text{Amino-acyl-AMP}} \cdot \underset{\text{(gly)}}{\text{Enzyme}} + \underset{\text{(gly)}}{\text{tRNA}} \rightleftarrows \underset{\text{(gly)}}{\text{Amino-acyl-tRNA}} + \underset{\text{(gly)}}{\text{Enzyme}} + \underset{\text{(gly)}}{\text{AMP}}$$

When transfer of the amino acid to the tRNA has been completed, the tRNA is said to be charged, that is, ready for participation in protein synthesis. It should be noted that the energy level of the acyl bond between the amino acid and AMP or between the amino acid and the tRNA is the same as that originally present in the equivalent bond of ATP. Consequently, the acyl bond is also a high energy bond, and its subsequent dissolution during polypeptide synthesis provides energy for peptide bond formation. The specificities of the enzymes which catalyze these reactions should also be emphasized. Each activating enzyme must recognize the R group of one specific kind of amino acid and no other. It must also recognize ATP, as well as a particular kind of tRNA.

It should be recalled that the ability to do so is a consequence of molecular structure.

In addition to regions which combine with activating enzymes, amino acids, or ribosomes, the folded structure of tRNA results in the presence of a specific loop, a required distance from and opposite to the two united ends of the molecule (Figures 15–7 and 15–8). This loop is known as the *anticodon loop* because it contains a set of three bases *(anticodon)* which must be matched with a set of three complementary bases of a codon of mRNA. The codon and anticodon are correctly aligned only if their respective bases can form hydrogen bonds with one another. Through the matching of successive codons with the anticodons of the tRNA's, different amino acids can be incorporated, in order, into a polypeptide. The specificity of base pairing, C with G and A with U, thus accounts for the accuracy with which proteins are assembled.

TRANSLATION OF mRNA INTO PROTEIN

Although the overall process of protein synthesis is known, most of the molecular mechanisms involved are not at all well defined. A good deal of our present knowledge has been gained from studies with bacteria whose comparatively simple organization is more amenable to analysis than is that of more complex higher forms. Since all cells, whether prokaryotic or eukaryotic, utilize the same general mechanism for synthesizing proteins, model systems developed in bacteria can be usefully applied to higher organisms, although the details of the process will unquestionably be different in the two groups. Cell-free systems have additionally been of great value in defining the respective roles of participants in the process. Although the following presentation is based primarily on studies with bacteria, the general flow of events is considered characteristic of all cells.

The current concept of the process of protein synthesis involves a series of steps beginning with the binding of the 5′ end of a mRNA to the smaller, 30 S subunit of the ribosome, the ribosome subunits being, at this time, separate from one another. The mRNA and the 30 S subunit are also joined by a particular kind of tRNA whose anticodon matches a codon of the mRNA and is held by hydrogen bonding. These three, the small ribosome subunit, the mRNA, and the tRNA, together form what is called an *initiation complex* (Figure 15–9a). Additional specific agents required for this association, for recognition, catalysis, or both, have been identified in bacteria and named factors 1, 2, and 3. Energy enabling the union is supplied by the hydrolysis of GTP whose phosphate bonds contain the same amount of energy as do those of ATP. Formation of the initiation complex is the first step in protein synthesis.

Let us examine this complex more closely. Although protein synthesis begins at the 5′ end of a mRNA, we cannot say that it begins with the first nucleotide or that this nucleotide is included in the initiation complex. On the contrary, the initiation complex seems to involve a codon some distance from the actual 5′ end of the mRNA. This codon is the initiating codon or "start" signal incorporated in the mRNA. The sequences preceding this codon are not

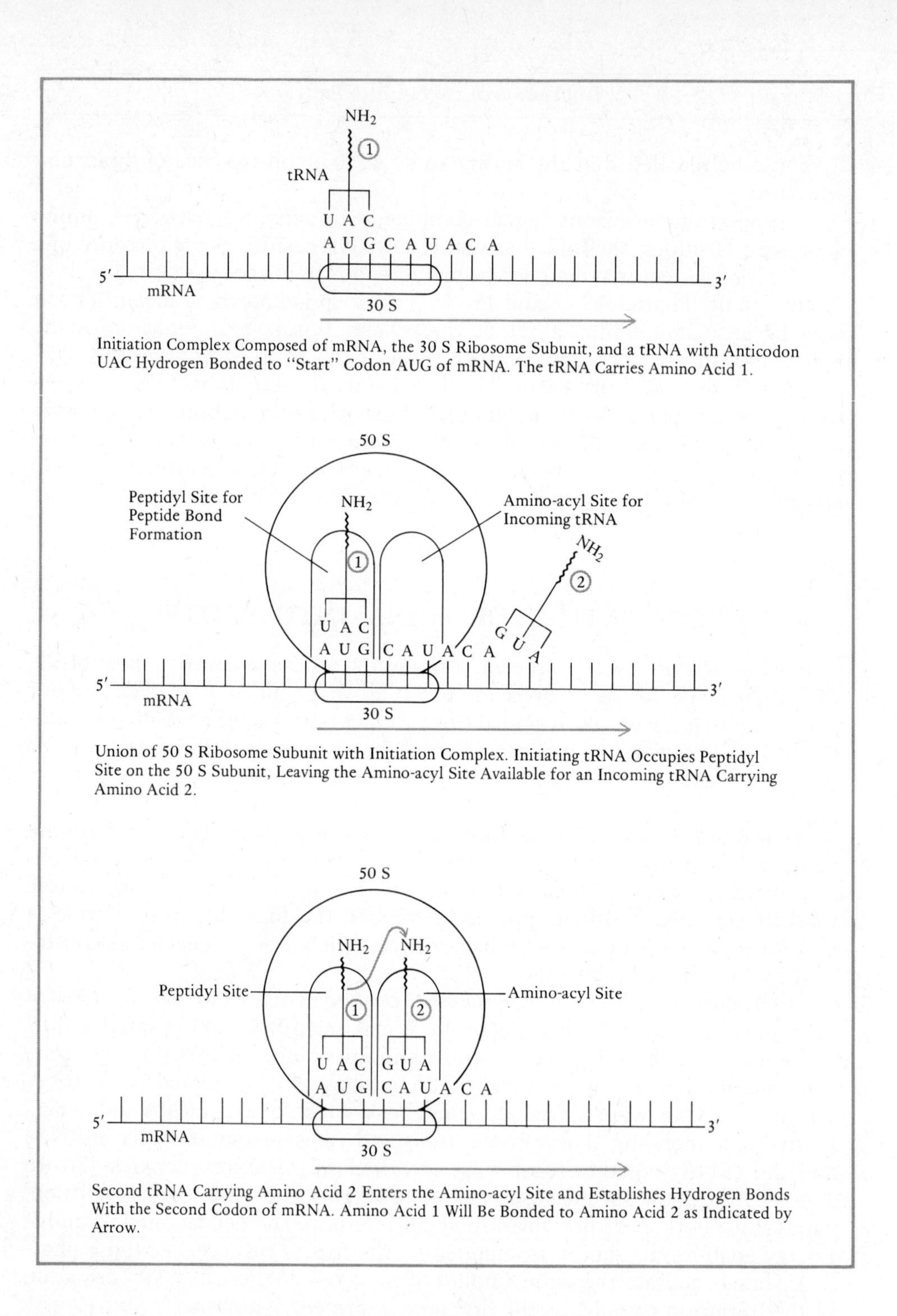

Initiation Complex Composed of mRNA, the 30 S Ribosome Subunit, and a tRNA with Anticodon UAC Hydrogen Bonded to "Start" Codon AUG of mRNA. The tRNA Carries Amino Acid 1.

Union of 50 S Ribosome Subunit with Initiation Complex. Initiating tRNA Occupies Peptidyl Site on the 50 S Subunit, Leaving the Amino-acyl Site Available for an Incoming tRNA Carrying Amino Acid 2.

Second tRNA Carrying Amino Acid 2 Enters the Amino-acyl Site and Establishes Hydrogen Bonds With the Second Codon of mRNA. Amino Acid 1 Will Be Bonded to Amino Acid 2 as Indicated by Arrow.

translated into protein and may be involved in ribosome attachment or recognition.

The particular tRNA involved in the initiation complex also deserves closer inspection, especially since it contains the anticodon for the message to start. In prokaryotes this tRNA has been found to transport N-formyl methionine, that is, the amino acid methionine modified by the addition of a

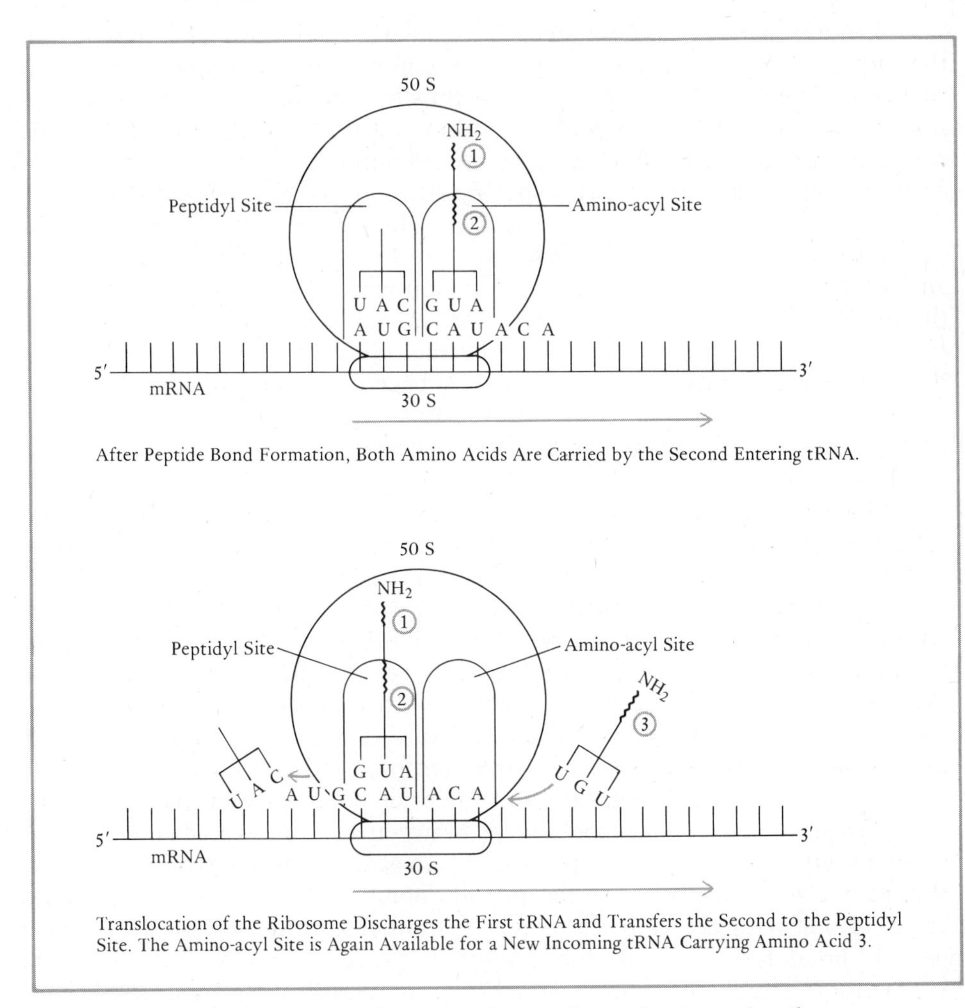

After Peptide Bond Formation, Both Amino Acids Are Carried by the Second Entering tRNA.

Translocation of the Ribosome Discharges the First tRNA and Transfers the Second to the Peptidyl Site. The Amino-acyl Site is Again Available for a New Incoming tRNA Carrying Amino Acid 3.

Figure 15–9. Sequential steps in polypeptide synthesis on the ribosome.

formyl (CHO) group. The anticodon of this tRNA is U-A-C, and thus the starting codon of mRNA must be the complementary sequence A-U-G. The situation is complicated by the fact that a slightly different tRNA which transports ordinary, unmodified methionine, possesses the same anticodon (UAC) as does the N-formyl methionine tRNA. Ordinary methionine is found incorporated into the interior of a polypeptide wherever called for, but N-formyl methionine is found only at the beginning. It is postulated that subtle structural differences between these tRNA's may account for their distinctive functions. It might be expected that since the first tRNA to be involved in translation carries N-formyl methionine, the polypeptide ultimately produced should also begin with N-formyl methionine. However, not all proteins contain this amino acid in the first position, and it has been found that after synthesis the formyl group may be removed or even the methionine itself cleaved from the polypeptide before the final active form is assumed.

The next and succeeding steps in translation begin with the addition of the larger, 50 S subunit to the initiation complex to form a complete 70 S ribosome. The two parts of the ribosome thus hold between them the ribbon of mRNA with its attached N-formyl methionine tRNA. At the joining of the ribosome subunits, this tRNA also becomes bound at its opposite, amino acid carrying end to the second of two specific sites on the large ribosome subunit. Energy for this binding is supplied by GTP. The two sites on the 50 S subunit can be visualized as lying side by side (Figure 15–9b). The first site, called the *amino-acyl site,* serves in facilitating codon-anticodon recognition between the mRNA and any incoming tRNA. For the very first codon, however, this function has already been carried out by the original initiation complex, and therefore the N-formyl methionine tRNA becomes bound to the second, or *peptidyl site* of the subunit, leaving the first site vacant. Once all of these interactions and arrangements have been completed, the system is ready to synthesize a polypeptide.

Since the first codon of the mRNA is occupied by the anticodon of the first tRNA, the next succeeding three bases of the messenger become available for base pairing. These three bases constitute the second codon which must be matched precisely with the anticodon sequence of a new, entering tRNA. When an appropriate tRNA is recognized, the codon and anticodon bases can pair with one another via hydrogen bonding, and the new tRNA can be admitted to the vacant amino-acyl site of the larger ribosome subunit. At this point both subunit sites are occupied by tRNA's, each of which carries an amino acid, and each of which is hydrogen bonded to the template mRNA (Figure 15–9c).

The next step is the formation of a peptide bond between the two amino acids. It will be recalled that the linkage between any amino acid and its tRNA involves only the carboxyl group of the amino acid, the amino group being free. We must also remember that the peptide bond is one that joins the amino group of one with the carboxyl group of another amino acid. To accomplish peptide bond formation on the ribosome, the first amino acid (N-formyl methionine) is removed from its attachment to its tRNA and transferred to the free —NH_2 terminus of the second amino acid. The first amino acid is thus placed "on top of" the second. The ensuing peptide bond thus joins the carboxyl group of the first amino acid with the amino group of the second amino acid (Figure 15–9d). The resulting compound is a dipeptide whose carboxyl end is still bonded to the second tRNA, but whose amino end is free. The reaction is catalyzed by an enzyme associated with the 50 S subunit and called *peptidyl transferase.* The energy for peptide bond formation is supplied by the dissolution of the amino-acyl bond between the first amino acid and its carrier tRNA, this energy having originally been donated by ATP.

Although thus far a dipeptide has been generated, continued synthesis requires that the next codon be made available and that the next tRNA be admitted to the amino-acyl site on the ribosome, this site being still occupied by the dipeptide-carrying tRNA. The problem is solved by a movement of the entire ribosome relative to the mRNA strand. Since the dipeptide-carrying tRNA is still hydrogen bonded to the template, this movement merely dislodges it from the first position, the amino-acyl site, on the ribosome and transfers it to the second position, the peptidyl site. Its entrance into this site displaces the tRNA whose amino acid has been discharged. The net result of movement is a release of the first tRNA, a transferral of the dipeptide-carrying tRNA to the peptidyl site, and a once more vacant amino-acyl site which can then receive a new tRNA (Figure 15–9e). This movement of the ribosome is not a random slippage along the mRNA. Instead, the ribosome is thought to move only "three paces forward," so that only the next codon is brought into alignment to be matched with an incoming tRNA. Ribosome movement, called *translocation,* is clearly a most complex event and very little is known concerning the mechanisms which bring it about. GTP, however, is once again involved as an energy donor.

The sequential formation of a polypeptide continues in the manner described above. A tRNA in the peptidyl site shifts its burden of growing polypeptide to the next succeeding tRNA, followed by translocation, exit of the discharged tRNA, and entrance of a new tRNA to base pair with a new codon at the amino-acyl site. Thus, the growing polypeptide is assumed, in turn, by each tRNA, with each successive amino acid being added, in effect, to the bottom of the stack. As the process continues, the mRNA is progressively translated, codon by codon, from the 5′ end to the 3′ end. The resulting polypeptide thus contains one amino acid per codon of mRNA, the translation of codons from the 5′ to the 3′ end corresponding to the sequence of amino acids from the amino to the carboxyl terminus of the polypeptide.

The occurrence of messengers which contain more bases than can be accounted for by the size of the product and the fact that mRNA of bacteria may contain information for more than one separate polypeptide suggests the presence of stop signals in the mRNA for polypeptide termination so that translation will cease at the proper codon. Three such stop signals have indeed been identified as the codons UAA, UGA, and UAG. These codons are read, not by a tRNA, but by protein release factors which appear to recognize a stop codon and subsequently cleave the acyl bond which holds the completed polypeptide to the last tRNA translated. As a result the polypeptide is released from the tRNA, and the ribosome dissociates into its separate subunits.

As described in Chapter 10, polypeptides do not usually occur as simple linear molecules. Instead, they are folded into specific configurations, the various twists and turns being held in place by sulfhydryl as well as by hydrogen bonds. The assumption of this final form is spontaneous and is determined by the properties of the amino acids built into the molecule. Even as translation proceeds, the polypeptide begins to fold according to the dictates of its structure so that when it is released from the ribosome, the final conformation which confers activity has already been achieved.

When one ribosome has completed its journey down the length of a mRNA, the messenger is not necessarily set free. On the contrary, the first ribosome is followed by a second, and a third, and so on, all of them proceeding along the template, one behind the other, with the result that many identical polypeptide molecules are produced. Such a linearly arranged cluster of ribosomes is called a *polyribosome* or *polysome* (Figure 15–10). It is thought that when the first ribosome has progressed a given distance beyond the initiation site of the template, a new initiation complex is formed. Thus a template can be translated many times in response to the requirements of the cell for a given

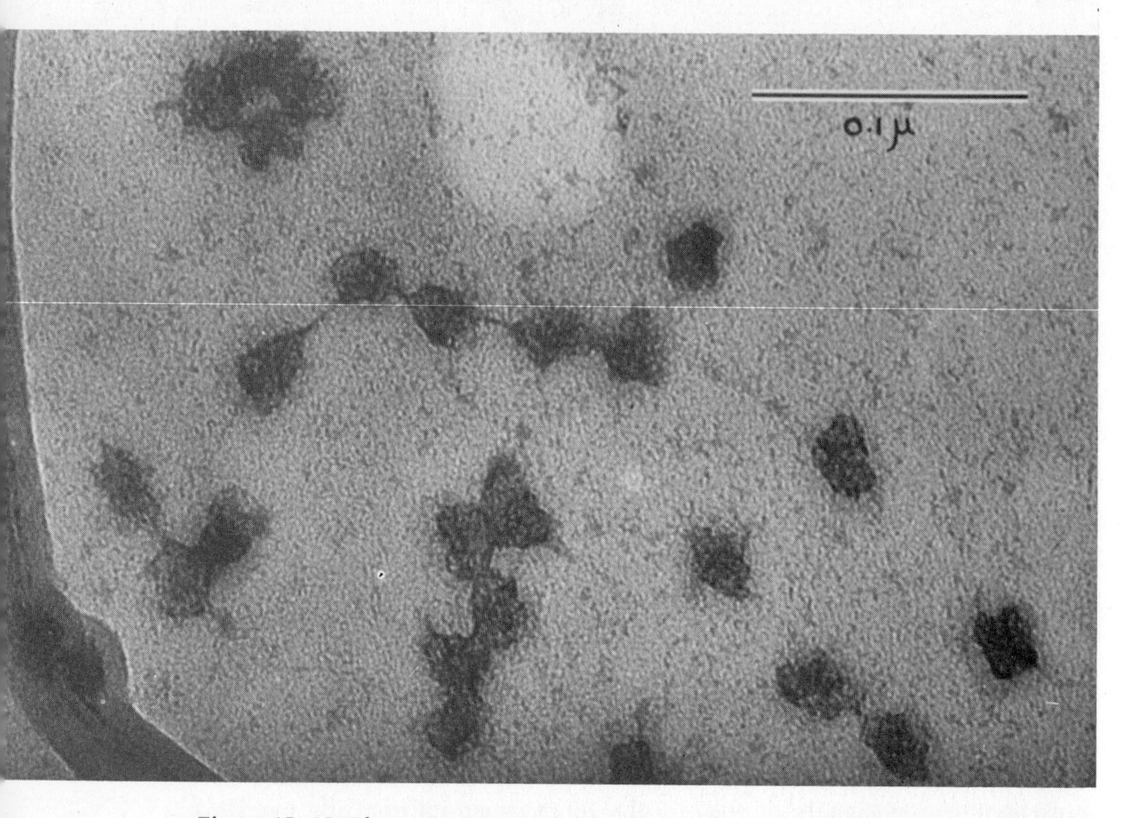

Figure 15–10. Electron micrograph of polyribosomes isolated from immature blood cells of the rabbit. Polysomes are ranged along thin strands of mRNA. The message in this case is presumably for the globin portion of hemoglobin. (280,300×) [Henry Slater, Farber Cancer Institute. From H. S. Slater et al., J. Mol. Biol., 7, 652–657, 1963.]

protein. Cells are most economical in their expenditure of energy however, and proteins for which there is no continuing need are not lavishly produced, a delicate balance between demand and supply being maintained at all times.

PROBLEMS

15–1. Distinguish between RNA polymerase, DNA polymerase, and reverse transcriptase with respect to template and product.

15–2. Why must adapter RNA molecules be postulated as participants in protein synthesis?

15–3. What function common to all tRNA's does the C-C-A sequence perform? By what steps is this function accomplished?

15–4. RNA with a base ratio of 2 A : 1 U : 1 C was enzymatically synthesized from the following DNA sequence:

```
T A T G T A T G T A T G T A T G T A T G   (strand 1)
| | | | | | | | | | | | | | | | | | | |
A T A C A T A C A T A C A T A C A T A C   (strand 2)
```

From which strand was the RNA transcribed?

15–5. Messenger RNA was transcribed *in vitro* from enzymatically synthesized DNA. The DNA was then denatured to the single-stranded state and the base ratio of each DNA chain was analyzed and compared with that of the mRNA. On the basis of the following ratios, which strand of the DNA served as the template for RNA synthesis?

	A	T OR U	G	C
DNA-1	18.5	22.0	27.0	32.5
DNA-2	22.1	18.4	32.7	26.8
mRNA	18.0	22.0	27.0	33.0

15–6. Suppose the following peptide chain was attached to a gly-tRNA during translation: met-leu-asp-glu-asn-cys-gly-tRNA. If the next incoming tRNA carries serine, where will serine be added to the chain?

15–7. Assume the following sequence of mRNA: U U C C A C G A C G G A C C U.
a. If this message is transcribed from left to right, which end is the 5′ terminus?
b. What is the base sequence from which this RNA was copied?
c. If this message is translated from left to right, what are the corresponding tRNA anticodons?
d. If UUC = leucine and CCU = proline, which amino acid will be at the carboxyl terminal of the resulting polypeptide?

REFERENCES

Crick, F. H. C., 1957. The structure of nucleic acids and their role in protein synthesis. *Biochem. Soc. Sympos., 45,* 25.

Ingram, V., 1972. *The Biosynthesis of Macromolecules,* 2nd ed. W. A. Benjamin, Menlo Park, Calif.

McConkey, E. H., 1974. Composition of mammalian ribosomal subunits: A reevaluation. *Proc. Nat. Acad. Sci., 71,* 1379.

Miller, O. L., and B. R. Beatty, 1969. Visualization of nucleolar genes. *Science, 164,* 955.

Miller, O. L., B. A. Hamkalo, and C. A. Thomas, 1970. Visualization of bacterial genes in action. *Science, 169,* 392.

Nomura, M., 1973. Assembly of bacterial ribosomes. *Science, 179,* 864.

Nomura, M., A. Tissières, and P. Lengyel (Eds.), 1974. *Ribosomes.* Cold Spring Harbor Laboratory, Cold Spring Harbor, N.Y.

Smith, J. D., 1972. Genetics of transfer RNA. *Ann. Rev. Genet., 6,* 235.

Söll, D., 1971. Enzymatic modification of transfer RNA. *Science, 173,* 293.

Watson, J. D., 1976. *Molecular Biology of the Gene,* 3rd ed. W. A. Benjamin, Menlo Park, Calif.

Chapter 16

THE GENETIC CODE AND ITS VALIDATION

We have assumed that a given base sequence of DNA constitutes a three letter, or triplet code which can be translated by the cell into a corresponding colinear polypeptide. The basis for this assumption can now be discussed.

THE TRIPLET CODE

The early predictions of A. L. Dounce and G. Gamow that the sequence of bases within DNA functioned as a code in the ordering of amino acids in the polypeptide prompted a number of questions concerning the nature of such a code. Such questions included the number of bases per amino acid "word" (coding ratio), the presence or absence of punctuation, and whether or not the code was overlapping, that is, whether one base was a part of two different adjacent words.

The question of overlapping was resolved through the observation that some mutant proteins, for example, hemoglobin variants, contain a change in only one amino acid. This finding indicated a nonoverlapping code, for overlapping symbols would be expected to alter at least two adjacent amino acids of a polypeptide.

With respect to the number of bases per word, it was clear that a single base could not specify an amino acid because about 20 amino acids commonly occur in proteins, but only 4 bases. A symbol made up of 2 bases is inadequate for the same reason, since a two-letter code involving 4 bases yields only 4^2 or

16 different combinations. However, a three-letter symbol will furnish 4^3 or 64 possible combinations, more than sufficient to specify 20 amino acids. For this reason, the idea of a triplet code was accepted as the most reasonable hypothesis, and the proposed coding unit consisting of three consecutive bases per amino acid was termed a *codon*. Such a code's 64 combinations, far in excess of the 20 needed, also suggested that more than one codon might stand for the same amino acid and that therefore the code might be "degenerate."

As a solution to the question of whether or not punctuation might be present within the code, Crick suggested that the translation of the code into protein might proceed from a fixed point. This concept imposes a "reading frame" on a sequence of bases. If a fixed point within the sequence constitutes the starting position for translation, all succeeding bases will fall into place as sets of three and no internal punctuation will be necessary.

Support for this hypothesis, as well as for the triplet code, was obtained in 1961 by Crick and associates in a series of experiments with the *rII* mutants of phage T4. Wild type phage was treated with a chemical mutagen, called proflavin, which acts on DNA to cause the insertion and deletion of bases. Among the mutations so induced was one located in the *B* cistron. It will be recalled that such a mutated phage can be identified by its ability to form plaques on *E. coli* strain B, but not on strain K-12. Crick found that when phage carrying the induced mutation were used to infect K-12 cells, some revertants to wild type appeared, as evidenced by plaque formation. However, these revertants seemed to have arisen not through a reversal of the original mutation, but through a second mutation which had occurred at a different, but nearby site within the *B* cistron. Evidence for an additional mutation was the slightly altered plaque phenotype when the progeny of revertants were grown on strain B and the fact that the second mutated site was separable from the first by recombination. It thus appeared that the effects of the first mutation, termed plus, were being suppressed by the presence of the second mutation, termed minus. Other minus-type mutations which could suppress the plus mutation were also found. Each was located in the vicinity of the original plus site, and in every case the combination of plus with minus led to the wild type phenotype. By repeating the treatment with proflavin, additional mutations within the *B* cistron were then recovered. Some of these, when combined with the preceding minus mutations, also gave rise to wild type phage. Since their behavior in combination with a minus mutant was the same as that of the original plus mutation, these newly induced sites were also called plus mutations. By making additional combinations and observing the resulting phenotype, it was found that when the genome contained either two plus or two minus mutations, the phenotype was mutant, and no plaques were formed on K-12 cells. However, the combination of three plus or three minus mutations yielded wild type phage.

Assuming a triplet code, read from a fixed starting point, Crick interpreted these results in the following way. He postulated that plus mutants were induced by the addition of a single nucleotide and minus mutants by the deletion of a single nucleotide. Applying the reading frame concept, he reasoned that a single plus mutation (addition) or a single minus mutation (deletion), or two plus or two minus mutations, would shift the reading frame from the original sequential order of three's from the point of mutation onward, thereby producing an abnormal inactive product. For example, if a sequence of triplets was 1 2 3 1 2 3 1 2 3 1 2 3 . . . , loss of one nucleotide (in intalics) would produce a shift in the bases comprising the succeeding triplets, changing the sequential order to 1 2 3 1 2 1 2 3 1 2 3 1 2 3 1 . . . . Such a shift would lead to an abnormal amino acid sequence.

The same type of change would result from the addition of a base to the sequence or the addition or deletion of two bases, but, if a plus mutation were followed by a nearby minus mutation, the original reading frame sequence would be restored for the balance of the cistron. The intervening region between the addition and deletion of a base would be abnormal, but if this region was short, the resulting polypeptide might not be greatly altered, and a phenotype approximating wild type could arise. In our example, if loss of the nucleotide is followed by gain of another nucleotide, 1 2 3 1 2 3 1 2 3 3 1 2 3 1 2 3 . . . , the resulting sequence will be: 1 2 3 1 2 1 2 3 3 1 2 3 1 2 3 . . . . Here, although the second and third triplets would be translated as two "wrong" amino acids, restoration of the reading frame beyond this point should produce, thereafter, the normal amino acid sequence. The finding that a combination of three plus or three minus mutations, when located close together, also produced a normal phenotype was entirely in accord with Crick's hypothesis.

These results provided strong support for a triplet code, as well as for the concept that this code contained no punctuation and was translated into protein from a fixed starting point. Although this hypothesis was derived on theoretical grounds and from the results of experiments with mutations of phage T4, studies on the mechanism of protein synthesis soon provided independent and conclusive proof of the triplet code.

An opportunity to analyze the code for proteins was provided through the synthesis of mRNA's of known base sequence which could be utilized in cell-free systems for the synthesis of protein. Such messengers were first produced by M. W. Nirenberg and J. H. Matthaei and by S. Ochoa through the use of an enzyme called polyribonucleotide phosphorylase which in the presence of magnesium can unite ribonucleotide diphosphates into polymers by forming the usual phosphodiester bonds between succeeding nucleotides with the release of inorganic phosphate. This reaction can be reversed when excess phosphate is added, resulting in a breakdown of the RNA polymer. The enzyme does not function as an RNA polymerase in the living cell, but instead acts to degrade RNA. However, in experimental systems the enzyme can be induced to construct a molecule of RNA whose base composition is a random reflection of the concentration of the various nucleotides present as substrates in the reaction mixture. Such synthetic polymers have provided elegant proof of the role of mRNA as well as establishing the correctness of the triplet code concept.

The first such polymer to be used in the study of protein synthesis was one constructed entirely of uracil nucleotides. It was thus a polyuracil or polyU RNA. When polyU was used as a messenger in a cell-free system containing bacterial ribosomes, tRNA, activating enzymes and other factors, as well as the 20 amino acids, a polypeptide was produced that consisted entirely of the amino acid phenylalanine. On the basis of a triplet code, this finding was interpreted to mean that the mRNA codon UUU specified phenylalanine.

Similarly, a polyC messenger, whose codons were all CCC, was found to direct the synthesis of a polypeptide composed entirely of proline, and a polyA messenger produced a product containing only lysine.

Various mixtures of nucleotides were then used to construct the synthetic messenger. Since random chance determined the incorporation of nucleotides into the RNA molecule, one could expect to distinguish classes of triplets within such a polymer whose frequency could be predicted by statistical means. For example, a polymer synthesized from a reaction mixture of $\frac{3}{4}$ adenine nucleotides (A) to $\frac{1}{4}$ cytosine nucleotides (C) would be expected to contain these classes of triplets:

CLASS	FREQUENCY	TRIPLET TYPE
3 A	$\frac{3}{4} \times \frac{3}{4} \times \frac{3}{4} = \frac{27}{64} = 0.42$	AAA
2 A : 1 C	$\frac{3}{4} \times \frac{3}{4} \times \frac{1}{4} = \frac{9}{64} = 0.14$	AAC, ACA, or CAA
1 A : 2 C	$\frac{3}{4} \times \frac{1}{4} \times \frac{1}{4} = \frac{3}{64} = 0.05$	ACC, CAC, or CCA
3 C	$\frac{1}{4} \times \frac{1}{4} \times \frac{1}{4} = \frac{1}{64} = 0.015$	CCC

If such a polymer were used as a messenger in protein synthesis, the resulting polypeptide would be expected to contain amino acids whose proportional presence could be correlated with the relative frequencies of the triplet classes. For example, the codon AAA is expected to occur approximately 3 times more frequently than a codon containing 2 A : 1 C, around 8 times more frequently than a codon of 1 A : 2 C, and about 28 times more often than the codon CCC. An amino acid exhibiting corresponding frequencies might be assigned to codon AAA. In general, this procedure allows amino acids to be assigned to codon classes, but does not distinguish individual members of each class because the order of the bases within the codons is not determined. In actual practice, however, quite accurate assignments can be made by choosing initial proportions, such as 5 : 1 or 10 : 1, which give relatively unequivocal results with respect to amino acid incorporation.

The actual order of bases within any triplet was finally deciphered by Nirenberg, P. Leder, and associates and by H. G. Khorana and coworkers. Nirenberg and Leder used a technique that relied on the specificity of binding between amino-acyl tRNA's, the ribosome, and mRNA. This binding is dictated by the mRNA, and it was found that a messenger composed of only three nucleotides was of sufficient length to initiate the complex, by itself strong supporting evidence for a three-letter code. Synthetic trinucleotides of known base order were then tested to determine which particular charged tRNA contained the specific complementary anticodon. By this technique it became possible to distinguish between codons such as AAC, ACA, and CAA and to determine the amino acid specified by each. That the code was indeed triplet in nature was demonstrated by the finding that only trinucleotides, and not dinucleotides, could effectively form an initiation complex with tRNA and ribosomes.

Khorana and associates utilized a different method to show that the ratio of nucleotides to amino acids was three, that all bases were read with no punctuation, and that the code was nonoverlapping. These workers synthesized RNA molecules by the chemical joining of dinucleotides to form a messenger of known sequence and polarity, that is, 5′ CU-CU-CU-CU-CU- . . . .

This RNA was then used *in vitro* to synthesize a polypeptide whose amino acid sequence could be compared with the known sequence of the messenger. In our example, the messenger codons would be CUC UCU CUC UCU . . . , leading to the synthesis of a polypeptide formed of an alternating sequence of leucine and serine, leu-ser-leu-ser-leu-ser- . . . (see Figure 16–1).

That the direction of translation proceeds from the 5′ to the 3′ end of the mRNA molecule has additionally been demonstrated with synthetic messengers of known sequence by P. Doty and by S. Ochoa and their co-workers. When such messengers are used for *in vitro* protein synthesis, the first amino acid at the NH_2 terminus of the resulting peptide is that specified by the first set of three bases located at the 5′ end of the synthetic messenger, the second amino acid is that specified by the second set of three bases, and so on.

By all of these means, assignments were made for the 64 possible codons of mRNA (Figure 16–1). Of the 64, 61 were matched by tRNA's, while 3 (UAG, UAA, and UGA) had no corresponding tRNA and did not specify an amino acid. These three are the stop signals for polypeptide termination and are often called nonsense codons. Microbiologists frequently use the colloquial (and confusing) terms amber for UAG, ochre for UAA, and opal for UGA.

The 64 codons of mRNA are referred to, collectively, as the *genetic code.* It must be remembered, however, that the complementary sequences of the tRNA's comprise a matching "anticodon code" and that these two codes to-

First (5′) Base	Second Base: U		C		A		G		Third (3′) Base
U	UUU	Phe	UCU	Ser	UAU	Tyr	UGU	Cys	U
U	UUC	Phe	UCC	Ser	UAC	Tyr	UGC	Cys	C
U	UUA	Leu	UCA	Ser	UAA	Stop	UGA	Stop	A
U	UUG	Leu	UCG	Ser	UAG	Stop	UGG	Trp	G
C	CUU	Leu	CUU	Pro	CAU	His	CGU	Arg	U
C	CUC	Leu	CCC	Pro	CAC	His	CGC	Arg	C
C	CUA	Leu	CCA	Pro	CAA	Gln	CGA	Arg	A
C	CUG	Leu	CCG	Pro	CAG	Gln	CGG	Arg	G
A	AUU	Ilu	ACU	Thr	AAU	Asn	AGU	Ser	U
A	AUC	Ilu	ACC	Thr	AAC	Asn	AGC	Ser	C
A	AUA	Ilu	ACA	Thr	AAA	Lys	AGA	Arg	A
A	AUG	Met Start	ACA ACG	Thr	AAG	Lys	AGG	Arg	G
G	GUU	Val	GCU	Ala	GAU	Asp	GGU	Gly	U
G	GUC	Val	GCC	Ala	GAC	Asp	GGC	Gly	C
G	GUA	Val	GCA	Ala	GAA	Glu	GGA	Gly	A
G	GUG	Val	GCG	Ala	GAG	Glu	GGG	Gly	G

Figure 16–1. The mRNA genetic code. (Amino acids are abbreviated as in Figure 10–9.)

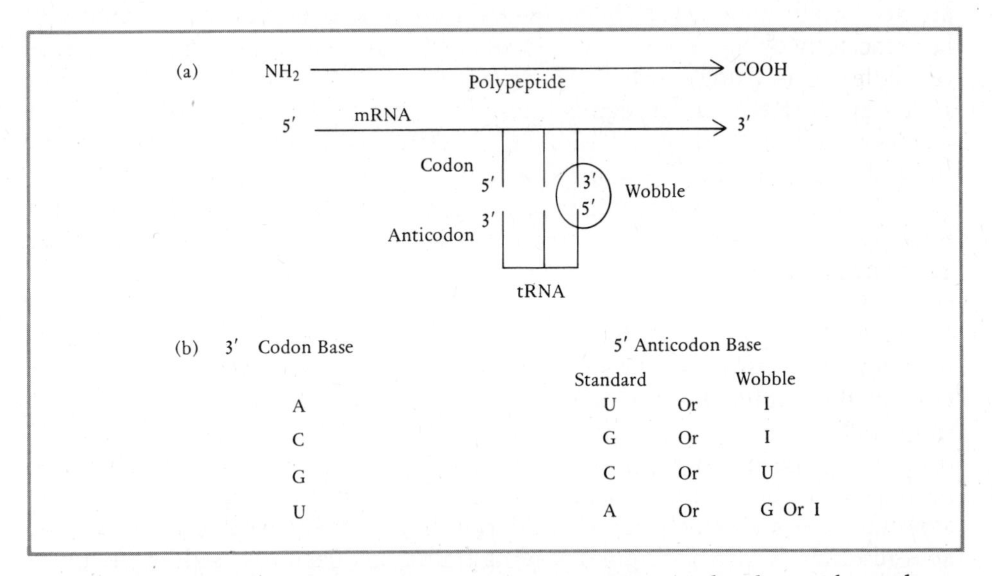

Figure 16–2. (a) Relationships between tRNA, mRNA, and polypeptide product. Arrows indicate direction of translation of mRNA and synthesis of the product. Note that the bases of the codon and anticodon are antiparallel in their orientation. (b) Possible standard and wobble base pairs at the third nucleotide position of the codon, between the 3′ base of the mRNA codon and the 5′ base of a tRNA. I is inosine, a rare base occasionally present in tRNA. Inosine is a derivative of adenine and has hydrogen bonding properties similar to guanine.

gether make sense in terms of amino acids and polypeptides only if they are read in the right direction. In visualizing the correspondence between codon and anticodon, we must include the fact that the two are antiparallel. Thus, the base at the 5′ end of the mRNA codon is the complement of the base at the 3′ end of the matching tRNA anticodon. These relationships are illustrated in Figure 16–2.

THE WOBBLE HYPOTHESIS

Examination of the genetic code will reveal that the same amino acid may be specified by more than one codon, a condition referred to as *degeneracy.* In view of this circumstance it is reasonable to inquire if particular tRNA's are produced for each of the alternative symbols for a given amino acid or if a single tRNA can recognize all such codons. In bacteria different tRNA's for the same amino acid have been identified, and since most organisms possess over 40 different kinds of tRNA, many of the alternative codons are presumably matched by complementary tRNA's. With respect to recognition of the same codon by different tRNA's, Crick has noted that the differences between the codons that specify the same amino acid lie primarily in the third base. This observation prompted his *wobble theory,* whereby it is postulated that hydrogen bonding between codon and anticodon is specific for the first two bases of the codon, but less precise for the third base (Figure 16–2). Thus, if the first two are accurately matched with the bases of the anticodon, a certain amount of latitude or "wobble" may be permitted at the third site. Experimental tests of this hypothesis, using synthetic trinucleotides of known sequence, have shown that a given tRNA can recognize more than one of the codons which specify the same amino acid. In addition, the alternative codons so recognized are those predicted on the basis of the wobble hypothesis.

Confirmation of the reading frame concept whereby translation begins at a fixed point in mRNA has been obtained by E. Terzaghi and coworkers. In phage T4 the normal amino acid sequence was determined for the T4 lysozymal enzyme which digests the host cell wall. This sequence was compared to that of the lysozyme produced by phage carrying two different proflavin-induced mutations, a minus mutation and a plus mutation, and it was found that the two enzymes differed by only five consecutive amino acid residues. Reasoning from the genetic code, it could be deduced that at one point an adenine nucleotide had been lost and that 15 nucleotides later a guanine nucleotide had been gained. The 15 nucleotides between the loss and the gain comprised five abnormal codons, which upon translation had yielded the five wrong amino acid residues. That the original reading frame had been restored by the insertion of guanine could be inferred from the fact that the amino acid sequence was normal throughout the remainder of the protein molecule.

Even more complete evidence for the validity of the genetic code has been obtained by W. Min Jou and associates in studies of the *E. coli* phage MS2. The genome of this virus is composed of single-stranded RNA consisting of around 3000 nucleotides and comprising three genes which specify a coat protein, a replicating enzyme, and a maturation protein. When phages of this type enter a host cell, the single-stranded genome immediately performs the role of mRNA, utilizing host ribosomes and tRNA in the process. In a correlated study of both product and messenger the sequence of the 129 amino acids of the coat protein was first determined. Then, the sequence of nucleotides comprising the RNA coat protein gene was analyzed by chemical means. Comparisons between the two indicated that each amino acid of the protein is matched by a specific codon within the gene, these codons conforming to those of the genetic code. Degeneracy of the code is present in that the same amino acid is specified by more than one codon, as predicted. An initiating codon (AUG) occurs at the beginning of the message which is translated via consecutive triplets without punctuation, and two stop signals (UAG, UAA) occur at the end (Figure 16–3). Extensive internal hydrogen bonding between complementary bases within the RNA produces a complex secondary structure which may, in itself, be an important regulator of translation.

MUTATION AND THE CODE

Although verification of the genetic code has been obtained primarily through *in vitro* biochemical studies, the presence of the triplet code can also be deduced from the effects of mutation on proteins of known amino acid sequence. For example, the majority of variant hemoglobins contain only one wrong amino acid in the entire molecule (see Table 10–2). By comparing the postulated triplet that codes for the correct residue with that specifying the

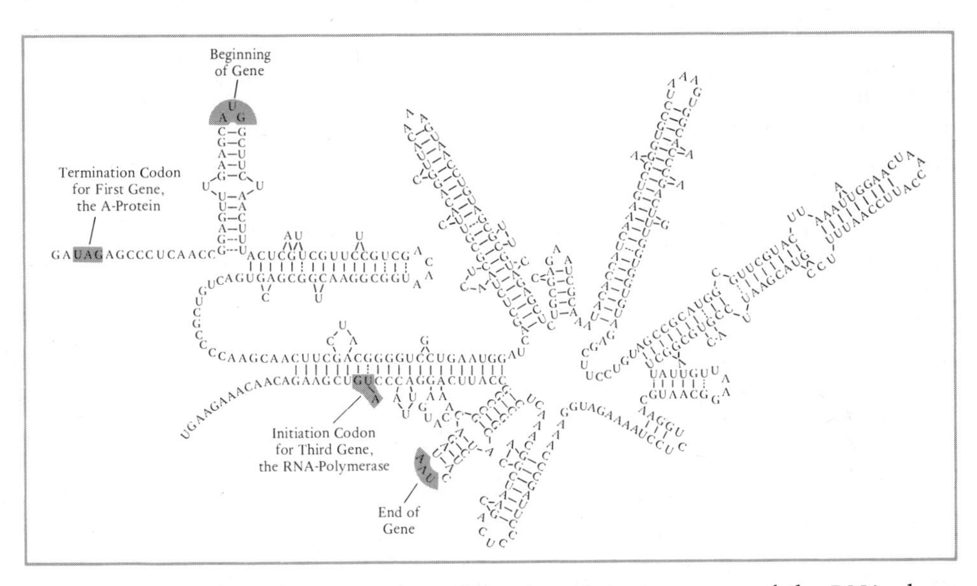

Figure 16–3. Three-dimensional configuration of the coat gene of the RNA phage MS2, showing the positions of the start and stop signals for translation. [Courtesy of Dr. Walter Fiers. From W. Min Jou et al., *Nature, 237,* 82–88, 1972.]

substituted residue, it is evident that the latter can be accounted for by a change in a single base. As an illustration, sickle cell hemoglobin contains valine at amino acid position 6. According to the genetic code, the codons for valine are GUU, GUC, GUA, or GUG. The normal β chain has, at position 6, glutamic acid, whose codons are GAA or GAG. By comparing the codons for these two amino acids, it can be seen that the substitution of U for A in the middle nucleotide of a glutamic acid codon will result in a codon for valine. Such a substitution in mRNA reflects a change from T to A in the corresponding genetic codon of the DNA sense strand used for transcription. If the genetic code were other than that postulated, such direct inferences as to the source of an amino acid substitution would not be possible.

Similar inferences have been drawn by C. Yanofsky and co-workers with respect to mutations affecting the A subunit of the enzyme tryptophan synthetase of *E. coli.* It will be recalled that amino acid substitutions within this polypeptide, when correlated with the position of mutated sites within the cistron, provided a demonstration of the colinearity of the gene with its product. Substitutions at the same position within the A polypeptide and which arise from mutations mapping at the same site within the *A* cistron can also be examined from the standpoint of the genetic code. At amino acid position 210 ten different substitutions have been identified, and all can be attributed to single base changes. For example, position 210 is normally occupied by glycine, whose mRNA codon is GGA. One mutation causes the substitution of arginine (AGA) for glycine, a different mutation substitutes valine (GUA), and a third mutation substitutes glutamic acid (GAA). It can be seen that in each case a single nucleotide substitution within the messenger will suffice to alter the GGA codon of glycine to that of arginine, valine, or glutamic acid.

In addition to analyzing the origin of amino acid substitutions in terms of changes within the mRNA codon, Yanofsky also performed crosses, by transduction, between cells carrying the above mutations. When cells carrying the arginine substitution at position 210 were crossed with cells containing either the valine or the glutamic acid substitution, some wild type progeny were recovered. Analysis of the A enzyme subunit present in these progeny indicated the restoration of glycine at amino acid position 210 (Figure 16–4). Assuming a triplet code, the origin of these wild type cells could be attributed only to recombination occurring between adjacent nucleotides within the DNA codon specifying amino acid 210. These findings, as well as the inferences drawn from the analysis of hemoglobin variants, can be used to clarify the nature of Benzer's muton and recon (see Chapter 14). The muton was proposed as the smallest unit whose mutation causes a change in phenotype, and the recon as the smallest unit involved in recombination. The results of Yanofsky's studies indicate that the muton can be defined as a single nucleotide and the recon as the distance between two adjacent nucleotides.

The probable effects of base substitutions on the subsequent activity of proteins can be predicted, at least in part, by examining the genetic code from the standpoint of the kinds of amino acids specified by particular codons. In this regard, the "meaning" of the second base of the codon must be emphasized. It will be recalled that the biological activity of a protein, such as an enzyme, is conferred by the conformation of the molecule. This conformation is such that hydrophilic groups are usually on the outside next to the aqueous environment and hydrophobic groups are inside, frequently lining the active site for catalysis. A significant change or reversal in position of hydrophilic versus hydrophobic residues results in changes in conformation as well as solubility, usually accompanied by loss of activity. If we examine the genetic code from this point of view, taking into consideration the importance of the

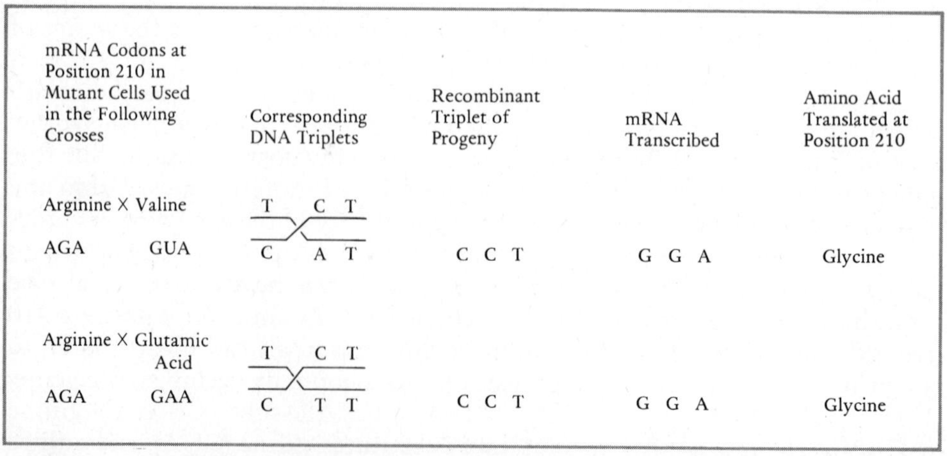

mRNA Codons at Position 210 in Mutant Cells Used in the Following Crosses	Corresponding DNA Triplets	Recombinant Triplet of Progeny	mRNA Transcribed	Amino Acid Translated at Position 210
Arginine × Valine	T C T			
AGA GUA	C A T	C C T	G G A	Glycine
Arginine × Glutamic Acid	T C T			
AGA GAA	C T T	C C T	G G A	Glycine

Figure 16–4. Yanofsky's experiment demonstrating recombination between single nucleotides within a DNA triplet coding for the amino acid in position 210 of the A subunit of tryptophan synthetase of *E. coli.*

TABLE 16–1
Amino Acids Classified According to the Second Base of Their Respective Codons

Second base U (nonpolar)	Second base C (nonpolar or neutral)	Second base A (polar)	Second base G (polar, except trp)
phenylalanine	serine	tyrosine	cysteine
leucine	proline	histidine	tryptophan
isoleucine	threonine	lysine	arginine
valine	alanine	asparagine	serine
methionine		glutamic acid	glycine
		glutamine	
		aspartic acid	

second base of a codon, the following observations can be made (Table 16–1). When the second base of the codon is U, the amino acid incorporated is nonpolar in nature and, after folding of the polypeptide, would be expected to end up inside the molecule. A base substitution in the first or third position of a codon whose second base is U will still result in the incorporation of a hydrophobic residue, compatible with the inside of the molecule, and, in all likelihood, some biological activity will be retained.

When the second base is C, the amino acid specified is either nonpolar or neutral, that is, it can fit into either a hydrophobic or a hydrophilic environment. In this instance a base substitution in the first or third position will cause incorporation of an amino acid that can be accomodated on the inside or on the surface of the polypeptide without causing significant conformational changes.

If the second base is A, the amino acids incorporated are polar and will be expected to occur on the outside of the protein. The same is true for most codons whose second base is G. Substitutions in the first or second positions of the codon in these instances will still result in the incorporation of a hydrophilic residue that can be situated on the outside of the polypeptide molecule. An exception should be noted in the case of the stop signals, UAA, UAG, and UGA, which possess either A or G in the second position. A base substitution in the first or third position that converts a codon from one that specifies an amino acid to one that signals stop will result in termination of polypeptide synthesis when the signal is encountered during translation. As a result, the polypeptide produced will be fragmentary, its length dependent on the position of the altered codon in the mRNA.

When all of these aspects are taken into account, it should be evident that the genetic code provides considerable flexibility. It should not be viewed merely as a code with extra words for some amino acids, for instead, it contains patterns which tend to conserve structure in the face of random mutation.

UNIVERSALITY OF THE CODE

Of profound significance is the question of whether or not the genetic code is universal in living organisms, that is, does the same codon specify the same amino-acyl tRNA in all forms. Evidence on this point, while not extensive, suggests that this is indeed the case. Synthetic messengers result in the same polypeptide when translated in cell-free systems by tRNA and ribosomes isolated from bacteria or from mammals. In addition, mammalian mRNA can be translated to mammalian protein by bacterial tRNA and ribosomes. When various viruses are used to infect cells of organisms widely different from their natural hosts, the viral mRNA can successfully produce its own protein by utilizing the translation machinery of its unusual host. Although some differences in codon-anticodon relationships are apparent in different species, in general, it may be said that the coding mechanism for the synthesis of proteins is universal. The question of its origin is unsolved, but its presence in a range of organisms that extends from bacteria to humans implies great antiquity.

PROBLEMS

16–1. The following codons appear in mRNA: 5′CAU GUG AAC UCA 3′. On the basis of the wobble hypothesis, give the corresponding anticodons of matching tRNA's, indicating the 3′ and 5′ terminals of each. (Ignore inosine as a possible base.)

16–2. The indicated amino acid substitutions occurred within the sequences of four proteins. Indicate the probable base changes that have occurred in each case and postulate the effects of such changes on the resulting polypeptide.

(1) . . . glu ser tyr lys *asn* glu thr arg ala . . . asn → ilu
(2) . . . glu gln *cys* tyr lys glu thr ala . . . cys → gly
(3) . . . ser his lys *glu* asn tyr arg . . . glu → asp
(4) . . . pro ala phe ilu *leu* met thr ser . . . leu → his

16–3. What are the advantages of a degenerate code?

REFERENCES

CRICK, F. H. C., 1966. Codon-anticodon pairing: the wobble hypothesis. *J. Mol. Biol., 19,* 548.

CRICK, F. H. C., 1967. The genetic code. *Proc. Roy. Soc. Lond. (B), 167,* 331.

GUEST, J. R., AND C. YANOFSKY, 1965. Amino acid replacements associated with reversion and recombination within a coding unit. *J. Mol. Biol., 12,* 793.

KHORANA, H. G., H. BÜCHI, H. GHOSH, N. GUPTA, T. M. JACOB, H. KÖSSEL, R. MORGAN, S. A. NARANG, E. OHTSUKA, AND R. D. WELLS, 1967. Polynucleotide synthesis and the genetic code. *Cold Spring Harbor Sympos., 31,* 39.

MARSHALL, R. E., C. T. CASKEY, AND M. NIRENBERG, 1967. Fine structure of RNA codewords recognized by bacterial, amphibian and mammalian transfer RNA. *Science, 155,* 820.

NIRENBERG, M. W., AND P. LEDER, 1964. RNA codewords and protein synthesis. *Science, 145,* 1399.

WEINSTEIN, I. B., 1963. Comparative studies on the genetic code. *Cold Spring Harbor Sympos., 28,* 279.

YANOFSKY, C., J. ITO, AND V. HORN, 1967. Amino acid replacements and the genetic code. *Cold Spring Harbor Sympos., 31,* 151.

Chapter 17

MUTATION

It has long been recognized that mutations are the source of all hereditary variation and provide the basis for evolution. In addition to their evolutionary significance, mutations are the means whereby the phenotype can be directly associated with the genotype. Only when inherited alternative phenotypes for a given character are recognizable, can the character be associated with a gene.

If we consider a mutation to be any heritable change within the genome, two major categories can be distinguished: chromosomal mutations and point mutations. *Chromosomal mutations* include major changes in structure, such as inversions, translocations, deletions, and duplications. These gross kinds of rearrangements, losses, and gains are cytologically visible in the chromosomes of higher organisms. *Point mutations* are defined as alterations of single genes and, as such, are below the limits of cytological detection. Since the nature and consequences of chromosomal mutations have been reviewed in Chapter 9, the present discussion will be restricted primarily to point mutations.

SOURCE AND FREQUENCY OF POINT MUTATIONS

A point mutation can arise spontaneously within any cell and at any time during the life cycle. If it occurs in a germ cell that later participates in fertilization, the mutation will be inherited, but if it occurs in somatic tissue, its presence and effects will be restricted to the individual in which it arises. An exception to this rule applies to plants that can be propagated asexually by grafting or by cuttings. Many interesting and economically desirable varieties originating as somatic mutations, (for example, the navel orange) have been preserved by these means.

The causes of spontaneous mutation are to a large extent unknown. The effects of cosmic rays, radiation from naturally occurring minerals, or chemicals present in the environment are insufficient to account for the spontaneous mutation rates observed. Although a number of cell constituents and by-products are mutagenic, the genome is ordinarily protected from the ill effects of its own metabolites by enzymes capable of destroying such products. As a specific example, the enzyme catalase, present in practically all cells, destroys peroxides formed during aerobic metabolism. The mutagenic nature of the peroxides and the protective function of catalase can be demonstrated by poisoning catalase with azide or cyanide. When the enzyme's action is altered, the accumulation of highly reactive peroxides is accompanied by a significant rise in the mutation rate.

A more likely source of spontaneous mutation is the accidental mispairing of bases during DNA replication. Mutant DNA polymerases are also implicated as causative agents. In addition, the presence of known mutator genes in some organisms suggests that the mutation of at least some loci is under genetic control (see below).

Since mutation is a recurring event, observations conducted over a period of time permit an estimate of the rate at which a given gene mutates. Studies of this type have shown that spontaneous mutation is rare, and that different genes, as well as different alleles of the same gene, mutate at different frequencies. A few examples will suffice to illustrate typical mutation rates (Table 17–1). Mutation rates for bacteria are ordinarily expressed as the number of mutations per cell per division, that is, per cell generation. The rate for higher

TABLE 17–1
Spontaneous Mutation Rates in Representative Organisms

Organism	Trait	Mutation rate*
E. coli	histidine auxotrophy	2×10^{-6}
(various sources)	streptomycin sensitivity	1×10^{-8}
	phage T1 resistance	$2–3 \times 10^{-8}$
Drosophila males	brown eyes	3×10^{-5}
(from Glass and	eyeless	6×10^{-5}
Ritterhoff, 1956)	yellow body	12×10^{-5}
Corn	colorless kernel	2×10^{-6}
(from Stadler, 1942)	shrunken kernel	1.2×10^{-6}
	waxy kernel	0×10^{-6}
Human	achondroplasia	1×10^{-5}
(from Vogel, 1964)	aniridia	2.9×10^{-6}
	retinoblastoma	$6–7 \times 10^{-6}$
	muscular dystrophy (Duchenne type)	9.2×10^{-5}

* Mutation rate for bacteria expressed as the number of mutations per cell per cell generation; for eukaryotes as the number of mutations per gamete per generation.

forms is given as the number of mutations per gamete per generation. Generation time in microorganisms differs vastly from that of higher organisms. If each cell, whether bacterial or eukaryotic, is subject to a low probability of acquiring a spontaneous mutation per cell division cycle, it might be expected that multicellular eukaryotes would accumulate numerous mutations in the course of a lifetime. Yet it can be seen from Table 17–1 that the estimated frequency of spontaneous mutation for a given gene is about the same order of magnitude for both single-celled bacteria and multicellular plants and animals. This finding suggests that eukaryotic cells possess enzyme systems capable of recognizing and repairing distortions and breaks within the genome with great efficiency. Experimental evidence supporting this hypothesis has been obtained with both plants and animals. From an evolutionary standpoint, it can also be argued that mutation rates are subject to selection, too high a rate leading to deleterious effects and too low a rate yielding insufficient variation for evolution. Such selection could operate at the molecular level through the effects of mutator (error prone) and antimutator (error free) DNA polymerases.

Since different genes mutate at different frequencies, it is pertinent to inquire if different sites within the same gene also mutate at different frequencies. Evidence on this point has been obtained primarily in phage T4 where Benzer has found that spontaneous mutation occurs nonrandomly with great frequency at several "hot spots," and not at all at "cold spots" (see Figure 14–7). The reasons why hot spots are more susceptible to mutation is unknown.

PREADAPTIVE NATURE OF MUTATIONS

That spontaneous mutations randomly affect any function or structure with no relationship to the possible adaptive nature of the change has long been observed in higher organisms. With microorganisms, however, the nonadaptive nature of mutation is not immediately evident. Bacterial cultures exhibit a remarkable ability to adjust to any new conditions of growth, such as the presence of a drug or a substrate not formerly utilized, and these kinds of adaptations are inherited by descendent cells. For this reason, many microbiologists of the 1930s and 1940s believed that bacteria mutated in response to some specific environmental factor or need, that is, the bacteria exhibited *postadaptive mutation.* This hypothesis was a restatement of Lamarck's proposal for the inheritance of acquired characteristics. Other microbiologists considered bacterial adaptation to result from *preadaptive mutation,* that is, from spontaneous mutations that occur in some cells of a culture before exposure to some specific environmental factor. Upon the introduction of that factor, such as a drug or new substrate, only the mutant cells are able to multiply, thus populating the culture with their descendents.

A test of the preadaptive hypothesis required a demonstration that mutant bacteria were already present in a culture prior to exposure to an agent which appeared to cause adaptation. One of the first such demonstrations was the fluctuation test devised by S. E. Luria and M. Delbrück in 1943. The phenotypic character used was resistance to phage T1. It was reasoned that if mutation was postadaptive, then every cell of a culture had an equal opportunity to undergo an adaptive mutation provoked by the phage, and if samples from separate cultures containing equal numbers of cells were plated with T1 particles, the same number of resistant colonies should be found on each plate. If, however, mutation to resistance was independent of the presence of phage T1, samples of separate cultures plated with phage should exhibit varying numbers of resistant colonies. Such variation, or fluctuation, was predicted on the basis that each cell had a small chance of undergoing mutation to resistance in any cell generation. If such a mutation occurred early in the growth of a culture, many resistant descendents would be produced and their presence in a sample plated with phage particles would give rise to a significant number of colonies. If mutation occurred in a later generation, however, fewer resistant descendents would be present in a sample and as a result fewer resistant colonies would be observed. When this test was performed, wide variations in the number of resistant colonies were evident.

A second experiment designed to demonstrate the preadaptive nature of mutation was performed by W. D. Newcombe. Bacterial cells sensitive to phage were incubated on nutrient agar until each had formed a colony. At this point the surface of one group of plates was rubbed with a spreader so as to redistribute all the cells, while colonies of a second group of plates were left undisturbed. Both groups were then sprayed with phage and after incubation the number of resistant colonies present in each set was scored. It was reasoned that if one of the original bacterial cells contained a preadaptive mutation for phage resistance, it would form a colony of resistant cells; if left undisturbed, only that colony would be scored after exposure to phage. However, if the cells of the resistant colony were first redistributed by spreading them over the surface of the agar, then each resistant cell would give rise to a separate colony after exposure to phage. Thus the number of cells on "spread" plates should be many times greater than the number present on the undisturbed plates, and this was the result obtained by Newcombe.

Although these experiments indicated that mutation was preadaptive in nature, many microbiologists did not accept these findings until 1952, when J. Lederberg and E. M. Lederberg demonstrated the presence of mutations in bacterial cells never exposed to the "adapting agent." These workers devised a technique called *replica plating* in which a dilute suspension of cells is plated on enriched medium and incubated long enough for each cell to give rise to a colony. At this time a sterile, velvet-covered cylinder, the same size as the petri plate, is pressed to the surface of the agar on which the colonies are growing and then consecutively pressed to the surface of a series of fresh plates, each marked with a reference point corresponding to one on the original plate. The fibers of the velvet pick up cells from the colonies present on the master plate, and the cylinder is oriented such that the location of cells on each successive plate exactly duplicates that of the master (Figure 17–1).

Lederberg and Lederberg cultured *E. coli* cells on nutrient agar for their master plate. Replicas of this plate were then transferred to plates containing nutrient agar plus phage T1. After incubation the positions of resistant colonies were noted, corresponding positions on the master plate were cut out, and the cells were removed and suspended. These cells were then used to prepare a second master plate from which replicas were once again transferred to plates

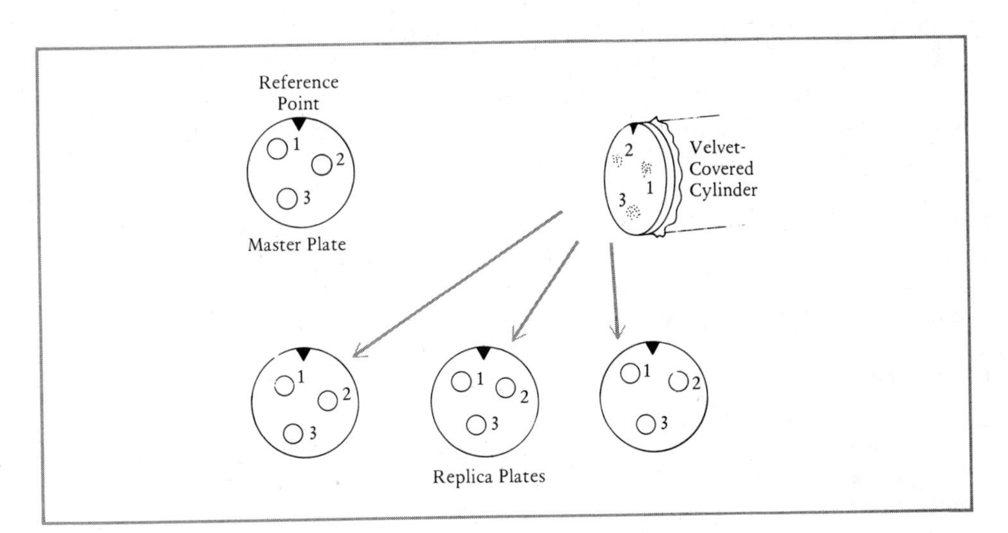

Figure 17–1. The replica plating technique.

containing phage, and the positions of resistant colonies noted. By repeating this procedure twice more, Lederberg and Lederberg were able to isolate a pure culture of cells resistant to phage T1. Since none of the cells present on any of the master plates had been exposed to T1 particles, the origin of mutations to resistance could not be a response to the presence of the virus. This indirect method of isolating mutations was used by Lederberg and Lederberg to show that resistance to streptomycin also originated by spontaneous preadaptive mutation. The results obtained in these experiments eliminated the last vestiges of Lamarckism and demonstrated that mutations in bacteria arise randomly, as in higher organisms.

MUTANT PHENOTYPES AND THEIR DETECTION

Most gene mutations are *forward mutations* in that they are changes from the wild type allele to a new allele or from one mutant allele to another. Far fewer in frequency are *reverse mutations,* whereby a mutant gene changes back to the wild type condition. Considering that spontaneous mutation is a rare, as well as a random event, the difference in overall frequency between the two is understandable. For example, the usual polypeptide is composed of over 300 amino acids, specified by over 900 bases, any one of which is subject to change. The probability that such a change will occur at a previously altered site and, in addition, will restore the wild type base sequence is extremely low. Nevertheless, true reversions to the wild type gene do occur. Reversions to the wild type phenotype are also brought about by *suppressor mutations* arising at a different site within the same gene or at another locus (see below).

The phenotypes produced by mutations are varied, the majority being recessive. All are recognized as departures in structure or physiology from the standard, normal, wild type condition. Since the wild type has been selected by evolution for adaptation to the present environment, most mutations beneficial to this adaptation will already have been incorporated into the genotype. Thus, random changes in the wild type genotype can be expected to alter, harmfully, the delicate balance between the organism and its environment. The fact that the vast majority of mutations are deleterious is therefore not surprising.

Common in higher organisms are *morphological mutations* that affect structure, appearance, or color. Although changes in these aspects stem from altered molecules, the association between phenotype and a specific biochemical lesion is usually difficult to make. Morphological mutations in fungi affect hyphal growth patterns and color. Such mutations in phage are recognized by plaque phenotype.

A second class, *biochemical mutations,* are those in which the altered phenotype can be traced to some specific metabolic defect, the nutritional mutations of fungi and bacteria being examples.

Lethal mutations constitute a third class and consist of those mutations that cause derangements so severe as to be incompatible with life. *Semilethal mutations* lower viability significantly, but do not cause the death of all individuals possessing them.

An additional class can be termed *conditional mutations,* in that the mutated gene may or may not be expressed depending on the environment. An illustration of this class is the temperature-sensitive mutation of the Himalayan rabbit (see Figure 4–11) in which the mutated gene is expressed at one temperature, but not at another. Conditional lethal mutations are another illustration. In these, growth or viability is possible only under defined, permissive conditions and lethality occurs with altered conditions of culture or environment.

Numerous experimental systems have been employed for the detection of mutations. Mutations in phage are identified by their ability to infect different strains of host cells and the types of plaques formed. Bacterial mutants resistant to antibiotics or other drugs can be recovered as colonies that grow on minimal medium containing the agent. Auxotrophic mutants are frequently selected through the use of penicillin which kills only growing and dividing cells. If a wild type bacterial population, suspected of containing auxotrophic mutations, is plated on minimal medium containing penicillin, only wild type cells will initiate rapid metabolism, thereby becoming susceptible to the lethal effects of the antibiotic. The remaining, viable auxotrophic cells can be washed and then grown on complete medium. Replica plating to medium supplemented with various amino acids or other substances then permits a determination of specific nutritional requirements or the selection of particular types of auxotrophs. The number of auxotrophic colonies as compared with the number of cells originally present in the population allows an estimation of the mutation rate. Prototrophic revertants can be obtained by plating auxotrophic strains on minimal medium and observing any presence of wild type colonies. Auxotrophic mutants of fungi are selected by similar methods (see Chapter 10).

In diploids the spontaneous mutation of a wild type locus to the recessive allele can be assayed by crosses of the homozygous dominant with the homozygous recessive. Any progeny exhibiting the recessive phenotype must originate through mutation occurring in a gamete of the dominant parent. This technique has been used to estimate the mutation rate at specific loci of corn,

Drosophila, mice, and other organisms. In mice special test strains containing seven different recessive mutations affecting a variety of external features have been developed to screen for spontaneous or induced mutations at these loci. Dominant lethal mutations in mice and other mammals can be detected by crossing treated males with untreated females. Dominant lethals usually cause death *in utero* and their presence will result in a significant decrease in fertility.

In *Drosophila* the occurrence of X-linked visible or lethal mutations, arising spontaneously or as a result of treatment with a mutagenic agent, can be estimated by several methods. Visible mutations can be identified by crossing treated males with attached-X females and examining the F_1 males arising via patroclinuous inheritance for morphological changes.

A determination of the frequency of X-linked lethal mutations can be carried out with the *ClB* or with the *Basc (Muller-5)* test stocks by methods devised by Muller. The *ClB* females contain on one X chromosome an inversion *(C)* for crossover suppression, a recessive lethal *(l)*, and the dominant Bar *(B)* duplication which serves as a chromosome marker. The other X chromosome is normal. As outlined in Figure 17–2, treated wild type males are crossed to *ClB* females. The resulting F_1 females are heterozygous for the *ClB* chromosome and for a treated X chromosome derived from the male parent. These females are then bred individually in separate cultures. Of the F_2 males, half die because of the lethal present in the *ClB* chromosome; if a new lethal was induced by treatment of the original male parent, all F_2 males will die. Thus the presence of a new lethal mutation can be determined by the simple observation of whether or not F_2 males are present. Since each separate culture represents a test of a single X chromosome, the number of cultures containing no males as a percent of all cultures provides an estimate of the X-linked lethal mutation rate. In addition, if F_2 males are present, they can be examined for visible mutations. The Muller-5 technique is similar. The stock contains inversions to

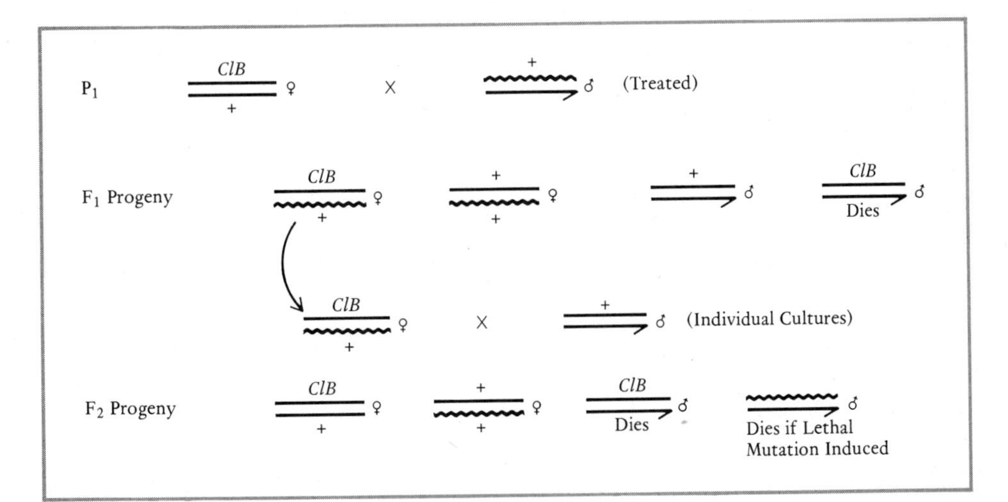

Figure 17–2. The *ClB* technique for the determination of X-linked lethal mutations in *Drosophila.*

prevent crossing over, Bar, and the recessive mutation apricot eye color *(ap)*. No lethals are present. Absence of wild type F_2 males indicates that an X-linked lethal mutation was induced (Figure 17–3).

To detect the presence in an autosome of a recessive visible or lethal mutation, it is necessary that the autosome be present in the homozygous condition. To this end, treated wild type males are crossed to females contain-

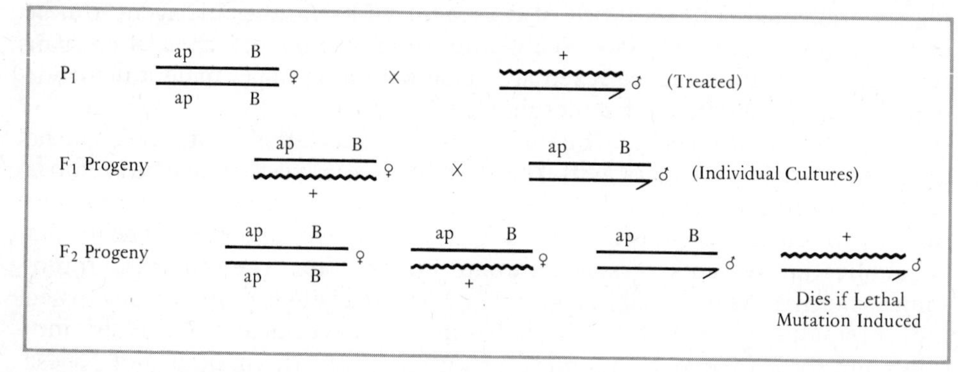

Figure 17–3. The Muller-5 (Basc) technique for the determination of X-linked lethal mutations in *Drosophila*.

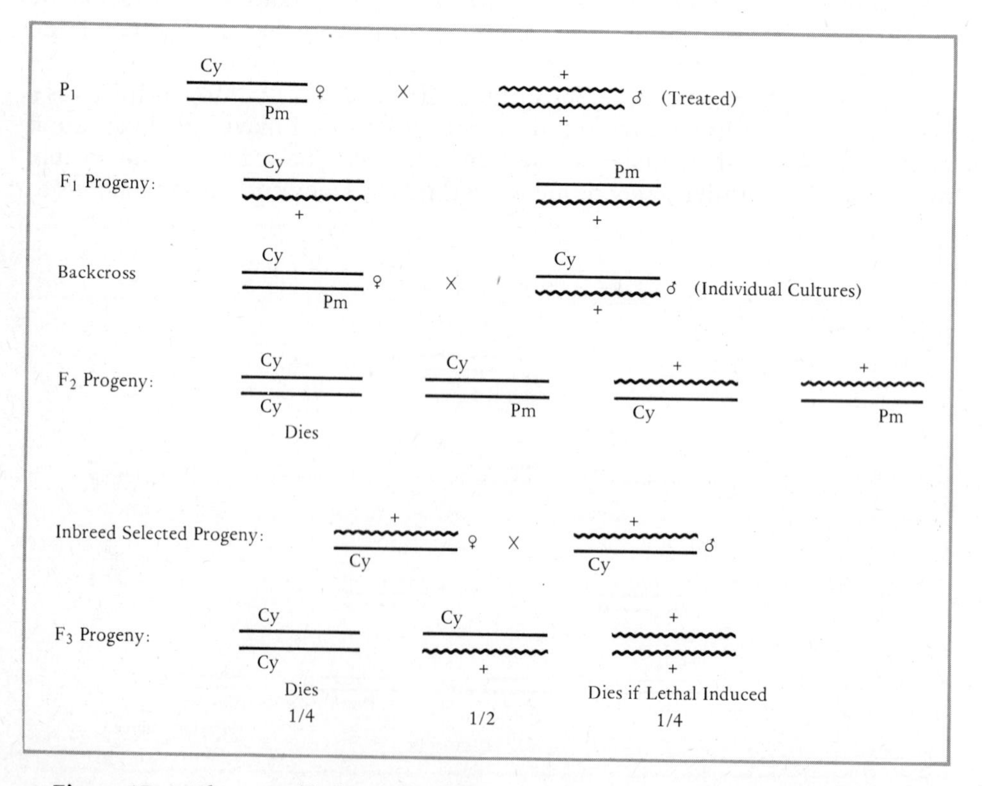

Figure 17–4. The use of balanced lethals to detect recessive autosomal lethal mutations in *Drosophila* (see also Figure 9–17).

ing balanced lethals on the autosome being tested, for example, the *Cy/Pm* stock outlined in Chapter 9, Figure 9–17. The F_1 males from this cross are then backcrossed separately to balanced lethal females. Appropriate F_2 progeny are then inbred to produce F_3 wild type individuals homozygous for the autosome under consideration (Figure 17–4). If a lethal mutation was induced in this autosome in the original P_1 treated male, wild type individuals will not appear in the F_3 progeny.

Spontaneous mutations in humans are recognized as inherited traits that appear in individuals whose family history contains no evidence of the prior existence of the trait. Human mutation rates are best estimated for dominant mutations expressed fully in the heterozygous condition. The reliability of such estimates also requires the consideration of other criteria. The trait should be rare, should not cause lethality before birth, and should be such that medical attention is uniformly required so that a record of the condition will be made. Traits listed in Table 17–1 conform to these criteria. Those interested in a compendium of human mutation rates should consult Vogel (1964, 1970).

MUTATION AT THE MOLECULAR LEVEL

All mutations stem from alterations in the base sequence of DNA. Such alterations consist of two general types: base substitution and frameshift.

The substitution of one base for another can occur anywhere within either chain of a DNA molecule. At replication the new base will demand a different partner for hydrogen bonding, thus leading to the incorporation of a new and different pair of bases into the daughter DNA molecule. Subsequent rounds of replication, followed by cell division, will insure the faithful reproduction of the altered double helix, as well as its distribution to daughter cells.

When a region containing a base substitution is called upon to serve as a template for transcription, the change in DNA base sequence will be reflected in the RNA synthesized from that region. If the mutant sequence is transcribed to a gene product which acts directly in the cell, such as rRNA or tRNA, the mutant product may not function normally during translation. For example, a mutant tRNA might transport an amino acid that did not correspond to its anticodon. If the base change has occurred in a structural gene that codes for a protein, the transcribed mRNA will contain an altered codon which may lead to an amino acid substitution in the protein product. If so, the mutation can be termed a *missense mutation,* because the meaning or sense of the message has been altered. On the other hand, a base substitution may change a codon from one that specifies an amino acid to one that signals stop. In this case the change is called a *nonsense mutation,* because premature termination of translation generally results in a meaningless nonfunctional polypeptide fragment.

As originally suggested by Crick, base changes within DNA may occur spontaneously as a result of *tautomerism.* Each of the four bases can exist in the usual, common form and also in a rare and unstable form, called a *tautomer,* which arises from a redistribution of hydrogen atoms in the molecule. Two examples are shown in Figure 17–5.

Tautomers cause base substitutions because the change to the rare state alters the hydrogen bonding properties of a base such that a rare purine

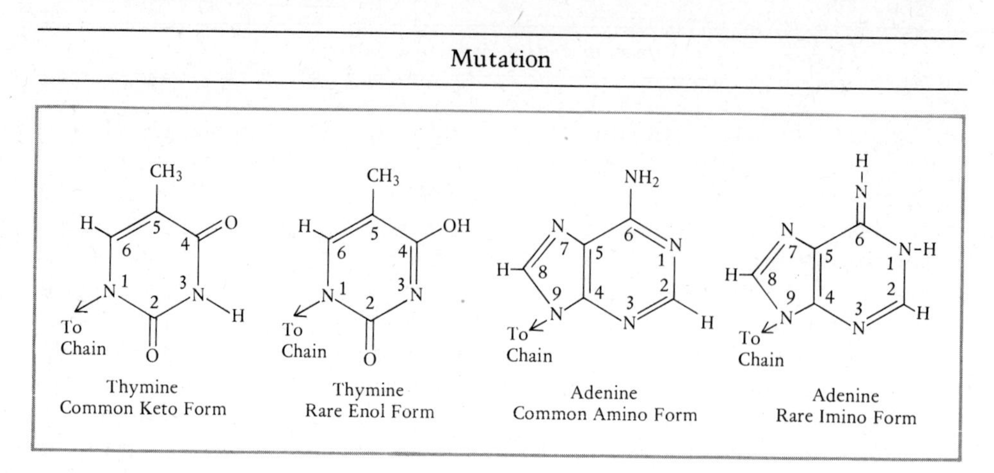

Figure 17–5. Common and rare tautomeric forms of thymine and adenine.

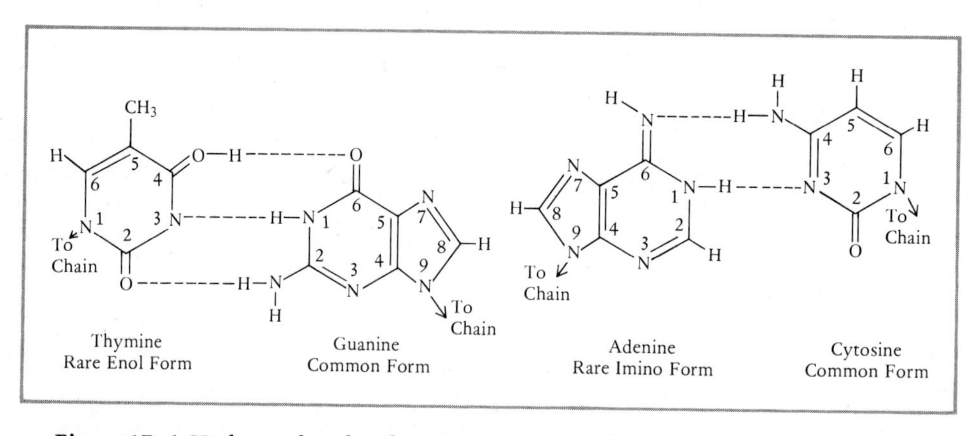

Figure 17–6. Hydrogen bonding between guanine and a rare tautomer of thymine and between cytosine and a rare tautomer of adenine.

tautomer will hydrogen bond with the wrong pyrimidine, and a rare pyrimidine tautomer will pair with the wrong purine (Figure 17–6). Thus, if any base of a template chain is in its rare tautomeric form at the time of replication, it will form hydrogen bonds with the wrong partner. Equally possible is the mistaken incorporation into a new chain of a nucleotide whose base is in the rearranged tautomeric form.

In either case, a base substitution will occur in the newly synthesized chain, and at the next cycle of duplication this substitution will be replicated to yield a different base pair (Figure 17–7). This kind of change is called a *transition.* With transitions pairing between a purine and a pyrimidine is preserved, but a substitution in the bases involved occurs, that is, cytosine for thymine, adenine for guanine, and vice versa. Transitions thus lead to changes in base pairs which can be summarized as A-T ⇄ G-C. Although such changes may be expected to occur with very low frequency, they may account for a portion of the spontaneous mutations observed. They are of additional interest because they can be induced by mutagenic chemicals.

Mutations originating from the gain or loss of nucleotides are called *frameshift mutations,* a name derived from their effects on the mRNA reading

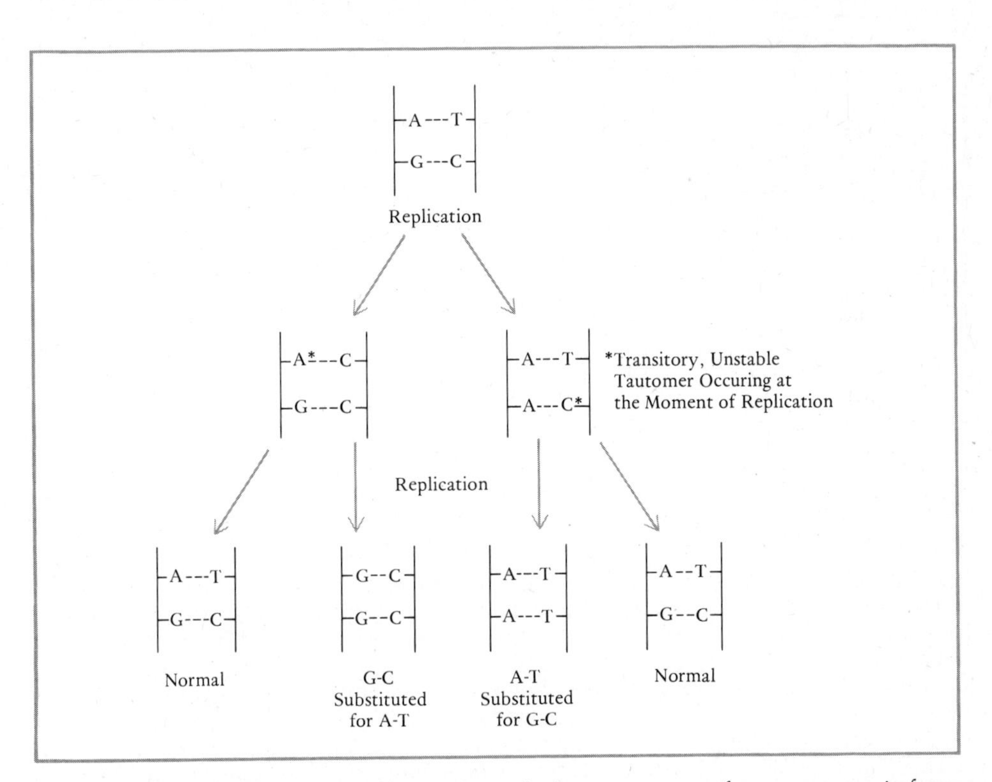

Figure 17–7. Base pair substitution through the occurrence of rare tautomeric forms in a template chain at replication.

frame. It will be recalled that mRNA is translated sequentially from a fixed point, successive codons being read from that point. As reviewed in Chapter 16, the loss or gain of a base within a codon of mRNA will cause a shift in the bases included in every succeeding triplet thereafter to the end of the message. As a result the polypeptide produced will contain a sequence of wrong amino acid residues starting with the codon where loss or addition occurred. Gain or loss of two, four, or five nucleotides also results in frameshifts. However, since codons consist of three bases, it can be predicted that the deletion or addition of a number of nucleotides divisible by three will result in a corresponding loss or gain of amino acids, followed by restoration of the normal reading frame. The activity of the resulting polypeptide will depend on whether or not the lost or added amino acid residues participate in function or lead to a significantly altered conformation.

Two interesting examples illustrating a base substitution and a frameshift can be cited from human hemoglobin. Two hemoglobin variants have been described in which the α chain contains additional amino acid residues beyond those normally present (Figure 17–8). One type, called Hb Constant Spring (HbCS) has 31 extra residues at the carboxyl terminal, for a total of 172, while Hb Wayne-1 (HbW1) contains 5 extra residues, a total of 146. The amino acid sequence of Hb Constant Spring is normal for the first 141 residues, indicating that mRNA codons are unchanged up to this point. Since the chain ordinarily

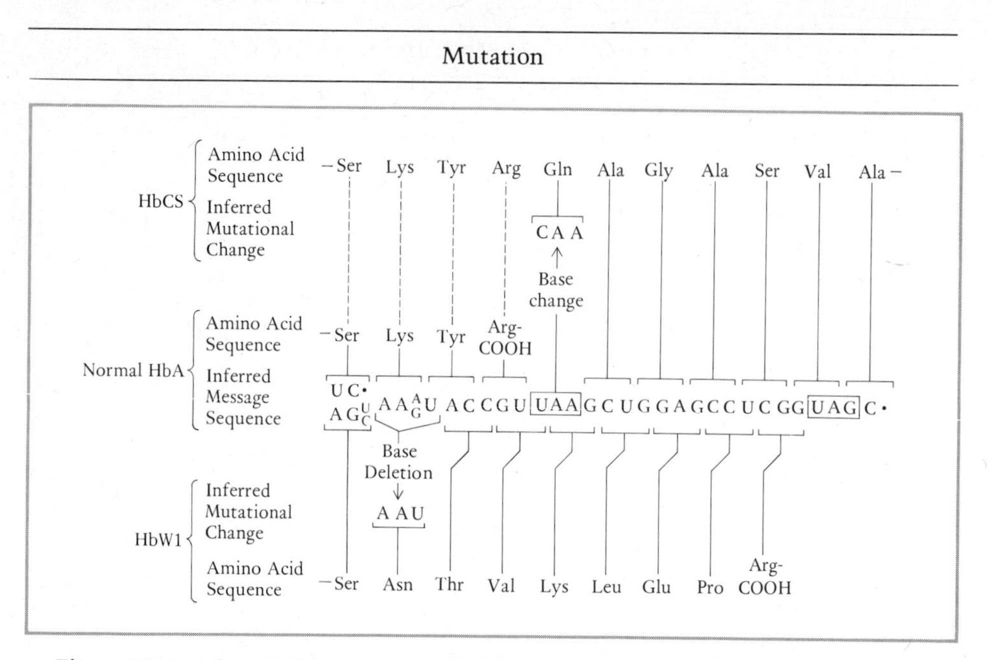

Figure 17–8. A frameshift mutation affecting human hemoglobin. The residues of the normal hemoglobin α chain (HbA) and two variant hemoglobins, Constant Spring (HbCS) and Wayne-1 (HbW1), are given. Both of the variant hemoglobins are longer than the normal HbA. Variant HbCS has 31 residues in addition to the normal sequence of 141 (only 7 of these extra residues are shown). Variant HbW1 differs from normal by 3 changed amino acids and 5 added residues. From these amino acid sequences one can infer the normal message sequence and the nature of the mutational events which lead to HbCS and HbWl. Probably termination signals (indicated in rectangles) are UAA for normal HbA and UAG for HbWl. [Reproduced, with permission, from J. R. Roth, *Ann. Rev. Genetics,* 8, 319–346, 1974. Copyright © 1974 by Annual Reviews Inc. All rights reserved.]

contains only 141 residues, corresponding to 141 sense codons, it can be inferred that the 142nd codon must normally be one that signals chain termination. It is postulatd that a base substitution has occurred in this codon to transform a stop signal (UAA) to a sense codon (CAA) which specifies glutamine. The presence of the extra 31 residues also indicates that the mRNA contains bases beyond those normally translated. Although the function of these bases is unknown, their sequence can be inferred from the additional amino acids present in HbCS. HbW1 appears to have arisen as a frameshift caused by the deletion of a nucleotide from a codon located prior to the normal termination of the α chain. This shift has resulted in a change of three amino acids and in the translation of the base sequence of mRNA into five additional residues.

SUPPRESSOR MUTATIONS

Suppressor mutations are those that partially or fully reverse the harmful effects of a previous mutational event. They can be classified into two major groups on the basis of their location within the genome: intragenic and intergenic.

Intragenic suppression involves the reversal of the effects of one mutation by a second mutation occurring within the same gene. For example, intragenic suppression frequently involves the restoration of the normal codon reading frame, previously altered by the deletion or insertion of a base. If the second and compensating gain or loss is located close to the first mutation site, it is likely that at least partial activity will be restored to the product of translation. Results of this nature were obtained by Crick in his study of plus and minus mutations in phage T4.

Intragenic suppression can also result from a number of other mutational events, for example, a second base substitution within the very same codon previously altered by mutation. This event could be a true reverse mutation at the original site so that the codon is restored or another base of the same codon could be changed resulting at translation in the incorporation of the same or a compatible amino acid. Lack of a normal initiation codon caused by mutation can also be compensated for by a second mutation which transforms another codon to an initiator. In bacteria, where instructions for several different polypeptides may be transcribed as a single, long message, absence of initiation codons or the presence of a stop signal at an early point will result in nontranslation or a significant decrease in the translation of more distal regions of the messenger RNA. A second mutation which creates by base substitution a new initiating codon or replaces a stop signal with a codon for an amino acid may result in proper translation of most of the message. A ribosomal or transfer RNA whose conformation was altered due to a previous mutational event can also be restored to partial or complete activity by a second intragenic mutation. All of these and other examples of intragenic suppression have been observed in various organisms, especially phage and bacteria.

Intergenic suppression involves suppression of the effects of one mutation by a second, totally different mutation which is located elsewhere in the genome. The second mutant, called a suppressor, may act directly or indirectly. *Direct suppression* is usually the result of a modification of some component of the protein synthesizing system which results in the misreading of mRNA. For example, the altered codon of a mutated mRNA may be read by a tRNA which is also mutated so that its specificity is changed. The change in tRNA specificity can result in the incorporation into a protein of a more compatible or even the correct amino acid. The gene for such a tRNA would be acting as a suppressor gene in this instance, for during translation of the mutant mRNA the altered tRNA might permit the synthesis of some normal or partially normal protein molecules which otherwise would not be produced at all.

A number of mutated tRNA's have been implicated in the mechanism of suppression. Some are undermethylated and some are thought to participate in inaccurate codon-anticodon base pairing with consequent loss of specificity. Even frameshift mutations have been found to be corrected by such outside suppression. In one instance in the bacterium *Salmonella* the modifying agent has been found to be a mutant tRNA with a four-letter anticodon which corrects for a frameshift caused by the gain of one nucleotide. Although a mutant tRNA might be expected to have deleterious effects on the structure of other proteins, such effects are largely counteracted through the presence of alternate tRNA's for most amino acids. Alterations in the ribosome components have also been postulated as the cause of direct suppression.

Indirect suppression can be attributed to suppressor genes whose products participate in a broad spectrum of cell function. For example, a change in the flow and concentration of metabolites may be brought about by the presence of an enzyme altered by a mutation. A higher concentration of one of these metabolites may be sufficient to stimulate to activity a protein originally responsible for the mutant phenotype, consequently causing partial to complete suppression of the abnormal phenotype. In this case a change in the cellular milieu allows the mutant protein to assume some activity.

Indirect suppression can also result from a variety of environmental changes: slight variations in pH, in the concentration of some ions which are protein activators, for example, zinc, or in the availability of modulating molecules, such as metabolites which function in feedback (allosteric) interactions. Tracing the cause of indirect suppression to its source can be a difficult and intricate undertaking. The mechanism is usually complex and is often exerted on multisubunit enzymes which ordinarily undergo significant conformation changes through interactions with ions and metabolites.

INDUCED MUTATION

In 1927 Muller reported the induction of mutations in *Drosophila* by X ray, and in 1928 L. J. Stadler obtained similar results with barley. These discoveries were of great importance, for up to that time the only mutations available for study were those arising spontaneously. Since then the use of ultraviolet (UV) radiation and a variety of chemicals to induce mutations has permitted the identification of numerous genes in all types of organisms.

Mutations Induced by Chemicals

Of the large number of organic and inorganic chemicals known to produce mutations, four general groups have been most thoroughly investigated in terms of their mode of action: (1) base analogues, (2) chemicals that act directly on DNA bases, (3) alkylating agents that remove purines from DNA, and (4) acridine dyes that cause the addition and subtraction of nucleotides from DNA.

Base analogues. Base analogues are mutagenic chemicals whose structure so mimics that of a naturally occurring base that they are incorporated by a cell into DNA. One such substance is 5-bromouracil (5-BU), which is similar to thymine (5-methyluracil) except that the methyl group of thymine is replaced by bromine. A similar compound is 5-bromodeoxyuridine (5-BUdR), which is similar to thymidine except for the bromine substitution. The presence of bromine in the thymine or thymidine molecule increases the frequency of tautomerism to the rare enol form, such that pairing of 5-BU with guanine, instead of adenine, is greatly increased (Figure 17–9). During replication 5-BU may be incorporated normally opposite adenine, but at a later duplication cycle may pair with guanine. In this case a transition from an A-T to a G-C base pair results. 5-BU can also be incorporated opposite guanine initially and pair with adenine at a later replication, in this case causing a G-C to an A-T transition. Thus, 5-BU can bring about transitions in either direction.

Another commonly used base analogue is 2-aminopurine (2-AP). This compound has hydrogen bonding properties similar to those of adenine, and therefore it normally forms base pairs with thymine. However, 2-AP shifts to a

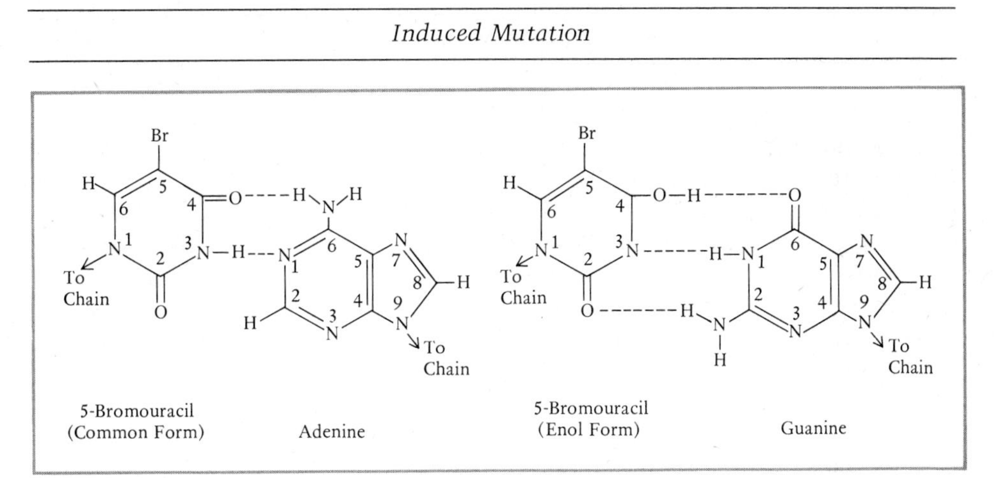

Figure 17–9. Hydrogen bonding properties of common and rare tautomers of 5-bromouracil (5-BU).

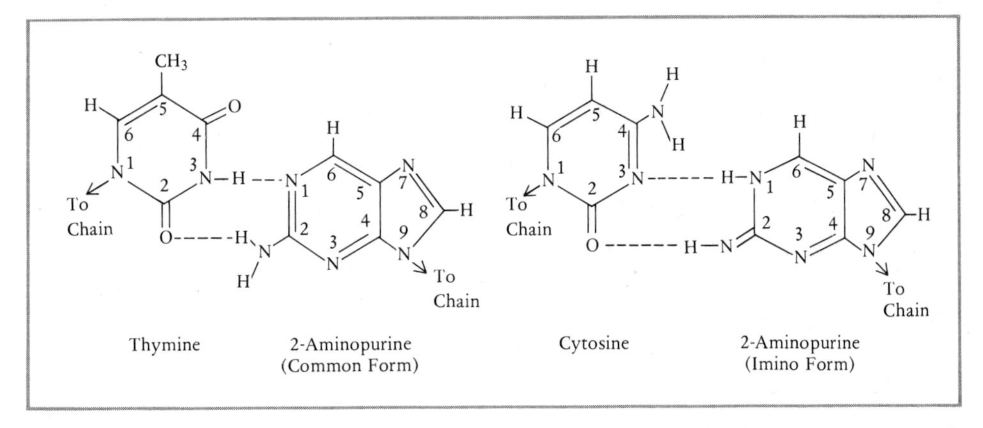

Figure 17–10. Hydrogen bonding properties of common and rare tautomers of 2-aminopurine (2-AP).

rare tautomer (imino form) more often than does adenine, and in this form it pairs with cytosine via two hydrogen bonds (Figure 17-10). Its presence in DNA can therefore be expected to cause transitions in either direction (A-T ⇄ G-C), as in the case of 5-BU.

Direct acting chemicals. Compounds of this category are not incorporated into DNA, but act directly on base structure. One such compound is nitrous acid (HNO_2), which removes amino groups from adenine, cytosine, and guanine by oxidative deamination. By this action adenine is changed to hypoxanthine, a compound that forms base pairs with cytosine, not thymine, and cytosine is changed to uracil, which pairs with adenine, not guanine (Figure 17–11). Guanine is deaminated to xanthine, but xanthine continues to form base pairs with cytosine, although by only two hydrogen bonds. Thymine and uracil are unchanged by treatment with nitrous acid since neither of these bases contains an amino group. Thus the mutagenic action of nitrous acid is exerted only through changes in adenine and cytosine.

The change of adenine to hypoxanthine causes the formation of a hypoxanthine-cytosine base pair at the first replication and a C-G base pair at the next replication. The result is a transition from A-T to G-C. The deamination of cytosine to uracil will result in a U-A base pair at replication and ultimately an A-T base pair, the overall consequence being a transition from G-C to A-T. Thus nitrous acid can induce transitions in both directions.

A somewhat similar compound, hydroxylamine (H_2NOH), appears to react specifically only with cytosine such that cytosine forms base pairs with

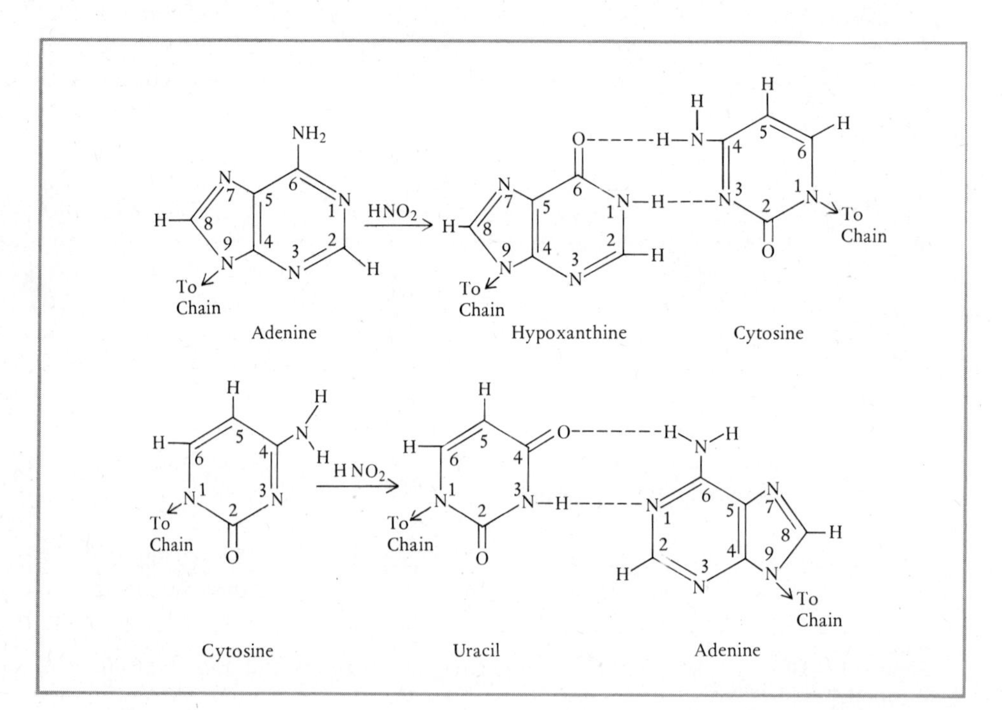

Figure 17–11. The action of nitrous acid (HNO_2) on adenine and cytosine.

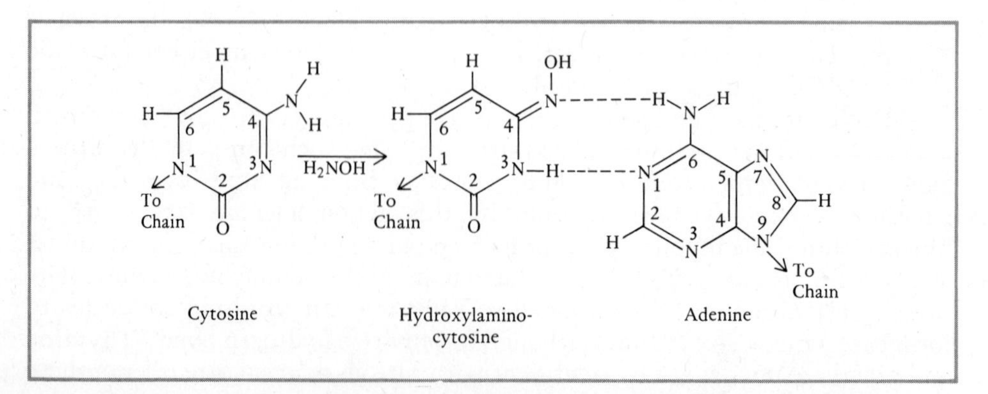

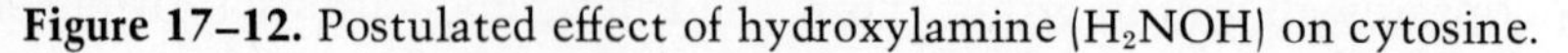

Figure 17–12. Postulated effect of hydroxylamine (H_2NOH) on cytosine.

adenine, not guanine (Figure 17–12). The result is a transition in one direction only, from G-C to A-T. Another potent mutagen, nitrosoguanidine, is also very effective in inducing transitions.

Alkylating agents. Alkylating agents include the nitrogen and sulfur mustards as well as the less toxic compounds ethyl ethanesulfonate (EES) and ethyl methanesulfonate (EMS). The frequently used EES and EMS act specifically on guanine, adding an ethyl or methyl group at position 7 and causing weakening and dissolution of the linkage to deoxyribose. As a result, guanine is lost from the DNA polymer. It is thought that the gap so formed may be filled at the next replication by any one of the four bases. If replaced by guanine, the normal G-C base pair will be restored; if replaced by adenine, an A-T transition will result. However, if replaced by cytosine, the original G-C base pair will be changed to a C-G pair, and if thymine is inserted, the G-C pair will become a T-A base pair. These two changes are *transversions* (Figure 17–13). In contrast to transitions, transversions are base changes whereby a purine is substituted for a pyrimidine or vice versa. In the present example a purine (G) is replaced by a pyrimidine (C or T). It is thought that transversion-type substitutions may also occur spontaneously, perhaps through accidental mispairing during replication. Examination of hemoglobin variants, where the normal DNA codon, inferred from the normal amino acid residue, can be compared with a substituted DNA codon, inferred from a wrong residue, indicates that transversions, as well as transitions, have arisen by spontaneous mutation.

If guanine is not lost from a DNA chain through the action of an alkylating agent, the presence of methyl or ethyl groups causes guanine to behave as a base analogue of adenine and pair with thymine, thus introducing a G-C to A-T transition.

Acridines. The acridine dyes, such as proflavin, are highly mutagenic for phage and probably for other organisms as well. They produce mutations at random sites which do not appear to coincide with those attacked by base analogues, nitrous acids, or alkylating agents. The action of the acridine dyes is related to their property of binding to DNA and becoming inserted between adjacent bases. If a parental template chain is distorted by an acridine molecule

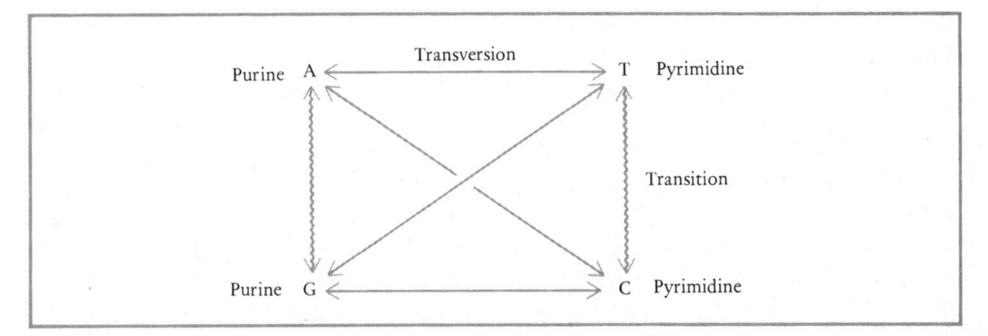

Figure 17–13. Transitions and transversions which can occur between the bases in DNA.

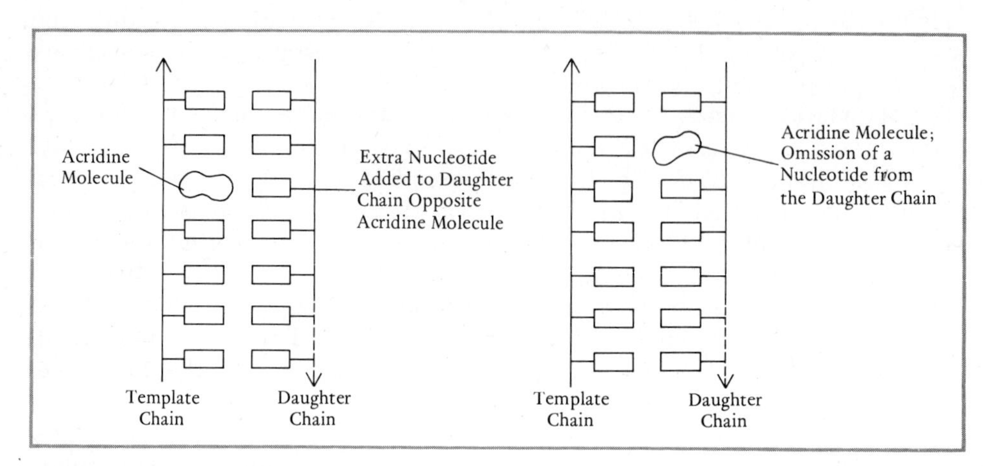

Figure 17–14. Addition and subtraction of nucleotides caused by the presence of acridine dyes during DNA replication.

at the time of replication, an additional nucleotide is added opposite the acridine molecule to the new polymer being synthesized (Figure 17–14). If the acridine molecule becomes inserted between nucleotides that are aligned for synthesis, the acridine will occupy a position opposite a base in the template strand, thereby preventing the incorporation of the complementary nucleotide in the daughter chain.

The insertions and deletions of nucleotides caused by acridines result in frameshift mutations. As expected, such mutations do not revert by treatment with base analogues, nitrous acid, hydroxylamine, or alkylating agents. It will be recalled, however, that reversions to wild type may be accomplished by a second, nearby suppressor mutation which restores the codon reading frame.

The action of all of these mutagenic chemicals has been most thoroughly investigated with phage. One approach has been to compare the map positions of mutations induced by these agents to determine if the agents' mutational "spectra" are the same. The results of these studies indicate that the positions of the mutations induced in each case are generally quite different, despite considerable overlap. In addition, the distribution on a genetic map of mutations arising spontaneously is significantly different from that of induced mutations, and this is particularly so for hot spots.

Other studies have been made in an attempt to gain evidence that the proposed chemical action of mutagenic substances actually occurs within DNA *in vivo.* In such experiments mutations are induced by one agent, followed by attempts to induce reversions of these mutations by the same or different agent. For example, nitrous acid should induce transitions in both directions, and mutations obtained by this agent should revert to wild type when this agent is applied once more. Hydroxylamine causes transitions in one direction only, and a second treatment with hydroxylamine should produce no reversions. However, mutations induced by hydroxylamine should revert with base analogues or nitrous acid, both of which cause transitions in either direction. In the case of alkylating agents that proportion of mutations predicted to be transversions should not revert when transition-causing agents are used. In general, experimental evidence appears to support the postulated actions of these mutagens on DNA. That some discrepancies exist is only to be expected, considering the complexity of cell and genome structure.

Ultraviolet Radiation

Ultraviolet light has been used for many years as a sterilizing agent because it kills bacteria. It also induces mutations through the production of photochemical changes in DNA. It is therefore not surprising that the most mutagenic wavelength, 2600 Å, is that which is most strongly absorbed by DNA. Because ultraviolet radiation has poor penetrating ability, it is useful as a mutagen only for dispersed cells or for thin layers of cells in culture.

Ultraviolet light damages DNA by causing the formation of dimers whereby adjacent pyrimidine bases become linked to one another by carbon to carbon bonds. Thymine-thymine dimers are most common, but cytosine-cytosine and cytosine-thymine unions are formed as well. Dimerization produces distortion of the DNA polymer, rupture of hydrogen bonds, and the blockage of DNA replication through the formation of cross linkages and is responsible for much of the killing effect of UV radiation.

Repair of this damage, if not too extensive, may occur in some cells, such as bacteria, by three methods. The first of these is called *photoreactivation* and requires exposure to an intense source of visible light in the blue range, a wavelength of 4000 Å being most effective. By this treatment, pyrimidine dimers can be split in place within the DNA polymer by a light-dependent photoreactivating enzyme, which has been found in bacteria and yeast cells.

The second repair mechanism, called *dark repair,* is characteristic of the cells of many organisms in addition to bacteria and is not dependent on visible light as an energy source. The process requires several steps (Figure 17–15). First, distorted regions within the genome are recognized by an endonuclease which produces a single strand break in the DNA backbone adjacent to a dimer. Then, an exonuclease attacks the free end, removing a single-stranded fragment containing the aberrant nucleotides. This action is followed by a local synthesis of DNA to fill in the gap. DNA polymerase, specifically DNA polymerase I in *E. coli,* catalyzes the synthesis using the opposing strand as a template. This type of replication results in no net increase in DNA and is therefore called unscheduled or nonconservative DNA replication. Final closure of the break is carried out by a polynucleotide ligase. Three *ultraviolet-resistance (uvr)* genes with widely separate loci are involved in dark repair in *E. coli.* These genes are *uvrA, uvrB,* and *uvrC.* The mutation of any one of them produces the same phenotype characterized by lack of endonuclease and the consequent inability to excise dimers. It also appears that after endonuclease action, both the excision and the replacement of nucleotides can be accomplished by DNA polymerase I.

In *E. coli* a third mechanism for repairing UV induced damage to DNA is called *postreplication recombination repair* and is associated with the activity of the *rec* loci, *recA, recB, recC,* and others. It will be recalled from Chapter 14 that mutations of these loci, and in particular *recA* mutations, result in loss of the ability to integrate donor cell DNA by recombination and also in extreme

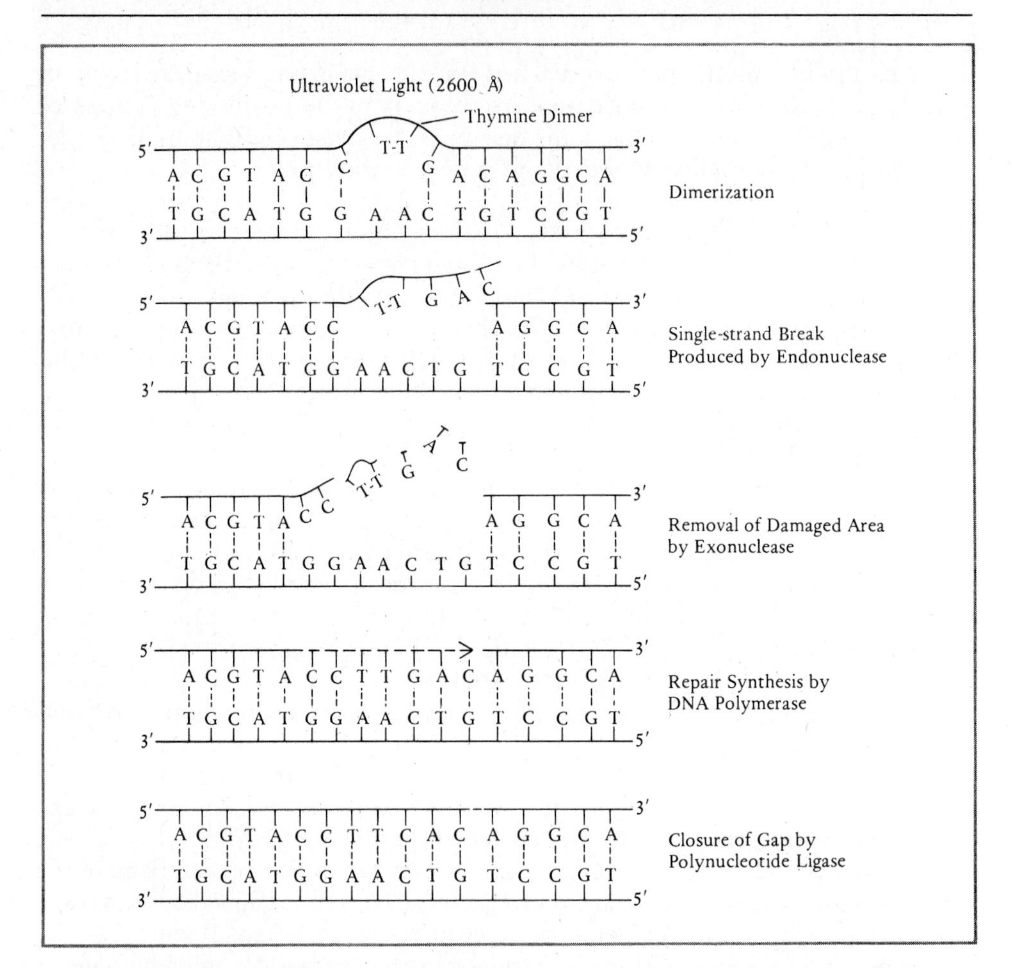

Figure 17–15. Diagrammatic representation of dark repair of a thymine dimer induced in DNA by ultraviolet radiation.

sensitivity to the effects of UV radiation. The repair mediated by these loci appears to occur after the replication of DNA which contains UV induced dimers. It has been shown that when such DNA undergoes replication, a gap is left in the newly synthesized strand opposite the site of a dimer present in the template strand. As a result, both strands of one daughter double helix are defective in that one contains dimers and the other single-strand gaps. The *rec* loci function by repairing the single-strand gaps, apparently through the excision and transfer of segments from one sister molecule to the other. This action restores the integrity of a chain opposite a dimer, and the regions from which segments were removed for patching purposes are then filled in by repair synthesis (Figure 17–16).

W. D. Rupp and associates have demonstrated that this process can occur apart from either photoreactivation or dark repair in that it is accomplished in the absence of visible light and in cells containing a *uvr* mutation. The *rec* loci may also be involved in the excision of dimers since it has been found that such excision occurs normally in the presence of a mutation called *polA*, which renders DNA polymerase I defective.

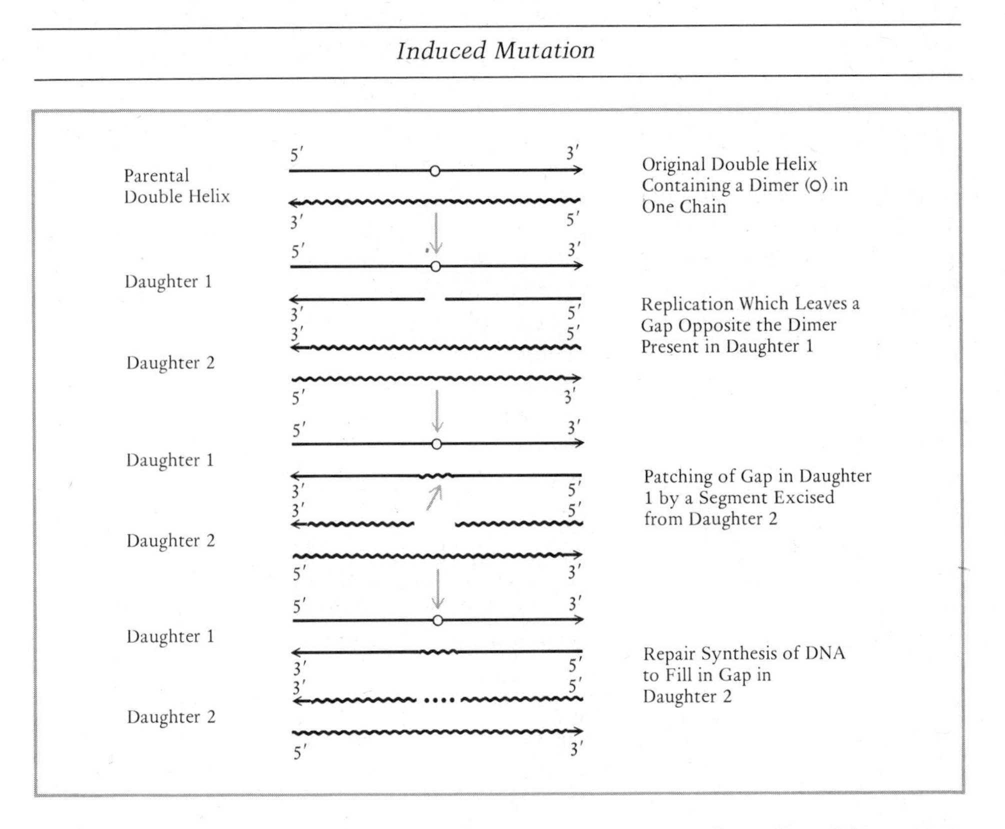

Figure 17–16. Postreplication recombination repair in *E. coli.* [Adapted from W. D. Rupp et al., *J. Mol. Biol.*, *61*, 25–44, 1971.]

In postreplication repair the transfer of portions of a chain from one molecule to another to fill in gaps results in recombination which is prone to error, and it has been proposed that such errors are the source of most ultraviolet induced mutations in *E. coli.* Support for this hypothesis is the finding that ultraviolet does not appear to induce mutations in strains containing a *recA* mutation.

Ultraviolet-sensitive mutations have also been identified in yeast and fungi that are recombination deficient. It appears that repair mechanisms similar to those of *E. coli* are widespread in occurrence. In humans a rare recessive defect (xeroderma pigmentosum) results in great sensitivity to sunlight and a predisposition to ulcerative and cancerous skin lesions. Cells from most xeroderma patients exhibit an almost complete absence of the first step in dimer excision repair and therefore appear to lack a specific UV-endonuclease. Cells from patients with another form of xeroderma (the XP variant) can carry out normal excision repair, but are unable to accomplish postreplication repair. It thus appears that more than one mechanism for the repair of damaged DNA exists in humans.

Ionizing Radiation

Ionizing, or high energy, radiation is electromagnetic radiation of short wavelength. It includes cosmic rays, X rays, particles emitted by radioactive elements and isotopes such as radium and cobalt^{60}, as well as protons and fast neutrons produced by atomic reactors. Whereas the kinds of base changes caused by chemical mutagens provide an insight into their mode of action, and the production of pyrimidine dimers and other photochemical changes in DNA by ultraviolet light has been documented, the molecular effects of ionizing radiation on DNA are much more obscure. The particles produced by high energy radiation are penetrating, and upon passage through cells and tissues their collisions with molecules such as water cause the expulsion of electrons. With the loss of one or more electrons a molecule becomes a chemically reactive ion, and the released electrons can also give rise to secondary ionizations. Thus the path of such particles through a cell leaves a wake of reactive ions and radicals, for example, peroxides, which in themselves are mutagenic. On the molecular level it is therefore impossible to determine whether a mutation induced by ionizing radiation arose directly through a collision of an ionizing particle with DNA itself or indirectly via the mutagenic effects of reactive ions and radicals.

The dosage of ionizing radiation is measured by the *roentgen unit (r)* which is defined by physicists as the amount of radiation that yields 2.08×10^9 ion pairs per cubic centimeter of air under standard conditions of temperature and pressure. In biological terms this amount of radiation produces 2 ionizations per cubic micron of tissue or water. Although more recent studies have sometimes utilized neutrons, β particles, or γ particles to induce mutations, most of the research with ionizing radiation has been conducted with X rays. The results of these studies have led to the following conclusions.

Irrespective of the wavelength used (0.1 Å to 10.0 Å), the number of lethal mutations induced by X rays is directly proportional to the dose in r units. Thus, when the lethal mutation rate is plotted against the dose in r units, a linear relationship is observed (Figure 17–17). Since this relationship holds

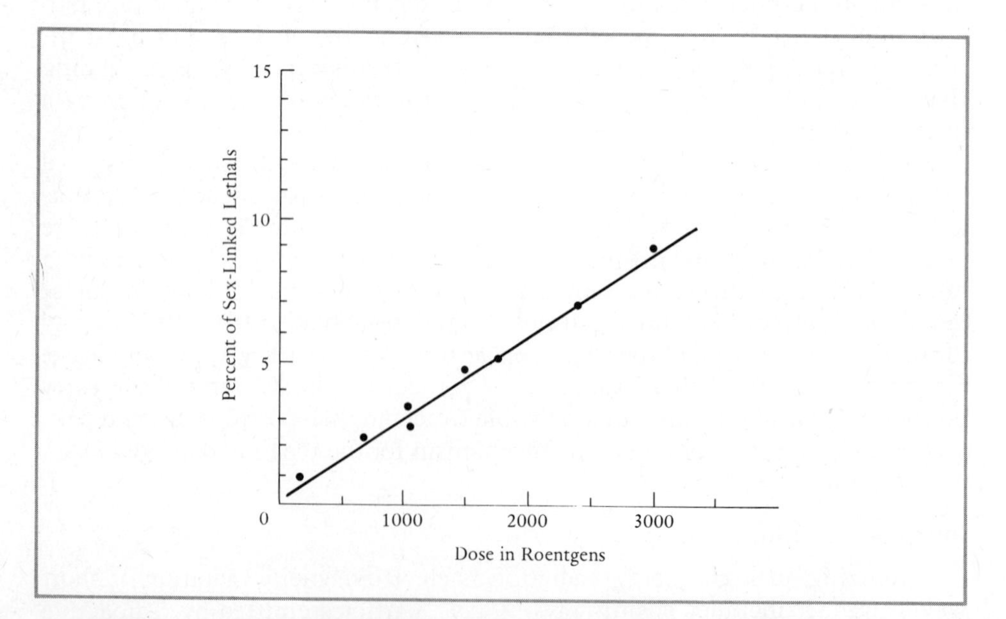

Figure 17–17. Relationship between X ray dosage and the sex-linked lethal mutation rate in *Drosophila.*

even for very low doses, it can be concluded that any amount of ionizing radiation is potentially mutagenic. At high doses a falling off from linearity occurs because more than one lethal mutation may be induced, and the methods of detection generally employed measure only a single phenotype, lethality. The techniques of measurement are those reviewed earlier in this chapter. Although Muller's original method for *Drosophila* employed the *ClB* test stock, the use of the Muller-5 method is now more general.

Another conclusion derived from studies of X ray induced mutation is that the linear relationship between dose in r units and the mutation frequency is independent of the intensity of the radiation. This means that a given dose delivered over an extended period of hours, days, weeks, or even years will produce the same number of mutations as the same dose delivered in a few minutes. In other words, a given amount of radiation will yield the same mutation frequency regardless of the rate at which the radiation is delivered. This finding indicates that the effect of ionizing radiation is cumulative. An exception to this generalization occurs in mice and probably all mammals. W. L. Russell has found that with mice, chronic low intensity radiation produces fewer mutations than a short, high intensity treatment delivering the same total amount of radiation. These results suggest the presence in mice of efficient repair mechanisms capable of healing damaged DNA, providing such damage is not too extensive at any given time.

The direct linear relationship between radiation dose and mutation rate, as well as the independence of mutation rate from the intensity of a given dose early prompted the so-called target theory. This theory proposed that the gene is a target "hit" by an ionizing particle, and a single hit was presumed to cause a single mutation. When chromosome aberrations resulting from X-radiation were studied, it was found that the production of cytologically recognizable single breaks in chromosomes, such as terminal deletions, were also independent of intensity and occurred at a rate directly proportional to dose. Such single breaks could also be considered a one-hit phenomenon resulting from a single ion cluster or trail. The relationship is complicated, however, by the fact that broken chromosomes may undergo healing within a relatively short time. Thus, the single breaks detected cytologically represent only those in which restitution has failed to occur.

Other types of chromosome aberrations also result from ionizing radiation. These include translocations, inversions, interstitial insertions and deletions, dicentrics, acentrics, and ring chromosomes. These aberrations can be recognized cytologically in favorable material and are frequently classified on the basis of whether chromosome breakage can be inferred to have occurred before replication (chromosome breaks) or in one or both of the two chromatids present after replication (chromatid and isochromatid breaks). In contrast to point mutations or single break or one-hit events, these kinds of rearrangements result from two or more separate breaks or hits, and their frequency is dependent on the intensity of the dose. A correlation of the number of such rearrangements with the rate at which a given dose of radiation is delivered indicates that their frequency rises exponentially as the square of the dose (Figure 17–18). This phenomenon can be interpreted in terms of the number of breaks present at any one time and their restitution by repair mechanisms.

When a dose of radiation is delivered at a high, intense rate, the number of ion clusters and the concentration of the associated mutagenic radicals that impinge on chromosomes is sufficient to bring about the simultaneous induction of two or more breaks. The subsequent healing of such multiple breaks can then lead to the rearrangements observed. As noted above, the actual number of chromosome breaks that are induced by any level of treatment is unknown since only those that fail to heal or that lead to cytologically visible rearrangements can be detected and scored.

The mutation rate obtained with ionizing radiation is influenced by environmental factors such as temperature and the concentration of oxygen present during irradiation. A higher temperature or a higher than normal concentration of oxygen increases mutation rate, and a lowering of the temperature or oxygen concentration decreases the rate. Varying sensitivities are also exhibited by different tissues or by cells in different phases of the cell cycle. *Drosophila* spermatogonia and oöcytes are less sensitive than are mature sperm, and cells in interphase are less subject to damage than are those whose chromosomes are condensed at metaphase. This latter finding underlies the use of X rays as a therapeutic agent in the treatment of malignant tumors containing rapidly dividing cells.

In summary, the action and effects of ionizing radiation are extremely complex. The actual agents which cause a mutation or rearrangement in any given instance are unknown, and the mutation rate is influenced by the differential sensitivity of cells, environmental effects, and the capacity of enzyme systems to recognize and repair distortions and breaks within DNA. Despite these uncertainties, ionizing radiation continues to provide a powerful tool for the analysis of the genome.

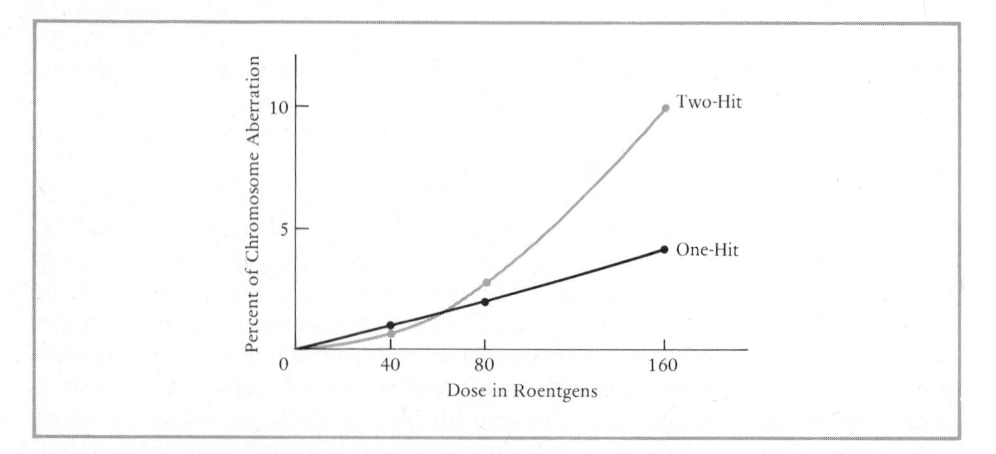

Figure 17–18. The relation between X-ray dosage and the frequency of one-hit and two-hit chromatid aberrations following irradiation at prophase. The one-hit aberrations tend to increase in direct proportion to dosage, but the two-hit types of aberrations tend to increase as the square of the dosage if the intensity is relatively high. [From K. Sax, *J. Cell. Comp. Physiol.*, *35* (suppl. 1), 71–81, 1948.]

MUTATOR AND MUTABLE GENES

Genes which promote mutations at other loci have been described in a number of organisms. In phage and bacteria the loci which exhibit mutator effects are those associated in some way with DNA replication, recombination, and repair. In phage T4 several temperature-sensitive mutations of gene *43*, which specifies DNA polymerase, strongly increase the mutation rate by causing transitions, transversions, and even frameshifts. Other mutations at this locus behave as antimutator genes by significantly decreasing the spontaneous mutation rate. Mutations of other genes of T4 also exert a mutating influence. Among these are mutants of gene *32*, which specifies a protein required for DNA replication, genes *42* and *td*, whose products function in the metabolism and synthesis of DNA precursors, and gene *px*, which participates in recombination and repair.

A number of mutator genes have also been identified in *E. coli*, and in some instances their general effects have been ascertained. One such gene, first described by H. P. Treffers and named *muT*, has been shown to cause A-T → C-G transversions. The locus of *muT* is adjacent to that for DNA polymerase II, and *muT*, presumably via its product, appears to be directly involved in DNA replication and is postulated to interact in some way with DNA polymerase III. Although the products of *muT* and other mutator genes of *E. coli* have not been isolated and the nature of their action is unknown, it again appears that all are involved with DNA replication, recombination, or repair.

In higher organisms mutator and mutable genes have been studied principally in maize and *Drosophila*. In maize several complex interactions between mutator genes and the loci which respond to their activity have been described. One such instance, reported by M. M. Rhoades, involves the dominant mutator gene Dotted *(Dt)*, located on chromosome 9. Dotted appears to induce the mutation of the third chromosome recessive allele *a* to a dominant A_1 allele. Since A_1 causes the production of purple pigment in both leaves and kernel, a mutation from *a* to A_1 is easily recognized. Both mutation frequency and the effects of gene dosage can be analyzed in the aleurone layer of the kernel, since this layer is derived from triploid tissue. When the recessive allele of Dotted is present in combination with gene *a*, that is, *dt/dt/dt a/a/a*, no pigment is present. However, when 1 recessive *dt* allele is replaced by the dominant allele *Dt*, around 7 small purple spots appear in the kernel, indicating that a mutation from *a* to A_1 has occurred in at least 7 different cells during the growth of the aleurone. If 2 *Dt* alleles are present, the number of spots increases to around 22, and if 3 are present, over 120 spots are evident. The biochemical basis for the sensitivity of the recessive *a* allele to the nonlinked *Dt* gene is unknown.

Other instances of mutable genes in maize involve transposable elements which appear to move from one position to another within the genome and cause the mutation or loss of genes next to which the element becomes inserted. These transposable agents have been termed controlling elements. Since their positions can be mapped, they appear to be integrated within the chromosome and therefore must be composed of DNA. Their activities parallel those of episomes, but their origin, whether foreign or intrinsic, is unknown, and none has been isolated biochemically as a separate entity. Several different transposable elements have been described. R. A. Brink has found such an agent, called modulator *(Mp)*, which, when located next to the dominant gene P^r (red pericarp, a layer of the kernel) causes instability and the apparent mutation of P^r to P^w (white pericarp). As a result, the kernel is variegated, red and white in appearance. Modulator is capable of autonomous transposition to a new site, and after it leaves its locus next to P^r, the instability of P^r ceases and no further color variegation is observed. Transposition of *Mp* from P^r to a site adjacent to the dominant gene *Wx* (starchy endosperm) has been observed. In this position *Mp* induces the mutation of *Wx* to the recessive allele *wx* (waxy endosperm).

More complex cases involving two transposable elements have been described by B. McClintock. One of these consists of a Dissociator *(Ds)* agent which causes instability, suppression of activity, and often chromosome breakage and gene loss at its site of integration. The Dissociator element induces these effects, however, only when an additional agent, called Activator *(Ac)*, is present, usually on a different chromosome. The Activator element appears to control not only the activity of *Ds*, but also the time and frequency of *Ds* action during development. Both agents undergo mutation and both are transposable. In a new site *Ds* again responds to signals from the Activator to suppress the function of the gene next to which it has integrated. The activity of these agents can be inferred from phenotypic changes in kernel color. For example, the dominant gene *I* prevents and the recessive allele *i* permits pigment formation. If *Ds* is located next to *I* in an aleurone genotype of *Iii*, loss of gene *I* or the suppression of its activity results in the formation of pigment in those cells in which such loss or suppression occurred, as well as in their descendents formed from mitosis. If *Ds* acts early, large colored sectors containing many pigmented cells will be produced, whereas if it acts late in development, only small spots formed by fewer pigmented cells will be evident. It has also been found that the presence of one activating element induces early action of *Ds* and consequently large colored areas in the kernel. With two activator agents the spots become smaller, and with three such agents only tiny dots are observed. Thus an increase in the dosage of *Ac* causes a delay in the action of *Ds*. Although McClintock has suggested that these controlling elements may be heterochromatic in nature, they are not essential components of the maize genome. Another more complex set of controlling elements, called the Suppressor-Mutator system, has also been described by McClintock.

M. M. Green has presented strong evidence for a controlling element integrated within the white eye locus of *Drosophila melanogaster*. The agent induces mutations along with a high frequency of cytologically visible deficiencies of varying length, some extending up to 20 salivary chromosome bands. Transpositions to the third chromosome of the portion of the white locus which includes the agent have also been identified. The agent, whatever its nature, is not a required component of the genome and, like the controlling elements of maize, exhibits characteristics in common with those of episomes and temperate viruses.

Other mutator genes are also known in *D. melanogaster.* One called *mu* and located on the third chromosome is active only in females where it induces both forward and reverse mutations. It also increases the frequency of primary nondisjunction. Of the sex-linked lethals induced by *mu,* approximately half appear to be deficiencies. In another species, *Drosophila robusta,* M. Levitan and associates have described the occurrence of a high rate of spontaneous chromosome aberrations, including inversions and translocations. The causative agent is unknown, but the trait is transmitted maternally, and only paternally derived chromosomes of both male and female progeny are affected.

One further phenomenon, quite different from those described above, is *paramutation.* Paramutation refers to the induction of a heritable change in one allele by another allele of the same gene which is located in the opposite homologue. Examples of paramutation have so far been found only in plants and have been analyzed in greatest detail with the *R* alleles of corn by Brink and associates. Presence of the allele R^r leads to full anthocyanin pigment production, while R^{st} produces stippled pigmentation. When R^r is combined with R^{st} in the heterozygous condition, different sectors of the plant body exhibit mosaicism for the degree of expression of R^r. When the R^r allele is extracted from such a heterozygote by an appropriate cross, it no longer produces full pigmentation, and this change in expression is inherited by succeeding generations. However, when a suppressed R^r allele is maintained as a homozygote, reversion part way back to full expression or reversion to complete expression may eventually occur. Thus the initial change in R^r elicited by association with the R^{st} allele does not appear to be a mutation in the usual sense, but rather a sustained change in expression. How such a change is brought about is unknown, but Brink has proposed that it arises through alterations in postulated regulatory sequences associated with the *R* locus. An understanding of paramutation will depend in large part upon a prior understanding of the organization and control of genetic units within the eukaryotic chromosome.

PROBLEMS

17–1. What are missense, nonsense, and frameshift mutations and what are the consequences of each?

17–2. If a drastic alteration occurred in the structure of one of the genes for 28 S rRNA, do you think that the translation of mRNA into protein would cease? If not, why not?

17–3. Explain the difference between a transition and a transversion and give an example of each.

17–4. What possible explanations can you offer for the reversion of a mutant to the wild type phenotype?

17–5. Compare the effects of nitrous acid, hydroxylamine, and 5-bromouracil on DNA.

17–6. If the α chain of adult hemoglobin contains 141 amino acids, how many bases, at a minimum, must the mRNA for this protein contain? What is the evidence that the mRNA contains more bases than can be accounted for by the number of amino acids in the α chain?

17–7. Distinguish between the three types of UV repair mechanisms that have been discovered in bacteria.

17–8. Which of the following mutagens is likely to produce the most harmful mutations and why: nitrous acid, base analogues, acridines, alkylating agents?

17–9. A dose of 1500 roentgens was given to two groups of *Drosophila* males. Group 1 received the dose over a five minute period, and group 2 over a five day period. How would these two groups compare with respect to lethal mutation rate?

17–10. What is the difference between intragenic and intergenic suppression? Give an illustration of each.

17–11. Suppose a deletion (in brackets), followed by an insertion (in italics) occurred in the following sequence of DNA: A T [G] G C A T T A A *A* T T A G A. What would be the resulting DNA sequence? If this strand were the sense strand for transcription of mRNA, which triplets would code for the same amino acid, in the same position as designated by the original sequence? Assume nondegeneracy of the code.

17–12. Although glycine normally occupies position 233 within the A subunit of tryptophan synthetase, Yanofsky has found three different mutations in which glycine has been replaced by cysteine, aspartic acid, or alanine. If the codons for cysteine, aspartic acid, and alanine are UGU, GAU, and GCU, respectively, and each mutation has involved a single base change, what is the codon for glycine? Characterize each change as a transition or a transversion.

17–13. Assume that the sequence of amino acids of a hypothetical wild type peptide is that given below and that the mRNA codons specifying this sequence are those indicated. Assume also that the same peptide is isolated from four different mutant strains and the amino acid sequence in each case is determined to be that shown. Referring to the genetic code (Figure 16–1) for appropriate codons, postulate the probable mutational events which resulted in the abnormal peptides.

Wild Type:	NH_2—ser	—leu	—arg	—gln	—glu	—asp	—val	—trp	—COOH
mRNA:	UCU	—CUU	—CGU	—CAA	—GAA	—GAU	—GUU	—UGG	
Mutant 1:	NH_2—ser	—leu	—arg	—gln	—gly	—asp	—val	—trp	—COOH
Mutant 2:	NH_2—ser	—leu	—val	—lys	—lys	—met	—phe	—COOH	
Mutant 3:	NH_2—ser	—leu	—val	—asn	—glu	—asp	—val	—trp	—COOH
Mutant 4:	NH_2—ser	—leu	—arg	—gln	—glu	—asp	—val	—COOH	

REFERENCES

Auerbach, C., and B. J. Kilbey, 1971. Mutation in eukaryotes. *Ann. Rev. Genet., 5,* 163.

BEIR (Biological Effects of Ionizing Radiation) Report, 1972. *The Effects on Populations of Exposure to Low Levels of Ionizing Radiation.* National Academy of Sciences—National Research Council, Washington, D.C.

Brink, R. A., 1973. Paramutation. *Ann. Rev. Genet., 7,* 129.

Cleaver, J. E., and D. Bootsma, 1975. Xeroderma pigmentosum: Biochemical and genetic characteristics. *Ann. Rev. Genet., 9,* 19.

Drake, J. W., 1970. *The Molecular Basis of Mutation.* Holden-Day, San Francisco.

Fincham, J. R. S., and G. R. K. Sastry, 1974. Controlling elements in maize. *Ann. Rev. Genet., 8,* 15.

Glass, B., and R. K. Ritterhoff, 1956. Spontaneous mutation rates at specific loci in *Drosophila* males and females. *Science, 124,* 314.

Hartman, P. E., and J. R. Roth, 1973. Mechanisms of suppression. *Adv. Genet., 17,* 1.

Hollaender, A. (Ed.), 1970. *Chemical Mutagens,* vols. 1, 2. 1974, vol. 3. Plenum Press, New York.

LEA, D. E., 1955. *Actions of Radiations on Living Cells,* 2nd ed. Cambridge University Press, Cambridge.

MCCLINTOCK, B., 1967. Genetic systems regulating gene expression during development. *Dev. Biol., 1* (Suppl.), 84.

PETERSON, P. A., 1970. Controlling elements and mutable loci in maize: Their relation to bacterial episomes. *Genetica, 41,* 33.

ROTH, J. R., 1974. Frameshift mutations. *Ann. Rev. Genet., 8,* 319.

STADLER, L. J., 1942. Some observations on gene variability and spontaneous mutation. *The Spragg Memorial Lectures* (Third Series). Michigan State College, East Lansing.

SUTTON, H. E., AND M. I. HARRIS (Eds.), 1972. *Mutagenic Effects of Environmental Contaminants.* Academic Press, New York.

VOGEL, F., 1964. Mutations in man. In *Genetics Today, Proceedings of the 11th International Congress of Genetics, 1963.* Pergamon Press, New York.

VOGEL, F., 1970. Spontaneous mutations in man. In F. Vogel and G. Röhrborn (Eds.), *Chemical Mutagenesis in Mammals and Man.* Springer-Verlag, New York.

WOLFF, S., 1967. Radiation genetics. *Ann. Rev. Genet., 1,* 221.

Chapter 18

GENE REGULATION IN BACTERIA AND BACTERIOPHAGES

THE existence of regulatory mechanisms for the control of gene expression has long been recognized. It is obvious that although cells contain the genetic capacity for the synthesis of an enormous number of different products, not all of these products are present at any given time, many being selectively evoked only upon occasion and in response to some environmental stimulus. In prokaryotes some enzymes are synthesized *constitutively* (that is, continuously), indicating that transcription of mRNA is constantly occurring. However, other enzymes are synthesized only when a need for their presence arises, and when this need has been satisfied, enzyme synthesis ceases. Transcription of mRNA in this case is evidently initiated only on demand and must therefore be subject to regulation. Although controls imposed at the level of the translation of mRNA into protein could inhibit enzyme synthesis, experimental evidence indicates that at least in prokaryotes and viruses transcription is the primary target of controlling mechanisms. The operation of these mechanisms has been intensively studied, and both negative and positive controls on transcription have been discovered. These controls are the subject of this chapter.

INDUCIBLE OPERONS

It will be recalled that in bacteria genes that specify enzymes belonging to the same metabolic pathway are characteristically clustered together in the genome and comprise a genetic unit called an *operon.* The members of an operon are transcribed coordinately as a single, long polycistronic mRNA molecule. One such operon in *E. coli,* called the *lactose* or *lac operon,* has provided a model system for the study of gene regulation. The *lac* operon contains three cistrons, *z, y,* and *a,* whose products are involved in the breakdown (catabolism) of the sugar lactose (Figure 18–1). Gene *z* specifies an enzyme, *β-galactosidase,* which converts lactose into glucose and galactose, while gene *y* determines the structure of a protein, *galactoside permease,* which facilitates the entrance of lactose into the cell. Gene *a* specifies an enzyme, *thiogalactoside acetylase,* whose precise function is poorly understood although it is presumed to be involved at some point in lactose utilization. Mapping experiments employing mutations of these three genes have demonstrated that the gene order is *z—y—a* (see Figure 13–14).

Normally, the synthesis of these three enzymes is not constitutive, and in the absence of lactose only a few molecules of each enzyme are present. However, when lactose is provided, all three are synthesized rapidly and simultaneously as a coordinated response to the presence of this substrate. Lactose thus acts to *induce* the production of the enzymes needed for its catabolism.

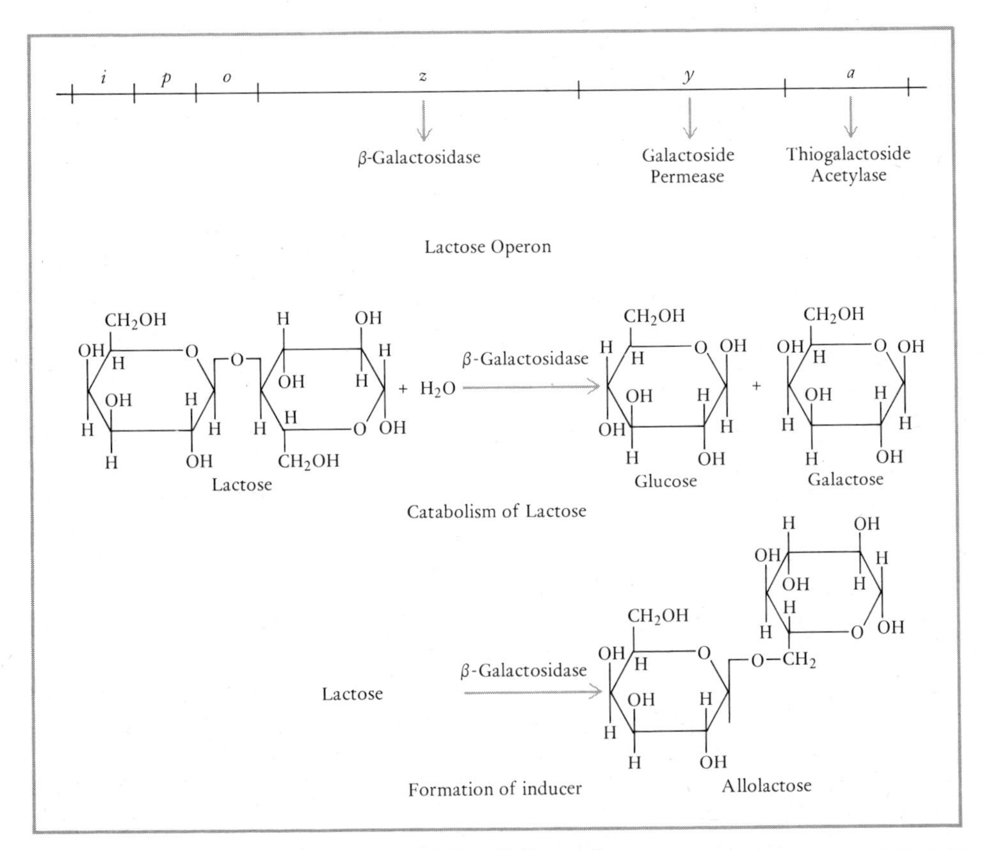

Figure 18–1. The lactose operon of *E. coli* (*i,* regulator; *p,* promotor; *o,* operator; *z, y* and *a,* structural genes) and reactions catalyzed by β-galactosidase, the enzyme specified by gene *z*.

Since all three enzymes are synthesized through the translation of a single polycistronic mRNA, it follows that the entire operon is responding as a unit to the presence of the inducer. It has been found that although lactose itself provides the basis for induction, the actual inducer is a compound called allolactose whose formation from lactose is also mediated by β-galactosidase (Figure 18–1).

A clue to the cause of this controlled response on the part of the *lac* operon was provided by a mutation in whose presence all three enzymes were produced constitutively regardless of the presence or absence of lactose. Since the enzymes themselves were normal, it was inferred that this mutation had occurred in a hitherto unidentified controlling locus which was given the name *operator (o)*. The mutation causing constitutive enzyme synthesis was called *operator-constitutive* (o^c). By mapping experiments the operator locus was found to be situated immediately adjacent to gene *z*, so that the overall gene order was $o—z—y—a$. That the operator was in fact a genetic site that regulated the transcription of *lac* genes was shown by experiments which employed the z^+ wild type allele and a β-galactosidase-deficient mutation z^-, along with the normal operator o^+ and the operator-constitutive mutation o^c. *E. coli* cells were made diploid for these genes through the introduction of an F' element. In the cis arrangement ($F'\ o^+\ z^+/o^c\ z^-$), where the wild type operator was physically adjacent to z^+, the phenotype was normal and β-galactosidase was produced inductively in response to lactose. However, in the trans arrangement ($F'\ o^c\ z^+/o^+\ z^-$), where the mutated operator was adjacent to z^+, β-galactosidase was produced constitutively. These results indicated that the operator affected the transcription of the z^+ cistron only when physically joined to this cistron, that is, the operator was *cis-dominant.*

Another clue to the mechanism controlling the *lac* operon was provided by the discovery of a gene called *i (inducer)*, whose locus was closely linked to the operon, but was separate from that of the operator, gene order being $i—o—z—y—a$. Mutations of the *i* locus also resulted in constitutive synthesis of *lac* enzymes, and a cell of genotype $i^-\ o^+\ z^+$ produced β-galactosidase in the absence of the inducer. The action of the *i* gene was clearly different from that of the operator, however. Partial diploids of the genotype $F'\ i^+\ o^+\ z^+/i^-\ o^c\ z^-$ or $F'\ i^+\ o^c\ z^-/i^-\ o^+\ z^+$ exhibited the normal inducible phenotype, and it was evident that the i^+ wild type allele could restore normal function in either the cis or trans arrangement, that is, whether or not it was physically adjacent to the wild type operator and z^+ gene. This implied that the *i* gene specified a product which could diffuse from the site of transcription and translation to an altogether different region of the genome and there influence the function of the wild type operator and z^+ gene.

To clarify the respective roles of these elements, F. Jacob and J. Monod in 1961 proposed that a distinction be made between *structural genes,* such as *z, y,* and *a,* that specify proteins required for cell metabolism, and *regulatory genes,* such as *i,* whose products participate in control mechanisms imposed on structural genes. The operator element was designated a *controlling site* which governed the transcription of physically adjacent structural genes. Jacob and Monod also proposed a theory to explain the interaction of these loci in the regulation of coordinate enzyme synthesis. According to this theory, the regulating *i* gene specified a *repressor protein* which, in the absence of the inducer

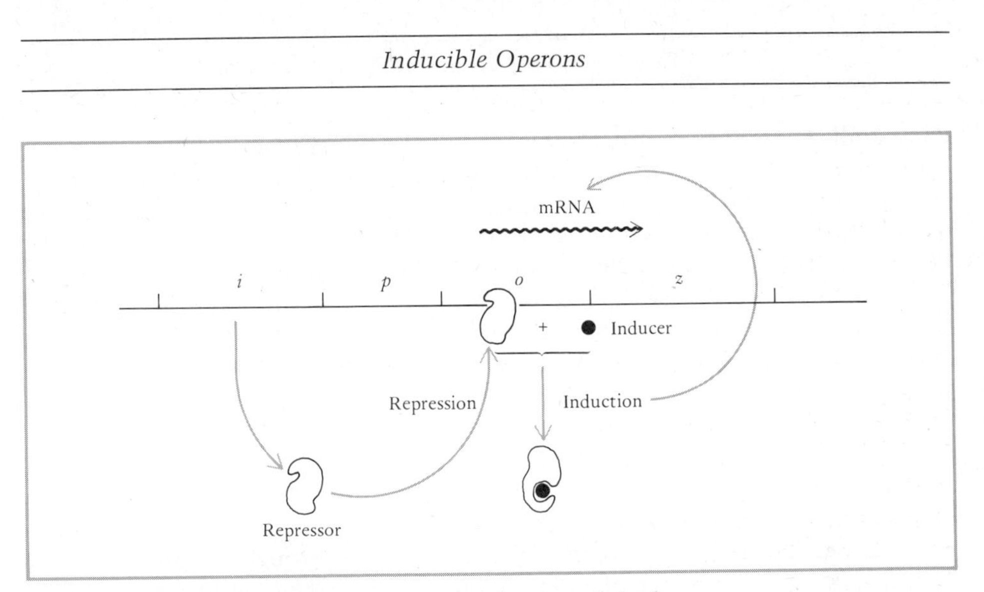

Figure 18–2. Repression and induction of the lactose operon.

(lactose), was bound to the operator, thereby inactivating the operator and preventing transcription of the three *lac* cistrons (Figure 18–2). Induction of transcription was explained as the result of the binding of the inducer to the repressor protein such that the repressor dissociated from the operator. Upon this release from repression, the operator would permit the transcription of the adjacent operon and coordinate enzyme synthesis would follow. The repressor protein was visualized as a molecule with two different, nonoverlapping binding sites, one for the operator and one for the inducer. Union of repressor with inducer would cause a change in the conformation of the repressor protein which rendered the binding site for the operator nonfunctional. However, when the supply of inducer was depleted through the activity of *lac* enzymes, dissociation of the inducer-repressor complex would occur permitting a reverse change in conformation so that the repressor could once more bind to the operator to shut down transcription. This kind of reversible change in conformation, called *allostery,* is well known and is exhibited by a large number of proteins.

In accord with their theory Jacob and Monod postulated that operator-constitutive mutations altered the operator to such an extent that affinity for the repressor molecule was lost, and in the absence of repression continuous transcription and enzyme synthesis would occur. Mutations at the *i* locus would be expected to result in changes in the protein repressor molecule such that it could no longer bind to the operator, and again constitutive enzyme synthesis would be observed. The action of an additional *i* mutation, i^S *(superrepressor),* in whose presence no enzyme production at all occurs, could also be reconciled with Jacob and Monod's hypothesis. It was proposed that this mutation caused an alteration of the repressor with loss of the inducer binding site with the result that the repressor molecule remained attached to the operator even when the inducer was present.

Jacob and Monod's theory for the regulation of the *lac* operon has since been fully substantiated and can be characterized as a *negative control* mechanism since the product of the regulatory gene acts to repress transcription. *Positive control* mechanisms can also be visualized whereby the product of a regulatory gene acts to stimulate transcription. The application of both types of control to a single operon would further ensure a measured and appropriate response to an environmental stimulus. Intensive studies of the *lac* operon have now demonstrated that while negative control is exerted upon the operator, positive control is imposed on yet another controlling element, the *promotor.*

The promotor *(p)* is the site of RNA polymerase attachment and it behaves as a cis-dominant locus. Mutations of the promotor cause a decrease in enzyme synthesis. The site has been mapped and has been found to lie between the loci of *i* and *o,* and thus the complete order of the *lac* operon and its controlling elements is *i—p—o—z—y—a* (Figure 18–1). Positive control of the *lac* operon via the promotor occurs as follows. Although lactose is a carbon source acceptable to *E. coli* cells, it is not as readily utilized as is glucose, and when both lactose and glucose are present, the *lac* operon is repressed. This repression, called *catabolite repression,* is mediated through the promotor and occurs despite the presence of a normal operator and the i^+ wild type regulator gene. The source of this repression has been traced to the interaction between an *activator protein (CAP,* for catabolite activator protein) and a small nucleotide, adenosine-3′, 5′-monophosphate (cyclic AMP, Figure 18–3). When cyclic AMP is present, the activator protein becomes bound to the promotor and the attachment of RNA polymerase is stimulated (Figure 18–4). However, when the level of cyclic AMP falls, the activator protein dissociates from the promotor and the attachment of RNA polymerase is inhibited. For reasons not well understood, the presence of glucose causes a decrease in the concentration of cyclic AMP, and this, in turn, results in the repression of transcription and *lac* enzyme synthesis.

Biochemical studies of the operator and promotor sites have permitted a determination of the base sequence of these regulatory elements (Dickson and others, 1973). The operator consists of two regions, the base sequences of these regions being mirror images of one another. The repressor protein which binds to the operator is composed of four subunits arranged as pairs such that their

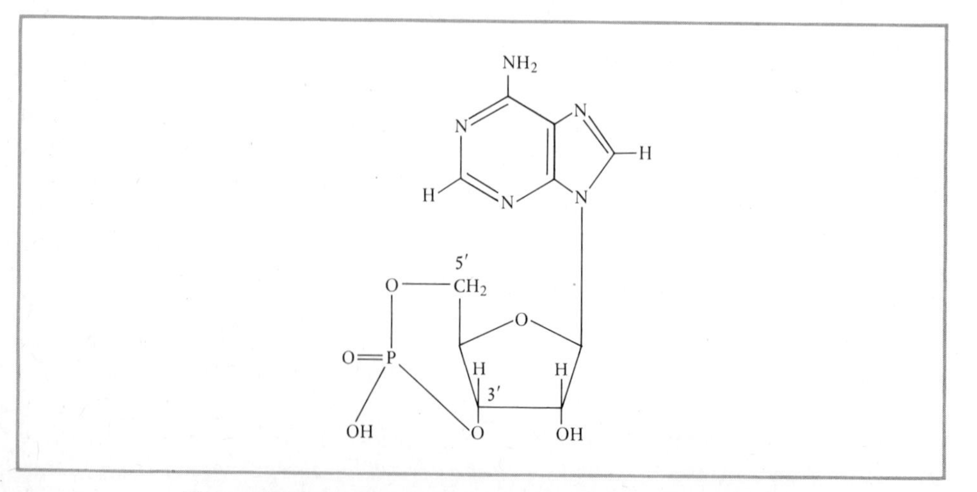

Figure 18–3. Cyclic AMP (adenosine-3′, 5′-phosphate).

symmetry matches the mirrored symmetry of the operator. The promotor site also consists of two regions, one for the binding of the activator protein (CAP) and one for the attachment of RNA polymerase (Figure 18–4). It is suggested that the complex formed between the DNA of the promotor and RNA polymerase serves to initiate strand separation and sense strand selection so that mRNA synthesis can begin when the start codon within the operator is reached.

Although the regulatory mechanisms of the *lac* operon have been those most intensively investigated, other inducible operons of *E. coli,* such as the galactose and arabinose operons, also exhibit dual systems of negative and positive control and catabolite repression. For example, the galactose *(gal)* operon consists of a promotor, an operator, three structural genes which specify enzymes for the catabolism of galactose, and an unlinked regulatory gene which codes for a repressor protein (Figure 18–5). This operon is regulated in the same manner as the *lac* operon. Negative control is exerted by the repressor protein which binds to the operator, and induction of enzyme synthesis is brought about by the union of inducer with repressor, accompanied by the release of the operator. Positive control to stimulate transcription is imposed by an activator protein (CAP) which binds to the promotor in the presence of cyclic AMP.

The arabinose *(ara)* operon (Figure 18–6) is somewhat different in that the regulatory gene product can function both as a repressor and as an activator. The regulatory gene *araC* specifies an allosteric protein which exists in two

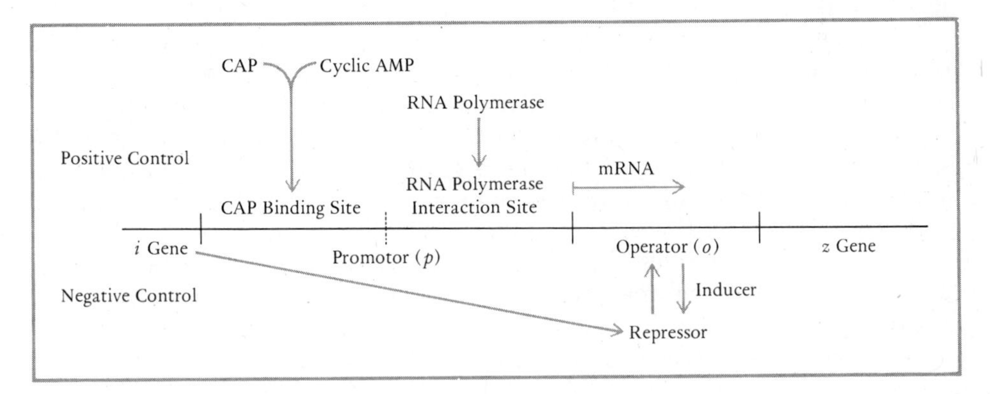

Figure 18–4. Positive and negative control of the lactose operon. [From R. C. Dickson et al., *Science, 187,* 27–35, 10 January 1975. Copyright © 1975 by the American Association for the Advancement of Science.]

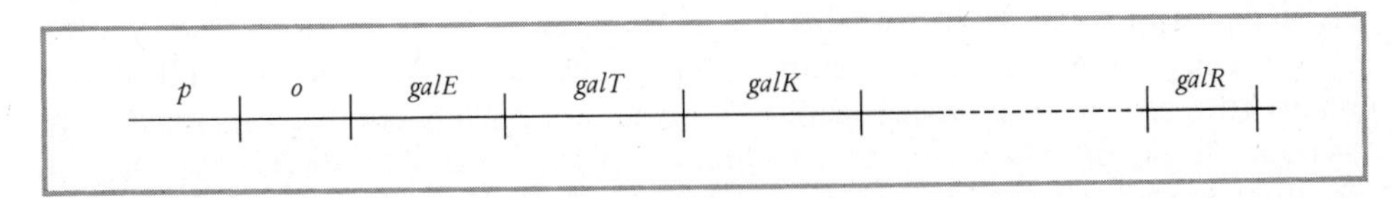

Figure 18–5. The galactose operon of *E. coli* (*p,* promotor; *o,* operator; *galE, galT,* and *galK,* structural genes); *galR* is an unlinked regulatory gene which codes for the repressor protein.

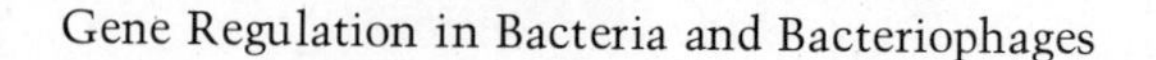

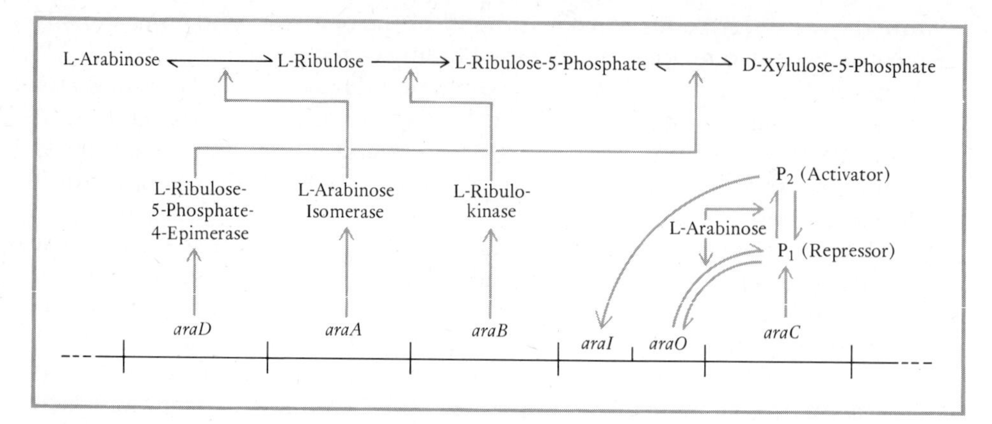

Figure 18–6. The arabinose operon of *E. coli* (*araI*, promotor; *araO*, operator; *araD*, *araA*, and *araB*, structural genes); *araC* is a regulator gene whose product can assume two different configurations, P_1 and P_2. [Reproduced, with permission, from E. Englesberg and G. Wilcox, *Ann. Rev. Genetics,* 8, 219–242, 1974. Copyright © 1974 by Annual Reviews Inc. All rights reserved.]

different configurations, P_1 and P_2. The shift back and forth between these two states is dependent on the concentration of the inducer L-arabinose. In the absence of L-arabinose the P_1 configuration is assumed, and the repressor exerts negative control over the operon by attaching to the operator to shut down transcription. In the presence of L-arabinose the repressor is removed from the operator, assumes the P_2 configuration, and then attaches to the promotor *(araI)* to positively stimulate transcription. Catabolite repression by glucose is also imposed on this operon, but in the absence of glucose a CAP-type protein in conjunction with cyclic AMP additionally stimulates transcription by binding to *araI.* This same locus probably also serves as the attachment region for RNA polymerase. Thus, *araI* may be a promotor containing three kinds of sequences: one for the P_2 activator, one for the CAP protein activator, and one for RNA polymerase.

From these examples it can be seen that inducible operons are a most efficient and economical method of handling sugars and other substrates not frequently encountered, for the enzymes needed for the catabolism of such substrates are not synthesized unless the substrate itself is present.

REPRESSIBLE OPERONS

The enzymes coded by genes of a repressible operon are usually involved in a synthetic pathway through which amino acids or other needed cell constituents are provided. In the absence of the end product of the pathway such operons are constitutive and continuously active in transcription. However, when the end product has reached a sufficiently high concentration, all enzymes of the pathway are coordinately repressed, and synthesis of the end product ceases. This system is admirably suited to supply a given product when needed, but avoids wasteful overproduction. The overall differences between inducible and repressible systems can be summarized as follows.

1. In inducible systems transcription occurs only in the presence of the substrate, and repression occurs in the absence of the substrate.

2. In repressible systems transcription is initiated in the absence of the end product, and repression occurs in the presence of the end product.

The elements involved in the control of a repressible operon consist of the structural genes specifying the enzymes of the pathway, an operator locus which controls transcription, a promotor locus for the attachment of RNA polymerase, and a regulatory gene which codes for a repressor protein. An additional participant in control is the end product of the reaction sequence.

Negative control occurs in the following manner (Figure 18–7). In the absence of end product the repressor protein is unable to bind to the operator. The operator therefore freely permits transcription which is followed by coordinate enzyme production and the synthesis of the end product. When a given concentration of end product is attained, this product assumes the role of *corepressor* and combines with the repressor protein, bringing about an allosteric change in conformation which transforms the repressor to the active state. The corepressor-repressor complex then binds to the operator and blocks transcription. With continuing utilization of the end product, its concentration declines whereupon it dissociates from the repressor, permitting a reverse shift in the conformation of the repressor and its removal from the operator. The operator is then free *(derepressed)* to initiate transcription anew.

An example of a repressible operon controlled in the manner described above is the tryptophan operon of *Salmonella typhimurium*. The *trp* operon includes five structural genes whose order is the same as the sequence of reactions for the synthesis of this amino acid (Figure 18–8; see also Figure 10–5).

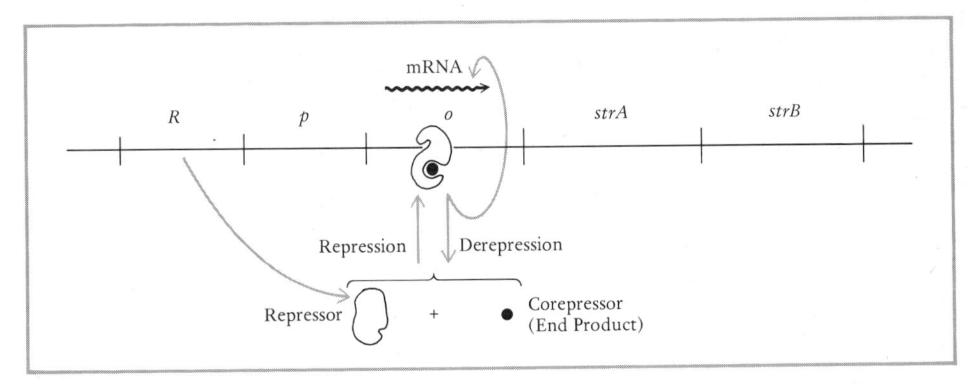

Figure 18–7. Negative control of a repressible operon. *R*, regulator gene whose product is a repressor; *p*, promotor; *o*, operator; *strA*, *strB*, structural genes *A* and *B*.

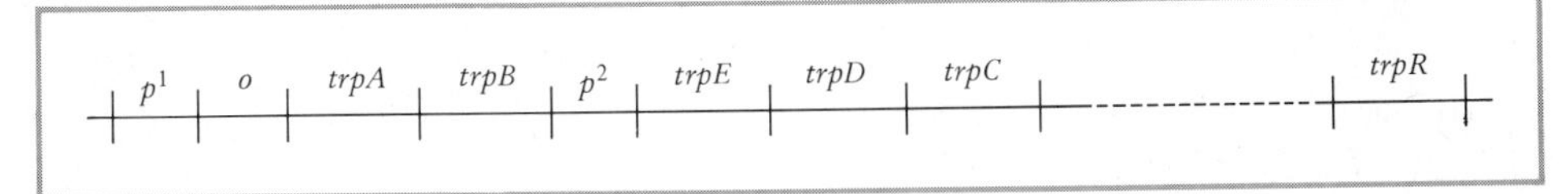

Figure 18–8. The tryptophan operon of *Salmonella typhimurium* (p^1, main promotor; *o*, operator; *trpA*, *trpB*, *trpE*, *trpD*, and *trpC*, structural genes; *trpR*, regulatory gene). A second internal promotor (p^2) is located between structural genes *trpB* and *trpE*.

The regulatory gene *(trpR)* which codes for the repressor is not closely linked to the operon. Transcription of all five structural genes is governed by a main promotor *(p^1)* which is located adjacent to the operator. In addition, a second internal promotor *(p^2)* has been found at the boundary between structural genes *B* and *E.* The activity of the main promotor is governed by the state of repression, but the second promotor functions to permit low level, but continuous transcription of the last three genes of the operon. With the exception of the use of different gene symbols, the *trp* operon of *E. coli* is essentially identical to that of *Salmonella.*

A second example of a repressible system is the histidine operon of *Salmonella.* The operon includes an operator and nine structural genes whose order is unrelated to the sequential steps in the pathway for histidine synthesis (Figure 18–9; see also Figure 10–6). The *his* operon differs from the *trp* operon in that no regulatory protein has yet been found and histidine does not act as corepressor. Instead the actual corepressor is *histidyl-tRNA* whose concentration determines the state of repression or derepression of the operon. Five separate genes with unrelated loci *(hisR, hisS, hisU, hisW,* and *hisT)* are involved in the specification and modification of his-tRNA, and a mutation in any one of these loci results in constitutive expression of the operon. *HisR* codes for his-tRNA, *hisS* specifies histidyl-tRNA synthetase, the activating enzyme, and the products of *hisU, hisW,* and *hisT* are involved in structural modification of his-tRNA. From this example, it can be seen that a variety of cellular components can participate in regulatory mechanisms.

Besides the controls imposed on the synthesis of the enzymes of an operon, an additional fine tuning of metabolism is provided by *feedback inhibition.* As described above, proteins frequently exhibit allosteric changes in conformation which alter their ability to combine with other molecules. In feedback inhibition a small molecule produced as the end product of a sequence of reactions combines with the first enzyme of the sequence and by doing so alters the conformation of the enzyme to render it inactive. As a result, the entire metabolic chain can be shut down. When the end product molecules are utilized by the cell and their concentration declines, dissociation from the enzyme occurs, and the metabolic chain is activated once more. Feedback inhibition imposed on enzyme activity and the controls imposed on enzyme synthesis together permit a most exquisite sensitivity to the changing metabolic needs of the cell.

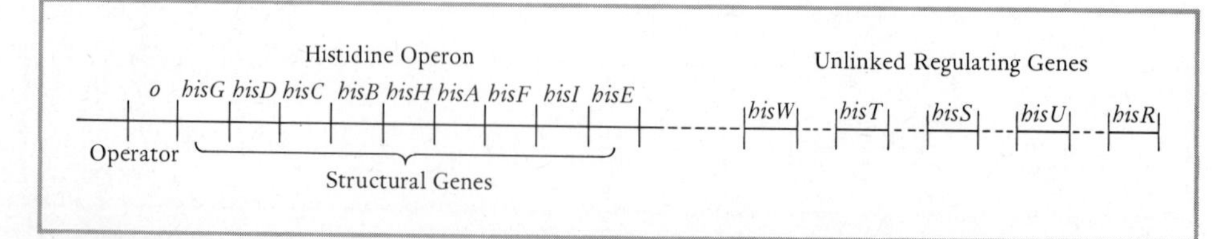

Figure 18–9. The histidine operon of *Salmonella* and its unlinked regulating genes. See also Figure 10–6.

POLAR MUTATIONS

Polar mutations were first observed in structural genes of the *lac* operon. For example, such a mutation in gene *z* not only prevented the synthesis of β-galactosidase, but also caused a decreased synthesis of permease (gene *y*) and acetylase (gene *a*). Similarly, a polar mutation in gene *y*, though it had no effect on β-galactosidase production, prevented the formation of permease and resulted in the decreased synthesis of acetylase. Mutations of this type thus exhibited *polarity* in that a mutation in an early gene of the sequence affected the transcription of distal genes of the sequence even though these genes were wild type.

Polar mutations have since been shown to be base substitutions or frameshifts which transform a sense codon to a nonsense codon (stop signal). Polarity has been interpreted in terms of the effect of such stop signals on coupled transcription and translation. It will be recalled that in bacteria mRNA is translated as fast as it is synthesized (see Figure 15–6). It is postulated that after RNA polymerase has encountered an early stop signal, dissociation of ribosomes from the mRNA may occur, leaving the message vulnerable to destruction by nucleases. Support for this hypothesis has been obtained. It has been shown that in polar mutants of *E. coli* transcription, translation, and the destruction of mRNA proceed together in the 5′ to 3′ direction. However, destruction of mRNA does not occur in the presence of a suppressor mutation causing loss of an endonuclease (ribonuclease V) which degrades RNA in the 5′ to 3′ direction.

Polar mutations of the type described above are seen only in connection with operons and they provide additional proof that the structural genes comprising an operon are transcribed as a single polycistronic mRNA molecule.

GENE REGULATION AND EXPRESSION IN BACTERIOPHAGES

The life cycle of a bacteriophage is invariably characterized by the sequential expression of viral genes. After infection of the host proteins and enzymes required for the termination of host activities and for initiating viral DNA replication and recombination are produced first (early phase), and proteins required for the construction of the viral capsule and for host cell lysis are produced somewhat later (late phase). Regulation of sequential gene expression is positive and is exerted primarily upon transcription. Typically, early acting genes are transcribed by host RNA polymerase, while the transcription of late acting genes is initiated by a phage coded protein. Operonlike clusters of cistrons with related functions are found among both early and late gene sequences, and polycistronic mRNA's have been identified.

A simple example of sequential gene expression is found in phage T7 of *E. coli*. After infection of the host, genes for early functions are transcribed into mRNA by *E. coli* RNA polymerase which, with the *E. coli* sigma subunit, recognizes the appropriate initiation site on the viral DNA molecule. Among these early genes is gene *1,* which codes for T7 RNA polymerase. Once synthesized, this enzyme then transcribes all other cistrons of the viral DNA. Thus in T7 (and also in T3) the programing of function is regulated by the sequential activities of two different RNA polymerases, each of which recognizes different promotor sites.

A more complex sequence of events is evident during infection of *E. coli* by phage T4. In T4 the early phase of the lytic cycle can be subdivided into immediate-early and delayed-early stages. Transcription of immediate-early genes is initiated within a minute after infection by *E. coli* polymerase and sigma, but transcription of delayed-early genes depends upon the presence of proteins produced in the immediate-early phase. One such protein appears to be a T4 sigma factor which replaces host sigma two minutes after infection. The presence of a new sigma, in conjunction with the host core enzyme, then permits the initiation of mRNA synthesis at different promotor sites. By twelve minutes after infection synthesis of mRNA from late genes begins, utilizing yet another T4 sigma factor coupled to the host core enzyme whose subunits have become modified by the action of viral enzymes. Thus, in contrast to T7, the host core enzyme (with some alterations) is utilized for the transcription of all viral genes, the temporal pattern of mRNA synthesis being at least partially controlled by the successive substitution of different sigma factors which recognize different promotor sites.

It has also been found that both DNA chains of the viral genome are used in transcription. Practically all immediate-early and delayed-early genes are transcribed from one strand in a leftward or counterclockwise direction, while late genes are transcribed predominantly from the opposite DNA strand and in a rightward or clockwise direction (Figure 18–10). The advantages of this arrangement are difficult to understand since the production of a mature virion still requires the expression of all genes regardless of the location of their sense sequences. This kind of strand switching for transcription purposes is also a feature of phage λ (see below).

With respect to operons the clustering of genes with related functions is a prominent feature of the T4 genome, and the discovery of polar mutations indicates that at least some mRNA is polycistronic. However, no regulatory genes or regulating proteins such as those found in bacteria have been identified as participants in the control of transcription.

One other interesting aspect of the T4 life cycle concerns the self-assembly of the constituent parts of the complex capsule and tail fibers. At least 50 different genes are involved at some point in the synthesis of these

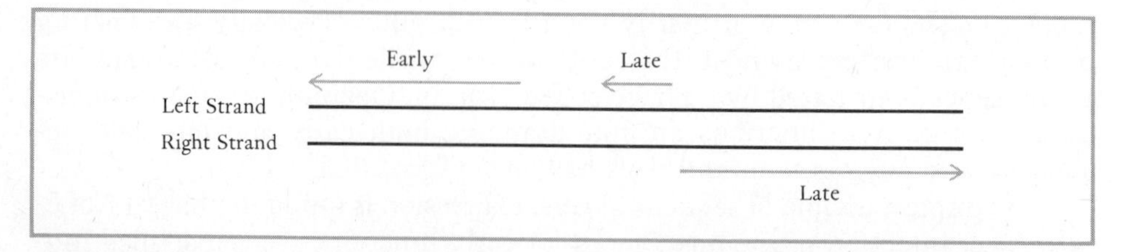

Figure 18–10. Overall pattern of transcription of early and late genes from opposite DNA strands of the T4 genome.

structures which are produced independently, but simultaneously during the late phase of the lytic cycle. All evidence indicates that construction of the final capsid is not dependent on sequential gene activity, but instead proceeds spontaneously in precisely ordered steps in which the attachment of one part depends on the prior molecular assembly and structure of a preceding part, that is, tails cannot be attached to heads until the head capsule is complete, and tail fibers cannot be attached to tails until the tail has been joined to the head (Figure 18–11). This kind of sequence suggests that the joining of parts causes modifications in structure which then permit the attachment of additional parts, and so on. A somewhat parallel situation occurs in the organization of cortical structures in ciliate protozoa (see Chapter 20).

Regulation of gene expression in phage λ and other temperate phages is more complex than that found in virulent viruses. Upon entrance into a non-

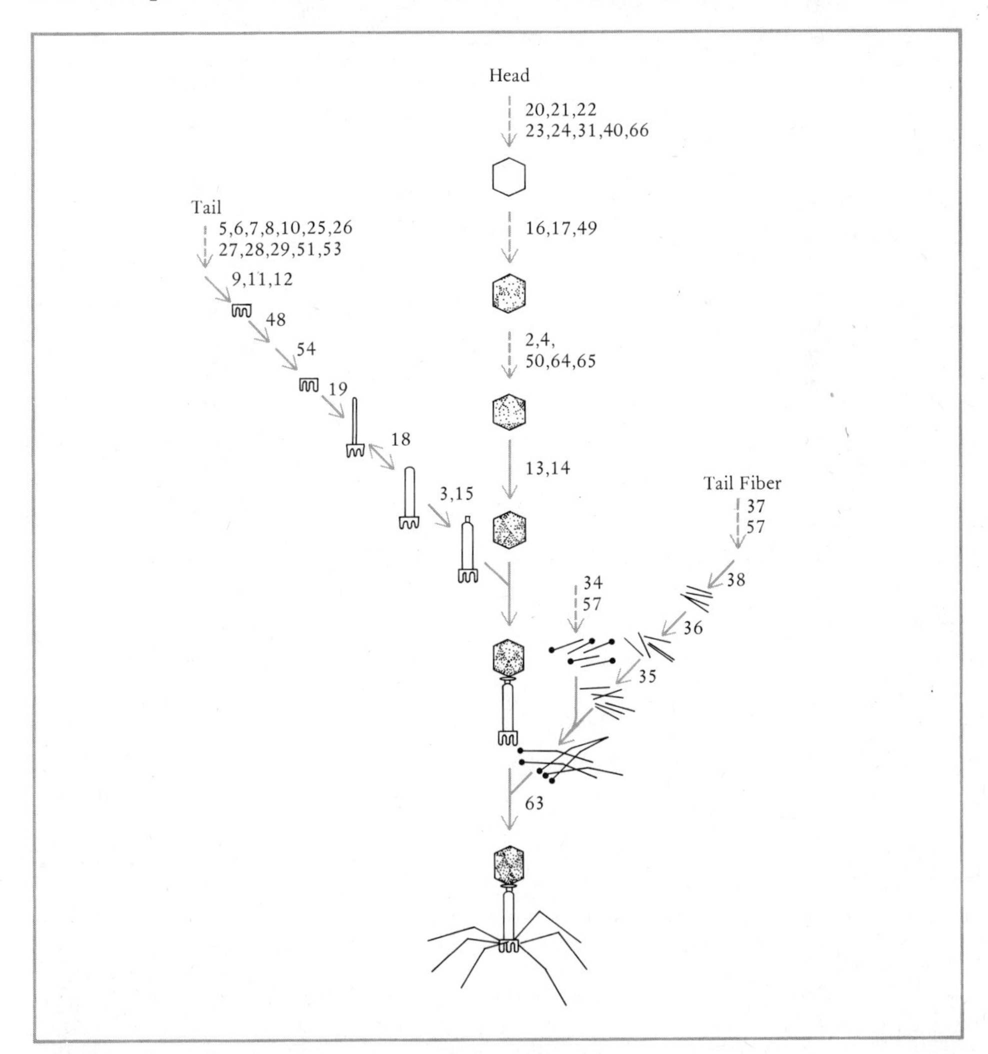

Figure 18–11. Morphogenesis in phage T4. [From W. B. Wood, in *Genetic Mechanisms of Development, Sympos. Soc. Dev. Biol., 31,* 29–46. Academic Press, New York, 1973.]

lysogenic cell, λ is confronted with a choice of two life-styles: *lysis,* whereby new particles are produced through lytic growth, or *lysogeny,* whereby the lytic cycle is suppressed and the viral DNA integrates within the host DNA as a prophage. The initial events preceding either pathway are alike. The linear λ genome circularizes through the union of the cohesive termini, and the immediate-early phase of development commences with the transcription of gene *N* by host RNA polymerase (Figure 18–12). Gene *N* protein then intro-

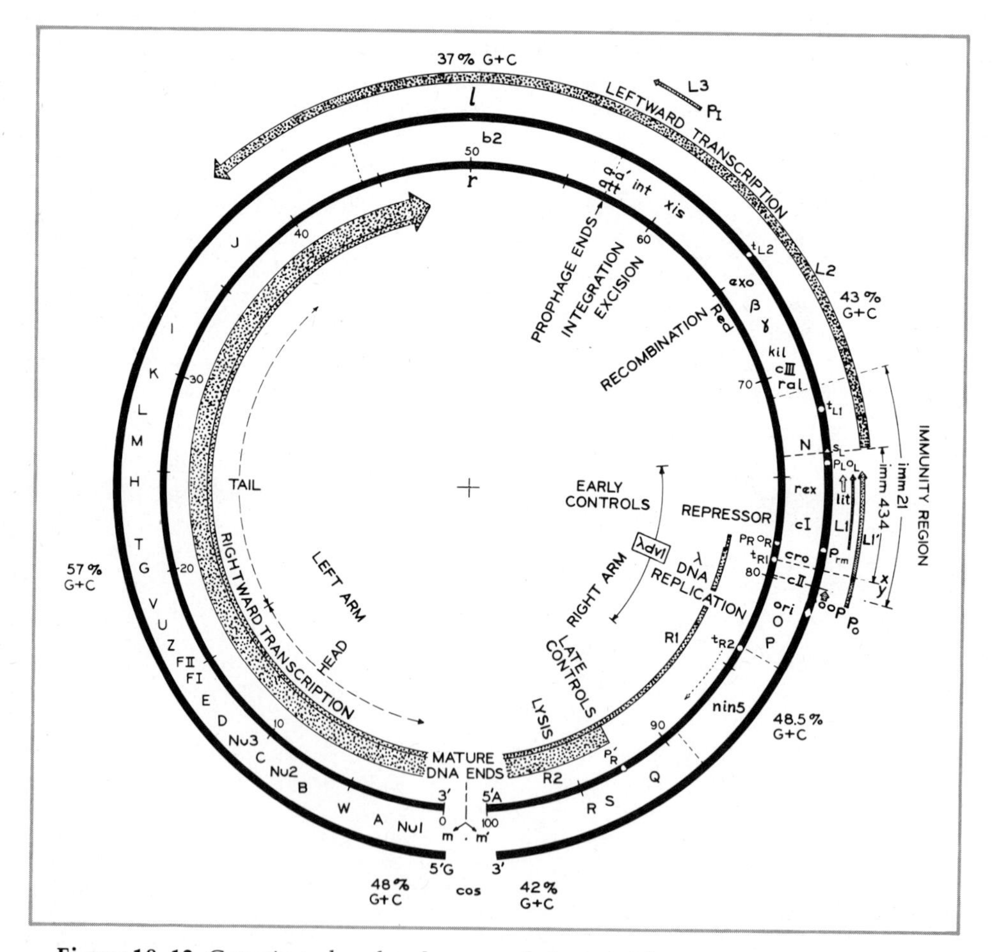

Figure 18–12. Genetic and molecular map of phage λ. The approximate base composition of each segment of λ DNA is indicated. The shaded arrows define the orientation of transcription of operons *L1, L1′, L2, L3, R1, R2, lit,* and *oop*. Within the immunity region P_RO_R and P_LO_L refer to the promotor and operator of operons *R1* and *L2,* respectively; P_{rm} is the promotor for operon *L1*. The complementary strands of DNA are referred to by *l* and *r;* cohesive termini by *cos*. Strand *l* is transcribed leftward (counterclockwise) and has a 5′-G at its left cohesive *(m)* terminus. Strand *r* is transcribed rightward (clockwise) and has a 5′-A at its right cohesive (*m′*) terminus. λ*dvl* refers to a fragment of λDNA which if present can replicate as an autonomous circular plasmid. [Courtesy of Dr. W. Szybalski and updated as of May 1977. Based on original figure by W. Szybalski et al., *J. Cell. Physiol., 74, Suppl. 1,* 33–70, 1969.]

duces the delayed-early stage by modifying host polymerase to stimulate the transcription of those genes required for replication, recombination, integration, and regulation. At this point development can proceed either to lysis or lysogeny. If gene *N* transcription is now repressed, lysogeny will occur, but if gene *N* continues to function, *N* protein will activate the transcription of gene *Q*. The late phase and lytic development then follow irreversibly, for once present, the product of *Q* stimulates the transcription of genes which specify head and tail proteins, as well as the enzyme needed for host cell lysis.

It can be seen that careful timing of the duration of gene *N* activity is pivotal in the choice between lysis or lysogeny. If the lysogenic pathway is followed, *N* protein must be evoked initially as a prerequisite for integration, but repression of gene *N* must then be imposed rapidly before late stage functions begin. If the lytic pathway is followed, repression of gene *N* must be prevented. The regulatory mechanisms involved in this choice are intricate and involve the interaction of repressor proteins with particular promotor-operator sites present in each DNA chain, both of which are used in transcription. One of these repressors, the product of gene c_I, inhibits the transcription of gene *N* and by doing so permits lysogeny to occur, while a different repressor inhibits the transcription of gene c_I and directs development towards lysis. The balance between these two repressors determines the final outcome. A detailed account of these interactions may be found in Davison (1973), Echols (1972), Herskowitz (1973), and Ptashne et al. (1976).

The complex control mechanisms exhibited by this small parasite permit its survival in the face of an uncertain supply of host cells. If host cells are numerous and growing rapidly, reproduction and the lytic cycle is adopted and the population of particles is increased, but if host cells are few and growth is stationary, lysogeny provides a route to survival. Prophage induction offers a means of escape from a host no longer able to maintain or replicate its genome. The choice among these alternative routes must depend not only on the regulatory mechanisms of the phage itself, but also on the physiological condition of the host.

THE OPERON MODEL AND EUKARYOTES

The operon concept provides an explanation for the control of coordinate enzyme synthesis in bacteria, but it does not appear to be applicable to eukaryotic organisms, since in eukaryotes related functions or enzymes belonging to the same pathway are almost always specified by genes with scattered loci. However, in some few instances in fungi the operon model has been invoked where mutations affecting the same pathway have been mapped to small regions of a linkage group. For example, the pyrimidine-3 *(pyr-3)* locus of *Neurospora* specifies two functions as demonstrated by complementation tests, and the discovery of polar mutations suggested an operon-type organization with mRNA containing information for two different enzymes. However, isolation of the protein product has indicated that both enzymatic functions are carried out by a single polypeptide, each function being relegated to a different region of the same molecule. In a similar case mutations affecting three enzymes of the histidine pathway of *Neurospora* map to the same locus *(his-3)*, and here again, isolation of the *his*-3 product has revealed a single polypeptide possessing three different enzymatic functions. Therefore, in both of these examples what initially appeared to be an operon-type cluster of genes has proven to be a single gene which specifies a single, but multifunctional product.

A third example from *Neurospora,* studied by N. H. Giles and M. E. Case, is more complex and concerns a pathway for the synthesis of a precursor for certain aromatic amino acids. Of the seven steps involved, five have been mapped to a small region (the *arom* region) of the second linkage group. Mutations causing a deficiency of one or another of these five enzymatic steps have permitted a linear ordering of these functions and polar mutations have also been identified, but whether there are five genes or only one gene whose product has five functions is not clear. All five enzymatic activities are carried out by a large aggregate (MW over 200,000) which is composed of two subaggregates, each of which also possesses the five functions. This subunit association can be viewed as one formed between separate polypeptides specified by separate genes or as one formed by polypeptides each with multiple functions. In the former case an operon-type organization would be indicated, and in the latter a single gene. A distinction between these possibilities must await further analysis of the enzyme aggregate. No controlling sites equivalent to operators have been found in the *arom* region or in any other similar regions of fungi, and it is probable that operons, at least in the bacterial form, are either rare or do not occur at all in eukaryotes.

PROBLEMS

18–1. Compare inducible and repressible pathways of metabolism.

18–2. What effects might be observed in the event of a nonpolar mutation in the *lac* operator; the *lac* promotor; gene *z;* gene *y;* gene *i;* structural gene for CAP protein.

18–3. If a polar mutation occurred in gene *galE* of the galactose operon, what effects would be anticipated?

18–4. Two independent mutations of gene *araC* are discovered. One mutation results in constitutive enzyme synthesis in the absence of L-arabinose, but in the presence of the second mutation all arabinose enzymes are repressed, even in the presence of L-arabinose. How would you interpret these phenotypes?

18–5. A mutation of gene *trpR* causes constitutive enzyme synthesis regardless of the cellular level of tryptophan. How would you explain this effect?

18–6. What is meant by the term cis-dominant? What examples can you cite? How are cis-trans tests carried out in bacteria?

18–7. What effects would be observed as the result of a missense versus a nonsense mutation in gene *z* of the *lac* operon?

18–8. Would enzyme synthesis be inducible or constitutive in:

(1) $i^+ o^+ z^+$	(4) $i^- o^c z^+$
(2) $i^- o^+ z^+$	(5) $i^+ o^+ z^+/i^- o^- z^+$
(3) $i^+ o^c z^+$	(6) $i^- o^+ z^+/i^+ o^c z^+$

REFERENCES

BECKWITH, J., AND P. ROSSOW, 1974. Analysis of genetic regulatory mechanisms. *Ann. Rev. Genet., 8,* 1.

BECKWITH, J., AND D. ZIPSER (Eds.), 1970. *The Lactose Operon.* Cold Spring Harbor Laboratory, Cold Spring Harbor, N.Y.

CALHOUN, D. H., AND G. W. HATFIELD, 1975. Autoregulation of gene expression. *Ann. Rev. Microbiol., 29,* 275.

DAVISON, J., 1973. Positive and negative control of transcription in bacteriophage. *Brit. Med. Bull., 29,* 208.

DICKSON, R. C., J. ABELSON, W. M. BARNES, AND W. S. REZNIKOFF, 1975. Genetic regulation: The *lac* control region. *Science, 187,* 27.

ECHOLS, H., 1972. Developmental pathways for temperate phage: Lysis vs lysogeny. *Ann. Rev. Genet., 6,* 157.

ENGLESBERG, E., AND G. WILCOX, 1974. Regulation: Positive control. *Ann. Rev. Genet., 8,* 219.

HERSKOWITZ, I., 1973. Control of gene expression in bacteriophage Lambda. *Ann. Rev. Genet., 7,* 289.

LEVINE, M., 1969. Phage morphogenesis. *Ann. Rev. Genet., 3,* 318.

PTASHNE, M., K. BLACKMAN, M. Z. HUMAYUN, A. JEFFREY, R. MAURER, B. MEYER, AND R. T. SAUER, 1976. Autoregulation and function of a repressor in bacteriophage Lambda. *Science, 194,* 156.

REZNIKOFF, W. S., 1972. The operon revisited. *Ann. Rev. Genet., 6,* 133.

STUDIER, W. F., 1972. Bacteriophage T7. *Science, 176,* 367.

Chapter 19

GENE REGULATION AND CHROMOSOME ORGANIZATION IN EUKARYOTES

CONTROL at the primary level of transcription in eukaryotes presents a problem in coordination of a magnitude and scale not encountered in bacteria, for the loci of genes which have related functions or which participate in the development and expression of a single trait are almost always noncontiguous and scattered through the linkage groups. Added to this problem are the further complexities presented by the two general types of transcriptional control distinguishable in higher organisms: *long-term regulation,* which leads to the assumption and maintenance of the morphological and biochemical specializations evident in different tissues and organs, and *short-term regulation,* which permits a response to immediate environmental stimuli. The underlying mechanisms which bring about coordinated transcription and both long- and short-term regulation are almost completely unknown, but some appreciation of the complexities of these processes can be gained by examining a series of examples which illustrate the kinds of control phenomena encountered in higher organisms.

DIFFERENTIATION AND THE CONSTANCY OF THE GENOME

In the higher eukaryotes, particularly the higher animals, fertilization is followed by a period of embryonic development during which the organ systems of the body are laid down and cells are differentiated into specialized types to form bone, muscle, kidney, liver, and so on. Differentiation is normally irreversible so that once a cell has been programed to form, for example, intestinal epithelium, that cell and all its descendents will assume the specialized structure and function of an intestinal epithelial cell and no other.

Two theories have been postulated to account for the extreme specializations and disparate functions exhibited by the cells of animals. (1) All genes except those expressed in a particular kind of differentiated cell are lost during the course of development, and (2) all genes originally present in the zygote are also present in all somatic cells, and the assumption of specialized functions reflects the imposition of regulating controls whereby particular genes are activated or repressed in the different tissues. Of these two explanations, the second has been shown to be correct, for proof has been obtained that the identical genome is present in all somatic cells, regardless of type.

The first demonstration of the constancy of the genome was carried out in 1952 by R. Briggs and T. J. King who successfully transplanted a nucleus from a specialized somatic cell into an enucleated egg of the frog *Rana pipiens* and found that the recipient egg could develop into a complete, normal organism, a process that requires the full complement of genes.

More recent experiments by J. B. Gurdon, utilizing the toad *Xenopus laevis,* have confirmed and extended these results. Unfertilized eggs obtained from female toads that have been treated with hormones are exposed to ultraviolet light to inactivate the egg nucleus. Each such egg is then injected with a single somatic nucleus obtained from a differentiated cell of brain, skin, intestine, or other organ. Although some of the resulting embryos are abnormal, others achieve complete development, metamorphose to the adult, and breed successfully. *Xenopus* is particularly useful for such studies since a mutation causing the deletion of the nucleolus organizer can be used as an identifying marker in interphase nuclei. Nuclei from individuals heterozygous for this mutation *(+/0-nuc)* contain only a single nucleolus as opposed to two nucleoli in +/+ individuals. Ordinarily, *+/0-nuc* nuclei are injected into eggs from +/+ females, and the presence of a single nucleolus in the cells of the developing embryo is proof that the cells were derived from mitotic divisions of the transplanted nucleus, rather than from an unkilled egg nucleus. The overall results of these studies have shown that somatic cells, no matter how specialized, contain all of the genes necessary for the formation and function of every organ and cell type characteristic of the species. In higher plants (carrot, tobacco) similar findings have been obtained, for complete adults which form seed have been grown from single phloem cells maintained in tissue culture. The process of differentiation must, therefore, be attributed to controls on transcription selectively applied to the identical genomes present in cells of the different tissue types. Another indication that transcriptional control is involved is the fact that less than 10 percent of the DNA present in the nucleus of a differentiated cell is active in RNA synthesis.

REGULATION AND THE CYTOPLASM

Experiments with nuclear transplantation have also demonstrated the controlling role of the cytoplasm in turning genes on and off. For example, if a nucleus from skin, which normally produces keratin, is transplanted to an egg, genes for keratin are no longer expressed, but as development proceeds, genes for hemoglobin, myosin, and other cell products are reactivated. Also if brain nuclei active in RNA synthesis are placed in a mature egg in which chromosome condensation has occurred and both RNA and DNA synthesis have ceased, the transplanted nuclei immediately stop RNA synthesis and their chromosomes also condense. If these brain nuclei are transplanted to an egg which has ceased RNA synthesis and completed meiosis, but is initiating DNA synthesis, the injected nuclei also cease RNA synthesis and initiate DNA synthesis. We must, therefore, conclude that the egg cytoplasm contains components capable of switching on or off the fundamental processes of chromosome replication and transcription.

An even more spectacular demonstration of this cytoplasmic capability can be seen in heterokaryons formed by the fusion of actively transcribing mouse (or human) fibroblast cells with mature chicken red blood cells in which all replication, transcription, and translation has ceased and whose nuclei are totally condensed and inactive. After fusion, which entails little incorporation of red cell cytoplasm, the condensed chick nuclei undergo enlargement, the chromatin becomes dispersed, RNA synthesis is initiated, and the nucleolus reappears. Shortly thereafter, chick-specific antigens can be detected on the surface of the heterokaryon, indicating that translation of chick-specific mRNA has occurred. These observations strongly suggest that the reactivation of the red cell nuclei and the renewed expression of the chick genome have been induced by controlling factors present in the fibroblast cytoplasm.

Cytoplasmic Control in Amphibian Eggs

Control by cytoplasmic regulating molecules has been intensively studied in the early developmental stages of amphibia, where it appears that materials contributed to the egg cytoplasm prior to fertilization play a directing role in the programing of early development (see also Chapter 20). Although the nature of these substances is unknown, they are produced under the direction of the maternal genotype and appear to assume a specific, but unequal distribution in the cytoplasm of the unfertilized egg. Because of this differential distribution, cleavage nuclei produced by mitotic divisions of the zygote are exposed to different kinds of regulating molecules which affect them in different ways. As a result, the cells comprising a stage as early as the blastula are already programmed with respect to their general fate in the new individual (that is, endoderm, ectoderm, and so on), and the subsequent movements and infoldings of cells which characterize later embryonic stages permit the further development and the defined expression of previously established potentials.

Two major kinds of maternal contributions to the egg have been identified in amphibia: informational RNA and ribosomal RNA.

Informational RNA. The primary oöcytes of amphibia and many other vertebrates undergo a period of intensive growth during a lengthened diplotene stage of the first meiotic prophase. At this stage homologous chromosomes have already undergone synapsis and crossing over and are held together only by chiasmata. During the growth phase these diplotene chromosomes attain giant size, becoming greatly extended up to lengths of 800 μm. Each chromosome consists of two closely associated sister chromatids, and linearly arranged along each are small, beadlike thickenings (chromomeres) from which extend paired loops, one from each chromatid. The overall appearance is hairy or brushlike, and therefore the name *lampbrush chromosome* has been applied (Figure 19–1). The lateral loops of the lampbrush chromosomes are coated with RNA complexed with protein, and it is inferred that the evident intense transcription taking place provides a store of informational RNA which will subsequently direct the course of early development (see Figure 15–2).

In amphibian embryos mRNA synthesis does not begin until the blastula stage has been attained, but protein synthesis begins almost immediately after fertilization. The messages being translated in the zygote are presumably those which were transcribed during the lampbrush stage and stored in stable form for future use upon activation of the egg by the sperm. Since initial programing of the egg occurs prior to mRNA synthesis by the embryo, it appears that some of the molecules synthesized during oöcyte growth provide for the early regulation of development. At the end of the growth phase the lateral loops of

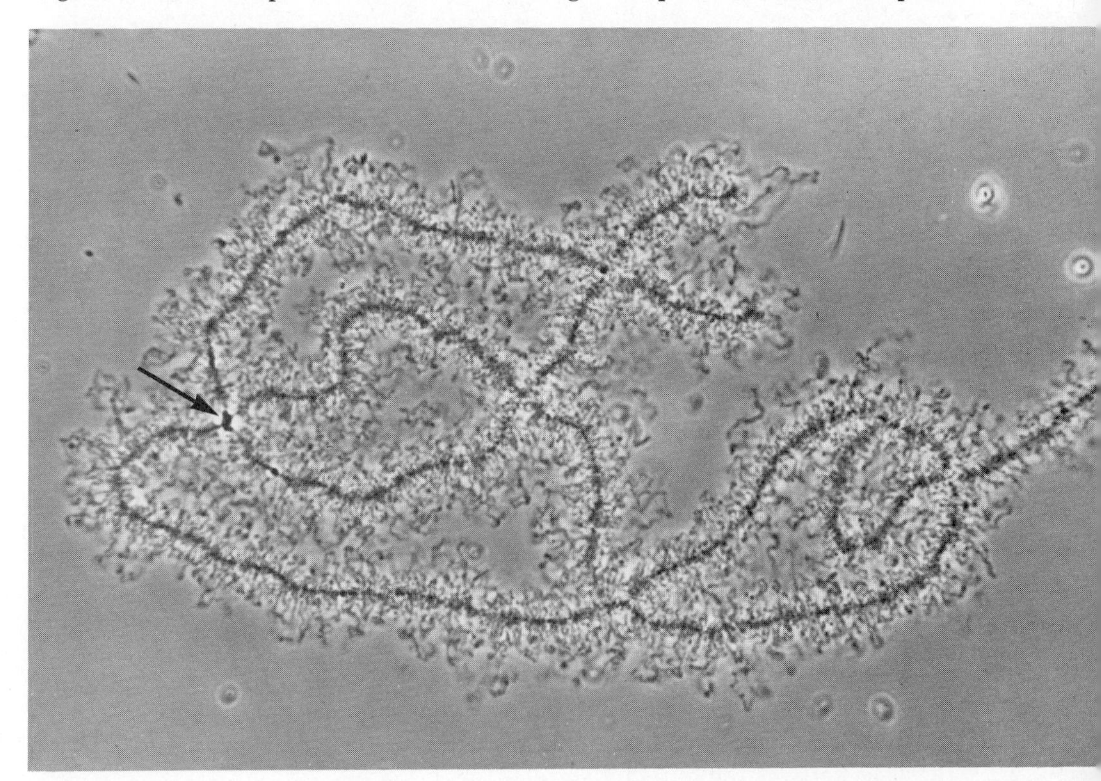

Figure 19–1. An isolated lampbrush chromosome of the newt, *Triturus,* composed of two homologues held together by three chiasmata and the fused centromeres (at arrow). Each homologue consists of two sister chromatids which give rise to paired loops at a series of chromomeres. (520×) [From J. G. Gall, In *Methods in Cell Physiology,* D. M. Prescott, Ed., Academic Press, New York, 1966.]

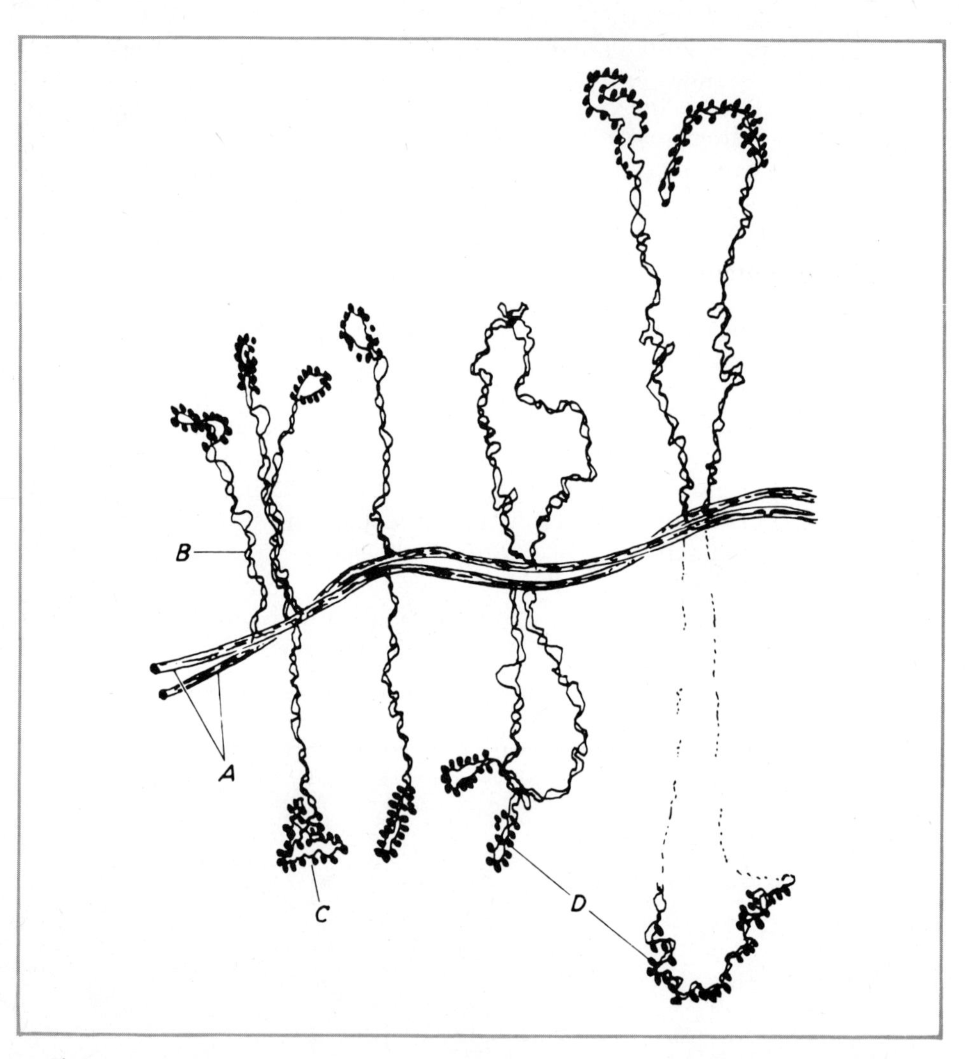

Figure 19–2. Diagrammatic interpretation of chromosome structure in the human primodial oöcyte, based on electron microscopy. For the sake of clarity, only a few lateral projections are shown. A, chromosome axis; B, coiled fibrils emerging from the axis and forming lateral projections; C, clusters of granules at the ends of the projections; D, loops formed by the reflection of the lateral projections back to the axis. [From T. G. Baker and L. L. Franchi, *Chromosoma, 22,* 358–377, 1967.]

the lampbrush chromosomes are retracted, RNA synthesis ceases, and chromosome condensation follows in preparation for the first meiotic division.

Lampbrush chromosomes have also been described in human primary oöcytes (Figure 19–2), but the role of the informational RNA produced during the diplotene stage is less clear, since mammalian zygotes begin mRNA synthesis as early as the four-cell stage.

Gene amplification and rRNA. In the amphibia a second major maternal contribution to the oöcyte consists of rRNA which is synthesized in enormous quantities during the growth phase. Since the embryo does not even initiate its own rRNA synthesis until the gastrula stage, and significant production does not occur until the hatching tadpole stage, the ribosomes present at fertilization must be sufficient to meet all cellular demands for protein synthesis prior to these stages. As a seeming response to this demand, the nucleolus organizer undergoes amplification to form up to 1500 additional micronucleoli which are present in the nucleus, free and unattached to the chromosomes (Figure 19–3). The additional nucleoli are circular double-stranded DNA molecules of varying size and contain a few to many templates for 18 S and 28 S RNA. These extra copies of ribosomal genes are formed prior to the diplotene stage, possibly by a rolling circle mechanism, and are intensely active in rRNA synthesis which is accompanied by ribosome formation.

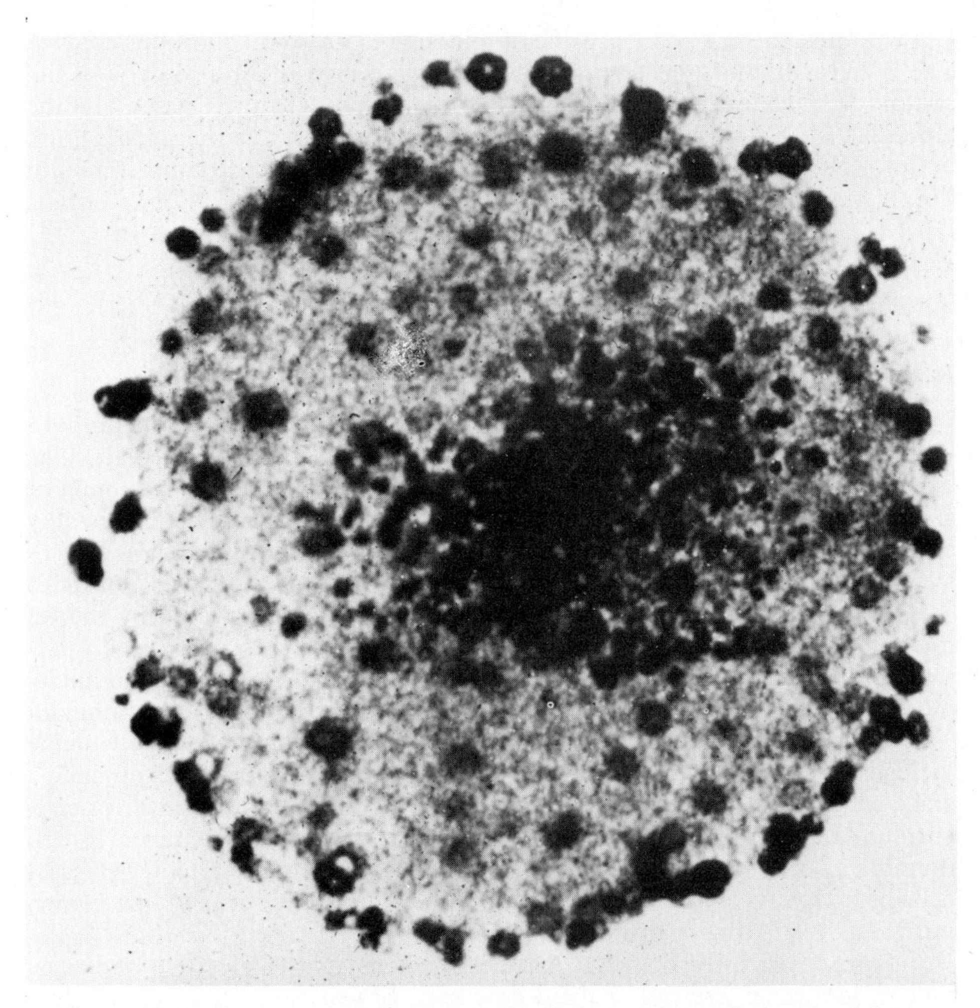

Figure 19–3. Photomicrograph of an isolated germinal vesicle (nucleus) from an oöcyte of *Xenopus laevis*, diameter about 400 μm. The deeply stained spots are some of the hundreds of additional nucleoli. [From D. W. Brown and I. B. Dawid, *Science, 160,* 272–280, 19 April 1968. Copyright © 1968 by the American Association for the Advancement of Science.]

When the egg is ready for ovulation rRNA synthesis ceases, and with the dissolution of the nuclear envelope at the first meiotic division the extra nucleoli are released to the cytoplasm and disappear. We should note that ribosomes also contain a 5 S RNA molecule which is transcribed coordinately with the 18 S and 28 S components. The genes for 5 S rRNA are not amplified, probably because they are highly reiterated, being present in approximately 20,000 copies per haploid genome (in *Xenopus*). This multiplicity would permit synthesis of the 5 S component to keep pace with the synthesis of 18 S and 28 S RNA. Amplification of rRNA genes occurs only during oögenesis and although the signals which initiate its onset are unknown, it represents a clear-cut example of genome regulation.

Amplification of specific DNA sequences is not limited to amphibia. In the giant polytene chromosomes of the salivary glands of sciarid midges, several chromosome bands are capable of releasing DNA in the form of micronucleoli, and a similar phenomenon has been described in footpad cells of the fleshfly *Sarcophaga.* Ribosomal genes are not involved in these cases, and the amplified DNA is thought to provide additional templates for the transcription of an RNA message needed in large quantities at a particular time in development.

EXAMPLES OF CONTROL MECHANISMS IN EUKARYOTES

Regulation in the Drosophila Egg

The early development of *Drosophila* is characterized by synchronous mitotic divisions which occur at approximately 10 minute intervals. The cleavage nuclei so formed are not contained in cells, but instead are found in small "islands" of cytoplasm which become distributed throughout the yolky egg. The early embryo can thus be considered a large multinucleate cell, or syncytium, in which the replication of chromosomes and the stages of mitosis are subject to a single, uniform timing mechanism, as evidenced by the remarkable synchrony with which they proceed through these stages.

At the end of the ninth cleavage division (512-nuclei stage), the nuclei migrate to the periphery of the egg, and the initially thin outer layer of cytoplasm broadens to a thick band. After three more synchronous mitoses have yielded over 4000 nuclei, cell membranes form between nuclei to produce a stage called the blastoderm which can be described as a single layer of cells surrounding the yolky interior (Figure 19–4). With the appearance of cell membranes the synchrony of division cycles ceases, and different groups of cells begin to divide at different rates as if they were now subject to their own individual regulating or timing devices.

Since practically no RNA synthesis occurs during cleavage and blastoderm formation, control of these processes, as well as later developmental events, is attributed primarily to substances synthesized under the direction of the maternal genotype. Prior to fertilization the *Drosophila* egg contains ribosomes, yolk, and other materials sufficient to support the embryo to the time of hatching. All of these substances are elaborated, not by the oöcyte itself, but by nurse cells which surround and pour their products into the egg across cytoplasmic bridges. Among these products are informational and regulating molecules of an unknown nature which appear to control the nuclear cycle during cleavage and to lead to blastoderm formation. We should note, however, that at least some control must be attributed to the genotype of the zygote, for nuclei of nullo-X individuals fail to form a complete blastoderm.

Experimental evidence has shown that the cells of the blastoderm are already determined bilaterally and in the anterior-posterior direction as to the specific structures they and their descendents will form, not only in the larva, but also in the adult insect. Accordingly, it appears that controlling molecules are spatially distributed in a precise pattern in the cytoplasm such that specific fates are imposed on cleavage nuclei that enter their area of influence. As a specific example, the posterior pole of the *Drosophila* egg contains granules as visible cytoplasmic inclusions; cleavage nuclei which happen to enter this region upon migration to the periphery of the egg are immediately enclosed in cell membranes, well in advance of cell membrane formation in all other regions of the blastoderm (Figure 19–4). Most of these cells, set aside before all others, give rise to the germ cells of the adult.

Structures of the adult insect, such as antennae, compound eyes, wings, legs, genitalia, and so on, are derived from discrete groups of cells called *imaginal discs* which are prominent features of the internal anatomy of the late stage larva and which undergo rapid growth and differentiation during pupation (Fig-

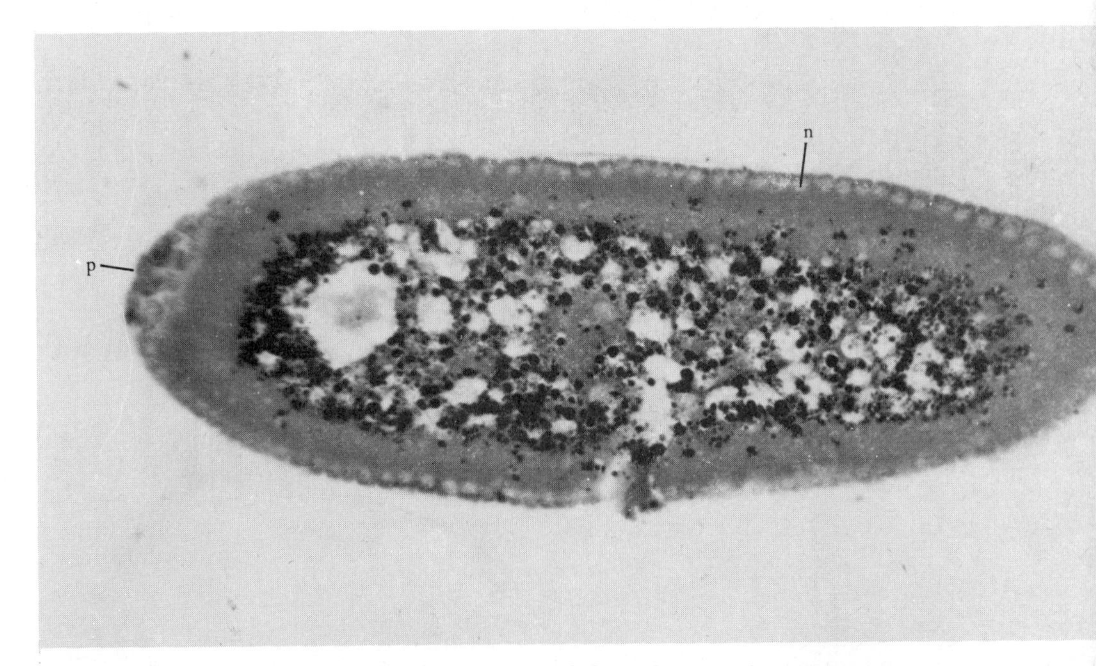

Figure 19–4. Longitudinal section of a *Drosophila* embryo in the blastoderm stage. The embryo consists of a thick layer of cytoplasm surrounding the yolk. Nuclei (n) have migrated to the periphery, but cell membranes between them have not yet appeared. The pole cells (p) from which the germ cells of the adult are derived are already formed posteriorly. (200×)

ure 19–5). The origin of these discs can be traced back to the primordial cells of the blastoderm, and it has even been possible to estimate the approximate number and relative locations of the blastodermal cells whose descendents will eventually give rise to adult body parts. Most such studies have employed gynandromorphs in which one chromosome of a pair, heterozygous for wild type and recessive marker genes, has been lost during mitosis at some stage of development. An example is the unstable ring X chromosome (see Chapter 7), whose loss in a female gives rise to X0 cells having a male phenotype and in which all genes of the remaining X chromosome are expressed.

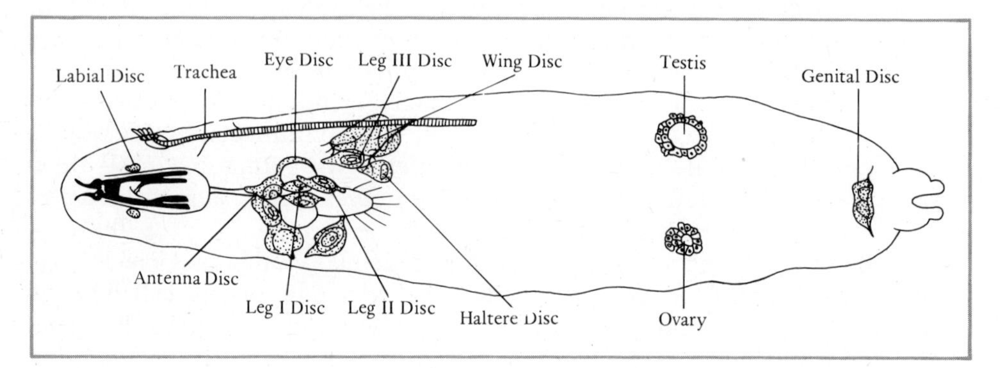

Figure 19–5. The locations of imaginal discs in the mature *Drosophila* larva. [From D. Bodenstein, in *Biology of Drosophila,* M. Demerec, Ed. John Wiley, New York, 1950.]

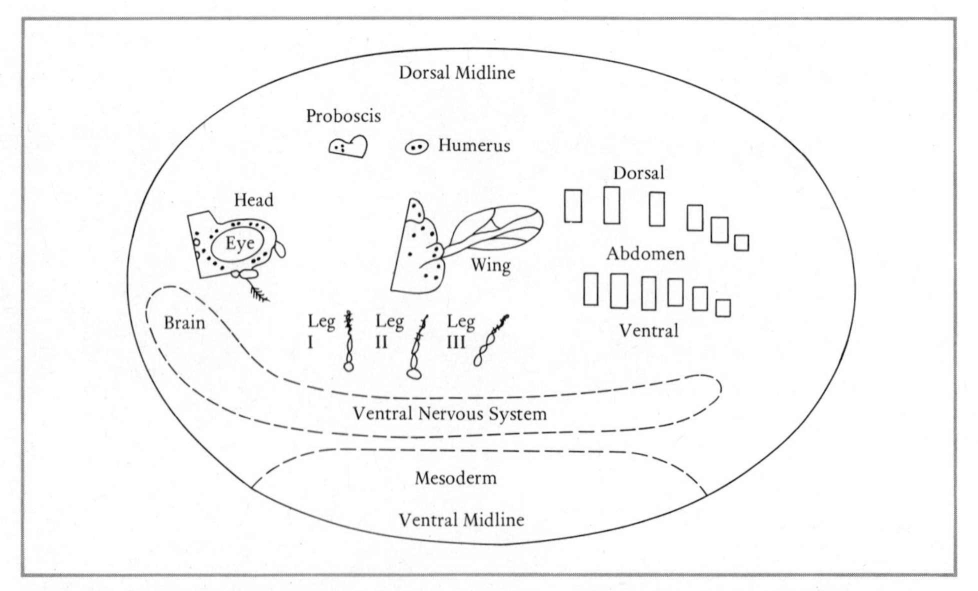

Figure 19–6. Pictorial sketch of the external parts of an adult fly derived from various sites on the blastoderm. The diagram represents only the right half of the fly; the left parts are formed from the other half of the blastoderm. Dotted lines indicate areas which give rise to the nervous system and the mesoderm. [From Y. Hotta and S. Benzer, *Nature, 240,* 527–535, 1972.]

The location on the blastoderm surface of cells destined to give rise to different adult structures is determined by observing how often a given mosaic region, arising as the result of chromosome loss, includes one or the other or both of two different adult parts, for example, wing and leg, eye and antenna, and so forth. If the two parts are frequently included within the same mosaic region, it can be concluded that the sites occupied by their respective antecedent cells were close together on the blastoderm surface, whereas if the two parts are rarely or not at all included in the same mosaic region, their sites must be far apart. By making such comparisons for many parts of the body, the relative positions occupied by cells programed to give rise to different structures can be estimated, and on the basis of such estimates a "fate map" of the surface of the blastoderm has been prepared (Figure 19–6).

It has also been shown that the cells of the imaginal discs derived from specific regions of the blastoderm are themselves fully determined to assume a defined role in the structure of the future adult organ. Although early regulation determines fate in terms of the kind of organ and its parts which will later develop, the final and detailed structure of any part depends on genotype. For example, from the blastoderm stage onward a given group of cells and their descendents might be destined to form a wing, but whether the wing is wild type, vestigial, scalloped, crossveinless, or some other phenotype, would depend on the activity of the particular alleles present in the genotype of the zygote.

We should note that blastoderm fate maps are also being developed for simple behavioral traits involving legs, wings, or other structures whose movements or functions are altered by mutation. Gynandromorphs are again used in these studies which attempt to correlate within a mosaic region the presence or absence of a trait, such as leg shaking, with the presence or absence of some recognizable body surface mutation whose locus is linked to that of the behavior mutation. The frequency with which one, but not the other trait occurs in the same structure can be used to estimate the distance apart on the blastoderm surface of their antecedent cells. As might be expected, characters such as leg shaking generally involve sites within the region of the blastoderm that give rise to the ventral nervous system. Study of the genetics and development of behavior patterns is only just beginning, and we can expect significant advances in this area in the future.

The programing imposed during embryogenesis is widely separated in time from the actual differentiation of an imaginal disc into a fully formed structure, an event which occurs much later, during the pupal period. Therefore, it is evident that early regulation involves sustained, long-term control which ensures the subsequent development of a given kind of structure while permanently repressing the potential to form any other.

The sustained control imposed upon the fate of imaginal discs has been demonstrated experimentally by E. Hadorn and others. An imaginal disc can be dissected from a larva and transplanted into the abdomen of an adult for culture. In this location the disc lives and grows, but does not undergo differentiation because adults do not produce ecdysone, the hormone that initiates pupation. Such discs can be transplanted repeatedly from adult to adult, and at any time their potential to form a given structure can be assayed by transfer to a larva. When the larva pupates and undergoes metamorphosis to the adult, the implanted disc likewise differentiates to an adult structure which can be dissected from the host and examined for realization of a previously established potential. Such studies have demonstrated that leg discs continue to differentiate into legs, eye discs into eyes, and so on, even after long culture in the abdomens of successive adults, indicating that controls imposed during early embryogenesis are still effective.

On occasion, however, after repeated transplantations an alteration in fate is observed, and a disc previously shown to develop into one kind of structure, for example, leg, develops instead into a different structure, such as wing. The cause of this change in fate is unknown, but we can infer that a shift in control has somehow occurred to activate genes for wing structures and repress those for leg structures. These sudden changes in potential, called *transdetermination,* are similar to those observed in the case of *homeotic mutations* which cause the transformation of one structure into another, embryologically homologous structure. For example, mutations at the bithorax *(bx)* locus of *Drosophila* cause the third thoracic segment to be structurally modified so as to resemble the second thoracic segment, and with certain combinations of *bx* mutations a four-winged, rather than a normal two-winged fly is produced (Figure 19–7). Another example is the mutant *aristopedia,* where a large bristlelike structure (arista) of the antenna is transformed into a leglike appendage.

Figure 19–7. Photograph of a *Drosophila* male containing a combination of bithorax mutations which cause the transformation of the metathorax into a mesothorax-type, wing-bearing segment. A four-winged fly is the result. [From E. B. Lewis, *Amer. Zoologist, 3,* 33–56, 1963.]

X Chromosome Inactivation

An outstanding example of regulation by the zygote, previously discussed in Chapter 7, is the inactivation of all but one of the X chromosomes of mammals during very early development. Since sex chromosome constitution is determined only at fertilization, the number of X chromosomes present in the zygote itself must be the decisive factor in this phenomenon. In addition, the mechanisms leading to inactivation are separate from those inducing morphogenesis, for somatic cells of many different types exhibit Barr bodies (see Figure 7–5). X chromosome inactivation is also selective with respect to germ line versus somatic cells, for both X chromosomes are active in the oöcytes of females, but in spermatocytes of males of many species, including nonmammals, the single X is condensed. We can therefore infer not only that cells of the germ line contain substances which distinguish them from all other cells of the organism, but these cells must also be specifically differentiated in response to the kinds of sex chromosomes present.

In mammals the condensation of one of the two X chromosomes of normal females or of all additional X chromosomes beyond one in males and females with abnormal sex chromosome complements is related to dosage compensation whereby the activity of X-linked loci is equalized in the two sexes (see Chapter 7). It will be recalled that dosage compensation also occurs in *Drosophila,* in this case by a decrease in transcriptional activity on the part of the two X chromosomes of females and an increase in this activity by the single X of males, control being exerted by X-linked and autosomal genes. The molecular nature of this evident control which adjusts the rate of RNA synthesis in the X chromosomes of the two sexes is unknown.

X chromosome inactivation is also involved in position effects which in mammals result from the transposition of autosomal genes to the X chromosome (see Chapter 9). Analysis of a number of such translocations in the mouse has permitted the tentative location of a controlling site present in the X chromosome from which the suppression of gene activity appears to extend along the introduced autosomal segment. Such spreading effects have also been observed in *Drosophila* when normally euchromatic genes are brought into the vicinity of broken heterochromatin, especially that associated with the constitutive heterochromatin adjacent to the centromere.

Y Chromosome Activity

Differential activity of the Y chromosome is also evidence for the precise switching on of genes essential for normal germ cell formation and function. In *Drosophila* species the Y chromosome does not determine sex and appears as a condensed inactive element in somatic cells. In primary spermatocytes, however, the Y chromosome engages in active RNA synthesis with the formation of lampbrush-type loops (Figure 19–8). The informational RNA so produced is evidently necessary for the normal differentiation of functional sperm, for X0 males are sterile.

In addition, when gross chromosomal aberrations are present in the paternal genotype, such that spermatids resulting from the meiotic divisions lack one or even most of the chromosomes, a condition ordinarily lethal, such

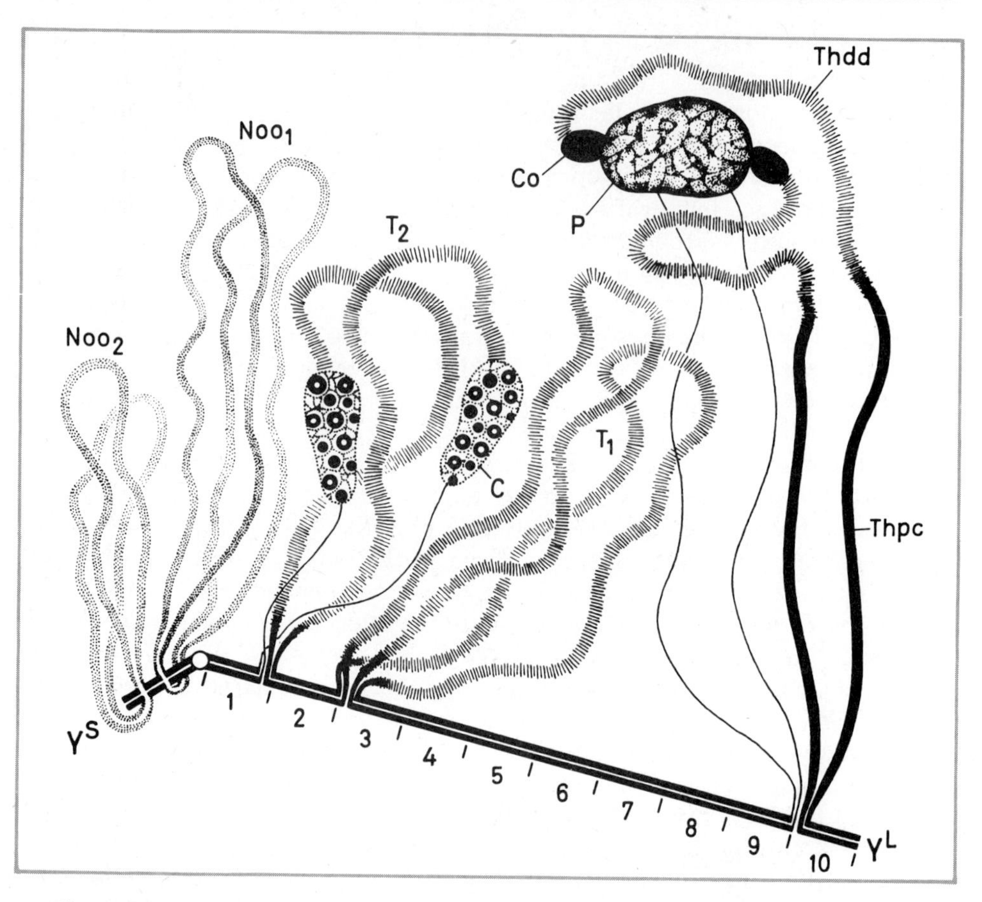

Figure 19–8. Diagram of the Y chromosome of *Drosophila hydei* in the lampbrush state in a spermatocyte, with loci of loop-forming sites. The loops have been variously named as nooses 1 and 2 (Noo_1, Noo_2); clubs (C); pseudonucleolus (P) and its cones (Co); threads with distal diffuse (Thdd) and proximal compact (Thpc) regions. Y^S and Y^L refer to the short and long arms of the Y chromosome. [From O. Hess, *Genetics, 56,* 283–295, 1964.]

spermatids will still undergo normal differentiation into fully functional sperm. This finding indicates that substances which direct the course of spermiogenesis are synthesized in the primary spermatocyte and distributed via the cytoplasm to the four products of meiosis. The intense transcriptional activity of the Y chromosome during growth of the primary spermatocyte suggests that at least some, and perhaps all, of these directing molecules are provided by Y-linked loci.

Gene Compensation and Magnification

In *Drosophila* the X and Y chromosomes each bear one of the two nucleolus organizers and from the percentage of DNA that hybridizes with rRNA, it is estimated that each nucleolus organizer contains between 200 and 250 genes that code for 18 S and 28 S rRNA, for a total of 400 to 500 such genes in normal males and females. It might be expected that in X0 males or in females where one X chromosome is deficient for the nucleolus organizer, only half the number of ribosomal genes would be present. However, it has been shown that the DNA isolated from such individuals contains nearly the same number of rRNA genes as does the DNA from wild type flies, indicating that a compensatory replication of these genes has occurred. Although the source of this *gene compensation* must be the normal X chromosome present in these individuals, the extra gene copies are not inherited by the next generation, for XX daughters and XY sons of nucleolus-deficient mothers exhibit the normal, not the amplified number of rRNA genes. It appears that the induced copies do not become integrated into the normal X chromosome. In addition, only the X chromosome engages in compensatory replication, since additional copies of rRNA genes are not found in individuals containing a normal Y plus a nucleolus-deleted X chromosome. Why the X and Y chromosomes differ in this ability is not known.

A second phenomenon, called *magnification,* also concerns rRNA genes of *Drosophila* and provides an explanation for the frequent reversions back to wild type of mutations called bobbed *(bb),* which are partial deletions of the nucleolus organizer. It has been found that when an X chromosome carrying *bb* is present with a Y chromosome also partially deficient for the nucleolus organizer, additional copies of rRNA genes are inherited as evidenced by the presence of both the wild type phenotype and the wild type rDNA content in descendent generations. Careful studies have shown that "magnified progeny," having more ribosomal genes than their fathers, are accompanied by a reciprocal class of lethal progeny having a significantly reduced rDNA content. K. Tartof has proposed that the simultaneous occurrence of magnification and reduction is the result of unequal crossing over between sister chromatids during the mitotic divisions which precede meiosis. It will be recalled (see Figure 9–11) that as a consequence of unequal crossing over, both duplication and deficiency chromosomes are produced. This explanation accounts for changes in the number of rRNA genes in germ line cells, but it does not provide an insight into the regulatory mechanism which initiates compensation. This mechanism, whatever its nature, is endowed with the ability to sense deficiencies in rRNA and to respond appropriately to make up such deficiencies in somatic cells.

Chromosome Puffs

As described in Chapter 9, the giant polytene chromosomes found in the larval salivary chromosomes of Dipterans are each composed of a pair of homologues which together may contain up to 1000 or more chromonemata held in precise parallel alignment. Since the transverse bands evident in these chromosomes are regions where the individual chromonemata are tightly coiled and condensed, the bands are considered to be the polytene equivalents of the chromomeres seen in single chromatids. Both cytological and genetic evidence indicate the presence of only one function, or complementation group, per band, that is, one function per chromomere, and if a single function is equated with a single gene, then the crossbands of the salivary chromosomes represent single gene sites. Studies of polytene chromosomes have demonstrated that the initiation of RNA synthesis by any band is accompanied by the swelling of the band to form a puff, the largest puffs being referred to as Balbiani rings (Figure 19–9). The puffed condition can be interpreted as the result of the uncoiling and extension into delicate loops of the previously compacted chromomeres which together form the band (Figure 19–10). Since polytene chromosomes contain numerous individual chromonemata, the uncoiling of all

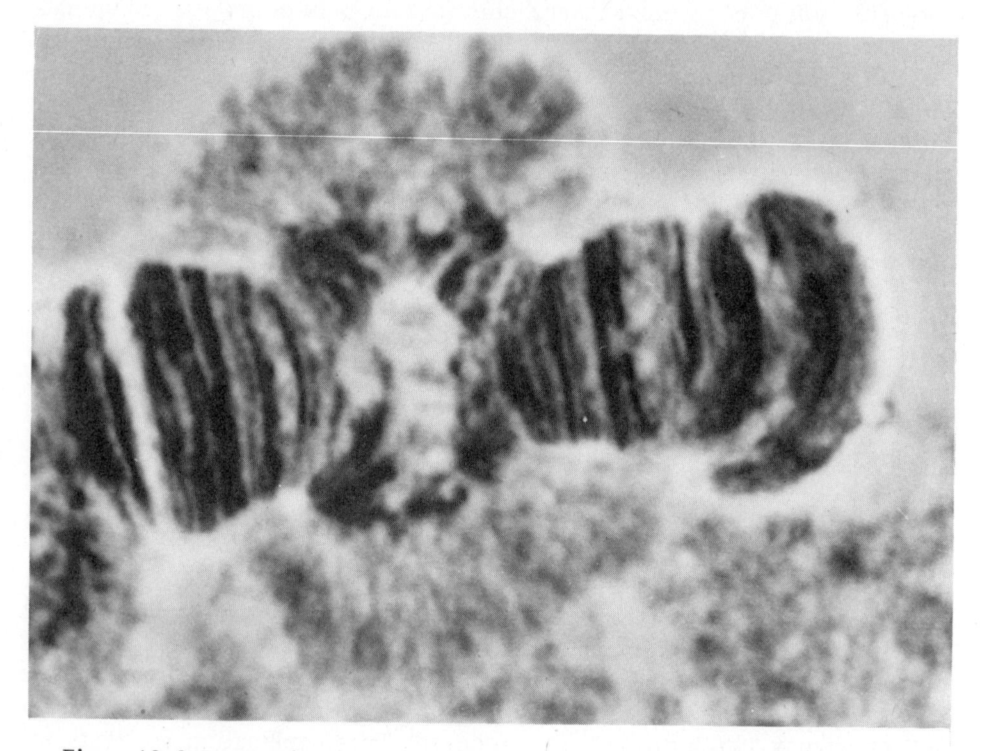

Figure 19–9. Proximal end of chromosome 4 from the salivary gland of *Chironomus pallidivittatus* with a large Balbiani ring (giant puff). Diameter of the puff is 25 mμ. [From U. Grossbach, *Cold Spring Harbor Sympos. Quant. Biol., 38,* 619–627, 1974.]

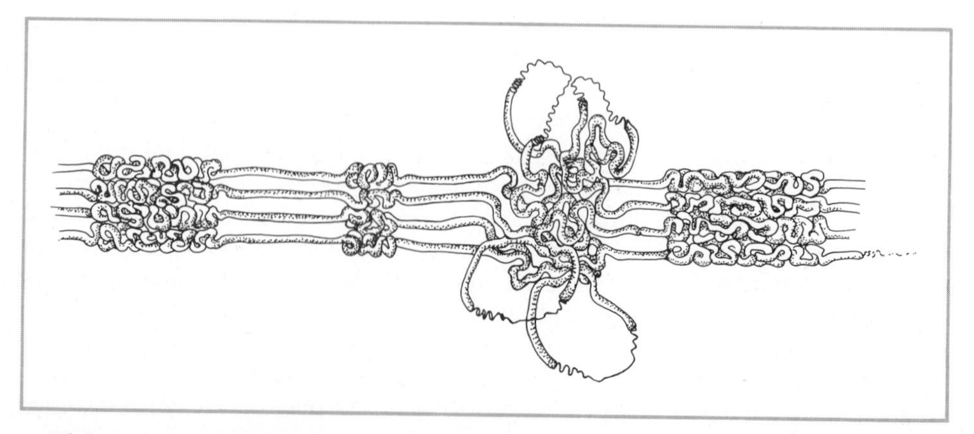

Figure 19–10. Model of a giant polytene chromosome illustrating a possible mode of puffing in one of the bands. [From E. J. DuPraw and P. M. M. Rae, *Nature, 212,* 598–600, 1966.]

of these gives rise to a greatly swollen region. The extension into loops for transcriptional activity is similar to that observed in the lampbrush chromosomes of amphibia.

Cytological evidence that puff formation is associated with gene activity has been obtained by W. Beerman in studies of two crossable species of midges, *Chironomus tentans* and *Chironomus pallidivittatus*. The salivary cytology of the two species differs in regard to a specific puff on the fourth chromosome. These species also exhibit a correlated difference with regard to the kind of secretions produced by one region of the salivary gland: *C. tentans*, without the puff, has a clear secretion and *C. pallidivittatus*, with the puff, has a granular secretion, clear versus granular being inherited as allelic states of a fourth chromosome gene. Although heterozygotes resulting from an interspecific cross produce a granular secretion, the two allelic states are manifested independently in the fourth chromosome, for one of the two homologues develops a puff, but the other does not. Since the banding pattern in the vicinity of the puff is the same in the two species, a difference in regulatory or controlling sites in the two species has been suggested in explanation.

The pattern of puffing in different polytene chromosomes is specific to the organ as well as to the stage in development, certain puffs being exhibited in salivary glands, others in midgut cells, and still others in rectum. Such patterns imply differential gene activity associated with the functioning of particular kinds of cells. In addition, puffs can be induced by changes in temperature or other environmental factors, and their appearance under these circumstances indicates the initiation of a physiological response to altered conditions of life. A dramatic example of such a response is seen when the molting hormone ecdysone is injected into *Chironomus* larvae, for within one half-hour, a large puff appears in one of the chromosomes, followed within another half-hour by a second large puff in a different chromosome. These are later joined by additional puffs which are initiated in other chromosomes of the complement. In *Drosophila* injection of the hormone has a similar effect in that it induces a new puffing pattern. In both cases the puffs elicited by hormone injection are the same as those which would normally appear at the onset of molting or pupation when ecdysone is present in the organism. The hormone thus acts to stimulate the initiation of transcription, either directly or indirectly, and in this respect it is similar to the hormones of vertebrates (see Table 19–1).

Isozymes

Isozymes are enzymes that exhibit an identical catalytic activity, but occur in different molecular forms. They may perform their function as single polypeptides (monomers) or as multimers whose subunits may be specified by the same or different genes. In humans the enzyme lactate dehydrogenase (LDH) is a good illustration. LDH is required for the reversible conversion of lactate to pyruvate and is present in all tissues. Two kinds of polypeptides, called A and B, occur in LDH and these are specified by two different, nonallelic genes. The active enzyme is composed of four subunits and consists of an apparently random association between the A and B polypeptides. Accordingly, the enzyme can occur in five different molecular forms composed of the following subunit associations: LDH 1 (B_4), LDH 2 (B_3A), LDH 3 (B_2A_2), LDH 4 (BA_3), and LDH 5 (A_4). Each of these isozymes is identifiable through the technique of electrophoresis. Since differing amounts of polypeptides A and B are produced by different tissues, the predominant molecular form characteristic of any tissue is specific to that tissue only (Figure 19–11). For example, heart muscle contains a high concentration of subunit B and much less of subunit A, and consequently the isozyme that predominates in heart muscle is the B_4 form or LDH 1. In contrast, the liver and skeletal muscle contain more of subunit A than of B, and thus the predominant molecular species in these tissues is A_4 or LDH 5.

An additional LDH polypeptide, called C, presumably the product of a third LDH gene, has also been identified, but its presence appears to be restricted to primary spermatocytes. Other enzymes, such as esterases, aldolases, and phosphoglucomutase have also been found in multiple molecular forms and the presence of polymorphic proteins and enzymes appears to be of widespread occurrence not only in humans, but also in other eukaryotic species examined for this phenomenon.

With respect to control mechanisms, we may ask, why is the ratio of subunit A to B in the neighborhood of 10 to 1 in skeletal muscle and liver, 1 to 20 in heart muscle, and around 1 to 10 in kidney? No answer based on experimental evidence exists for this question, but we can at least postulate that some sort of regulating mechanism associated with the differentiated state is responsible for maintaining a controlled transcription of genes A and B which is specific to each tissue. We can also infer that the presence of different isozymes for the same function is a reflection of subtle biochemical distinctions which exist between these tissues. That the transcription of subunit C is repressed in all but primary spermatocytes is additional evidence that the kinds of enzyme subunits present in a given cell are dependent on precise transcriptional control.

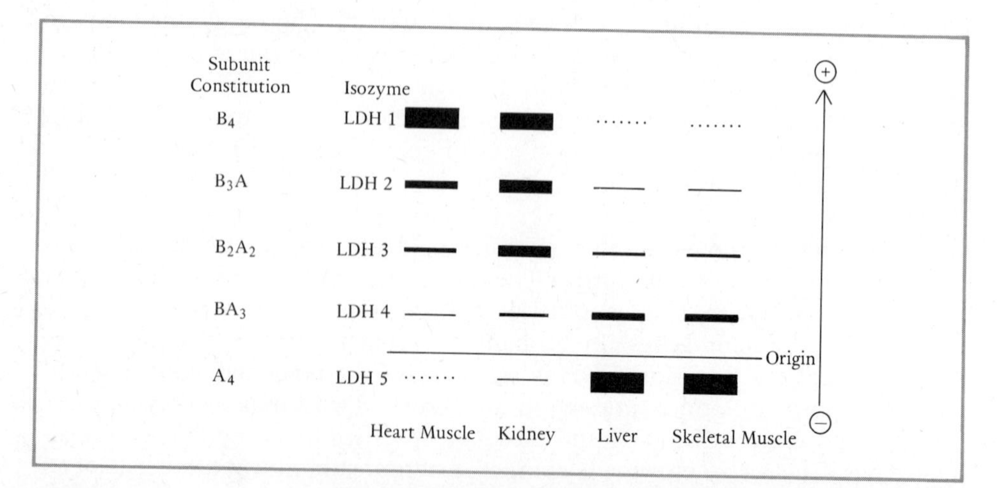

Figure 19–11. Diagram of electrophoretic patterns of lactate dehydrogenase isozymes in heart muscle, kidney, liver, and skeletal muscle. [From H. Harris, *The Principles of Human Biochemical Genetics,* 1st ed., North-Holland, Amsterdam, 1970.]

Immunoglobulins

The immune system of vertebrates, undoubtedly one of the most highly specialized and complex systems of the body, functions in the recognition and destruction of foreign macromolecules (antigens). This function is carried out by lymphocytes which are thought to be programed in advance to respond to a specific antigen before the antigen itself has been encountered. Upon meeting the antigen to which a cell has been precommitted, the lymphocyte is stimulated to undergo division to form a clone of cells, all of whom respond to the same antigen.

Two types of lymphocytes are involved in the immune response: T cells and B cells (Figure 19–12). In humans both types are thought to originate in the

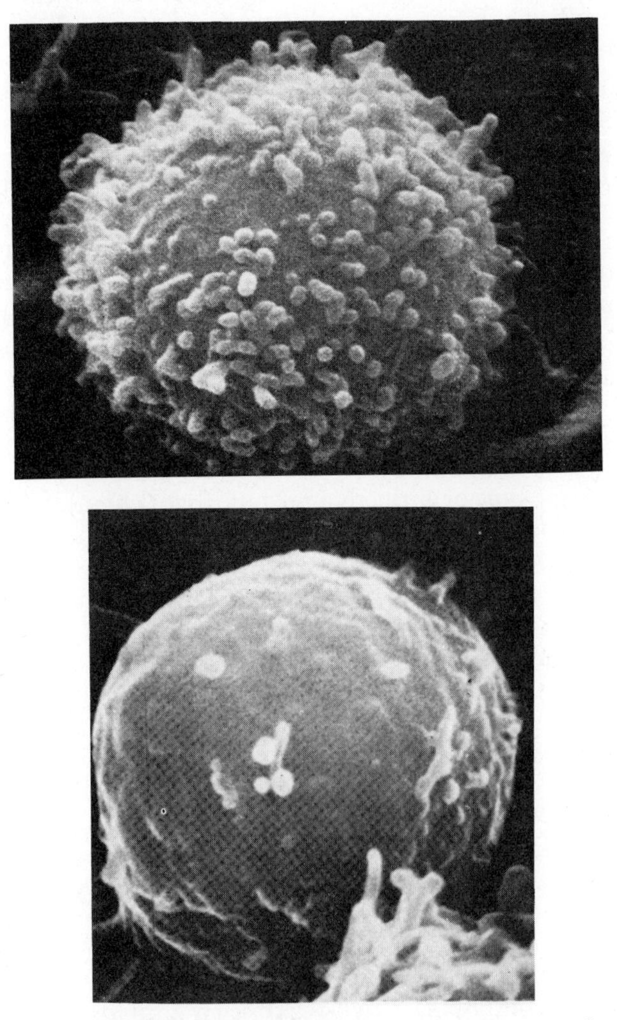

Figure 19–12. Scanning electron micrographs of human B and T lymphocytes. Above, typical normal B lymphocyte with complex surface villi; below, normal T lymphocyte having a relatively smooth surface with few villi. (14,000×) [Courtesy of Dr. Aaron Polliak, M.D., Hadassah University Hospital, Jerusalem, Israel. From A. Polliak et al., *J. Expt'l. Med.*, *138*, 607–624, 1973.]

bone marrow. T cells, so called because they undergo further differentiation in the thymus, do not secrete antibodies (immunoglobulins), but instead produce poorly understood substances, called lymphokines, some of which inhibit the movement of other cells. Under certain circumstances activated T cells can function as "killers" to destroy newly arisen malignant tumor cells. In general, T cells constitute a cellular response to infection.

The other type of lymphocyte, the B cell, is named after a central lymphoid organ called the bursa of Fabricius, which is located near the rectum in birds. In chickens B cells differentiate in the bursa, but mammals do not possess this organ, and differentiation is thought to occur in bone marrow. Although individual B lymphocytes are programed to respond to a particular antigen by secreting a corresponding antibody, the production of antibodies does not occur until the antigen has been encountered. Activation of a B cell requires not only the antigen itself, but also a T cell which somehow acts as a helper in the process. Once triggered, a B cell undergoes proliferation to form a clone of antibody secreting cells (plasma cells), as well as "memory" cells which can be triggered to form plasma cells if the antigen is encountered again in the future. B cells constitute what is called a humoral response to foreign macromolecules, for their descendents secrete the antibodies present in blood and body fluids.

Different species of antibodies may be produced in response to a single antigen, and the same species of antibody may bind to different, but related antigens. These capabilities permit the immune system to respond to an enormous number of foreign macromolecules, even to those synthesized in the laboratory and not encountered in nature. Since the antibodies present in blood serum comprise a spectrum of different types, each present in low concentration, it is difficult to obtain enough of any one kind of antibody for analysis. However, in multiple myeloma, a cancer of antibody secreting cells, large quantities of a single type of antibody molecule (light chain) are excreted in the urine (Bence Jones protein), and their study has provided much information on structure and function.

Immunoglobulins are proteins and in humans they can be classified into three major groups, immunoglobulins G (IgG), M (IgM), and A (IgA), and two minor groups, immunoglobulins D (IgD) and E (IgE). Using IgG as the example, an antibody molecule is composed of four polypeptides held together by sulfhydryl bonds (Figure 19–13). Two of the component polypeptides are identical heavy chains (440 residues), while the other two are identical and shorter light chains (214 residues). The functional subunit of any antibody molecule is a light chain–heavy chain pair.

One family of heavy chains and two families of light chains, kappa (κ) and lambda (λ), are recognized in humans. The genes which specify the members of each of these families are located on separate autosomes and are thought to comprise a closely linked, tandem array of numerous similar, but not identical genes. As a further complexity, in any antibody molecule each heavy chain and each light chain is composed of two regions, an amino terminal or *variable region* and a carboxyl terminal or *constant region*. The terms variable and constant are relative, in that so-called constant regions exhibit much less diversity than do the variable regions, for it is the variable regions that confer upon an antibody molecule the ability to recognize one of the thousands of different antigens to which the organism may be exposed.

It is now thought that the constant and variable regions of each kind of chain (heavy, κ, and λ) are specified by different genes, and if so, an immediate problem arises. Each chain is a single polypeptide, and if two portions of the same molecule are encoded by different genes, then some means must exist whereby either the genes are joined to transcribe a single mRNA or two separate mRNAs are joined prior to translation or two separate proteins are translated and then joined prior to assuming the functional state. Through sequence analysis it has been found that a single mRNA serves as the template for the translation of a complete light chain, including both the variable and constant regions, indicating that the joining, at least in this case, has occurred at either the RNA or DNA level. No means has yet been found to distinguish between these two alternatives, but immunologists favor the idea of joining at the DNA level, and a number of models which attempt to explain how this might be accomplished have been proposed. Some models invoke somatic recombination, others somatic translocation, and still others the excision and integration of episome-like DNA segments, but as yet there is no unifying theory which satisfactorily accounts for antibody diversity.

Although each of the multigene families appears to contain one to relatively few genes for the constant region, the number of genes for the variable region is unknown. One theory postulates that each kind of variable region that an organism can produce is specified by a separate gene, thousands of these genes being present in the germ line. The origin of these genes is attributed to gene duplication occurring during the course of evolution. Another theory proposes the existence of a more limited number of immunoglobulin genes that can perhaps undergo somatic mutation to produce gene variants or may engage in some combinational process, such as those suggested above, to produce the antibody diversity observed. Since no way has been found to identify and map individual genes for the variable regions, a choice between these hypotheses, or some combination thereof, cannot as yet be made. Despite these uncertainties,

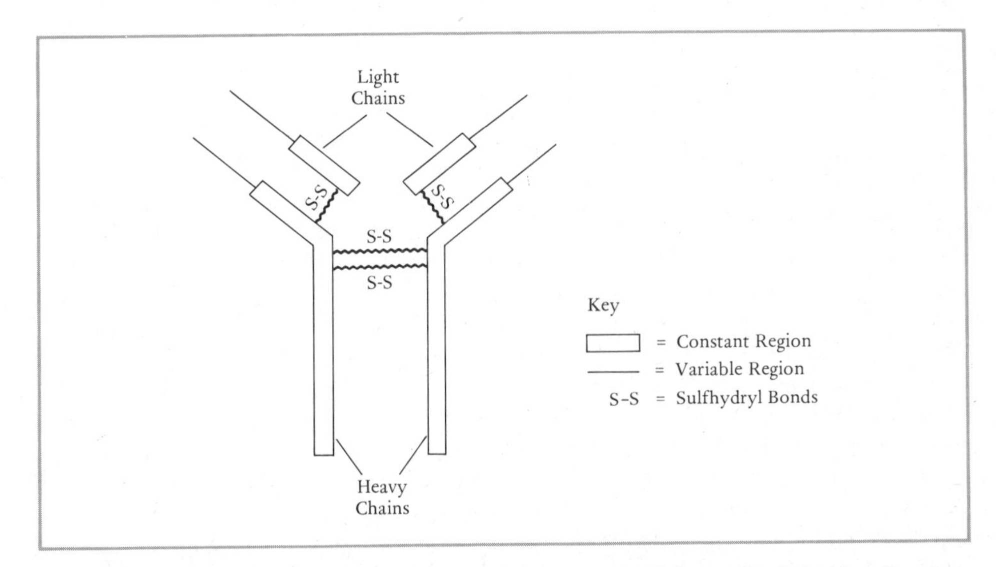

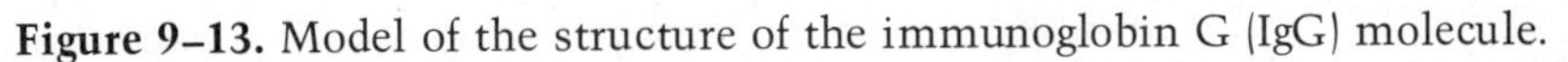
Figure 9–13. Model of the structure of the immunoglobin G (IgG) molecule.

it is nevertheless clear that the regulation of antibody synthesis involves the selection of only those specific sequences whose expression is required, along with the suppression of all other immunoglobulin genes within any T or B cell. How this programing is accomplished is a fascinating riddle, suggesting the presence of sophisticated systems of information storage, retrieval, and control.

POSSIBLE REGULATORS OF GENE ACTIVITY

The obvious presence of gene regulatory mechanisms in eukaryotic cells has stimulated the search for possible controlling molecules. Since most transcription occurs in interphase, particular attention has been directed toward chromosomal activity in this stage. It will be recalled that during interphase the nucleoprotein fiber which constitutes the chromosome ordinarily becomes greatly extended and is referred to as *chromatin* (see Figure 6–12). Chromatin can be isolated in purified form from a variety of sources and its transcriptional capacity can then be tested *in vitro.* Such studies have shown that the kinds of RNA synthesized by chromatin extracted from different tissues of the same plant or animal vary, indicating that selective control of transcription has been imposed in these tissues. The proteins bound to the DNA of chromatin can be removed, and the capacity of naked DNA to synthesize RNA can also be assayed in an *in vitro* system. The results of such experiments have shown that the ability of chromatin to serve as a template for RNA synthesis is only 5 to 10 percent that of DNA alone, suggesting that the proteins normally complexed with DNA perform a regulatory role. Analysis of chromatin has revealed that in addition to enzymes, such as DNA and RNA polymerases, ligases, endonucleases, and others, chromatin contains two classes of chromosomal proteins, *histones* and *nonhistones,* along with a small amount of RNA.

Histones

Histones are small proteins (MW 10,000 to 25,000) that contain an abundance of basic amino acids and are therefore positively charged under physiological conditions. They have been classified into five major groups according to the proportion of lysine and arginine they contain: H1 (very lysine rich), H2A and H2B (slightly lysine rich), and H3 and H4 (arginine rich). The amino acid sequence of most of these fractions has been determined for a number of different organisms, and while some fractions are highly variable (H1), others (H3 and H4) exhibit great evolutionary stability suggesting a universal role related to chromosome structure and function. Histones are present in a 1 : 1 ratio with DNA. Their synthesis is coupled to that of DNA, occurs during the S phase of the cell cycle, and experimental evidence indicates that during DNA replication newly synthesized histones rapidly associate with the newly synthesized strand of DNA.

Histones are the only proteins known to be encoded by clustered, reiterated genes in all eukaryotic species so far examined. Analysis of these genes has been made possible by the isolation of histone mRNA's from sea urchin embryos, where the rapid synchronous nuclear divisions are accompanied by significant histone synthesis. Hybridization experiments between such mRNA's and single-stranded DNA of the same or different species has permitted estimates of the number of copies per haploid genome of each kind of histone gene present. For example, sea urchins, depending on species, have 400 to 1200 copies, *Drosophila* and most mammals 50 to 100 copies, and humans 10 to 20 copies of each histone gene. These genes appear to be clustered in a complex tandem arrangement in which structural gene sequences are separated by spacer DNA. Evidence for clustering can be seen in *Drosophila* in that sea urchin histone mRNA hybridizes with a region on the left arm of chromosome 2, which contains three to five bands in the salivary chromosomes.

The function of histones with respect to regulation is associated primarily with the repression of genetic activity, as suggested by E. Stedman and E. Stedman in 1950. In general, histones show no tissue specificity and do not appear to vary, quantitatively or qualitatively, during the development and differentiation of specialized cell types. Three exceptions to this generalization are known, however. In chicken red blood cells an additional histone is synthesized whose function appears to be associated with the highly repressed state of the mature red cell nucleus in avian species. In spermatocytes special histones are produced and are thought to aid in the extreme condensation of the genome which occurs during sperm differentiation. In fish small basic proteins, called *protamines,* replace the usual somatic histones during sperm development. Outside of these exceptions, the histones present in all cells, whether metabolically active or inactive and from zygote to adult, appear alike, and for this reason histones are thought to function more generally in the repression of transcription, rather than as specific gene activators.

Their role in repression is intimately associated with their role as structural proteins of the chromosome, for it has become evident that interactions between the different types of histones are involved in the folding and coiling of the DNA fiber. Recent evidence suggests that the basic unit of coiling is the *nucleosome* or *nu body.* This structure is thought to be a spherical body composed of the DNA fiber coiled about and complexed with histones. It is proposed that such bodies are spaced at regular intervals corresponding to a repeat of about 200 base pairs and are separated by histone-poor regions, the DNA fiber and its associated histones presenting a "beads on a string" appearance. Bodies of this nature have been seen with the electron microscope. The molecular arrangements between the DNA and the histones of nu bodies are unknown; also unknown is the mechanism by which higher orders of coiling might be imposed on nu bodies.

Although much remains to be learned, it is clear that the packaging of the DNA may determine whether or not a given sequence is accessible for transcription. If a control region or promotor site must be exposed and available in order that transcription of a structural gene be initiated, any conformation or folding of the DNA fiber which prevents this exposure will repress gene expression. In this regard it is significant that removal of histones from chromatin causes a marked increase in the capacity of the DNA to serve as a template for RNA synthesis. We can also surmise that position effects may be due, at least in part, to the insertion of genes into regions of the genome, such as condensed heterochromatin or an inactivated X chromosome, where the packaging pattern causes repression. We can also visualize that one function of a gene-regulating molecule may be to produce alterations in DNA-histone relationships in different tissues such that promotor or controlling sites for specific structural genes are uncovered.

Evidence for the existence of such sites has been inferred in instances where a mutation causes a change in the rate of synthesis, rather than a change in the polypeptide being synthesized. For example, in β-thalassemia, a hemolytic anemia of humans common in Mediterranean countries, synthesis of the β chain of hemoglobin is greatly decreased or absent, although the α and δ chains are produced normally. The β-thalassemia locus is closely linked to that of the β chain structural gene, and it has been proposed that reduced synthesis of the β chain is caused by a mutation in a promotor or initiation site. In *Drosophila* A. Chovnick and associates have recently identified a cis-acting controlling site associated with the rosy eye color cistron, which is the structural gene for the enzyme xanthine dehydrogenase. Fine structure mapping has located this site immediately adjacent to the rosy cistron. Although we can extrapolate this finding to other structural genes, the question still remains as to what sorts of molecules uncover, activate, or bind to these controlling sites such that transcription is initiated. No firm answers to this question are as yet available, but the evidence thus far points to the nonhistone proteins and possibly to RNA as possible candidates for this role.

Nonhistone Chromosomal Proteins

In its broader sense the title nonhistone proteins includes every kind of protein, except histone, found in isolated chromatin. Although it is recognized that a number of different components are present, the term nonhistone protein is frequently used in a narrower sense to refer specifically to those proteins that appear to activate the transcription of particular mRNA's, and this definition will be used here.

The nonhistone proteins are sometimes called acidic nuclear proteins because of their high content of acidic amino acids, such as aspartic and glutamic acids, which confer upon the molecule a negative charge under physiological conditions. Unlike the histones, acidic nuclear proteins are highly varied, possess tissue specificity, and exhibit electrophoretic patterns of great complexity which change both quantitatively and qualitatively during the cell cycle and during the course of development and differentiation in various cell types. They are produced throughout the cell cycle, peak synthesis occurring during the G_1 phase, and some evidence suggests that synthesis at this time is a necessary prerequisite for subsequent DNA and RNA synthesis. In addition, these proteins have been found to reverse histone inhibition of DNA-dependent RNA synthesis. Overall, these characteristics make the nonhistone proteins prime candidates for gene-regulating molecules.

The ability of these proteins to regulate transcription has been assessed by a number of techniques, among them *chromatin reconstitution.* Purified chromatin can be dissociated into DNA, histone, and nonhistone proteins, and these components can then be recombined to reconstitute chromatin. By interchanging histones, nonhistones and DNA derived from different tissues, it is possible to assess the role of each in stimulating the transcription of some specific type of mRNA which can itself be identified. The kind of mRNA chosen for assay is ordinarily one which is produced in large quantities in some particular tissue; a favorite choice is mRNA for the globin portion of hemoglobin, synthesized in large amounts by erythropoietic tissue. Identification of specific mRNA transcripts is accomplished by use of a molecular probe.

Purified mRNA, proven by *in vitro* translation to contain information for a specific protein, such as globin, is utilized as a template for the *in vitro* synthesis of complementary DNA (cDNA) through the use of reverse transcriptase (RNA-directed DNA polymerase). The cDNA produced by this technique will hybridize specifically only with globin mRNA and thus can be used to single out and identify this particular transcript from among any others that might also be produced in an experiment.

J. Paul and R. S. Gilmour have utilized these methods to demonstrate that nonhistone proteins can determine the kind of mRNA transcribed from chromatin. These workers used mouse fetal liver (globin producing) and mouse brain (non-globin producing) and established that globin-specific mRNA was produced by both native and reconstituted chromatin from fetal liver, but was not synthesized by chromatin isolated from brain. However, when brain chromatin was dissociated and reconstituted with nonhistone proteins derived from fetal liver, globin-specific mRNA was transcribed, whereas interchanging either the DNA or the histones had no such effect. Similar experiments by G. Stein and associates have demonstrated that mRNA for histone is produced by chromatin isolated from cells in the S phase, but not by chromatin derived from cells in the G_1 phase of the cell cycle. However, when DNA from G_1 chromatin and nonhistone proteins from S phase chromatin are combined, histone-specific mRNA can be identified among the transcripts.

These and other experiments have provided convincing evidence that nonhistone proteins can determine transcriptional specificity although the molecular mechanisms responsible for this action are unknown. Perhaps related to their proposed role in gene activation is the fact that nonhistone proteins may become phosphorylated by the addition of phosphate to the OH groups of serine and threonine residues, and a correlation between the phosphorylation of these proteins and gene activation has been observed. For example, increased phosphorylation of these proteins is associated with the stimulation to transcription which accompanies the treatment of prostate cells with testosterone or mammary gland with prolactin. In contrast, phosphorylation decreases when DNA and RNA synthesis ceases, as in maturing avian red blood cells. Changes in the phosphorylation of nonhistone proteins are also observed during the cell cycle. The function of the added phosphate groups is unknown, but it has been suggested that their presence aids in the removal of histones blocking promotor sites for transcription.

Although the nonhistone proteins appear to be intimately involved in the activation of transcription, it should be noted that a variety of small RNA molecules are also associated with chromatin and are frequently contained within nonhistone protein fractions. The function of these RNA's is unknown, but they have been assigned a role in gene regulation by a number of investigators.

COORDINATED TRANSCRIPTION AND THE ORGANIZATION OF THE GENOME

Although the examples of gene regulation cited previously in this chapter permit an insight into specific instances of controlled gene activity, they fail to convey an appreciation of the overall magnitude of the task of coordination which is carried out so smoothly in eukaryotic cells. For a more concrete idea as to what controlling molecules must accomplish to permit an appropriate cellular response to a stimulus, attention is directed to Table 19–1 wherein are listed some of the responses by the uterus to the administration of estrogen. Reflection as to what might be involved in each of these responses will bring to mind numerous metabolic pathways, a multitude of different enzymes, and the synthesis of a wide variety of cell constituents. We must add to this complexity the fact that the genetic loci necessarily involved are mostly noncontiguous, yet are activated almost simultaneously. In view of this amazing feat of integration, it is reasonable to ask what kind of structural framework makes possible these coordinated responses. No definite answer to this question can as yet be given, but it is probable that the genetic regulatory mechanisms of eukaryotes reflect an underlying pattern of organization present in the chromosome as well as in the total functional gene itself. Some clues as to the nature of such a pattern have recently become available.

One such clue has been provided by studies of the rate of reassociation of single-stranded DNA fragments. In this technique DNA is isolated and purified, sheared into uniformly sized fragments approximately 500 base pairs long, and denatured to the single-stranded state. Under standardized conditions it is then allowed to renature, or anneal, to form the double-stranded state once more. Successful annealing depends on random collision between a pair of complementary base sequences, and the probability that such a collision will occur is greatly increased if many such complementary sequences are present, that is, their rate of reassociation will be dependent on their concentration. For example, if two complementary sequences are present in very low numbers, the chance that a random collision between them will occur is also very low and, therefore, reassociation of these sequences will require an extended period of time. However, if the two complementary sequences are present in high

TABLE 19–1
Some Effects of Estrogen on Uterine Cells

Increase in total cell protein
Increase in transport of amino acids into cell
Increase in protein synthesis activity per unit amount of polyribosomes
Increased synthesis of new ribosomes
Alteration of amounts of nuclear protein to nucleus
Increased amount of polyribosomes per cell
Increase in nucleolar mass and number
Increase in activity of two RNA polymerases
Increase in synthesis of contractile proteins
Imbibition of water
Increased synthesis of many phospholipids
Increased *de novo* synthesis of purines (dependent on new enzyme synthesis)
Alteration in membrane excitability
Alteration in glucose metabolism
Increase in synthesis of various mucopolysaccharides

Source: R. J. Britten and E. H. Davidson, *Science, 165,* 349–357, 25 July 1969.

numbers, frequent random collisions will occur between them and their reassociation will be rapid.

When DNA from eukaryotes of all types is assayed by this method for reassociation rate, three major classes of DNA sequences are usually apparent. One class renatures with extreme rapidity, indicating the presence of a sequence having a high degree of reiteration; a second class reassociates at an intermediate rate, indicating the presence of moderate repetition frequency; and a third class reassociates so slowly that it can be inferred that only one to a few of these sequences are present. The proportion of the genome represented in these classes can also be estimated (Figure 19–14).

An example of a high degree of reiteration can be seen in the mouse in which approximately 10 percent of the DNA is composed of a sequence of 240 base pairs which are repeated a million times. In other organisms this highly repetitive fraction may consist of up to a million repeats of a short sequence of 2 to 10 base pairs. In the guinea pig the repeating unit is a sequence of 6 bases, C C C T A A (and the complementary G G G A T T sequence), and in some crabs 30 percent of the genome is composed of the alternating bases A T A T A T and so on.

In some cases, for example, mouse and crab, the base composition of the fragments containing highly repetitive sequences differs sufficiently from that of the bulk of the DNA to confer a different buoyant density in a CsCl gradient. Density gradient centrifugation results in the presence of a small satellite band having either a lower density (high A-T) or a higher density (high G-C) than the

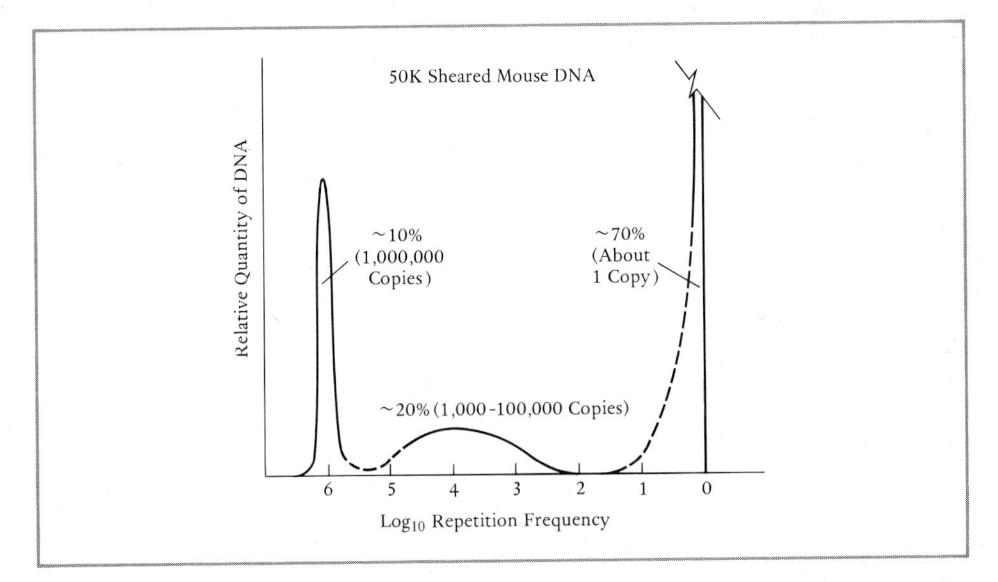

Figure 9–14. Pictorial representation of the frequency of nucleotide sequences in the DNA of the mouse; relative quantity of DNA plotted against the logarithm of the repetition frequencies. Dashed segments of the curve represent regions of uncertainty. The graph is not a direct description of measurements but illustrates indications of knowledge at the time. [From R. J. Britten and D. E. Kohne, *Science, 161,* 529–540, 9 August 1968. Copyright © 1968 by the American Association for the Advancement of Science.]

main band which contains the rest of the DNA. In other cases, for example, green monkey, the base composition of fragments containing highly repetitive sequences does not differ significantly from that of the rest of the DNA and therefore no satellite band containing this fraction is observed. It should be noted that other satellite bands besides those containing highly repetitive base sequences may also be present; some of these represent mitochondrial or plastid DNA which frequently has a base composition different from that of nuclear DNA.

The location in the chromosomes of the highly repetitive fraction has been determined for a number of organisms by the technique of *in situ* hybridization developed by M. L. Pardue and J. G. Gall. Chromosomes of a cytological preparation are treated to remove protein and RNA, after which the DNA is denatured to the single-stranded state. The DNA of the chromosomes is then allowed to anneal with satellite DNA labeled with a radioactive isotope or with labeled RNA copies of satellite DNA. (RNA copies, cRNA, are produced *in vitro* with the use of RNA polymerase along with satellite DNA as the template.) The autoradiographs so prepared can be examined to determine the chromosomal site of hybridization. In mammals, including mouse, guinea pig, hamster, horse, calf, and human, the highly repetitive fraction has been localized to the constitutive heterochromatin adjacent to the centromeres of all chromosomes except the Y, where it is found mainly in the long arm (Figure 19–15). In *Drosophila* the centromeric heterochromatin is again the site exhibiting the greatest degree of annealing.

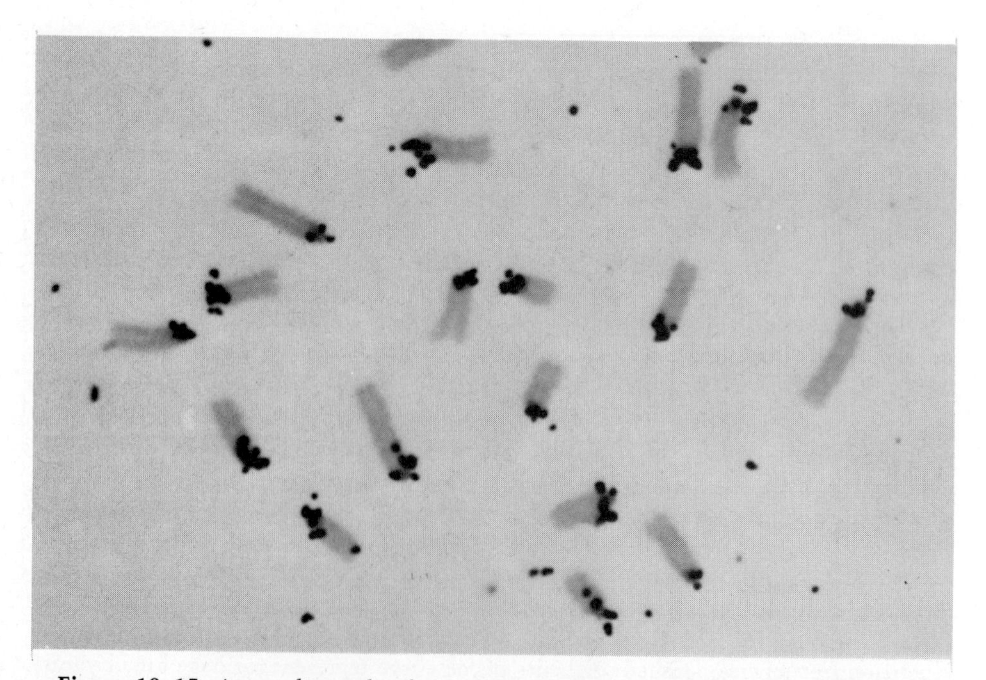

Figure 19–15. Autoradiograph of a mouse tissue culture preparation after *in situ* cytological hybridization with radioactive RNA copied *in vitro* from mouse satellite DNA. The RNA has bound to the centromeric heterochromatin of the chromosomes. (2000×). [Courtesy of M. L. Pardue and J. G. Gall.]

The function of these repeated sequences, which on the average make up 5 to 10 percent of the genome, is unknown, for they do not appear to code for protein and are not transcribed. Nevertheless, the fact that they are present and conserved points to some utility, perhaps structural in nature.

The moderately repetitive fraction of DNA sequences identified in renaturation experiments appears to consist of families of related sequences whose members are similar, but not identical to one another. These sequences are scattered throughout the genome. It has been estimated that in *Drosophila* around 4500 such families are present, a number that is in the same range as the total number of bands (5000) present in the salivary chromosomes. Moderately repetitive sequences appear to be transcribed and are thought to contribute to those RNA's present in the nucleus, but absent from the cytoplasm, and for whom a regulatory role has been proposed. In mammals 15 to 20 percent of mRNA has been found to be complementary to moderately repetitive sequences.

The third class of DNA identifiable in reassociation experiments is single copy DNA which comprises 70 percent of the genome of the mouse (Figure 19–14) and around 55 percent of that of the calf. These unique sequences are dispersed throughout the genome and are thought to represent structural genes.

The overall pattern of organization which emerges from these studies is one in which highly repetitive sequences are located mostly around the centromere, while the arms of the chromosomes contain unique single copy sequences interspersed with moderately repetitive families containing similar, but not identical sequences. To this very tentative model must be added the interpretations which have resulted from genetic studies of complementation and those which have arisen from the examination of mRNA's.

B. H. Judd and co-workers have made an extensive analysis of the complementation groups present within a specific region of the X chromosome of *Drosophila* and have demonstrated the presence of only one complementation group, that is, one functional cistron, per salivary chromosome band. A somewhat similar analysis for the fourth chromosome has also indicated a one gene–one band relationship. This proposal is in accord with Callan's hypothesis that each chromomere of the amphibian chromosome is the locus of a single gene. If the one chromomere (or one band) to one gene relationship is accepted, a paradox immediately arises. The DNA content of the average salivary chromosome band is around 30,000 nucleotide pairs which is 30 times greater than the amount needed to code for an average sized polypeptide of 300 amino acids. If it is assumed that the lampbrush loops of amphibian chromosomes represent the unwinding of the DNA of a single chromomere, as interpreted by Callan, then the amphibian chromomere contains an even greater amount of DNA beyond that necessary to encode a single protein. In an attempt to resolve this dilemma, Callan has postulated that the DNA of the lampbrush loops represents tandem copies or slaves of a master structural gene. His proposal requires that the slaves be somehow matched against the master and, if necessary, corrected for any mutations at every meiosis in order to maintain uniformity of sequence. In addition, Callan's hypothesis requires that the function of crossing over be restricted to the single master gene. This hypothesis is not supported by crossover data which, at least in *Drosophila*, indicate that the amount of crossing over is directly related to the DNA content of a salivary band, and this would not be the case if each band contained only one master copy capable of crossing over. Furthermore, data from reassociation experiments suggest that there is not enough repetitive DNA in the genome to permit the presence of identical slave copies in each chromomere or band.

Other attempts by C. A. Thomas and associates to ascertain the nature of the DNA within a chromomere have involved the shearing of *Drosophila* DNA, followed by treatment with an exonuclease to produce single-stranded ends which anneal to form circles. (A previous example of the annealing of single-stranded ends of a linear DNA molecule to form a circle was described for the bacteriophage λ; see Figure 12–6.) Since annealing would not occur without the presence of complementary bases, the formation of circles is interpreted to indicate the presence of serially repetitive segments within the genome. J. Bonner and J. R. Wu have proposed that the middle repetitive sequences of *Drosophila* consist of separate families, each containing sequences of around 125 base pairs repeated an average of 30 to 35 times. These workers estimate that the number of such families is approximately 4500 and therefore within the range of the total number of bands in the salivary chromosomes. They have proposed that a salivary band is composed of approximately 35 unique single copy sequences, each about 750 base pairs in length, which are separated from one another by the identical 125 base pair sequences, all of which belong to one family of the middle repetitive group. The hypothesis of Bonner and Wu thus proposes that a salivary band contains not only some sort of spacer repetitive DNA, but also a series of up to 35 unique sequences. If these unique sequences are considered to be separate functional genes, the hypothesis would be inconsistent with genetic data that clearly demonstrates the one band–one function relationship. However, if the many unique sequences were to code for controlling molecules plus a structural gene, this inconsistency could be resolved.

Some information as to the nature of a gene can be gained by comparing the RNA transcripts present in the nucleus with those in the cytoplasm associated with the polysomes. It will be recalled from Chapter 15 that giant RNA molecules, called heterogeneous nuclear RNA (hnRNA) are found in eukaryotic nuclei. In *Chironomus* B. Daneholt has identified a 75 S nuclear transcript which can be shown by *in situ* hybridization and autoradiography to anneal to a specific puffed band in the salivary chromosomes (Figure 19–16). Estimates of the amount of DNA in the puff versus the amount of RNA in the transcript indicate that the entire band is transcribed as the 75 S molecule. The transcript appears to contain repetitive sequences, but their nature is not known. The entire 75 S transcript has also been identified in the cytoplasm which is quite unusual, since in most eukaryotes the message which reaches the cytoplasm is considerably smaller than the original transcript.

Studies by G. P. Georgiev of globin mRNA produced by immature erythrocytes has revealed that the initial transcript (pre-mRNA) is a giant precursor molecule containing at the 3′ end no more than one or two informational sequences for protein; mRNA is thus restricted to the 3′ end of the transcript. Also at the 3′ end (and perhaps at the 5′ end as well) is a short repetitive sequence. During the preparation and processing of the transcript prior to transport to the cytoplasm, a length of around 200 adenine nucleotides, the poly(A) sequence, is added enzymatically to the 3′ end, the short reiterated sequence at this end perhaps serving as a signal for this addition or as a terminator in translation. We should note that the function of the poly(A) sequence which is characteristically found in mRNA has been associated with the stability and longevity of the message once it reaches the cytoplasm. Other portions of the pre-mRNA transcript appear to contain a number of sequences, some of them repetitive and complementary to one another. These do not code for protein, but instead are cleaved from the molecule and remain within the nucleus. It has been estimated that the mRNA which finally reaches the cytoplasmic polysomes represents only a tenth of the pre-mRNA initially transcribed. The function of the RNA which remains behind is unknown, but a role in gene regulation has been proposed by a number of investigators.

In summary, the model of chromosome and gene organization currently emerging is one that views the chromomere as the equivalent of the cistron. It appears that most of the DNA of the chromomere is transcribed and that ordinarily the transcript contains only one to a few sequences for protein at the 3′ end, the rest of the molecule containing regulating sequences which may ensure proper processing of the transcript or may act to switch on other genes or may serve both functions. A model of gene regulation for eukaryotes which incorporates many of these features has been proposed by R. J. Britten and E. H.

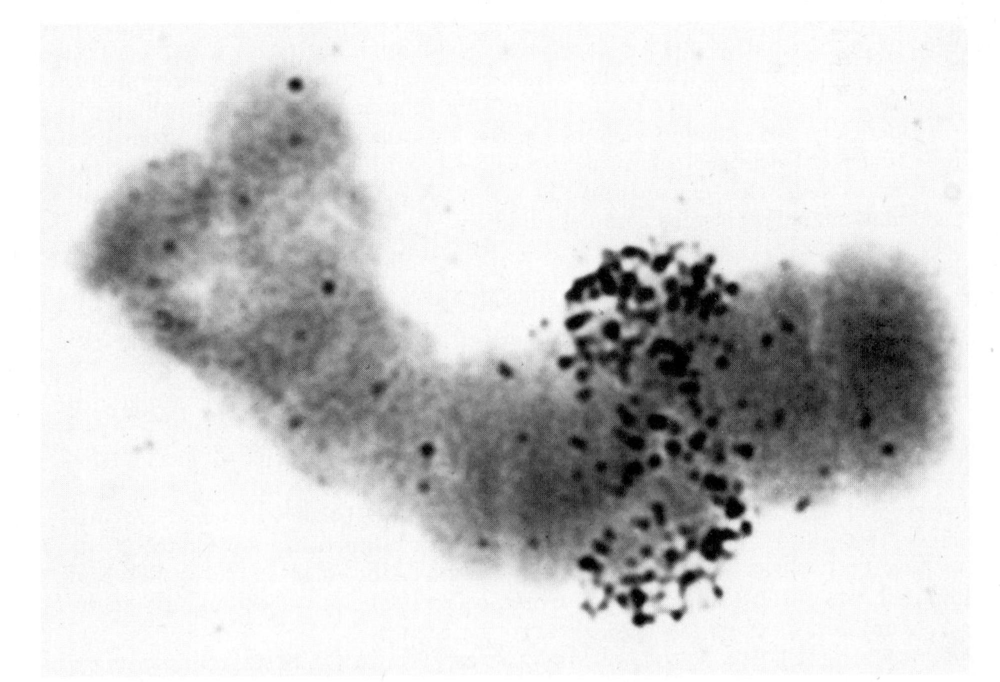

Figure 19–16. *In situ* hybridization between the DNA of the giant puff of chromosome 4 of *Chironomus* and the corresponding 75 S mRNA transcript present in the nuclear sap. (900×) [From B. Lambert, *Cold Spring Harbor Sympos. Quant. Biol.*, *38*, 637–644, 1974.]

Davidson. The model ascribes a major role in structural gene activation to RNA molecules transcribed from multiply present regulating genes. These, in turn, are subject to stimulation by some agent, such as a hormone. This model has provoked much discussion and speculation, and those interested in eukaryotic gene regulation will find much of value in Britten and Davidson's paper (1969).

PROBLEMS

19–1. In an individual heterozygous for a variant (A*) of the A subunit of lactate dehydrogenase, how many different kinds of isozymes are possible? What subunit combinations would occur in each?

19–2. Persistence of fetal hemoglobin (HbF) occurs in a hereditary anemia called high HbF, and neither the β nor the δ chains are produced in homozygotes. The locus of the high HbF mutation is closely linked to the β chain structural gene. Criticize the hypothesis that the high HbF mutation represents a change in an operator and that the β and δ genes belong to a single operon. Are there alternative explanations?

19–3. Design an experiment to show that the pole cells of the *Drosophila* blastoderm give rise to germ cells of the adult.

19–4. Enumerate some of the ways by which a variegated or mosaic individual can arise.

19–5. What evidence in addition to the analysis of mRNA's indicates the presence in mRNA of a sequence of bases at the 3′ end beyond those normally translated into protein?

19–6. In nuclear transplantation experiments with amphibian eggs, a number of abnormal embryos whose cells contain abnormal chromosome sets are observed. Does this finding alter the hypothesis that the nuclei of somatic cells are genetically the same? In view of the fact that the egg cytoplasm automatically undergoes cleavage to form the two-cell stage within two hours after a nucleus is implanted, what explanation can you suggest to account for the presence of abnormal chromosome complements?

REFERENCES

Ashburner, M., 1970. Function and structure of polytene chromosomes during insect development. *Adv. Insect Physiol., 7,* 1.

Berendes, H. D., 1973. Synthetic activity of polytene chromosomes. *Int. Rev. Cytol., 35,* 61.

Britten, R. J., and E. H. Davidson, 1969. Gene regulation for higher cells: A theory. *Science, 169,* 349.

Britten, R. J., and D. E. Kohne, 1968. Repeated sequences in DNA. *Science, 161,* 529.

Callan, H. G., 1963. The nature of lampbrush chromosomes. *Int. Rev. Cytol., 15,* 1.

Davidson, E. H., 1968. *Gene Activity in Early Development.* Academic Press, New York.

Elgin, S. C. R., and H. Weintraub, 1975. Chromosomal proteins and chromatin structure. *Ann. Rev. Biochem., 44,* 725.

Fudenberg, H. H., J. R. L. Pink, D. P. Stiles, and A. Wang, 1972. *Basic Immunogenetics.* Oxford University Press, New York.

Gally, J. A., and G. M. Edelman, 1972. The genetic control of immunoglobulin synthesis. *Ann. Rev. Genet., 6,* 1.

Gurdon, J. B., 1973. Nuclear transplantation and regulation of cell processes. *Brit. Med. Bull., 29,* 259.

Hess, O., and G. F. Meyer, 1968. Genetic activities of the Y chromosome in *Drosophila* during spermatogenesis. *Adv. Genet., 14,* 171.

Hildemann, W. H., 1973. Genetics of immune responsiveness. *Ann. Rev. Genet., 7,* 19.

Hood, L., J. H. Campbell, and S. C. R. Elgin, 1975. The organization, expression, and evolution of antibody genes and other multigene families. *Ann. Rev. Genet., 9,* 305.

Hotta, Y., and S. Benzer, 1972. Mapping of behaviour in *Drosophila* mosaics. *Nature, 240,* 527.

Lefevere, G., Jr., 1974. The relationship between genes and polytene chromosome bands. *Ann. Rev. Genet., 8,* 51.

Lewin, B., 1975. Units of transcription and translation: The relationship between heterogeneous nuclear RNA and messenger RNA. *Cell, 4,* 11.

Mintz, B., 1974. Gene control of mammalian differentiation. *Ann. Rev. Genet., 8,* 411.

Nesbitt, M. N., and S. M. Gartler, 1971. The applications of genetic mosaicism to developmental problems. *Ann. Rev. Genet., 5,* 143.

Postlethwait, J. H., and H. A. Schneiderman, 1973. Developmental genetics of *Drosophila* imaginal discs. *Ann. Rev. Genet., 7,* 381.

Tartof, K., 1975. Redundant genes. *Ann. Rev. Genet., 9,* 355.

Ursprung, H., 1967. Developmental genetics. *Ann. Rev. Genet., 1,* 139.

Wright, T. R. F., 1970. The genetics of embryogenesis in *Drosophila. Adv. Genet., 15,* 262.

Chapter 20

INHERITANCE THROUGH THE CYTOPLASM

ALTHOUGH our discussion up to this point has been concerned almost exclusively with nuclear genes, these are not the only entities involved in inheritance, for numerous instances have been found in which a cytoplasmic agent may exert a decisive influence upon the phenotype of one or more succeeding generations. This influence may stem from the presence of stored products of nuclear genes or from the activities of autonomous cytoplasmic elements. Indeed, in some cases even preformed patterns of structural organization can be transmitted to descendent cells. The role of the cytoplasm, as distinct from that of the nucleus, is difficult to assess, but it is nevertheless clear that some characters are transmitted apart from chromosomal genes. The evidence for cytoplasmic inheritance is reviewed in this chapter.

MATERNAL EFFECTS

Maternal effects refer to those exerted on a developing organism by substances, such as mRNA, that have been included in the egg cytoplasm prior to fertilization. Since such inclusions are synthesized during oögenesis, they reflect the genotype of the mother and not that of the zygote. These effects are exclusively maternal in origin because essentially no cytoplasm is contributed to the zygote by sperm or pollen.

When maternal effects are present, the results of reciprocal crosses differ. As an illustration, in the flour moth, *Ephestia,* black eyes is determined by the dominant nuclear gene *A* and red eyes by the recessive allele *a,* and these phenotypes are expressed in young larvae. In a testcross $Aa \times aa$, where the male parent carries the dominant allele, the expected Mendelian ratio of $\frac{1}{2}$ black-eyed *(Aa)* and $\frac{1}{2}$ red-eyed *(aa)* larvae results. However, in the reciprocal cross $Aa \times aa$, where the female parent carries the dominant allele, all young larvae are black-eyed, regardless of genotype. This dark color is not permanent, however, for as growth proceeds, the eye color of individuals genotypically *aa* gradually lightens to red. It has been shown that the black color initially present is due to a pigment precursor (kyneurinine) included in the cytoplasm of eggs produced by black-eyed mothers. Since *aa* individuals do not synthesize this precursor, their eye color changes as the precursor is used up and is not replaced.

A second and classic example of a maternal effect concerns the direction of coiling of the shell of the snail *Limnea peregra* (Figure 20–1). Right handed (dextral) coiling is determined by a dominant gene *D* and left handed (sinistral) coiling by the recessive allele *d.* The orientation of the shell, and indeed of the entire body, is determined at the first cleavage division by the plane of the mitotic spindle. Apparently substances which establish the symmetry of the future zygote are included in the egg cytoplasm prior to fertilization, and these exert a directing influence at the first cleavage division. Since coiling, once

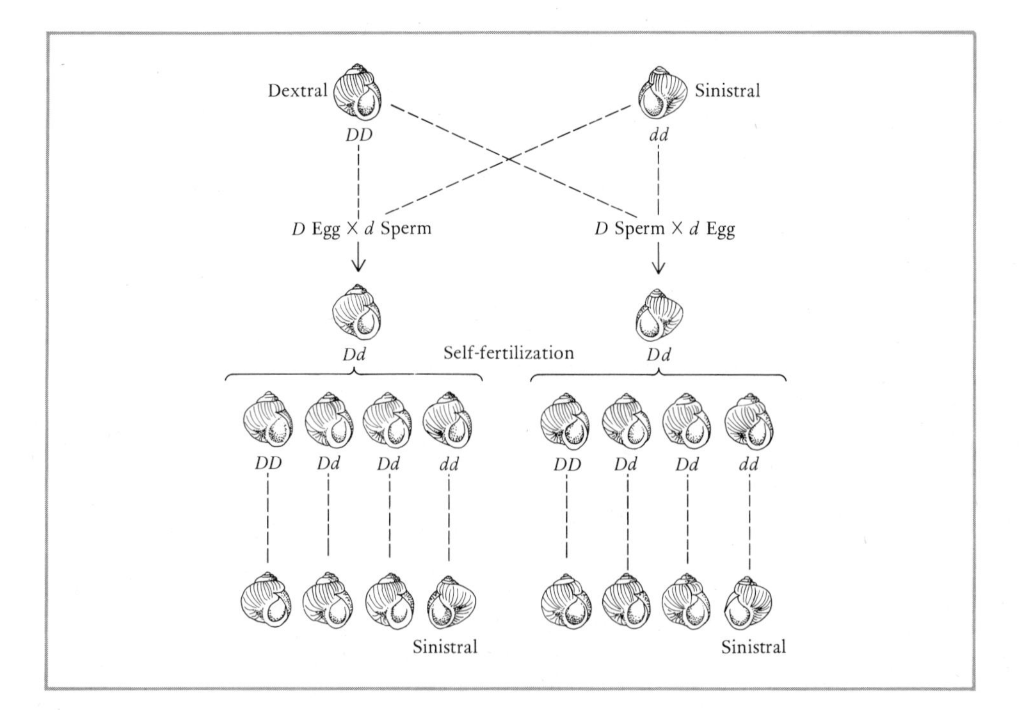

Figure 20–1. Inheritance of coiling in *Limnea peregra.* [Reprinted with permission of Macmillan Publishing Co., Inc. from *Genetics,* 2nd ed., by Monroe Strickberger. Copyright © 1976 Monroe W. Strickberger.]

established, is a permanent character, the effect lasts for the lifetime of the individual. In the next following generation the genotype of the maternal parent again determines the direction of coiling of all offspring, regardless of whether she herself is dextral or sinistral. *Limnea* is hermaphroditic and capable of self-fertilization or cross fertilization. In either case the genotype of the individual producing the eggs determines the coiling phenotype of the offspring. For example, a *Dd* female, whose mother was *dd*, will herself be sinistral, but will produce all dextral progeny because of the presence of the dominant gene *D* in her genotype.

A third example is a mutation of *Drosophila subobscura* called granddaughterless *(gs)* which results in the sterility of all progeny of *gs/gs* females. The character is expressed only via the egg cytoplasm and is independent of the genotype of the male parent. Sterility is caused by the absence or rudimentary nature of the reproductive organs. These organs arise from primordial germ cells formed by the migration of cleavage nuclei into a specialized region of the cytoplasm at the posterior end of the egg (see Figure 19–4). Evidently, eggs produced by *gs/gs* females lack a necessary cytoplasmic substance required for the formation of a normal gonad.

All of these and other instances of maternal effects can be considered examples of delayed Mendelian inheritance in that they stem from the activities of nuclear genes whose effects are expressed in the next following generation.

CYTOPLASMIC INHERITANCE

Cytoplasmic or extrachromosomal inheritance differs from the maternal effects described above in that the phenotype produced is not limited to the immediate progeny, but instead is inherited indefinitely by succeeding generations. The entities responsible for cytoplasmic inheritance exist apart from nuclear genes and appear capable of autonomous self-replication and of independent transmission to daughter cells. Since they are not associated with the apparatus of cell division, they do not segregate in a regular, predictable fashion at germ cell formation and thus give rise to non-Mendelian ratios. In higher organisms maternal inheritance through the cytoplasm of the egg (rarely through pollen or sperm) is a general characteristic and consequently the results of reciprocal crosses usually differ. Even when they do not differ, the varying non-Mendelian ratios observed indicate an origin other than nuclear genes.

Cytoplasmic inheritance in fungi can be recognized by the transmission of a character from one strain to another via heterokaryosis. It will be recalled that cells of a heterokaryon contain within a united cytoplasm the separate nuclei of each parental strain. Since these nuclei do not fuse, no opportunity is afforded for recombination, and therefore the uninucleate asexual spores formed by the heterokaryon give rise to homokaryons of each parental genotype, neither of which has been altered by residence in a common cytoplasm. Thus, if a phenotypic character is transmitted from one parental strain to the other by heterokaryosis, a cytoplasmic, as opposed to a nuclear, origin is inferred.

The kinds of characters transmitted through the cytoplasm are varied. As an illustration, P. Michaelis has shown that several attributes of plant growth and morphology are passed on for many generations through the egg cytoplasm of the willow herb *Epilobium*. Michaelis crossed two strains of this plant obtained from different geographical regions of Europe. Eggs of strain 1, the initial female parent, were fertilized by pollen of strain 2, the initial male parent, and the F_1 plants then served as females in a backcross which again used strain 2 pollen. These backcrosses were continued for 24 generations, the plants arising in each generation providing the eggs, and strain 2 providing the pollen. Assuming that the pollen contributed no cytoplasm to the zygote, it was reasoned that any cytoplasmic determinants present in the initial strain 1 female would be transmitted to each succeeding generation via the egg. At the same time the chromosomes of strain 1 would be replaced through continuous backcrossing by those of strain 2 plants. In the F_{24} generation Michaelis found that backcross progeny, despite their strain 2 genotype, still exhibited continuing differences from strain 2 plants with respect to male fertility, flower shape, growth pattern, and other characters. Since the chromosomes of F_{24} plants were derived from strain 2, such characters could be attributed only to cytoplasmic determinants transmitted for 24 generations through the egg. The cytoplasmic entities responsible for the inheritance of these character differences have not been identified in *Epilobium*, but in other instances cytoplasmic inheritance can be associated with specific cell organelles, chloroplasts and mitochondria.

Chloroplasts

One of the first recognized instances of cytoplasmic inheritance was described by Correns in 1909 and concerned green and white leaf variegation in four-o'clocks *(Mirabilis)*. Correns found that in one strain, regardless of whether the pollen was obtained from flowers on green, white, or variegated shoots, flowers borne on green shoots produced only green progeny and those on white-leaved stems yielded only white progeny. Flowers of variegated shoots gave rise to green, white, and variegated offspring, but in no constant and predictable ratios. Correns concluded that the character of green, white, or variegated leaves was inherited maternally through the cytoplasm. Since the green pigment of plants is associated with the chloroplast, inheritance through this organelle can be postulated as an explanation.

As another instance, E. Baur found in certain strains of geraniums *(Pelargonium)* that green plants crossed to those with white-margined leaves gave rise to green, white, or white-margined progeny, but in non-Mendelian ratios. Although the results of reciprocal crosses were the same, the varying, inconstant proportions of the different phenotypes exhibited by progeny suggested cytoplasmic inheritance. Baur postulated that plastids were contributed to the zygote by the pollen as well as by the egg, and the presence of plastids in male germinal cells of *Pelargonium* now supports this proposal.

A third example, somewhat similar to that described by Correns for four-o'clocks, has been analyzed by Rhoades in corn. A kind of green and white leaf striping, called *iojap*, is determined by the recessive nuclear gene *ij*. In crosses of *ij/ij* with green strains, where the maternal parent is green, all progeny are also green. However when the maternal parent is *ij/ij*, the progeny are green, white, or striped. When self-fertilized or in testcrosses to green, the green F_1's from this cross produce only green offspring, but the striped F_1's, used as the maternal parent, continue to produce green, white, and striped progeny in non-Mendelian and highly variable ratios. Rhoades has concluded that the striped iojap character results from the action of a nuclear gene on the chloroplast, and that once the character has been initiated, it is thereafter inherited maternally through the altered proplastid present in the egg cytoplasm.

It is now well established that plastids possess a genome of their own which enables them to perform a genetic function separate from that of nuclear chromosomes. Chloroplasts contain circular, double-stranded DNA molecules which resemble the genophores of bacteria in that chromosomal proteins, such as histones, are absent. Replication of the chloroplast genome is semiconservative and is mediated by an organelle-specific DNA polymerase. Chloroplast DNA molecules measure 40 μm in *Euglena* and 43 μm in corn, and it is estimated that this length is sufficient to code for approximately 300 average sized polypeptides. Each chloroplast contains several to many of these DNA molecules, there being about 20 per single plastid in higher plants.

Isolated chloroplasts exhibit active protein synthesis. They contain their own ribosomes which are similar to those of prokaryotes in that they are 70 S and composed of 16 S and 23 S subunits. Chloroplast ribosomes are also sensitive to the same substances (for example, chloramphenicol) that inhibit prokaryotic ribosomes, but are little affected by specific inhibitors (for example, cycloheximide) of the 80 S ribosomes of the surrounding cytoplasm. In addition to ribosomes chloroplasts also contain their own tRNA's and amino acid activating enzymes, and experimental evidence suggests that most of the RNA present in these organelles has been transcribed from chloroplast DNA by chloroplast RNA polymerase.

The chloroplast is bounded by a double membrane and at least some of the proteins of the inner of these two membranes are specified by genes of the organelle. The source of other proteins and enzymes of the plastid is much less certain, however, for many plastid characters are determined by nuclear genes. It has also been found that some enzymes or enzyme subunits characteristic of the chloroplast are synthesized in the cytoplasm and subsequently moved into the organelle. Despite all of these complexities, it is nonetheless clear that chloroplasts constitute a genetic and protein synthesizing system separate from that of the cell which contains them.

The genetics of eukaryotic cell organelles is much more difficult to analyze than is the genetics of Mendelizing nuclear genes. Until recently few well-defined phenotypic markers that could be directly associated with the chloroplast were available. Even with the use of such markers, the phenotype of the organelle itself cannot be scored, but only that of the cell which contains it. Adding to these problems is the fact that most plant cells contain more than one chloroplast, and each chloroplast contains several DNA molecules. Despite these difficulties, considerable progress has been made, particularly with the unicellular alga *Chlamydomonas,* which contains only a single chloroplast.

Chlamydomonas is haploid and the sexual cycle involves the fusion of cells of opposite mating types, mt^+ (♀) and mt^- (♂), mating type being determined by a single pair of nuclear genes (see Figure 6–26). Chloroplast genetics began with a report by Ruth Sager in 1954 that a mutation to streptomycin resistance *(sm_2)* in *Chlamydomonas* exhibited non-Mendelian inheritance. Crosses of wild type sensitive mt^+ cells with resistant mt^- cells yielded all sensitive progeny, whereas the reciprocal cross of resistant mt^+ with wild type sensitive mt^- produced essentially all resistant progeny. Although nuclear mutations that conferred a similar resistance were known, it appeared that in this case the inheritance of resistance to streptomycin was uniparental and transmitted only by the mt^+ parent. A number of other mutations that also exhibit uniparental inheritance through the mt^+ parent have since been identified. These include temperature-sensitive mutations, a nutritional requirement for acetate, and resistance to various antibiotics. They have been associated with the chloroplast genome because their pattern of inheritance follows that of the plastid which, in *Chlamydomonas,* is also transmitted to progeny almost exclusively through the mt^+ parent. In addition, mutations producing resistance to certain antibiotics result in a detectable alteration in chloroplast ribosomes.

Studies of the sexual cycle of *Chlamydomonas* have demonstrated that the cytoplasm and nuclei of cells of opposite mating types undergo fusion to form the diploid zygote. The two chloroplasts contributed one from each parent cell also fuse, but, for unknown reasons, only the chloroplast DNA of the mt^+ parent is ordinarily retained and transmitted via daughter chloroplasts to the products of meiosis and their descendents. The retention of the mt^+ chloroplast genome is not absolute, however, since very rarely chloroplast characters derived from the mt^- parent or from both parents appear in descendent cells. Sager and Ramanis have found that by irradiating mt^+ cells with low doses of ultraviolet before mating followed by photoreactivation, almost 100 percent biparental inheritance of plastid characters can be obtained. This technique has permitted an analysis of the segregation and recombination of chloroplast genes. For example, when this procedure is applied to the cross streptomycin-resistant $mt^+ \times$ wild type sensitive mt^-, the characters of resistance or sensitivity to this antibiotic begin to segregate from one another during the mitotic divisions which follow the completion of meiosis. As a result, cells that carry one or the other of these determinants are produced and these subsequently give rise to sensitive or resistant clones.

When two different chloroplast factors are used in a cross, along with ultraviolet irradiation of the mt^+ parent, recombinant phenotypes can be detected in some of the cells produced by the mitotic divisions which follow meiosis. As an illustration, a cross between two different and distinguishable acetate mutants, $ac_1 \times ac_2$, has been found to yield after meiosis segregant cells carrying ac_1 or ac_2, as well as some cells that are wild type or doubly mutant, and these latter two types must originate from recombination between parental type chloroplast DNA molecules. Sager and associates have performed crosses with eight different chloroplast mutants, and on the basis of the number of recombinants produced in the first few mitotic divisions after meiosis, they have prepared a circular linkage map on which all eight mutations can be positioned.

Although these and other studies provide a good beginning to an understanding of chloroplast inheritance, no conclusive information is yet available as to the time or mechanism of recombination. In addition, numerous nuclear genes play a role in the control of chloroplast development and function, and distinguishing between the effects of nuclear and chloroplast genes presents a most complex and formidable problem.

Mitochondria

Mitochondria are eukaryotic cell organelles which, like chloroplasts, are enclosed by a double membrane and constitute a discrete genetic and protein synthesizing system apart from the nucleus and cytoplasm (Figure 20–2). The

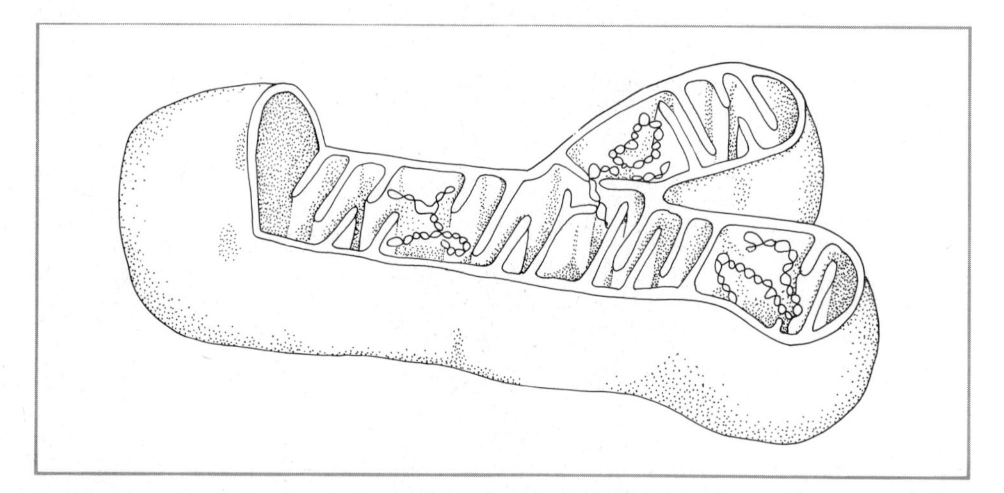

Figure 20–2. Three-dimensional representation of the possible arrangement of circular DNA within the mitochondrion. A branched mitochondrion typical of many cell types emphasizes the polymorphous structure of these organelles. The number of DNA molecules is variable, and these may be attached to portions of the membranes and coiled or at least folded inside the matrix compartments. [From M. M. K. Nass, *Science, 165,* 25–35, 4 July 1969. Copyright © 1969 by the American Association for the Advancement of Science.]

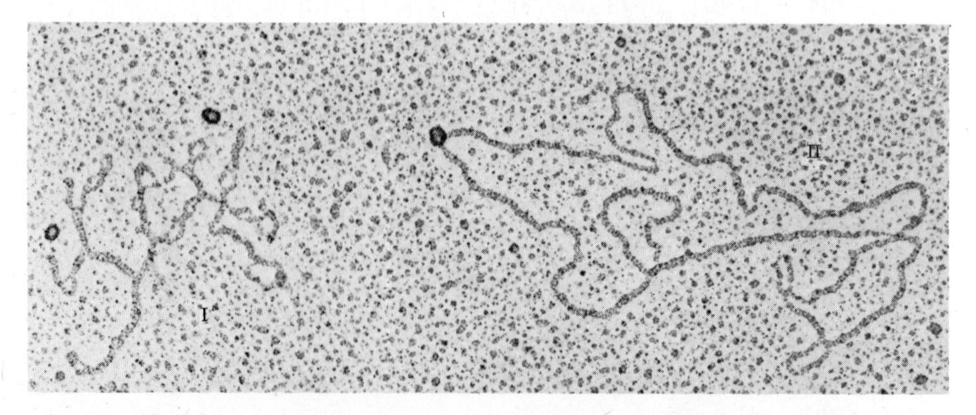

Figure 20–3. Electron micrograph of circular molecules of mitochondrial DNA isolated from *Drosophila eggs*. Left, twisted intact circle; right, open nicked circle. (53,000×). [From M. L. Polan et al., *J. Cell Biol., 56,* 580–589, 1973.]

double-stranded DNA molecules found in mitochondria are usually circular in form and of varying contour length depending on species. They are approximately 20 μm long in *Neurospora,* 25 μm in yeast, 30 μm in higher plants, and only 5 μm in metazoan animals (Figure 20–3). Although replication is semiconservative, it does not occur at the same time as chromosome duplication. Typically, numerous mitochondria are present in a eukaryotic cell, each mitochondrion containing more than one and sometimes several copies of the genome.

Mitochondria possess ribosomes and tRNA's different from those found in the cytoplasm or in the chloroplast. The mitochondrial ribosomes of animal cells are quite small, sedimenting at 55 S, but considerably larger ribosomes, 73 S, have been isolated from the mitochondria of fungi. The mitochondrial ribosomes of both plants and animals are similar to those of chloroplasts in that they are sensitive to substances that inhibit prokaryotic ribosomes, but not to those that affect the 80 S ribosomes of the cytoplasm. Approximately 10 to 15 tRNA's have been identified in the mitochondria of the toad *Xenopus* and 20 or more have been found in yeast.

Mitochondria are the site of cellular respiration, and they contain the enzymes of the electron transport chain through which most of the ATP of the cell is generated. Experimental evidence indicates that at least some of these enzymes or their subunits are specified by mitochondrial genes. Mitochondrial mutations (see below) frequently result in respiratory deficiencies which in some instances have been traced to a lack of cytochrome oxidase (cytochromes a, a_3), cytochrome b, and cytochrome c_1, all of which are components of the electron transport chain.

These complex enzymes, bound to the inner membrane, appear to be of dual origin, however. For example, seven different subunit polypeptides have been identified as comprising the cytochrome oxidase of *Neurospora,* but only one of these polypeptides is synthesized on mitochondrial ribosomes, the others being produced on the 80 S ribosomes of the cytoplasm. These findings suggest that information for the complete structure of cytochrome oxidase is encoded partly in the nucleus and partly in the mitochondrion, and similar findings have been reported for cytochrome b. However, F. Sherman and coworkers have demonstrated that in yeast another enzyme of the chain, cytochrome c, is specified by a nuclear, not a mitochondrial gene, and this appears to be the case in higher organisms as well. The small size of the mitochondrial genome, especially that of animal mitochondria, does not provide the capacity to code for more than a minimum number of proteins, and it is postulated that the majority of mitochondrial components are specified by nuclear genes and synthesized on cytoplasmic ribosomes. An intimate interrelationship thus exists between nuclear and mitochondrial loci with respect to the structure and function of this organelle.

Mitochondrial characters have been studied most intensively in bakers yeast, *Saccharomyces cerevisiae* (see Figure 6–27). The first such character was described by Ephrussi in 1956 and named *petite* because of the small size of the colony produced by mutant cells. Numerous different petites have since been identified and all are characterized by respiratory deficiencies stemming from lesions in the electron transport chain. Loss of the ability to carry out aerobic metabolism is not lethal to yeast cells, for they can survive by glycolysis alone as long as a fermentable energy source, such as glucose, is supplied in the medium. By this means both haploid and diploid petites can be maintained as vegetative colonies. Spore formation cannot be carried out, however, either by diploid petite cells or by a zygote resulting from a cross between haploid petites.

Three different kinds of petites are known. The first type, *segregational petites,* can be attributed to nuclear gene mutations because expected 1 : 1 Mendelian segregations are obtained in the ascospores arising from crosses of a petite of this kind with wild type. A second type is called a *neutral petite.* From a cross of a neutral petite with wild type, all progeny and their descendents are wild type, whether they are diploid vegetative cells produced by mitotic divisions of the zygote or whether they are ascospores arising from meiosis. In other words, as a result of such a cross the petite character disappears completely from both immediate and subsequent descendents. This behavior points to a cytoplasmic, rather than a nuclear origin, and continuing studies have demonstrated that many neutral petites lack mitochondrial DNA. The results of crosses between neutral petites and wild type can thus be attributed to the inheritance by progeny of normal mitochondria contributed by the wild type parent to the cytoplasm of the zygote.

The third type of petite is called a *suppressive petite.* In a cross with wild type varying fractions of diploid vegetative progeny of the zygote exhibit the petite character. The proportion of respiratory deficient progeny depends on the particular suppressive petite mutant used in the cross, and with some mutants up to 99 percent of the progeny are phenotypically petite and unable to respire aerobically. If the zygote from such a cross is induced by experimental means to undergo meiosis, non-Mendelian ratios are obtained, numerous asci containing only spores that give rise to petites. Suppressive petites are characterized by the loss of cytochromes a, a_3, b, and c_1, and biochemical analysis has indicated that the mitochondrial DNA of such petites has undergone gross losses and alterations in sequence.

Mitochondrial mutations in yeast that confer resistance to various antibiotics have also been described, and when such mutants are used in crosses, recombinant phenotypes can be recognized. In addition, the buoyant density, determined by ultracentrifugation, of mitochondrial DNA extracted from zygotes is intermediate as compared with the densities of the mitochondrial DNA obtained from parental types used in a cross of a suppressive petite with wild type. This finding indicates that physical exchanges between parental mitochondrial DNA molecules have occurred. Electron micrographs of young zygotes have demonstrated that after parental cell fusion mitochondria lose their structure, but that somewhat later they redifferentiate. It is postulated that during this period of reorganization frequent recombination occurs between the mitochondrial genomes contributed by the two parents of the cross.

Cytoplasmic mutations, called *poky,* have been identified in *Neurospora.* They are similar to the petite mutations of yeast in that the cytochromes of the electron transport chain are abnormal. When reciprocal crosses can be made, the poky phenotype is transmitted only by the larger female (protoperithecial) parent which contributes virtually all of the zygote cytoplasm. Thus, in the cross poky ♀ × normal ♂ all progeny are poky, while in the reciprocal cross normal ♀ × poky ♂ all progeny are normal. When heterokaryons are made between poky and nonpoky strains, poky homokaryons can be isolated whose nuclear genotype is normal and unchanged. All of these results indicate cytoplasmic inheritance of the poky character. It must be noted, however, that Mendelizing nuclear genes which affect respiration have also been identified in *Neurospora.*

Mitochondrial mutations have been reported in *Paramecium* and *Chlamydomonas,* as well as in yeast and *Neurospora.* It has been suggested that some inherited conditions in humans (myopathies), characterized by abnormal numbers of mitochondria or by giant mitochondria and for which a chromosomal or Mendelian basis has not been found, may in fact be mitochondrial mutations.

Origin of Eukaryotic Cell Organelles

Much attention has been directed toward the theory that mitochondria and chloroplasts arose, not gradually in the course of evolution, but as prokaryotic symbionts of primitive eukaryotic cells. Bacteria-like aerobic prokaryotes are proposed as ancestors of mitochondria, and blue-green algal-like prokaryotes as ancestors of choloroplasts. The association between a primitive eukaryotic cell and either or both types of prokaryotic cells could have occurred, perhaps many times, either fortuitously or by capture, and the presence of numerous chromosomal genes which specify organelle proteins is presumed to have resulted through the gradual transfer of such genes from the symbiont to the eukaryotic cell nucleus. Support for this theory is derived from the striking similarities which exist between modern prokaryotic cells and mitochondria and chloroplasts. In both, DNA molecules are generally circular, are not enclosed in a nuclear envelope, and histones are absent. Furthermore, ribosomes are 70 S or smaller and are sensitive to specific inhibitors which do not affect eukaryotic ribosomes. However, this theory of the endosymbiotic origin of organelles presupposes the anaerobic nature of primitive eukaryotic cells, and it is difficult to imagine how such cells could have survived in competition with aerobic and photosynthetic prokaryotes.

An opposing view postulates that primitive eukaryotic cells arose from the common ancestral type that also produced bacteria and blue-green algae. According to this hypothesis, both mitochondria and chloroplasts constitute features inherited from an aerobic and photosynthetic ancestor.

THE INHERITANCE OF PREFORMED STRUCTURE

Although it is clear that the majority of molecular constituents of a eukaryotic cell are derived from information contained within DNA and transcribed to RNA, the organization and assembly of these constituents and the way in which they are positioned to form a given cytoplasmic structure may be independent of the genes which specify the molecules themselves.

Such appears to be the case in protozoa. Ciliate protozoans have a surface pellicle of gelled cytoplasm which contains an elaborate pattern of cilia, basal bodies, and connecting fibers, and this cortical organization is perpetuated from one generation to the next. T. M. Sonneborn and coworkers have shown that in *Paramecium aurelia* if a small piece of pellicle is removed, rotated, and implanted back into the pellicle once more, the cilia on the inverted patch beat in a direction opposite to the cilia of the rest of the animal. Moreover, the inverted rows of cilia are inherited by all descendents, even through 800 cell generations and numerous sexually produced generations. Thus, once initiated, a specific pattern of macromolecular organization is inherited without change, apparently determined by the preexisting state of this region of the pellicle, and not by the action of nuclear genes.

Similar findings with respect to the inheritance of pellicle organization have been reported for *Tetrahymena* by D. L. Nanney. The gullet region of *Paramecium* is also self-reproducing, for after loss or major injury no restitution occurs in descendent cells despite the presence in these cells of a normal nuclear genome. Again, preexisting molecular organization determines the assembly of a new, reproduced structure.

In multicellular organisms the flatworm, *Stenostomum,* exhibits a somewhat similar cytoplasmic control of structure, for abnormal individuals produced by surgical intervention perpetuate their changed structure in asexually produced progeny. It is intriguing to speculate that mechanisms of this type may play a role in the development of eggs of higher organisms, but whether or not they do is unknown.

INFECTIOUS INHERITANCE IN EUKARYOTES

A number of microorganisms have adopted a symbiotic or infectious relationship with various eukaryotes whereby they are transmitted to the progeny of their hosts. One of the better known examples is a small bacterium found in the cytoplasm of certain strains of *Paramecium aurelia.* The presence of this symbiont converts such strains to "killers," for they release into the culture medium a substance toxic to sensitive strains of paramecia. The condition was first described by Sonneborn, who found that presence of the symbiont, named *kappa,* was dependent upon a dominant nuclear gene *K,* killers being either *KK* or *Kk* and paramecia unable to maintain kappa being *kk.* Since killing does not occur during sexual union (conjugation) between killer and sensitive cells, the inheritance of the ability to harbor kappa particles could be determined. Although kappa-free strains of paramecia exhibit sensitivity to the toxin regardless of genotype, the presence of kappa bacteria within the host cytoplasm confers immunity. Besides kappa, viruslike particles also appear in killer paramecia and it has been proposed that these are temperate bacteriophages released from dead, lysed kappa bacteria. Thus the virus is a parasite of kappa which in turn is a symbiont of *Paramecium.* The killing of sensitive paramecia has been attributed to a toxic protein formed during the lytic cycle of the bacteriophage. At least eight other bacterial symbionts also occur in *Paramecium.* These produce a variety of effects on the host and, in most instances, their maintenance appears to depend on the presence of a host gene.

Infectious inheritance has also been reported in *Drosophila.* One of the first such cases, described by P. L'Héritier and G. Teissier, concerns the phenotype of carbon dioxide (CO_2) sensitivity. Normal flies recover immediately from anesthesia with CO_2 and show no aftereffects, whereas CO_2-sensitive strains are permanently paralyzed by exposure to this gas. The character is transmitted by sensitive females to all progeny, but only rarely do sensitive males pass the trait to offspring, and if so, these offspring, while themselves sensitive to CO_2, usually fail to contribute the character to the next generation. The causative agent is a virus named sigma which can be transmitted through the germ line of females, but only infrequently through spermatozoa. Sigma is infectious and extracts of sensitive flies injected into resistant adults transform them into sensitives after an initial latent period. A number of mutations of sigma have been described which alter the infectivity of the particle. In addition, several *Drosophila* genes are known to promote resistance to infection.

A trait called sex-ratio *(SR)*, which is found in a number of *Drosophila* species, has also been traced to the effects of a hereditary symbiont, in this case, a spirochete. The trait shows maternal transmission through the cytoplasm of the egg. *SR* females produce virtually all daughters since the presence of the agent causes the death of male zygotes during early developmental stages. Normal flies can be infected by injecting hemolymph or tissue extracts obtained from *SR* individuals. As in the case of *sigma*, maintenance of the spirochete is dependent on *Drosophila* nuclear genes.

Numerous other instances of the inheritance of a symbiotic or infectious particle could be cited, for example, the controlling elements of higher plants and *Drosophila* which in many respects resemble episomal agents of foreign origin. Although none of these many particles or elements are normal components of the host cell or organism, their presence results in a recognizable phenotype and they are inherited by daughter cells and in many cases by subsequent generations.

Tumor Viruses

Both DNA and RNA tumor viruses have been isolated from a variety of vertebrates including primates. The first activity undertaken by the virus upon entering a host cell is the replication of its genome, usually by a virion associated polymerase. Double-stranded DNA viruses produce double-stranded DNA replicates, but tumor viruses with an RNA genome first form RNA-DNA hybrid molecules from which DNA-DNA molecules are synthesized. The enzyme required for this synthesis is a DNA-dependent RNA polymerase (reverse transcriptase) which is introduced to the host along with the viral RNA. In the form of double-stranded DNA molecules, both the DNA and RNA tumor viruses may be covalently inserted into chromosomes of the host as a provirus, or prophage, and in this form the viral genome undergoes replication along with the chromosomes and is transmitted to daughter cells. In the presence of the prophage a cell may undergo a number of phenotypic changes with respect to morphology, motility, metabolism, cell surface characters, and rate of multiplication and invasiveness, which are collectively referred to as transformation. Although some of these changes may be initiated by a response of the cell to the presence of the parasite, others are thought to originate from the activity of viral gene specified products. For example, temperature-sensitive mutants isolated from some strains of mammalian viruses (for example, polyoma) are unable to effect transformation at a restrictive temperature and this observation suggests that at least some aspects of transformation are governed by viral genes.

RNA tumor viruses have been positively associated with the origin of mouse leukemia and mammary cancer and evidence for their presence in human lymphomas, leukemias, breast cancer, and other tumors has been brought forth by a number of investigators. A detailed description of the RNA and DNA tumor viruses may be found in the reviews by Baltimore (1971), Huebner and Todaro (1969), and Temin (1974).

TRANSMISSIBLE PLASMIDS OF BACTERIA

Plasmids are hereditary elements of bacteria present in the cytoplasm apart from the bacterial genophore. They are capable of autonomous replication and can promote their own transfer to other cells by conjugation. The best known example is the *F* factor of *E. coli* which can be considered an episome as well as a plasmid since it integrates stably within the bacterial genophore (see Chapter 13). Numerous other plasmids have been described, most of them unable to achieve integration in the bacterial genophore. These plasmids are transferred from cell to cell via pili. In some cases the pilus is similar to that formed in the presence of the sex factor, but other types of pili also occur. These are distinguishable from one another and from *F*-type pili by antigenic properties and the ability or inability to adsorb certain phages (see Figure 13–8).

Plasmid DNA occurs as double-stranded closed circles. Although replication is autonomous, control by host genes is evident in that only a limited number of replicates (usually one or two) is permitted, and mutations of host genes can affect plasmid duplication. During conjugation transfer of plasmid DNA is achieved by the same processes utilized by the sex factor (see Figure 13–9). Transfer of the host genophore occurs only rarely since most autonomous plasmids fail to integrate within the host genophore. Two types of plasmids are of general occurrence: colicin factors and R factors.

Colicins are bacteriocidal proteins produced by many different strains of enteric bacteria. Synthesis of these substances is controlled by genes carried on colicin plasmids, called *Col factors,* and bacterial cells which harbor such plasmids are termed colicinogenic. A number of colicins have been identified, each specific for a given sensitive strain of bacteria. Some Col factors require the concurrent presence of the sex factor for transfer during conjugation, but others may promote their own transfer with great efficiency in the absence of *F.* In addition, Col factors may enter a recipient cell via transduction.

R factors are plasmids that contain two different types of genetic determinants: genes that confer resistance to a variety of drugs and genes that control the replication and transfer of the plasmid itself. This latter component is called the *resistance transfer factor (RTF).* R factors promote the formation of their own pili by which they are transferred from cell to cell within the same or different species. Since R-type plasmids contain genes conferring resistance to numerous antibiotics, they are of great importance from the standpoint of public health.

Plasmids and Recombinant DNA

It has been discovered that after treatment with calcium chloride ($CaCl_2$) *E. coli* cells can be "transformed" by purified DNA isolated from plasmids, as well as from such viruses as λ. Plasmid DNA introduced in this fashion appears as closed circles within the host cytoplasm and is biologically functional. If R factor DNA is used, transformed cells become drug resistant, and the R factor DNA is capable of replication and transfer to new cells. This technique for the introduction of exogenous DNA into *E. coli* has been exploited in conjunction with the use of *restriction enzymes.*

Restriction enzymes are specified by genes present in the bacterial genophore as well as by some loci within plasmids. They act as endonucleases and function to protect the cell from invading viruses by producing single-strand cuts between specific nucleotides of DNA. As a result, foreign DNA can be fragmented and subjected to degradation by exonucleases. At the same time the genophore of the cell is protected from destruction by the activities of additional enzymes, called *methylases,* which add methyl groups specifically to those nucleotides subject to the action of restriction enzymes. The presence of the methyl group prevents cleavage of the DNA and provides a means whereby "self" DNA can be distinguished from foreign DNA. The tight coordination of the action of these two types of enzymes has suggested to some investigators that in the intact cell both activities may be carried out by one complex enzyme aggregate.

The specificity of restriction enzymes has been subjected to intense investigation. One such enzyme, EcoRI, has been found to recognize a specific complementary sequence of six nucleotide pairs and to produce cuts between the same two nucleotides within each DNA chain, as shown below (arrows indicate the cuts).

5′ . . . . G↓A A T T C . . . . 3′
3′ . . . . C T T A A↑G . . . . 5′

A different restriction enzyme, EcoRII, recognizes a complementary sequence of five bases and again cleaves each DNA chain at equivalent sites.

5′ . . . . ↓C C A G G . . . . 3′
3′ . . . . G G T C C↑ . . . . 5′

Since many similar sequences can be expected to occur within the extremely long DNA molecule which forms the genome of a phage or a plasmid, the action of these enzymes serves to cut the DNA into a number of segments, each of which has single-stranded complementary termini. Such termini are cohesive and can form hydrogen bonds with the complementary termini of any other fragment.

This tendency of cohesive ends to anneal has been used *in vitro* to produce various types of artifically recombined DNA which could not be generated by the natural means of genetic recombination. Such experiments employ purified restriction enzyme, isolated plasmid DNA, and donor DNA obtained from some source such as a virus or even a eukaryote. After cleavage of the DNA, the fragments of donor and plasmid DNA are allowed to anneal, whereupon they form circles held together by hydrogen bonds formed between the cohesive termini (Figure 20–4). Upon treatment with polynucleotide ligase, the circles become covalently closed and can be used to transform bacterial cells.

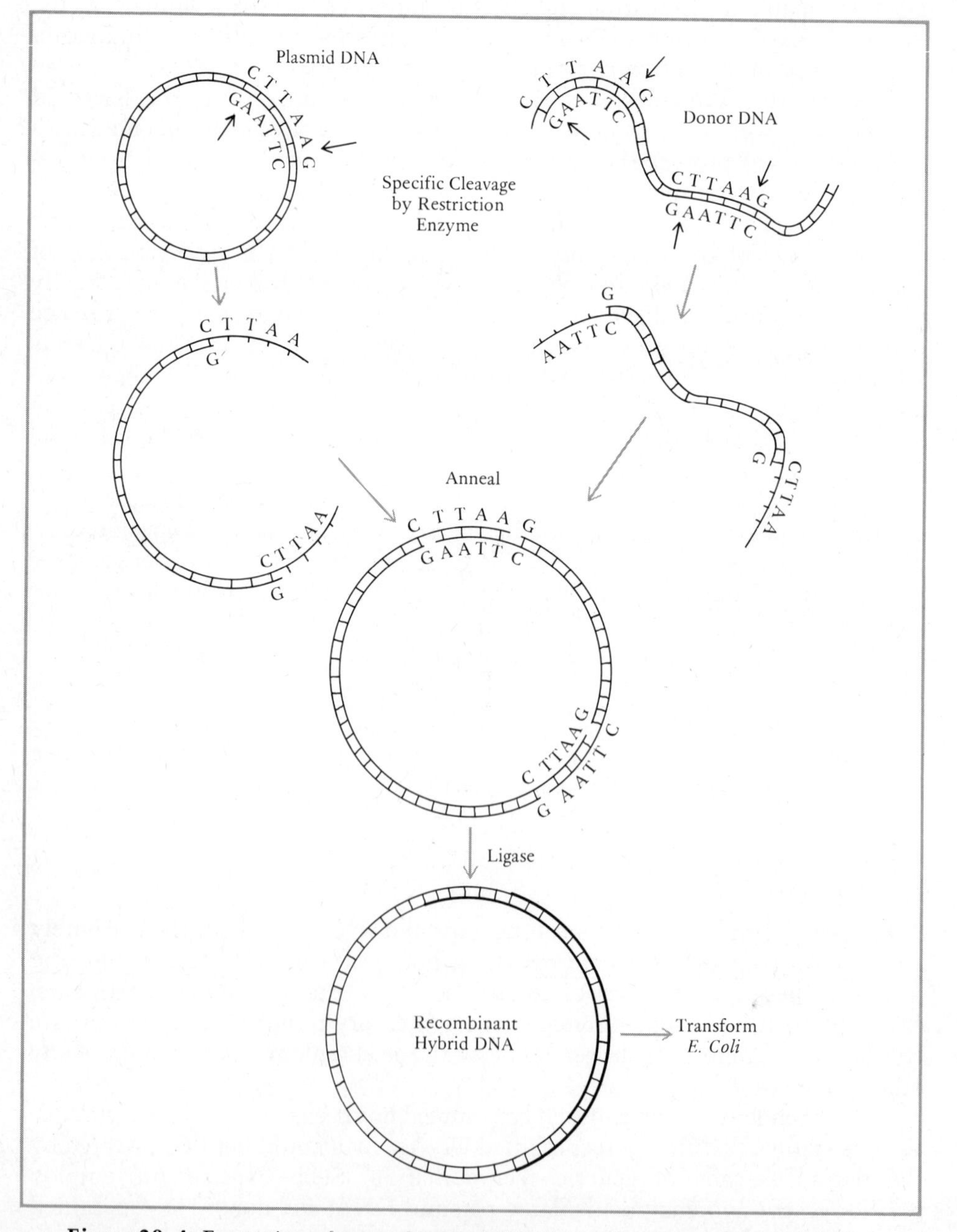

Figure 20–4. Formation of recombinant DNA from plasmid and donor DNA.

Such reconstituted, hybrid plasmids are capable of biological function within the host, for the expression of both donor and plasmid genes can be demonstrated. By these techniques DNA from a variety of bacterial species as well as eukaryotic species can be introduced and maintained in *E. coli.* For example, ribosomal DNA from the toad *Xenopus* has been successfully inserted into plasmid DNA and used to transform *E. coli.* After entrance, the hybrid plasmid was capable of replication and transmission, and the activity of *Xenopus* rDNA was demonstrated by the production of rRNA which was complementary to, and hybridized with *Xenopus* rDNA. A similar experiment utilizing *Drosophila* rDNA has also been performed.

These techniques for recombining DNA offer a means of producing new combinations of genes not found in nature. Such "new species" hold a potential for great benefits and also for disaster, and the wisdom of carrying out these experiments has been the subject of much debate. On one hand, it is visualized that bacteria capable of synthesizing a wide variety of medically important substances such as hormones and antibiotics might be produced. On the other hand, there is the danger that accidental escape of *E. coli* cells containing recombinant DNA might transmit resistance to antibiotics to pathogenic organisms. In addition, genes derived from pathogens could inadvertantly be transferred to *E. coli.* The insertion into a plasmid or bacterial virus of random pieces of eukaryotic DNA could also lead to the spread of tumor viruses contained within the donor DNA as prophages. A major hazard is the use of *E. coli* itself, for this species is ubiquitous in distribution and lives within the human intestine. Although the K-12 strain used in these experiments is attenuated and considered unable to compete successfully with organisms normally present in the intestine, there is no guarantee that recombinant plasmids contained in accidentally ingested cells would not be transferred to other, more viable strains or species. In summary, it is clear that if the benefits envisioned are to be realized and the hazards avoided, great care in the containment of "engineered" bacteria and viruses must be practiced.

PROBLEMS

20–1. What criteria can be used to distinguish extrachromosomal (cytoplasmic) inheritance from chromosomal inheritance?

20–2. Suppose that in mice muscle fiber degeneration occurs in strain 1, but not in strain 2. Crosses of strain 1 females with strain 2 males produce F_1's all of whom show muscle degeneration. F_1's of crosses of strain 2 females with strain 1 males are normal. What criteria, crosses or procedures would you use to determine if the condition is due to a sex-linked nuclear gene, cytoplasmic inheritance, or maternal influence through the placenta or milk?

20–3. In *Aspergillus* a sexually fertile strain forms white conidia, in contrast to the green conidia formed by a sexually sterile strain. When heterokaryons are made between these two strains, the conidia produced are either green or white, but all give rise to fertile homokaryons. How would you explain the inheritance of spore color and sterility?

20–4. In a given cross fertilizing plant pale green leaves are produced in the homozygous presence of the recessive nuclear gene *p*. In another strain homozygous for the dominant allele *P* pale green leaves are produced due to a chloroplast mutation.

The phenotypes of the two strains are indistinguishable from one another. If you were given plants of each type, what crosses would you make to distinguish those carrying the nuclear mutation from those carrying the chloroplast mutation? (Assume that no plastids are contributed to the zygote by the pollen.)

20–5. In *Chlamydomonas* resistance to streptomycin may be controlled by a chromosomal or by a chloroplast gene. If a strain carrying both nuclear and chloroplast genes for resistance is crossed to a sensitive strain, what results are expected in the haploid zoospores formed by meiosis when,

a. the mt^+ parent is resistant?

b. the mt^+ parent is sensitive?

20–6. In yeast a haploid segregational petite, otherwise normal, was crossed to a neutral petite, also otherwise normal. If the diploid zygote undergoes meiosis to produce ascospores, what proportion are expected to be petite in phenotype?

20–7. The snail *Limnea* can undergo self-fertilization as well as cross fertilization. A self-fertilizing snail produces all dextral progeny. When some of these progeny undergo self-fertilization, one-fourth of the offspring are sinistral. What was the genotype of the P_1 snail?

20–8. In sugar beets tissue from a male-sterile strain was grafted onto a male-fertile strain which produced seeds, all of which developed into male-sterile plants. What explanation can you offer for this result?

REFERENCES

BALTIMORE, D., 1971. Expression of animal virus genomes. *Bact. Rev., 35,* 235.

BERG, P., D. BALTIMORE, H. W. BOYER, S. N. COHEN, R. W. DAVIS, D. S. HOGNESS, D. NATHANS, AND N. D. ZINDER, 1974. Potential biohazards of recombinant DNA molecules. *Proc. Nat. Acad. Sci., 71,* 2593.

BOYER, H. W., 1971. DNA restriction and modification mechanisms in bacteria. *Ann. Rev. Microbiol., 25,* 153.

GILLHAM, N. W., 1974. Genetic analysis of the chloroplast and mitochondrial genomes. *Ann. Rev. Genet., 8,* 347.

HUEBNER, R. J., AND G. J. TODARO, 1969. Oncogenes of RNA tumor viruses as determinants of cancer. *Proc. Nat. Acad. Sci., 64,* 1087.

MAHLER, H., 1973. Biogenetic autonomy of mitochondria. *Crit. Rev. Biochem., 1,* 381.

MEYNELL, G. G., 1972. *Bacterial Plasmids.* Macmillan, London.

NANNEY, D. L., 1968. Cortical patterns in cellular morphogenesis. *Science, 160,* 496.

PREER, J. R., JR., 1971. Extrachromosomal inheritance. *Ann. Rev. Genet., 5,* 361.

PREER, J. R., JR., L. B. PREER, AND A. JURAND, 1974. Kappa and other endosymbionts in *Paramecium aurelia. Bact. Rev., 38,* 113.

RAFF, R. A., AND H. R. MAHLER, 1972. The nonsymbiotic origin of mitochondria. *Science, 177,* 575.

RAVEN, P. H., 1970. A multiple origin for plastids and symbionts. *Science, 169,* 641.

ROODYN, D. B., AND D. WILKIE, 1968. *The Biogenesis of Mitochondria.* Methuen, London.

SAGER, R., 1972. *Cytoplasmic Genes and Organelles.* Academic Press, New York.

SONNEBORN, T. M., 1970. Gene action in development. *Proc. Roy. Soc. Lond. (B), 176,* 347.

TEMIN, H. M., 1974. On the origin of RNA tumor viruses. *Ann. Rev. Genet., 8,* 155.

UZZELL, T., AND C. SPOLSKY, 1974. Mitochondria and plastids as endosymbionts: A revival of special creation? *Amer. Sci., 62,* 334.

WILKIE, D., 1970. Reproduction of mitochondria and chloroplasts. *Sympos. Soc. Gen. Microbiol., 20,* 381.

Chapter 21

GENE FREQUENCIES IN NATURAL POPULATIONS

Up to this point we have been concerned primarily with individual crosses or individual examples of gene expression and have paid little attention to whether or not a given mutation or alteration in genotype would persist in a natural population or, if it did so, would have any impact on the character and genetic structure of the population leading to evolutionary change. A consideration of these matters requires that we shift our focus to the larger group, since evolution occurs in populations and not in individuals.

The branch of evolutionary biology which seeks to analyze the forces leading to evolution is population genetics. It should be understood from the outset that population genetics is a different kind of genetics than that described elsewhere in this book, for it is primarily theoretical, rather than experimental. Using the rules of Mendelian inheritance and probability, population geneticists attempt to construct mathematical models to describe the effects of mutation, natural selection, and other factors on the genetic composition of populations. The theoretical foundations of the science were laid by S. S. Chetverikov, J. B. S. Haldane, R. A. Fisher, and S. Wright, and their work has since been extended by others to embrace complex mathematical formulations. Our review of population genetics is, of necessity, limited to the more elementary considerations.

CHARACTERISTICS OF POPULATIONS

A population can be defined as a community of sexually interbreeding organisms. Theoretically, such a definition could include a whole species having a wide geographic range, but potentially capable of breeding with one another. A more reasonable definition for our purposes, however, would be a community of individuals in which free access to other members of the community as potential mates is not restricted by distance or other geographic limitation.

This situation is a radical departure from the laboratory condition. In the laboratory matings and genotypes are controlled, and the frequencies of the alleles under consideration are selected in advance according to the dictates of the experiment. For example, if an experiment begins with an outcross between individuals homozygous for different alleles *(AA × aa),* the input or frequency of each allele is 50 percent of the total genes at the locus. In contrast, the members of natural populations may be homozygous or heterozygous for a pair of alleles and if mating is random, all types of crosses will occur. In addition, the alleles of a given gene are usually present in widely different frequencies from those deliberately chosen in the laboratory, for one allele may be present in 99 percent of individuals and the other in only 1 percent. Under these conditions the typical Mendelian ratios obtained as the result of controlled experiments cannot be expected.

If we were to inspect the members of a natural population, we would find that while all were superficially alike, variations among individuals would be evident in morphological characteristics such as color, stature, or size of some body part and in biochemical attributes such as blood groups, isozymes, and so on. This phenotypic variation could be ascribed in part to genotypic variation and to the presence in different individuals of different alleles of one or more genes, such allelic states having arisen by mutation. This inherent variation present in natural populations is the raw material for evolution.

Populations are not static entities. They undergo growth and their identity or overall character is subject to the progressive changes in time and space known as evolution. If the presence of a given gene (or genes) endows its carriers with a phenotype conferring greater reproductive success than that enjoyed by the bearers of an alternate allele, then the carriers of the favorable gene will contribute more progeny to the next and succeeding generations. As a consequence, over the course of time, the favorable allele and the phenotype which it confers will increase in frequency in the population at the expense of the less favorable allele, which will decrease in frequency. Such progressive changes in the frequencies with which different alleles occur in a population constitute evolution and are the basis for the progressive alterations in the phenotypes of organisms which have occurred over geologic time and are preserved in the fossil record. This concept of ongoing change as the result of the differential reproduction of favorable variations is the cornerstone of Darwin's theory of evolution.

If we wished to analyze the genetic constitution of a population at any given time, we would begin by determining the frequency with which the different allelic states of one or more genes occurred. This could be done by examining a random segment of the population, large enough to reduce sampling errors, for a distinctive phenotype caused by the gene being studied. Of the total genes of the locus present in the sample, the proportion that were the allele in question would represent the gene frequency of that allele. As an illustration we can use the autosomal locus for peptidase A, an enzyme which hydrolyzes dipeptides and is present in red blood cells and most other tissues of humans. Two codominant alleles, A_1 and A_2, specify the enzymes A1 and A2, the presence of either being detectable by electrophoresis. Heterozygotes contain a hybrid enzyme, A1A2, also distinguishable by electrophoresis, so that all three phenotypes and genotypes are readily identifiable. Although the A_2 allele is extremely rare in some parts of the world, it is fairly common in Africa. In our hypothetical example, suppose we assayed the blood of 500 people chosen at random and found that of these 405 or 81 percent were A_1A_1, 90 or 18 percent were A_1A_2, and 5 or 1 percent were A_2A_2. Codominance results in the same phenotypic and genotypic ratios, namely, 405 : 90 : 5 (or .81 : .18 : .01). Obviously, this is not a classical 1 : 2 : 1 Mendelian ratio, and a moment's reflection will tell us why. The 1 : 2 : 1 ratio could be obtained only from a cross of heterozygotes, that is, $A_1A_2 \times A_1A_2$, and in our population all types of crosses take place, such as $A_1A_1 \times A_2A_2$, $A_1A_1 \times A_1A_1$, and so on, and thus our sample represents the overall result of every kind of mating that can occur.

Our next step would be a determination of the proportion of the total genes at the peptidase A locus of our sample that are A_1 or A_2. Since we have surveyed 500 persons and each is diploid, the total number of peptidase genes in our sample is 1000. Of these 1000, the 405 A_1A_1 persons contributed 810 A_1 alleles ($405 \times 2 = 810$), the 90 A_1A_2 persons contributed 90 A_1 and 90 A_2 alleles, and the 5 A_2A_2 persons contributed 10 A_2 alleles (Table 21–1). By dividing the total number of each of the alleles by the number of alleles in the sample, we can readily obtain the proportion or frequency of each gene in the sample. To generalize, we can let N = the number of diploid individuals in the sample, $2N$ = the total number of genes, N_1 = the number of A_1 alleles, N_2 = the number of A_2 alleles, and p_1 and p_2 = the frequency of A_1 and A_2 alleles, respectively. Thus,

$$p_1 = \frac{N_1}{2N} \quad \text{and} \quad p_2 = \frac{N_2}{2N}$$

TABLE 21–1
Determination of Allele Frequency in a Sample Population

Number of persons	Genotypes	Number of A_1 alleles	Number of A_2 alleles
405	A_1A_1	810	—
90	A_1A_2	90	90
5	A_2A_2	—	10
500		900	100
		Frequency of A_1 = 900/1000 = 0.9	Frequency of A_2 = 100/1000 = 0.1

In our example the frequency of gene $A_1 = p_1 = 0.9$ and the frequency of gene $A_2 = p_2 = 0.1$. Note that these frequencies sum to 1.0 ($0.9 + 0.1 = 1.0$), because together they represent 100 percent of the genes at this locus in our sample. Since the sample of 500 persons was randomly chosen, we are safe in assuming that these frequencies are representative of the population itself. A gene frequency can also be viewed as the probability that any particular gene, selected at random from the population, will be the allele in question. Since progressive changes in gene frequency cause evolution, a gene frequency is also an evolutionary quantity possessed by a population at some given time.

HOW GENES COMBINE IN POPULATIONS

The basic theorem describing how genes combine in populations was derived independently from Mendelian principles by G. H. Hardy, an English mathematician, and by W. Weinberg, a German physician, in 1908. The Hardy-Weinberg theorem assumes the following conditions.

1. Individuals within a population mate randomly. Therefore, the probability of any two types mating is the product of the probabilities of the two types involved.

2. The frequency with which any allele is present in the gametes (eggs or sperm) is the same as the frequency of the allele in the population itself.

3. The offspring produced by random mating have equally viable phenotypes, that is, are equally likely to survive and be scored. The genotypes are also equally fecund.

4. All factors which could produce a change in gene frequency are absent. These factors include mutation, selection, random genetic drift, and the influx of new individuals from a different population.

5. The population is infinitely large, or at least so large that sampling errors due to chance can be disregarded.

With the above assumptions all factors leading to change are eliminated, and therefore the allele frequencies of the parental generation become the determinants of the allele frequencies of the next succeeding generation. To demonstrate this fact, it is helpful to explore the concept of a *gene pool.* Since the population is mating at random, we can ignore individuals and consider instead that all members of the population are contributing their alleles to a large pool of gametes and that the various combinations of gametes to form zygotes is dependent on chance. Using the previous example of the peptidase A_1 and A_2 alleles, where the frequency of allele $A_1 = p_1 = 0.9$ and the frequency of allele $A_2 = p_2 = 0.1$, the probability of an A_1 gamete meeting another A_1 gamete and forming an A_1A_1 zygote is the product of the frequencies with which these kinds of gametes occur in the gene pool. Thus, $0.9 \times 0.9 = 0.81$, that is, 81 percent of the zygotes of the next generation should be A_1A_1.

Inspection of Figure 21–1 will demonstrate that on the basis of predictions made from the frequencies of gametes in the gene pool, the proportions of genotypes and phenotypes of the next generation will be the same as those of the parental generation. In addition, since the frequencies of alleles A_1 and A_2 in the new generation have not changed, it can be seen that random mating has established an equilibrium between these gene frequencies and genotypes after one generation. Such an equilibrium is called a Hardy-Weinberg equilibrium.

The gene pool concept makes predictions on the assumption of random union of gametes which are identical to those that would be predicted from an analysis of all of the possible matings that could occur in the population. These random matings and their frequencies are shown in the checkerboard of Figure 21–2, and the data from this checkerboard are summarized in Table 21–2. It can be seen that the results of this analysis are identical with those previously derived from gene pool considerations.

The derivation of the Hardy-Weinberg equilibrium can be generalized by the use of symbols. Assuming there are no other alleles, if p_1 = the frequency of allele A_1 and p_2 = the frequency of allele A_2 in the parental generation, then $p_1 + p_2 = 1$, and gametes in the frequencies of p_1 and p_2 will be formed. Random union between these gametes will then produce the following offspring.

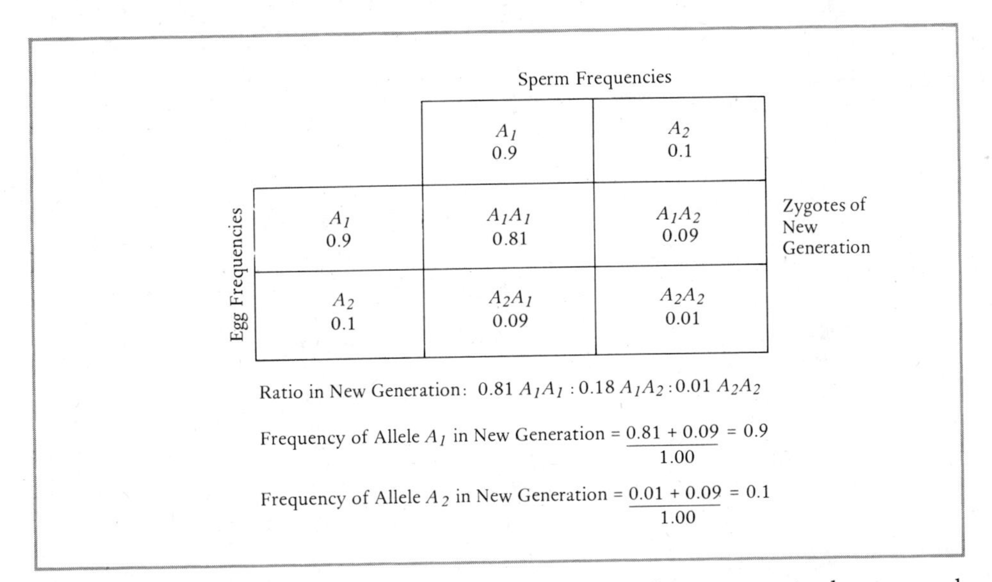

Figure 21–1. Results of random combinations of A_1 and A_2 gametes in the gene pool where the frequency of allele A_1 = 0.9 and that of allele A_2 = 0.1.

Genotypes of Females	Genotypes of Males: A_1A_1 (0.81)	A_1A_2 (0.18)	A_2A_2 (0.01)
A_1A_1 (0.81)	0.6561	0.1458	0.0081
A_1A_2 (0.18)	0.1458	0.0324	0.0018
A_2A_2 (0.01)	0.0081	0.0018	0.0001

Figure 21–2. Random matings and their frequencies, where A_1A_1 = 0.81, A_1A_2 = 0.18, and A_2A_2 = 0.01.

TABLE 21–2
Frequencies of Progeny Resulting from the Matings Shown in Figure 21–2

Matings	Mating frequency	Resulting progeny types	Frequencies of progeny types		
			A_1A_1	A_1A_2	A_2A_2
$A_1A_1 \times A_1A_1$ 0.81×0.81	0.6561	all A_1A_1	0.6561	—	—
$A_1A_1 \times A_1A_2$ 0.81×0.18	0.1458	$1/2A_1A_1, 1/2A_1A_2$	0.0729	0.0729	—
$A_1A_1 \times A_2A_2$ 0.81×0.01	0.0081	all A_1A_2	—	0.0081	—
$A_1A_2 \times A_1A_1$ 0.18×0.81	0.1458	$1/2A_1A_1, 1/2A_1A_2$	0.0729	0.0729	—
$A_1A_2 \times A_1A_2$ 0.18×0.18	0.0324	$1/4A_1A_1, 1/2A_1A_2, 1/4A_2A_2$	0.0081	0.0162	0.0081
$A_1A_2 \times A_2A_2$ 0.18×0.01	0.0018	$1/2A_1A_2, 1/2A_2A_2$	—	0.0009	0.0009
$A_2A_2 \times A_1A_1$ 0.01×0.81	0.0081	all A_1A_2	—	0.0081	—
$A_2A_2 \times A_1A_2$ 0.01×0.18	0.0018	$1/2A_1A_2, 1/2A_2A_2$	—	0.0009	0.0009
$A_2A_2 \times A_2A_2$ 0.01×0.01	0.0001	all A_2A_2	—	—	0.0001
Totals	1.000		0.81	0.18	0.01

Frequency of allele $A_1 = 0.9$
Frequency of allele $A_2 = 0.1$

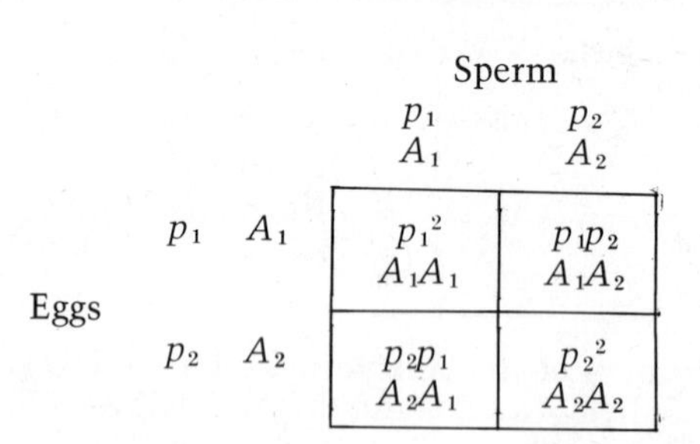

From the above, the genotype frequencies of the offspring can be expressed as:

$$p_1^2 = \text{the frequency of the homozygous class } A_1A_1$$
$$2p_1p_2 = \text{the frequency of heterozygotes } A_1A_2 + A_2A_1$$
$$p_2^2 = \text{the frequency of the homozygous class } A_2A_2$$

Overall, the distribution of genotypes in the progeny is thus: $p_1^2 + 2p_1p_2 + p_2^2 = 1$, which is the expansion of the binomial $(p_1 + p_2)^2$. The frequency of the A_1 and A_2 alleles in the progeny can be derived from these genotype frequencies as follows.

The frequency of gene A_1 is $p_1 = N_1/2N$ and the values of N_1 and $2N$ are:

$$N_1 = p_1^2 + p_1p_2$$
$$2N = p_1^2 + 2p_1p_2 + p_2^2$$

Substitution of these values in the equation gives us

$$p_1 = \frac{p_1^2 + p_1p_2}{p_1^2 + 2p_1p_2 + p_2^2} = \frac{p_1(p_1 + p_2)}{(p_1 + p_2)(p_1 + p_2)} = \frac{p_1}{(p_1 + p_2)} = p_1 \text{ (since } p_1 + p_2 = 1).$$

Similarly, the frequency of allele $A_2 = p_2 = N_2/2N$ or

$$p_2 = \frac{p_2^2 + p_1p_2}{p_1^2 + 2p_1p_2 + p_2^2} = \frac{p_2}{(p_1 + p_2)} = p_2$$

It can be seen from the above that the frequencies of genes A_1 and A_2 in the new generation are the same as they were in the parental generation. In addition, a Hardy-Weinberg equilibrium in which genotype frequencies are distributed as $p_1^2 + 2p_1p_2 + p_2^2$ has been established in one generation of random mating and will be maintained as long as random mating continues and other factors leading to change are absent.

The example we have used is one in which all three genotypic classes can be recognized by their phenotypes, the alleles A_1 and A_2 being codominant. However, when dominance is present, only two phenotypic classes are evident, the dominant class which includes both homozygotes and heterozygotes and the homozygous recessive class. In this case, and assuming random mating and a Hardy-Weinberg equilibrium, the frequency of the two alleles can be determined with the use of the terms of the expanded binomial $(p_1 + p_2)^2$. Since the term p_2^2 represents the frequency of the homozygous recessive class in the population, the square root of p_2^2 is the frequency of the recessive allele. Thus, if the frequency of recessives in a population is, for example, 0.04, then the frequency of the recessive allele is $\sqrt{0.04}$ or 0.2. Also, since $p_1 + p_2 = 1$, the frequency of $p_1 = 1 - p_2$. In this example the frequency of the dominant allele would be $1.0 - 0.2 = 0.8$. The expected frequency of heterozygotes can then be determined by substituting these values in the term $2p_1p_2$ of the binomial expansion. In this case, $2p_1p_2 = 2(0.8 \times 0.2) = 0.32$. By these calculations, we would expect that if 4 percent of the population was homozygous for the recessive allele, then 64 percent should be homozygous for the dominant allele and 32 percent should be heterozygotes. If it were possible to distinguish homozygous dominants from heterozygotes, it could be determined whether or not two alleles were in equilibrium by comparing the expected with the actual frequencies of genotypes, and if some doubt arose, the chi-square test could be applied. We should note that if a χ^2 test were used to determine the probability that the actual gene frequencies conformed to Hardy-Weinberg expectations, only one degree of freedom (rather than two) is available, since one parameter (p_2) is estimated from the data.

The Hardy-Weinberg theorem can also be applied to the case of multiple alleles. For example, with the three alleles A_1, A_2, and A_3 in a population in equilibrium, if the frequency of $A_1 = p$, that of $A_2 = q$, and that of $A_3 = r$, the trinomial $p + q + r = 1$ can be used. The expansion of this trinomial gives the following genotypes and their frequencies.

$$\begin{matrix} A_1A_1 & & A_1A_2 & & A_1A_3 & & A_2A_2 & & A_2A_3 & & A_3A_3 & \\ p^2 & + & 2pq & + & 2pr & + & q^2 & + & 2qr & + & r^2 & = 1 \end{matrix}$$

If each homozygous genotype, p^2, q^2, and r^2, is recognizable by phenotype, then the respective square roots of these values provide the required gene frequencies, or if the frequencies of two of the alleles are known, the third can be obtained by difference, since $p + q + r = 1$.

In the case where the alleles of two different gene loci are under consideration and the population is in equilibrium, the frequencies of the various genotypes obtained separately for each pair of alleles can be multiplied together to arrive at the frequency of any overall class. As an illustration, suppose that alleles *A* and *a* occur in the frequencies of $A = p_1 = 0.8$, $a = p_2 = 0.2$, and $p_1 + p_2 = 1$; and that alleles *B* and *b* occur in the frequencies of $B = q_1 = 0.6$, $b = q_2 = 0.4$, and $q_1 + q_2 = 1$. With these values, the frequency of any composite genotypic class can be determined, as in the following examples.

$$\text{genotypic class } AABB = (p_1^2)(q_1^2) = (0.8)^2(0.6)^2$$
$$\text{genotypic class } Aabb = (2p_1p_2)(q_2^2) = (2)(0.8)(0.2) \times (0.4)^2$$
$$\text{genotypic class } AaBb = (2p_1p_2)(2q_1q_2) = (2)(0.8)(0.2) \times (2)(0.6)(0.4)$$

These examples assume a population already in equilibrium. It should be noted that single autosomal loci reach equilibrium in one generation of random mating, but because of the effects of linkage, many generations of random mating may be required before the frequencies of genes at two or more loci reach the steady state.

GENE FREQUENCIES AND SEX

The above discussion has dealt with autosomal genes without consideration of the sex of individuals in the population, on the assumption that the frequency of any gene is the same in both sexes. If the frequency is equal in the two sexes, an equilibrium is attained in one generation of random mating. However, if the initial frequency of an autosomal gene differs in males and females, two generations, rather than one, will be required to establish the equilibrium. The first generation serves to equalize the frequency in the two sexes, producing an average gene frequency, and the second generation then initiates the Hardy-Weinberg equilibrium.

In contrast to autosomal genes, sex-linked genes are not equally present in the two sexes, for females have two X chromosomes, while males have only one. If the initial frequency of a sex-linked gene is unlike in males and females, the approach to equilibrium is gradual, requiring a number of generations (Figure 21–3). Equilibrium cannot be reached in one or two generations because sons inherit the X chromosome only from mothers, and therefore the allele frequency of mothers will determine that of the sons. However, daughters receive an X chromosome from each parent, and therefore the allele frequency in the daughters will be the average of those found in the parents, male and female. The gradual approach to equilibrium for a sex-linked gene is illustrated in Figure 21–3, where it can be seen that the gene frequency of sons is the same as that of their mothers, while that of daughters lies halfway between the respective gene frequencies of the parent generation.

Even when sex-linked alleles have not attained equilibrium, it is possible to determine their average frequency for an entire population of males and females, providing allowance is made for the fact that $\frac{2}{3}$ of the X chromosomes of the population are found in females and only $\frac{1}{3}$ in males. Thus, the average frequency (p_1) of one allele can be represented as

$$p_1 = (\tfrac{1}{3}\overset{\text{♂}}{p_1} + \tfrac{2}{3}\overset{\text{♀}}{p_1})/3$$

and the average frequency of the other allele (p_2) can be determined in the same way:

$$p_2 = (\tfrac{1}{3}\overset{\text{♂}}{p_2} + \tfrac{2}{3}\overset{\text{♀}}{p_2})/3$$

Once equilibrium has been attained, with random mating and equal numbers of both sexes, the genotype frequencies of females will conform to the Hardy-Weinberg distribution characteristic of autosomal genes. Because two X chromosomes are present, three genotypes occur in females, *AA*, *Aa*, and *aa*, and if $A = p_1$ and $a = p_2$, the genotype frequencies can be represented as

$$\begin{matrix} AA & & Aa & & aa \\ p_1^2 & + & 2p_1p_2 & + & p_2^2 = 1 \end{matrix}$$

In males only one X chromosome is present, and therefore genotype frequencies are represented as

$$\begin{matrix} A & & a \\ p_1 & + & p_2 = 1 \end{matrix}$$

If the population is in equilibrium for alleles *A* and *a*, then the allele frequency values for p_1 and p_2 will be the same in both sexes, and $p_1 + p_2 = 1$ in both

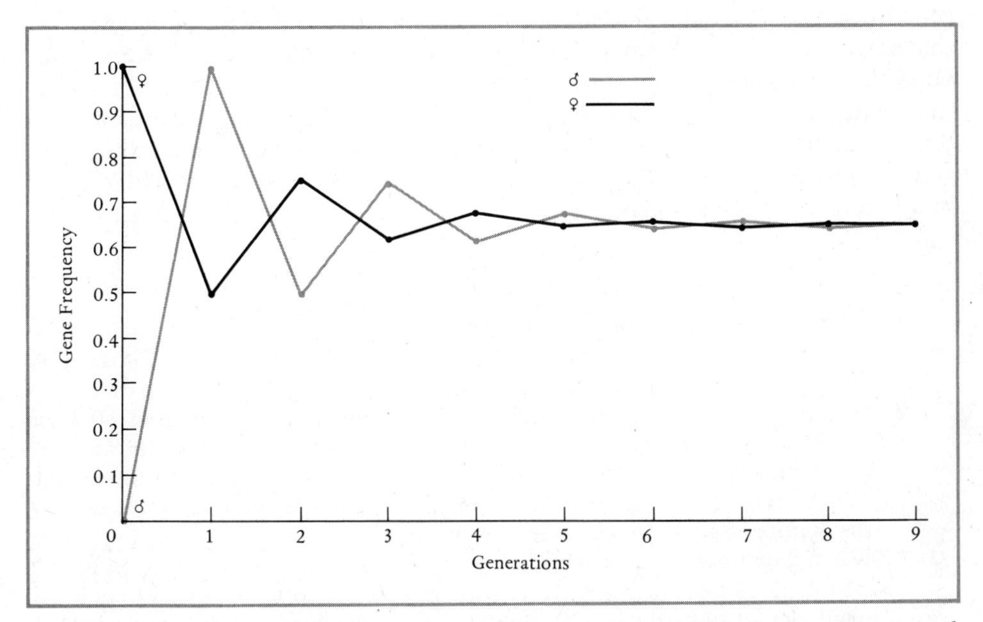

Figure 21–3. The gradual approach to equilibrium of a sex-linked gene whose initial frequency is 1.0 in females and 0 in males.

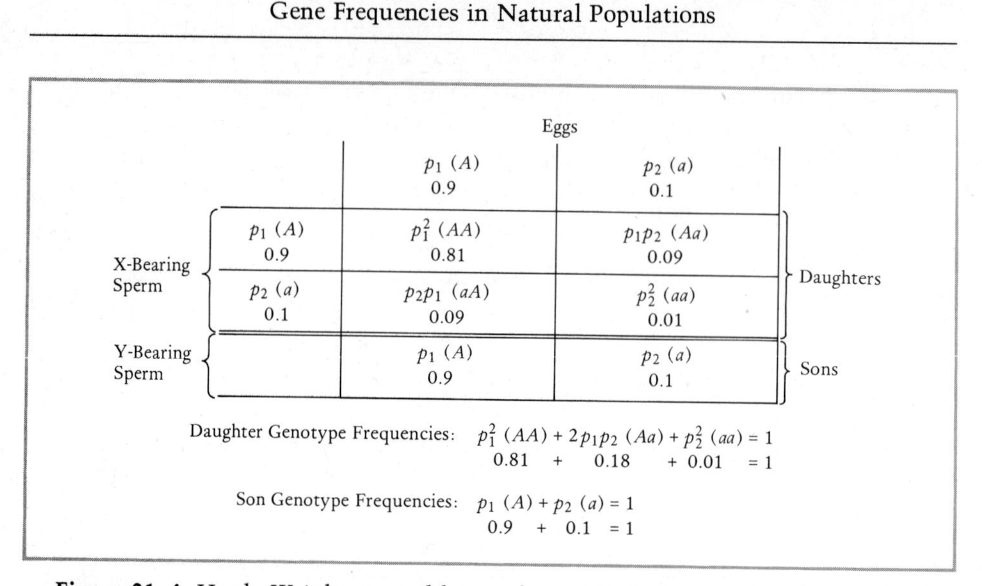

Figure 21–4. Hardy-Weinberg equilibrium for sex-linked genes *A* and *a*, where the frequency of $A = p_1 = 0.9$, and the frequency of $a = p_2 = 0.1$.

males and females. Such an equilibrium is shown in Figure 21–4. In this example notice that the frequency of males containing allele *a*, and therefore expressing "trait a" (0.1), is much greater than the frequency of homozygous *aa* females (0.01). This accounts for the fact that many more males than females exhibit sex-linked recessive conditions, such as hemophilia and colorblindness.

The preceding discussion has been concerned with methods by which gene frequencies can be calculated and with the characteristics of populations that are in equilibrium. A population in a Hardy-Weinberg equilibrium maintains the status quo; it is a static population in which evolution is not occurring. Although such an idealized population would rarely if ever be encountered in nature, it provides a theoretical yardstick against which changes in gene frequency can be measured. For this reason, it is a basic point of departure in studies of population genetics.

PROBLEMS

21–1. The M and N blood groups are determined by two codominant alleles. In a random sample of 200 persons, 128 were found to belong to group M. Assuming random mating, what is the frequency of the *N* allele?

21–2. Within a randomly mating population of mammals, the homozygous presence of the recessive allele *a* causes white spotting in the fur, while the presence of the dominant allele *A* causes solid color. A survey was made, and of 137 tagged, 4 were spotted. What is the frequency in the population of the recessive gene, of the dominant allele, and of heterozygotes?

21–3. Two different populations, both in Hardy-Weinberg equilibrium, exhibit different frequencies of the recessive gene *d*, as opposed to the dominant allele *D*. Population A consists of 100 individuals of which 25 are *dd*. Population B consists of 100 individuals of which 47 are *dd*.

a. If these two groups merge, so as to form a combined population of 200 individuals, what will be the frequency of gene d in the combined population?

b. If the combined population now mates randomly, what will be the frequency of gene d in the next generation?

c. If the next generation consists of 400 individuals, how many of these are expected to be heterozygous Dd?

21–4. The following genotype frequencies were found in random samples of 100 individuals from each of four different populations. Which of these populations is not in equilibrium?

	A^1A^1	A^1A^2	A^2A^2
(1)	.36	.48	.16
(2)	.7291	.3204	.0064
(3)	.5184	.3456	.0576
(4)	.0289	.2822	.6889

21–5. The codominant alleles B_1, B_2, and B_3 occur in a randomly breeding population with the respective frequencies of 0.1, 0.3, and 0.6. What are the frequencies of the different genotypes?

21–6. Assume that in a hypothetical population in equilibrium the frequency of individuals of blood group M is 0.64 and the frequency of those showing the dominant secretor trait is 0.91.

a. What should be the frequency of nonsecreting individuals of blood group N?

b. What should be the frequency of genotype $MN\ Ss$?

c. What should be the frequency of blood group M secretors?

21–7. In a certain species of plant where red and white flowers are determined by a single pair of alleles, R (red) and r (white), a grower counted 160 white flowering plants in a field of 1000. If random fertilization is occurring, what proportion of the fertilizations are expected to involve an R-bearing gamete with an r-bearing gamete?

21–8. Colorblindness is sex linked. In a human population at equilibrium the frequency of colorblind females is 40 per 1000 females.

a. What is the frequency of colorblind males?

b. What proportion of the females of the population would be heterozygotes?

21–9. In a given human population in the Philippines, if approximately 16 percent of the males and 9 percent of the females are colorblind, what is the average frequency of the gene for colorblindness in the population?

21–10. A blood bank collected 10,000 pints of blood from citizens of a city over the course of a year. Blood typing indicated that 3,200 pints were type A, 1,500 pints were type B, 400 were type AB, and 4,900 were type O. Assuming a Hardy-Weinberg equilibrium, what are the frequencies of genes I^A, I^B, and I^O among these citizens?

REFERENCES

CAVALLI-SFORZA, L. L., AND W. F. BODMER, 1971. *The Genetics of Human Populations.* Freeman, San Francisco.

CROW, J. F., 1976. *Genetics Notes,* 7th ed. Burgess, Minneapolis, Minn.

FALCONER, D. S., 1960. *Introduction to Quantitative Genetics.* Ronald Press, New York.

LI, C. C., 1955. *Population Genetics.* University of Chicago Press, Chicago.

LI, C. C., 1967. Genetic equilibrium under selection. *Biometrics, 23,* 397.

METTLER, L. E., AND T. G. GREGG, 1969. *Population Genetics and Evolution* (Foundations of Modern Genetics Series). Prentice-Hall, Englewood Cliffs, N.J.

Chapter 22

CHANGES IN GENE FREQUENCIES

We have seen that, once achieved, an equilibrium between the frequencies of two or more alleles in a population will continue indefinitely, providing the initial assumptions made for the population hold. These assumptions include random mating, genotypes with equal viability and fecundity, absence of selection and mutation, and infinite size. A change in any one of these assumed conditions can disturb the Hardy-Weinberg equilibrium. In this chapter we will examine the results of a violation of each assumption.

DEPARTURES FROM RANDOM MATING

A violation of the assumption of random mating occurs when individuals mate more frequently with others like themselves than would be expected on the basis of chance. This is called *positive assortative mating.* For example, humans may choose mates on the basis of physical beauty, height, intelligence, and so on, and to the extent that such characteristics have a genetic basis, such assortative mating would tend to increase the frequency of homozygosis of genes associated with the character. *Negative assortative mating* where individuals mate preferentially with those unlike themselves would increase heterozygosis. Another obvious departure from random mating is *inbreeding,* where individuals tend to carry identical genes descended from a common ancestor. The most extreme form of inbreeding is *self-fertilization,* found in many plant species.

The effects of self-fertilization can be illustrated by example. If we start with a heterozygote, *Aa* in genotype, then the first generation will consist of $\frac{1}{4}$ *AA*, $\frac{1}{2}$ *Aa*, and $\frac{1}{4}$ *aa*. Thus half of the F_1 population consists of homozygotes, *AA* and *aa*. Since the condition of self-fertilization precludes outcrosses, *AA* individuals will produce only *AA* offspring and *aa* individuals will produce only *aa* offspring. The other half of the F_1 population consists of heterozygotes, *Aa*, and these individuals upon self-fertilization will produce $\frac{1}{4}$ *AA*, $\frac{1}{2}$ *Aa*, and $\frac{1}{4}$ *aa*. The F_1 *Aa* heterozygotes thus contribute $\frac{1}{2} \times \frac{1}{4} = \frac{1}{8}$ *AA* to the F_2 *AA* class, and $\frac{1}{2} \times \frac{1}{4} = \frac{1}{8}$ *aa* to the F_2 *aa* class, and only $\frac{1}{2} \times \frac{1}{2} = \frac{1}{4}$ of the F_2 generation will be heterozygotes. Therefore, of the total F_2 population, each of the homozygous classes will contain $\frac{1}{4}$ (contributed by self-fertilization of homozygotes) plus $\frac{1}{8}$ (contributed by self-fertilization of heterozygotes), or $\frac{3}{8}$ of the total, while F_2 heterozygotes will comprise only $\frac{1}{4}$ of the total population. Since heterozygotes represented $\frac{1}{2}$ of the F_1 population, but are now only $\frac{1}{4}$ of the F_2 population, it is evident that their frequency has been halved, and this halving in their frequency continues in each successive generation of self-fertilization (Table 22–1).

After a large to infinite number of generations of self-fertilization, the frequency of heterozygotes will decrease essentially to zero, while the homozygous classes *AA* and *aa* will each come to comprise half of the population. Thus the overall effect of self-fertilization is to increase homozygosis and decrease heterozygosis. Note, however, that *the frequency of the genes A and a is unchanged and only the frequency of genotypes is altered.* Of course, self-fertilization is the most extreme form of inbreeding, but less intensive inbreeding produces a similar, if less pronounced, effect on the frequency of heterozygotes.

A measure of the decrease in heterozygosity due to inbreeding is provided by S. Wright's coefficient of inbreeding, called *F.* In the preceding example of self-fertilization, the value of *F* in the F_2 generation would be $\frac{1}{2}$, since $\frac{1}{2}$ of the heterozygosis originally present in the first generation has been lost due to

TABLE 22–1
The Distribution of Genotypes Resulting from Self-Fertilization Carried Out over Four Generations, Beginning With an *Aa* Individual

	Genotypes		
Generation	*AA*	*Aa*	*aa*
0		1	
1	$\frac{1}{4}$	$\frac{1}{2}$	$\frac{1}{4}$
2	$\frac{1}{4} + \frac{1}{8} = \frac{3}{8}$	$\frac{1}{4}$	$\frac{1}{4} + \frac{1}{8} = \frac{3}{8}$
3	$\frac{3}{8} + \frac{1}{16} = \frac{7}{16}$	$\frac{1}{8}$	$\frac{3}{8} + \frac{1}{16} = \frac{7}{16}$
4	$\frac{7}{16} + \frac{1}{32} = \frac{15}{32}$	$\frac{1}{16}$	$\frac{7}{16} + \frac{1}{32} = \frac{15}{32}$
∞	$\frac{1}{2}$	0	$\frac{1}{2}$

self-fertilization. In the third generation heterozygotes are only $\frac{1}{8}$ of the population, compared to the original first generation value of $\frac{1}{2}$. Thus $\frac{3}{4}$ of the original heterozygosity has been lost and therefore in the F_3 generation $F = \frac{3}{4}$. With increasing generations of self-fertilization, the value of F will continue to rise until, with an infinite number of generations, heterozygotes are eliminated altogether, $F = 1$, and the population consists entirely of the homozygotes AA and aa. With random mating $F = 0$, because with the absence of inbreeding the frequency of heterozygotes in the population does not decrease. In the case where some intermediate or average amount of inbreeding is occurring, the genotypes of a population can be expressed as follows. According to the Hardy-Weinberg equilibrium, the frequency of AA homozygotes is p_1^2, that of aa homozygotes is p_2^2, and that of heterozygotes is $2p_1p_2$. However, with the introduction of inbreeding the frequency of heterozygotes will be reduced by an amount $2p_1p_2F$, and half of this amount (p_1p_2F) will be added to one homozygous class and half to the other homozygous class. Therefore the frequency of genotypes in the next generation will be:

$$\underset{AA}{p_1^2 + p_1p_2F} + \underset{Aa}{2p_1p_2 - 2p_1p_2F} + \underset{aa}{p_2^2 + p_1p_2F}$$

Although in the above example F was used to express the average amount of inbreeding in one generation of a population, F is more commonly employed as the probability that two genes in an individual are *identical by descent*, that is, they are replicates of one gene present in an ancestor. Note that as used here identity has a different meaning from homozygosis. A diploid individual may contain two like alleles which through random mating between unrelated individuals are combined to yield the homozygous state. In this case the two like alleles are independent. On the other hand, the two like alleles in a homozygote may be actual replicates of one single allele present in some ancestor, and in this case they are identical.

The probability of identity, F, can be computed from a pedigree, as follows. Let us consider the simple case of a child of two half sibs (two individuals with a parent in common). The individuals in the pedigree would be the common ancestor (A), the son (S), the daughter (D), and the individual in question (I).

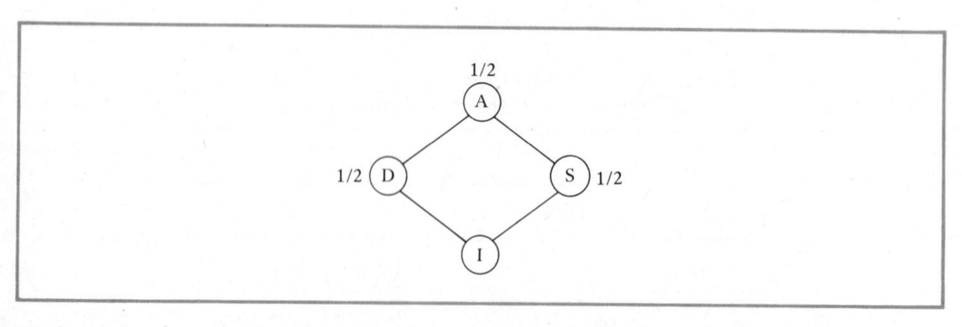

At each point (D, A, and S) in the pedigree we may ask, what is the probability that the gametes carry an identical gene? The probability of this event at any juncture in the pedigree is $\frac{1}{2}$, that is, one for each of the three junctures in the pedigree between the child (I) and the common ancestor (A) and then back to the child. Thus in this instance, $F = (\frac{1}{2})^3 = \frac{1}{8}$.

In the above example only one common ancestor is involved, but if more than one common ancestor must be considered, identity by descent of two genes in an individual could be due to any of these ancestors (although only one can actually be involved). Therefore, the probability of identity through all possible pathways of descent must be summed to arrive at a value for F. In the first illustration in Figure 22–1 two common ancestors are involved, and thus two possible pathways for the identity of a pair of genes by descent in individual I are possible. Here $F = (\frac{1}{2})^3 + (\frac{1}{2})^3 = \frac{1}{4}$. In first cousin marriages the two pathways each involve five junctures in the pedigree, and therefore $F = (\frac{1}{2})^5 + (\frac{1}{2})^5 = \frac{1}{16}$. We can generalize the computation of F to

$$F = \Sigma (\tfrac{1}{2})^n$$

where n is the number of junctures in a pathway of descent.

It should be noted that sometimes the ancestor is itself inbred and, if so, is more likely to produce identity in descendents, since the ancestor may possess a pair of genes that are themselves identical by descent. The probability that two identical genes are present in the ancestor is the inbreeding coefficient of the ancestor (F_A). Taking the possibility of an inbred ancestor into consideration, the inbreeding coefficient can additionally be generalized as

$$F = \Sigma (\tfrac{1}{2})^n (1 + F_A)$$

As previously stated, the effects of inbreeding are exerted on genotypes and not on gene frequencies. In addition, the effects of inbreeding may be

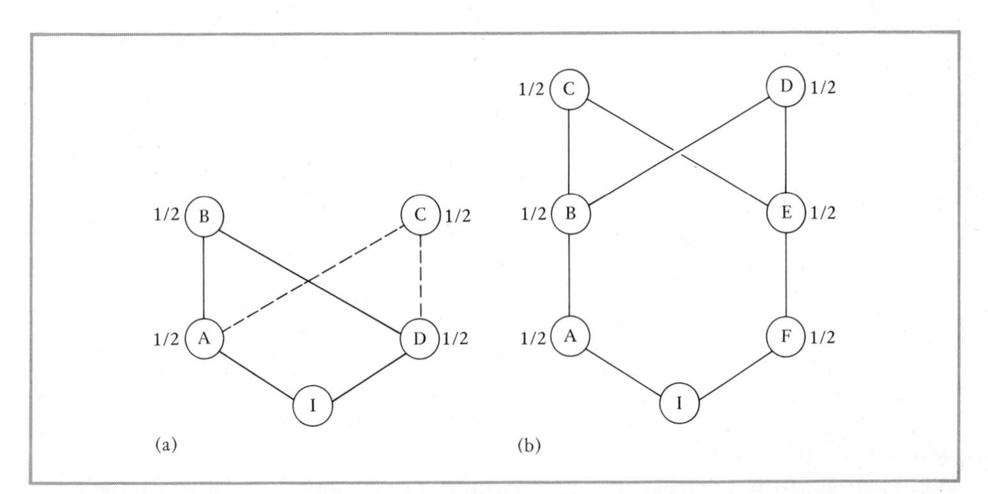

Figure 22–1. Pedigrees illustrating pathways whereby a gene may become identical by descent in an individual (I). (a) Two pathways are possible, A-B-D (solid lines) and A-C-D (dash lines). $F = (\frac{1}{2})^3 + (\frac{1}{2})^3 = \frac{1}{4}$. (b) Individual I is the offspring of first cousins (A and F) having common grandparents (C and D). Pathways of descent are A-B-C-E-F, $(\frac{1}{2})^5$, and A-B-D-E-F, $(\frac{1}{2})^5$. $F = (\frac{1}{2})^5 + (\frac{1}{2})^5 = \frac{1}{16}$.

considered from two points of view: in terms of the genes of an individual or in terms of a particular gene in a population.

With respect to the individual, inbreeding will cause F percent of the genes of that individual to become identical by descent, but *which* genes are involved is not known.

For a particular gene in a population we can consider that F percent of the individuals will be homozygous for a pair of genes made identical by descent, and $1 - F$ percent will consist of homozygotes and heterozygotes independently generated through random breeding. Thus, if p_1 = gene A and p_2 = gene a, the frequency of genotypes in the population will be

$$\begin{array}{ccc} AA & Aa & aa \\ p_1^2(1-F) + p_1F & 2p_1p_2(1-F) & p_2^2(1-F) + p_2F \end{array}$$

The above expression is entirely equivalent to the formula previously derived, wherein F was treated as the decrease in heterozygosity, that is,

$$\begin{array}{ccc} AA & Aa & aa \\ p_1^2 + p_1p_2F & 2p_1p_2 - 2p_1p_2F & p_2^2 + p_1p_2F \end{array}$$

Note that with both expressions when $F = 0$, random mating is occurring and the proportions of genotypes revert to the Hardy-Weinberg equilibrium.

$$\begin{array}{ccc} AA & Aa & aa \\ p_1^2 & 2p_1p_2 & p_2^2 \end{array}$$

As previously discussed, inbreeding results in an equal contribution of former heterozygotes to each of the homozygous classes. This effect of increasing homozygosis is most evident in the case of rare deleterious genes. For example, in humans the approximate frequency of the gene for phenylketonuria (PKU) is $p_2 = 0.01$, whereas the frequency of the normal allele is $p_1 = 0.99$ (let us use $p_1 \cong 1.0$ for convenience). Here, when $F = 0$, PKU homozygotes $= p_2^2 = (.01)^2 = 1 \times 10^{-4}$; but when $F = \frac{1}{16}$ (as in first cousin marriages), PKU homozygotes $= p_2^2 + p_1p_2F = (.01)^2 + (1)(.01)(\frac{1}{16}) \cong 7 \times 10^{-4}$. From this it can be seen that the inbreeding resulting from a marriage of first cousins has produced a sevenfold increase in the probability of homozygosis for PKU. Similar computations can be carried out for other deleterious genes of humans where an estimate of the frequency of a given gene is available.

INTRODUCTION OF NEW ALLELES THROUGH MUTATION AND MIGRATION

The phenotypic variation observed among the individuals in a population can be attributed in part to the effects of the environment (see Chapter 4) and in part to the fact that every member of the population, with the exception of identical twins, possesses a different genotype. Table 2–1 should be consulted as a reminder of the genotypic, and therefore phenotypic variability made possible by Mendelian segregation and independent assortment, and to this genetic variability must be added that contributed by recombination. Although the shuffling of genes which occurs in each generation is capable of producing enormous numbers of new allele combinations, only some of which can be realized even in a large population, it does not produce new alleles themselves, for new alleles arise only by mutation.

The origin of novelty through mutation must be distinguished from the establishment and maintenance of that novelty within a population, and therefore we can inquire if mutation alone is a significant factor in the progressive gene frequency changes that lead to evolution. To this end, let us consider whether or not a single unique mutation would persist in a large population when the mutation is neutral, that is, is neither advantageous nor disadvantageous, thus eliminating selection as a force in its preservation or extinction. If the mutation occurs in a gamete that participates in fertilization, a heterozygote will be produced. If the heterozygote leaves no progeny, the new mutation will immediately be lost. If the heterozygote leaves progeny, only $\frac{1}{2}$ are expected to carry the new gene, so that the probability of extinction is $(\frac{1}{2})^n$ where n is the number of offspring. With every generation chance again plays a major role, and the probability that the unique neutral mutation will survive and become established in the population is negligible. Even selectively advantageous mutations can be lost in early generations before selection acts in their favor.

Mutation to a given allele is not a unique event, however, for mutation is recurrent, and even though mutation rates are low, in the order of 10^{-5} to 10^{-6} (see Table 17–1), over the course of time more than one opportunity for the survival of a given mutation will be afforded. Additionally, mutation occurs in both the forward and reverse directions, though usually at different rates for each. Thus, considering the neutral alleles A and a, allele A can mutate to allele a at one rate, at the same time that allele a is mutating back to A at a different, and usually lower rate. Whenever A mutates to a, a decrease in the frequency of A and an increase in the frequency of a will result; likewise, the reverse mutation of a to A will cause a decrease in the frequency of a and an increase in the frequency of A. Theoretically, at some time in the history of the population an equilibrium will be established in which the mutation of A to a will be balanced by the mutation of a to A. The probable frequencies of these alleles when such a state of mutational equilibrium has been reached can be estimated in the following way. Let u = the rate of mutation from A to a, v = the rate of mutation from a to A, p_2 = the frequency of gene a, and $1 - p_2$ = the frequency of gene A. An increase in the frequency of A can be expressed as vp_2 and a decrease in the frequency of gene a as $u(1 - p_2)$. Thus, when forward mutation of $A \rightarrow a$ balances reverse mutation of $a \rightarrow A$

$$\begin{aligned} vp_2 &= u(1 - p_2) \\ vp_2 &= u - up_2 \\ vp_2 + up_2 &= u \\ p_2(v + u) &= u \\ p_2 &= \frac{u}{v + u} \end{aligned}$$

Although a theoretical equilibrium between forward and reverse mutation can be derived as above, the low mutation rates observed and the odds against the survival and establishment of a mutation in a large population combine to make mutation pressure, by itself, a weak and minimal force in gene frequency change. The major contribution of mutation to evolution is thus the formation of new alleles. The preservation of these alleles within a population depends primarily on selection.

A second means by which new alleles can enter the gene pool of a population is migration, for an influx of individuals from another population having different gene frequencies can introduce new alleles and augment the variation originally present. If two adjacent populations engage in frequent exchange of their respective members, the gene frequencies of the two populations are likely to be similar. However, if the populations are separated by geographic barriers of some type, adaptation to their respective environments will generally result in the presence of different allele frequencies in the two groups. The entrance of migrants from one group into the second group and their interbreeding with members of the recipient population may introduce a number of uncommon alleles sufficient to change the original gene frequencies.

It is possible to estimate the change, due to migration, in the frequency of an allele as follows. Let p_0 = the frequency of gene A in a recipient population before the influx of migrants, P = its frequency in migrants, M = the percent of the population who are migrants, and $1 - M$ = the percent of the population who are recipients. After one generation of interbreeding, the frequency of gene A (p_1) will be

$$p_1 = p_0(1 - M) + PM$$

In other words, the frequency of gene A will be the average of the proportional input from recipients and migrants.

The change in gene frequency will be the difference between the starting frequency of the recipients, p_0, and the frequency after one generation, p_1, or

$$\Delta p = p_0 - p_1$$

We can also consider that if M percent of the genes of a migrant population are continuously introduced into a recipient population having a different gene frequency, the difference in gene frequency between the two populations will be reduced by the proportion $1 - M$ each generation. After n generations, the difference between the two populations will be

$$p_n - P = (1 - M)^n(p_0 - P)$$

Assuming that the migration rate is constant and that the original gene frequencies are known, use of this formula permits an estimate of the rate of admixture of two populations between which gene flow is occurring.

SELECTION

It can readily be observed that the reproductive potential of the members of any population far exceeds the finite carrying capacity of the environment, and thus population size is limited by the availability of food supply, space, or appropriate ecological niches. Not all of the individuals produced will survive the rigors of the environment, and those who do can be considered more "fit" than those who do not. In evolutionary terms fitness involves more than survival; it requires fecundity as well. An organism that survives but does not reproduce is equivalent to a genetically caused death, that is, neither casualties to selection nor effectively sterile individuals pass their genes to the next generation. Evolution by natural selection can thus be viewed in terms of the differing reproductive capacities of members of a population, that capacity involving not only survival to reproductive age, but more important the transmission of genes to the next and future generations.

A consideration of fitness is the first step in an assessment of the effects of selection on changes in gene frequency. As indicated above, fitness has two components: differential viability, which involves an unequal probability of the survival of genotypes, and differential fecundity, which involves disproportionate quantities of gametes carrying an allele. The effects of these two components are multiplicative. To include both survivorship and fecundity in our considerations, we can use the concept of a *cohort,* whereby a group of individuals is followed from one point in the life cycle (for example, birth) to the same point (birth) in the next generation. Thus we start with a number of newborn individuals, follow them to reproductive age, and score the number of newborns they produce. (We should note that the cohort receives only half the credit for the children produced, the other half belonging to the marriage partner.) As a specific example, suppose we use three cohorts of 100 individuals each, of genotypes *AA, Aa,* and *aa,* respectively, and follow them from birth scoring the number of living newborns they produce by some given age (Table 22–2). The *absolute fitness* of a genotype is a measure calculated by multiplying the proportion surviving to reproductive age (viability) by the number of offspring per survivor (fecundity). In other terms, it is the average number of offspring per cohort. Thus, for genotype *AA* the proportion surviving is 0.8 ($\frac{80}{100}$) and the average number of offspring per survivor is 3. The absolute fitness is therefore 0.8×3 or 2.4. Similarly, the absolute fitness of genotypes *Aa* and *aa* is computed as 1.2 and 0.6, respectively.

Although the fitness of a genotype can be stated in absolute terms, it is more usually expressed in relative terms and in comparison to an optimum genotype. In our example genotype *AA* is selected as the optimum genotype against which the others are to be compared. In the data of Table 22–2 genotypes *Aa* and *aa* can be assigned a *relative fitness* (w) of 0.5 and 0.25, respectively, as compared to genotype *AA*.

TABLE 22–2
Relative Fitness of Three Cohorts of Genotypes *AA, Aa,* and *aa*

Genotype	Total number in cohort	Number surviving to reproduce	Number of living offspring	Average number of offspring per survivor	Average number of offspring per cohort (absolute fitness)	Relative fitness (w)
AA	100	80	240	3.0	$0.8 \times 3.0 = 2.4$	$2.4/2.4 = 1.0$
Aa	100	60	120	2.0	$0.6 \times 2.0 = 1.2$	$1.2/2.4 = 0.5$
aa	100	40	60	1.5	$0.4 \times 1.5 = 0.6$	$0.6/2.4 = 0.25$

The difference between the fitness of a given genotype and the optimum is called a *selection coefficient (s)*. Thus, $s = 1 - w$ and $w = 1 - s$. The selection coefficient of genotype *Aa* in our example would be $1.0 - 0.5 = 0.5$, and the selection coefficient for genotype *aa* would be $1.0 - 0.25 = 0.75$. Both genotype fitnesses are below 1.0 and we consider these selection coefficients to be positive. If a selection coefficient is positive, the genotype is at a selective disadvantage, while if the selection coefficient is negative, the genotype can be considered selectively advantageous, that is, better fitted than the genotype chosen as the standard of comparison. In our example both genotypes *Aa* and *aa* compare unfavorably with the optimum, and therefore we would expect selection to operate against them in proportion to their relative fitness. To summarize, differences in fitness result in selection which in turn leads to changes in gene frequency.

The effects of differences in fitness on gene frequency are illustrated in Table 22–3 and can be explained as follows. We can begin with a large, randomly breeding population in which the frequency of gene $A = p_1$, that of gene $a = p_2$, and $p_1 + p_2 = 1$. The frequencies of genotypes *AA, Aa,* and *aa* in the population exhibit the Hardy-Weinberg proportions of p_1^2 $(AA) + 2p_1p_2$ $(Aa) + p_2^2$ $(aa) = 1$, these genotypes contributing gametes to the gene pool in the proportions of *A*-bearing gametes $= p_1^2 + p_1p_2$ and *a*-bearing gametes $= p_2^2 + p_1p_2$.

Having established the initial conditions for the population, suppose we now impose selection, designating the relative fitness *(w)* of genotype *AA* as w_{11}, that of *Aa* as w_{12}, and that of *aa* as w_{22}. As a result of selection, the gametes forming the first generation will be present in the proportions of

$$A\text{-bearing gametes} = p_1^2w_{11} + p_1p_2w_{12}$$
$$a\text{-bearing gametes} = p_1p_2w_{12} + p_2^2w_{22}$$

Random union of these gametes will then produce the following genotype frequencies.

$$p_1^2w_{11} + 2p_1p_2w_{12} + p_2^2w_{22} = \overline{w}$$

TABLE 22–3
Gene and Genotype Frequency Before and After Selection in a Randomly Breeding Population

Genotype	Gene frequency before selection	Genotype frequency before selection	Relative fitness	Genotype frequency after selection	Relative proportions of genotypes after selection	New gene frequency	Genotype frequency *before* the next round of selection
AA	p_1	p_1^2	w_{11}	$p_1^2w_{11}$	$p_1^2w_{11}/\overline{w}$	p_1'	$p_1'^2$
Aa		$2p_1p_2$	w_{12}	$2p_1p_2w_{12}$	$2p_1p_2w_{12}/\overline{w}$		$2p_1'p_2'$
aa	p_2	p_2^2	w_{22}	$p_2^2w_{22}$	$p_2^2w_{22}/\overline{w}$	p_2'	$p_2'^2$
Total	1	1		$\overline{w}$	1	1	1

The term $\overline{w}$ is called the *mean fitness* of the selected population; it represents the genotype frequencies weighted by their relative fitnesses. When both sides of the equation are divided by $\overline{w}$, the sum of the relative proportions of the three genotypes in the population after selection is 1.0. These proportions are:

$$\begin{array}{ccccccc} AA & & Aa & & aa & & \\ \dfrac{p_1^2 w_{11}}{\overline{w}} & + & \dfrac{2p_1p_2w_{12}}{\overline{w}} & + & \dfrac{p_2^2 w_{22}}{\overline{w}} & = & 1 \end{array}$$

Using primes to indicate the new gene frequencies in the first generation after selection

$$\text{gene } A = p_1' = \frac{p_1^2 w_{11} + p_1p_2w_{12}}{\overline{w}}$$

$$\text{gene } a = p_2' = \frac{p_2^2 w_{22} + p_1p_2w_{12}}{\overline{w}}$$

$$p_1' + p_2' = 1$$

These gene frequencies will be those involved in the next round of mating and selection. With recurring selection, new genotype frequencies and new values for p_1 and p_2 would arise in each subsequent generation.

A formula expressing the change in the frequency of a particular gene due to selection can be derived as follows. In our example the starting frequency of gene a before selection was p_2 and its frequency after one cycle of selection was p_2'. Accordingly, the change in the frequency of gene a is: $\Delta p_2 = p_2' - p_2$. Since the frequency of gene a after selection was derived above as

$$a = p_2' = \frac{p_2^2 w_{22} + p_1p_2w_{12}}{\overline{w}} = \frac{p_2(p_2w_{22} + p_1w_{12})}{\overline{w}}$$

we can substitute

$$\Delta p_2 = \frac{p_2(p_2w_{22} + p_1w_{12})}{\overline{w}} - p_2$$

$$\Delta p_2 = \frac{p_2(p_2w_{22} + p_1w_{12}) - p_2\overline{w}}{\overline{w}}$$

and since

$$\overline{w} = p_1^2 w_{11} + 2p_1p_2w_{12} + p_2^2 w_{22}$$

we can substitute:

$$\Delta p_2 = \frac{\underset{①}{p_2}(p_2w_{22} + p_1w_{12}) - p_2(p_1^2w_{11} + \underset{②}{2p_1p_2w_{12}} + p_2^2w_{22})}{\overline{w}}$$

$$\Delta p_2 = p_2' - p_2 = \frac{p_1p_2[p_2(w_{22} - w_{12}) + p_1(w_{12} - w_{11})]}{\underset{③}{\overline{w}}}$$

The meaning of this long equation for gene frequency change can be better understood if we use a device originated by Mettler and Gregg whereby the component parts of the equation (shown by the numbered brackets) are examined separately.*

Part 1. The change in gene frequency under selection is proportional to p_1p_2. This change will be greatest when the value of $p_1 = p_2 = \frac{1}{2}$, that is, at intermediate gene frequencies.

Part 2. The change in gene frequency is inversely proportional to the average fitness of the population, and as the average fitness gradually improves, the rate of change in the frequency of a gene slows down.

Part 3. The change in gene frequency reflects the effects of gene substitution. For allele *A* a change from *Aa* to *AA* has a selective effect $(w_{11} - w_{12})$ which is weighted by the frequency of the allele that remains (p_1 of *A*). The change from *aa* to *Aa* also has a selective effect $(w_{12} - w_{22})$ weighted by the frequency of the allele that remains (p_2 of *a*).

The formula derived above can be used to quantitate gene frequency changes if values for the actual frequency of genes *A* or *a* are obtained by counting a representative sample of the population.

To estimate a change in allele frequency under selection, an additional factor to be considered is dominance, for the presence of partial or complete dominance influences the relative fitness of a genotype. Suppose we consider the case of partial selection against the dominant phenotype where genotypes *AA* and *Aa* have a relative fitness of $w_{11} = w_{12} = 1 - s$ and genotype *aa* has a relative fitness of 1. To use the gene frequency change formula, let us use 1 for w_{22}, and $1 - s$ for w_{11} and w_{12}, p_2 for the frequency of gene *a*, and $1 - p_2$ for the frequency of gene *A*. With these substitutions the formula for change in the frequency of gene *a* (p_2) in one generation of selection is

$$\Delta p_2 = \frac{(1-p_2)p_2\{p_2[1-(1-s)] + (1-p_2)[(1-s)-(1-s)]\}}{1 - 2(1-p_2)p_2s - (1-p_2)^2s}$$

$$\Delta p_2 = \frac{sp_2^2(1-p_2)}{1 - s(1-p_2^2)}$$

Here, the change in the frequency of gene *a* is positive, and therefore an increase in the frequency of allele *a* would occur.

When both dominant phenotypes are sterile or lethal, the selection coefficients in both cases are 1, and the relative fitnesses are 0. In this circumstance

$$\Delta p_2 = \frac{p_2^2(1-p_2)}{1 - 1 + p_2^2} = \frac{p_2^2(1-p_2)}{p_2^2} = 1 - p_2$$

In the above it can be seen that the change in the frequency of allele *a* is equal to the frequency of allele *A* $(1 - p_2)$, and therefore the dominant allele, as well as the genotypes *AA* and *Aa* are completely eliminated from the population in one generation. As a result, the recessive allele is fixed, and all members of the population become *aa* in genotype after one generation of selection.

Let us next consider the case where partial selection is exerted against the recessive genotype *aa*, while genotypes *AA* and *Aa* each have a relative fitness of 1. If $a = p_2$, $A = 1 - p_2$, $w_{11} = w_{12} = 1$, and $w_{22} = 1 - s$, the formula for

* L. E. Mettler and T. G. Gregg, *Population Genetics and Evolution*, Prentice-Hall, Englewood Cliffs, N.J., 1969.

change in the frequency of gene *a* is

$$\Delta p_2 = \frac{(1 - p_2)p_2\{p_2[(1 - s) - 1] + (1 - p_2)(1 - 1)\}}{1 - sp_2^2}$$

$$\Delta p_2 = \frac{-sp_2^2(1 - p_2)}{1 - sp_2^2}$$

Here, the change in the frequency of gene *a* is negative, and therefore gene *a* would progressively decrease in frequency until finally it would be lost from the population altogether.

If individuals *aa* in genotype are lethal or sterile, complete selection would be exerted against them. In this case, $s = 1$, and the gene frequency change formula would be

$$\Delta p_2 = \frac{-p_2^2(1 - p_2)}{1 - p_2^2} = \frac{-p_2^2(1 - p_2)}{(1 + p_2)(1 - p_2)} = \frac{-p_2^2}{1 + p_2}$$

Here, selection would eliminate all *aa* individuals as contributors to the gene pool, so that the frequency of gene *a,* if initially moderately high, would decrease rapidly. However, as the frequency of *aa* individuals decreased, the rate of change in the frequency of gene *a* would also decrease because fewer and fewer individuals would be exposed to the forces of selection, heterozygotes carrying the gene being protected by the relative fitness of 1. Neither natural selection nor deliberate eugenic practices can totally eliminate a gene from a population because the equilibrium frequency between loss and addition is such that genes lost by selection are replaced by mutation.

Types of Selection

More than one type of selection can be recognized, depending on the relative fitness of the genotypes involved. To describe modes of selection we can borrow the terms, *directional, stabilizing,* and *disruptive,* which were used by Kenneth Mather with reference to selection for quantitative traits.

In the example above, genotype *AA* was the optimum, and selection would bring about an increase in the frequency of allele *A* and a corresponding decrease in the frequency of allele *a.* If the situation is reversed such that genotype *aa* exhibits the highest degree of fitness, the frequency of gene *a* would increase at the expense of that of gene *A.* Selection which favors one or the other homozygous genotype is called *directional selection,* and it can be summarized by diagram.

$$\underset{w_{11} \geq W_{12} > W_{22}}{A \longleftarrow} \qquad \text{or} \qquad \underset{W_{11} < W_{12} \leq W_{22}}{\longrightarrow a}$$

With directional selection one allele is replaced by a "better" allele, the process being one of gene substitution. During the course of gene substitution the average fitness of the population increases. The difference between the average fitness of the population and 1 can be considered the cost of natural selection, that is, the price the population pays for evolving. This price is often called the *substitutional genetic load.* Assuming a fitness of 1 when all individuals are *AA,* the fitness of a population evolving in this direction would be reduced at any time by selection directed against heterozygotes (s_1) and by that directed against homozygous recessives (s_2). Thus, the substitutional genetic load for replacement of allele a by A could be expressed as

$$\text{load} = 1 - (1 - 2p_1p_2s_1 - p_2^2s_2) = 2p_1p_2s_1 + p_2^2s_2$$

Stabilizing selection is selection directed against both homozygous genotypes and in favor of the heterozygote. It can be diagramed as

$$\xrightarrow{\quad\quad} Aa \xleftarrow{\quad\quad}$$
$$w_{11} < \quad w_{12} \quad > w_{22}$$

In this case Aa is the best adapted genotype, but it can be maintained only through the segregation of inferior homozygotes, and therefore an equilibrium between the frequencies of genes A and a will, in the course of time, be established. To examine this further, let p_1 = the frequency of gene A and p_2 = the frequency of gene a, the relative fitness of genotype Aa being 1 or $w_{12} = 1$. Also, if w_{11} = the relative fitness of genotype AA and s = the selection coefficient for genotype AA, then $1 - w_{11} = s$ or $w_{11} = 1 - s$. Similarly, w_{22} = the relative fitness of genotype aa, t = the selection coefficient for genotype aa, and $1 - w_{22} = t$ or $w_{22} = 1 - t$. Substituting these values, the equilibrium gene frequencies, $\hat{p}_1$ and $\hat{p}_2$, can be computed from the gene frequency change formula by setting it equal to 0 (that is, no change). Thus

$$0 = \Delta p_1 = \frac{\hat{p}_1\hat{p}_2\{\hat{p}_1(1 - s - 1) + [(1 - \hat{p}_1)(1 - 1 + t)]\}}{\overline{w}}$$

$$0 = \Delta p_1 = -\hat{p}_1 s + t - \hat{p}_1 t$$

and

$$\hat{p}_1 = \frac{t}{s + t} \quad \text{and} \quad \hat{p}_2 = \frac{s}{s + t}$$

The gene frequency equilibrium associated with stabilizing selection assures the presence in the population of each allele, so that the advantageous heterozygote is continually generated. As an extreme example, we can recall the true breeding balanced lethal condition in *Drosophila* (see Figure 9–17) where both homozygous classes die, the only viable members of each generation being heterozygotes.

Selection directed against heterozygotes and in favor of both homozygous genotypes is called *disruptive selection*. It can be diagramed as

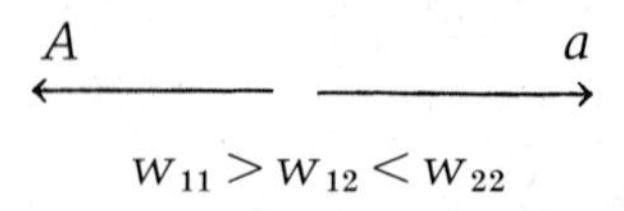

With disruptive selection a gene frequency equilibrium may be established as in the case of stabilizing selection where $\hat{p}_1 = t/(s + t)$ and $\hat{p}_2 = s/(s + t)$. However, the equilibrium is unstable, and any chance deviation from equilibrium frequencies can cause directional selection leading to fixation of one or the other of the two alleles involved. Thus, if $p_1 > t/(s + t)$, the first allele, in this case A, will become fixed, and if $p_1 < t/(s + t)$, the alternate allele, in this case a, will become fixed.

Equilibrium Between Mutation and Selection

Although selection has been analyzed as a separate process, it does not occur apart from mutation, for while selection is removing an allele from a population, mutation is continually replacing it. These two opposing forces eventually balance out at some time in the history of a population so that the number of alleles added by mutation is the same as the number eliminated by selection, that is, an equilibrium between mutation and selection is established.

To illustrate, let us assume that recessive allele a is selectively disadvantageous and that allele A mutates to a at the rate u per allele per generation. Then at equilibrium the recessive allele a is being removed at the rate sp_2^2, and at the same time replacement is occurring at a rate $up_1^2 \cong u$, since $p_1 \cong 1$. Thus, at equilibrium

$$s\hat{p}_2^2 \cong u,\ \hat{p}_2^2 \cong \frac{u}{s} \text{ and } \hat{p}_2 \cong \sqrt{\frac{u}{s}}$$

When $s = 1$, as in lethal or sterile genotypes, $\hat{p}_2 \cong \sqrt{u}$. In more concrete terms, if the mutation rate of a lethal gene were 10^{-5}, the equilibrium gene frequency would be about 3×10^{-3} or 0.003. For less deleterious recessives, where $s = 0.1$, $\hat{p}_2 \cong 10^{-2} \cong 0.01$. In humans or other organisms where better means are not available, mutant phenotype frequencies can be used to obtain a rough estimate of the mutation rate of a recessive lethal.

We should note that our discussion has not taken reverse mutation $(a \rightarrow A)$ into account. It will be recalled that reverse mutation occurs with a significantly lower frequency than forward mutation and therefore can probably be ignored.

There are many situations encountered in nature where two or more alleles coexist within a population at frequencies higher than can be explained by mutational equilibria, a phenomenon termed *polymorphism* by E. B. Ford. One mechanism which can account for the maintenance of this condition is stabilizing selection where two alleles coexist with frequencies of $\hat{p}_1 = t/(s + t)$ and $\hat{p}_2 = s/(s + t)$, respectively. Such a condition is known as balanced polymorphism, an example being sickle cell hemoglobin. Polymorphism is also exhibited as a result of disruptive selection and may be a transient phenomenon during directional selection. Since polymorphism is best illustrated by example, further discussion of this generally widespread condition will be deferred to Chapter 23.

RANDOM GENETIC DRIFT

One of the initial assumptions made for a Hardy-Weinberg population is infinite size, wherein the allele frequencies in a new generation are identical with those of the parental generation. Actual populations are not infinite in size, however, and thus the alleles present in a descendent generation represent a random, but not necessarily identical sample of the parental group. In addition, sample gene frequencies will be binomially distributed with a mean gene frequency the same as in the parental generation (see Figure 3–2 for examples of binomial distributions.) Although sampling errors decrease with increasing population size, in a finite population there will always be some chance differences between the gene frequencies of the parental and the new generation. The most likely gene frequency in a sample is the parental, but other gene frequencies (including 0 and 1) are possible even though the probability is small. Thus, there exists a small, but non-zero possibility that a gene will be fixed by chance alone. The process of gene frequency change among populations due to random sampling is called *random genetic drift.*

Suppose we examine the fate of a single allele in a population of constant finite size. If the population contains N individuals, there will be $2N$ alleles in the sample of gametes forming the next generation. If the starting allele frequency at time 0 (t_0) is p_0, then, in the next generation the probabilities for the frequency of the allele will be binomially distributed. Although the most probable gene frequency in the succeeding generation is the parental gene frequency, some other gene frequency than the parental is often just as likely. Since a Hardy-Weinberg population has no "memory" of a previous gene frequency, the new gene frequency in the first generation will be the starting determinant of the gene frequency of the next generation. A second and independent sampling event will then occur from which will arise the gene frequency of generation 2. Each successive sampling is thus independent of all previous events and depends solely on the gene frequency of the previous generation (Figure 22–2). The process can be called a "random walk," and in any particular population the walk ends when the gene frequency strikes a boundary and becomes fixed at either $p = 1$ or $p = 0$. Although fixation may require a long time in large populations, one or the other of these two alternatives is the ultimate fate of a gene in all finite populations (in the absence of mutation).

As an analogy, we can consider the movement of a tipsy man staggering forward along a sidewalk which is bounded on either side by a ditch, one ditch representing $p = 1$, the other $p = 0$. As our man progresses, his possible direction at any time will be binomially distributed with a maximum probability that he will move in the same direction as his previous step. Although he may or may not move in this direction, whatever step he takes will then initiate a new binomial distribution for the direction of his next succeeding step, and eventually he will stray into either one ditch or the other.

We can also examine the process of random drift from the standpoint of a large number of populations of the same size starting at the same gene frequency. At generation 1, their gene frequencies will be binomially distributed with a few populations fixed at 0 and 1. With successive generations more and more of the populations will become fixed at the boundaries, while those remaining unfixed and therefore free to vary in gene frequency will decrease in number. These are represented in Figure 22–2(b) by the curve between the boundaries which with an increasing number of generations becomes flatter and flatter. Ultimately all of the populations become fixed, p_0 of them at 1 and $1 - p_0$ of them at 0. At this point, if the overall gene frequency in all of these populations together were determined, no change would be observed. However, the variation originally present *within* these populations would have been converted to variation *between* these populations. Random genetic drift occurring in small ancestral populations is thought to account for some of the differ-

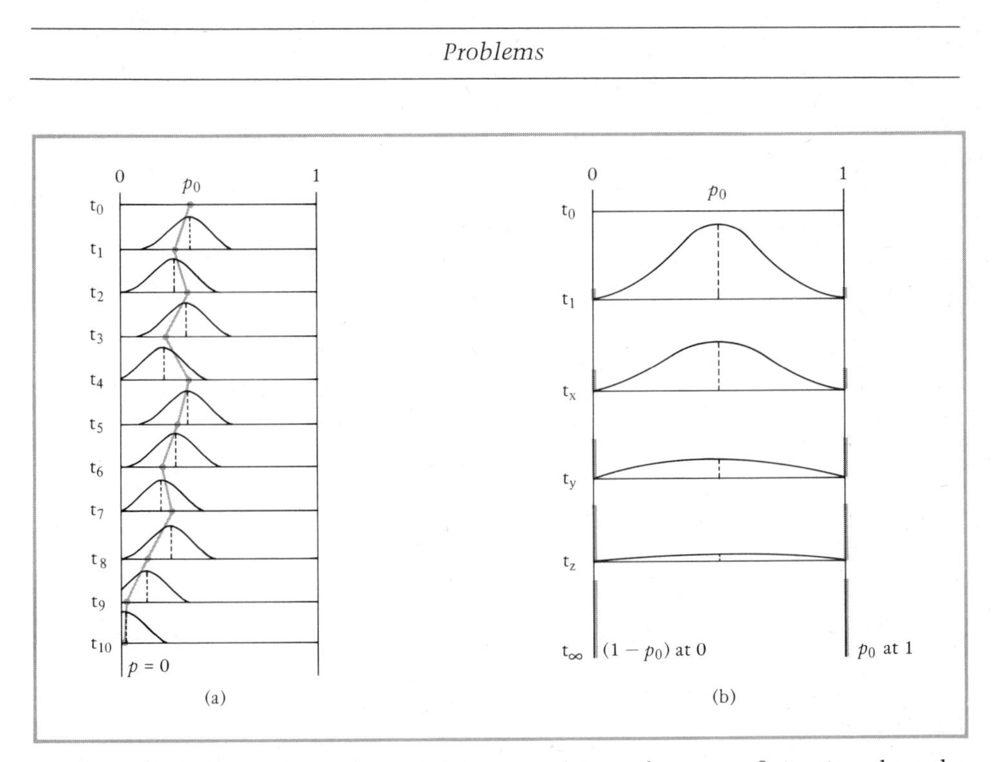

Figure 22–2. (a) Random genetic drift in a population of constant finite size where the starting gene frequency is p_0 at time t_0. The probabilities for the frequency of gene p in each generation are binomially distributed. Although in any sample the most likely frequency is the same as the mean frequency of the parental generation, other frequencies are possible through chance alone, leading ultimately to fixation, in this case, loss of allele p. (b) Random genetic drift in a large number of populations of the same size, starting with the same gene frequency. At t_1 the gene frequencies of these populations are binomially distributed with a few populations fixed at 0 or 1. With succeeding generations and as more and more populations become fixed, fewer remain free to vary in gene frequency and therefore the curve representing such groups becomes flatter. Finally p_0 of these populations is fixed at 1 and $(1 - p_0)$ is fixed at 0. (These diagrams courtesy of W. F. Duggleby).

ences between human population subgroups in the frequencies of some alleles such as those determining blood types.

This chapter has outlined the ways by which gene frequencies are changed and some of the mathematical models which provide the basis for quantitating such change. We have seen that while mutation and migration can introduce new alleles into a population, their preservation or loss depends largely upon the fitness which they confer. Selection imposed by the environment upon the phenotypes present in a population is mainly responsible for shaping the characteristics of the group, and to this force must be added the chance fixation of alleles through random genetic drift. In the following chapter, we will examine some examples of these processes.

PROBLEMS

22–1. If barley plants, initially heterozygous for genes *C* and *c* were allowed to self-fertilize, what would be the overall frequency of genes *C* and *c* in the population after five generations of self-fertilization?

22–2. In the following pedigree determine the pathways of descent and compute the value of F for individual I.

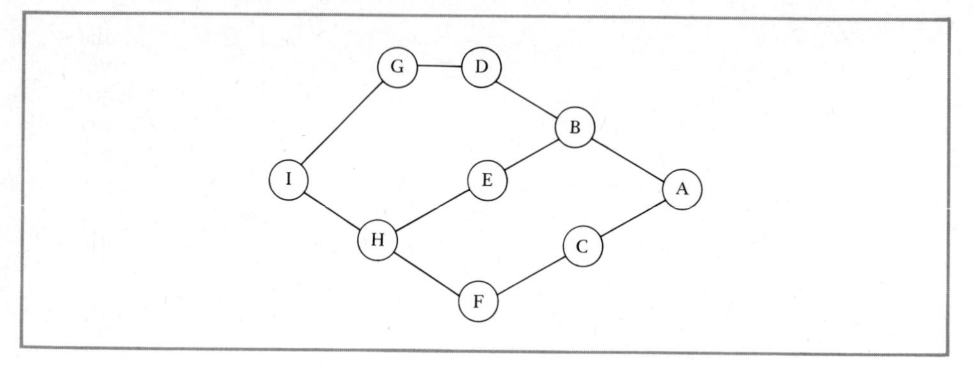

22–3. If the inbreeding coefficient of individual C of Figure 22–1(b) is $\frac{1}{4}$, calculate the value of F for individual I.

22–4. Assume that the frequency of the gene for aniridia in humans is approximately 1.6×10^{-3} and the frequency for the normal condition is approximately 1.0.

a. What is the probability that a homozygote for aniridia will be produced by a marriage between unrelated individuals?

b. What is the probability that a homozygote will be produced by a marriage between first cousins?

22–5. If the mutation rate of the wild type allele D to the recessive allele d is 6×10^{-5} and the reverse mutation rate of d to D is 1×10^{-5}, in the absence of selection what will be the equilibrium frequency of gene d under mutation in a randomly breeding population?

22–6. In an isolated island population of 75 individuals the frequency of gene A is 0.8. A migrant group of 25 individuals wherein the frequency of A is 0.2 joins the natives to form a combined population of 100 members. As a result of one generation of random breeding, what would be the change in the frequency of gene A?

22–7. In a randomly breeding population the frequency of gene d is 0.2. If the relative fitness of genotype $DD = Dd = 1$ and that of genotype $dd = 0.3$, what would be the frequencies of genes d and D after one generation of selection?

22–8. A large random sample of the initial population of problem 22–7 was taken.

a. What proportion of the sample would be expected to show the dominant phenotype?

b. If a sample were taken from the population one generation after selection, but *before* selection was again imposed, what proportion would show the dominant phenotype?

22–9. In a randomly breeding population a drastic alteration in the environment occurs with the result that genotype aa has a relative fitness of 0 compared with genotypes AA and Aa, each of which has a fitness of 1. If the starting frequency of gene a is 0.4, what would be the frequency of gene A one generation after the environmental change?

22–10. A large random sample of the population of problem 22–9 was taken prior to environmental change.

a. What proportion of the sample would be expected to be aa?

b. If selection against *aa* individuals occurs only after hatching from the egg, what proportion of the eggs laid by individuals one generation after environmental change would be *aa* in genotype?

22–11. In a randomly breeding population of butterflies individuals with black and orange spotted wings blend with their surroundings and have a relative fitness of 1; those with solid orange wings are more obvious to predators and have a relative fitness of 0.5. The black and orange phenotype is due to the homozygous presence of recessive gene *b*, while solid color is produced in the presence of the dominant allele *B*. If the initial frequency of gene *b* is 0.7, what are the frequencies of genes *B* and *b* after one generation of predation, that is, selection?

22–12. A random sample of 500 butterflies was collected from the initial population of problem 22–11.

a. How many would be expected to be solid colored?

b. If a second sample of 500 was taken from this population *after* predation, what proportion of the sample would be solid colored?

22–13. Suppose an acute viral disease is introduced into a population of squirrels and that those homozygous for gene *r* are resistant, while those of genotypes *RR* and *Rr* are susceptible and suffer 100 percent mortality. If the initial frequency of gene *r* is 0.65, what is its frequency (barring mutation) one generation after the introduction of this disease?

22–14. If the mutation rate of gene *B* to allele *b* is 6×10^{-5}, and *bb* homozygotes have a relative fitness of 0.2 as compared to the dominant phenotype, what will be the frequency of gene *b* when an equilibrium between mutation and selection is reached?

22–15. If the rate at which the normal gene mutates to the recessive lethal allele for retinoblastoma is 6×10^{-6}, what is the frequency of the lethal allele at equilibrium?

22–16. If, in a randomly breeding population, the relative fitness of genotype $Aa = 1$, that of $AA = 0.4$, and that of $aa = 0.8$, in the absence of mutation what would be the frequencies of these alleles at equilibrium?

REFERENCES

Crow, J. F., 1976. *Genetics Notes,* 7th ed. Burgess, Minneapolis, Minn.

Falconer, D. S., 1960. *Introduction to Quantitative Genetics.* Ronald Press, New York.

Li, C. C., 1955. *Population Genetics.* University of Chicago Press, Chicago.

Li, C. C., 1967. Genetic equilibrium under selection. *Biometrics, 23,* 397.

Mettler, L. E., and T. G. Gregg, 1969. *Population Genetics and Evolution* (Foundations of Modern Genetics Series). Prentice-Hall, Englewood Cliffs, N.J.

Chapter 23

POPULATION CHANGE AND EVOLUTION

THE initial examples used by Darwin to illustrate the power of selection in changing the characteristics of a species were taken from domesticated plants and animals. Darwin himself was a pigeon fancier and in *The Origin of Species* he discourses at length upon the numerous varieties of pigeons developed by breeders and traces the origin of these varieties to the wild rock pigeon. Selection in pigeons and other domesticated species is directed by human design, however, and not by adaptation to the natural environment, and the traits which have been selected for have been primarily quantitative, rather than qualitative.

It will be recalled that quantitative traits are those in which the phenotype varies continuously from one extreme to the other, no separate classes ordinarily being distinguishable. Quantitative traits include such aspects as height, weight, quantity of some component such as carbohydrate, fat, or protein, number of eggs laid, growth rate, and so on. These attributes must be scored by measurement, and the expression of the trait within a sample can then be represented by a histogram or by a curve and analyzed by statistical means (see Chapter 5). Many genes, called polygenes, are involved in the determination of these traits, their separate and slight contributions to the character being obscured by the effects of the environment. Since separate phenotypic classes cannot usually be identified, the number of genes contributing to a quantitative character must be estimated indirectly, rather than determined directly through Mendelian ratios. Quantitative traits are of great economic importance, and selection practiced by animal and plant breeders is responsible for the development of the numerous breeds of domestic animals, the strains of food crops of improved vigor and yield, and the many varieties of ornamental shrubs and flowers.

DIRECTIONAL SELECTION

The types of selection previously described for qualitative traits, that is, directional, stabilizing, and disruptive, also pertain and, indeed, were first applied to quantitative traits (Figure 23–1). Directional selection is ordinarily practiced in breeding programs directed toward the improvement of a strain. In such programs only the best individuals are used as parents of the next generation. Although the progeny exhibit a mean which is better than that of the entire parental generation, they do not ordinarily achieve the mean of the selected parents. This indicates that the superiority of the parents is partly due to environmental effects, for if it were completely genetic in origin, the progeny mean would coincide with the parental mean. "Superiority" is judged by the plant or animal breeder in terms of the trait being selected for, and in the case of artificial selection the percentage of parental superiority achieved by the progeny is called *heritability* or, more strictly, *realized heritability.* Heritability can be considered a measure of the approximate degree to which a trait is genetically determined (see Table 5–4).

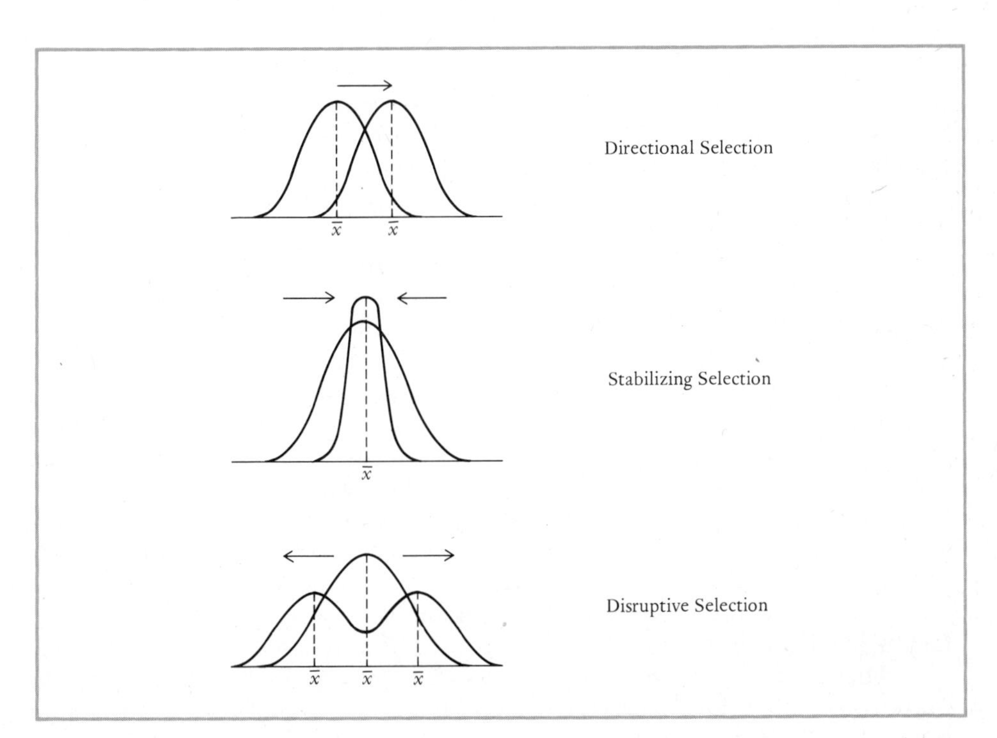

Figure 23–1. Three modes of selection for quantitative traits. The means $(\bar{x})$ are indicated by dotted lines.

In domestic species almost any quantitative trait can be altered by directional selection continued over many generations. As an example, I. M. Lerner reported that selection applied over the years 1933 to 1965 for the number of eggs laid per year by White Leghorn chickens resulted in an increased average production per hen of from 125.6 eggs per year in 1933 to 249.6 eggs per year in 1965. Equally effective has been the artificial selection for high and low protein and oil content in corn. Starting with an initial strain with an average of around 10.9 percent protein and 4.7 percent oil, 50 generations of selection for high levels of these components resulted in corn containing 19.4 percent protein and 15.4 percent oil, while selection for low levels yielded corn containing 4.9 percent protein and 1 percent oil. These are only two of many examples that could be cited and similar experiments performed with laboratory species have produced analogous results (see below).

Quantitative Heterosis

Programs of directional selection, of necessity, include at least some degree of inbreeding between individuals deemed superior in regard to the trait under consideration, and it will be recalled that inbreeding results in an increase in homozygosity and a decrease in heterozygosity. Because detrimental genes originally concealed in heterozygotes by dominant alleles become homozygous and are expressed in the phenotype, the usual net result of inbreeding is a decline in all components of fitness, such as vigor, yield, and survival. Inbreeding also decreases the variability of a strain by decreasing the number of alleles still free to undergo segregation. Although highly inbred homozygous lines may themselves be weak, the F_1 hybrids formed by crossing two such lines are more vigorous and exhibit a higher yield than either parental strain. In addition, the F_1 hybrids are characteristically uniform in phenotype because of their uniform heterozygosity. This hybrid vigor with its accompanying phenotypic uniformity is called *quantitative heterosis,* and advantage has been taken of this phenomenon by plant breeders in the production of superior crops, notably hybrid corn.

The usual method of producing hybrid corn is to begin by crossing two inbred lines by artificial pollination to produce F_1 hybrid seed. This seed is formed on stalks of the inbred strain serving as the female parent in the cross. However, since the parental strain is weak, the amount of seed set is insufficient for large scale use. To circumvent this difficulty, the F_1 hybrid seed is grown, and the resulting plants are crossed either to another inbred line or to another F_1 hybrid. Since the seed from this cross develops on an F_1 hybrid exhibiting heterosis, large ears and a high yield of seed result. It is this double cross seed which is sold to farmers, and when planted, it develops into a uniform, high yielding crop. The seed from this crop is ordinarily all used for grain and not saved to produce another generation, because with random mating via wind pollination the progeny will be inbred, no longer uniform in phenotype, and yield will decline. Thus, hybrid seed must be purchased anew each season.

The source of quantitative heterosis can be attributed not only to the masking of deleterious recessives by dominant alleles, but also to numerous other subtle factors which are not well understood. These probably operate at the molecular level through a variety of mechanisms, for example, by increasing the efficiency of enzyme-substrate interactions or increasing the permeability of cells. Indeed, it has even been demonstrated that the mitochondria of hybrids also exhibit heterosis in terms of increased oxygen uptake and ATP production. An understanding of the basis for heterosis will depend to a large extent on greater knowledge concerning the effects of those genes which determine quantitative traits.

Directional Selection in Experimental Populations

Artificial selection for quantitative traits can readily be carried out in the laboratory using noninbred strains of experimental organisms. *Drosophila* species, in particular, have been favorite subjects in such controlled studies, and as in domestic animals it appears that practically any quantitative trait in these insects can be changed by selection. For example, body and wing size, bristle number, rate of development, fecundity, resistance to insecticides, behavioral patterns, and even the frequency of recombination in different chromosome regions have been successfully modified by selection. Two such examples can be considered in greater detail.

In one set of experiments Kenneth Mather and associates have demonstrated directional changes in the number of bristles on the ventral surface of the fourth and fifth abdominal segments of *D. melanogaster* adults. The starting population exhibited a mean of 36 bristles. After 20 generations of intense artificial selection to produce a "high line," the average bristle number was increased to 56; after 30 generations of selection to produce a "low line," the mean bristle number was reduced to 25 to 30.

In addition to successful selection for what would seem to be a trivial character, it was found that correlated changes occurred in a number of other phenotypic features, for example, viability and fertility were lowered, and alterations were observed in pigmentation, eye shape, and number of sperm storage organs (spermathecae) in females. The correlated changes can be interpreted in terms of the separate effects of numerous genes associated with the quantitative trait of bristle number. Genes affecting this character are found at several loci on all chromosomes, and the intense selection practiced in the high and low lines would tend to eliminate alleles having an opposite effect than that desired. Thus, the lines would become increasingly homozygous, thereby uncovering detrimental genes whose effects had previously been concealed in the heterozygous condition.

We must also consider that genes whose cumulative effects regulate bristle number may also have pleiotropic effects on other and perhaps more important parts of the body. For example, a gene specifying an enzyme whose activity was rate limiting at some stage of development would be expected to produce wide ranging effects on many organ systems besides the integument and its associated structures. Intense selection for a seemingly trivial trait without regard for correlated phenotypic effects, such as fertility, would hardly occur in nature, but a laboratory demonstration of its effects indicates that the genome must evolve as a whole and that it is the fitness of the entire phenotype, not just a single character, that is subject to natural selection. In these experiments, selection was subsequently terminated in some high and low lines, and it was found that abdominal bristle number gradually returned toward the

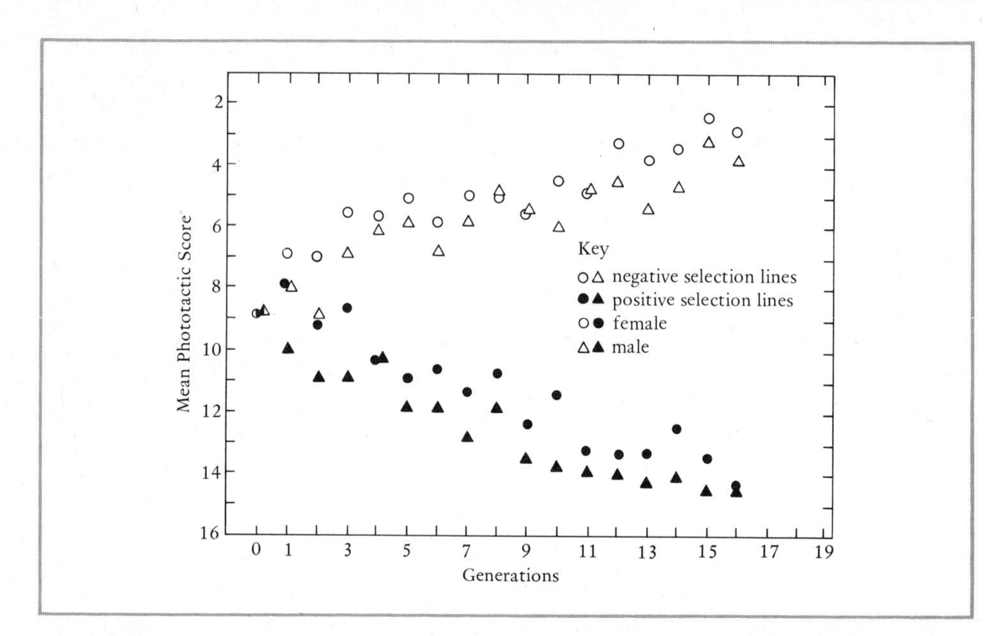

Figure 23–2. Directional selection of phototaxis. [From Th. Dobzansky and B. Spassky, *Proc. Roy. Soc. London, B, 168,* 27–47, 1967.]

original mean value. We can interpret this observation as the reestablishment by natural selection of more optimal genotypes leading to better adapted phenotypes, one of whose aspects involved bristle number.

A second and interesting example of directional selection for a quantitative trait may be taken from the studies of Th. Dobzhansky and B. Spassky, where selection was applied for positive and negative responses to light on the part of *D. pseudoobscura.* In these experiments 300 males and 300 females in each generation were run through a maze containing 15 choices for a dark or light passage, and those with high scores for either light or dark were utilized as parents of the next generation. After 16 generations of selection separate light and dark lines were obtained (Figure 23–2). At this point, selection was relaxed and parents were chosen regardless of their light preferences. With the cessation of selection for phototaxis each line tended to return to the norm of the original population, a finding similar to that obtained above with abdominal bristle number. Similar observations have long been recorded by plant and animal breeders, for it has often been noted that loss of a selected trait may occur when selection for the trait is discontinued. We can view such loss in terms of artificial selection having resulted in genotypes which, while producing a desired trait, are less well adapted overall. Accordingly, it is not hard to understand why the products of intense artificial selection rarely survive under natural conditions in the wild.

Directional Selection in Nature

The best recorded example of directional selection in nature is the industrial melanism of moths which has accompanied the industrial revolution in

Britain, as well as in the rest of Europe and North America. The moth *Biston betularia* had long been observed in Britain, as a peppered form, light gray in color with a scattering of black spots. Since during the day the moths often rest on gray lichen-covered tree trunks, their color pattern provides protective camouflage against bird predators. The peppered variety was the only form known for this species until a black variant (called *carbonaria*) was recorded in Manchester, England, in 1848; by 1895 it constituted about 98 percent of the members of the species in the Manchester region (Figure 23–3). During this period industrialization caused a dramatic change in the environment in this and other areas of Britain. Factories using soft coal produced tons of soot which descended on the surrounding countryside blackening the trees and killing the lichens. Thus, when the light-colored, peppered variety rested on tree trunks in these industrialized regions, it was highly visible to birds, while the melanic *carbonaria* form blended with the blackened tree trunk and was less subject to predation. That the increase in the *carbonaria* variety in industrialized regions is due to environmental change brought about by industrial pollution has been

Figure 23–3. *Biston betularia* the Peppered Moth, and its black form *carbonaria,* (a) at rest on lichened tree trunk in unpolluted countryside, Dorset, England; (b) at rest on soot covered Oak trunk near Birmingham, England. [From the experiments of H. B. D. Kettlewell, University of Oxford.]

experimentally demonstrated by Kettlewell, Ford, and others. Kettlewell released equal numbers of marked light and dark forms in polluted and unpolluted areas, and by the use of attractants (female pheromones) was able to recapture marked individuals. Many more marked melanic than light forms were recaptured in industrial areas, while just the reverse was true in unpolluted districts. That bird predation was responsible for these results was proven by using high speed photography of birds perceiving and eating moths resting on trees. In unpolluted areas melanic forms were highly visible and most often chosen, while in polluted regions the ancestral peppered variety suffered the greatest predation. Recently, the frequency of melanic forms in industrial areas has declined as a consequence of smoke abatement programs.

Industrial melanism in *Biston* is not a polygenic trait, for breeding experiments have shown that the *carbonaria* phenotype is due to a single dominant gene. The locus may not always have been dominant, however, for recent melanics are darker than those preserved in early collections. Enhanced darkness of the modern *carbonaria* form is thought to be due to selection for modifiers of dominance. It has also been found that the melanism present in one population may not behave as a complete dominant in crosses with peppered forms from a different population, suggesting that the alleles responsible for the dark form may not be identical in these groups. A variant called *insularia,* which is not as dark as *carbonaria,* has been found in regions peripheral to those with intense pollution, and crosses of *insularia* with *carbonaria* indicate the presence of separate loci for melanism. Over 100 cases of industrial melanism have been recorded for other species of moths in Britain, Europe, and North America. Although the progress of natural selection is usually too slow to be witnessed even within several lifetimes, the historical record along with modern observations has permitted documentation of the event in this instance.

STABILIZING SELECTION

The term stabilizing selection has been applied to both qualitative and quantitative traits. In the case of single gene loci the heterozygous state is best adapted and selection is exerted upon the two homozygous classes. With quantitative traits the degree of expression of the trait approximates a normal distribution wherein the mean is favored as the optimal form and selection is imposed upon the two extremes (Figure 23–1). In this case the best adapted genotypes are those that are highly heterozygous.

Stabilizing Selection in Experimental Populations

In an experiment with *D. melanogaster,* S. Polivanov studied selection against stubble *(Sb),* a third chromosome dominant mutation with a recessive lethal effect. Four experimental populations were initiated by heterozygotes *(Sb/+)*, the gene frequency being $p_1 = p_2 = 0.05$ in each case. In founders of two of these populations the wild type allele of *Sb* was contained in a third chromosome descended from a single ancestor and thus these populations can be termed monochromosomal. In the other two populations the founders contained wild type third chromosomes derived from a number of different ancestors, that is, they were polychromosomal. The experimental populations were maintained in cages into which fresh medium could be introduced when needed and from which larval or adult samples could be removed from time to time for a determination of gene frequency. In Figure 23–4 it can be seen that the frequency of *Sb* declined in all four populations, but much more quickly than predicted in the polychromosomal groups and much less quickly than predicted in the monochromosomal groups. In the monochromosomal populations the heterozygote *Sb*/+ enjoyed a temporary advantage over +/+ individuals for some generations. This unexpected advantage was attributed to the masking by genes of the *Sb*-bearing chromosome of subvital alleles on the wild type chromosome. Since the wild type chromosome had been derived from a single ancestor, +/+ individuals were of necessity homozygous for such genes, a condition causing lowered fitness as compared to heterozygotes. The replacement of subvital genes by their wild type counterparts eventually occurred through crossing over, but since the *Sb* chromosome contains a large inversion, the process required some time. During this time heterozygotes displayed an advantage, but after replacement had occurred, the +/+ genotype was strongly favored by selection and the frequency of *Sb* decreased rapidly.

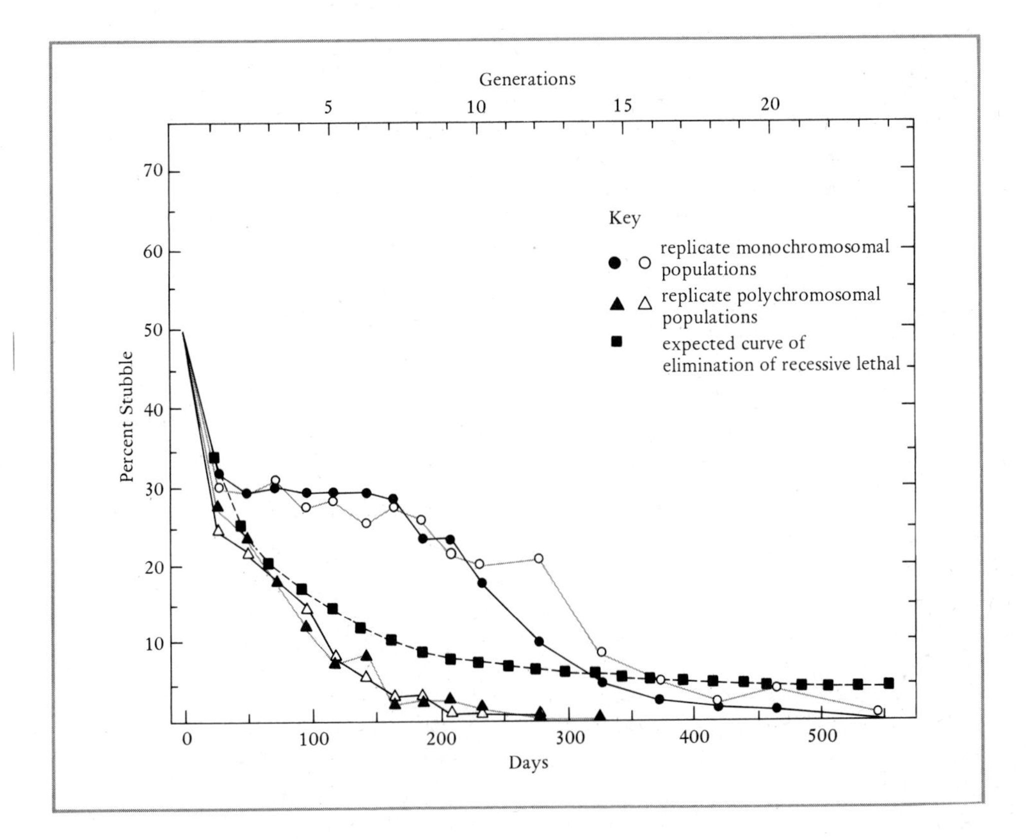

Figure 23–4. Changes in frequencies of Stubble in monochromosomal and polychromosomal populations. [From S. Polivanov, *Genetics, 50,* 81–100, 1964.]

In polychromosomal populations no such heterozygote advantage was present. The wild type chromosomes of these populations were derived from several sources, and combinations between them immediately rendered the +/+ genotype superior in fitness to that of the heterozygote. As a result, directional selection in favor of +/+ individuals occurred continuously.

Stabilizing Selection in Natural Populations

The best documented example of stabilizing selection in nature is that involving sickle cell hemoglobin (HbS), where heterozygotes enjoy a selective advantage in malarial regions of Africa. The homozygous condition of this mutation (HbS/HbS) is essentially lethal in early childhood due to severe hemolytic anemia. Homozygous normal individuals (HbA/HbA) are more susceptible to infection by the protozoan parasite *Plasmodium falciparum,* the causative agent of the most severe type of malaria. Serious to fatal infection by this parasite is more common in such persons than in heterozygotes (HbS/HbA).

In experimental studies of African volunteers injected with the parasite (and subsequently cured), heterozygotes exhibited a significantly reduced parasite count as compared to normals. This resistance can in part be ascribed to the occurrence of red cell sickling caused by low oxygen tension in capillaries, sickled cells, including those containing parasites, being more subject to hemolysis and removal from the bloodstream. A more effective immunity resulting from prior infections may additionally contribute to the resistance shown by heterozygous individuals. There is also some evidence for an increased fecundity on the part of heterozygotes in some populations, perhaps due to a lower rate of spontaneous abortion resulting from a decreased frequency of malarial infection of the placenta.

In African populations heterozygotes become more frequent with age as would be expected when such individuals enjoy a selective advantage. Assuming the selective disadvantage of genotype HbA/HbA to be $s = 0.15$ and the selection coefficient of HbS/HbS to be $t = 1$, then the equilibrium frequency of the HbS allele would be: $p_S = s/(s + t) = .15/(.15 + 1) = 0.13$, which is approximately that observed. It should be noted that outside of malarial regions heterozygotes are at a selective disadvantage and their numbers decrease with age as compared to homozygous normal individuals.

There is some indication that another abnormal hemoglobin, HbC, is replacing HbS in Africa. Heterozygotes for HbC are also more resistant to parasitic infection and this allele has an advantage in that homozygotes, while anemic, do not die in early childhood. Other mutant hemoglobins associated with malaria have also been identified, such as HbE in Thailand and other areas of Southeast Asia, HbO in Indonesia, and HbD in India (Figure 23–5).

Besides the hemoglobin variants, other human molecular diseases related to malaria resistance include the thalassemias of Mediterranean regions and glucose-6-phosphate dehydrogenase deficiency (G-6-PD). G-6-PD deficiency is sex linked; in heterozygous females normal and G-6-PD deficient red blood cells occur in approximately equal numbers due to random inactivation of one of the X chromosomes. L. Luzzato and coworkers have shown that in such females G-6-PD deficient cells are much less frequently parasitized than are

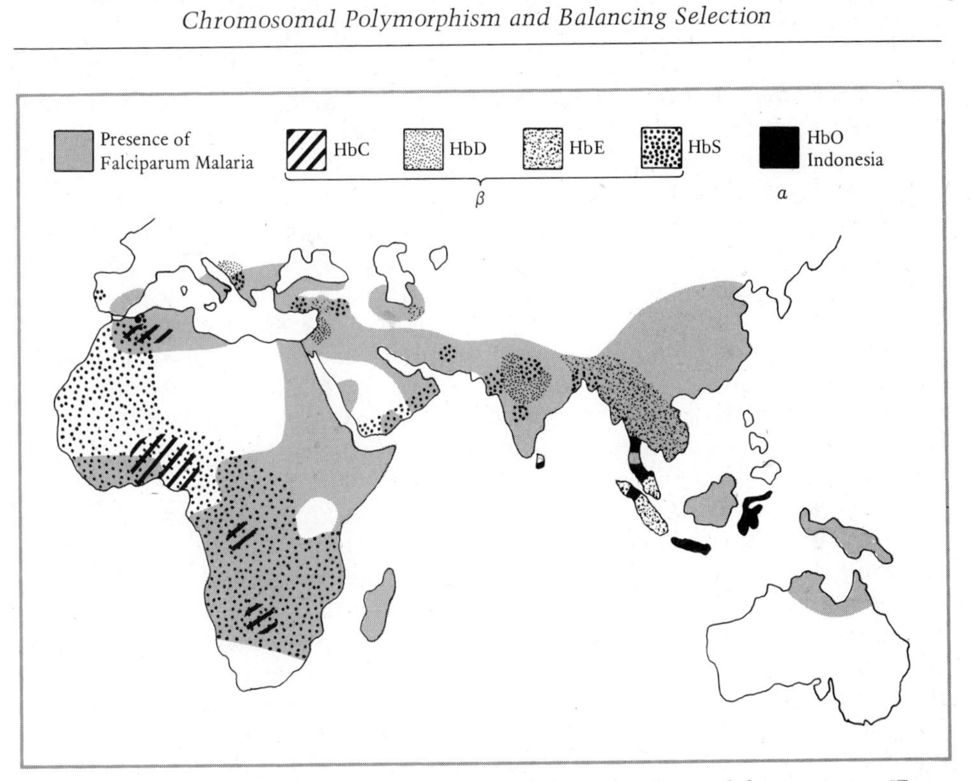

Figure 23–5. Map showing the distribution of the major hemoglobin variants. [From *Evolutionary Biology* by Stanley N. Salthe. Copyright © 1972 by Holt, Rinehart and Winston. Reprinted by permission of Holt, Rinehart and Winston.]

cell having normal enzyme activity. This finding could account for the selective advantage enjoyed in regions where malaria is endemic by heterozygotes over homozygous normal females or hemizygous normal males.

CHROMOSOMAL POLYMORPHISM AND BALANCING SELECTION

Polymorphism refers to the presence of several different forms of alleles or phenotypes, the rarest of these existing in a population with a higher frequency than can be accounted for by mutation. All types of polymorphism—morphological, cytological, and biochemical—occur in natural populations; of these, let us consider chromosomal polymorphism in the form of inversions.

An aberration such as an inversion can be viewed as a chromosomal mutation which arises as a chance, nonrecurring event. We must also recall that inversions can be termed crossover suppressors because, when heterozygous, recombination usually leads to the production of acentric and dicentric chromosomes or to duplication and deficiency gametes (see Chapter 9). Since recombination generally results in inviable zygotes, the genes present within an inverted chromosome segment tend to be inherited together as a group or gene complex. If such an inversion happens to involve alleles whose presence together in the same chromosome segment confers better adaptation to the environment, such an inversion will be advantageous, for loss of this particular association of alleles through recombination will be prevented. For this reason, an advantageous grouping of genes within an inversion is sometimes called a supergene, because the entire gene complex is inherited as a whole.

Dobzhansky and coworkers carried out detailed studies of particular third chromosome inversions in natural populations of *D. pseudoobscura.* The frequencies of these inversions vary seasonally and geographically, as well as within different regions of the local environment. Sampling of wild populations resulted in the identification of a number of cytologically distinguishable inversions which occur widely and are present in different frequencies in various regions of southwestern United States and northern Mexico. These common inversions were named for their prevalence, for example, Standard (ST) or for the localities in which they were first found, for example, Chiricahua (CH), Arrowhead (AR), and Pikes Peak (PP). In wild populations living at different elevations in the Sierra Nevada mountains of California, it was found that the respective frequencies of different inversions changed with altitude. For example, the prevalence of the ST chromosome decreased with increasing altitude, dropping from a frequency of 46 percent at 850 feet to 10 percent at 10,000 feet. Analysis of samples captured at different seasons of the year (spring, summer, and fall) from a locality on Mt. San Jacinto indicated the presence of cyclic fluctuations in the frequencies of different inversions (Figure 23–6).

These results suggest that the different gene complexes contained within the inversions have differing adaptive values which are correlated with seasonal and local environmental changes. Samples of these populations brought into the laboratory and maintained in cages have been subjected to experimental environments where temperature, humidity, or food supply were variables, and it has been found that individuals heterozygous for two different inversions are usually more viable and therefore more frequent in number than those homozygous for either inversion. This finding confirms early observations made in the field that the majority of individuals captured are similarly

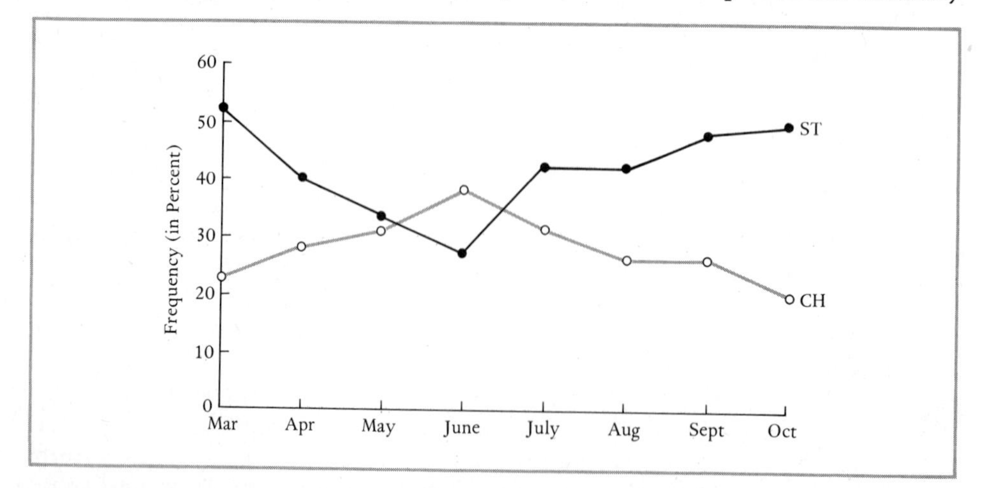

Figure 23–6. Seasonal changes in the frequencies of third chromosomes with two different inversions (ST and CH) in a population of *Drosophila pseudoobscura* from California. [From Th. Dobzansky, *Genetics of the Evolutionary Process,* p. 135, Columbia University Press, New York, 1970.]

heterozygous. Dobzhansky has concluded that the inversions found in high frequency in natural populations represent supergenes, that is, groups of advantageous genes whose collective association is preserved by crossover suppression. The chromosomal polymorphism present in these populations thus confers a better average fitness through a kind of balanced chromosomal heterosis in which individuals heterozygous for two different supergenes experience greater versatility in coping with a changing environment.

DISRUPTIVE SELECTION

Disruptive or diversifying selection is selection exerted in favor of the extremes and against the mean or intermediate phenotype. Disruptive selection may be associated with the adaptation of a species to a heterogeneous environment in which a variety of ecological niches are present. In such an environment a certain amount of genetic flexibility might be required which could lead to polymorphism and the presence of more than one favorable phenotype. In the case of single genes the equilibrium between alleles in a population undergoing disruptive selection is unstable rather than balanced, and in the absence of gene flow fixation of alleles within population subgroups may follow.

Equivalents of natural environmental heterogeneity are difficult to establish, and thus laboratory experiments utilize a presumed homogeneous environment and apply a rigorous program of disruptive selection for quantitative, rather than single-gene traits. In such experiments members of extreme classes are chosen as parents of each generation and all other individuals are discarded. The studies of J. M. Thoday and colleagues are an example. These workers used the quantitative trait of sternopleural bristle number in *D. melanogaster.* Initial populations were subjected to selection for high and low bristle number. Thereafter, males from high lines with the highest bristle number were mated with females with a high bristle number but derived from low lines. The reciprocal cross was carried out as well, that is, males from low lines with the lowest bristle number were mated with females from high lines with the lowest bristle number. These crosses ensured a 50 percent gene flow between the high and low populations, but only the extreme individuals were used as parents, all other classes of individuals being discarded. Application of disruptive selection for continuing generations resulted in an increased difference in mean bristle number between the high and low lines, this divergence being maintained as a genetic polymorphism (Figure 23–7).

Additional experiments by J. B. Gibson and Thoday using the same trait resulted not only in divergence between these lines, but also in reproductive isolation because of the loss of fecundity. The genetic conditions leading to this result, though not understood, must have been unusual. It is unlikely that the intensity of selection imposed experimentally would occur in nature, and thus whether or not disruptive selection could cause divergence to the point of isolation in a population in the wild is debatable.

In natural populations disruptive selection is thought to be the major force in the evolution of Batesian mimicry (named after H. W. Bates, an English naturalist). In this type of mimicry a palatable organism adopts the coloration

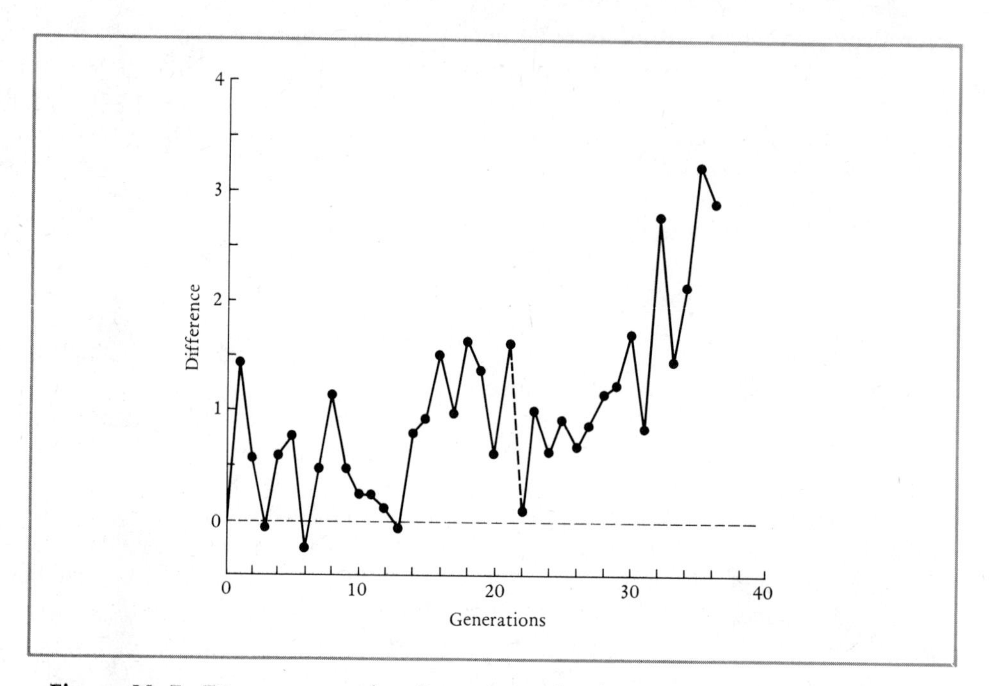

Figure 23–7. Divergence under disruptive selection. Mean differences in bristle number per fly between high-bristle number and low-bristle number lines. [From J. M. Thoday and T. B. Boam, *Heredity, 13,* 205–218, 1959.]

of a distasteful species, thus gaining protection from predators who select their prey visually. The predator, after some trials, learns to avoid the form of a noxious species, and if the edible mimic is sufficiently similar to the unpalatable model, the predator will also avoid the edible species. One stringent requirement in this relationship is that members of the distasteful species far outnumber those of the mimetic species, for otherwise the predator will not be sufficiently conditioned to learn avoidance.

The best documented example of Batesian mimicry is the African swallowtail butterfly, *Papilio dardanus,* studied by B. C. Clarke and P. M. Sheppard (Figure 23–8). In this species the females mimic several different noxious models, while the males are nonmimetic. This difference between the sexes is possible because females select their mates visually, but males are attracted to females by species-specific odors, regardless of coloration. The phenotypic polymorphism of the females whereby the patterns of several model species are adopted is thought to have arisen by disruptive selection, more than one phenotype being selectively advantageous. It is postulated that the color patterns resulting from the expression of major alleles have been perfected through the selection of modifying genes specific to each locality or subpopulation of *P. dardanus.* When hybrids are formed experimentally between members of such subgroups, the hybrid phenotype is intermediate and less like that of the noxious model, and thus any hybrids formed in nature are likely to be more subject to predation and therefore less well adapted than either parental type.

Figure 23–8. Batesian mimicry. 1, female, and 7, male, of *Papilio dardanus* from a nonmimicking population in Madagascar. 2–6, unpalatable model species; 8–12, *P. dardanus* mimics of these models; 13–16 imperfect mimics formed by hybridization with members of a population not possessing the particular mimic. [From P. M. Sheppard, *Cold Spring Harbor Sympos. Quant. Biol., 24,* 131–140, 1959.]

FREQUENCY-DEPENDENT SELECTION

Frequency-dependent selection refers to the situation where a phenotype is selected against when common and selected for when rare. The circumstances of Batesian mimicry provide an example. In the case where unpalatable models are rare and mimics are more common than nonmimics, the mimics will suffer heavier predation than nonmimetic individuals. However, in the case where models are common and nonmimics are more common than mimics, the nonmimics will suffer heavier predation. Although selection pressure is influenced by the availability of models, it is exerted on mimics or nonmimics depending on their relative frequency in the population.

A laboratory demonstration of frequency-dependent selection has been carried out by K. Kojima and colleagues using the trait of the enzyme esterase-6 of *D. melanogaster.* Two different alleles of the esterase-6 locus specify two electrophoretic variants of the enzyme, one fast (F) and one slow (S). In a large laboratory population the frequencies of these alleles were allowed to reach the equilibrium values of F = 0.3, and S = 0.7, whereupon additional populations, homozygous for the fast or slow alleles, were initiated. By crosses between these lines and counts of the number of eggs laid by FF, FS, and SS females, it was found that despite its high frequency in the equilibrium population, individuals SS in genotype were the least fecund of the three genotypes, that is, SS < FF < FS.

This finding suggested that differences in viability were involved in the establishment of the equilibrium gene frequencies. To analyze this possibility, additional populations were established in which the starting gene frequencies were known. This permitted a calculation of the expected genotype frequencies from the Hardy-Weinberg equation and thus comparisons between the expected and the actual frequencies of genotypes could be made by sampling the populations at intervals. By this method estimates of the relative viability of genotypes and the frequency of the F and S alleles could be obtained. It was found that when a population having a high frequency of the fast allele was initiated, the homozygous genotype FF was less frequent and therefore less viable than predicted on the basis of Hardy-Weinberg considerations and that the frequency of the F allele declined continuously until, after 30 generations, the equilibrium value of 0.3 was reached. However, when a population having a frequency of the fast allele below the equilibrium value was initiated, the FF genotype was more frequent than anticipated and the frequency of the F allele increased continuously until again the equilibrium value of 0.3 was achieved (Figure 23–9). Similar findings were obtained for the slow allele, that is, it was selected for when initially lower and selected against when initially higher than the equilibrium frequency of the original base population.

Estimates of the relative fitness of the three genotypes, FF, FS, and SS, made by sampling the experimental populations at intervals, revealed that only when the original equilibrium frequencies were reached was fitness approximately equal to 1.0 for all three genotypes. Since at this point the viability of all genotypes and phenotypes was equal, it could be concluded that the F and S alleles were selectively neutral at equilibrium. These results have prompted the proposal that frequency-dependent selection resulting in neutrality may be

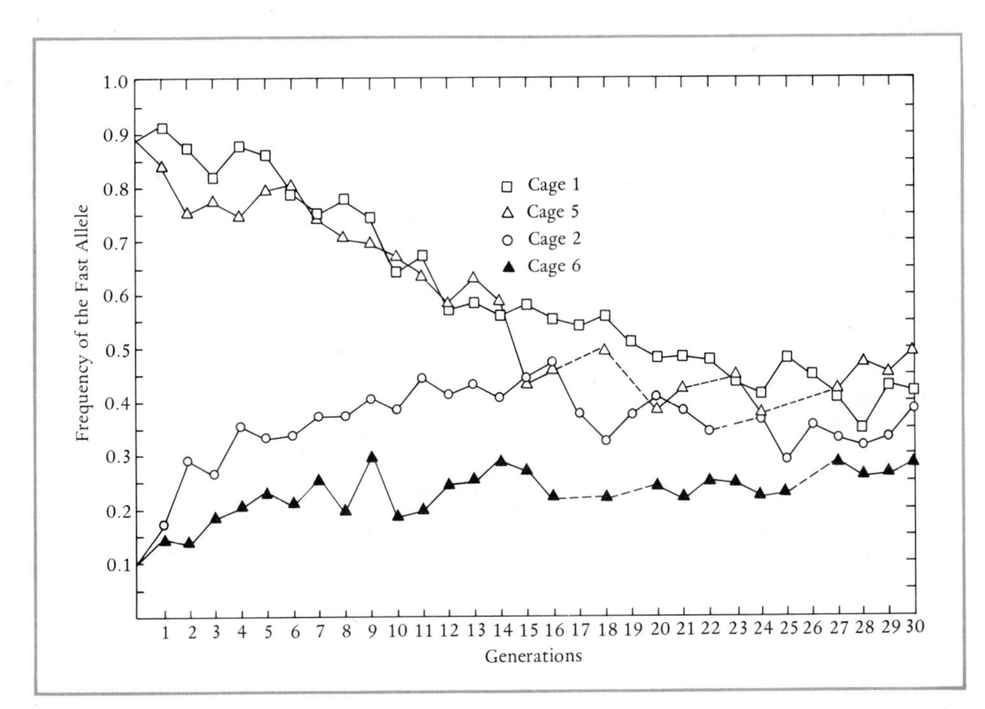

Figure 23–9. Frequency-dependent selection. Allele frequency changes at the esterase-6 locus of *D. melanogaster.* [From K. Yarborough and K. Kojima, *Genetics, 57,* 677–686, 1967.]

involved in the maintenance of the numerous genetic polymorphisms found in nature.

Frequency-dependent selection has also been described by C. Petit and by L. Ehrman in observations of mating in several species of *Drosophila.* When females are presented with two types of males in equal numbers in an observation chamber, neither kind of male has a mating advantage, but when one type of male is rare and the other type numerous, the rare type is significantly more successful in copulation. As a result, the alleles of such males are represented in greater frequency in the next generation than the alleles of competitors. The difference between rare and other males can be almost anything, for example, geographic origin, presence of mutant genes or chromosome inversions, or even development in a slightly altered culture environment. These observations demonstrate that fitness can be conferred in a variety of ways and by a variety of subtle factors.

RANDOM GENETIC DRIFT

It will be recalled that in a finite population the frequency of a gene in a progeny generation will vary by chance according to a binomial distribution. The most likely event is that the gene frequency of the progeny will be the same as the mean of the parental generation, but other frequencies, including 0 and 1, can also occur by chance. Therefore, there exists in all populations, and especially in small ones, the possibility that a gene will be fixed by chance alone.

Experiments designed to test this possibility were carried out by W. Kerr and S. Wright with 96 separate laboratory populations of *D. melanogaster,* in each of which the starting frequency of the gene for forked bristles and that of its wild type allele was 0.5. In these experiments each generation was produced by randomly selected males and females, 4 of each sex, and after 16 generations it was found that the wild type allele had become fixed in 41 and the forked allele in 29 of these small populations. Although selection was not absent in that wild type individuals possessed greater fitness than forked individuals, it was evident that chance had played a major role in gene fixation and in the consequent loss of variation.

In large populations natural selection is the most important cause of gene frequency change and genetic drift is of minor significance. Occasionally, however, a large population may be subjected to a "bottleneck" in the form of an ecological catastrophe, such as a flood, during which the population is drastically reduced in size. The survivors of such an event may not constitute a representative sample of the original population and, consequently, a considerable amount of the genetic variation originally present may be lost by chance. In addition, the survivors will be the founders of a new population, and while this initial population is still small, chance fluctuations in gene frequency and fixation by drift may be significant. Similarly, marginal populations at the boundaries of a species range are more subject to random drift than the larger populations in the center of a range. For example, if a small migrant group of founders establishes an outpost at the borders of a range, random drift may initially play a role in fixing the genetic variation upon which selection is imposed.

The interplay between the forces of selection and drift in large and in small founder populations has been studied by Dobzhansky and Pavlovsky. These workers used as the trait being followed two different third chromosome inversions, Arrowhead (AR) and Pikes Peak (PP), which had previously been found in natural populations of *D. pseudoobscura.* With the foreknowledge that AR/PP heterozygotes were selectively favored, a starting population in which the frequencies of the two inversions were equal was established. Random samples from this population were used to initiate 10 large populations starting with 4000 flies each and 10 small populations starting with 20 founders each. The small populations increased within two generations to the initial size of the large populations, and it was considered that at this size selection was exerted with equal force in all experimental groups. The groups were sampled to determine the frequency of the two inversions after 4 months and after 17 months with results illustrated in Figure 23–10. It can be seen that at an early stage in the history of these populations (4 months), the change in the frequency of the PP inversion was much greater in the small founder populations than in the large populations, due primarily to random genetic drift. Thereafter selection undoubtedly played a greater role in both groups in the determination of the ending frequencies of the two inversions.

Examples of genetic drift can be found in natural populations and particularly in human populations. B. Glass and colleagues have studied the members

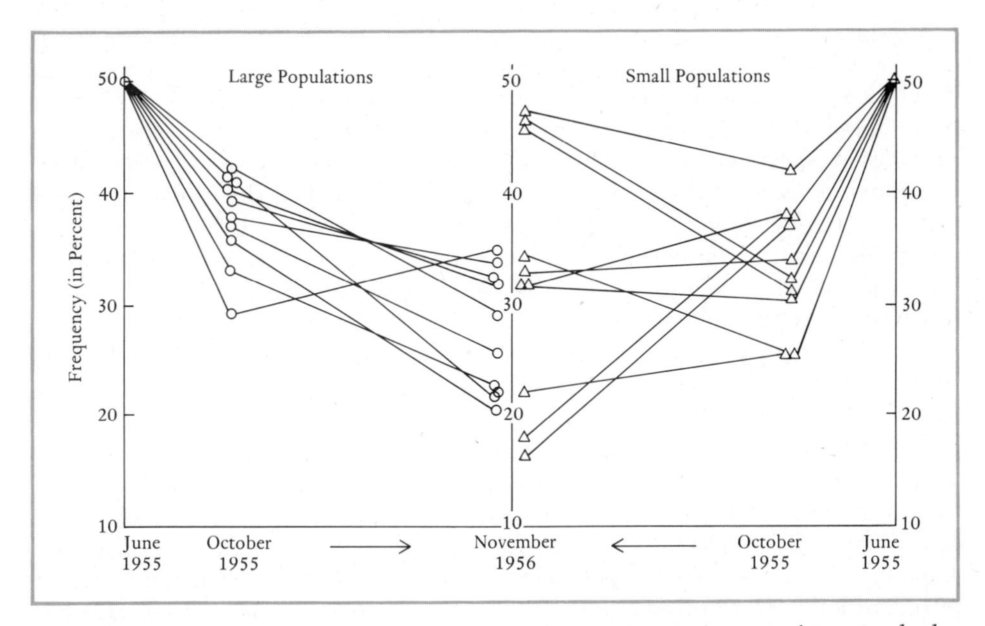

Figure 23–10. Random genetic drift in large and small populations of *D. pseudoobscura.* The frequencies of PP chromosomes in 20 replicate experimental populations were sampled at 4 and 17 months. [From Th. Dobzhansky and O. Pavlovsky, *Evolution, 11,* 311–319, 1957.]

of a small religious sect, called Dunkers, living in Pennsylvania. The forebears of this group emigrated from Germany in the eighteenth century and their descendents have maintained their identity by intermarriage. Within this self-isolated group the frequencies of the ABO blood group alleles differ considerably not only from the parental population in Germany, but also from that of the surrounding population in Pennsylvania. For example, 60 percent of the Dunkers were found to be blood group A, whereas only 40 to 45 percent of the U.S. or German populations are of this blood type; also group B individuals are virtually absent in the Dunker population, but comprise 10 to 15 percent of the U.S. and German populations. These and other differences in allele frequency between the Dunkers and other populations suggest that random drift has occurred during the history of this sect in the United States.

The blood groups fround in American Indians are another example of genetic drift. Human populations generally possess all three alleles, I^A, I^B and I^O, but most American Indians belong to blood group O and lack alleles I^A and I^B. However, in two tribes, the Blood and the Blackfeet, allele I^A is present in very high frequency. It has been suggested that the prehistoric migrants that crossed the Bering Strait to the North American continent were few in number and heterogeneous as to blood type and that the loss of I^A and I^B in most Indian groups is due to random genetic drift.

VARIATION IN NATURAL POPULATIONS

Although natural selection imposed by the innumerable facets of the environment provides the mechanism for the population changes that lead to evolution, selection can be effective only in proportion to the amount of genetic variation present. A population containing little genetic diversity and much homozygosity may be doomed to extinction, for continuing adaptation to a constantly changing environment demands a ready store of variation to provide a wide range of phenotypes upon which selection can act. This direct relationship between the amount of variation present and the ultimate degree of fitness achieved by a population was early recognized by R. A. Fisher. Fisher developed a mathematical expression called the *fundamental theorem of natural selection* which demonstrates that the rate of selection is proportional to the amount of genetic variation present in a population.

Until recently, however, the question of just how much genetic variation is actually present has been difficult to answer. Early workers attempted to discover the extent to which single gene mutations, such as those known in laboratory strains of *Drosophila,* were present in wild populations. By the use of balanced lethal marker stocks (see Figure 17–4), chromosomes derived from natural populations were made homozygous and the phenotypic effects of recessive mutations observed. It was found that such mutations were indeed present, but in low frequency. In similar studies estimates of the number of genes causing semilethality, lethality, or sterility have also been made. For example, Dobzhansky and Spassky have tested a large number of homozygous and heterozygous combinations of second, third, and fourth chromosomes of *D. pseudoobscura* captured in the wild. Their results (Figure 23–11) indicate that the viability of individuals carrying such chromosomes in the homozygous condition ranges all the way from lethality to supervitality, using heterozygotes as the standard of comparison. Overall, these workers found that the average viability of the tested homozygotes was well below the mean viability of heterozygotes, which suggests that the randomly mating individuals of the populations sampled carry a large number of deleterious genes concealed in the heterozygous condition.

N. E. Morton, J. F. Crow, and H. J. Muller have attempted to estimate the number of concealed deleterious and lethal mutations present in human populations. Their estimates are based on studies of the frequency of early deaths in children of related parents (primarily first cousins) versus those of unrelated parents. By examining church records for a rural French population of a past generation, these workers found that deaths prior to maturity occurred in approximately 13 percent of the children of unrelated parents, but in 25 percent of those whose parents were first cousins. The excess mortality of the children of related parents was thus 12 percent ($0.25 - 0.13$). It will be recalled that the inbreeding coefficient for first cousin marriages is $\frac{1}{16}$, which means that $\frac{1}{16}$ of all of the genes of an offspring of such a marriage will be made homozygous, although we do not know which ones. We can reason, however, that if an increase in mortality of 12 percent (0.12) arises when $\frac{1}{16}$ of the loci are homozygous, then an increase in mortality of 192 to 200 percent (16×0.12) should occur with complete homozygosis. Morton, Crow, and Muller have interpreted these data in terms of lethal equivalents, meaning a single lethal gene or several subvital genes whose cumulative effects cause death when homozygous. Since in this example the increase in mortality through complete homozygosis is calculated to be around 200 percent (2.00), it can be estimated that each gamete must have carried two lethal equivalents which were rendered homozygous by inbreeding. From this it follows that on the average each diploid parent must have contained four lethal equivalents concealed in the heterozygous condition.

The extent of chromosomal polymorphism, that is, the presence of inversions or other structural changes, has also been assessed by cytological studies of a variety of organisms and has been found to be widespread. The many different third chromosome inversions found by Dobzhansky and coworkers in natural populations of *D. pseudoobscura* are examples of this phenomenon. As

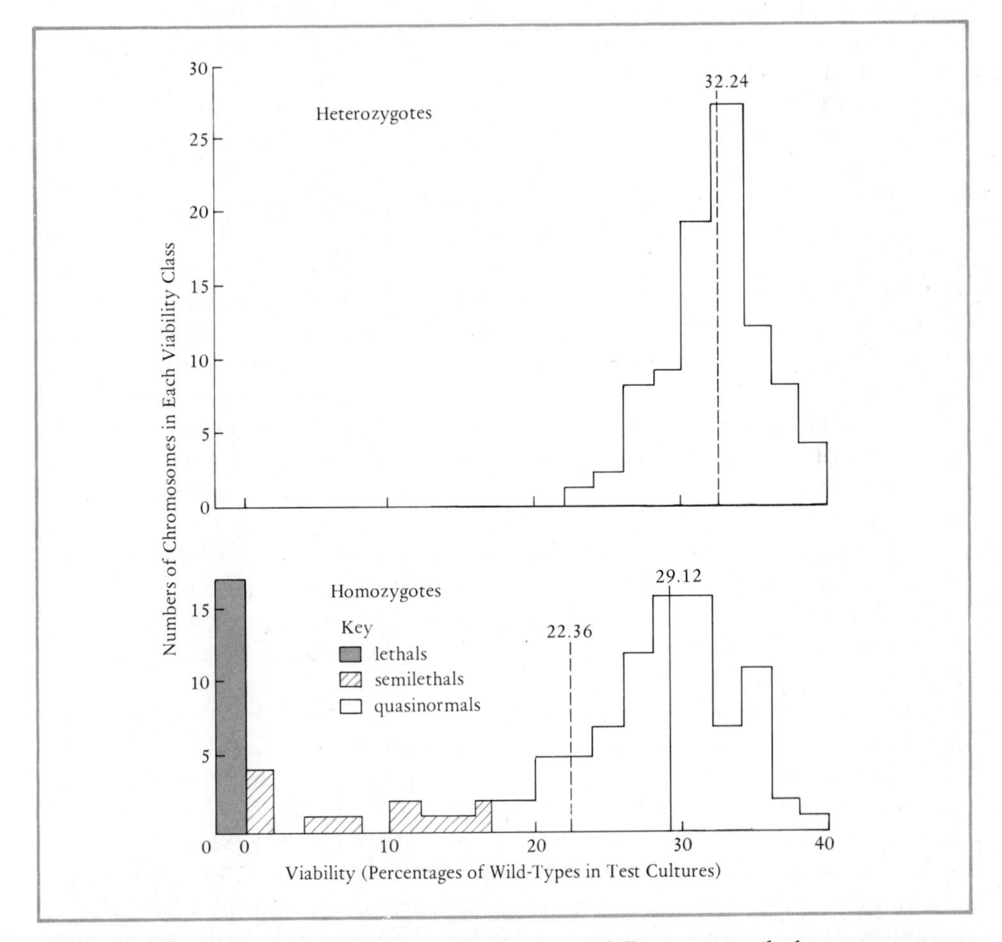

Figure 23–11. The influence of homozygosis of different second chromosomes on viability of *D. pseudoobscura* from Arizona. The numbers of second chromosomes giving various viabilities when homozygous and heterozygous are shown. The homozygotes are divided into three classes: lethals (solid), semilethals (diagonal hatching), and quasinormals (open). The mean values of the heterozygotes and all homozygotes (dashed lines) and of the quasinormals alone (solid lines) are given above the lines. [From Lawrence E. Mettler and Thomas G. Gregg, *Population Genetics and Evolution*, © 1969, p. 144. Reprinted by permission of Prentice-Hall, Inc., Englewood Cliffs, New Jersey. Drawn from data presented by Th. Dobzhansky and B. Spassky, *Genetics, 48,* 1467–1487, 1963.]

previously mentioned, such inversions are thought to maintain a group of linked genes which are inherited together as a supergene, conferring better fitness in different environments or at different seasons of the year.

For quantitative traits, the practice of inbreeding provides a means of estimating variation. When a normally outbreeding species is continuously inbred, numerous different strains homozygous for different alleles can be isolated. Since most mutations are deleterious, the lines developed by inbreeding usually exhibit a decrease in overall fitness which is correlated with the value of the inbreeding coefficient *F*. This decreased viability is called *inbreeding depression* and the extent to which it occurs with respect to a given trait is a measure of the variation for that trait originally present.

All of these and other studies, while demonstrating the pervasive presence of variation, failed to provide any means for estimating quantitatively the total amount of variation in the genome. This problem is being resolved, however, by the application of starch gel electrophoresis to the enzymes and other proteins present in wild populations. It will be recalled that a 1 : 1 relationship exists between a structural gene and its corresponding polypeptide and that gene mutations are reflected in an altered amino acid sequence of the product. Further, the electrophoretic mobility of a protein depends upon its charge (see the appendix to Chapter 10), and if an amino acid substitution which alters the charge on the molecule has occurred, the mutant protein will move more slowly or more rapidly in an electric field than the normal unchanged protein. By this means variant proteins indicating the presence of variant alleles can be detected. R. C. Lewontin and J. L. Hubby estimate that only around half of the possible amino acid substitutions will result in such a change in charge and therefore only half of the single gene variation actually present can be identified by electrophoresis.

The technique of electrophoresis was first applied in 1966 to serum enzymes of humans by H. Harris and to unbiased samples of proteins from wild populations of *Drosophila* by F. M. Johnson and coworkers and by Hubby and Lewontin. It has since been extended to many other species. Some examples of human enzyme polymorphisms are given in Table 23–1. The amount of variation present in natural populations and detected by this means is truly astounding. Minimum estimates suggest that 30 percent or more of all gene loci are polymorphic and that 5 to 20 percent of the loci of sexually reproducing organisms may be heterozygous. If we take the not unreasonable figure of 10,000 as the total number of gene pairs in *Drosophila,* then up to 2,000 of these would be heterozygous in an outbred individual; if the figure of 20,000 gene pairs is taken as an estimate for humans, then on the average around 4,000 of these would be heterozygous. Clearly, this amount of variation provides an enormous potential for evolutionary change.

The high degree of polymorphism discovered in the genomes of sexually reproducing forms might suggest a corresponding degree of phenotypic polymorphism, but this expectation is not realized. Most members of a natural population exhibit a generally uniform appearance, and in explanation it must be remembered that much of the genetic variation present is concealed in the heterozygous condition. Thus, the wild type phenotype results from numerous different genotypes and no one optimum phenotype or genotype can be singled out as the best for a species. Although extreme phenotypes arise through genotypes generated by independent assortment and recombination, most individuals, being heterozygous, conform to an adaptive norm, that is, they appear to be normal or well adapted to the environment with only slight differences from one another.

The range of phenotypes present in a population can be viewed in the form of a normal distribution with the mean representing the fitness of heterozygotes. B. Wallace and C. Madden have utilized this concept to define the adaptive norm of a population as those individuals whose phenotypes are included within two standard deviations above or below the mean fitness of heterozygotes. This definition can be used quantitatively in assessing an actual population once the mean fitness of a representative sample of heterozygotes has been determined.

Uniformity in phenotype, despite variety in genotype, stems from the stability of developmental events. The physiology of most higher organisms is sufficiently flexible to provide a buffer against the effects of a changing environment, for example, the maintenance of a constant body temperature in the face of fluctuating environmental temperatures. Developmental processes exhibit the same kinds of flexible adjustments to changing conditions. The phenomenon, called *homeostasis,* permits development to proceed more or less normally under a wide variety of environmental conditions. C. H. Waddington has proposed that, in addition to homeostasis, a process termed *canalization* occurs by which major regulatory genes initially set the course of development

TABLE 23–1.
Enzyme Polymorphisms in Humans, Classed According to the Reported Presence of Quantitative Differences in Enzyme Activity Between the Common Phenotypes

Polymorphisms where quantitative differences between the common phenotypes have been reported	Polymorphisms where as yet no quantitative differences have been reported
glucose-6-phosphate dehydrogenase	adenosine deaminase
red-cell acid phosphatase	peptidase D (prolidase)
phosphogluconate dehydrogenase	pancreatic amylase
adenylate kinase	pepsinogen
placental alkaline phosphatase	glutamate-oxalate transaminase
peptidase A	phosphoglucomutase
peptidase C	locus PGM_1
galactose-1 phosphate uridyl transferase	locus PGM_3
glutathione reductase	
liver acetyl transferase	
red cell NADase	
salivary amylase	
serum cholinesterase	
locus E_1	
locus E_2	
alcohol dehydrogenase	
locus ADH_2	
locus ADH_3	

Source: H. Harris, 1971, *J. Med. Genetics, 8,* 444–452, 1971. (C. A. Clarke, Ed.)

which then leads to a generally similar result, the adaptive norm. Selection for canalizing genes would thus stabilize developmental processes such that certain basic traits, for example, four legs, two eyes, would appear in all normal individuals in practically all environments. Although it might be concluded that a trait resulting from a canalized process would be refractive to selection, this is not necessarily true. For example, the number of whiskers in the mouse appears to be fixed at 19 due to prior canalization during development. However, in the Tabby mutation, whisker number is reduced and variable and if through appropriate crosses non-Tabby mice derived from a Tabby background are obtained, the imposition of selection for whisker number results in mice with an increased or decreased number of whiskers. These results indicate that variation for whisker number, while present, is normally not expressed due to the action of canalizing genes which initiate the standard or normal developmental pattern.

GENETIC LOAD

Genetic load is the relative deficiency in viability or fecundity or both viability and fecundity as compared to that which would occur if the population were all of the optimum phenotype. Thus, genetic load is the difference between the actual fitness and an optimum fitness, this difference being caused by the presence of suboptimal genetic variants, concealed or not, within the population. Three kinds of genetic loads will be described: mutational, segregational, and substitutional.

Mutational load is the result of the presence of rare deleterious genes maintained in a population by mutational-selectional equilibria. It will be recalled from Chapter 22 that at equilibrium the number of new mutations of this type that arise in each generation is equal to the number removed by selection, or $\hat{p}_2 \cong \sqrt{u}$. The mutations that contribute most to the mutational load are those that cause lethality, gross abnormality, or sterility.

Segregational load is the loss in average fitness of a population caused by the segregation of the less fit homozygous genotypes. Under stabilizing selection the heterozygote is the favored genotype, but the formation of heterozygotes is necessarily accompanied by the generation of the less well-adapted homozygous genotypes whose lowered viability and fecundity cause the segregational load. Because selection acts to preserve the heterozygous genotype, neither allele, even if one or both are lethal, is eliminated from the population, and if the frequencies of the two alleles are at equilibrium in the absence of environmental change, the same segregational load will arise in each generation. As an example, in malarial regions of Africa the loss or lowered fecundity of HbS and HbA homozygotes as compared to heterozygotes constitutes the segregational load for that locus.

Substitutional load results from directional selection where one allele is being replaced by a new allele whose effects confer greater fitness. The substitutional load thus consists of the relative number of deaths and lack of births which the population must endure in order to progress from the original to the new state of fitness. During the period of replacement selection will favor the new allele, and the proportion of the population containing the old allele will, by comparison, be less well adapted. The loss of these individuals through reduced viability, death, or lessened fecundity will place a substitutional load upon the population. Sickle cell anemia can again be used as an example. In the United States where falciparum malaria is absent HbA homozygotes possess the best adapted phenotype and directional selection in favor of the normal allele is occurring. As a result, the lowered fecundity and viability of HbS/HbA heterozygotes and HbS homozygotes, as compared to normal individuals, contribute a substitutional load to the black population of the United States.

The substitutional genetic load has been called the price of evolution, and J. B. S. Haldane has made some theoretical computations of the extent of this price in terms of reduced fitness or loss of individuals. He has estimated that with a selection intensity of 0.1, approximately 300 generations will be required to replace a gene at a cost in members lost by death or sterility of 30 times the average number of individuals present in a single generation. Thus, according to these calculations, approximately 10 percent of a population would be lost per generation if gene replacement occurred over a time span of 300 generations.

Because of the enormous amount of protein, and therefore genic, polymorphism recently revealed at the molecular level by immunological and electrophoretic methods, these computations pose a quandary. Evolution requires the concurrent replacement of many genes, and to the substitutional load borne by a population must be added the mutational load as well as the segregational load contributed by stabilizing selection. It would seem that no population could withstand such a drain on its members, especially if the effects of individual subvital genes were independent and therefore multiplicative. The question, then, is how are these numerous genic polymorphisms maintained without incurring a genetic load so heavy as to cause the extinction of a population? Two different hypotheses attempt to resolve this problem.

The *selection hypothesis* proposes that the observed polymorphisms are each due to heterozygote advantage and are maintained in a population by various types of stabilizing or balancing selection. Examples include balanced polymorphisms, such as sickle cell heterozygotes, as well as instances where one or another allele or chromosomal inversion is favored in different seasons, regions, or stages of the life cycle or in different sexes. Frequency-dependent selection also serves to maintain polymorphism and at equilibrium the alleles involved are selectively neutral, contributing no genetic load.

Proponents of the selection theory suggest that the primary cause of the extensive heterozygosity observed in nature is heterosis at hundreds or even thousands of gene loci, the cumulative effects conferring fitness. J. A. Sved, J. L. King, R. D. Milkman, and others propose that the genetic load should not be estimated on the assumption that each locus contributes to this load independently and that the individual effects are multiplicative. In view of the known interaction between genes they suggest, instead, that the effects of most loci are slight as well as additive and that the values of the sums of these effects exhibit a normal distribution in a population. Since selection scans the entire organism and not single gene loci, only those individuals that fall below a certain threshold of fitness, that is, those with fewer heterotic loci than a minimum number required for average fitness, are culled from the population. The members of a population are thus divided into two classes, those above and those below a given threshold, and under these circumstances it is estimated that perhaps 1000 selectively maintained polymorphisms, each conferring an advantage to heterozygotes of around 1 percent, could persist in a population with only minimal effects on genetic load.

An opposing view, sometimes called the *neutralist theory,* is held by J. F. Crow, K. Kimura, T. Ohta, T. H. Jukes and others. These workers suggest that the majority of the thousands of genic polymorphisms discovered at the biochemical level have no significant effect on fitness and are therefore selectively neutral, making no contribution to the genetic load. They propose that adaptation does indeed occur through natural selection, but that selection acts primarily to remove deleterious mutations from a population. Proponents of the neutralist theory point out that while mutations which result in the substitution in a protein of a compatible amino acid are detectable by sophisticated methodology, they may not be detected by natural selection because the function of the protein will remain essentially unchanged. Under these circumstances the frequencies of polymorphic genes conferring no advantage or disadvantage will be determined primarily by random genetic drift rather than by selective processes.

A realistic assessment of the relative contributions of selection and drift to the presence of protein polymorphisms cannot yet be made because of the paucity of evidence with respect to the selective advantages or disadvantages of given enzyme or protein variants, the number of genes undergoing directional selection at any time, the interactions between gene products, or the linkage relationships between genes being selected for or against in some environmental setting. Some polymorphisms are undeniably maintained by selection in certain environments, notably sickle cell heterozygotes in Africa. Also, the frequency of some enzyme variants is greater than that explainable by chance, suggesting the operation of selective processes. However, whether or not selection can account for the maintenance of all polymorphisms is most uncertain and the likelihood that some of these are neutral and present in a population through random drift cannot at the present time be refuted.

SPECIATION

In outbreeding species no two individuals, except identical twins, have the same genotype, and therefore pure isogenic lines or "pure races" do not exist in nature. What do exist are Mendelian populations each containing a spectrum of genotypes. Such populations are usually dispersed in space and since every environment poses somewhat different problems, adaptive changes in gene frequencies which permit better fitness in a particular environment occur through selection. Examples of such adaptations are seen in melanic varieties of the moth *Biston* or in the different kinds of chromosomal inversions of *D. pseudoobscura* which vary in frequency with altitude or geographic region. In these instances the observed adaptive changes can be used to characterize each subpopulation, but since interbreeding between individuals of these populations occurs in nature with the production of viable and fertile hybrids, the members of these populations can be considered to belong to the same species. A species is thus defined in dynamic, not descriptive, terms as one or more populations comprising individuals between whom gene flow occurs in nature and whose hybrid offspring have a fitness comparable to that of their parents. As the reverse of this definition, different species are those whose gene pools have become so distinct from one another that either interbreeding between their members does not occur in nature or hybrid offspring resulting from interbreeding are inviable or sterile. Speciation is thus characterized by the progressive divergence of gene pools to the extent that gene flow between them is no longer possible.

In most instances the process of speciation has as an initial prerequisite the separation of populations by some kind of geographic barrier so that gene flow between them is prevented. Such isolated groups are called *allopatric* populations. Because of mountains, bodies of water, deserts, or other inhospitable terrain or conditions, the members of allopatric populations cannot meet and breed with individuals from outside their group. As a result and with continued isolation, the allopatric population becomes more and more adapted to its immediate environment and its gene pool diverges from that of the ancestral type. Eventually the genetic differences between the allopatric and other populations become sufficiently great that if hybrids were formed between them, the hybrids would be poorly adapted. The divergence achieved at this point between an allopatric and an ancestral population is considered to be the first stage in geographic speciation by evolutionary biologists and is called the allopatric stage. If the geographic barrier is somehow removed or otherwise overcome so that contact with the ancestral population is regained, that is, if the two populations become *sympatric,* the second or sympatric stage may occur. Speciation would be carried to completion if natural selection favors the maintenance and strengthening of reproductive isolation between members of the two groups.

The process of speciation is lengthy and the divergence of gene pools due to the accumulation of genetic differences is very gradual. Although the entire process is far beyond our observational abilities, it has been possible to survey incipient species that appear to be on the verge of reproductive isolation. Dobzhansky and Spassky have found that crosses between such semispecies of *Drosophila* occur with difficulty and that females ordinarily reject males of a different group due to differences in mating behavior. These incipient species have also been shown to differ with regard to chromosomal polymorphisms, some of which are restricted to one semispecies only. The use of starch gel electrophoresis to assess genetic differences between incipient species has additionally revealed that in *Drosophila,* fishes, amphibians, and mammals, around 20 percent of the genes are changed during the allopatric stage, but little further change occurs during the sympatric stage.

Although geographic speciation is the most usual process by which new species arise, some less common modes also exist. In the higher plants "instant speciation" can occur through the formation of allopolyploids (see Chapter 9). It will be recalled that allopolyploidy (amphidiploidy) is initiated by hybridization between closely related species living in the same environment. Hybridization is followed by chromosome doubling, perhaps in a zygote, so that the genome contains two complete diploid chromosome complements, one from each original species. Such a plant is fertile, produces gametes carrying a haploid complement from each parental species, and if capable of self-fertilization, will produce abundant seeds. Allopolyploids are reproductively isolated from both parental species because crosses with either will result in zygotes having unbalanced chromosome numbers.

An additional, but probably rare mode of sympatric speciation is that which may occur as the result of disruptive selection. It will be recalled that reproductive isolation was achieved by Thoday in experimental populations of *Drosophila* by this means. Although the intensity of this type of selection in nature is presumably much less than that employed by Thoday, theoretically, at least, disruptive selection is capable of producing subgroups within a population so specialized for their own ecological niches that they become reproductively isolated from one another.

Reproductive isolating mechanisms can be divided into two major classes: prezygotic and postzygotic. *Prezygotic isolating mechanisms* include any aspects of the ecology or physiology of two species that prevent or interfere with mating and the union of gametes. For example, two species may live in different areas within the same general environment or they may mate or flower at different seasons; courtship behavior in one species of animal may be inappropriate for another so that little attraction occurs between the sexes, or insects necessary for pollination may not be attracted by one species or the other; the structure of floral parts may prevent pollination or structural differences in genitalia, particularly in insects, may prevent gamete transfer; pollen or sperm may be inviable when introduced to the stigmas of flowers or the reproductive tract of females of the opposite species.

Postzygotic isolating mechanisms include those leading to the inviability or sterility of F_1 hybrids. An additional mechanism, called *hybrid breakdown,* is the production of sterile or otherwise unfit F_2 progeny or backcross progeny.

In most cases, gene flow between two species is effectively prevented by more than one of these mechanisms, and reproductive isolation once achieved is irreversible.

REFERENCES

BODMER, W. F., AND L. L. CAVALLI-SFORZA, 1976. *Genetics, Evolution, and Man.* Freeman, San Francisco.

CAVALLI-SFORZA, L. L., AND W. F. BODMER, 1971. *The Genetics of Human Populations.* Freeman, San Francisco.

DOBZHANSKY, T., 1951. *Genetics and the Origin of Species,* 3rd ed. Columbia University Press, New York.

DOBZHANSKY, T., 1970. *Genetics of the Evolutionary Process.* Columbia University Press, New York.

FISHER, R. A., 1930. *The Genetical Theory of Natural Selection.* Oxford University Press (Clarendon Press), New York. (Reprinted in paperback by Dover Publications, New York, 1958).

FORD, E. B., 1975. *Ecological Genetics,* 4th ed. Chapman & Hall, London.

GRANT, V., 1963. *The Origin of Adaptations.* Columbia University Press, New York.

KETTLEWELL, B., 1973. *The Evolution of Melanism.* Oxford University Press (Clarendon Press), New York.

KIMURA, M., AND T. OHTA, 1971. *Theoretical Aspects of Population Genetics.* Princeton University Press, Princeton, N.J.

LEVINS, R., 1968. *Evolution in Changing Environments.* Princeton University Press, Princeton, N.J.

LEWONTIN, R. C., 1974. *The Genetic Basis of Evolutionary Change.* Columbia University Press, New York.

MANGELSDORF, P. C., 1974. *Corn, Its Origin, Evolution and Improvement.* Harvard University Press, Cambridge, Mass.

MAYR, E., 1963. *Animal Species and Evolution.* Harvard University Press, Cambridge, Mass.

METTLER, L. E., AND T. G. GREGG, 1969. *Population Genetics and Evolution* (Foundations of Modern Genetics Series). Prentice-Hall, Englewood Cliffs, N.J.

RENDEL, J. M., 1967. *Canalization and Gene Control.* Academic Press, New York.

SALTHE, S. N., 1972. *Evolutionary Biology.* Holt, Rinehart and Winston, New York.

STEBBINS, G. L., JR., 1950. *Variation and Evolution in Plants.* Columbia University Press, New York.

STEBBINS, G. L., JR., 1971. *Chromosome Evolution in Higher Plants.* Arnold Press, London.

WADDINGTON, C. H., 1962. *New Patterns in Genetics and Development.* Columbia University Press, New York.

WALLACE, B., 1970. *Genetic Load.* Prentice-Hall, Englewood Cliffs, N.J.

WHITE, M. J. D., 1973. *Animal Cytology and Evolution,* 3rd ed. Cambridge University Press, Cambridge.

WRIGHT, S., 1969. *Evolution and the Genetics of Populations,* vol. 2. University of Chicago Press, Chicago.

Chapter 24

MOLECULAR EVOLUTION

A variety of techniques, such as DNA hybridization, gel electrophoresis, immunology, and the sequencing of proteins, are currently being employed to study phylogenetic relationships between various groups of plants and animals and to determine the specific effects on function caused by amino acid substitutions in structural proteins. The use of these techniques has provided remarkable insights into the evolutionary process.

DNA hybridization has been used for estimates of the degree of similarity, or complementarity, between different groups of organisms. The technique involves the hybridization of single-stranded DNA's derived from different species, and it will be recalled that the extent of annealing can be evaluated quite rigorously, since the mismatching of bases or the presence of noncomplementary single-stranded regions is reflected in changes in absorbance at 260 nm and in a lower melting temperature (see the appendix to Chapter 11).

B. J. McCarthy and coworkers have carried out extensive hybridization experiments with DNA's derived from human, chimpanzee, monkey, sheep, cow, pig, mouse, hamster, chicken, salmon, and others and have found that the DNA's of animals belonging to the same taxonomic group, such as a family, are more similar to one another than are those obtained from animals of a different family. Comparison of DNA's of animals from higher taxonomic groups gives the same kind of results, that is, DNA's of animals of the same order or class are more similar than those of different orders or classes. These findings conform to expectations and correlate in general with the amount of time that is estimated to have elapsed since the various groups diverged during the course of evolution.

Differences between closely related species have also been estimated by hybridization techniques with somewhat surprising results. In experiments with DNA's derived from *D. melanogaster* and *D. simulans,* species that are crossable with difficulty, yielding sterile hybrids, C. D. Laird and B. J. McCarthy have found that approximately 80 percent of the DNA sequences are held in common in these two species, but that 20 percent are so unlike that stable duplexes are not formed. Similar experiments comparing DNA from these two species with that of the more distantly related species, *D. funebris,* has revealed that only 25 percent of the sequences are held in common. These findings indicate that speciation involves not only gene mutations and alterations in allele frequency, but also major changes in the organization and structure of the genome.

PROTEIN EVOLUTION

The amino acid sequencing of homologous proteins found in different groups of organisms is providing a most powerful tool for the study of evolutionary relationships. The protein best studied from the standpoint of evolution is cytochrome c, a component of the mitochondrial electron transport chain. Cytochrome c is an ancient protein thought to have originated around 1.2 billion years ago in some remote ancestor common to the eukaryotes. It is composed of approximately the same number of amino acids in all species (104 in vertebrates). The positions within the polypeptide of over half of these amino acids are the same in all forms examined and in an *in vitro* system the cytochromes of all species can react with cytochrome oxidase derived from any other species. These characteristics indicate that cytochrome c is homologous both in origin and in function.

The complete sequence of amino acids of cytochrome c has been determined for over 40 different eukaryotic species from fungi to humans. Comparisons of these sequences reveal differences that correlate well with evolutionary relationships based on anatomical or other considerations. For example, about 10 amino acid substitutions separate primates from other mammals, 19 separate higher vertebrates from fishes, 27 separate vertebrates from insects, and 47 separate plants and animals from fungi.

One most interesting finding from these studies is that the amino acid substitutions which have been preserved over the course of evolution are not randomly situated within the molecule. When detailed comparisons are made between the amino acid sequences of cytochrome c from different species, it is evident that positions 70 through 80 are invariant in all forms, suggesting that these sites are crucial for normal function and that any change therein has been eliminated by natural selection. These positions have now been shown to contain the binding site for cytochrome oxidase and also to be associated with the iron-containing heme prosthetic group (see Figure 24—3). Other sites also appear to be related to function in that, although substitutions occur, the amino acids found at these sites are restricted to those having the same properties. At a few sites more radical substitutions are evidently permissible, for amino acids with unlike properties may be found at such positions in different eukaryotic organisms.

Since cytochrome c is a homologous protein, the genes which specify cytochrome c in all modern forms must have descended from an ancestral gene present at some time in the remote past. During the course of evolution mutational changes in this gene have occurred and these are reflected in the observed amino acid substitutions. Although mutation is a random process, the distribution of amino acid substitutions within the molecule is not random, suggesting that those substitutions retained in different modern species either contribute to fitness in some way or are selectively neutral.

The sequencing of homologous proteins from a variety of species has also permitted the construction of evolutionary trees based on the number of amino acid changes that separate related organisms. One such tree, which utilizes differences in the α chain of vertebrate hemoglobin, is illustrated in Figure 24–1. The numbers which appear within this tree indicate the number of differences in sequence among the animals studied, and the arrangement of the branches of the tree is based on phylogenetic relationships.

Since paleontologists can infer from the fossil record the approximate time that has elapsed since the divergence of these animals from a common ancestral form, it is possible to correlate amino acid changes with time and thus to

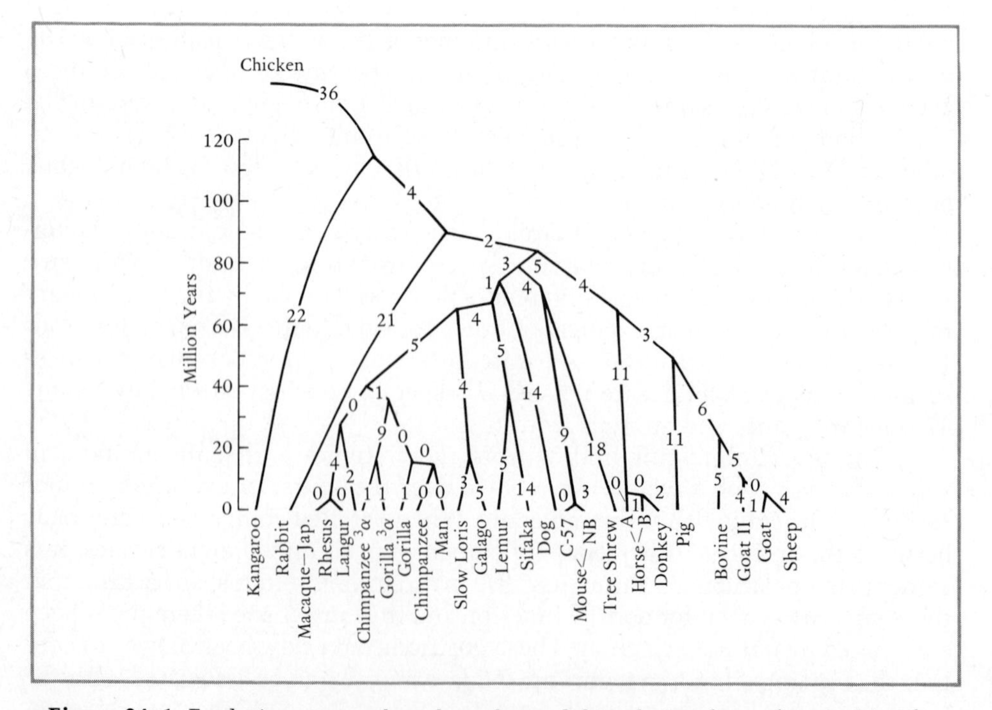

Figure 24–1. Evolutionary tree based on the α globin chain of vertebrates. Numbers within the tree are numbers of amino acid substitutions. [From M. Goodman et al., *J. Molec. Evol., 3,* 1–48, 1974.]

estimate the rate of protein evolution. When different proteins are compared (Figure 24–2), it can be seen that the rate of evolution is fairly constant for any one given protein, but that it varies widely for different proteins. This variation is associated with function.

The fastest rate of change is exhibited by fibrinopeptides. These are short peptide sequences within the fibrinogen molecule which at the time of blood clotting are enzymatically excised from fibrinogen to permit the transformation of fibrinogen to fibrin. It would appear that as long as the particular amino acid residues recognized by the excising enzyme are present, the rest of the

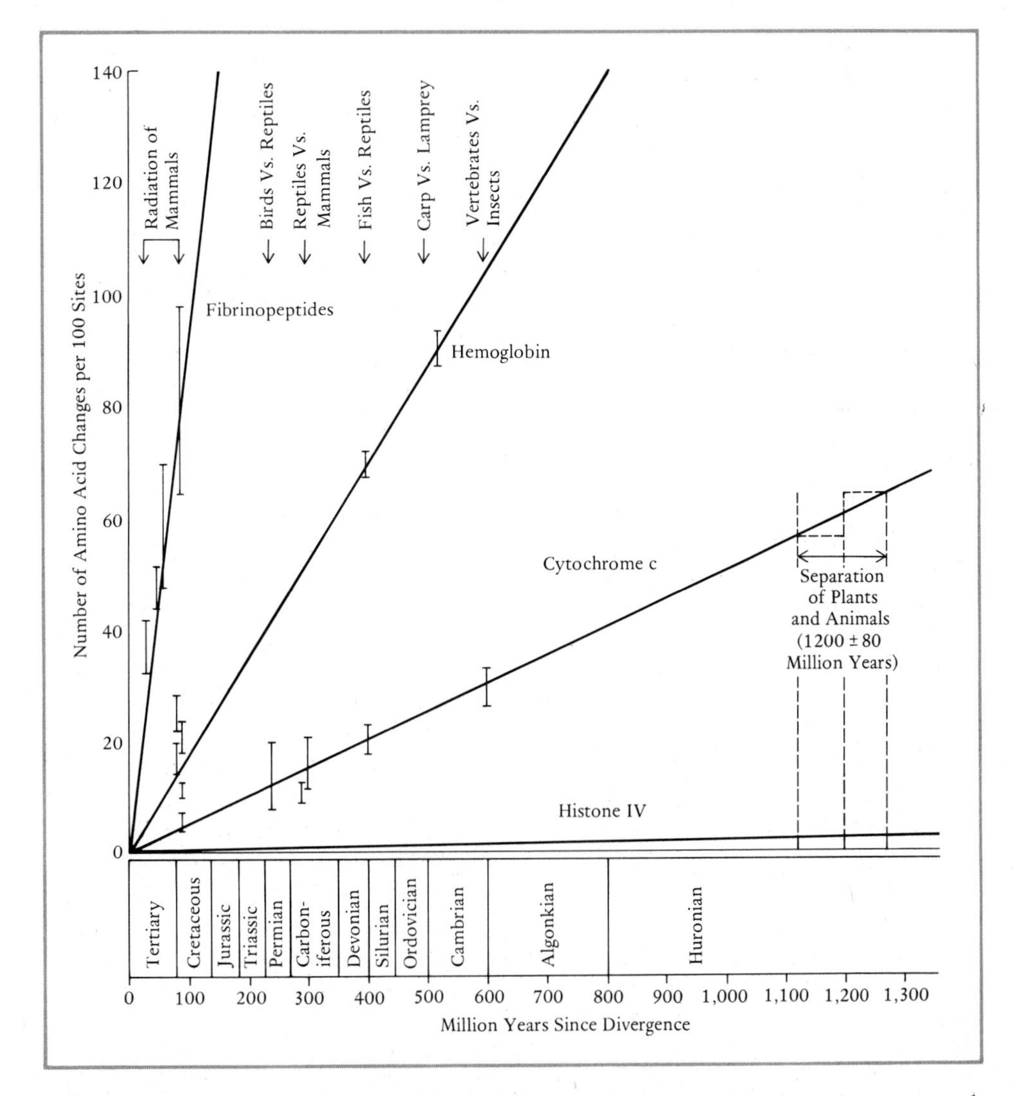

Figure 24–2. Rates of evolution of fibrinopeptides, hemoglobin, cytochrome c and histone IV. Vertical bars indicate mean errors in amino acid differences. [From R. E. Dickerson, *Scientific American, 226,* 58–72, April, 1972.]

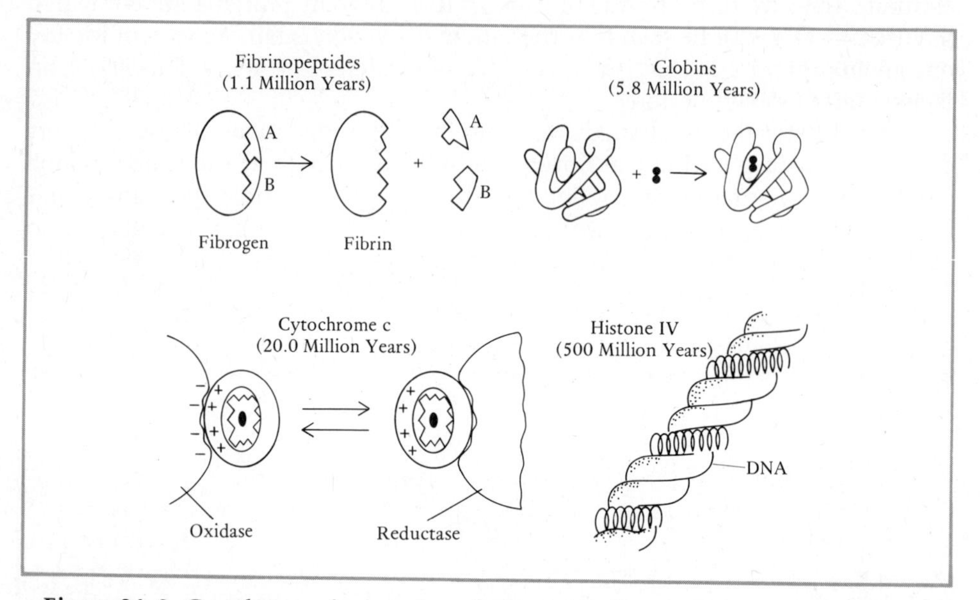

Figure 24–3. Correlation of rates of evolution with the complexity of molecular interactions. The discarded fibrinopeptides A and B have the fastest rate and exhibit an average of a 1 percent change in amino acid sequence per 1.1 million years. The same change for hemoglobin requires 5.8 million years, for cytochrome c, 20 million years, and for histone IV 500 million years. These rates reflect the functional constraints placed upon these molecules. [From R. E. Dickerson, *J. Molec. Evol., 1,* 26–45, 1971.]

fibrinopeptide molecule can tolerate a variety of amino acid substitutions without impaired function (Figure 24–3).

The hemoglobin molecule has a slower rate of evolution and we can therefore judge that it can withstand fewer changes and still maintain normal function. It will be recalled that the four polypeptide chains of hemoglobin must form associations with one another as well as with the heme portions of the molecule and these requirements restrict the number of random changes which can be accomodated without serious impairment of function. Even greater functional constraints are placed upon cytochrome c, and the slowest rate of evolution found thus far is that of histone IV. This histone is a structural protein bound to the DNA of all eukaryotic chromosomes and is thought to participate in the control and packaging of DNA (see Chapter 19). Since only two amino acid substitutions have been found in this protein between organisms as diverse as peas and the calf, we must conclude that the selection imposed upon histone IV must be stringent indeed.

It has been proposed that the evolution of proteins proceeds randomly and at a constant rate. If so, differences in the structural proteins of modern species could be used as evolutionary clocks to estimate the time when these species diverged from a common ancestor. Unfortunately, the changes in proteins over evolutionary time are highly heterogeneous and cannot be interpreted as random variations from a constant rate. Nevertheless, when many proteins are taken together, they do exhibit a fairly constant rate of change and therefore provide most valuable adjuncts to the fossil record. It must be noted, however, that the rates of protein evolution do not reflect the higher rates at which changes in morphology, physiology, or chromosome structure have occurred. To account for this difference, it has been proposed that mutations in regulating genes, rather than in structural genes, may be primarily responsible for significant evolutionary advances.

GENE DUPLICATION

The presence of duplications within the genome was first recognized through cytological studies of the giant salivary chromosomes of *Drosophila* species where such duplications are usually observed as tandem repeats of one or more bands. Salivary analysis of the Bar mutation in *D. melanogaster* coupled with genetic studies provided evidence that Bar was a duplication which had arisen by unequal crossing over (see Figures 9–10 and 9–11). The multiple cistrons for rRNA and tRNA substantiate the presence of duplications, and studies of the reassociation kinetics of single-stranded DNA indicate that all eukaryotes possess sequences with a high repetition frequency (see Figure 19–14). Gene duplications, probably arising by unequal crossing over, are thus ubiquitous among eukaryotes and provide for increases as well as decreases in genome size. The evolutionary advantages of an increase in gene number are obvious. Duplicated loci can evolve independently, permitting the assumption of new functions and new specializations, and this process has probably been responsible for the development of the intricate regulatory mechanisms so evident in eukaryotes.

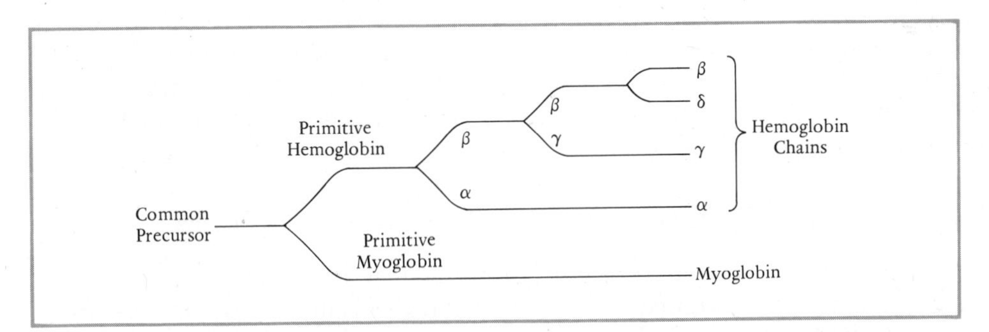

Figure 24–4. Evolutionary tree showing the probable order of gene duplications which gave rise to the genes for human globins. The most ancient duplication and divergence produced the primitive myoglobin and hemoglobin genes. Duplication of the hemoglobin gene early in vertebrate evolution gave rise to primitive α and β chains, which, with further evolution (especially of the β gene), developed the ability to form tetramers. The β gene line underwent two further duplications. The first yielded the γ gene, expressed during fetal life, and the β gene, expressed in children and adults; the second yielded the two closely linked adult type genes which produce the major β component and the minor δ component. Other very recent duplications have occurred in both the α and γ lines; the two α loci produce identical proteins, while the two γ loci produce chains differing in one residue. Insufficient evidence is available to place the fetal ϵ and ζ genes on the tree with certainty; however, the ϵ chain appears to be most similar to the β chain and the ζ chain to the γ chain. [From *An Introduction to Human Genetics,* Second Edition by H. Eldon Sutton. Copyright © 1975, 1965 by Holt, Rinehart and Winston. Reprinted by permission of Holt, Rinehart and Winston.]

Analysis of the amino acid sequences of a number of proteins gives evidence that their loci originated through gene duplication. The globins are a good example. These proteins include the polypeptide chains of the various vertebrate hemoglobins found at different stages of development (see Table 10–1 and Figure 10–12). On the basis that the most closely related amino acid sequences will be those whose specifying genes have most recently diverged, V. M. Ingram as well as E. Zuckerkandl and L. Pauling have proposed the following evolutionary history of the globin genes (Figure 24–4). Originally an ancestral gene was present which coded for a monomeric myoglobinlike molecule. About 650 million years ago, this gene underwent duplication to give rise to two myoglobinlike genes. The product of one of these was preserved in the monomeric form, ultimately evolving into the modern myglobin molecule which functions in the transport of oxygen in muscle. The other myoglobinlike gene evolved to a primitive hemoglobin gene whose product was capable of forming dimers. Duplication of the primitive hemoglobin gene then produced the primitive α and β genes. Two more duplications of the β gene then gave rise first to the γ gene and subsequently to the δ gene.

With regard to specialized function in humans, the α chain is present in all hemoglobins, embryonic, fetal, and adult, while the γ chain is normally limited to fetal stages where its greater affinity for oxygen indicates specialization for this stage of life. The β chain occurs in the 90 percent of adult hemoglobin that is HbA ($\alpha_2 \beta_2$), while the δ chain is found only in the minor adult hemoglobin HbA^2. Hemoglobin A^2 and its contained δ chain appear to be relatively unimportant, suggesting that little specialization for function on the part of the δ gene has yet taken place.

Although the globins comprise one of the best illustrations of gene duplication, another well-analyzed example is that of bovine trypsinogen and chymotrypsinogen which have 50 percent of their sequences in common. In addition, the numerous genes coding for the light and heavy chains of the immunoglobulins are also postulated to have arisen by the duplication of previously existing loci.

EVOLUTION OF BIOLOGICAL SYSTEMS

Living organisms arise only from preexisting organisms, and on the assumption that life originated on earth through natural processes, scientists have long sought to understand how life came into being. Informed estimates place the age of the earth at around 4.5 to 4.7 billion years, and the oldest rocks have been dated at around 3.3 to 3.6 billion years. The earliest known fossils are blue-green algal mats estimated to have existed approximately 3.2 billion years ago. They have been found in sedimentary rocks, and geologic evidence indicates that these organisms lived in the shallow seas which then covered most of the earth. Sometime prior to their existence, life originated, and although the occurrence of such an event might seem to be highly improbable, given enough time, improbable events become probable and even inevitable.

Experiments which attempt to duplicate the conditions of the prebiotic earth have demonstrated that most of the kinds of organic chemicals necessary for life can be formed from modest inputs of energy, for example, an electric spark mimicking lightning. The kinds of compounds produced include sugars, amino acids, fatty acids, purines, peptides and polypeptides, and it is proposed that after long abiotic synthesis, the seas contained a "soup" of many such organic chemicals. How the transition from nonliving to living occurred is conjectural, although a number of plausible hypotheses have been proposed.

One interesting aspect which points to a single successful ancestral proto-life system is the presence in all living forms of particular asymmetric molecules which exhibit optical activity when placed in a beam of plane-polarized light. Thus levomolecules rotate such light in one direction and dextromolecules rotate such light in the opposite direction. Although inorganic synthesis produces both optical isomers, all living forms are characterized by the presence of dextrosugars and levoamino acids. The advantage of having only one optical isomer lies in the fact that it permits the construction of orderly polymers which would not be possible if both right- and left-handed molecules were utilized. However it arose, the ability to select and incorporate only one of two available optical isomers would lead to greater efficiency and make possible the formation of more complex molecules, and such a capability would confer a selective advantage upon any proto-life system.

The present mechanism for the synthesis of proteins consists of a stored code in the form of DNA, a labile messenger RNA, and a system for the translation of the message into protein using rRNA and tRNA along with a multitude of enzymes. The origin of such a complex system is difficult to visualize. It is easy to conceive that a group of polypeptides could carry out some catalytic activity, but without some form of information storage it would be difficult to maintain the same type of primitive "metabolism" over time. C. R. Woese has proposed that both polynucleotides and polypeptides were present in the earliest proto-life systems, for if they were both present, the polynucleotides could assume the role of templates for the construction of polypeptides which in turn could catalyze the synthesis of polynucleotides. As to which came first, DNA or RNA, no answer can be given, although most scientists favor the idea that RNA formed the earliest code. One reason is that, theoretically, polypeptides could be synthesized without DNA, that is, a messenger-type RNA would initially suffice. A second reason is that modern living organisms synthesize deoxyribotides from previously synthesized ribotides and this pathway may represent the ancestral mode of origin.

Woese, Crick, and Cohen have suggested that the earliest systems comprised a group of "statistical" proteins and a primitive kind of rRNA, the two working in concert to produce more of one another. Transfer RNA is visualized as evolving from this system and leading to greater efficiency in polypeptide synthesis. Also evolving from this system might be mRNA-type molecules which could perform a coding function, the presence of such molecules leading to the production of a greater variety of proteins and to the enhanced efficiency and precision of their synthesis. Messenger RNA is notably labile, however, and a system which by chance acquired the ability to make DNA would be at a great selective advantage, for DNA provides a means whereby coded information can be stored in stable form. With the acquisition of a stable code quantitative control over polypeptide synthesis becomes possible, and this in turn permits the evolution of more advanced and complex systems.

Certain aspects of the code provide additional evidence that all living forms have descended from a single ancestral system. For example, the same code is universally used by modern organisms. Furthermore, the degeneracy of the code and the nonrandom pattern of codon assignments which it contains and which tend to lead to compatible amino acid substitutions could only have arisen through natural selection acting on a single ancestral system. If two or more such systems had evolved into present-day forms, not only their codes, but also their respective patterns of codon assignments would be different, but, as it is, the code and its patterns are the same for all.

It will also be recalled that of the three bases of a codon, the first two are more important (see Chapter 16). This observation has led some biologists to propose that a doublet code preceded the evolution of a triplet code and that degeneracy appeared only when this transition was accomplished. On the other hand, a triplet code could well have been present initially with the first two bases serving as discriminating agents, and degeneracy could have arisen as the result of increasing sophistication of the system. Crick has proposed that relatively few amino acids were specified by the earliest codes and that more were utilized with increasing size of the code. Further, the new amino acids added were likely to be those whose properties would not disrupt the ongoing system. Crick also suggests that with the increasing complexity and numbers of proteins specified, the rate of addition of new amino acids to the system would decrease until ultimately no more could be added without affecting the functioning of those proteins already present. Crick's hypothesis is that this point of diminishing returns was reached when 20 amino acids were encoded, and the fact that modern forms contain the same 20 amino acids is further evidence of common ancestry.

With the evolution of a stable code which could be replicated and a method for the translation of the code into protein, we can envision that a further step in evolution resulted in some means whereby any buds from the parent system could be assured of receiving at least one copy of the code. The acquisition of such a mechanism would produce an organism that could be called a primitive prokaryote. The earliest prokaryotes must have been anaerobic since no free oxygen was present. However, with the evolution of algal forms capable of photosynthesis, oxygen was increasingly added to the atmosphere. Survival in an oxygen environment requires the presence of enzymes, such as catalase, to destroy the harmful peroxides produced in the presence of free oxygen, and it is likely that such enzymes, as well as the ability to carry out aerobic respiration, arose with the development of photosynthetic capability.

Because of great similarities in metabolic pathways, it can be inferred that eukaryotes arose from prokaryotes through the evolution of chromosomes, nuclear and other internal cellular membranes, mitochondria, plastids, and a mitotic apparatus to assure orderly and precise cell division. As previously discussed (Chapter 20), according to one view, the cell organelles of eukaryotes represent ancient prokaryotic symbionts, while the opposing view holds that these structures were inherited from aerobic, photosynthetic ancestral forms.

The origin of eukaryotes is placed at 1.2 to 1.4 billion years ago and from their modest beginnings has stemmed all of the varied plant and animal life so profusely generated over geologic time. From the standpoint of evolution, one of the most significant advances associated with the development of eukaryotes was the introduction of sexual reproduction with its accompanying process of genetic recombination. As we have seen, sexual reproduction and recombination permit enormous variation in genotype in each generation, whereas in prokaryotes variation allowing adaptation to the environment arises primarily through mutation. Reliance on mutation as the major source of genetic variety requires that large populations be produced within a very short generation time, and this requirement cannot ordinarily be met by other than unicellular forms reproducing by fission. In contrast, when genetic variation can be introduced by sexual reproduction, generation times can be lengthened and the acquisition of a complex, multicellular physical structure becomes possible.

Throughout the history of biological systems natural selection has played a creative role, producing forms ever more adapted to their conditions of life, whether these be a primal sea or a modern ecological niche. The rules of natural selection are simple. They all serve one purpose: bearers of genotypes that confer greater fitness to the environment survive and reproduce more often than do those of lesser fitness. These rules are nowhere better expressed than by Darwin in the last paragraph of *The Origin of Species.*

> It is interesting to contemplate an entangled bank, clothed with many plants of many kinds, with birds singing on the bushes, with various insects flitting about, and with worms crawling through the damp earth, and to reflect that these elaborately constructed forms, so different from each other, and dependent on each other in so complex a manner, have all been produced by laws acting around us. These laws, taken in the largest sense, being Growth with Reproduction; Inheritance which is almost implied by reproduction; Variability from the indirect and direct action of the external conditions of life, and from use and disuse; a Ratio of Increase so high as to lead to a Struggle for Life, and as a consequence to Natural Selection, entailing Divergence of Character and the Extinction of less-improved forms. Thus, from the war of nature, from famine and death, the most exalted object which we are capable of conceiving, namely, the production of the higher animals, directly follows. There is grandeur in this view of life, with its several powers, having been originally breathed into a few forms or into one; and that, whilst this planet has gone cycling on according to the fixed law of gravity, from so simple a beginning endless forms most beautiful and most wonderful have been, and are being, evolved.

REFERENCES

CRICK, F. H. C., 1968. The origin of the genetic code. *J. Mol. Biol., 38,* 367.

DARWIN, C., 1859. *On the Origin of Species,* 1st ed. (Paperback facsimile reprinted by Atheneum, New York, 1967.)

FOX, S. W., AND K. DOSE, 1972. *Molecular Evolution and the Origin of Life.* Freeman, San Francisco.

JUKES, T. H., 1966. *Molecules and Evolution.* Columbia University Press, New York.

MILLER, S. L., AND L. E. ORGEL, 1974. *The Origins of Life on Earth.* Prentice-Hall, Englewood Cliffs, N.J.

OHNO, S., 1970. *Evolution by Gene Duplication.* Springer-Verlag, New York.

SALTHE, S. N., 1972. *Evolutionary Biology.* Holt, Rinehart and Winston, New York.

WOESE, C. R., 1967. *The Genetic Code: The Molecular Basis for Gene Expression.* Harper & Row, New York.

ZUCKERKANDL, E., 1965. The evolution of hemoglobin. *Sci. Amer., 212,* 110–118.

ANSWERS TO PROBLEMS

CHAPTER 1

1–1. a. tan b. $\frac{3}{4}$ tan, $\frac{1}{4}$ ebony c. $\frac{1}{2}$ ebony, $\frac{1}{2}$ tan

1–2. a. red

b. $RR \times RR$ gives all RR, or $RR \times Rr$ gives $\frac{1}{2}$ RR and $\frac{1}{2}$ Rr

$Rr \times Rr$ gives $\frac{1}{4}$ RR, $\frac{1}{2}$ Rr, $\frac{1}{4}$ rr

$RR \times rr$ gives all Rr

$rr \times rr$ gives all rr

$Rr \times rr$ gives $\frac{1}{2}$ Rr and $\frac{1}{2}$ rr

1–3. $\frac{1}{4}$

1–4. a. solid, dominant; half-white, recessive

b. $SS \times ss$ c. $\frac{1}{4}$ SS, $\frac{1}{2}$ Ss, $\frac{1}{4}$ ss d. Ss e. $\frac{1}{2}$

1–5. a. deafness recessive

b. Sell dogs B and C, since each carries the gene for deafness in the heterozygous condition.

1–6. a. short hair dominant

b. Using a (angora) and A (short hair):

(1) $aa \times aa$, progeny all aa

(2) $aa \times AA$, progeny all Aa

(3) $Aa \times aa$, progeny $\frac{1}{2}$ Aa and $\frac{1}{2}$ aa

c. $\frac{3}{4}$ short haired, $\frac{1}{4}$ angora

d. all kittens short haired

1–7. a. yellow dominant b. $\frac{1}{4}$ white c. $\frac{1}{2}$, i.e., those that are heterozygous d. $\frac{1}{4}$, i.e., those homozygous for yellow e. 2500

1–8. a. Yy (yellow) $\times$ yy (white) b. Yy and yy

c. Yy gives $\frac{1}{4}$ YY, $\frac{1}{2}$ Yy, $\frac{1}{4}$ yy ($\frac{3}{4}$ yellow, $\frac{1}{4}$ white); yy gives all white F_2

d. self-fertilization of F_1 Yy gives $\frac{1}{4}$ YY and $\frac{1}{4}$ yy F_2 homozygotes; self-fertilization of F_1 yy gives all yy F_2 homozygotes

1–9. a. Son's wife has attached ear lobes

b. recessive trait

1–10. The child was adopted, illegitimate, or (much less likely) had inherited a gene which had mutated to the dominant condition.

1–11. a. dominant

b. I 1, Aa; I 2, 3, 4, aa

II 1, AA; II 2, 5, Aa; II 3, 4, 6, 7, aa

c. $\frac{1}{2}$

d. III 5, 6, 7, 8, 10, aa; III 9, 11, Aa

e. $\frac{1}{2}$ f. $\frac{1}{2}$

1–12. a. recessive

b. I 1, aa; I 2, Aa

II 1, probably AA; II 2, Aa; II 3, aa; II 4, probably AA

III 2, 3, Aa; III 6, 7, Aa

c. testcross, outcross, inbreed

CHAPTER 2

2–1. a. *Ee* b. *AA*

2–2. a. $\frac{9}{16}$ green, starchy; $\frac{3}{16}$ orange, starchy; $\frac{3}{16}$ green, waxy; $\frac{1}{16}$ orange, waxy
b. $\frac{1}{4}$ *Oror Wxwx*, green, starchy; $\frac{1}{4}$ *Oror wxwx*, green, waxy; $\frac{1}{4}$ *oror Wxwx*, orange, starchy; $\frac{1}{4}$ *oror wxwx*, orange, waxy

2–3. a. (1) *OrOr Wxwx* (2) *OrOr wxwx* (3) *oror Wxwx*
(4) *oror WxWx* (5) *OrOr WxWx* (6) *Oror WxWx*
(7) *Oror wxwx* (8) *Oror Wxwx*
b. $\frac{1}{2}$ *or Wx*, $\frac{1}{2}$ *or wx*
c. $\frac{1}{4}$ *Or Wx*, $\frac{1}{4}$ *Or wx*, $\frac{1}{4}$ *or Wx*, $\frac{1}{4}$ *or wx*

2–4. a. P_1: *ff cc* × *FF CC*; F_1: *Ff Cc*
b. 18 continuous flowering, crinkled; 6 continuous flowering, noncrinkled; 54 seasonal, crinkled; 18 seasonal, noncrinkled
c. 24 d. 12 e. 6

2–5. *Tt Bb*

2–6. a. barred, white: *Bb ww*; non-barred, orange: *bb Ww* b. each class $\frac{1}{4}$ of the total c. $\frac{3}{16}$ d. $\frac{9}{16}$ e. $\frac{3}{16}$

2–7. a. $\frac{1}{2}$ b. all progeny B

2–8. a. $(\frac{1}{4})^4$ or $\frac{1}{256}$ *AABBCCDD* in genotype; $(\frac{3}{4})^4$ or $\frac{81}{256}$ A, B, C, D in phenotype
b. $(\frac{1}{4})^4$ or $\frac{1}{256}$ *aabbccdd* in phenotype and in genotype
c. 2^4 or 16 types of gametes
d. 2^4 or 16 different phenotypes
e. 3^4 or 81 different genotypes

2–9. a. 2^3 or 8 phenotypes b. 3^3 or 27 genotypes

2–10. a. $\frac{1}{4}$ b. $\frac{1}{4}$ c. $\frac{1}{8}$ d. 16 phenotypes
e. 16 genotypes

2–11. a. *Ee Dd* female × *Ee dd* male
b. $\frac{1}{4}$ *ED*, $\frac{1}{4}$ *Ed*, $\frac{1}{4}$ *eD*, $\frac{1}{4}$ *ed*
c. $\frac{1}{2}$ *Ed*, $\frac{1}{2}$ *ed*
d. 3 tan : 1 ebony
e. 1 long : 1 dumpy
f. *Ee Dd*, *EE Dd*, tan, long; *Ee dd*, *EE dd*, tan, dumpy; *eedd*, ebony, dumpy; *ee Dd*, ebony, long

2–12. a. $\frac{1}{3}$ homozygous tan b. $\frac{2}{3}$ heterozygous tan
c. testcross to ebony homozygotes

2–13. a. both traits recessive
b. I 1, *AAbb*; I 2, *aaBB*
c. II 1, *AAbb*; II 2, *AaBb*; II 6, *AaBb*; II 7, *aaBB*; III 5, *Aabb*; III 6, *aaBb*; IV 1, 4, *AbBb*; IV 2, *aaBb*; IV 3, 6, *Aabb*; IV 5, *aabb*.

2–14. a. Assume both traits dominant. Could trait 1 be recessive?
b. I 1, *aabb*; I 2, *aaBb*
II 1, *AaBB*; II 2, *aabb*; II 3, *aaBb*; II 4, *Aabb*;
III 1, 3, 4, 7 *aaBb*; III 2, 5, 10, *AaBb*;
III 6, 8, *aabb*; III 9, *Aabb*
c. $\frac{3}{4}$ will show trait 1, $\frac{3}{4}$ will show trait 2, $\frac{9}{16}$ will show both traits
d. $\frac{1}{2}$ will show trait 1, $\frac{1}{2}$ will show trait 2, $\frac{1}{4}$ will show both traits

2–15. a. trait 1 recessive, trait 2 dominant
b. I 1, *aabb*; I 2, *AAbb*; II 1, *AABb*; II 2, *Aabb*;
II 5, *Aabb*; II 6, *aabb*; III 7, *aabb*; III 8, *Aabb*
c. $\frac{1}{2}$ d. if individual III 6 is heterozygous for trait 1 *(Aa)*

CHAPTER 3

3–1. a. $\frac{1}{4}$ b. $\frac{1}{4}$ c. $\frac{1}{2}$ d. all gametes carry *c*

3–2. a. $\frac{1}{2}$ b. $\frac{1}{4} \times \frac{1}{2}$ or $\frac{1}{8}$ c. $\frac{1}{2} \times \frac{3}{4} \times \frac{1}{2}$ or $\frac{3}{16}$ d. $\frac{1}{2} \times \frac{1}{2} \times \frac{1}{2}$ or $\frac{1}{8}$ e. $\frac{1}{2} \times \frac{1}{4} \times \frac{1}{2}$ or $\frac{1}{16}$

3–3. a. all gametes b. $\frac{1}{2}$ c. $\frac{1}{8}$ d. $\frac{1}{8}$ e. $\frac{3}{16}$

3–4. a. $(a + b)^4$

b. a^4, all children normal; $6a^2b^2$, two normal and two abnormal; $4ab^3$, one normal and three abnormal

c. $\frac{1}{16}$; $\frac{3}{8}$; $\frac{1}{4}$

3–5. a. $\frac{3}{64}$ b. $\frac{9}{64}$

3–6. a. $\frac{1}{4}$

b. No, prediction is the same.

c. Yes, prediction should be based on three children and would be $\frac{1}{8}$.

3–7. player 1: $\chi^2 = 13.68$, $P = <0.01$; data do not fit ratio
player 2: $\chi^2 = 1.76$, $P = >0.5$; data fit ratio well
player 3: $\chi^2 = 9.84$, $P = <0.02$; data do not fit ratio
player 4: $\chi^2 = 5.16$, $P = >0.05$; data fit ratio minimally

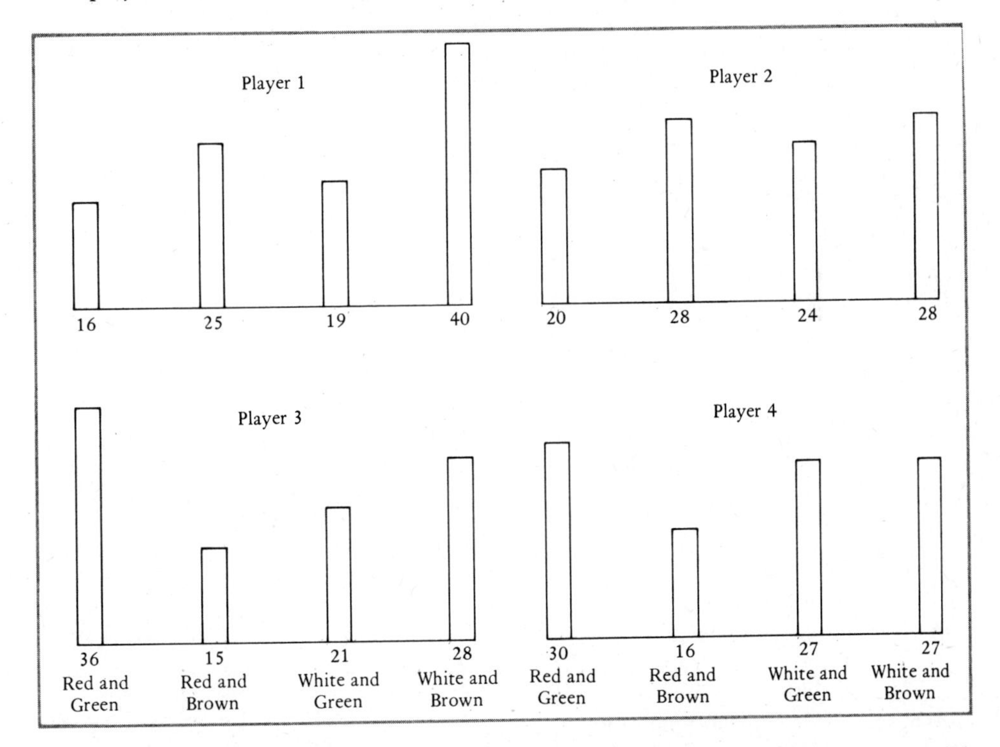

3–8. *P* values obtained for samples 1 through 5 are: (1) $P = <0.02$; (2) $P = >0.8$; (3) $P = >0.7$; (4) $P = >0.5$; (5) $P = <0.05$. The data of samples 2, 3 and 4 fit the 9 : 3 : 3 : 1 ratio. Significant deviations from this ratio occur in the data of samples 1 and 5, where in both cases *P* is below 0.05. If the data of samples 2, 3 and 4 are pooled, $P = >0.95$.

3–9. The genotypes of the parents of problem 2–11 are *EeDd* × *Eedd*. From this cross we expect a 3 : 1 ratio for tan versus ebony and a 1 : 1 ratio for normal versus dumpy wings. Since $P = >0.5$ in each case, these data fit the expected ratios.

3–10. Association of the condition with the male sex is highly significant. Assuming a 1 : 1 ratio, χ^2 is 22.86 and *P* is much less than 0.01, indicating that this ratio is inapplicable to the data.

CHAPTER 4

4–1. a. genes are allelic
b. one pair of genes exhibiting intermediate dominance; homozygotes, black *(BB)* and splashed *(bb)*; heterozygotes, blue *(Bb)*
c. 1. $BB \times bb$ gives all Bb
2. $bb \times bb$ gives all bb
3. $bb \times Bb$ gives $\frac{1}{2}$ bb, $\frac{1}{2}$ Bb
4. $BB \times Bb$ gives $\frac{1}{2}$ BB, $\frac{1}{2}$ Bb
5. $Bb \times Bb$ gives $\frac{1}{4}$ BB, $\frac{1}{2}$ Bb, $\frac{1}{4}$ bb

4–2. a. two pairs of genes, *B* and *b*, as in problem 4–1, and *W* and *w*. *W* is a dominant gene permitting color; *w*, when homozygous, blocks color formation.
b. P_1 *BBWW* (black) × *bbww* (white)
c. F_1 *BbWw* (blue)
d. Black: *BBWW, BBWw*; blue: *BbWW, BbWw*; splashed: *bbWW, bbWw*; white: *BBww, Bbww, bbww*

4–3. a. Phenotypic ratio is 9:3:4, estimated as follows. The sum of the classes is 82 (45 + 16 + 21); $\frac{1}{16}$ of the progeny $(\frac{82}{16})$ is 5 individuals. Thus, $\frac{9}{16} = 45$, $\frac{3}{16} = 15$, and $\frac{4}{16} = 20$ individuals, in agreement with the data.
b. two gene pairs
c. *BBCC* (black) × *bbcc* (white)
d. F_1 *BbCc*
e. F_2 white: *BBcc*, $\frac{1}{16}$ of total or $\frac{1}{4}$ of white progeny
Bbcc, $\frac{2}{16}$ of total or $\frac{1}{2}$ of white progeny
bbcc, $\frac{1}{16}$ of total or $\frac{1}{4}$ of white progeny

4–4. 1. *BbCC* (black) × *bbcc* (white)
2. *BBCc* (black) × *BBcc* (white)
3. *BbCc* (black) × *bbcc* (white)

4–5. a. Phenotypic ratio is 9:6:1, obtained as explained in answer to problem 4–3.
b. two pairs of genes
c. Since there are two gene pairs, members of the largest class ($\frac{9}{16}$ brown) must carry two different dominant genes. The tan individuals are $\frac{6}{16}$ (or $\frac{3}{16} + \frac{3}{16}$) of the total and must carry one or the other dominant gene, but not both. White individuals ($\frac{1}{16}$) must be homozygous for both recessive alleles. Thus, assume two different dominant genes, B^1 and B^2, and their respective recessive alleles, b^1 and b^2.
P_1: $B^1B^1b^2b^2$ (tan) × $b^1b^1B^2B^2$ (tan)
F_1: $B^1b^1B^2b^2$ (brown)
F_2 tan: $B^1B^1b^2b^2$, $B^1b^1b^2b^2$, $b^1b^1B^2B^2$, $b^1b^1B^2b^2$
F_2 white: $b^1b^1b^2b^2$

4–6. a. 12:3:1 phenotypic ratio
b. red offspring: *RRYY, RRYy, RRyy, RrYY, RrYy, Rryy*
c. yellow offspring: *rrYy, rrYY*
d. Only those homozygous for red *(RR)* will produce red progeny exclusively upon selfing. These are *RRYY, RRYy,* and *RRyy*; they comprise $\frac{4}{12}$ or $\frac{1}{3}$ of the F_1 red individuals.
e. Only those homozygous for the dominant gene *Y* will produce yellow progeny exclusively upon selfing. These are *rrYY*; they comprise $\frac{1}{3}$ of the F_1 yellow individuals.

4–7. a. *CCWW* × *ccWw*
b. F_1: $\frac{1}{2}$ *CcWw*, $\frac{1}{2}$ *CcWW*
c. *CcWw* × *CcWw* gives white, black, and Siamese; *CcWw* × *CcWW* and *CcWW* × *CcWW* gives all white F_2

4–8. Variable penetrance is present since not all carriers of the gene *W* have blue eyes and are deaf. Variation in the degree of expression (expressivity) is also evident in that only one eye may be blue. Pleiotropism is also involved since the gene *W* affects not only coat color and eye color, but hearing as well.

4–9. Let S = straight hair and s = curly hair; P = black tongue and p = pink tongue.
a. Parents: *SSpp* (male) × *ssPP* (female)
b. F_1: *SsPp*, wavy hair, gray tongue

4–10. $\frac{1}{16}$ *SSPP*, straight, black; $\frac{2}{16}$ *SSPp*, straight, gray; $\frac{1}{16}$ *SSpp*, straight, pink; $\frac{2}{16}$ *SsPP*, wavy, black, $\frac{4}{16}$ *SsPp*, wavy, gray; $\frac{2}{16}$ *Sspp*, wavy, pink; $\frac{1}{16}$ *ssPP*, curly, black; $\frac{2}{16}$ *ssPp*, curly, gray; $\frac{1}{16}$ *sspp*, curly, pink

4–11. Docked tail is a dominant trait (gene *D*), as opposed to long tail (gene *d*). Since in the cross docked X docked some long-tailed offspring were produced, both parents must have been heterozygous (*Dd*). The ratio from this cross is 2 : 1, suggesting that *DD* individuals are inviable and that the gene *D* is lethal when homozygous.

4–12. a. *SsDdpp* × *ssDdPp* b. $\frac{3}{4}$ c. $\frac{1}{2}$
d. $\frac{1}{2}$ e. $\frac{1}{4}$ f. $\frac{1}{3}$ g. $\frac{1}{6}$

4–13. multiple allelic series with A dominant over B, C, and D; B dominant over C and D; C and D with intermediate dominance

4–14. Let A = agouti, a = nonagouti, and A^Y = yellow.
a. 1. $AA \times AA$
2. $AA \times aa$
3. $AA \times AA^Y$
4. $AA^Y \times aa$
b. yellow dominant over both agouti and nonagouti; agouti dominant over nonagouti
c. all alleles d. yellow lethal when homozygous

4–15. a. all T/t^0, short tail b. 6

4–16. a. curly and stubble both dominant
b. Using *Cy* for curly, *Sb* for stubble, and + for the respective wild type alleles, parental genotypes are *Cy*/+ +/+ × +/+ *Sb*/+.
c. *Cy*/+ +/+, curly, long bristle; *Cy*/+ *Sb*/+, curly, stubble; +/+ +/+, normal wing, long bristle; +/+ *Sb*/+, normal wing, stubble

4–17. curly and stubble both lethal when homozygous

4–18. $\frac{1}{3}$

4–19. $\frac{7}{16}$ inviable, $\frac{9}{16}$ viable; of these, $\frac{7}{9}$ hairy, $\frac{2}{9}$ hairless, and $\frac{4}{9}$ *H*/+ *Su* – *H*/+

4–20. child AB adopted

4–21. $\frac{3}{8}$

4–22. a. $\frac{1}{2}$ I^AI^O, $\frac{1}{2}$ I^OI^O
b. $\frac{1}{4}$ I^AI^A, $\frac{1}{4}$ I^AI^O, $\frac{1}{4}$ I^BI^A, $\frac{1}{4}$ I^BI^O
c. $\frac{1}{2}$ I^BI^A, $\frac{1}{2}$ I^BI^O
d. $\frac{1}{4}$ I^AI^A, $\frac{1}{2}$ I^AI^B, $\frac{1}{4}$ I^BI^B

4–23. a. I^AI^O b. I^BI^O c. I^AI^O

4–24. a. 1. $r'r'$ or *dCe/dCe*
2. R^1R^1 or *DCe/DCe*; R^1r' or *DCe/dCe*
3. R^zR^2 or *DCE/DcE*; R^zr'' or *DCE/dcE*; R^2r^y or *DcE/dCE*
4. $r''r''$ or *dcE/dCE*
b. genotypes 2 and 3 classed as Rh^+.

4–25. no, because the woman is Rh^+

4–26. a. woman Rh^-/Rh^-; her mother Rh^-/Rh^-; her father Rh^+/Rh^-
b. $\frac{1}{2}$

4–27. a. $\frac{1}{4}$ b. $\frac{1}{16}$ c. $\frac{3}{32}$ d. $\frac{1}{128}$

4–28. Individuals excluded as the father are: b on the basis of blood groups O and N, c on the basis of blood group S, d on the basis of blood group A and nonsecretor.

CHAPTER 5

5–1. a. A normal distribution is observed and a normal curve can be drawn.

b. The mean ($\bar{x}$) is 67.2 inches, σ is 2.6 inches, and standard error of the mean is 0.08.

c. $\pm 1\sigma$ gives a range from 64.6 to 69.7 inches and includes 68 percent of the soldiers; $\pm 2\sigma$ gives a range from 62 to 72 inches and includes 95.4 percent of the soldiers; $\pm 3\sigma$ gives a range from 59.5 to 74.9 inches and includes 99.7 percent of the soldiers. The sample is representative of French soldiers of 1871.

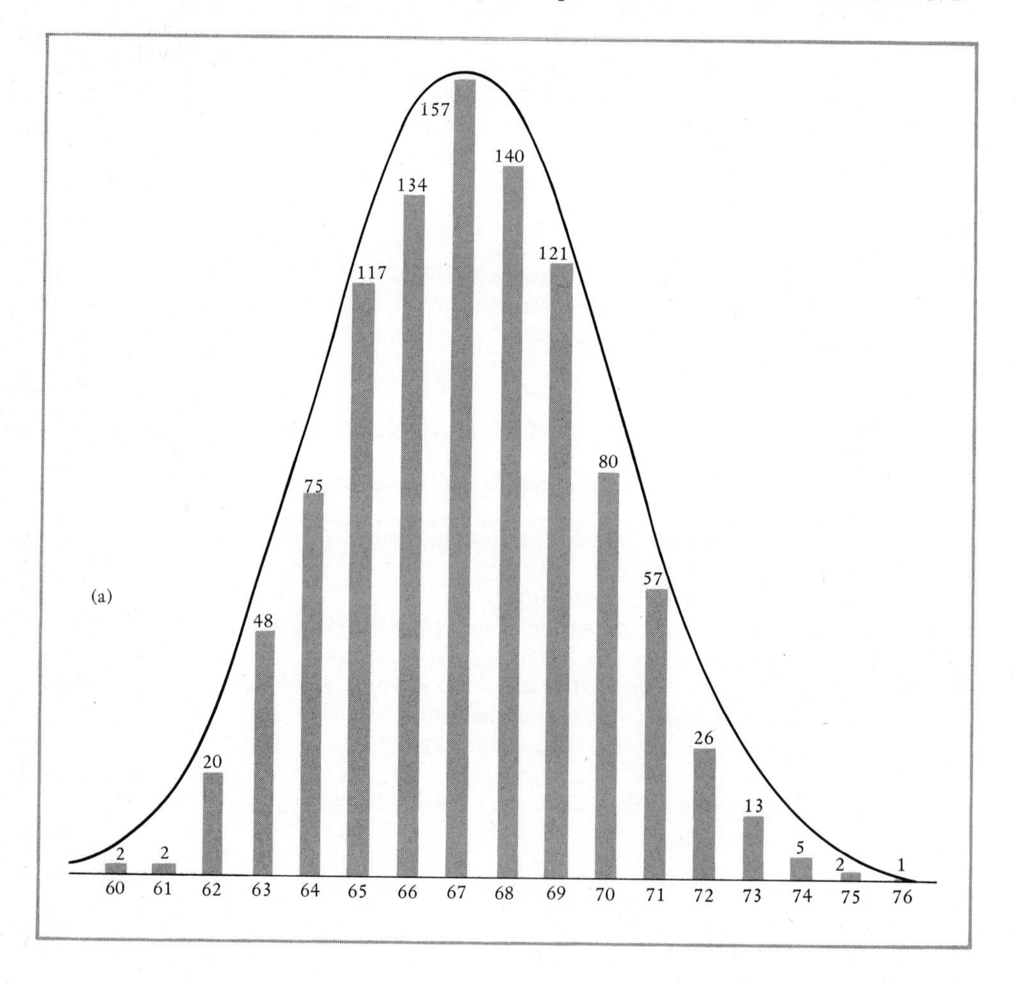

5–2. a. P_1: $\bar{x}$ = 22.9 mm, σ = 1.03 mm, $s_{\bar{x}}$ = 0.07
P_2: $\bar{x}$ = 45.9 mm, σ = 1.03 mm, $s_{\bar{x}}$ = 0.07
F_1: $\bar{x}$ = 34 mm, σ = 1.34 mm, $s_{\bar{x}}$ = 0.08
F_2: $\bar{x}$ = 34 mm, σ = 1.93 mm, $s_{\bar{x}}$ = 0.08

b. The difference between the P_1 and P_2 varieties is very significant and far above 3 times the standard error of the difference in means (s_D).

c. The difference between the means of the F_1 and F_2 progeny is not significant.

d. There are probably 4 pairs of genes. With 4 gene pairs, either homozygote is expected with a frequency of 1 in 256. In this case, individuals falling within the range of the parental varieties number 2 in 542, or 1 in 271.

5–3. a. strain A, *aabbcc*; strain B, *AABBCC*
b. F_1 would be 225 gm *(AaBbCc)*
c. 150 gm; 175 gm; 200 gm; 225 gm; 250 gm; 275 gm; 300 gm
d. 256

5–4. a. Yes, the difference is significant for all three genotypes.
b. No.
c. Yes, the difference is very significant.
d. Increasing temperature causes an increase in the size of the spots produced in the presence of allele sp^1, but decreases the size of spots produced by allele sp^2.
e. Proteins, particularly enzymes, are highly sensitive to temperature changes.

CHAPTER 6

6–2. a. 28 b. 7 c. 7 d. 14
e. 28 f. 7 g. 42 h. 21
i. 28 j. 14

6–3. a. 76 b. 152 c. 152 d. 76 e. 152
f. 38 g. 76 h. 38

6–4. a. $(\frac{1}{2})^7$ b. $(\frac{1}{2})^7$ c. $(\frac{1}{2})^7 + (\frac{1}{2})^7$

6–5. $(\frac{1}{2})^4$

6–6. Because of independent assortment, the gametes produced by the mule contain chromosomes derived from both horse and donkey and such mixed chromosome complements lead to inviability. Only very rarely, by chance, would a gamete be formed that contained a complete set of chromosomes from one species or the other.

6–7. a. 10 and 6 b. 5, 5, 3, 3 c. $n + 1$ and $n - 1$

6–8. 4, 4, 3, 5

6–9. $\frac{1}{2}\ p^+$ and $\frac{1}{2}\ p^-$

6–10. a. $p^+p^+p^+p^+p^-p^-p^-p^-$ (or reverse)
b. 4 orders: $p^+p^+p^-p^-p^-p^-p^+p^+$, $p^-p^-p^+p^+p^+p^+p^-p^-$, $p^+p^+p^-p^-p^+p^+p^-p^-$, and $p^-p^-p^+p^+p^-p^-p^+p^+$.

6–11. $\frac{1}{4}p^+q^+$, $\frac{1}{4}p^-q^+$, $\frac{1}{4}p^+q^-$, $\frac{1}{4}p^-q^-$

6–12. *Neurospora, Chlamydomonas,* and similar organisms are haploid and dominance or recessiveness can be observed only in the diploid condition.

CHAPTER 7

7–1. Carnation is sex linked, short wing is autosomal.
a. female: *c/c +/s* b. male: +/Y *s/s*
c. progeny males: *c/* Y *+/s* and *c/* Y *s/s*
progeny females: *c/+ s/+* and *c/+ s/s*

7–2. Reduced eyes is sex linked and dominant.

7–3. parental male: +/Y +/*j*; parental female: +/*p* +/*j*

7–4. $\frac{1}{2}$ white (*w*/Y *st*/+ and *w*/Y *st/st*), $\frac{1}{4}$ wild type (+/Y *st*/+), $\frac{1}{4}$ scarlet (+/Y *st/st*)

7–5. parental female: +/*v st/st*; parental male: *v*/Y *st*/+

7–6. Three hypotheses are possible: (1) black is an autosomal recessive; *b/b* ♀ × *b*/+ ♂ yields a 1 : 1 ratio; (2) black is an autosomal dominant; *B*/+ ♀ × +/+ ♂ yields a 1 : 1 ratio; (3) black is a sex-linked dominant; *B*/+ ♀ × +/Y ♂ yields a 1 : 1 ratio. Note: Black *cannot* be a sex-linked recessive.

7–7. a. Short bristle *(S)* is dominant and sex linked, and the data indicate that it could not be recessive.
b. Black body could be either dominant *(B)* or recessive *(b)*, but in either case it is autosomal.
c. If black is recessive, genotypes are *S*/+ +/*b* × +/Y *b*/*b*. If black is dominant, genotypes are *S*/+ +/+ × +/Y *B*/+.
d. Absence of short-bristled males indicates lethality when this gene is hemizygous. Note that the number of male progeny is only half that of the female progeny.

7–8. a. Females (1) and (2) differ.
b. Brown mutant in (1) is sex linked, dominant, and lethal homozygous or hemizygous; brown mutant in (2) is autosomal, dominant, and lethal homozygous.

7–9. a. $\frac{3}{8}$ b. $\frac{2}{3}$ c. $\frac{1}{4}$ d. $\frac{2}{3}$

7–10. In birds females are XY and males are XX.
males: $\frac{1}{2}$ *B*/*b s*/*S*, barred, nonsilky; $\frac{1}{2}$ *B*/*b*, *s*/*s*, barred, silky
females: $\frac{1}{2}$ *b*/Y *s*/*S*, black, nonsilky; $\frac{1}{2}$ *b*/Y *s*/*s*, black, silky

7–11. Gene *M* is located on the Y chromosome.

7–12. The probability for a daughter is $\frac{1}{2}$ and the probability for colorblindness is $\frac{1}{2}$. Thus, the probability for a colorblind daughter is $\frac{1}{2} \times \frac{1}{2}$ or $\frac{1}{4}$.

7–13. The mother was heterozygous for colorblindness; nondisjunction of the X chromatids, each bearing gene *c*, occurred at the second meiotic division to produce an egg of sex genotype X^cX^c. Fertilization by a Y-bearing sperm resulted in a son.

7–14.

	Drosophila	*Human*
a.	superfemale	female
b.	male	male
c.	female	male (Kleinfelter)
d.	male	female (Turner)
e.	male	male
f.	intersex	male
g.	female	male
h.	intersex	female
i.	supermale	male

7–15. a. trait 1 dominant and autosomal; trait 2 recessive and sex linked
b. II 1, *A*/*a B*/Y; II 2, *a*/*a B*/*B*; II 3, *a*/*a B*/Y; II 4, *a*/*a B*/*b*;
III 5, *A*/*a B*/*B*; III 6, *a*/*a B*/Y; IV 1, *a*/*a b*/Y; IV 2, *A*/*a B*/*b*;
V 1, *A*/*a b*/*b*; V 2, *a*/*a B*/ Y; V 3, *a*/*a b*/ Y; V 4, *A*/*a B*/*b*; V 5, *a*/*a B*/*b*; V 6, *A*/*a B*/Y; V 7, *A*/*a b*/Y

7–16. a. trait recessive and sex linked
b. I 1, *A*/Y; I 2, *A*/*a*
II 1, *a*/Y; II 2, *A*/*a*; II 8, *A*/*A*
III 6, *a*/*a*; III 10, *A*/*a*; III 11, *A*/Y

7–17. a. trait dominant and sex linked
b. I 1, *A*/Y; I 2, *a*/*a*; I 4, *A*/*a*
II 1, *A*/*a*; II 6, *a*/Y; II 7, *A*/*a*
III 1, *A*/*a*; III 2, *A*/Y; III 6, *a*/*a*; III 7, *a*/Y

7–18. a. trait 1 dominant and sex linked; trait 2 recessive and autosomal
b. I 1, *a*/Y *b*/*b*; I 2, *A*/*a B*/*B*; I 4, *a*/*a b*/*b*
II 1, *a*/*a b*/*b*; II 2, *A*/Y *B*/*b*; II 5, *a*/Y *B*/*b*; II 6, *a*/*a B*/*b*
III 1, *a*/Y *b*/*b*; III 6, *A*/*a b*/*b*; III 10, *a*/*a b*/*b*

7–19. a. trait 1 recessive and sex linked; trait 2 recessive and autosomal

b. I 3, *a*/Y *B*/*B*; I 4, *A*/*A b*/*b*;
II 2, *A*/Y *B*/*B*; II 3, *A*/*a B*/*b*; II 7, *A*/*a B*/*b*; II 8, *A*/Y *B*/*B*; III 3, *A*/*a B*/*b*; III 4, *a*/Y *B*/*b*;
IV 2, *a*/Y *B*/*b* or *a*/Y *B*/*B*; IV 3, *a*/*a b*/*b*; IV 4, *A*/Y *b*/*b*

7–20. 2 X : 2 Y : 1 A sperm could originate through normal disjunction of the autosomes, but nondisjunction of the X and Y chromosomes at both the first and second meiotic divisions. 4 X : 1 A eggs could originate similarly, by normal disjunction of autosomes and nondisjunction of the X chromosomes at the first and second meiotic divisions. 2 X : 1 A sperm could originate by normal disjunction of autosomes and nondisjunction of X chromatids at the second meiotic division.

7–21. 1. Nondisjunction at the first meiotic division will produce a secondary oöcyte lacking the X chromosome; such an oöcyte will yield an ovum lacking the X.
2. Nondisjunction at the second meiotic division, where both X chromatids enter the polar body, will result in an ovum without the X chromosome.

7–22. Duchenne muscular dystrophy would be found only in females homozygous for this mutation. Since males carrying the gene do not live to reproduce, no such females are formed.

7–23. The mosaic phenotype results from random inactivation of one of the two X chromosomes. Albinism occurs in those cells in which the X chromosome bearing the normal allele is inactivated; pigmented areas consist of cells in which the X carrying the mutant allele has been inactivated.

CHAPTER 8

8–1. a. The data indicate two groups of linked genes: *F-A-E-D* and *C-G-B*.
b. Recombination frequencies are: *F-A*, 10 percent; *A-E*, 20 percent; *E-D*, 10 percent; *F-D*, 40 percent; *C-G*, 20 percent; *G-B*, 5 percent; *C-B*, 25 percent

8–2. a. All loci are linked.
b. parental genotypes: *ACB*/*acb* and *acb*/*acb*
c. map distances: *A* —— 20 —— *C* —— 15 —— *B*

8–3. a. 50 percent, with results indistinguishable from independent assortment
b. 25 percent

8–4. a. c^1 and c^2 are not allelic b. c^1 and c^2 are linked

8–5. The male was the heterozygous parent and the female was *cn vg*/*cn vg* because *cn* and *vg* progeny were produced in a 1 : 1 ratio, indicating that no crossing over had taken place. Recall that crossing over does not occur in *Drosophila* males.

8–6. The genes *S* and *Y* are linked; the ratio is 3 : 1 indicating complete linkage, that is, no crossing over. If not linked, a 9 : 3 : 3 : 1 ratio would have been obtained.

8–7. a. F_1 females: *fa* +/+ *sn* (wild type); F_1 males: *fa* +/Y (facet)
b. From the data the crossover frequency between *fa* and *sn* is 18 percent. Therefore, crossover eggs will be + + (9 percent) and *fa sn* (9 percent), and noncrossover eggs will be *fa* + (41 percent) and + *sn* (41 percent). Union of eggs with *fa* +, X-bearing sperm will produce females in the phenotypic proportions of 50 percent wild type and 50 percent facet. Union of eggs with Y-bearing sperm will produce males in the phenotypic proportions of 41 percent facet, 41 percent singed, 9 percent facet, singed, and 9 percent wild type.

8–8. Genotype of parental females: *p r* +/+ + *q*
gene order and map distances:

p ← 7 → *r* ← 14 → *q*
← 21 →

coefficient of coincidence: 0.7

8–9. Genotype of parental females: *b* + + *c*/+ *d a* +
gene order and map distances:

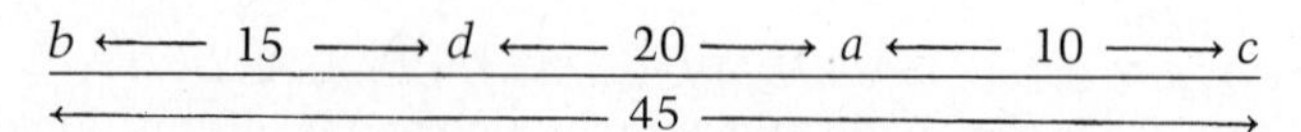

8–10. To solve this problem, the number of double crossovers expected and realized must be determined first. There are 10 map units between genes *w* and *s* and 20 map units between *s* and *t*. Therefore, $0.1 \times 0.2 = 0.02$ frequency of doubles. Since the progeny total 500, the expected doubles are $0.02 \times 500 = 10$ individuals. The coefficient of coincidence is 0.6, however, and therefore actual doubles will be $0.6 \times 10 = 6$ individuals.

The parental genotype is *w* + *t*/+ *s* +. Single exchanges between *w* and *s* produce the classes *w s* + and + + *t*. Their frequency is $0.1 \times 500 = 50$ individuals from which we must subtract the 6 double crossovers, giving a total of 44 individuals for these two genotypes together. Since only half are expected to be + + *t*, 22 + + *t* are predicted. Single crossovers between *s* and *t* produce classes *w* + + and + *s t*. Their frequency is $0.2 \times 500 = 100 - 6 = 94$ in all. Only half will be + *s t*, and therefore this class should include 47 individuals.

Genotype + *s* + is a parental type. By subtracting the total of all crossover classes from the number of progeny, the number of parental type individuals can be determined. Total crossovers are $6 + 44 + 94 = 144$ total; $500 - 144 = 356$ parental types. Half of these (178) are + *s t*, the other half *w* + *t*.

8–11. a. 16 individuals b. 4 individuals

8–12. 10,000

8–13. Forty percent of the gametes of *ds mp*/+ + parent will carry both *ds* and *mp*; 10 percent of the gametes produced by the *ds* +/+ *mp* parent will carry both recessives. Therefore, $0.4 \times 0.1 = 0.04$ or 4 percent of the progeny will be homozygous for these genes.

8–14. The cross is *a* +/+ *b* × *a* +/+ *b*. The frequency of crossovers will be 8 percent and that of parentals will be 92 percent. Thus, both pollen and eggs will occur in the proportions of 0.46 *a* +, 0.46 + *b*, 0.04 *a b*, 0.04 + +, and union between these gametes will produce approximately 24.8 percent individuals + *b* in phenotype. Proof can be obtained by using a checkerboard, weighting the gametes as above, and summing the frequencies of + *b* phenotypic classes.

8–15. The smallest class of offspring from such a cross is expected to be the double recessive *a b*/*a b*. This genotype can be produced only by recombination, and its frequency is the product of the frequencies of *a b* gametes produced equally by both sexes. The frequency of the smallest class is 0.008 (0.8 percent) and therefore the frequency of *a b* gametes whose union produces this class is $\sqrt{.008}$ or approximately 0.09. Since the two types of crossover gametes, *a b* and *A B*, are produced in equal frequency, the total recombination frequency will be $2 \times 0.09 = 0.18$ or 18 percent recombination. Proof can be obtained through a checkerboard, using 0.09 for the frequency of each type of crossover gamete (*a b* and *A B*) and 0.41 for the frequency of each parental type gamete (*A b* and *a B*) and determining the frequencies of the phenotypic classes in the progeny.

8–16. a. *a* +/*a b* and + +/+ *b* or *a* +/+ *b* and + +/*a b*
b. *a* +/*a* + and + *b*/+ *b*, or *a* +/+ *b* and + *b*/*a* +
c. Daughter cells can be *a* +/+ + and *a b*/+ *b*, or *a* +/+ *b* and *a b*/+ +

8–17. a. genes *b* and *c* linked
b. 5 map units
c. gene *a* 5 map units from its centromere

8–18. a. genes *x* and *y* linked

b. gene and centromere locations and map distances:

$$x \longleftarrow 1 \longrightarrow \overset{}{\underset{4}{o}} \longleftarrow 3 \longrightarrow y$$

(x to y: ⟵ 4 ⟶)

8–19.

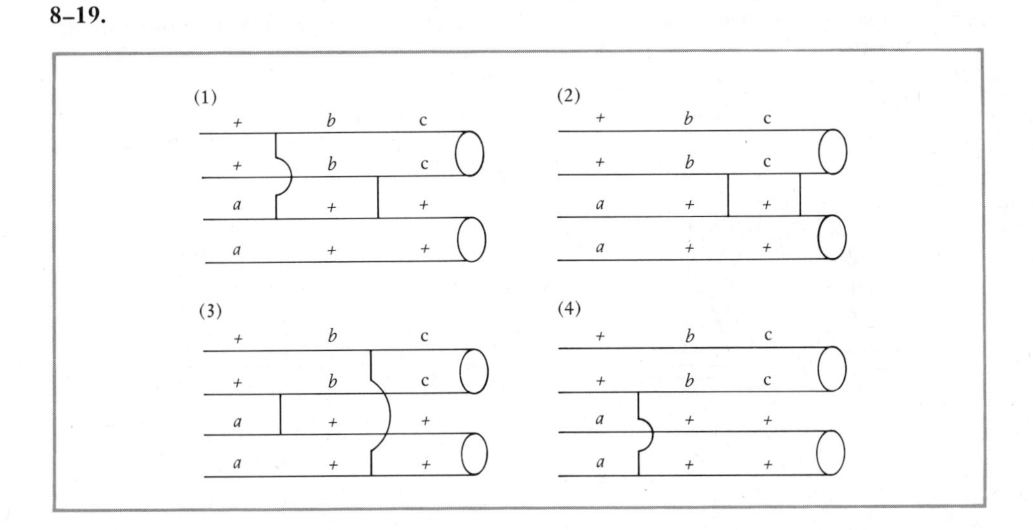

8–20. 1. *a* + *c*

2. + *b* +

3. *a b c*

4. + + +

CHAPTER 9

9–1. Examine meiotic cells cytologically. If an autopolyploid, complex symaptic associations involving four chromosomes (eight chromatids) will be evident. If an allopolyploid, tetrads composed of two chromosomes (four chromatids) each should occur.

9–2. Tetraploid males could be X0 in sex chromosome constitution, having 40 autosomes and 2 X chromosomes, as compared to females with 40 autosomes and 4 X chromosomes.

9–3. a. Because of centromere homology, the translocated chromosome, 14.21, will segregate from the normal chromosome 14; normal chromosome 21 can pass to either pole. As a result, the following four egg types can be formed: (1) eggs containing the normal chromosomes 14 and 21; (2) eggs containing the translocated chromosome 14.21; (3) eggs containing normal chromosome 14; (4) eggs containing normal chromosome 21 along with the translocated chromosome 14.21.

b. $\frac{1}{4}$. Egg type 4 above fertilized by a normal sperm will produce a trisomic zygote of genotype 14/14.21 21/21.

9–4. Six kinds of gametes could be produced: AA^1, A^2, AA^2, A^1, A^1A^2, A.

9–5. a. $\frac{1}{2}$ will be dipolid.

b. $\frac{1}{3}$ will be *Bbb*.

9–6. Let us use (1) for chromosome *Abc*, (2) for chromosome *aBc*, and (3) for chromosome *abC*. Examining the data, it can be seen that: the segregation of (1) vs. (2)

and (3) = 45 percent; the segregation of (3) vs. (1) and (2) = 45 percent; the segregation of (2) vs. (1) and (3) = 10 percent. Summing, (1) and (3) are found in separate progeny 90 percent of the time; (1) and (2) are found in separate progeny 45 + 10 = 55 percent of the time; and (2) and (3) are found in separate progeny 45 + 10 = 55 percent of the time. Therefore, chromosome (1), *Abc*, and chromosome (3), *abC*, were most frequently involved in pairing and separating from one another.

9–7.

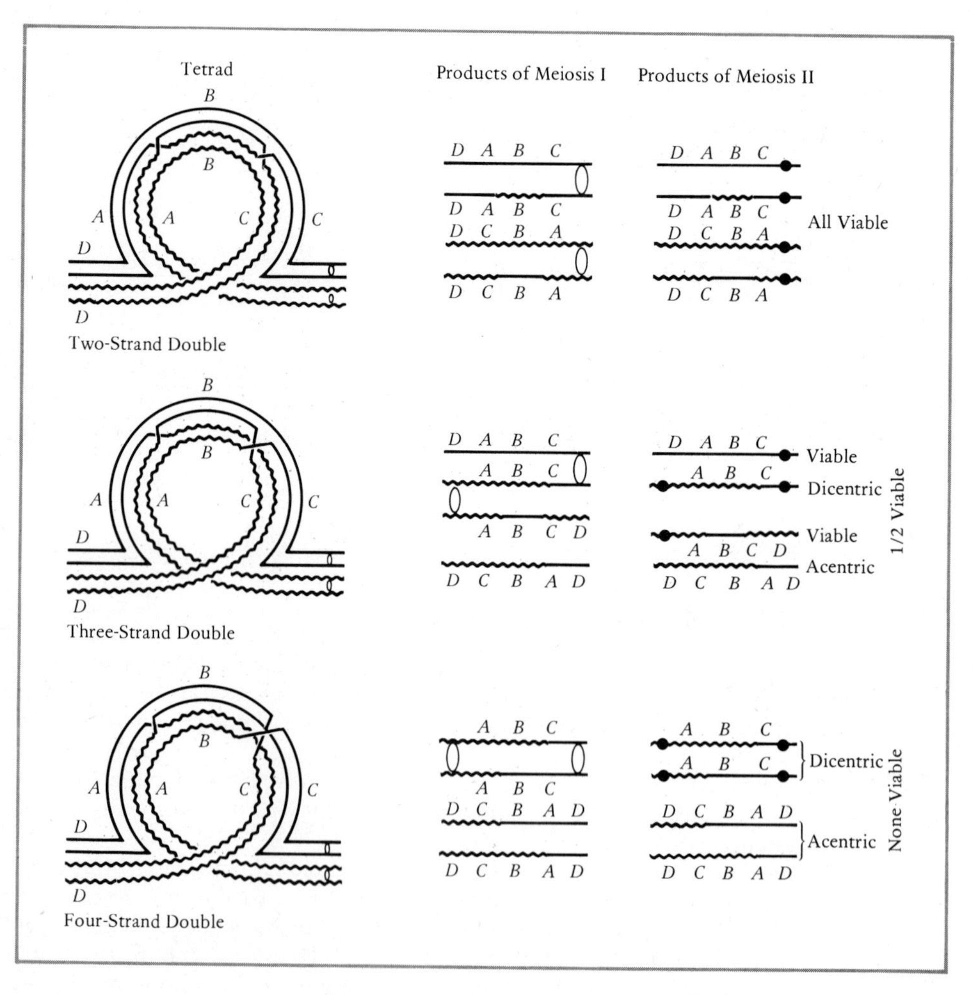

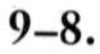
9–8.

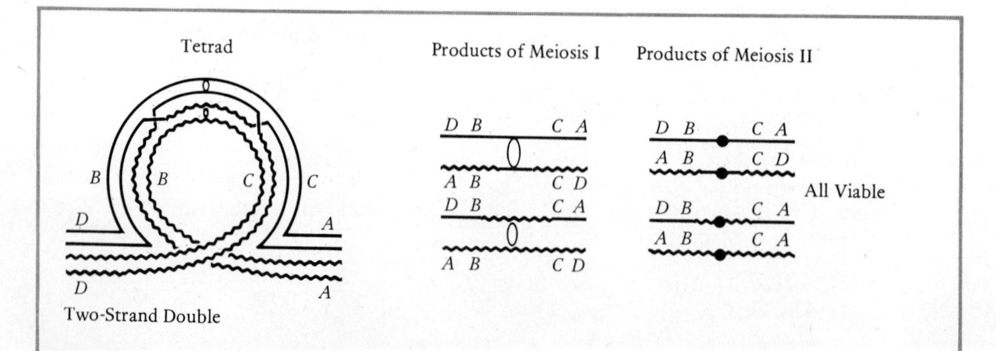

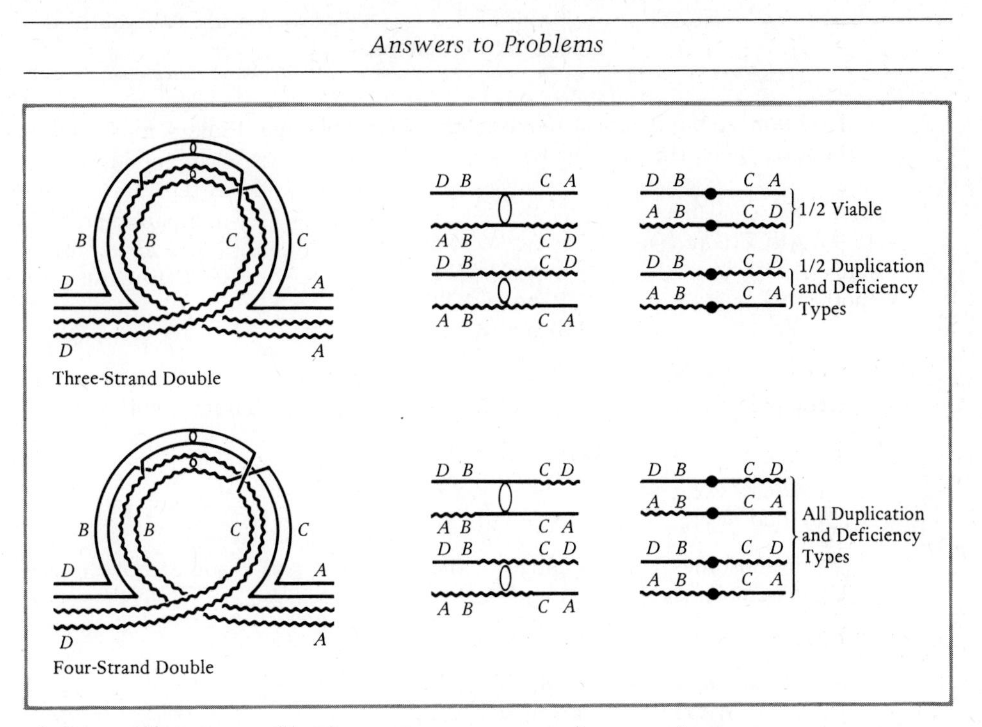

9–9. a. All gametes of both parents carry a complete set of genes.
b. All gametes of the homozygous parent and half the gametes of the heterozygous parent will carry a complete set of genes, giving a total of $\frac{3}{4}$ or 75 percent of all gametes produced.

9–10. a. inversion involving *m* and *f*
b. inversion involving *w* and *m*

9–11. a.

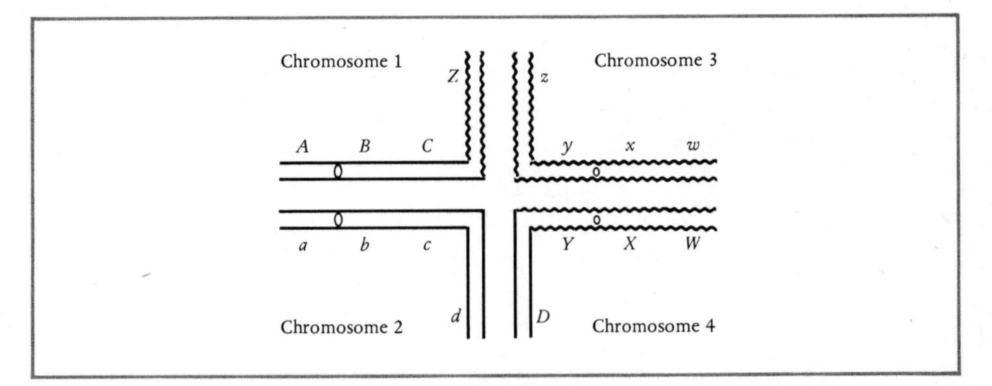

b. Crossover between genes *W* and *X:* If chromosomes 1 and 3 move to the same pole and 2 and 4 move to the opposite pole, the gametes will be:

From 1 and 3	From 2 and 4
(1) *ABCZ, wxyz*	(3) *abcd, wXYD*
(2) *ABCZ, Wxyz*	(4) *abcd, WXYD*

If chromosomes 1 and 4 move to the same pole and 2 and 3 move to the opposite pole, the gametes will be:

From 1 and 4	From 2 and 3
(1) *ABCZ, wXYD*	(3) *abcd, wxyz*
(2) *ABCZ, WXYD*	(4) *abcd, Wxyz*

If chromosomes 1 and 2 move to the same pole and 3 and 4 move to the opposite pole, the gametes will be:

From 1 and 2	From 3 and 4
(1) *ABCZ, abcd*	(1) *wxyz, wXYD*
	(2) *wxyz, WXYD*
	(3) *Wxyz, wXYD*
	(4) *Wxyz, WXYD*

c. Crossover between genes *B* and *C:* If chromosomes 1 and 3 move to the same pole and 2 and 4 move to the opposite pole, the gametes will be:

From 1 and 3	From 2 and 4
(1) *ABCZ, wxyz*	(3) *abCZ, WXYD*
(2) *ABcd, wxyz*	(4) *abcd, WXYD*

If chromosomes 1 and 4 move to the same pole and 2 and 3 move to the opposite pole, the gametes will be:

From 1 and 4	From 2 and 3
(1) *ABCZ, WXYD*	(3) *abCZ, wxyz*
(2) *ABcd, WXYD*	(4) *abcd, wxyz*

If chromosomes 1 and 2 move to the same pole and 3 and 4 move to the opposite pole, the gametes will be:

From 1 and 2	From 3 and 4
(1) *ABCZ, abCZ*	(1) *wxyz, WXYD*
(2) *ABCZ, abcd*	
(3) *ABcd, abCZ*	
(4) *ABcd, abcd*	

9–12. a. translocations
b. (1) chromosomes 2 and 3
(2) chromosomes 2 and 4
(3) chromosomes 3 and 4
(4) chromosomes 2, 3, and 4

9–13. gene *p*, bands A1 or A2; gene *rt*, band A3; gene *m*, bands B1 or B2; gene *q*, band B5

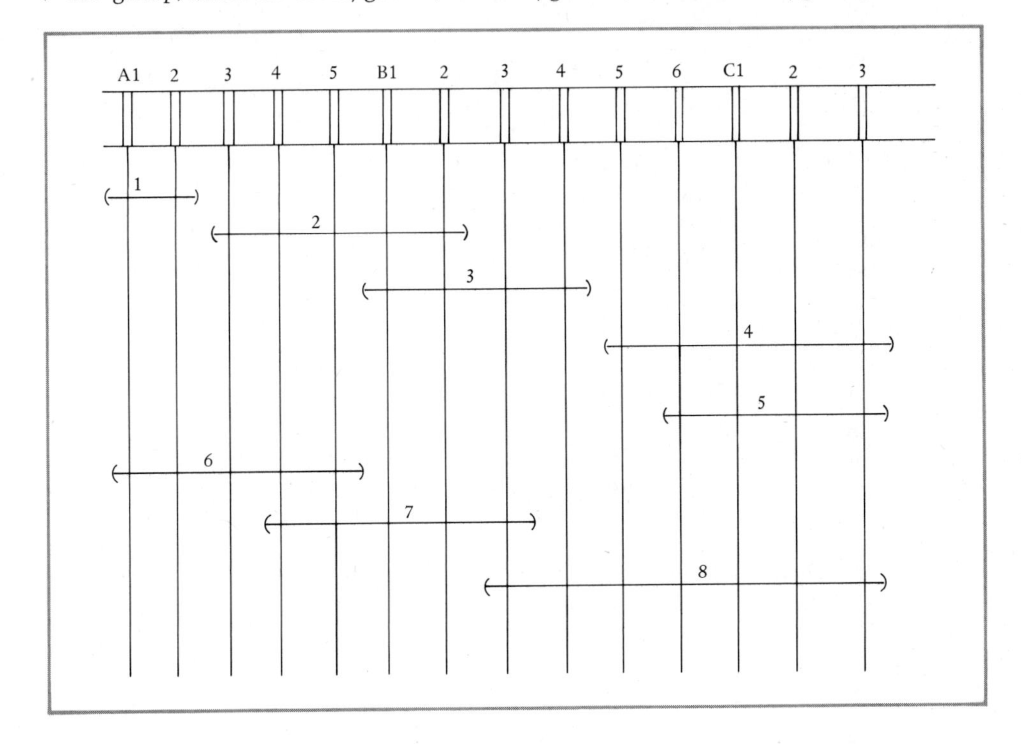

CHAPTER 10

10–1. Conformation is likely to undergo a drastic change if the cysteine residue normally forms a sulfhydryl bond with another cysteine residue located elsewhere in the molecule, since such bonds establish specific folds or loops required for function or enzyme recognition.

10–2. In PKU homozygotes one of the pathways leading to the formation of tyrosine is blocked, and since tyrosine is the precursor of melanin, melanin production and therefore pigmentation may be decreased.

10–3. precursor —gene 1→ intermediate —gene 2→ pigment

10–4. precursor —*A*→ monomethyl compound —*B*→ dimethyl compound ——→ choline

10–5. No, a branched metabolic pathway could be involved so that two separate end products were required for growth.

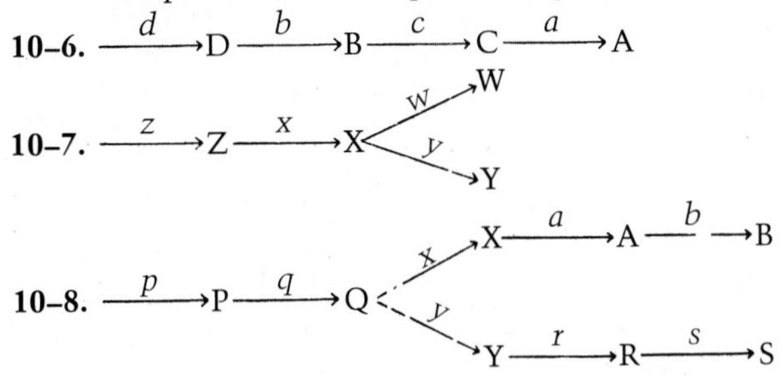

10–9. 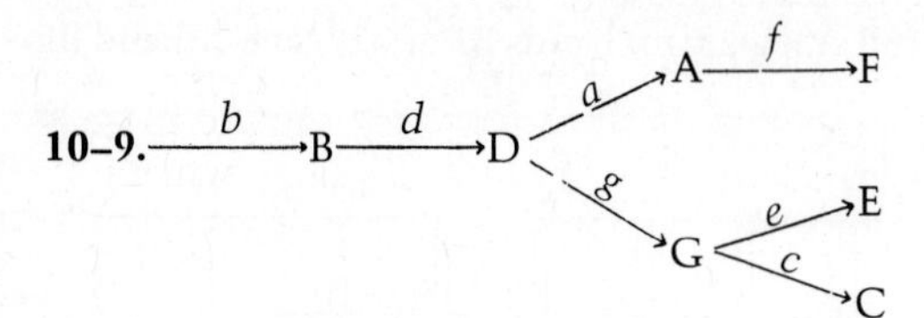

10–10. a. aa, aa′, a′a′
b. 3 electrophoretic bands
c. aaaa, aaaa′, aaa′a′, aa′a′a′, a′a′a′a′
d. 5 electrophoretic bands

10–11. a. deletion

10–12. a. (1) $\alpha_2\beta_2$, $\alpha_2\beta_2^S$, $\alpha_2\delta_2$
(2) $\alpha_2\beta_2^S$, $\alpha_2\beta_2^C$, $\alpha_2\delta_2$
(3) $\alpha_2\beta_2$, $\alpha_2\beta_2^S$, $\alpha_2^E\beta_2$, $\alpha_2^E\beta_2^S$, $\alpha_2\delta_2$, $\alpha_2^E\delta_2$
(4) $\alpha_2\beta_2$, $\alpha_2^E\beta_2$, $\alpha_2\delta_2$, $\alpha_2\delta_2^X$, $\alpha_2^E\delta_2$, $\alpha_2^E\delta_2^X$
(5) $\alpha_2\beta_2^S$, $\alpha_2\beta_2^C$, $\alpha_2^E\beta_2^S$, $\alpha_2^E\beta_2^C$, $\alpha_2\delta_2$, $\alpha_2\delta_2^X$, $\alpha_2^E\delta_2$, $\alpha_2^E\delta_2^X$

b. In genotypes 1 and 2 HbF is normal because the α chain is normal ($\alpha_2\gamma_2^A$ and $\alpha_2\gamma_2^G$). Normal HbF molecules in 3, 4, and 5 are $\alpha_2\gamma_2^A$ and $\alpha_2\gamma_2^G$. Variant HbF molecules in 3, 4, and 5 are $\alpha_2^E\gamma_2^A$ and $\alpha_2^E\gamma_2^G$. Neither the β nor the δ chains are found in HbF.

10–14. Five functional groups are represented. Mutants 8, 7, and 1 are alleles, as are mutants 6 and 10. Multisite mutants are 2 and 9, as shown below:

```
  8
  7                       10
  1         3              6          4          5
 ____     ____           ____       ____       ____
               2                          9
 ______________________________    ________________
```

10–15.

	1	2	3	4	5	6	7	8	9	10	11	12	13	14	15
1	0	+	+	+	0	0	0	+	+	+	+	+	+	+	+
2		0	+	0	0	+	+	+	0	+	+	+	+	+	+
3			0	+	+	+	+	+	+	+	+	0	+	+	0
4				0	0	+	+	+	0	+	+	+	+	+	+
5					0	0	0	0	0	+	0	0	+	0	+
6						0	0	+	+	+	+	+	+	+	+
7							0	0	+	+	+	+	+	+	+
8								0	+	+	+	+	+	+	+
9									0	+	+	+	+	+	+
10										0	+	+	0	+	+
11											0	0	+	0	+
12												0	+	0	0
13													0	+	+
14														0	+
15															0

CHAPTER 11

11–1. a. Bases at the 3′ end of the opposite strand will be complementary to those at the 5′ end of the polymer given. Therefore, the first 5 bases will be 3′ G-T-G-C-A.

b. 5′ T-A-A-G-C-T-G-G-A-A

c. A-T = 60 percent; G-C = 40 percent

d. The result of semiconservative replication:

```
             3' A-T-T-C-G-A-C-C-T-T-A-T-T-A-C-T-G-C-A-C 5'
                | | | | | | | | | | | | | | | | | | | |
             5' T-A-A-G-C-T-G-G-A-A-T-A-A-T-G-A-C-G-T-G 3'
daughter ---^
                             and
strands ---v
             3' A-T-T-C-G-A-C-C-T-T-A-T-T-A-C-T-G-C-A-C 5'
                | | | | | | | | | | | | | | | | | | | |
             5' T-A-A-G-C-T-G-G-A-A-T-A-A-T-G-A-C-G-T-G 3'
```

11–2. a. (1) 15 percent C, 15 percent G, 35 percent T, 35 percent A
(2) 12 percent G, 12 percent C, 38 percent T, 38 percent A
(3) 35 percent T, 35 percent A, 15 percent C, 15 percent G
(4) 28 percent A, 28 percent T, 22 percent G, 22 percent C

b. Yes, samples 1 and 3

11–3. (1) single-stranded DNA (2) double-stranded DNA
(3) single-stranded RNA (4) double-stranded RNA

11–4. (1) double-stranded DNA (2) single-stranded RNA
(3) single-stranded DNA

11–5. The second sample has a higher G-C content.

11–6. a. 1 million b. 3.4 mm

11–7. a. over 111.7 million b. Over 2.5 billion

11–8. a. both labeled due to the mechanism of semiconservative replication

b. Daughter cells could contain the first chromosome labeled, second unlabeled in one cell and the first chromosome unlabeled, the second labeled in the other cell; or they could contain both chromosomes labeled in one cell and both chromosomes unlabeled in the other cell.

11–9. a. All chromosomes of the daughter cells from the first mitosis would be labeled. b. $(\frac{1}{2})^{46}$

11–10. From labeled guanine, labeled adenine, cytosine, and thymine monophosphates would be found. From labeled cytosine, labeled guanine and adenine monophosphates would be found. From labeled thymine, labeled adenine, guanine and thymine monophosphates would be found. From labeled adenine, labeled cytosine, thymine, guanine and adenine monophosphates would be found.

11–11. From adenine labeling, A-G and A-T; from thymine labeling, T-A; from cytosine labeling, C-G; from guanine labeling, G-A and G-C.

CHAPTER 12

12–2. Grow each mutant at 25° during early stages and at 37° during late stages. With this treatment the mutant defective in DNA replication should form plaques, but the lysozyme-defective mutant should be unable to do so. With the reverse experiment, that is, 37° at early stages and 25° at late stages, only the lysozyme-defective mutant should form plaques.

12–5. Three different functions are represented. Phages 1, 3, and 4 are mutant for the same function; phages 2 and 5 are each mutant for different functions.

12–7. $s \leftarrow 8 \rightarrow f \leftarrow 2 \rightarrow t$
$\leftarrow 10 \rightarrow$

CHAPTER 13

13–1. genes *b* and *c* linked and on the same fragment
13–2. transformation linkage distance: 0.45 or 45 percent
13–3. $thr \leftarrow .14 \rightarrow tyr \leftarrow .23 \rightarrow his$
$\leftarrow .30 \rightarrow$
13–4.

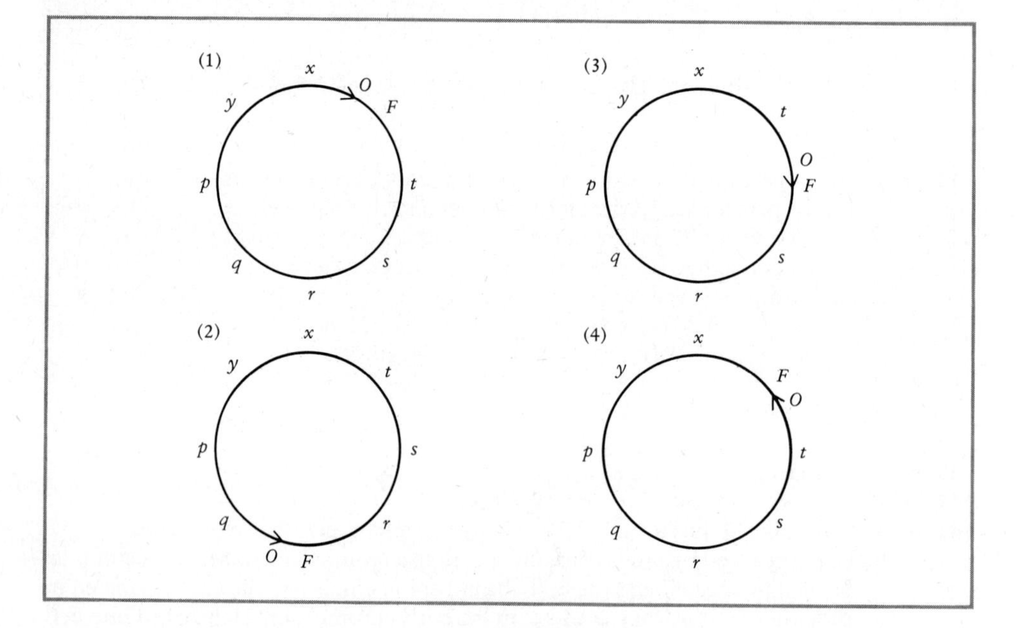

13–5. frequency of recombinants: 0.018 (1.8 percent)
13–6. We can infer the presence of a prophage inserted after gene c^+; when transferred to recipient, zygotic induction occurred with resulting lysis of the F^- cell.
13–7. Reciprocal crosses of Set 1: minimal medium is used and recombinants prototrophic for all genes are scored.
13–8. gene order: *leu—cys—tyr*
13–9. total recombinants = 42 + 36 = 78; total trandsuctants = 474 + 42 + 36 = 552; linkage distance = $\frac{78}{552}$ = 0.14 or 14 percent recombination between these genes
13–10. colony 1: $thr^+\ leu^-\ arg^+\ his^-\ str^R$
colony 2: $thr^-\ leu^-\ arg^+\ his^+\ str^R$
colony 3: $thr^+\ leu^+\ arg^-\ his^+\ str^R$
colony 4: $thr^-\ leu^+\ arg^+\ his^-\ str^R$
colony 5: $thr^-\ leu^+\ arg^+\ his^+\ str^R$

CHAPTER 14

14–1. a. $s/s \times r/r$
b. A mutant wing phenotype would indicate allelism.
c. $s/r \times s/s$ or $s/r \times r/r$

d. The appearance of a rare wild type recombinant indicates that *s* and *r* occupied different sites within the cistron.
e. 10,000

14–2. a. 0.04 map units b. 400 nucleotides

14–3. Enzyme A probably functions as a single polypeptide; enzyme B is multimeric, composed of a minimum of two subunits. Partial function of enzyme B in the heterokaryon of mutants 3 and 4 is an example of intragenic complementation.

14–4. order of mutant sites (left to right): 4 . . . 1 . . . 3 . . . 2

14–5. order of mutant sites (left to right): 3 . . . 7 . . . 2 . . . 10 . . . 8 . . . 4 . . . 1 . . . 5 . . . 6 . . . 9

14–6. (1) +/*a* and +/+
(2) *a*/*a* and +/*a*
(3) +/+ and (*a*/+)/*a*
(4) (*a*/+)/+ and +/*a*
(5) +/(*a*/+) and *a*/(+/*a*)

CHAPTER 15

15–4. strand 1

15–5. DNA-2

15–6. Serine will be added to glycine.

15–7. a. The 5′ terminus is the left end, starting U U C.
b. A A G G T G C T G C C T G G A
c. anticodons: AAG GUG CUG CCU GGA
d. Proline will be at the carboxyl terminus.

CHAPTER 16

16–1. codon CAU: anticodons 3′ GUA 5′ and 3′ GUG 5′
codon GUG: anticodons 3′ CAC 5′ and 3′ CAU 5′
codon AAC: anticodon 3′ UUG 5′
codon UCA: anticodon 3′ AGU 5′

16–2. (1) Probable single base change: AAU (asn) to AUU (ilu) or AAC (asn) to AUC (ilu). A change in the second base of the codon from A to U has occurred and has resulted in the substitution of a nonpolar residue in the midst of polar and/or neutral residues. This change can be expected to alter the solubility of the molecule.
(2) Probable single base change: UGU (cys) to GGU (gly) or UGC (cys) to GGC (gly). No change in the second base of the codon has occurred, but mutation has resulted in the substitution of glycine for cysteine. Although solubility might not be altered, the loss of cysteine would lead to the loss of sulfhydryl bonding with another cysteine residue located elsewhere in the molecule. This will cause the unfolding of at least one portion of the polypeptide chain and the loss of biological activity.
(3) Probable single base change: either GAA (glu) to GAU or GAC (asp), or GAG (glu) to GAU or GAC (asp). No change in the second base of the codon has occurred and mutation has resulted in the substitution of a compatible amino acid. It is likely that little change in biological activity will be observed.
(4) Probable single base change: CUU (leu) to CAU (his) or CUC (leu) to CAC (his). A change in the second base of the codon from U to A has occurred

and has resulted in the substitution of a polar amino acid (his) for a nonpolar amino acid (leu) located within a group of other nonpolar residues. This change is likely to alter an internal site of catalysis leading to loss of activity.

CHAPTER 17

17–1. Missense mutations result in amino acid substitutions which may or may not be compatible with activity. Nonsense mutations cause the substitution of a terminating codon (UAG, UAA, UGA) for a sense codon. The resulting polypeptide could be either fragmentary or shorter than normal. Frameshifts are caused by the addition or deletion of bases and result in a different sequence of amino acids from the point of mutation onwards, as in the case of Hb Wayne-1.

17–2. Translation would not cease since numerous genes for rRNA are present in the genome. The mutation of one of these would probably not interfere with protein synthesis.

17–3. Transitions cause the substitution of one purine (or pyrimidine) for another, for example, G-C ⇄ A-T. Transversions cause the substitution of a purine for a pyrimidine and vice versa, for example, T-A ⇄ A-T.

17–4. Intragenic mutation within the same codon, either restoring the original amino acid or resulting in the presence of a compatible amino acid; intragenic mutation within the same cistron, such as one that restores the normal reading frame; intergenic direct suppression, such as an alteration in some component directly involved in protein synthesis, for example, tRNA; intergenic indirect suppression, by an alteration in the cellular milieu.

17–6. The minimum number of bases required to code for 141 amino acids is 423. That mRNA for the α chain of hemoglobin contains more than 423 bases is indicated by the lengthened amino acid sequence of Hb Constant Spring and Hb Wayne-1.

17–8. Nitrous acid and base analogues produce transitions, and alkylating agents may produce transversions as well. The severity of the effects in any case would depend on the kind of amino acid substitution that occurred; also, if a base change resulted in a stop signal, a fragmentary polypeptide lacking any activity could result. Acridines are likely to cause the most harmful mutations because they produce frameshifts which alter all succeeding codons.

17–9. The lethal mutation rate would be the same in the two experimental groups.

17–11. The original sequence is ATG GCA TTA ATT AGA, the new sequence is ATG CAT TAA ATT AGA. Therefore, triplets 1, 4, and 5 would still code for the same amino acid.

17–12. The codon for glycine is GGU.
Gly → cys, *G*GU → *U*GU is a transversion.
Gly → asp, G*G*U → G*A*U is a transition.
Gly → ala, G*G*U → G*C*U is a transversion.

17–13.
1. single base substitution in codon 5, changing GAA (glu) to GGA (gly)
2. deletion of the first base (C) in codon 3, resulting in a frameshift with a new coden sequence of NH_2-UCU-CUU-GUC-AAG-AAG-AUG-UUU-GG-COOH
3. Deletion of the first base (C) of codon 3 and insertion of a base (C) between codons 4 and 5 of the original sequence, resulting in a new codon sequence of NH_2-UCU-CUU-GUC-AAC-GAA-GAU-GUU-UGG-COOH
4. base substitution in codon 8 transforming UGG (trp) to UGA (termination)

CHAPTER 18

18–2. A mutation in the *lac* operator would probably result in constitutive enzyme synthesis in the absence of inducer. A mutation in the promotor would probably result in no synthesis or reduced synthesis of *lac* enzymes in the presence of inducer. A mutation in gene *z* could result in an abnormal form of the enzyme β-galactosidase and in the consequent inability to utilize lactose. A mutation in gene *y* would result in reduced uptake of lactose into the cell. A mutation in gene *i* could cause either constitutive enzyme synthesis in the absence of inducer because of the inability of the repressor protein to bind to the operator or lack of enzyme synthesis because of the inability of the repressor protein to combine with the inducer. A mutation in the structural gene for CAP protein would result in the absence of positive control of the operon.

18–3. Since gene *galE* is the first structural gene of the operon, a polar mutation in this gene would result in reduced synthesis of all other enzymes of the galactose pathway.

18–4. Repression of enzyme synthesis in the presence of arabinose would result if the regulatory protein coded by gene *araC* was altered such that (1) it was incapable of interacting with the inducer and therefore remained bound to the operator or (2) it dissociated from the operator but was unable to assume the P_2 configuration and bind to the promotor.

18–5. Gene *trpR* codes for the regulatory protein of the operon. The mutation has probably resulted in an alteration of this protein such that it can no longer combine with the corepressor tryptophan.

18–7. If the missense mutation resulted in the substitution of a compatible amino acid, β-galactosidase might be active in lactose utilization; if an incompatible amino acid was substituted, β-galactosidase would likely be inactive. If a nonsense mutation occurred, the enzyme would probably be incomplete in amino acid sequence and therefore inactive.

18–8. (1) inducible (2) constitutive
(3) constitutive (4) constitutive
(5) inducible (6) inducible

CHAPTER 19

19–1. 15 possible isozymes could occur:

LDH 1	B_4
LDH 2	B_3A
	B_3A^*
LDH 3	B_2A_2
	$B_2A_2^*$
	B_2AA^*
LDH 4	BA_3
	BA_2A^*
	BAA_2^*
	BA_3^*
LDH 5	A_4
	A_3A^*
	$A_2A_2^*$
	AA_3^*
	A_4^*

19–2. The β and δ genes do not appear to be components of an operon because in the hereditary disease β-thalassemia only the β, and not the δ, chain is affected. Thus, these genes could not be controlled by the same operator. As an alternative explanation, high HbF could be caused by a deletion or rearrangement affecting both the β and δ loci.

19–3. Extirpate or irradiate only the pole cells and examine late stage larvae, pupae, and adults for the presence or absence of germinal tissue.

19–4. Mosaics can arise in a variety of ways, such as X chromosome inactivation in female mammals heterozygous for sex-linked genes, position effects caused by translocation or inversion, somatic crossing over, chromosomal nondisjunction in somatic cells, or even the accidental failure to include a lagging chromosome in a daughter nucleus.

19–5. The abnormal hemoglobins Constant Spring and Wayne-1 provide evidence for the presence in mRNA of additional bases beyond those normally translated.

19–6. The finding of abnormal embryos does not alter the hypothesis in any way. The aberrant chromosome complements in abnormal embryos can be explained as follows. After activation the egg cleaves whether or not chromosome replication in the injected nuclei has been completed. Somatic cells require 7 hours for genome replication and on the average divide only once in every 1 to 2 days. If by chance an implanted nucleus is one in which chromosome replication has not been completed, the cleavage division imposed by the egg cytoplasm will disrupt the normal diploid number and lead to aneuploidy and hyperploidy.

CHAPTER 20

20–3. Spore color is controlled by a nuclear gene, and sterility is determined by a cytoplasmic factor.

20–4. Use reciprocal crosses between the two types of plants. When a plant carrying the chloroplast mutation is the female parent, all progeny should be pale, but when it is used as the male parent, all progeny should be green.

20–5. a. all progeny resistant
b. $\frac{1}{2}$ resistant, $\frac{1}{2}$ sensitive

20–6. $\frac{1}{2}$ petite, $\frac{1}{2}$ wild type

20–7. P_1 parent was heterozygous for dextral and sinistral coiling *(D/d).*

20–8. Sterility is probably due to a virus brought in by the male-sterile graft.

CHAPTER 21

21–1. frequency of allele $N = 0.2$

21–2. frequency of the recessive allele (spotted) = 0.17
frequency of the dominant allele (solid) = 0.83
frequency of heterozygotes = 0.28

21–3. a. frequency of allele *d* in the combined population = 0.6
b. 0.6
c. 192 persons (48 percent of 400)

21–4. populations 2 and 3 not in equilibrium

21–5. $B_1B_1 = 0.01$; $B_1B_2 = 0.06$; $B_1B_3 = 0.12$
$B_2B_2 = 0.09$; $B_2B_3 = 0.36$; $B_3B_3 = 0.36$

21–6. a. frequency of nonsecretors = 0.09, frequency of blood group N individuals = 0.04; therefore, frequency of persons N, nonsecretor = 0.0036.
b. frequency of *MN* = 0.32, frequency of *Ss* = 0.42; therefore, frequency of genotype *MNSs* = 0.1344
c. frequency of *MM* = 0.64, frequency of *SS* = 0.49, frequency of *Ss* = 0.42, giving a total of 0.91 secretors; therefore, frequency of *M*, secretors = $.64 \times .91 = 0.5824$.

21–7. 0.48 or 48 percent

21–8. a. frequency of colorblind males = 0.2,

b. frequency of heterozygous females = 0.32

21–9. 0.11 or 11 percent

21–10. $I^A = 0.2$, $I^B = 0.1$, $I^O = 0.7$

CHAPTER 22

22–1. allele $C = 0.5$, allele $c = 0.5$

22–2. paths (common ancestor italicized): G D *B* E H = $(\frac{1}{2})^5 = \frac{4}{128}$

G D B *A* C F H = $(\frac{1}{2})^7 = \frac{1}{128}$

Sum: $\frac{5}{128} = 0.039 = F_I$

22–3. $F_I = 0.07$

22–4. a. Probability of aniridia in a child of unrelated parents is:

$$p_2^2 = (1.6 \times 10^{-3})^2 \cong 2.5 \times 10^{-6}$$

b. Probability of aniridia in a child of first cousins is:

$$p_2^2 + p_1 p_2 F = (2.5 \times 10^{-6}) + (1)(1.6 \times 10^{-3})(\tfrac{1}{16}) \cong 1 \times 10^{-4}$$

22–5. The equilibrium frequency of gene d (p_2) is computed as:

$$p_2 = \frac{u}{v+u} = \frac{6 \times 10^{-5}}{(1 \times 10^{-5}) + (6 \times 10^{-5})} = 0.85$$

22–6. The starting frequency of gene A in the native population $= p_0 = 0.8$ and the frequency of gene A in migrants $= P$. After one generation the frequency of gene A (p_1) will be:

$$p_1 = p_0(1 - M) + PM = .8(1 - .25) + .2(.25) = 0.65$$

The change in the frequency of gene A will be:

$$p_0 - p_1 = 0.8 - 0.65 = 0.15$$

22–7. After one generation of selection, the frequency of gene D is:

$$p_1 = \frac{p_1^2 w_{11} + p_1 p_2 w_{12}}{\bar{w}}$$

and the frequency of gene d is:

$$p_2 = \frac{p_2^2 w_{22} + p_1 p_2 w_{12}}{\bar{w}}$$

$\bar{w}$ is computed as:

$$\bar{w} = p_1^2 w_{11} + 2p_1 p_2 w_{12} + p_2^2 w_{22}$$
$$\bar{w} = (.8)^2(1) + 2(.8)(.2)(1) + (.2)^2(.3) = 0.972$$

Substituting:

$$p_1 = \frac{(.8)^2(1) + (.8)(.2)(1)}{.972} = 0.823 \text{ gene } D$$

$$p_2 = \frac{(.2)^2(.3) + (.8)(.2)(1)}{.972} = 0.177 \text{ gene } d$$

Alternately, the gene frequency change formula can be used to compute the change in the frequency of gene d (p_2) and from this the frequencies of genes D and d in the population after one generation of selection can be determined.

22–8. At generation 0, $p_1^2 + 2p_1p_2$ or $(.8)^2 + (2)(.8)(.2) = 0.96$ show the dominant phenotype. After selection, $p_1' = 0.823$ and $p_2' = 0.177$. Thus, $p_1'^2 + 2p_1'p_2'$ or

$(.823)^2 + (2)(.823)(.177) = 0.969$ or 97 percent would show the dominant phenotype.

22–9. Since genotype *aa* has a fitness of 0 and $s = 1$, the change in the frequency of gene *a* is:

$$p_2 = \frac{-p_2^2}{1 + p_2} = \frac{-(.4)^2}{1 + .4} = -0.114$$

The new frequency of gene *a* will be $0.4 - 0.114 = 0.286$. Since the frequency of gene $A = p_1 = 1 - p_2$, the new frequency of gene *A* is $1 - 0.286 = 0.714$. Alternately, the new gene frequencies can be computed as in problem 22–7.

22–10. a. Prior to environmental change 16 percent of the sample should be *aa*.
b. After selection 8.2 percent of the eggs should be *aa* in genotype.

22–11. One generation after selection the frequency of gene *B* is 0.202 and that of gene *b* is 0.798. The solution is computed as in problem 22–7.

22–12. a. In the first sample 51 percent or approximately 255 individuals should be solid colored.
b. In the second sample 34 percent or approximately 170 individuals should be solid colored. To solve the second part of the problem it is first necessary to compute the value of $\bar{w}$. This value is then used in the solution of the formula.

$$\frac{\overset{AA}{p_1^2 w_{11}} + \overset{Aa}{2p_1p_2w_{12}} + \overset{aa}{p_2^2 w_{22}}}{\bar{w}} = 1$$

The sum of the frequencies of genotypes *AA* and *Aa* is the frequency of solid-colored individuals.

22–13. The frequency of gene *r* would be 1.0.

22–14. Since the fitness of genotype $bb = 0.2$, the selection coefficient $s = 0.8$.

$$\hat{p}_2 = \sqrt{\frac{u}{s}} = \sqrt{\frac{6 \times 10^{-5}}{.8}} \cong 8.6 \times 10^{-3}$$

22–15. The frequency of the lethal allele at equilibrium is:

$$\hat{p}_2 = \sqrt{u} = \sqrt{6 \times 10^{-6}} \cong 2 \times 10^{-3}$$

22–16. If the fitness of genotype $AA = 0.4$, its selection coefficient $s = 0.6$; if the fitness of genotype $aa = 0.8$, its selection coefficient, $t = 0.2$. At equilibrium, the frequency of gene *A* would be:

$$\hat{p}_1 = \frac{t}{s + t} = \frac{.2}{.6 + .2} = 0.25$$

Since $1 - p_1 = p_2$, the frequency of gene *a* would be $1 - 0.25 = 0.75$.

INDEX

Index

79 80 81 9 8 7 6 5